PHOTOSTABILIZATION OF POLYMERS

Principles and Applications

PHOTOSTABILIZATION OF POLYMERS

Principles and Applications

J. F. RABEK

Department of Polymer Technology,
The Royal Institute of Technology,
Stockholm, Sweden

ELSEVIER APPLIED SCIENCE
LONDON and NEW YORK

ELSEVIER SCIENCE PUBLISHERS LTD
Crown House, Linton Road, Barking, Essex IG11 8JU, England

Sole distributor in the USA and Canada
ELSEVIER SCIENCE PUBLISHING CO., INC.,
655 Avenue of the Americas, New York, NY 10010, USA

WITH 98 TABLES AND 157 ILLUSTRATIONS

British Library Cataloguing in Publication Data
Rabek, J. F. (Jan F.)
Photostabilization of polymers.
1. Polymers. Stabilization
I. Title
668.4′1

ISBN 1-85166-408-4

Library of Congress Cataloging-in-Publication Data
Rabek, J. F.
Photostabilization of polymers: principles and applications/
J. F. Rabek.
p. cm.
Includes bibliographical references.
ISBN 1-85166-408-4
1. Polymers—Deterioration. 2. Photochemistry. I. Title.
QD381.9.P56R34 1990
620.1′920422—dc20 89-16993
CIP

Printed in Northern Ireland by The Universities Press (Belfast) Ltd.

Dedicated to my aunt Irena Gawronska, chemist, her family Adam and Mary and their lovely children Jan, Anna, Ronan and Paula

'None of us is as good as all of us'

People involved in polymer stabilization

Preface

During the last two decades, the production of polymers and plastics has been increasing rapidly. In spite of developing new polymers and polymeric materials, only 40–60 are used commercially on a large scale. It has been estimated that half of the annual production of polymers is employed outdoors. Increasing the stability of polymers and plastics towards heat, light, atmospheric oxygen and other environmental agents and weathering conditions has always been a very important problem. The photochemical instability of most of polymers limits them to outdoor application, where they are photodegraded fast over periods ranging from months to a few years.

To the despair of technologists and consumers alike, photodegradation and environmental ageing of polymers occur much faster than can be expected from knowledge collected in laboratories. In many cases, improved methods of preparation and purification of both monomers and polymers yield products of better quality and higher resistance to heat and light. However, without stabilization of polymers by application of antioxidants (to decrease thermal oxidative degradation) and photostabilizers (to decrease photo-oxidative degradation) it would be impossible to employ polymers and plastics in everyday use. At best, the choice of suitable stabilizers is at present governed not only by the data collected in practical outdoor or indoor ageing of polymers and plastics, but also by the collected knowledge of thermal oxidative and photo-oxidative degradation mechanisms, stabilization mechanisms, interactions of additives (synergistic and antagonistic effects), etc. The main task has been to bridge the gap between the theory of thermal degradation and/or photodegradation of polymers and their practical stabilization.

This book represents the author's attempt, based on his 20 years' experience in research on photodegradation and photostabilization, to collect and generalize a number of available data on the photostabilization of polymers. The space limitation and the tremendous number of publications in the past decade have made a detailed presentation of all important results and data difficult. The author apologizes to those whose work has not been quoted or widely presented in this book. Because many published results are very often contradictory, it has been difficult to present a fully critical review of collected knowledge, without antagonizing the authors. For that reason, all available theories, mechanisms and different suggestions have been presented together, and only practice can evaluate which of them are valid.

Very different chemical nomenclature of antioxidants and photo-stabilizers is sometimes used for the same compounds; therefore in this book the nomenclature employed is the same as was used by authors in their original publications. However, all available chemical nomenclature has been collected in tables and it should be very easy to find one's way through the jungle of different names.

This is an effort to integrate this vitally needed information on photostabilization of polymers in a single book written as a monograph of the subject. This book should be a useful aid to all people from the polymer and plastics industries, but also to graduate or advanced undergraduate students.

The following figures and tables were reproduced by permission of John Wiley & Sons, Inc: Figs 1.8, 1.9, 2.12, 2.14, 3.1, 3.6, 4.2–4.4, 4.14, 5.3, 5.4, 5.9, 5.24, 5.30, 5.31, 5.34, 5.37, 6.4, 6.5, 8.4, 9.1, 9.2, 9.8, 9.9, 9.16, 9.17, 9.20, 9.23–9.25, 9.30; Tables 2.5, 3.1, 3.3–3.6, 3.11, 4.5, 4.7, 4.10, 5.2–5.4, 5.10, 6.7, 7.3, 7.5.

In conclusion, I should like to express my gratitude to members of my family for their patience while this book was being prepared, having to forego my company during innumerable evenings, weekends and holidays.

J. F. RABEK

Contents

Chapter 3 *Antioxidants*

Chapter 4 *Pigments*

Chapter 5 *Photostabilizers*

Chapter 9 *Experimental Methods in Polymer Degradation and Stabilization*

1

Introduction to the Photo-oxidative Degradation Mechanisms of Polymers

1.1 DEFINITIONS OF PHOTOSTABILITY OF POLYMERIC MATERIALS

Polymers currently used on a large scale may be divided into three groups depending on their resistance to photodegradation (photo-oxidative degradation):

(i) Highly photostable polymers, that are commonly used without photostabilizer added, e.g. polytetrafluoroethylene and poly(methyl methacrylates).
(ii) Moderately photostable polymers that can be used without photostabilizer, e.g. poly(ethylene terephthalate) and polycarbonates.
(iii) Poorly photostable polymers which need extensive stabilization for outdoor uses, e.g. polyolefins, poly(vinyl chloride), polystyrene, polyamides, polyurethanes, rubbers, coatings and cellulose.

The definition of *long-term photostability* of polymers and plastics differs depending on their application. For the packaging industry 'long-term' means 1–4 years. Gradual photodegradation after use decreases the litter problem.[164] For the building and machine industries long-term properties of 10–40 years are required.

Oxidative degradation of unstabilized polymers in atmospheres containing even parts-per-million levels of oxygen occurs for prolonged periods above 80°C, and over the course of months or years at ambient temperature. Oxidation of polymers occurs during synthesis, storage and processing.

UV radiation of polymers causes photo-oxidative degradation which

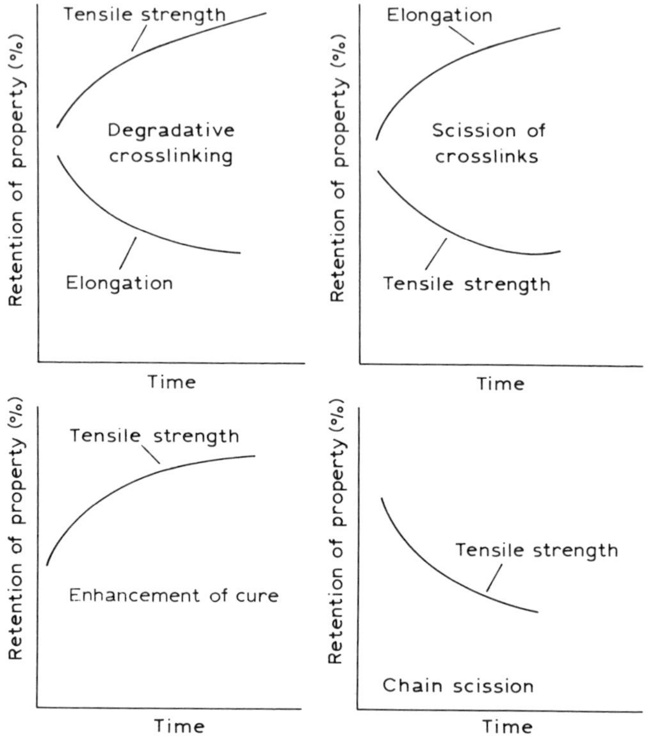

Fig. 1.1. Typical effects of various types of degradation processes on physical properties of polymers. Reprinted with permission from Ref. 780, © 1983 American Chemical Society.

results in cracking, chalking, colour changes, and the loss of physical properties such as impact strength, tensile strength, elongation etc. (Fig. 1.1).

Many reviews have been published on polymer photo-oxidation[32,35,139,140,201,241,250,435,523,745,936,1019,1056] and books on polymer oxidation,[710,777,894,900,934,1000,1135] degradation[375,485,611,961,986] and photodegradation, photo-oxidative degradation and photostabilization.[80,283,882,919,934,1135]

1.2 GENERAL MECHANISM OF POLYMER PHOTO-OXIDATION

Photo-oxidative degradation of polymers, which includes such processes as chain scission, crosslinking and secondary oxidative reactions, occurs by the free radical mechanism (similar to thermal oxidation) in the following steps:

Reaction	Step	No.
$\text{Polymer} \longrightarrow P\cdot + P\cdot$	Initiation	(1.1)
$P\cdot + O_2 \longrightarrow POO\cdot$	Chain propagation	(1.2)
$POO\cdot + PH \longrightarrow POOH + P\cdot$	Chain propagation	(1.3)
$POOH \longrightarrow PO\cdot + \cdot OH$	Chain branching	(1.4)
$PH + \cdot OH \longrightarrow P\cdot + H_2O$	Chain branching	(1.5)
$PO\cdot \longrightarrow$ Chain scission reactions	Chain branching	(1.6)
$POO\cdot + POO\cdot$, $POO\cdot + P\cdot$, $P\cdot + P\cdot$ $\longrightarrow$ Crosslinking reactions to non-radical products	Termination	(1.7)

where PH = polymer,
P• = polymer alkyl radical,
PO• = polymer oxy radical (polymer alkoxy radical),
POO• = polymer peroxy radical (polymer alkylperoxy radical),
POOH = polymer hydroperoxide,
HO• = hydroxy radical.

This mechanism applies to both thermal and photo-oxidation of the polymer; however, the initiation steps differ in each case:

—in thermal degradation, initiation results from the thermal dissociation of chemical bonds;

—in photodegradation, initiation results from different photophysical processes such as formation of electronically excited species, energy transfer processes or direct photodissociation of chemical bonds.

1.3 PHOTOINITIATION OF POLYMER DEGRADATION

Photoinitiation of polymer degradation can be caused by the following reactions:

(i) External impurities (RR′) with chromophoric groups present in a polymer and/or internal impurities in the form of chromophoric groups attached to polymer chains absorb light and produce low-molecular-weight radicals (R•) and/or polymeric radicals (P•):

$$RR' + h\nu \longrightarrow R\cdot + R'\cdot \quad (1.8)$$

$$\text{Polymer} + h\nu \longrightarrow P\cdot + R\cdot \text{ (and/or } P\cdot) \quad (1.9)$$

(ii) Energy absorbed by a given chromophoric group can be transferred to another group (energy-transfer process), which further dissociates into free radicals.

(iii) Energy absorbed by a given chromophoric group can be accumulated at a given bond (by an energy migration process), which further dissociate into free radicals.

Saturated polyolefins should not be expected to absorb beyond 200 nm. The absorption in the atmospheric sunlight spectrum around 300 nm has commonly been attributed to the low concentrations of ketone and/or hydroperoxide groups introduced into commercial polymers during processing. Wavelengths of maximum sensitivity of several polymers are shown in Table 1.1.

Saturated hydrocarbon molecules, which alone do not absorb UV radiation, can exhibit charge-transfer absorption in the presence of oxygen.[27,285] In a similar way polystyrene forms contact complexes with oxygen under high pressure.[926,927] These complexes have extended absorption towards longer wavelengths and can be photolysed into free radicals; however, this process has low efficiency because of its small absorption coefficients and quantum yields.

In polymers the distribution of chromophoric groups is non-homogeneous. The local concentration of chromophoric groups can be very high. Because of the interaction with the polymer main chain and side chain, the motion of a chromophoric group is retarded. Interac-

Table 1.1
Wavelength of maximum sensitivity of various polymers[1042]

Polymer	*Wavelength (nm)*
Polyesters (various formulations)	325
Polystyrenes	318
Polyethylene	300
Polypropylene (non-heat stabilized)	310
Poly(vinyl chloride)	310
Polyvinyl chloride copolymer	322
with vinyl acetate	364
Polycarbonate	295
Poly(methyl methacrylate)	290–315
Polyformaldehyde	300–320

tion between chromophore groups and other groups by either intrapolymer or interpolymer association depends on the structure of the polymer matrix. Below and above the glass transition temperature (T_g), the molecular motion is critically different and therefore the photophysical processes and photochemical reactions require diffusion or activation free-volume changes.

There has been much controversy in the literature about the relative importance of the various possible mechanisms for initiation (involving ketone groups, hydroperoxide, catalyst residue, singlet oxygen), but the controversy is practically irrelevant if initiation does not greatly influence the course, rate or extent of photo-oxidation.[70,240,263,284,919]

An important role may be played by ozone, which in the reactions with polymers may cause formation of different oxidized species.[944]

1.4 CHAIN PROPAGATION

The key reaction in the propagation sequences is the formation of polymer peroxy radicals (POO•) by reaction of polymer alkyl radicals (P•) with oxygen:

$$P\cdot + O_2 \longrightarrow POO\cdot \qquad (1.10)$$

This reaction is very fast, but diffusion-controlled.[346,1165,1166]

The next propagation step is the abstraction of a hydrogen atom by the polymer peroxy radical (POO•) to generate a new polymer alkyl radical (P•) and polymer hydroperoxide (POOH):[238,436,707]

$$POO\cdot + PH \longrightarrow P\cdot + POOH \qquad (1.11)$$

Hydrogen abstraction occurs principally from the tertiary carbon atoms:

$$POO\cdot + \text{—}CH_2\text{—}\underset{\underset{H}{|}}{\overset{\overset{R}{|}}{C}}\text{—}CH_2\text{—} \longrightarrow POOH + \text{—}CH_2\text{—}\underset{\bullet}{\overset{\overset{R}{|}}{C}}\text{—}CH_2\text{—} \qquad (1.12)$$

However, it may also occur from the secondary carbon atoms in methylene groups:

$$POO\cdot + \text{—}CH_2\text{—}CH_2\text{—} \longrightarrow POOH + \text{—}CH_2\text{—}\dot{C}H\text{—} \qquad (1.13)$$

Hydrogen abstraction reactions may also occur intramolecularly if a favourable stereochemical arrangement can be formed:

(1.14)

In the solid-state polymer most of the peroxy radicals formed as a result of hydroperoxide photolysis recombine (terminate) after only a few propagation steps (secondary cage recombination). The small proportion of peroxy radicals escaping from the secondary cage propagate with very long kinetic chain lengths to produce the bulk of photoinitiating hydroperoxides so that even relatively inefficient radical scavenging of these intermediates can result in a significant reduction in the rate of hydroperoxide formation.[437]

1.5 CHAIN BRANCHING

Chain branching by thermolysis and/or photolysis of polymer hydroperoxides (POOH) results in the formation of very reactive polymer oxy radicals (PO•) and hydroxy radicals (HO•):

$$\mathrm{POOH} \xrightarrow{\Delta\ (\text{or } h\nu)} \mathrm{PO\bullet} + \mathrm{\bullet OH} \tag{1.15}$$

Polymer oxy radicals (PO•) and very mobile hydroxy radicals (HO•) abstract hydrogen from the same or from a nearby polymer (PH) chain:

$$\mathrm{PO\bullet} + \mathrm{PH} \longrightarrow \mathrm{POH} + \mathrm{P\bullet} \tag{1.16}$$

$$\mathrm{HO\bullet} + \mathrm{PH} \longrightarrow \mathrm{P\bullet} + \mathrm{H_2O} \tag{1.17}$$

Polymer oxy radicals (PO•) undergo a number of other reactions including:

(i) *β*-Scission reactions which result in fragmentation of the polymer chain together with formation of end carbonyl (or end aldehyde) groups and end alkyl radicals:

$$-\mathrm{CH_2}-\underset{\underset{\bullet}{\mathrm{O}}}{\overset{\mathrm{R}}{\mathrm{C}}}-\mathrm{CH_2}-\overset{\mathrm{R}}{\mathrm{CH}}- \longrightarrow -\mathrm{CH_2}-\underset{\mathrm{O}}{\overset{\mathrm{R}}{\mathrm{C}}}\!=\; + \;\bullet\mathrm{CH_2}-\overset{\mathrm{R}}{\mathrm{CH}}- \tag{1.18}$$

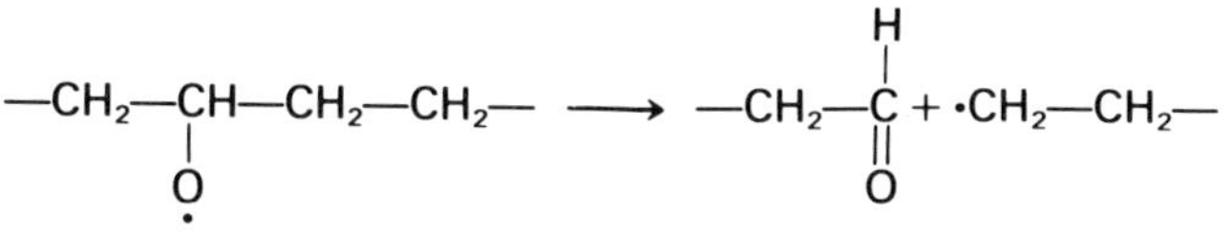

(1.19)

(ii) Formation of in-chain ketone groups:

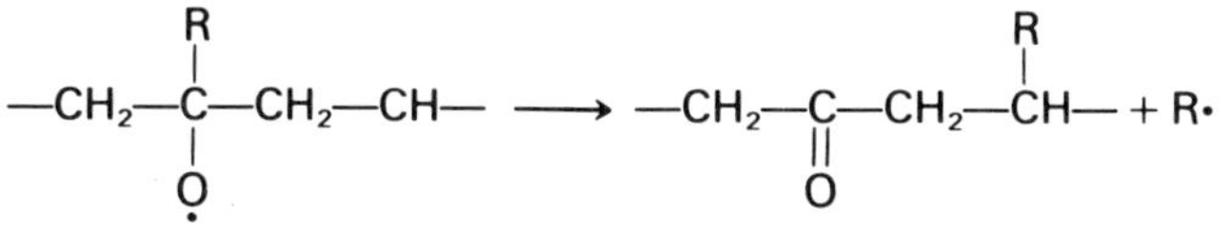

(1.20)

(iii) Induced hydroperoxide decomposition:

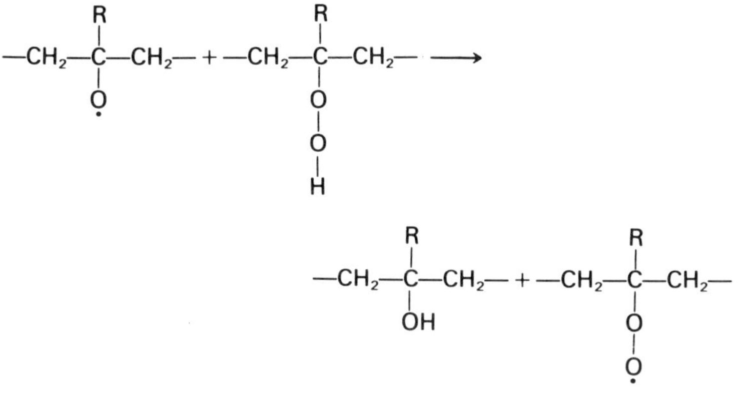

(1.21)

(iv) Crosslinking (termination):

$$PO\cdot + P\cdot \longrightarrow POP \quad (1.22)$$

$$PO\cdot + PO\cdot \longrightarrow POOP \quad (1.23)$$

$$PO\cdot + POO\cdot \longrightarrow POOOP \longrightarrow POP + O_2 \quad (1.24)$$

$$PO\cdot + \cdot OH \longrightarrow POOH \quad (1.25)$$

These processes are dependent on the chemical and physical structure of the irradiated polymers.

1.6 PHOTO-REACTIONS OF CARBONYL GROUPS

Carbonyl groups present in a virgin polymer or formed during photo-oxidative degradation of a polymer are photolysed by:[80,330,914,919,934]

(i) *Norrish Type I* reaction, during which carbon monoxide and two macroradicals are formed:

$$-CH_2-\underset{\displaystyle \overset{\|}{O}}{C}-CH_2- \xrightarrow{h\nu} -CH_2\cdot + CO + \cdot CH_2- \qquad (1.26)$$

(ii) *Norrish Type II* reaction, a non-radical, intramolecular process which occurs with the formation of a six-membered cyclic intermediate. Abstraction of a hydrogen from the γ-carbon results in its subsequent decomposition into an unsaturated polymer chain end, and a polymer chain with an end carbonyl group:

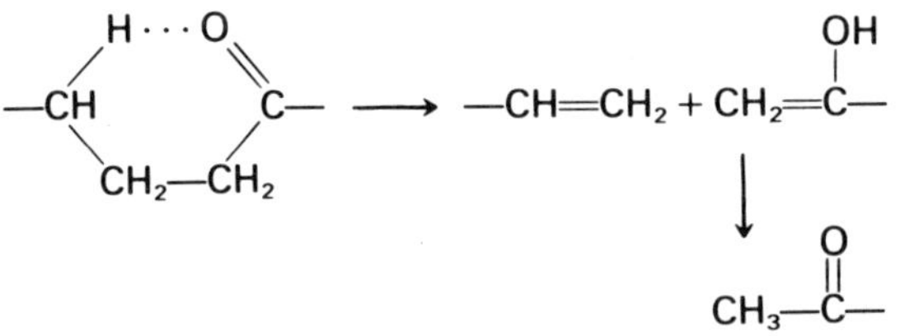

(1.27)

(1.28)

In polymers, the Type II reaction may also involve an intramolecular process with a longer chain segment.

1.7 HYDROGEN ABSTRACTION

Depending on the polymer chains structure, the hydrogen atoms can be abstracted from the tertiary, secondary and even primary C—H sites.

The highest level of tertiary C—H sites in commercial and laboratory synthesized polyethylene samples is ~3/100 carbons. Several studies have shown that branched polyethylenes oxidize more rapidly than their linear analogues because of the presence of the tertiary C—H site at each branch point.[343,592] However, the intrinsic rate of oxidation of polyethylene and the composition of its oxidation products are virtually independent of the presence of backbone branching, whether short or medium length, at levels found in commercial materials.[137,231,392,1054,1112]

Hydrogen abstraction by polymer alkoxy (PO•) and hydroxy (HO•) radicals from the photocleavage of hydroperoxy groups in polypropylene occurs with 50% probability at methyl groups. This means that up to 25% of the polymer peroxy radical (POO•) pairs produced in the initiation process may be pairs of primary peroxy radicals and the remaining 75% would be pairs between secondary, tertiary and mixed peroxy radicals. Due to the high radical concentration within such pairs of immediately adjacent radicals, the termination processes would be favoured over the slow hydrogen abstraction reaction of these radicals.

The intramolecular hydrogen atom abstraction by a polymer peroxy radical (POO•) can take place unless six-membered (or larger) rings are formed in the transition state. It is obvious that polymer segment conformation will determine the probability of formation of a transition complex of the optimum structure. Since structural relaxation is slow in a polymer matrix and the local mobility depends on the local conformation of a segment of macromolecule, the kinetics of intramolecular reactions will be influenced by the segment conformation of the macromolecule.

1.8 PHOTODECOMPOSITION OF HYDROPEROXIDE GROUPS

The light quanta produced by solar irradiation (Fig. 1.2) are energetically sufficient to cleave PO—OH and also P—OOH, but hardly POO—H bonds, which have the dissociation energies 42 kcal mol^{-1} (PO—OH), 70 kcal mol^{-1} (P—OOH) and 90 kcal mol^{-1} (POO—H).[171,172] The large difference in bond dissociation energy between PO—OH and P—OOH means that reaction with formation of PO• and •OH radicals will predominate during light irradiation.[691,1107,1161]

Hydroperoxide groups are transparent at wavelength >340 nm and they have very low molar absorptivity (molar extinction coefficient) ($\varepsilon = 10$–$150\ mol^{-1}\ cm^{-1}$) at wavelength $\lambda = 340$ nm. the O—O bond has no low-lying stable excited state, and the potential energy surfaces of the first excited states are dissociative.[385] The quantum yield of the photocleavage hydroperoxide groups in the near-ultraviolet is close to 1·0.

The photolysis of hydroperoxide groups under solar irradiation is a slow process. The average lifetime of an —OOH group in 10-μm polypropylene film under constant UV irradiation is ~25 h, equivalent to roughly 4–5 days in the solar radiation.[241,248]

The most probable mechanism of photodecomposition of OOH

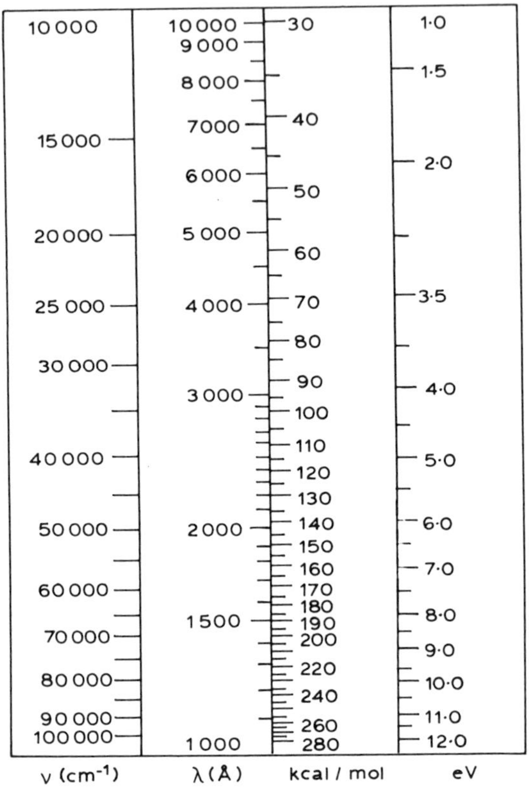

Fig. 1.2. Dependence of photon energy upon wavelength and upon frequency.

groups occurs via an energy transfer process from the excited carbonyl or aromatic hydrocarbon groups (donors) to hydroperoxy groups (acceptors):[54,108,198,199, 201,393,919,934]

$$\text{(donor)}^* + \text{—OOH(acceptor)} \longrightarrow \text{donor} + (\text{—OOH})^* \quad (1.29)$$

$$(\text{—OOH})^* \longrightarrow \text{—O}\cdot + \cdot\text{OH} \quad (1.30)$$

Experimental data obtained for the rate of disappearance of hydroperoxy groups during the ketone-sensitized photolysis of *tert*-butylhydroperoxide and of *cis*-polyisoprene hydroperoxide suggest that a significant portion of energy transferred from the excited carbonyl groups to the hydroperoxide groups results in direct scission of the peroxy linkage by an energy transfer mechanism.[506,507]

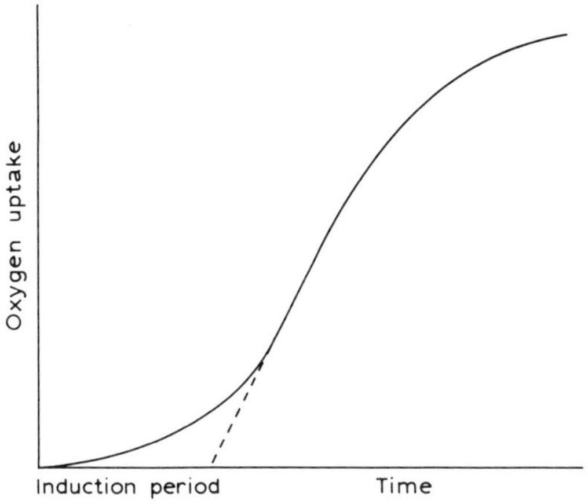

Fig. 1.3. Typical oxygen uptake curve.

In this mechanism localization of hydroperoxy groups in a polymer matrix play an important role. In consequence hydroperoxides can be divided into two groups:[198,200,201]

(i) Active hydroperoxides which are formed in the neighbourhood of phenyl groups (a sphere surrounding the phenyl groups). They are photolysed by an energy transfer process in a collision complex formed between excited phenyl groups (donors) and vibrationally excited hydroperoxides (acceptors).
(ii) Inactive hydroperoxides resulting from the propagation of polymer peroxy radicals outside a sphere, which avoids photodecomposition.

After an induction period, at the beginning of irradiation, the active hydroperoxides reach a steady concentration, while the inactive hydroperoxides continue to increase slowly, but regularly, with irradiation time (Fig. 1.3).

1.9 FORMATION OF ACIDS AND PERACIDS

There are several ways in which polymer acyl radicals $(\mathrm{P{-}\underset{\underset{\displaystyle O}{\|}}{C}\cdot})$ can be formed:

(i) Reaction of polymer end aldehyde groups (formed from the

β-scission reaction of polymer oxy radicals (PO•), with polymer peroxy radicals (POO•):

$$\text{P—CHO} + \text{POO}\bullet \longrightarrow \text{P—}\overset{\text{O}}{\overset{\|}{\text{C}}}\bullet + \text{POOH} \qquad (1.31)$$

(ii) Norrish Type I reaction:

$$\text{—CH}_2\text{—}\overset{\text{O}}{\overset{\|}{\text{C}}}\text{—CH}_2\text{—} \longrightarrow \text{—CH}_2\text{—}\overset{\text{O}}{\overset{\|}{\text{C}}}\bullet + \bullet\text{CH}_2\text{—} \qquad (1.32)$$

Polymer acyl radicals (**1.1**) are further oxidized to polymer peracid radicals (**1.2**):

$$\underset{(\mathbf{1.1})}{\text{—CH}_2\text{—}\overset{\text{O}}{\overset{\|}{\text{C}}}\bullet} + \text{O}_2 \longrightarrow \underset{(\mathbf{1.2})}{\text{—CH}_2\text{—}\overset{\text{O}}{\overset{\|}{\text{C}}}\text{—OO}\bullet} \qquad (1.33)$$

which can further abstract hydrogen and form polymer peracids (**1.3**):

$$\text{—CH}_2\text{—}\overset{\text{O}}{\overset{\|}{\text{C}}}\text{—OO}\bullet + \text{PH} \longrightarrow \underset{(\mathbf{1.3})}{\text{—CH}_2\text{—}\overset{\text{O}}{\overset{\|}{\text{C}}}\text{—OOH}} \qquad (1.34)$$

The cleavage of polymer peracids yields polymer carboxy radicals (**1.4**) and hydroxy radicals (HO•):

$$\text{—CH}_2\text{—}\overset{\text{O}}{\overset{\|}{\text{C}}}\text{—OOH} \longrightarrow \underset{(\mathbf{1.4})}{\text{CH}_2\text{—}\overset{\text{O}}{\overset{\|}{\text{C}}}\text{—O}\bullet} + \bullet\text{OH} \qquad (1.35)$$

The polymer carboxy radical abstracts hydrogen and produces polymer with the end carboxylic group:

$$\text{—CH}_2\text{—}\overset{\text{O}}{\overset{\|}{\text{C}}}\text{—O}\bullet + \text{PH} \longrightarrow \text{—CH}_2\text{—}\overset{\text{O}}{\overset{\|}{\text{C}}}\text{—OH} + \text{P}\bullet \qquad (1.36)$$

(iii) Polymer peracid radicals ($\text{P—}\overset{\text{O}}{\overset{\|}{\text{C}}}\text{—OO}\bullet$) and freely mobile hydroperoxy radicals (HOO•) can also be formed from the cage recombination of two polymer peroxy radicals (POO•):[396]

$$\underset{\text{cage}}{[\text{POO}\bullet + \bullet\text{OOP}]} \longrightarrow \text{P—}\overset{\text{O}}{\overset{\|}{\text{C}}}\text{—OO}\bullet + \bullet\text{OOH} \qquad (1.37)$$

The freely mobile HOO• radical escapes relatively quickly from the cage, leaving the polymer peracid radical, which can abstract hydrogen from surrounding C—H sites:

$$\mathrm{P{-}\overset{\overset{\displaystyle O}{\|}}{C}{-}OO\cdot + PH \longrightarrow P{-}\overset{\overset{\displaystyle O}{\|}}{C}{-}OOH + P\cdot} \qquad (1.38)$$

As a result, an isolated polymer may react with oxygen to produce a polymer peroxy radical (POO•).

Each hydroperoxy radical (HOO•) leaving a radical pair can probably initiate the formation of a propagating polymer peroxy radical (POO•). In reality, this might not be the case, since freely mobile hydroperoxy radicals (HOO•) are expected to disappear quite quickly by recombination with other species before chain initiation can take place.

The number of initiated photo-oxidation chains can be reduced by the inhibition of peracid radicals at the time of their formation. Scavenging of propagation polymer peroxy radicals reduces the kinetic chain length of photo-oxidation.

1.10 FORMATION OF UNSATURATED GROUPS

During photo-oxidative degradation of polymers, e.g. polyethylene, formation of unsaturated groups (—CH=CH—) has been observed. The presence of these groups can be detected by UV absorption spectroscopy, e.g. tetracene (310 nm), trienyl (325 nm), pentacene (342 nm), tetraenyl (360 nm) and pentaenyl (394 nm).[440]

The α,β-unsaturated ketones (**1.5**) and aldehydes (**1.6**) which are formed

$$\mathrm{{-}CH_2{-}CH{=}CH{-}\overset{\overset{\displaystyle O}{\|}}{C}{-}} \qquad \mathrm{{-}CH_2{-}CH{=}CH{-}\overset{\overset{\displaystyle O}{\|}}{C}{-}H}$$

(1.5) **(1.6)**

may undergo further isomerization to give β,γ-unsaturated carbonyl groups which then photolyse by the Norrish Type I and II processes:[35,36]

$$\underset{(1685\ \mathrm{cm^{-1}})}{\mathrm{{-}CH_2{-}CH{=}CH{-}\overset{\overset{\displaystyle O}{\|}}{C}{-}}} \longrightarrow \underset{(1720,\ 1732\ \mathrm{cm^{-1}})}{\mathrm{{-}CH{=}CH{-}CH_2{-}\overset{\overset{\displaystyle O}{\|}}{C}{-}}} \qquad (1.39)$$

Unsaturated groups, e.g. in polypropylene, have an influence on the induction period during the photo-oxidation process.[56]

1.11 TERMINATION REACTIONS

The termination of the polymer radicals occurs by bimolecular recombination:

$$\mathrm{P\cdot + P\cdot \longrightarrow P{-}P} \tag{1.40}$$

$$\mathrm{P\cdot + PO\cdot \longrightarrow POP} \tag{1.41}$$

$$\mathrm{PO\cdot + PO\cdot \longrightarrow POOP} \tag{1.42}$$

$$\mathrm{PO\cdot + POO\cdot \longrightarrow POOOP\ (or\ POP + O_2)} \tag{1.43}$$

$$\mathrm{POO\cdot + POO\cdot \longrightarrow POO{-}OOP\ (or\ POOP + O_2)} \tag{1.44}$$

$$\mathrm{P\cdot + \cdot OH \longrightarrow POH} \tag{1.45}$$

$$\mathrm{PO\cdot + \cdot OH \longrightarrow POOH} \tag{1.46}$$

$$\mathrm{POO\cdot + \cdot OH \longrightarrow POOOH\ (or\ POH + O_2)} \tag{1.47}$$

$$\mathrm{HO\cdot + \cdot OH \longrightarrow H_2O_2\ (or\ H_2O + \tfrac{1}{2}O_2)} \tag{1.48}$$

When the oxygen pressure is high (atmospheric pressure), the termination reaction almost exclusively occurs by reaction (1.44). At low oxygen pressure other termination reactions take place to some extent.

Several factors influence recombination reactions, such as:[919]

(i) The cage effect.
(ii) Effect of steric control.
(iii) Mutual diffusion of reacting radicals. In solid polymers the recombination of polymer peroxy radicals (POO•) is controlled by the rate of their encounter with each other and is influenced by the intensity of molecular motion.[377,434]
(iv) Structural parameters of the polymer matrix.[433]
(v) Molecular-dynamical parameters of the polymer matrix.

If the polymer peroxy radicals are in the neighbouring positions, they can recombine to form stable cyclic peroxides (**1.7**) or epoxides (**1.8**):

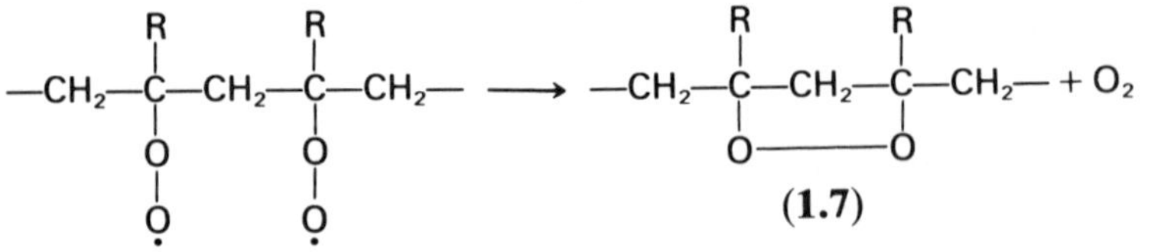

(1.49)

(**1.7**)

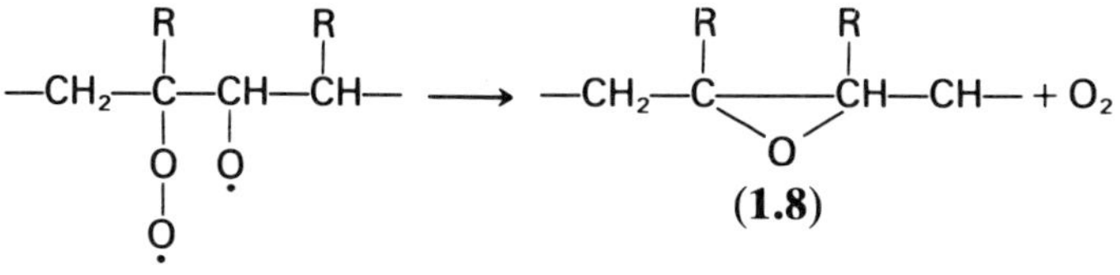

(1.8) (1.50)

Some of the termination reactions cause crosslinking, which creates a brittle polymer network.

1.12 PHENYL RING OPENING PHOTO-REACTIONS

Phenyl rings in some polymers such as polystyrene can be photo-oxidized with formation of hydroperoxy groups as a result of ring-opening reactions:[921,934]

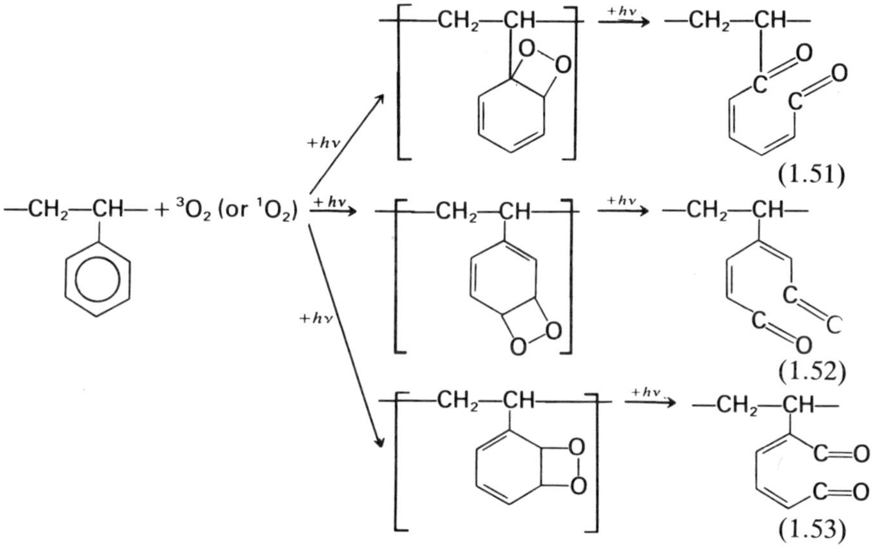

(1.51)

(1.52)

(1.53)

These proposed reactions were experimentally proved by the results obtained from IR-FTIR, NMR and ESCA spectroscopy.[726,727,888]

1.13 KINETICS OF PHOTO-OXIDATIVE DEGRADATION OF POLYMERS

Photo-oxidation kinetics of polymers are generally interpreted in terms of free-radical reactions (1.1)–(1.7). The determination of rate constants for these reactions has been the subject of extensive studies and considerable disagreement still exists as to their absolute values (particularly in solid polymers).[448,606,630,914, 940,1126,1184]

In the new approach to the photo-oxidation kinetics of polymers a number of new factors have been taken into careful consideration:

(i) A kinetic treatment developed for the liquid phase (solutions) is not completely valid for solid polymers, where radical mobility is very restricted. Pair recombination within cages of polymer peroxy radicals (POO•) is considered to be the main difference between liquid- and solid-state photo-oxidation processes:[235,236,346,437]

- —in liquid-state photo-oxidation of low molecular species, diffusion quickly randomizes radical populations;
- —in the solid-state process, the polymer peroxy radicals (POO•) will only separate by slow segmental diffusion.

(ii) Secondary cage recombination of peroxy radicals.[437] In a solid polymer, a pair of polymer peroxy radicals (POO•) is trapped in the polymer matrix. When a radical pair, produced by photo-initiation, escapes the initial cage, the probability of its recombination remains high even after several propagation steps. This phenomenon, known as *secondary cage recombination,* has a pronounced effect on the kinetics of oxidation (and on the mechanism of photostabilization by radical scavengers, cf. Section 3.21). It also has a pronounced effect on the distribution of kinetic chain lengths in the oxidation process.

(iii) Structural parameters of a polymer matrix. The structural microscopic heterogeneity of solid samples has important kinetic implications even if the polymer is completely amorphous. The micro-non-uniformities in reactant distribution through the polymer matrix may:

- —differ greatly from the mean value, and the rate constants determined experimentally may not agree with true values;
- —give rise to non-uniform kinetics, i.e. reaction will be faster in the zones of higher local concentrations.

(iv) Molecular-dynamical parameters of a polymer matrix.[377,940] Polymer materials have to withstand:

- —internal stresses, which are formed during the processing of a polymer and/or during ageing;

—external stresses, which originate during service.

The kinetics of photo-oxidation processes occurring in unstressed and stressed polymers will be different. In the stressed polymer all the structural and physical parameters of polymer matrix which affect chemical kinetics are changed.

A number of studies on polymer photo-oxidation have been based on free-radical oxidation of model compounds in solution.[376,378,387,914]

Investigations on low molecular models of the polymer are much more advantageous in order to obtain mechanistic information. However, there is a great risk in that special observations and findings relevant to the model might be overestimated. Due to the restricted mobility of the polymer chain, it is difficult to transfer kinetic results collected from measurements in solution to semicrystalline solid polymers.

1.14 ROLE OF POLYMER MORPHOLOGY IN PHOTO-OXIDATION

Polymer morphology (amorphous and semicrystalline regions) play an evident role in the oxidation and/or photo-oxidation reactions.

Oxidation of semicrystalline polymers such as polyolefins (polyethylene and polypropylene) is generally considered to occur within the amorphous region which can be treated as a boundary phase of the neighbouring crystalline regions. Tie molecules which connect crystallites through amorphous regions are scissioned in the oxidation process, resulting in a decrease of elongation and other physical properties. At later stages of oxidation when many chains in the amorphous phase and also at the crystalline boundary are destroyed, samples exhibit brittleness upon external stress.[564]

Because polyolefins are semicrystalline, quantitative studies of such heterogeneous polymers are very difficult. On the contrary, ethylene–propylene copolymers are completely amorphous, and can be selected as models for the investigation of the photo-oxidation of polyolefins in general.

A number of interesting conclusions have been drawn from the study of photo-oxidative degradation of ethylene–propylene copolymers at 310 nm in the presence of air:[453–456]

(i) The Norrish Type II reaction of ketones is the initial chain scission process.

(ii) The photolysis of hydroperoxides does not contribute to main chain scission through β-scission of alkoxy radicals.
(iii) As the local concentration of oxidation products (hydroperoxides and ketones) progressively increases, the production of carboxylic acid tends to become the dominant mechanism for chain scission. This proceeds through the photolysis of hydroperoxides hydrogen-bonded to ketones.
(iv) The photo-oxidation of ethylene–propylene copolymers is a very short chain reaction since the probabilities for propagation, reaction (1.54), and termination, reaction (1.55), are about the same:

$$\text{POO}\cdot + \text{PH} \longrightarrow \text{POOH} + \text{P}\cdot \qquad (1.54)$$

$$\text{POO}\cdot + \text{>CHOO}\cdot \longrightarrow \text{POH} + \text{>C=O} + \text{O}_2 \qquad (1.55)$$

Longer chains have greater probability for oxidative attack and chain rupture than shorter chains. The statistically greater scissioning of longer chains has a greater impact on changes in weight-average molecular weight ($\bar{M}_w$) than on the number-average molecular weight ($\bar{M}_n$). Chain scissioning affects not only molecular weight, but also mechanical properties, such as elongation.

1.15 ROLE OF POLYMER MANUFACTURING

The process by which a polymer is manufactured has a large influence over its subsequent behaviour during oxidation.[290] The fact that the formulations are generally more efficient in the Ziegler polymer is due to its lower unsaturation content. The higher unsaturation content of the Philips polymer causes a faster initial rate of oxidation, which consumes the antioxidant more rapidly than in the case of the Ziegler sample.

1.16 ROLE OF METALS OR METALLIC COMPOUNDS IN THERMAL AND/OR PHOTO-OXIDATIVE DEGRADATION OF POLYMERS

Metals or metallic compounds in polymers originate from:

(i) Catalyst used in the polymerization process. For example, commercial isotactic polypropylene is polymerized from a

Table 1.2
Metals and chlorine contents in polypropylene samples determined by X-ray fluorescence spectroscopy[60]

Sample	*Al (ppm)*	*Ti (ppm)*	*Fe (ppm)*	*Cl (ppm)*	*Antioxidant[a] (content 1% by wt)*
Diluent phase	140	25	1	35	0·1
Gas phase	230	50	6	240	0·22
Gas phase	200	65	1	220	0·5

[a] 2,6-Di-*tert.*-butyl-4-phenol.

heterogeneous organoaluminium–titanium complex (Ziegler–Natta process), or less frequently from metallic oxides of chromium, vanadium, or molybdenum bonded to an inert support (e.g. the Philips process). The transition metal ion contents vary in different commercial samples (Table 1.2).

(ii) Processing equipment and containers. The extrusion, milling, chopping, and compounding steps involved in polymer processing can all introduce traces or even particles of metals such as iron, nickel or chromium into the polymer.

(iii) Pigments (ZnO, TiO_2, Fe_2O_3), stabilizers (metal chelates), and vulcanization promotors.

(iv) Contact with copper wires in cable insulators.

(v) Corrosion of metal joints with polymer particles, machinery parts, containers, etc. (cf. Section 1.16.1).

(vi) Atmospheric pollution, water pollution, etc.

The role of these metal and/or metallic compounds in polymer photo-oxidative degradation has been reviewed,[864,865,914,919,934] and depends on:

(i) the nature of the polymer;

(ii) environmental conditions to which the polymer has been exposed;

(iii) the nature of the metal (transition or non-transition);

(iv) the valency of the metal;

(v) the anion or ligand of the metallic compound;

(vi) the spin state of the transition metal and the symmetry of its ligand field.[1104]

Residual titanium and aluminium catalysts increase photolysis of additives such as antioxidants and photostabilizers.[60,95]

1.16.1 Metallic Corrosion

It is an obvious fact that most of the polymers (plastics) used in practice are in direct contact with metals, which corrode in the environment surrounding them.

Metallic corrosion is the surface wastage that occurs when metals are exposed to reactive environments. At temperatures above 200°C there is usually significant reaction of most metals in dry air, and the rate and extent of reaction progressively increase, either as the temperature is raised or the air is contaminated by other gases.

At temperatures where water is liquid, the predominating corrosion process is electrochemical. Even in moist air, where there is no bulk water present, a very thin film of water may develop as a result of hydration of an initially chemically formed solid film of oxide, sulphide or carbonate. This thin film of water acts as the solvent and connecting electrolyte needed for electrochemical corrosion. Metal first dissolves as ions, and solid products such as rust are products of subsequent reaction.

At room temperature the progress of electrochemical corrosion depends on the nature of any oxidizing reactants present, pH, concentration and specimen geometry. Mechanical stress accelerates corrosion processes (corrosion fatigue).

Metallic corrosion is a result of oxidation and reduction reactions of metals. Oxidation of metals occurs with an increase in valence state:

$$Fe + \tfrac{1}{2}O_2 \longrightarrow FeO \text{ (increase from Fe(0) to Fe(II))} \qquad (1.56)$$

$$2\,FeO + \tfrac{1}{2}O_2 \longrightarrow Fe_2O_3 \text{ (increase from Fe(II) to Fe(III))} \qquad (1.57)$$

Reaction in the reverse direction, corresponding to a decrease in valency of the metal, is called reduction.

Oxygen is itself reduced in oxidation reactions:

$$\tfrac{1}{2}O_2 + 2e \text{ (from Fe)} \longrightarrow O^{2-} \text{ (in FeO lattice)} \qquad (1.58)$$

where e represents an electron. A general definition is that *reduction* consumes electrons and *oxidation* liberates electrons. The oxidation process of metals can be represented as:

$$Fe \longrightarrow Fe^{2+} \text{ (in FeO lattice)} + 2e \text{ (remaining in Fe)} \qquad (1.59)$$

whereas simultaneous ionic oxidation and reduction is represented as:

$$Fe \longrightarrow Fe^{2+} + 2e \qquad (1.60)$$

$$\tfrac{1}{2}O_2 + 2e \longrightarrow O^{2-} \qquad (1.61)$$

$$Fe^{2+} \longrightarrow Fe^{3+} + e \qquad (1.62)$$

$$\tfrac{1}{2}O_2 + 2e \longrightarrow O_2^- \qquad (1.63)$$

These reactions can be considered as *electron-transfer processes,* in which various atoms or molecules ionize or ions become discharged. Electron transfer processes play a very important role in the photoactivity of pigments such as titanium oxide (TiO_2) (cf. Section 4.2) and zinc oxide (ZnO) (cf. Section 4.3) and have an important impact on the photostability of polymers.

1.16.2 Reactions of Transition Metals

Transition metals participate in a number of important reactions, such as:

(i) Formation of complexes with hydroperoxy groups, which are further decomposed into free radicals and ions:[59,184,250,865]

$$Fe^{3+} + ROOH \rightleftharpoons \underset{\text{complex I}}{(Fe^{3+} \cdots ROOH)} \qquad (1.64)$$

$$(Fe^{3+} \cdots ROOH) \longrightarrow Fe^{2+} + ROO\cdot + H^+ \qquad (1.65)$$

$$Fe^{2+} + ROOH \rightleftharpoons \underset{\text{complex II}}{(Fe^{2+} \cdots ROOH)} \qquad (1.66)$$

$$(Fe^{2+} \cdots ROOH) \longrightarrow Fe^{3+} + RO\cdot + OH^- \qquad (1.67)$$

Transition metals accelerate photodecomposition of hydrogen peroxide.[736]

(ii) Activation of oxygen. Transition metals may form a charge-transfer complex with oxygen, which can further react with a polymer:

$$M^{n+} + O_2 \longrightarrow \underset{\text{CT complex}}{(M^{n+} \cdots O_2)} \qquad (1.68)$$

$$(M^{n+} \cdots O_2) + PH \longrightarrow M^{n+} + P\cdot + HO_2\cdot \qquad (1.69)$$

(iii) Formation of oxygen radical ion:[624,865]

$$M^{n+} + O_2 \longrightarrow M^{(n+1)+} + O_2^{\overline{\cdot}} \qquad (1.70)$$

$$O_2^{\overline{\cdot}} + H^+ \longrightarrow HO_2\cdot \qquad (1.71)$$

Transition metal impurities, both from catalyst residues and from contamination picked up during processing, have been shown to be present in polyolefin samples.[291,309,387,947] The accelerating effect of the presence of such species on the auto-oxidation of liquid hydrocarbons and polymers has been demonstrated.[57,669,1118] In polypropylene metal impurities play a more significant role than in polyethylene, probably due to the higher concentration of these moieties in this polymer.[59]

1.17 ROLE OF PROCESSING IN THERMAL OXIDATION OF POLYMERS

Most polymers have to be processed into plastic materials used for the final production of commercially available articles.[1192] Plastic materials normally include polymer and a few per cent of one or more different additives such as antioxidants, photostabilizers, pigments, dyes, lubricants, fillers, etc. Prior to practical application, polymers are usually granulated and further mixed with other polymers (polymer blends), additives, etc. Such mixtures are further melted in an extruder in the processing step and converted to the final product.

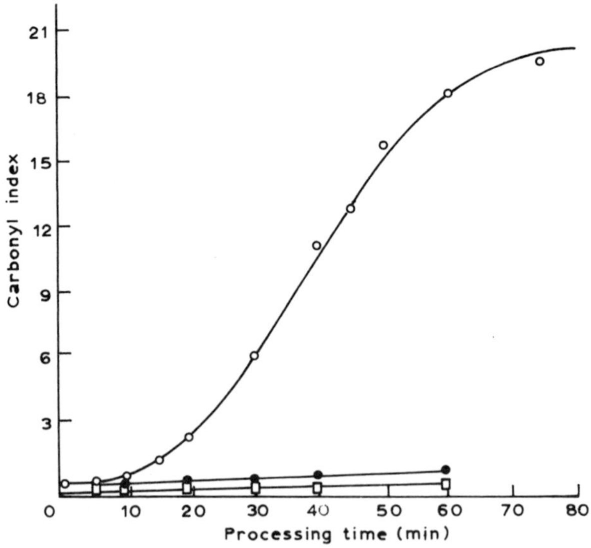

Fig. 1.4. The effect of thermal processing at 150°C on the carbonyl index of low-density polyethylene (LDPE): ○, in the presence of air (open chamber); ●, in the presence of a limited amount of air (closed chamber); □, in argon. Reprinted with permission from Ref. 261, © 1977 Pergamon Press PLC.

The temperatures employed in the processing are generally between 150 and 250°C depending on polymer type, technology and the shape of the finished article. Sometimes very high temperatures, up to 300°C, are required for certain processing technologies like extrusion coating or melt spinning of fibres, or for injection moulding of parts with a very complicated geometry and shape. At these high processing temperatures (150–350°C) polymers are partially thermally oxidized. In order to protect polymers against thermal oxidation, it is necessary to add thermostabilizers and/or antioxidants.

The main processes observed during the thermal oxidation of polymers are formation of carbonyl groups (Fig. 1.4) and hydroperoxy groups (Fig. 1.5).[134,261,1007,1008] The further thermal treatment of the processed polymer samples increases thermal oxidation processes (Figs. 1.6 and 1.7).

Figures 1.5 and 1.7 show that the maximum rate of initial carbonyl formation in polyethylene (LDPE) observed in Fig. 1.6 is associated

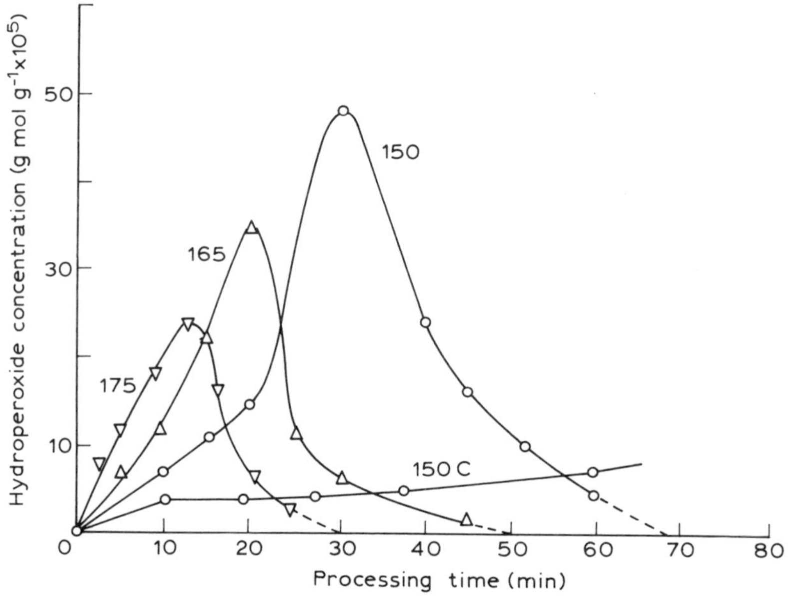

Fig. 1.5. The effect of processing temperature and time on peroxides concentration in low-density polyethylene (LDPE). The numbers on the curves are processing temperatures (in °C); C indicates a closed mixer. Reprinted with permission from Ref. 261, © 1977 Pergamon Press PLC.

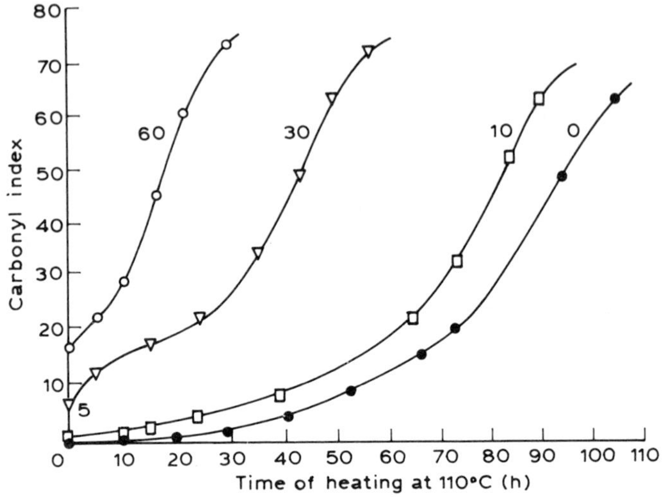

Fig. 1.6. The effect of processing at 150°C in air on the thermal oxidative stability of low-density polyethylene (LDPE) films. The numbers on the curves are processing times in minutes. Reprinted with permission from Ref. 261, © 1977 Pergamon Press PLC.

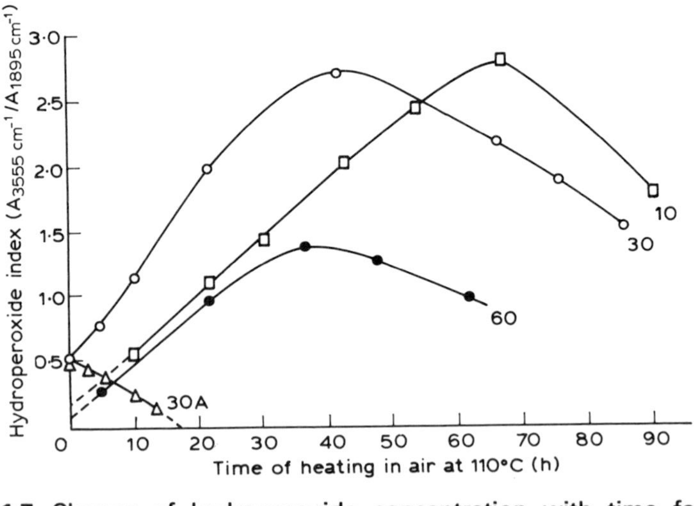

Fig. 1.7. Change of hydroperoxide concentration with time for low-density polyethylene (LDPE) films in air at 110°C. The numbers on the curves are processing times at 150°C in air; A indicates heated in argon. Reprinted with permission from Ref. 261, © 1977 Pergamon Press PLC.

with a higher initial hydroperoxide concentration (Fig. 1.5) and a higher rate of hydroperoxide formation during subsequent thermal oxidation (Fig. 1.7). Figures 1.5 and 1.7 also show that the hydroperoxide concentration rises to a maximum and then decays with heating time both in the melt and the solid phase and that the maximum concentration achieved increases with decreasing temperature. In the absence of oxygen, the hydroperoxide concentration decayed to zero in less than 20 h at 110°C.[261] The half-life of polyethylene hydroperoxide is 6·4 h at 100°C.[564]

Polymers may undergo quite different physical changes during processing in a mixer, depending on their chemical structure and on the temperature.[256,478,979,1008,1010,1017] For example, polyethylene undergoes crosslinking reactions, leading to a decrease in its Melt Flow Index,[261,470] whereas polypropylene under similar conditions undergoes chain scission, leading to an increase in the Melt Flow Index.[470,545] In both cases, the initial reaction occurring is chain scission due to shearing forces acting on the polymer when its viscosity is high.

Polymer processing conditions are sometimes called the *processing history* of a polymer and this is very important information for the choice of suitable antioxidants.

1.18 MECHANISMS INVOLVED IN THE PHOTOLYSIS OF THE THERMALLY OXIDIZED POLYOLEFINS DURING PROCESSING

All commercial polyolefins always contain a small amount of olefinic unsaturation of which vinyl is the most important in high-density polyethylene (HDPE) and vinylidene in low-density polyethylene (LDPE).

Polyethylene:

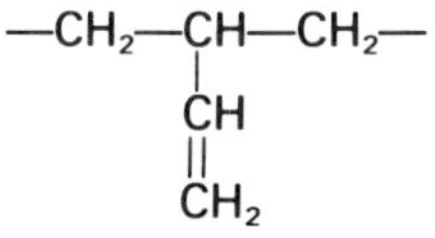

$CH_2{=}CH{-}CH_2{-}CH_2{-}$

Vinyl groups (mainly in HDPE)

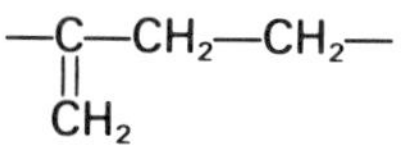

Vinylidene groups (mainly in LDPE)

Polypropylene: 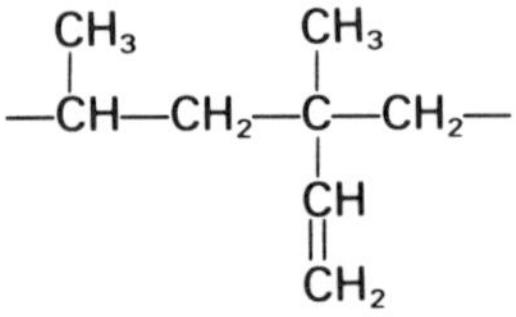 or 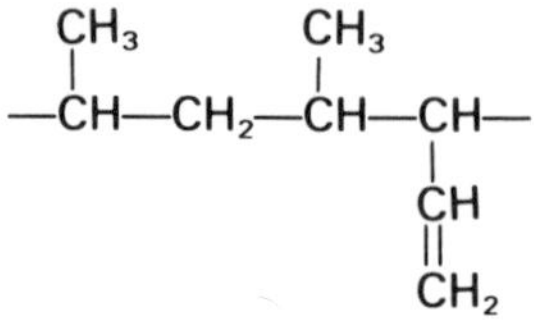

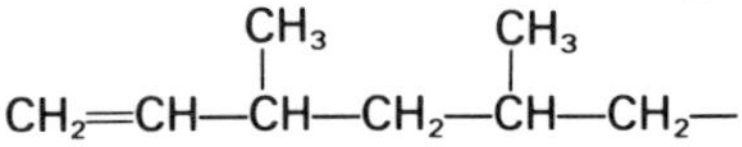

Vinyl groups

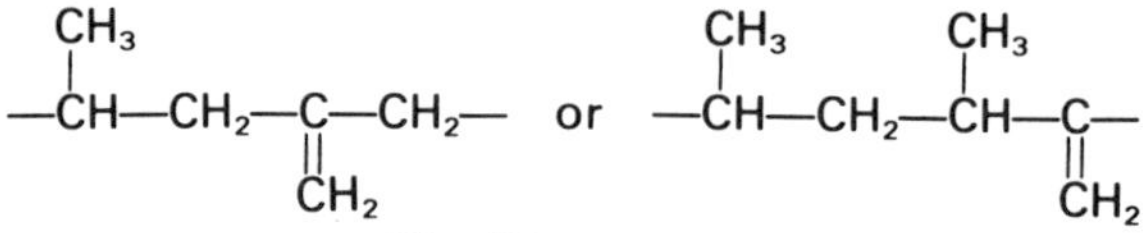

Vinylidene groups

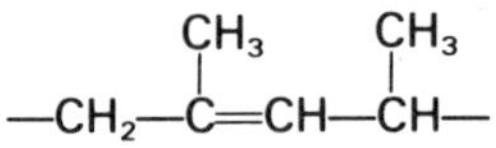

Unsaturated groups in the backbone

These groups, like unsaturation in polydiene polymers, become involved in oxidative reactions during a typical processing operation:[1008]

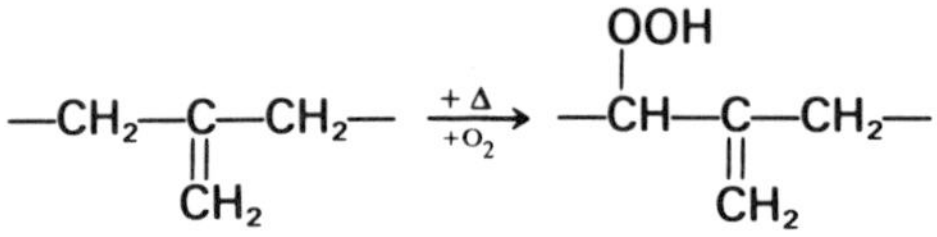

(1.9) (1.10)

(1.72)

Allylic hydroperoxides (**1.10**) formed from thermal oxidation of vinylidene groups (**1.9**) are the most important photoinitiating groups initially present in thermally treated polyolefins. The thermal decomposition and/or photodecomposition of hydroperoxy groups leads to formation of ketonic carbonyl groups (Fig. 1.8) by:[291,1008]

(i) A chain scission reaction, during which end aldehyde groups (**1.11**) are formed:

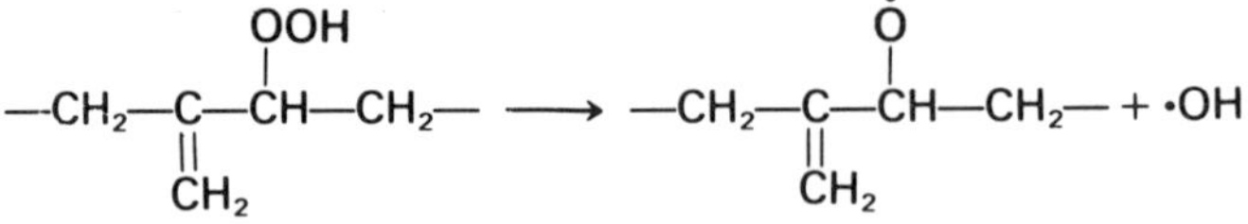

(1.73)

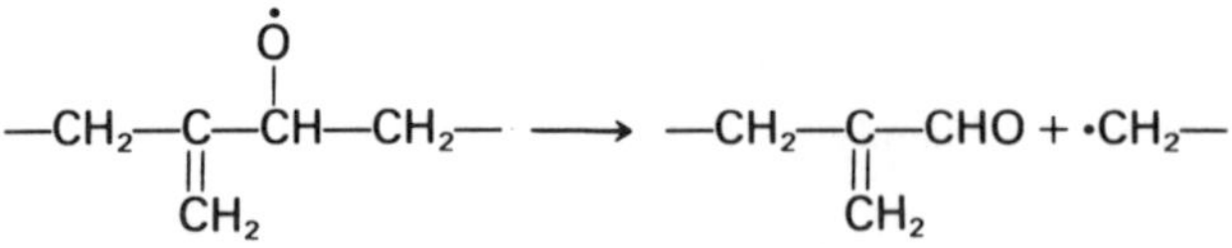

(1.74)

(1.11)

(ii) Carbonyl group formation:

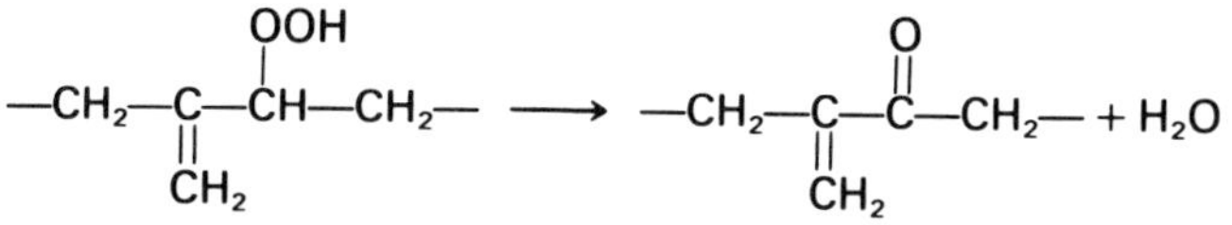

(1.75)

The hydroperoxide photolysis and chain scission are the most important initiating processes occurring in the photo-oxidation of polyolefins.

(iii) A crosslinking reaction, which occurs only during UV irradiation, but not in thermal oxidative degradation, and leads to a very rapid decrease of vinylidene groups (Fig. 1.8).

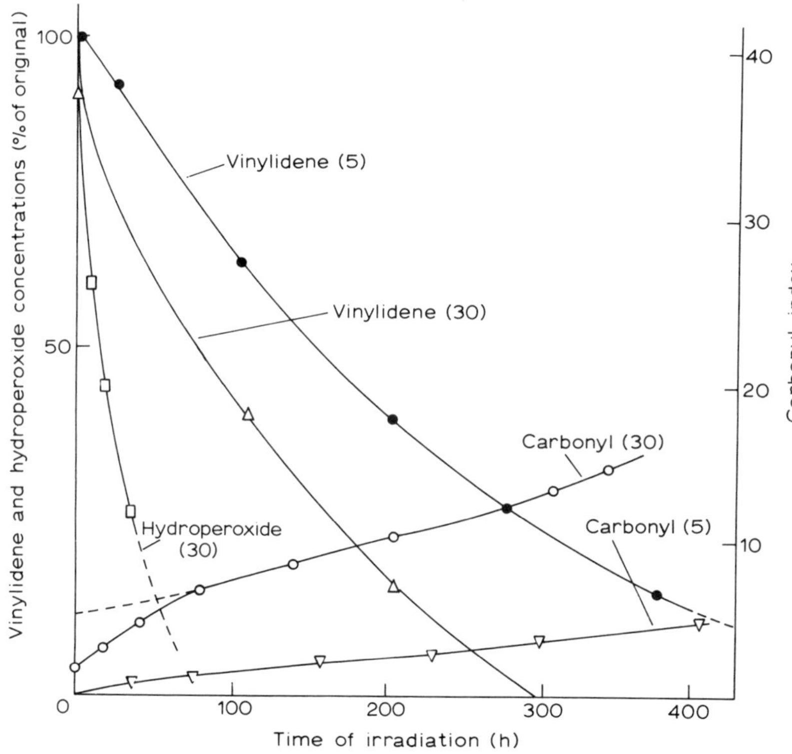

Fig. 1.8. Kinetics of formation (carbonyl) and disappearance (hydroperoxide and vinylidene) of groups in low-density polyethylene (LDPE) during UV irradiation. The numbers on the curves are processing times in minutes.[1008]

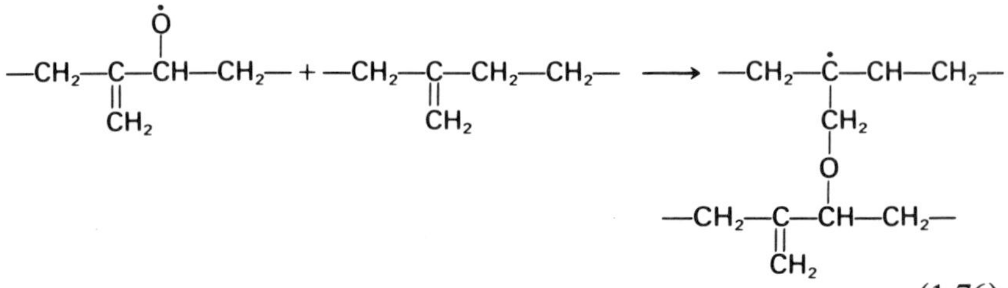

(1.76)

This process is associated with the formation of insoluble gel and an increase of polymer density (Fig. 1.9).

Chain scission and crosslinking reactions are closely associated with changes in mechanical properties (elongation at break) during the early stages of photo-oxidation (Fig. 1.9).

Increase of the processing time will increase the rates of photo-oxidation of a polymer up to the time of maximum hydroperoxide concentration.[134] In Fig. 1.10 are shown kinetic photo-oxidation curves (carbonyl index) for unprocessed polyethylene (LDPE) (0 h) and repeatedly processed (30, 60 h) samples. The photolysis of the hydro-peroxide groups is the major reaction leading to the rapid increase in

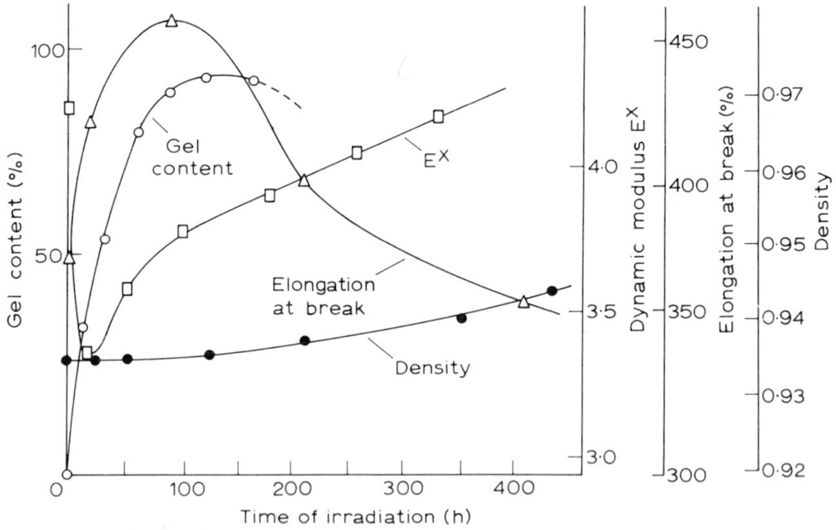

Fig. 1.9. Effect of UV irradiation on the mechanical and physical properties of low-density polyethylene (LDPE).[1008]

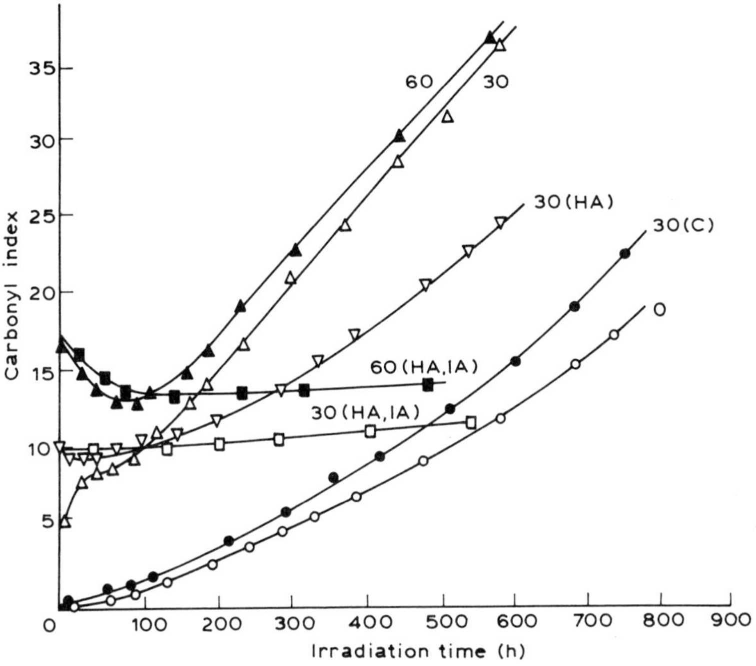

Fig. 1.10. Change in carbonyl index of low-density polyethylene (LDPE) films during UV irradiation. The numbers on the curves indicate processing times at 150°C in air: HA, heated in argon; IA, irradiated in argon; C, closed mixer. Reprinted with permission from Ref. 261, © 1977 Pergamon Press PLC.

carbonyl concentration. The modes and rates of carbonyl group formation under thermal oxidative conditions (Fig. 1.6) and under photo-oxidative conditions (Fig. 1.10) are similar. These results show that hydroperoxides are the most important initiators in normally processed polymers during the early stages of photo-oxidation. Carbonyl initiation is important in polymers which have been subjected to enhanced oxidation and during the later stages of normally processed polymers.

Processing time has an effect on the time to embrittlement. The time to embrittlement of polyolefins during photo-oxidation is not directly related to the concentration of carbonyl groups formed (Fig. 1.11).[1007]

Oxidation during processing has a remarkable effect on the photostability of polyolefins.[584,693,779]

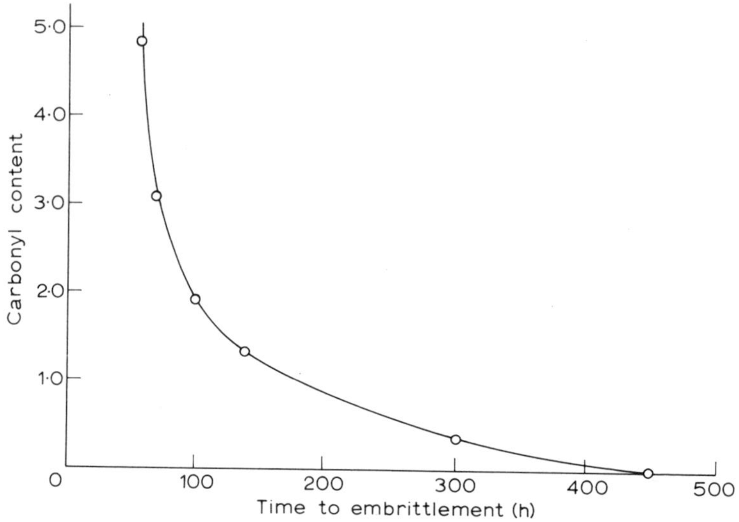

Fig. 1.11. Relationship between carbonyl content and embrittlement time of high-density polyethylene (HDPE). Reprinted with permission from Ref. 1007, © 1976 American Chemical Society.

1.19 PHYSICAL PROCESSES IN THE PHOTO-OXIDATION OF POLYMERS

1.19.1 Role of Diffusion in a Polymer Matrix

The diffusion of small molecules (vapours and liquids) occurs by movement of these small molecules from one 'hole' to another.

The term *holes* indicates, from the thermodynamic point of view, the probability of creating either sorption sites for the penetrant molecules or diffusion channels in the polymer through which the penetrant can move from one site to another.[315,324,639]

Diffusion requires a concentration gradient of the penetrant in the polymer, and flux occurs from regions of high concentration to regions of low concentration.

The *diffusion flux* (J) of a species is proportional to the concentration gradient measured normal to the unit area of cross-section through which a species diffuses (*Fick's First Law*) (also called *Case I Diffusion*) and is given by:[316,323,959]

$$J = -D\frac{\partial c}{\partial x} \tag{1.77}$$

where D = diffusion coefficient,
c = concentration of penetrant molecules (diffusant),
x = the space coordinate measured normal to the section,
$\frac{\partial c}{\partial x}$ = concentration gradient across a thickness x.

The *diffusion coefficient* (D) is a function of the concentration of diffusant (penetration molecule) (*Fick's Second Law*) and for Cartesian coordinates is given by:

$$\frac{\partial c}{\partial x} = D\left[\frac{\partial^2 c}{\partial x^2} + \frac{\partial^2 c}{\partial y^2} + \frac{\partial^2 c}{\partial z^2}\right] \tag{1.78}$$

Polymers whose diffusion characteristics can be described by Fick's First and Second Laws are said to follow *Fickian diffusion behaviour.*

Fickian diffusion behaviour is evidenced by a linear relationship between the sorption (or desorption) of penetrant and the square root of the time of diffusion ($\sqrt{t}$) (Fig. 1.12).

Non-Fickian diffusion behaviour (also called *Case II diffusion*) is an anomalous diffusion which cannot be described in terms of Fickian

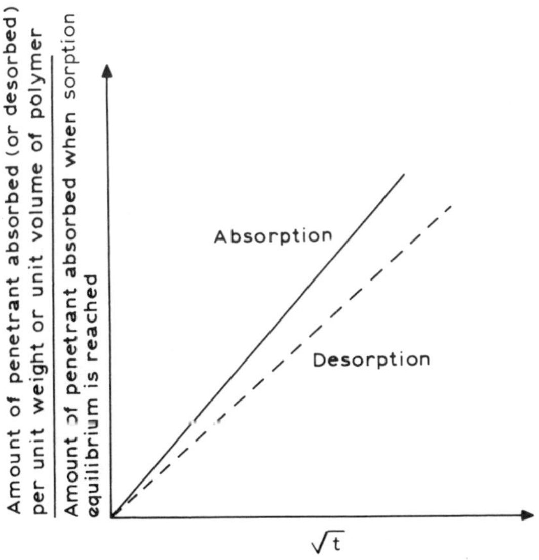

Fig. 1.12. Absorption and desorption curves of a given penetrant by a given polymer (Fickian diffusion behaviour).

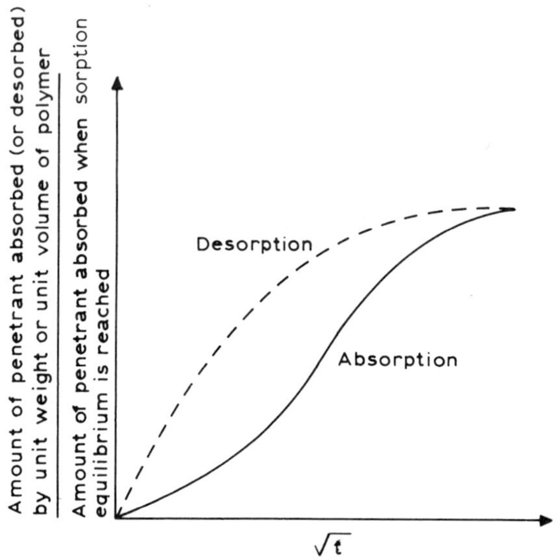

Fig. 1.13. Absorption and desorption curves of a given penetrant by a given polymer (non-Fickian diffusion behaviour).

laws.[1159] It can exist in polymers:

(i) below the glass transition temperature (T_g) or
(ii) above the glass transition temperature in crosslinked and crystalline polymers.

Anomalous diffusion behaviour is evidenced by significant deviations from the linear relationship between absorption of penetrant and the square root of the time of diffusion ($\sqrt{t}$) (Fig. 1.13). The rates of absorption and desorption are not equal in non-Fickian diffusion.

Both types of diffusion, i.e. Fickian and non-Fickian (Case II), can be superimposed. Some penetrants will show Fickian behaviour in a given polymer while others will show non-Fickian (Case II) behaviour. By increasing the crosslink density of a given polymer, the diffusion tends to move from Fickian type to non-Fickian (Case II) type.

Oxygen diffuses only into amorphous regions of a polymer making them more susceptible towards photo-oxidative degradation. Consequently, stabilization processes also occur mainly in the amorphous phase.[182,226,420,939,973,974]

1.19.2 Mathematical Models of Diffusion-Controlled Oxidation

A number of mathematical models to describe diffusion-controlled oxidation of polymers have been constructed. They generally assume the following facts:

(i) The rate of oxygen absorption is diffusion-controlled from the film surface in thin films; however, the rate is linearly proportional to the film thickness.[183,195]

(ii) The diffusion effects inherent in a polymer film are generally expressed by a differential equation according to Fick's Laws of diffusion. The solution to this equation is determined by applying suitable initial and final boundary conditions.[183,195,323,389,612,1158]

(iii) Oxygen diffusing into a solid polymer film through both surfaces is being consumed by a first-order or pseudo-first-order chemical reaction.[346,389,848]

In a thin polymer film diffusion of oxygen is not a limiting factor on oxidation reactions, whereas in a thick film diffusion-controlled oxidation is symmetrical about the midpoint (Fig. 1.14).

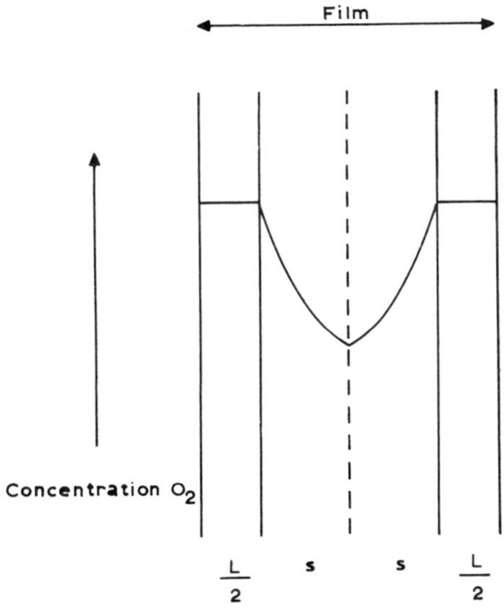

Fig. 1.14. Schematic representation of oxygen concentration in a degrading polymer film. *L* is the thickness in which diffusion of oxygen is not a limiting factor in the oxidation reaction, and *s* is the diffusion-controlled zone.[389]

For any point in the thick polymer film equation (1.79) is valid:

$$\frac{\partial C(x, t)}{\partial t} = D\frac{\partial^2 C(x, t)}{\partial x^2} - kC(x, t) \tag{1.79}$$

where $C(x, t)$ = concentration of oxygen at a given depth x and a given time t;
x = direction defined as perpendicular to the facets of the film;
t = time of reaction;
k = first-order rate constant;
D = oxygen diffusion constant in a polymer film.

Solution of eqn (1.79), assuming the following boundary conditions:[389]

$$C = C_0 \text{ at } x = 0 \text{ and } t \geqslant 0 \tag{1.80}$$

$$C = C_0 \text{ at all } x \text{ and } t = 0 \tag{1.81}$$

$$C = 0 \text{ at } x = \infty \text{ and } t > 0 \tag{1.82}$$

after final simplification gives the equation

$$C(x, t) = C_0 \exp(-x\sqrt{k/D}) \tag{1.83}$$

Figure 1.15 shows the variation of oxygen concentration with depth into the film obtained from the approximate solution for $C(x, t)$ in eqn (1.83).

This mathematical model is further developed in the original paper.[389]

1.19.3 Effect of Diffusion on the Bimolecular Reactions of Free Radicals Formed in a Polymer Matrix

Diffusion in a polymer matrix affects the rates of all the bimolecular reactions to some degree. The following types of such bimolecular reactions may occur.[346]

(i) Between small molecules and/or radicals, e.g. addition of oxygen to small radicals such as methyl ($CH_3\cdot$) or acyl ($CH_3CO\cdot$) radicals.

(ii) Between small molecules and polymer radicals, e.g. addition of oxygen to polymer alkyl radicals ($P\cdot$), or addition of water molecules to polymer peroxy radicals ($POO\cdot$) with formation of

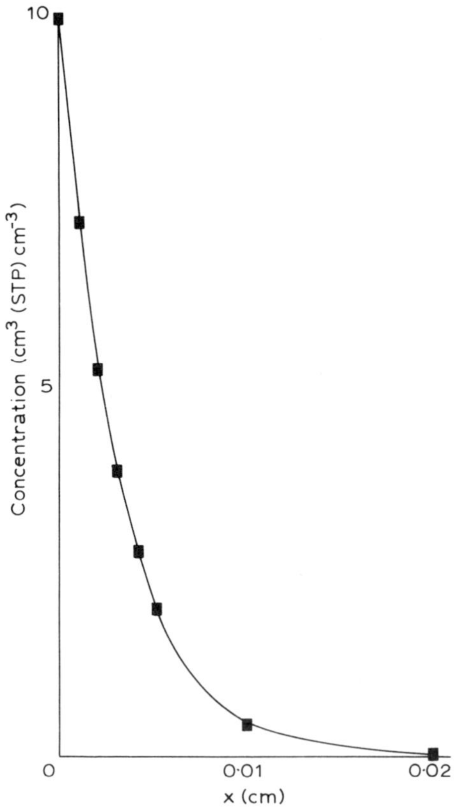

Fig. 1.15. Approximate solution for $C(x, t)$ in eqn (1.83); variation of oxygen with depth into film.[389]

solvated radicals (complex):

$$POO\cdot + nH_2O \longrightarrow (POO\cdot \cdots nH_2O) \quad (1.84)$$

Such solvated radicals have a lower activity than a precursor.[745]

(iii) Between small radicals and polymer radicals, e.g. recombination of polymer alkyl (P•), polymer oxy (PO•) and/or polymer peroxy (POO•) radicals with hydroxyl radicals (HO•). The polymer matrix reduces the bimolecular collision reaction rates by a factor of about 10^{-2} relative to fluid solution rates, due to the reduced mobility of the chains.

(iv) Between small radicals and macromolecules which involve hydrogen abstraction from the polymer backbone (and/or side groups). In this case all small radicals will have C—H bonds in the immediate vicinity, there is no dependence on diffusion and rates would be similar to fluid solution rates.
(v) Between all available polymer radicals. In a polymer matrix the mobility of polymer radicals is very restricted. The bimolecular rates are reduced by a factor of about 10^{-4} relative to fluid solution rates.
(vi) Between chromophoric groups, e.g. carbonyl (donor) and hydroperoxide (acceptor) groups which participate in the energy transfer process.

In a low-viscosity medium, reactants and products mix to a uniform concentration by rapid diffusion. However, when reactions occur in a highly viscous medium this randomization process becomes inefficient and highly localized concentration develop.

1.19.4 Effect of Oxygen Permeability

The *permeation* of a gas through a polymer film can be described in terms of the gas or vapour dissolving at one surface of the film, diffusing through the film under a concentration gradient, and evaporating from the other surface at the low-concentration, i.e. low-pressure, side.[315]

If the constant pressure difference is maintained across the film, the gas will diffuse through the film at a constant rate. Under steady-state conditions the permeability process is described by Fick's First Law:[959]

$$J = -D\frac{\partial c}{\partial x} \tag{1.77}$$

where J = flux (i.e. rate of gas transportation) or the amount of gas diffusing through unit area of the film in unit time:

$$J = \frac{\text{Quantity of gas}}{\text{Film area} \times \text{Time}} \tag{1.85}$$

D = diffusion coefficient,

$\frac{\partial c}{\partial x}$ = concentration gradient across a thickness x.

After integration of eqn (1.77) across the total thickness of the film (l), the amount of gas diffusing through the polymer film (J) is given by:

$$J = \frac{D(c_1 - c_2)}{l} \tag{1.86}$$

where c_1 and c_2 are the concentration of gas in the two surfaces.

Gas concentrations are normally measured in terms of the pressure (p) of the gas which is at equilibrium with the film. According to *Henry's Law*:

$$c = Sp \tag{1.87}$$

where c = gas concentration,
S = *solubility coefficient* of the gas in the polymer film and expressed as:

$$S = \frac{\text{Quantity of gas}}{\text{Mass of polymer} \times \text{Differential pressure of gas}} \tag{1.88}$$

p = gas pressure.

Hence

$$J = \frac{DS(p_1 - p_2)}{l} \tag{1.89}$$

The *permeability constant* (P) is defined as the product

$$P = DS \tag{1.90}$$

It gives a measure of the quantity of gas which can flow through a polymer film:

$$P = \frac{\text{Quantity of gas} \times \text{Thickness of film}}{\text{Film area} \times \text{Time} \times \text{Differential pressure of gas}} \tag{1.91}$$

Permeability of oxygen will be low in polyolefins, because of its low solubility in the polymer matrix; hence the progress of oxidation will be slower inside a thick sample, since only a small flow of oxygen can reach this part of the sample and maintain reaction.[643,788,940]

Solubility (S) of oxygen in semicrystalline polyolefins depends on the content of amorphous phase and can be expressed by the following equation:[788–790]

$$S = \alpha_a S_a \tag{1.92}$$

where α_a = fraction of the amorphous phase,
S_a = oxygen solubility in the amorphous phase.

Table 1.3
Oxygen solubility in different polymers (at 25°C)[801]

Polymer	*Solubility (ppm)*	*References*
Amorphous polyethylene	110	788, 790
Natural rubber	160	1120
Polybutadiene	140	1120
Butyl rubber	175	1120
Poly(vinyl acetate)	80	778

Oxygen solubility and oxygen diffusion coefficients are given in Tables 1.3 and 1.4, respectively.

The *activation energy for permeation* (E_P) is given by the following equation:

$$E_P = E_D + \Delta H_S \tag{1.93}$$

where E_D = activation energy for diffusion,
ΔH_S = heat of solution.

Table 1.5 gives some values of activation energies for permeation and diffusion of oxygen in different polymers. The following conclusions can be drawn from this table:

(i) Activation energies for permeation (E_P) and diffusion (E_D) for

Table 1.4
Oxygen diffusion coefficients in different polymers[801]

Polymer	*D*(10^{-7} *cm*^{2} *s*$^{-1}$)	*References*
Polyethylene, density 0·964	1·70	789
Polyethylene, density 0·914	4·60	789
Polyethylene, density 0·915	5·4	790
Polyethylene, density 0·931	2·5	790
Polyethylene, density 0·967	1·6	790
Poly(vinyl acetate)	0·4	778
Poly(methyl methacrylate)	3·8 (20°C)	746
Natural rubber	1·58	1120
Polybutadiene	1·5	1120
Butyl rubber	0·081	1120
Polystyrene	1·1	324, 639
Poly(ethylene terephthalate)	0·036	324, 639

Table 1.5
Activation energies of diffusion and permeation, and heat of solution, for oxygen in different polymers (between 25 and 50°C)[801]

Polymer	E_P (kJ mol^{-1})	E_D (kJ mol^{-1})	H_S (kJ mol^{-1})	*References*
Polyethylene, density 0·964	35·1	36·8	−1·7	789
Polyethylene, density 0·954	36·8	—	—	828
Polyethylene, density 0·922	41·4	—	—	828
Polyethylene, density 0·914	42·6	40·1	2·5	789
Polyethylene, density 0·907	47·6	—	—	828
Polypropylene	46	—	—	1049
Poly(vinyl acetate)	56·0	60·6	−4·6	778
Natural rubber	31·4	34·7	−3·3	1121
Butyl rubber	44·7	49·7	−5·0	1121

oxygen in the polymer matrices depend on

(a) the type of polymer,
(b) the polymer density (cf. polyethylene).

(ii) The heat of diffusion is very small and often negative. In the last case it means that the solubility will decrease with increasing temperature.
(iii) Permeation of surface layers should control oxidation of a thick sample. However, the limiting thickness for oxidation will be smaller as the temperature increases, due to faster oxidation kinetics.

Oxygen permeability in a polymer matrix evidently decides the kinetics of oxidation or photo-oxidation. The typical oxidation profile depends on:[179,183,327,426, 585,593]

(i) Sample thickness.
(ii) Oxygen permeability, i.e. diffusion and solubility.
(iii) Oxygen pressure.
(iv) Mode of sample irradiation (one- or two-sided exposure).
(v) Percentage of crystalline phase (in semicrystalline polymers).
(vi) Tensile stresses. In glassy amorphous polymers and semicrystalline polymers, the effect of tensile stresses on permeability is complex:

(a) Within the elastic region, the sorption of low-molecular compounds increases with stress as a consequence of an increase in the free volume fraction.

(b) In the plastic region the sorption of low-molecular compounds markedly decreases with an increase of stress, because the density of the amorphous component rises in this case.

(vii) The presence of cracks, pinholes and voids leads to changes of the oxidation profile and also to rather catastrophic loss of barrier properties against permeability of oxygen.

Oxidation profiles are symmetrical with the respect to the centre of the sample stack; however, photoirradiation from one side of a film makes them unsymmetrical. At the very beginning of the reaction the concentration of oxidation products (in a low-density polyethylene) is higher in the bulk of the film than on the upper surface layer. With increase of irradiation time the situation is reversed, with an increase of oxidation product concentration on the surface.[466]

1.19.5 Role of Crystallinity, Orientation and Crosslinking in Diffusion, Solubility and Permeation in Polymers

The presence of crystalline domains in a polymer matrix has two effects on sorption and diffusion behaviour:[959]

(i) At temperatures well below the melting point (T_m), crystalline regions are generally inaccessible to oxygen and to most penetrants. Hence, they act as excluded volumes for the sorption process and as impermeable barriers for the diffusion process.

(ii) Crystalline domains require penetrant migration around them which increases the average pathlength relative to the nominal dimensions of the sample (*tortuosity effect*).

Diffusion coefficients decrease approximately linearly with crosslink density or the reciprocal of the molecular weight between crosslinks at low to moderate degrees of crosslinking. At higher densities the decrease is nonlinear.

The solubility is relatively unaffected except at very high degrees of crosslinking or when the penetrant significantly swells the polymer.

The restraints of crosslinking on the segmental mobility of the polymer make the diffusion process more dependent on the size and shape of penetrant molecules. The decrease in inherent segmental mobility makes the diffusion process more concentration-dependent.

The orientation of the polymer matrix introduces an asymmetrical structure such that diffusion rates vary relative to the axis of orientation. Generally, the rate of penetration parallel to the axis is less than the rate in a randomly oriented sample, which is less than the rate perpendicular to the axis.

Deformation of elastomers does not appreciably affect diffusion and solubility coefficients until elongation results in crystallization. The effects of orientation of semicrystalline polymers depend on the initial degree of crystallinity.

2

Introduction to the Oxidative and Photo- Stabilization of Polymers

2.1 ANTIOXIDANTS

Antioxidants are chemical compounds which protect polymers and plastics against thermal and/or photo- oxidative processes which occur in thermal or photo- degradation or during natural ageing. Antioxidants which decrease photo-oxidation processes are also called *photo-antioxidants*.

Antioxidants are generally classified into two groups, according to their protection mechanism:

(i) Kinetic chain-breaking antioxidants (chain terminators, chain scavengers). They have capability to scavenge some or even all available low-molecular radicals (R•, RO• ROO•, HO•, etc.) and polymeric radicals (P•, PO•, POO•) by a process called a chain-breaking electron donor mechanism.

(ii) Peroxide decomposers, which decompose hydroperoxy groups (HOO—) present in a polymer.

Antioxidants play an important role in life processes in biological systems and also in food industries (protection of food against oxidation).[1018]

Antioxidants and their stabilization mechanisms have been the subject of a number of books,[905,919,1000] and review articles.[101,906,907,911,991,1001–1003,1012,1016,1018,1021,1022,1079] The present list of commercially available antioxidants is given in Appendix 1.[1196]

The effectiveness of antioxidants in polymers is determined by three independent parameters:[1003,1033]

(i) Intrinsic antioxidant activity. This is a measure of the efficiency of the antioxidant, measured as oxygen absorption (uptake) in a model substrate in which it is completely soluble and from which it cannot be physically lost.

In general, members of a homologous series of antioxidants which contain the same functional groups (e.g. hindered phenol) have a very similar activity on a molar basis.[1033]

(ii) Antioxidant solubility in the polymer. Additives are in general less soluble in polymers than in lower-molecular-weight analogues. At concentrations greater than their equilibrium concentration they diffuse out of the polymer, when they may be lost by volatilization or they may deposit (bloom) on the surface.[179,270,271]

(iii) Antioxidant substantivity in the polymer. *Substantivity* is defined as the ability of the antioxidant to remain in the polymer, particularly under high-temperature conditions in a frequently changing air stream or with solvent leaching of the antioxidant from the surface layers.[1013]

Substantivity has to be related to the following parameters:[179]

(a) The rate of diffusion of the antioxidant, since this determines the rate at which the surface is replenished.
(b) The rate of loss of antioxidant from the surface. In an air stream, this depends on the partial pressure of the antioxidant. In a solvent it depends on relative solubility in the solvent and in the polymer and the rate of replacement of the solvent in contact with the surface.
(c) The ability of the solvent to swell the polymer. Swelling effectively increases the diffusion coefficient of the antioxidant and hence its rate of removal.

Increasing the molecular weight of antioxidants generally decreases the molecular mobility as well as the volatility, and which factor dominates depends on the thickness of the sample.[179,180,182] In practice, increasing the molecular weight of antioxidants has a remarkable effect on improving the stabilizing activity.

All antioxidants in a polymer are slowly consumed by reactions with radical and peroxy compounds, but in practice small or large amounts of antioxidants are lost from the polymer by precipitation at the surface[186] and evaporation[537,771] into the atmosphere rather than serving as a radical scavenger or peroxide decomposer. These phenomena are complicated and depend on the solubility and mobility of the antioxidants in a polymer, and the volatility of the antioxidants. A number of articles have been published on these characteristics of antioxidants.[180,225,957,984,1074,1151]

A special group of antioxidants are antiozonants (which are not discussed in this book). The main function of antiozonants, specially important for rubbers, is their capability of migrating to the surface of the polymer where they react with ozone and are slowly converted to ozone-impermeable films.[684,943]

2.2 PHOTOSTABILIZERS

The *photostabilization* of polymers involves the inhibition or retardation of photochemical processes (mainly photo-oxidation) in polymers and plastics by a reduction in the rate of photoinitiation and/or a reduction in the kinetic chain length of the propagation stage of the photo-oxidation mechanism.

Photostabilizers (*ultraviolet stabilizers, light stabilizers*) are additives to plastics and other polymer materials which prevent the photochemical destructive processes and reactions caused by ultraviolet light (radiation) present in sunlight and various kinds of artificial light sources.

Photo-antioxidants (common *antioxidants*) are additives to plastics and other polymer materials which prevent the oxidative and/or photo-oxidative destructive reactions (cf. Section 2.1).

Many reviews on thermal[102,103,126,241,299,482,531,540,699,841,1001,1007,1012,1016,1017,1019,1020] and photo- stabilization[34,35,52,241,279,313,341,479,541,548,552,652,1008,1117,1154,1155] and books on polymer stabilization[283,375,428,483,485,530,532,611,685,842,934,1043,1057,1076,1135] have been published. The present list of commercially available photostabilizers is given in Appendix 2.[1196]

Synthesis of new photostabilizers, and formulations for effective photostabilization of commercially available polymers, are continuously published in hundreds of patents and technical reports every year. On the other hand, a number of academic publications and scientific papers from research institutions on photostabilization mechanisms and activity of photostabilizers is highly fragmented and often contradictory. Various research groups or individuals have studied narrow aspects of photodegradation and photostabilization in different experimental conditions using different commercial samples of the same polymers. As a consequence, the results often differ and a mosaic of collected data can give only a general image of these very complicated reactions.

Photostabilizers can be classified according to their ability to act as:[481,502,697,980]

(i) Absorbers to reduce the number of photons absorbed by internal and external chromophores present in a polymer (UV absorbers and light screeners).
(ii) Compounds which may deactivate excited states (singlet and/or triplet) of chromophoric groups present in a polymer (quenchers).
(iii) Compounds which can decompose hydroperoxide groups before they are photolysed by absorbed photons (decomposers).
(iv) Compounds which can react with free radicals and thus interrupt degradative chain processes (free-radical scavengers or free-radical traps).
(v) Compounds which can react with or deactivate singlet oxygen ($^1O_2$) (singlet oxygen quenchers).

For example, the stabilization of polyolefins by energy transfer mechanisms is not effective. Peroxy radical trapping is a far more effective mechanism of stabilization than energy transfer. The optimum photoprotection for polyethylene films would be obtained with a stabilizer which effectively scavenges peroxy radicals and also decomposes hydroperoxide. The optimum stabilizer would be a molecularly dispersed additive in very low concentration which could trap peroxy radicals. An additive which decomposes hydroperoxides would also be effective but would require higher concentrations.

A photostabilizer in general should be:

(i) Effective over a long period of time.
(ii) Not liable to volatilize, be leached out, or removed from the plastic material.
(iii) Distributed in the polymeric matrix where it is most needed, particularly on the surface of the materials.
(iv) Compatible within the amorphous fraction of the polymer.

2.3 STABILIZERS IN A POLYMER MATRIX

The following factors are responsible for the effectiveness of stabilizers (antioxidants and/or photostabilizers):

(i) The intrinsic stabilizer behaviour.
(ii) The permeance of the stabilizer in the polymer matrix.

For certain polymer–stabilizer systems, the effectiveness of the stabilizer depends on its concentration in the polymer, which decreases during processing and long-term use due to chemical reactions and physical loss from the polymer.[738,908,1124]

Significant amounts of stabilizers are often lost from polymeric materials because of volatilization during fabrication or because of exudation, leaching, or solvent extraction during use. This problem is expecially severe with the semicrystalline polymers which have relatively small amorphous fractions and with articles having a high surface-to-volume ratio, such as films or fibres.

In semicrystalline materials, the stabilizers are substantially excluded from the crystalline part and the spherulites of such polymers. Low-molecular-weight photostabilizers (even oligomeric photostabilizers) readily diffuse through the amorphous regions.

A decrease in stabilizer concentration during processing and long-term storage and use is a consequence of the following physical processes:[738,801,823]

(i) distribution;
(ii) compatibility;
(iii) volatility;
(iv) extractibility.

Distribution, compatibility and mobility of stabilizers in a polymer matrix are functions of:

(i) molecular weight;
(ii) molecular size;
(iii) polarity.

For example, in polyolefins, the compatability of stabilizers is relatively low, because most stabilizers (antioxidants and photostabilizers) are aromatic compounds of relatively high polarity which are not very compatible with the paraffin-like hydrocarbon polymers.

To increase compatibility and decrease the loss of stabilizers, a number of high-molecular-weight products having several stabilizing groups in the same molecule (hybrid stabilizers) have been synthesized.

The high mobility of low-molecular-weight stabilizers is not essential for the effectiveness of these compounds in polymeric materials.

Polymerizable stabilizers incorporated into polymer chains have been found to be especially effective in long-term exposure. Sometimes a flexible, rather short spacer group between the polymer backbone chain and the active stabilizer moiety is necessary.[1132,1133]

Most commercially available stabilizers were developed empirically, and it is still difficult to predict accurately their activity in polymers under practical conditions.

2.4 DISTRIBUTION OF STABILIZERS IN A POLYMER MATRIX

During crystallization of a molten polymer containing stabilizers, partition of solutes between the solid and liquid phases will give rise to inhomogeneous distribution of additives, which are concentrated in the interlamellar amorphous regions at the growing edges of spherulites.[226] Figure 2.1 shows the distribution of 2,2-thiobis(4-*tert.*-octophenolato)-*n*-butylamine nickel(II) (Cyasorb UV 1084) (**5.86**) in polypropylene partly crystallized at 130°C and quenched. These results

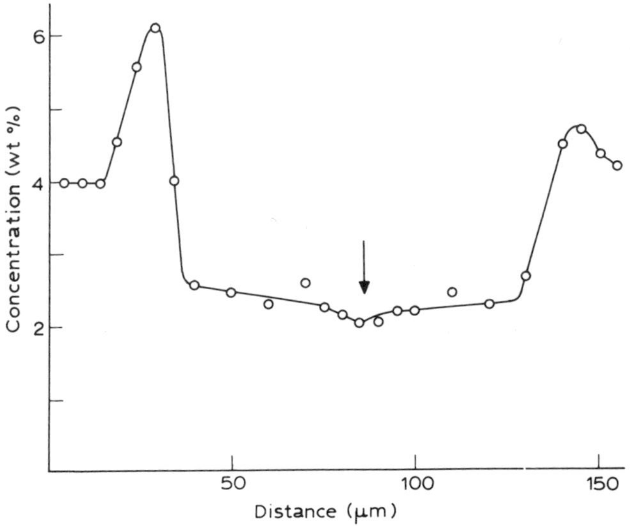

Fig. 2.1. Nickel distribution for 2,2-thiobis(4-*tert.*-octophenolato)-*n*-butylamine nickel(II) (Cyasorb UV 1084; **5.86**) (0·4 wt%) in polypropylene partly crystallized at 130°C and quenched. Reprinted with permission from Ref. 226, © 1978 Butterworth & Co. (Publishers) Ltd.

indicate that a growing spherulite pushes ahead of itself a wave of stabilizer; however, some part of it is incorporated into the crystal phase.

The 2-(2-hydroxy-3,5-di-*tert.*-amylphenyl)benzotriazole (Tinuvin 328) (**5.47**) is uniformly distributed in polyethylene melt, whereas in the solid polymer it is distributed non-uniformly, depending to a great extent on the crystallization conditions. The amorphous phase can dissolve this stabilizer at a concentration of 3 wt%, but it is rejected by the crystalline phase.[420]

In polymers which contain stabilizers grafted into macromolecules, the grafted chains or groups are located mainly in the amorphous phase. The grafting reaction does not disrupt the crystallites, at least for a low graft content (about 15–30% grafted monomer).

2.4.1 Compatibility of Stabilizers with Polymers

In stabilized polymers physical interaction occurs between the polymer matrix and low-molecular-weight stabilizers that is known as *compatibility*.

Compatibility in polymer systems cannot be defined unequivocally. It has its origin in the mixing of two liquids. When liquids are mixed to form a homogeneous and single-phase mixture, they are said to be *compatible*. From thermodynamic, kinetic and mechanical considerations, a homogeneous single-phase polymer–additives mixture is improbable. Compatibility in such a system is a representation of how close it can approach a single-phase mixture and is a relative measure of the degree of homogeneity of the system.

It has been found that photostabilizers and other additives in amorphous regions in isotactic polypropylene are distributed very non-uniformly.[595]

Some additives have a poor compatibility with polymers, resulting in dissolution of only a portion of the added stabilizer if it is used at the relatively high concentrations employed in practice.[270,271,367] By improving solubility in a polymer matrix it is possible to improve the light stability of the polymer. Addition of surfactant (hydroxyethylated cetyl alcohol) improves the compatibility of additives with the polymer matrix and considerably improves the light stability.

The more similar are the chemical structures and physical properties of the stabilizer and polymer matrix, the better is the compatibility. Most polymers such as polyolefins are totally nonpolar materials, whereas antioxidants and stabilizers are more or less polar compounds. In order to improve the compatibility of such systems, stabilizers have a long alkyl chain attached (for example, compare Tables 3.2 and 6.1).

In general, stabilizers containing more polar groups and fewer alkyl substituents are less compatible with non-polar polymers. However, the presence of polar groups in the stabilizer may improve the compatibility for the polar polyolefins, such as copolymers of ethylene with acid or ester monomers (e.g. acrylic acid, maleic anhydride, vinyl acetate, ethyl acetate), due to interactions between the copolymer and stabilizer.

Polymer stabilization by grafting of additives may overcome the problem of incompatibility, because the stabilizing groups are chemically linked to the macromolecules and cannot be rejected by the polymer matrix.

Grafter polyolefins with a high content of stabilizing groups, obtained by polyolefin grafting with monomeric stabilizers, may be used as stabilizer masterbatches by blending into the parent polyolefins.[823,824] In such blends compatibility is better than in blends of low-molecular-weight stabilizer in polyolefin at the same concentration of stabilizing groups. In blends with the parent polyolefin, the grafted polyolefin with stabilizer acts as a *Compatibilizer* to improve compatibility between the polymer and the homopolymeric and monomeric stabilizers.

2.4.2 Solubility of Additives in a Polymer Matrix

The solubility of an additive in a polymer matrix is determined by the free energy change associated with the transfer of the additive from its equilibrium state as a pure material into the polymer matrix.[180]

The following factors affect the solubility of additive molecules in a polymer matrix:[799,801]

(i) Temperature. The solubility coefficient (S) depends on temperature, increasing with temperature (Fig. 2.2) according to eqn (2.1):[180,799]

$$S = S_0 \exp(-\Delta H_S/RT) \qquad (2.1)$$

where S_0 = pre-exponential factor,
ΔH_S = enthalpy or heat of solution,
R = universal gas constant,
T = temperature.

Solubility increases with increasing temperature.

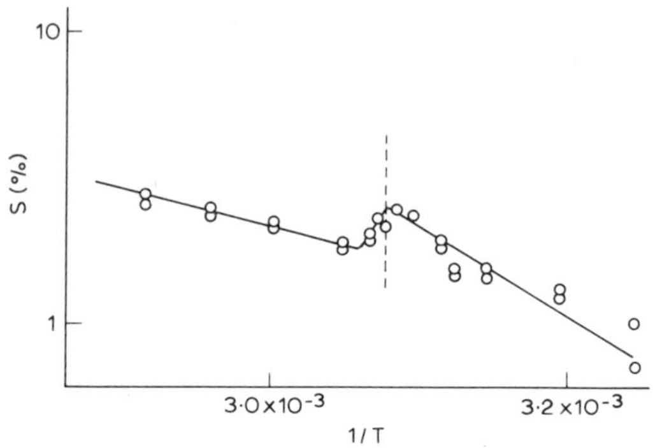

Fig. 2.2. Solubility of octadecyl-3-(3,5-di-*tert.*-butyl-4-hydroxyphenyl-propionate) (Irganox 1076; **3.21**) in low-density polyethylene (LDPE).[315]

(ii) Additive melting temperature. A number of additives show a discontinuity in the Arrhenius plot at the melting point (T_m) (Fig. 2.2). The discontinuity does not take place at the melting point of the additive but, as for diffusion, slightly above it. ΔH_S has different values above and below the discontinuity. In Table 2.1 are given solubility parameters and ΔH_S for a series of stabilizers in polyethylene.[800]

(iii) Polymer melting temperature (T_m) and polymer glass transition temperature (T_g). Mobility of macromolecules in a polymer matrix affects the solubility of additives.

(iv) Structure and size of the additive. The solubility of the amorphous additive (in amorphous polymer) is greater than that of the crystalline form. In nonpolar polymers the solubility of polar additives is limited and once the solubility is exceeded, particularly in localized regions, the molecule will tend to form insoluble molecular aggregates which consequently will be extruded from the bulk of a polymer matrix to its surface during processing. Generally the solubility of additives decreases with increasing molecular weight and size of additive (Fig. 2.3).

(v) Polymer structure. Solubility decreases as the density of the polymer matrix (Fig. 2.4) and the content of crystalline phase increase.[23,180,226]

(vi) Orientation. Solubility of additives increases with a rapid stretching of the polymer sample. A rapid orientation of the

Table 2.1
Solubility parameters (S) for some additives in polyethylene[315]

Additive	*Structure*	*Trade name*	*Molecular weight*	*Molecular length* (*Å*)	*Solubility coefficient* S (*ppm*)	*Enthalpy or heat of solution* ΔH_S (*kJ mol*$^{-1}$)
3,3′-Di-*tert*.-butyl-2,2′-dihydroxy-5,5′-dimethyldiphenylmethane (**3.36**)	See Table 3.2	Plastanox 2246	340	13.3	$2{\cdot}2 \times 10^{7}$	15·9
5,5′-Di-*tert*.-butyl-4,4′-dihydroxy-dimethyldiphenyl sulphide		Santonox	358	13·9	$1{\cdot}3 \times 10^{15}$	80·7
3,3′-Di-*tert*.-butyl-2,2′-dihydroxy-5,5′-diethyldiphenylmethane (**3.37**)	See Table 3.2	Plastanox 425	368	13·3	$9{\cdot}0 \times 10^{7}$	26·8
3,3′-Di(methylcyclohexyl)-2,2′-dihydroxy-5,5′-dimethyldiphenylmethane		Nonox WSP	420	15·0	$8{\cdot}1 \times 10^{8}$	32·8
3,5,3′,5′-Tetra-*tert*.-butyl-4,4-dihydroxydiphenylmethane		Ionox 220	424	13·3	$1{\cdot}6 \times 10^{9}$	38·5
Octadecyl-3-(3,5-di-*tert*.-butyl-4-hydroxyphenyl)propionate (**3.21**)	See Table 3.2	Irganox 1076	531	35·1	$1{\cdot}0 \times 10^{11}$	45·5
					$9{\cdot}3 \times 10^{14}$	72·6
1,1,3-Tri(2-*tert*.-butyl-4-hydroxy-6-methylphenyl)butane (**3.38**)	See Table 3.2	Topanol CA	544	13·4	$2{\cdot}0 \times 10^{10}$	47·2
2,2′-Thiodiethylenebis[3-(3,5-di-*tert*.-butyl-4-hydroxyphenyl)propionate] (**3.27**)	See Table 3.2	Irganox 1035	643	29·0	$1{\cdot}0 \times 10^{13}$	61·4
					$6{\cdot}1 \times 10^{10}$	46·8
1,3,5-Tri(3,5-di-*tert*.-butyl-4-hydroxy)-benzyl-2,4,6-trimethylbenzene		Ionox 330	774	14·8	$1{\cdot}1 \times 10^{7}$	28·0
Pentaerythrityl tetrakis[3-(3,5-di-*tert*.-butyl-4-hydroxyphenyl)propionate] (**3.32**)	See Table 3.2	Irganox 1010	1178	25·9	$4{\cdot}0 \times 10^{12}$	62·3
Didodecanyl 3,3′-thiobispropanoate (**3.56**)	See Table 3.7	Cyanox LTPD	514	43·0	$1{\cdot}6 \times 10^{9}$	53·0
					$1{\cdot}4 \times 10^{2}$	10·7
Dioctadecanyl 3,3′-thiobispropanoate (**3.59**)	See Table 3.7	Cyanox STPD	682	50·5	$1{\cdot}0 \times 10^{5}$	29·2
					$4{\cdot}4 \times 10^{1}$	7·7
2-(2-Hydroxy-3,5-di-*tert*.-butylphenyl)-5-chloro-2*H*-benzotriazole (**5.46**)	See Table 5.7	Tinuvin 327	358	—	$3{\cdot}9 \times 10^{5}$	10·9

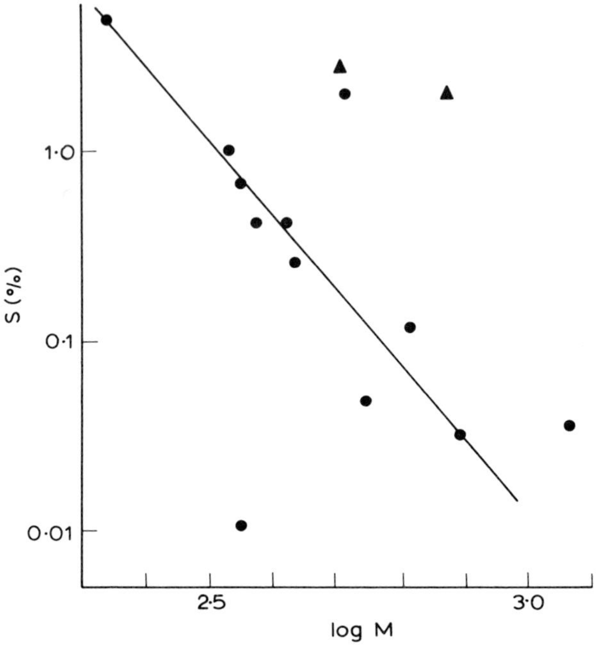

Fig. 2.3. Solubility at 50°C in low-density polyethylene (LDPE) as a function of molecular weight: ▲, thiodipropionic esters; ●, hindered phenolic antioxidants.[315]

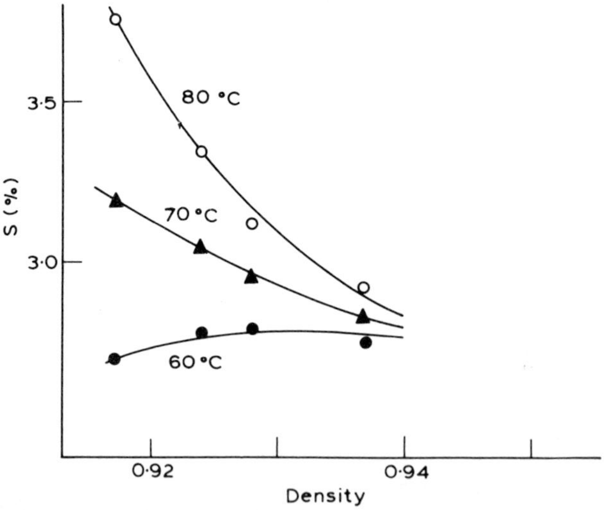

Fig. 2.4. Solubility of octadecyl-3-(3,5-di-*tert*.-butyl-4-hydroxyphenyl-propionate) (Irganox 1076; **3.21**) in various quenched polyethylenes at different temperatures: ●, 60°C; ▲, 70°C; and ○, 80°C.[315]

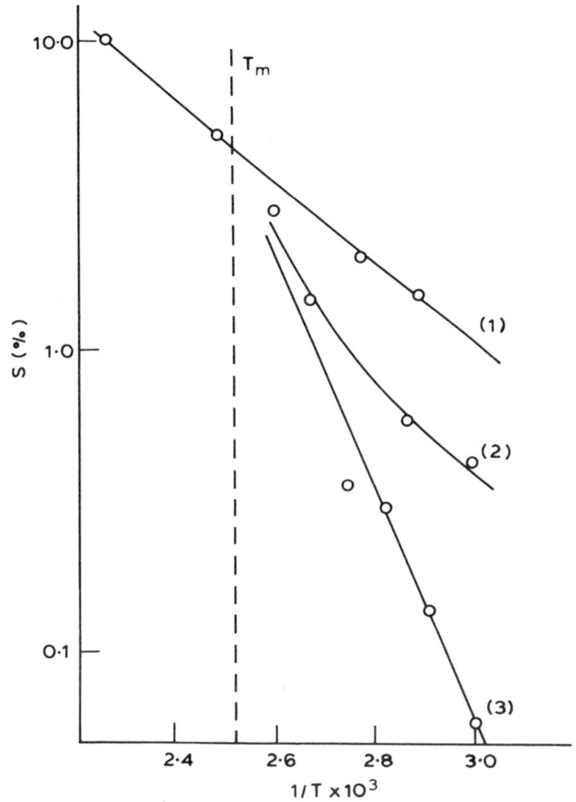

Fig. 2.5. Solubility of pentaerythrityl tetrakis[3-(3′,5′-di-*tert*.-butyl-4′-hydroxyphenyl)propionate] (**3.32**; Irganox 1010) with temperature: curves (1), (2) and (3) are results obtained by different authors.[315]

crystalline phase leads to an increase in free volume in the amorphous phase.

It is worth noting that measured values of solubilities of additives in a polymer vary according to the methods used (Fig. 2.5 and Table 2.2).

2.4.3 Diffusion of Additives in a Polymer Matrix

The following factors affect the diffusion of additives in a polymer matrix:[801]

(i) Temperature. The diffusion coefficient (D) depends on temperature (increasing with temperature) according to the

Table 2.2
Solubility parameters (*S*) of antioxidants in polyolefins at two temperatures[315]

Antioxidant	*Structure*	*Trade name*	*Polyolefin*	*S*(%) *at 25°C*	*S*(%) *at 70°C*	ΔH_S *(kJ mol^{-1})*	*References*
1,1,3-Tri(2-*tert.*-butyl-4-hydroxy-6-methylphenyl)butane (**3.38**)	See Table 3.2	Topanol CA	LDPE[a]	0·009	0·12	50 ± 1	180
				0·011	0·125	47·2	799
Pentaerythrityl tetrakis-3-(3,5-di-*tert.*-butyl-4-hydroxyphenyl)-propionate (**3.32**)	See Table 3.2	Irganox 1010	LDPE	0·00006	0·012	100	956
				—	0·026	62	180
				0·020	0·24	39 ± 4	419
				0·003	0·14	62·3	799
			Atactic polypropylene	0·25	1·5	24·4	419, 420
			Isotactic polypropylene	0·15	0·40	32·6	180

[a] LDPE, Low-density polyethylene.

equation:

$$D = D_0 \exp(-E_D/RT) \quad (2.2)$$

where D_0 = pre-exponential constant,
E_D = activation energy for diffusion,
R = universal gas constant,
T = temperature.

A typical Arrhenius plot for the diffusion coefficient of 1,1,3-tri(2-*tert.*-butyl-4-hydroxy-6-methylphenyl)butane (Topanol CA; **3.38**) is shown in Fig. 2.6. Diffusion coefficients for different additives in molten polypropylene at different temperatures are shown in Table 2.3.[568,799,956]

(ii) Diffusant melting temperature. A number of additives show a discontinuity in the Arrhenius plot at the melting point (T_m) (Fig. 2.7). The activation energy is lower at temperatures above the discontinuity. Hence, there are different activation energies, $E_D(hT)$, $E_D(lT)$, for high and low temperatures.

(iii) Polymer melting temperature (T_m) and polymer glass transition temperature (T_g). It is obvious that mobility of macro-

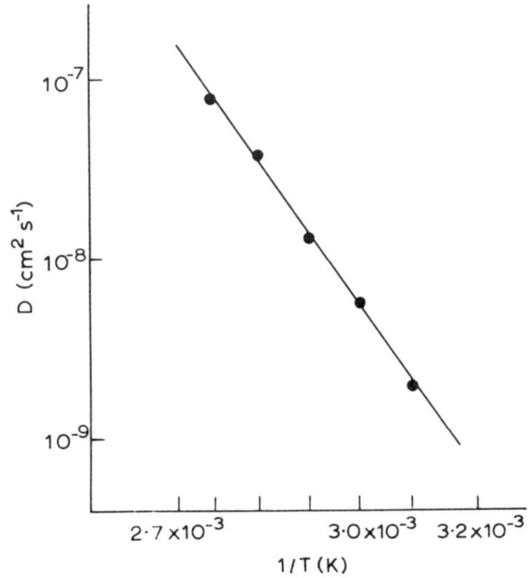

Fig. 2.6. Temperature dependence of the diffusion coefficient (D) of 1,1,3-tri(2-*tert.*-butyl-4-hydroxy-6-methylphenyl)butane (Topanol CA; **3.38**) in low-density polyethylene (LDPE).[315]

Table 2.3
Observed values of additive diffusion coefficients in molten polypropylene[a]

Additive	*Structure*	*Molecular weight*	*Diffusion coefficients (cm^2 s)*		
			120°C	*125°C*	*130°C*
2,6-Di-*tert.*-butyl-4-methoxyphenol (Topanol 354)		236	$1{\cdot}0 \times 10^{-7}$	—	$1{\cdot}5 \times 10^{-7}$
2,2′-Methylenebis(4-methyl-6-*tert.*-butylphenol) (CAO-5)		324	$>2{\cdot}5 \times 10^{-7}$	—	$>5{\cdot}0 \times 10^{-7}$
2-Hydroxy-4-*n*-octoxybenzophenone (UV531) (**5.30**)	See Table 5.5	326	—	$2{\cdot}0 \times 10^{-7}$	—
2,5-Di(5-*tert.*-butyl-2-benzoxazolyl)thiophene (Cyasorb Uvitex OB)		430	$6{\cdot}0 \times 10^{-8}$	$7{\cdot}0 \times 10^{-8}$	$>1{\cdot}0 \times 10^{-7}$
1,3,5-Tri(3,5-di-*tert.*-butyl-4-hydroxy)benzyl-2,4,6-trimethylbenzene (Ionox 330)	See Table 3.2	768	$2{\cdot}0 \times 10^{-8}$	$2{\cdot}0 \times 10^{-8}$	$1{\cdot}0 \times 10^{-8}$
1,3,5-Tri(2,6-di-*tert.*-butyl-4-methylphenyl) isocyanurate (Goodrite 3114) (**3.31**)	See Table 3.2	777	$1{\cdot}0 \times 10^{-8}$	—	—
1,3,5-Tri(ethyl[3,5-di-*tert.*-butyl-4-hydroxy-phenyl]propionate) isocyanurate (Goodrite 3125)		999	$<1{\cdot}0 \times 10^{-8}$	—	—
Tetrakis[methylene(3,5-di-*tert.*-butyl-4-hydroxyhydrocinnamate)]methane (Irganox 1010) (**3.32**)	See Table 3.2	1176	$<1{\cdot}0 \times 10^{-8}$	—	—

[a] Reprinted with permission from Ref. 226, © 1980 Butterworth & Co. (Publishers) Ltd.

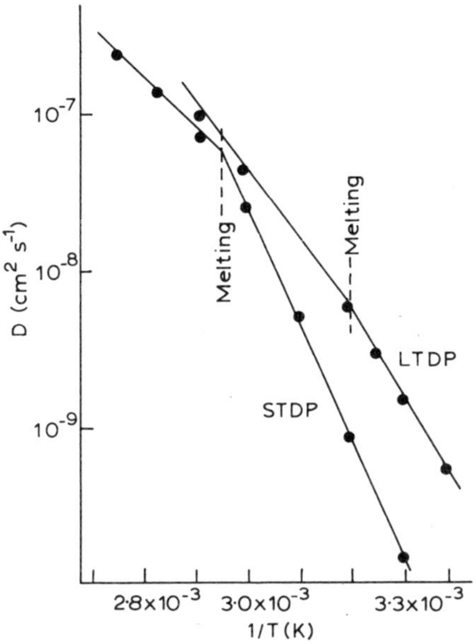

Fig. 2.7. Temperature dependence of the diffusion coefficient (*D*) for dilaurylthiodipropionate Cyanox LTDP; (**3.56**) and distearylthiodipropionate Cyanox STDP; (**3.59**) in low-density polyethylene (LDPE).[315]

molecules in a polymer matrix should affect the kinetics of diffusion.

(iv) Size of diffusing molecule. The diffusion coefficient decreases with increasing size of the diffusing molecules. The diffusion of 2-hydroxy-4-alkoxybenzophenone in polypropylene decreases as the chain length increases from methyl to octyl and dodecyl.[608]

(v) Polymer structure. The diffusion coefficient decreases as the density of a polymer matrix and content of the crystalline phase increases.

(vi) Factors which decrease the segmental mobility (crystallization, chain stiffness, and crosslinking) decrease the diffusion rate.

(vii) Factors which increase the fractional free volume, e.g. deformation. Elastic strain in a direction normal to crystalline lamellae will increase the thickness and the specific volume of the amorphous layers between the lamellae.

(viii) Factors affecting processes of immobilization of the penetrant in the polymer matrix.

In conclusion, it may be stated that the kinetics of diffusion of stabilizers in a polymer matrix depend on parameters related both to the structure of the diffusant and to that of the polymer.

The concentration of added stabilizers decreases during ageing (Fig. 2.8). As soon as the stabilizer concentration becomes too low, oxidation appears (Fig. 2.9). Two antioxidants such as octadecyl-3-(3,5-di-*tert.*-butyl-4-hydroxyphenyl)propionate (Irganox 1076) (**3.21**) and 1,1,3-tri(2-*tert.*-butyl-4-hydroxy-6-methylphenyl)butane (Topanol CA; **3.38**) have practically the same diffusion coefficients but very different solubilities. This results in protection times which are very different. On the other hand, Topanol CA and 2,2′-thiodiethylene-bis[3-(3,5-di-*tert.*-butyl-4-hydroxyphenyl)propionate)] (Irganox 1035;

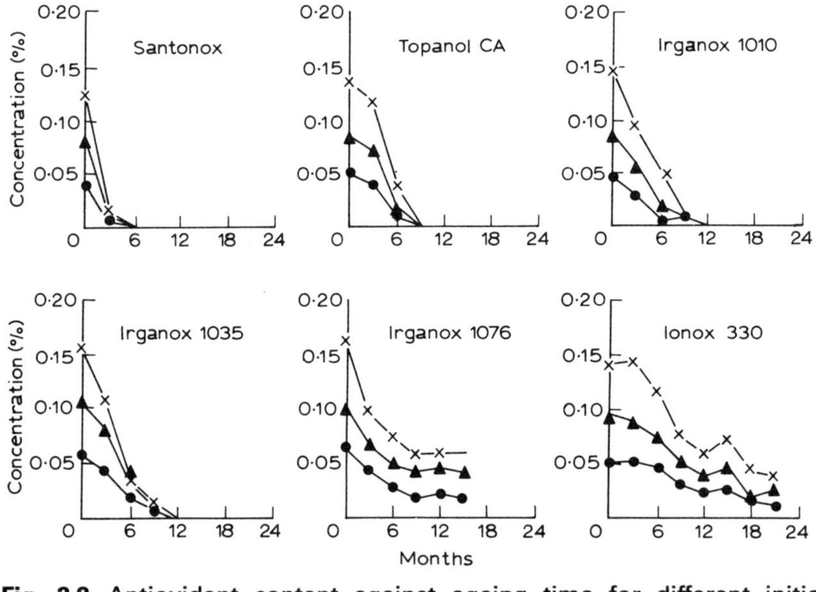

Fig. 2.8. Antioxidant content against ageing time for different initial loadings: ×, initially 1500 ppm; ▲, initially 1000 ppm; and ●, initially 500 ppm.[315] The antioxidants were as follows. Santonox, 5,5′-di-*tert.*-butyl-4,4′-dihydroxydimethyldiphenylsulphide; Topanol CA, 1,1,3-tri(2′-methyl-4′-hydroxy-5′-*tert.*-butylphenyl)butane (**3.38**); Irganox 1010, pentaerythrityl tetrakis[3-(3′,5′-di-*tert.*-butyl-4′-hydroxyphenyl)propionate] (**3.32**); Irganox 1035, 2,2′-thiodiethylenebis[3-(3,5-di-*tert.*-butyl-4-hydroxyphenyl)propionate] (**3.27**); Irganox 1076, octadecyl 3-(3,5-di-*tert.*-butyl-4-hydroxyphenyl)propionate (**3.21**); and Ionox 330, 1,3,5-tris(3,5-di-*tert.*-butyl-4-hydroxy)benzyl-2,4,6-trimethylbenzene .

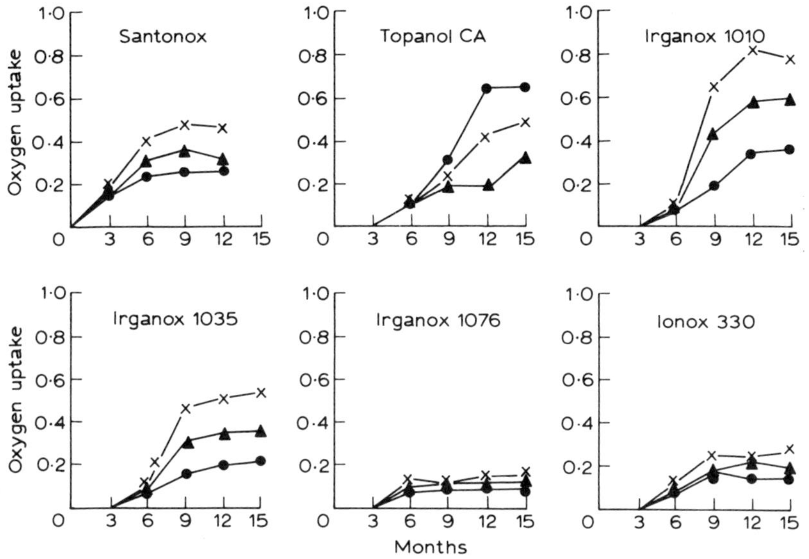

Fig. 2.9. Oxidation level (relative units) against ageing time for different initial antioxidant concentrations (for names cf. Fig. 2.8): ×, initially 1500 ppm; ▲, initially 1000 ppm; ●, initially 500 ppm.[315]

3.27) have diffusion coefficients differing by 0·2, but their similar solubilities lead to similar protection times.[802]

2.4.4 Volatility of Stabilizers

Many low-molecular-weight stabilizers are more or less volatile and they can escape during the processing step (cf. Section 1.17), storage and use. Consequently, their concentration in a polymer matrix continuously decreases.

For example, compounding of phenyl salicylate (MW = 211) and polypropylene at 325°C in 5 min caused 61% of the stabilizer added to volatilize. Only 20% of a higher-molecular-weight aromatic salicylate (MW = 500) and less than 5% of some salicylates (MW = 325–450) volatilized during compounding.[845]

Loss of additives is controlled either by the rate of their loss from the surface or by the rate of their migration through the bulk to reach the surface or by some combination of these parameters.[179,180,225] The rate of loss of additive from a polymer surface can be predicted from the volatility of the pure additive and its solubility in the polymer, whilst the rate of migration of the additive to the surface to replace lost material is controlled by its diffusion coefficient.

The shape of the final product may determine the relative importance of diffusion and surface loss. Thus, high surface area:volume ratios, and high temperatures, increase the importance of surface loss by volatilization and leaching, whereas in thick sections the rate of diffusion may be rate-determining.[179]

The rate of evaporation of additives per unit surface area of the polymer is proportional to the concentration of additives at the polymer surface.[225] When the concentration in the bulk polymer is homogeneous, the rate of evaporation per unit area of the polymer (V) ($\mathrm{g\,cm^{-2}\,min^{-1}}$) is given by:

$$V = ck \tag{2.3}$$

where c = concentration of additives in the bulk polymer (ppm),
k = rate constant related to the volatility of the pure additive and the solubility of the additive in the polymer ($\mathrm{g\,cm^{-2}\,min^{-1}}$).

The rate of decrease due to evaporation of the additive in polymer films is obtained by dividing the rate of evaporation (V) by the mass per unit surface area of the film:

$$-\frac{\mathrm{d}c}{\mathrm{d}t} = \frac{V}{l\rho} = \frac{cK}{l\rho} \tag{2.4}$$

where l = thickness of the polymer film (cm),
ρ = density of the polymer film ($\mathrm{g\,cm^{-3}}$).

Loss of additives (particularly antioxidants) from polymers and plastics is increasing in importance in food contact applications and in medical uses of polymers. It is also very important at increased working-temperatures of polymers, e.g. on tyres on motorways.[972]

2.4.5 Mechanisms of Moisture Penetration in a Polymer Matrix

Moisture penetration into a polymer matrix occurs by:[760]

(i) Diffusion of water molecules.
(ii) Microcracks and capillarity (the latter only in composites). Each of these becomes active only after specific damage occurs to the polymeric material:

(a) Transport of moisture by microcracks involves both flow and storage of water in microcracks or other forms of micro-damage which results from environmental effects or from service conditions.

(b) The capillarity mechanism involves flow of water molecules along the fibre–matrix interface followed by diffusion from the interface into the bulk resin. It is not active unless debonding between the fibres and the matrix has occurred, often as a result of water attack at the interface.

The activation of these mechanisms is distinguished by increasing both the rate and the maximum capacity of moisture absorption.

2.4.6 Extractability of Stabilizers from Polymers

A number of plastic products and articles come into contact with water (rain), organic solvents, oils, fuels, detergents or other liquids during both indoor and outdoor uses. These liquids may extract the stabilizers from the polymer matrix. Extractability depends on:[628,820]

(i) The polymer type (e.g. chemical structure, physical properties, molecular weight, crystallinity).
(ii) The chemical structure and molecular weight of the stabilizer.
(iii) The nature of the extraction liquids.
(iv) Extraction conditions (e.g. time, temperature).
(v) Working conditions (e.g. mechanical stress).

The extraction mechanism is diffusion controlled. The stabilizer which has migrated to the surface layer disappears rapidly by contact with leaching liquids. The rate of extraction of stabilizers therefore depends strongly on their solubility in the liquids. Thus organic solvents are in general much better leaching agents than is water. Solvents may also penetrate the polymer matrix and participate in the transportation of stabilizers to the surface. The permanence of stabilizers in polymers is often more affected by extraction than by volatilization.

The extractability problem is most severe for articles with a high surface area : volume ratio such as fibres and film. Some examples can illustrate this problem:[1080]

(a) Perchloroethylene cleaning (dry cleaning) of polyester and polyamide fibres stabilized with 2,2′-dihydroxy-4,4′-dimethoxy-

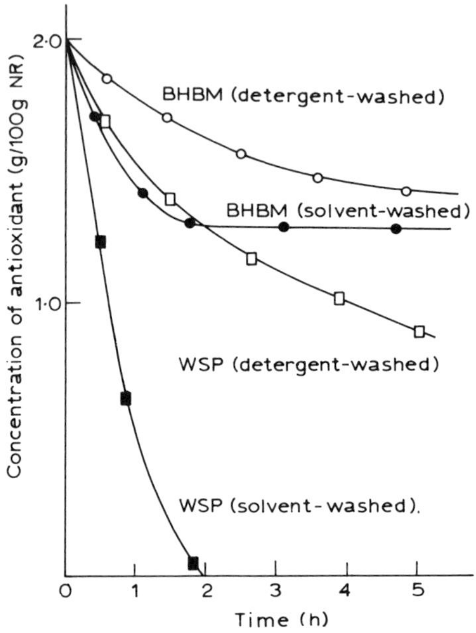

Fig. 2.10. Loss of antioxidants during cold solvent extraction (80:20 ether/toluene) and leaching by boiling detergent (2% tide). BHBM is 3,5-di-*tert.*-butyl-4-hydroxybenzyl mercaptan (**3.44**) and WSP is 2,2′-methylenebis[(6-phenyl-1-methyl)-4-methylphenol]. Reprinted with permission from Ref. 674, © 1978 Pergamon Press PLC.

benzophenone extracts as much as 20% of the stabilizer from the fibres.

(b) Resistance of antioxidants to 200 h of aqueous extraction at 60°C from films of low-density polyethylene increases with increasing stabilizer molecular weight.[537]

(c) Cold extraction or leaching by boiling detergent removes most of the antioxidants from rubber vulcanizates (Fig. 2.10).[674] Antioxidants should be specially resistant to solvent leaching (aqueous detergents and dry-cleaning solvents).[673,742,1013]

Formation of a thin surface layer of the stabilizer on the surface of the polymer improves the effectiveness of an additive but abrasion of the polymer or washing leads to removal of the protective layer and thus to a reduction in the effectiveness of incompatible stabilizers.

Washing with water[537] or dry cleaning in organic solvents[739] decreases the content of low-molecular-weight stabilizers in the polymer.

Physical properties such as volatility, extractability and diffusion rates of stabilizers decrease with increasing molecular weight of these additives.[557,897,984] Thus, a molecular weight of about 500 is enough to ensure adequate stabilizer permanence from the viewpoint of volatility, but not from that of extractability. Only highly polymeric stabilizers are resistant to extraction, although even such polymeric stabilizers, with molecular weights exceeding 15000–20000, may still be involved in diffusion processes. However, in blends of polymers and polymeric stabilizers the diffusion coefficients are considerably lower.

A decrease in stabilizer content may also result from migration of stabilizers to the polymer surface due to incompatibility.[363]

Polymers stabilized by grafting are completely resistant to extraction. Even organic solvents cannot extract the stabilizing groups from the polymer matrix because they are chemically bound to the polymer chains.

2.5 QUENCHERS

A number of commercially available photostabilizers act as quenchers of the excited singlet and/or triplet states. The most important group of such photostabilizers are nickel chelates (cf. Section 5.10).

The quenching reaction may occur over:[508,919,934]

(i) short distances (10–15 Å), in the electron exchange energy transfer (by the Dexter mechanism), or
(ii) long distances (50–150 Å), in the resonance–excitation energy transfer (the dipole–dipole or Förster mechanism).

Both mechanisms of quenching can be represented by a simple reaction, where excited *donor* (D*) (an excited chromophoric group in a polymer, which can be responsible for the initiation of photodegradation) is deactivated by an *acceptor* (*quencher*) (A or Q) molecule:

$$D^* + A \longrightarrow D + A^* \tag{2.5}$$

$$A^* \longrightarrow A + \text{energy in the form of emission or heat} \tag{2.6}$$

The quenching process is successful only if the quencher molecule is within quenching distance of the excited (singlet and/or triplet) state of the chromophoric group within the latter's lifetime. High diffusion constants in a polymer matrix, i.e. good mobility of quencher and/or excited chromophoric group and long lifetime of excited chromophore group, may therefore enlarge the apparent sphere of action of the quencher. In other words, it may lower the concentration of quencher necessary for a certain effect to be observed. The action sphere can be calculated for each specific case, e.g. a freely diffusing quencher of molecular weight 500 and a fixed excited chromophoric group with an active diameter of 5 Å.[541] In Fig. 2.11 the quencher concentration is plotted as a function of the mean square displacement ($\bar{x}^2$), i.e. the concentration needed to allow quenching within the lifetime (τ_s) in a polymer matrix characterized by a diffusion coefficient (D). Each curve represents a specific distance (R_Q) at which the energy transfer becomes operative.

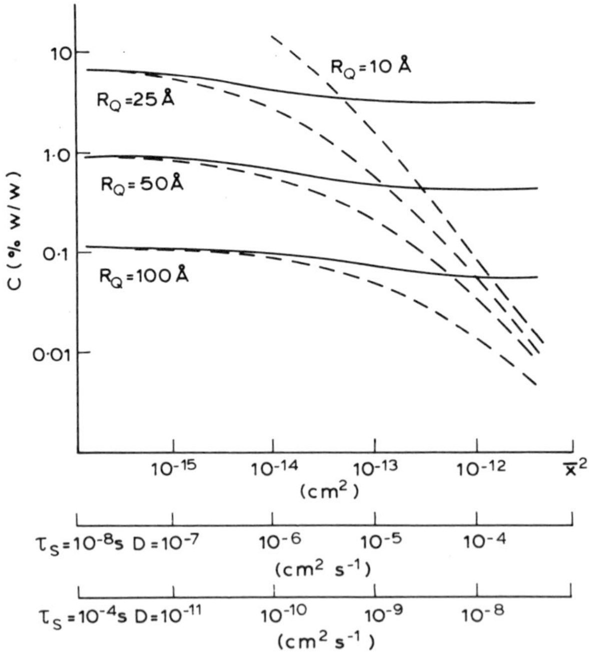

Fig. 2.11. Concentration of quencher versus mean displacement. Reprinted with permission from Ref. 541, © 1973 Pergamon Press PLC.

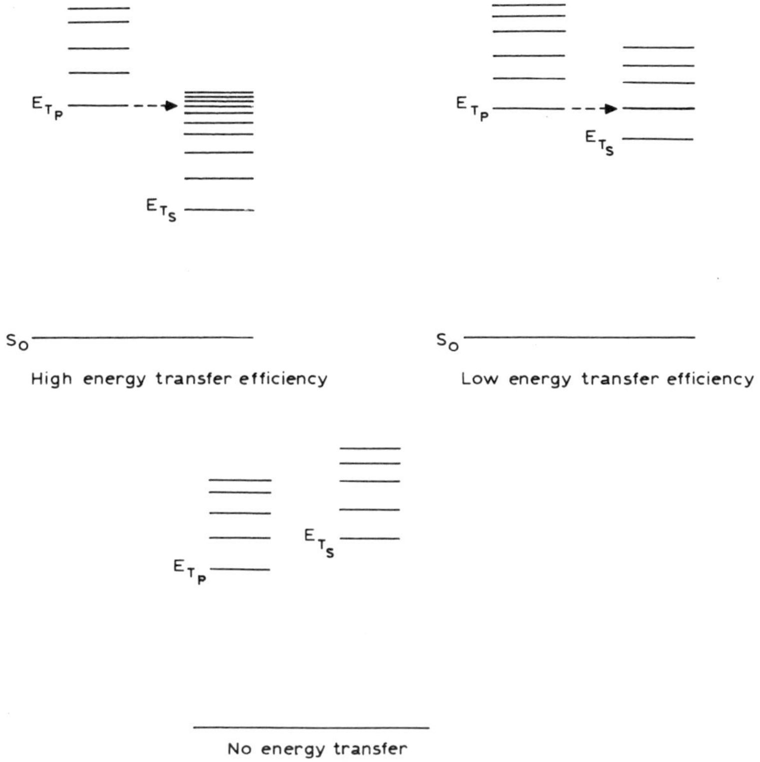

Fig. 2.12. Diagram of energy transfer between polymer and stabilizer for different triplet energy levels: E_{T_p}, polymer and E_{T_s}, stabilizer.

The broken lines give the best possible case (unit probability), i.e. the greatest enlargement of the quencher action sphere, which results if each quencher is surrounded by excited chromophoric groups. Under these circumstances any movement of the quencher, independent of direction, would lead to a successful encounter.

The full lines reflect a situation in which the concentration of excited chromophoric groups is equal to or even lower than the quencher concentration. In this case the direction of the quencher diffusional movement—towards or away from the excited chromophoric group—starts to play a role.

The concentrations presented in Fig. 2.12 can be considered from the point of view of photostabilizer concentrations actually used in

Table 2.4
Examples of triplet energies (E_T) of some commercially available photostabilizers[a]

Chemical name	*Structure*	*Commercial name*	E_T (*kcal mol*$^{-1}$)
Phenyl ester of benzoic acid	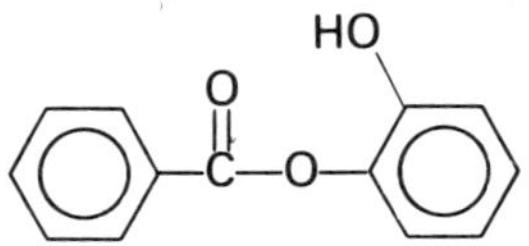		80·8
Nickel chelate	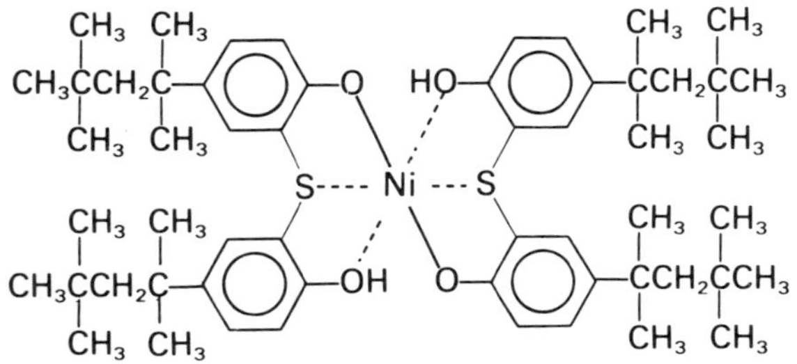	Ferro AM 101	77·8
o-Hydroxybenzophenone	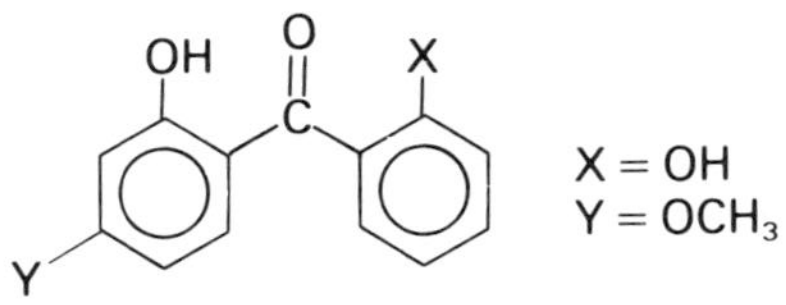	Cyasorb UV24	67·3
o-Hydroxybenzotriazoles	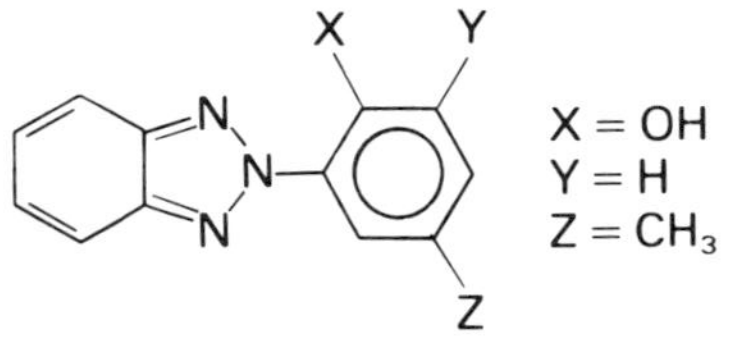	Tinuvin P	63·0

[a] Reprinted with permission from Ref. 166, © 1974 Pergamon Press, PLC.

practice, which range from 0·1 to 0·5% w/w. This figure demonstrates that at practical additive levels only quenchers with an operational mode effective at or above 50 Å can be expected to deactivate excited states of chromophoric groups efficiently. In other words, only long-range energy transfer can be expected to contribute to deactivation of excited states with the usual half-life of singlets and triplets.

Photostabilization of polymers by the energy transfer mechanism depends on the relative energy levels of the triplet of the polymer and the photostabilizer (Fig. 2.12). The triplet energies of some commercial photostabilizers are listed in Table 2.4.[166] This table shows that the commercial photostabilizers exhibit a wide range of triplet energies, which are highest for salicylate and nickel chelate, but are lowest for hydroxybenzophenones and hydroxybenzotriazoles.

Photodegradation and photo-crosslinking of poly[bis(4-benzoylphenoxy)]phosphazene which originate from the triplet state of the benzophenone moiety appended to the phosphazene polymer backbone can be stabilized by energy transfer processes from free naphthalene and naphthalene attached via oxygen to the cyclophosphazene ring in hexakis(β-naphthoxycyclophosphazene) (**2.1**).[472]

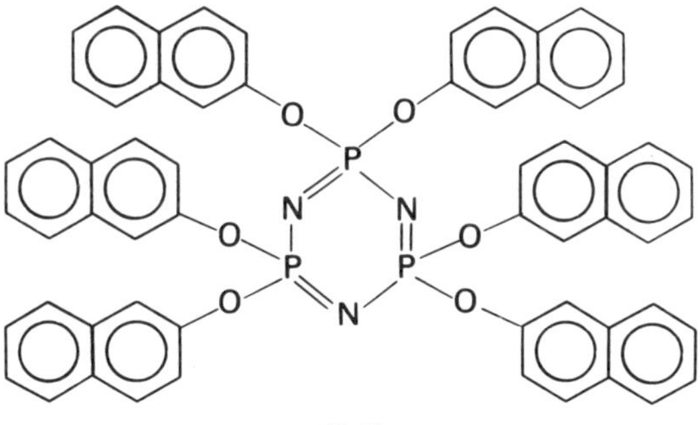

(2.1)

2.6 SINGLET OXYGEN DEACTIVATORS

Single oxygen ($^1O_2$) generated by chemical or phyisical methods reacts with polymers containing unsaturated bonds, mainly with polydienes, by the *ene*-type reaction which leads to the formation of hydroperoxides:[744,847,918,922]

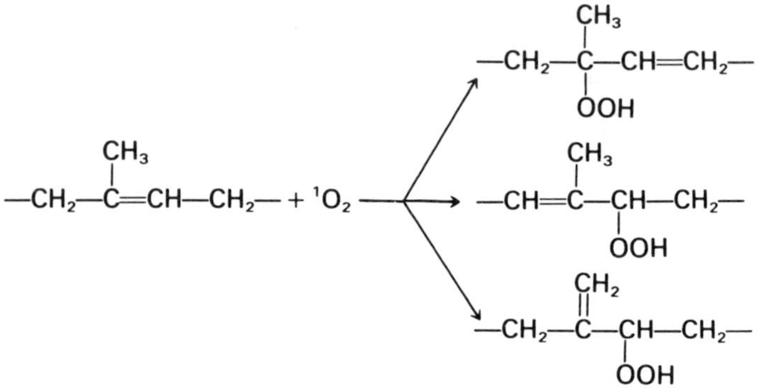

(2.7)

(2.8)

(2.9)

In order to protect polymers against the ene-type reaction, it is logical first to deactivate singlet oxygen ($^1O_2$) by adding to the polymer singlet oxygen quenchers (Q), which may react by two mechanisms:

(i) Chemical quenching:

$$Q + {}^1O_2 \xrightarrow{k_1} QO_2 \tag{2.10}$$

(ii) Physical quenching:

$$Q + {}^1O_2 \xrightarrow{k_2} Q + O_2 \tag{2.11}$$

The rate constant of quenching (k_q), depending on the reaction, can be: $k_q = k_1$, or $k_q = k_2$, or $k_q = k_1 + k_2$. The determination of k_q is described in Section 9.13.

One of the most effective $^1O_2$ quenchers is 1,3-diphenylisobenzofuran (referred to also as 2,5-diphenyl-3,4-isobenzofuran) (**2.2**), which reacts with $^1O_2$ to form the ozonide of 1,2-diphenyl-3,4-benzocyclodiene (**2.3**):[692]

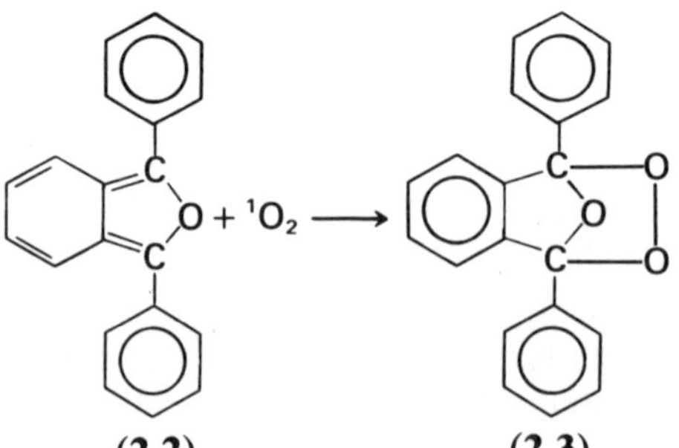

(2.12)

(**2.2**) (**2.3**)

The rate of quenching (k_q) for 1,3-diphenylisobenzofuran is 8×10^8 litres $mol^{-1} s^{-1}$. The 1,3-diphenylisobenzofuran has been tested as an effective deactivator of $^1O_2$ in the protection of polybutadiene.[693]

The β-carotene (**2.4**) and other carotenoids deactivate $^1O_2$ by physical and chemical quenching.[413–415,432,920]

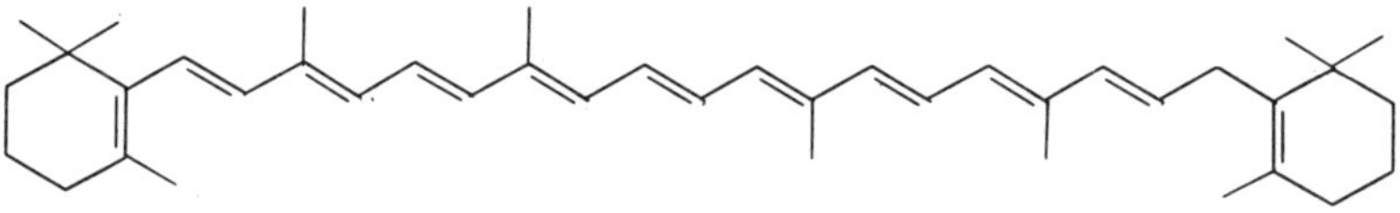

(2.4)

The rate constant of quenching (k_q) for β-carotene is 3×10^{10} litres $mol^{-1} s^{-1}$.[415] Caroteneoides are also very effective deactivators of $^1O_2$ in the protection of polybutadiene.

Hindered phenols are capable of both reacting with and quenching physically singlet oxygen ($^1O_2$), depending on their substituents (R_1, R_2 and R_3) and the reaction conditions (cf. Section 3.1.1).

2.7 PHOTOSTABILITY OF ADDITIVES

Most antioxidants and photostabilizers are photostable in polymer matrix in the absence of oxygen, but they are destroyed by photo-oxidative processes, probably by peroxy radicals (ROO• and/or POO•).[242] Several photostabilizers are almost completely destroyed well before the end of their photo-oxidative lifetime of the polymer film. In Figs 2.13 and 2.14 are shown changes of absorption spectra of some photostabilizers under UV irradiation.

The effect of UV irradiation for a number of antioxidants and photostabilizers is shown in Table 2.5.

Residues of metal catalysts used for polymerization, e.g. titanium and aluminium, increase the photolysis of phenolic antioxidants[60] and 3,5-di-*tert.*-butyl-4-hydroxybenzoate light stabilizer.[95]

2.8 SYNERGISM AND ANTAGONISM BETWEEN ADDITIVES

The mixture of additives is called *synergistic* if its effectiveness (observed effect) is greater than would be expected from the additivity law (i.e. the total sum of the effects of each additive individually).

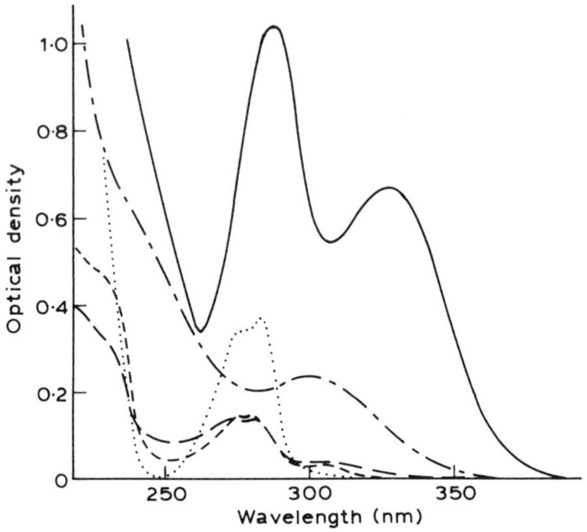

Fig. 2.13. Absorption spectra of stabilizers (0·5 wt%) in polypropylene films (40 μm): ···, 2,6-di-*tert.*-butyl-4-methylphenol; ——, octadecyl 3-(3,5-di-*tert.*-butyl-4-hydroxyphenyl)propionate (Irganox 1076; **3.21**); – –, 2-hydroxy-4-*n*-dodecyloxybenzophenone (Chimassorb 125; **5.31**); ———, 2,2′-thiobis(4-*tert.*-octylphenolato)-*n*-butylamine nickel(II) (Cyasorb UV 1084; **5.63**); ---, bis(*o*-butyl-3,5-di-*tert.*-butyl-4-hydroxybenzyl phosphonate) nickel(II) (Irgastab 2002; **5.67**).[242]

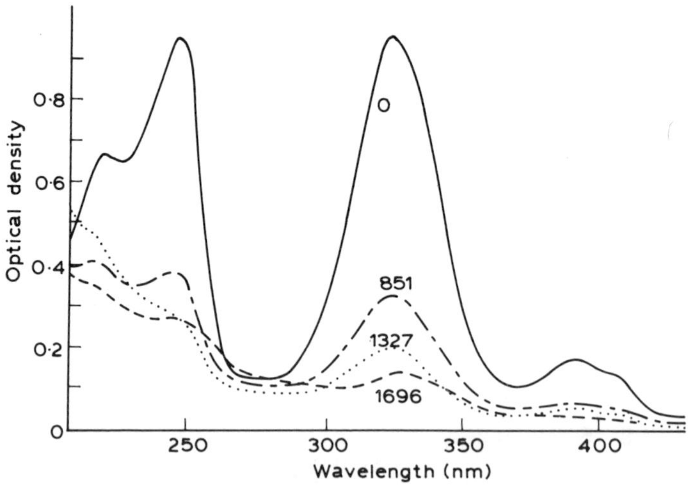

Fig. 2.14. Changes of absorption spectra of dibutyldithiocarbamate nickel(II) (**5.66**) (0·1 wt%) in polypropylene films (40 μm) during UV irradiation in the xenon-arc Weatherometer.[242] The numbers on the curves are irradiation times in hours.

Table 2.5
Effects of irradiation of phostabilizers[242]

Stabilizer	Structure	Photoresistance in PPH			Photoresistance in iso-octane[a,b]			
		Concn (wt %)	*Under N_2 τ_{N_2} (h)*	*Under air τ_{O_2} (h)*	*Concn (wt %)*	*τ_0 (h)*	*τ_0/τ_{ox} in t-octyl OOH*[c]	*τ_0/τ_{ox} in t-BuOO-t-Bu*[d]
2,6-Di-*tert.*-butyl-4-methylphenol		0·5	110	17	0·072	19	4·6	1·4
Octadecyl 3-(3,5-di-*tert.*-butyl-4-hydroxyphenyl)propionate (**3.21**)	See Table 3.2	0·5	450	200	0·082	21	4·5	1·2
2-Hydroxy-4-dodecyloxybenzophenone (**5.31**)	See Table 5.5	0·5	>6 000	1 300	0·030	40	0·99	0·95
2,4-Di-*tert.*-butylphenyl-4′-hydroxy-3′,5′-di-*tert.*-butylbenzoate (**5.7**)		0·5	>5 000[e]	1 100[e]	0·040	9	2·3	1·3
1,2,2,6-Tetramethyl-4-hydroxypiperidine		0·1	>120	50	0·063	>100	N.D.[f]	(τ_{ox} = 32 h)
2,2,2,6,6-Tetramethyl-4-hydroxypiperidine oxy radical (**6.30**)		0·1	~300	50	0·067	33	1·1	1·8
2,2′-Thiobis(4-*tert.*-octylphenolato)-*n*-butylamine Ni(II) (**5.63**)	See Table 5.9	0·5	≫2 000	700	0·040	60	2·8	0·95
Dibutyldithiocarbamate Ni(II) (**5.66**)	See Table 5.9	0·1	~6 000	500	0·018	120	N.D.[f]	20
Bis(*o*-butyl-3,5-di-*tert.*-butyl-4-hydroxybenzylphosphonate) Ni(II) (**5.67**)	See Table 5.9	0·5	~2 000	400	0·046	22	3.1	1·3

[a] τ_0 = time for the stabilizer to photolyse to half its initial level in iso-octane under N_2.
[b] τ_{ox} = half-life in the presence of the peroxide or hydroperoxide.
[c] *t*-Octyl hydroperoxide concentration $2{\cdot}0 \times 10^{-3}$ M.
[d] Di-*t*-butyl peroxide concentration $1{\cdot}2 \times 10^{-3}$ M.
[e] Stability of the photo-Fries product.
[f] Not determined because the stabilizer is a hydroperoxide decomposer.

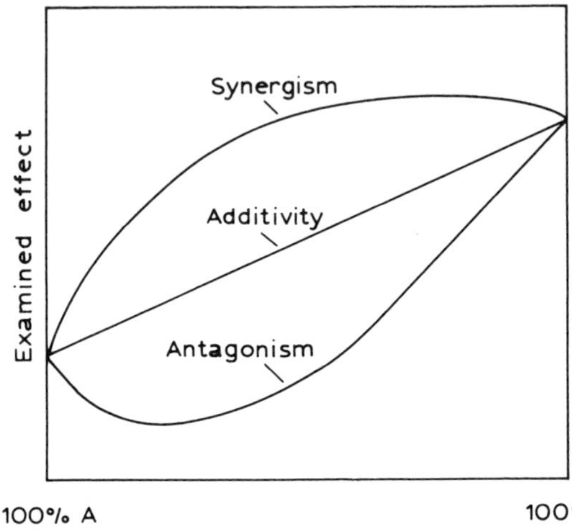

Fig. 2.15. Synergism and antagonism effects.

In the opposite case, when the overall effect is weaker, *antagonism* between the additives occurs. Both effects are shown in Fig. 2.15.

The different definitions encountered in practice are shown in Table 2.6.

A typical synergistic curve obtained from mixtures of calcium bis(3,5-di-*tert.*-butyl-4-hydroxybenzylmonoethylphosphonate) (Irganox 1425; **3.34**) and distearyl pentaerythrityl diphosphate (Weston 618; **3.113**) is shown in Fig. 2.16.[51]

Table 2.6
Definition of terms associated with combination of additives[216]

Description of effect	*Additive concentration*		*Effect*
	[*X*]	[*Y*]	
	1	—	x
	—	1	y
	2	—	$2x$
	—	2	$2y$
Additive effect	1	1	$x + y$
Antagonism	1	1	$<(x + y)$
Synergism	1	1	$>(x + y)$
Practical synergism	1	1	$>(x + y)$ and $>2x$ and $>2y$

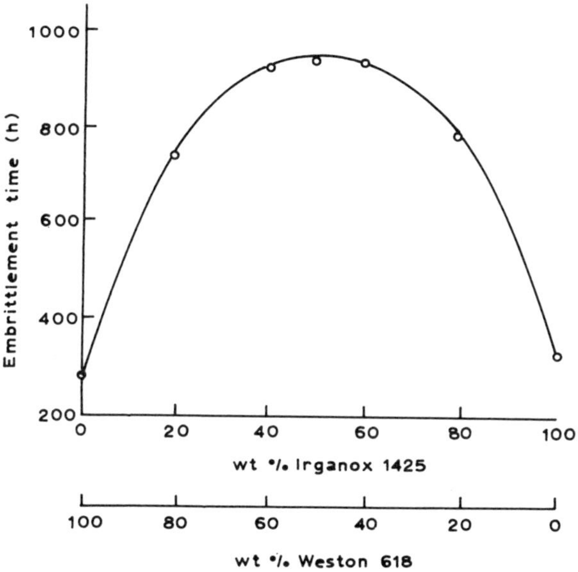

Fig. 2.16. Synergism between calcium bis(3,5-di-*tert.*-butyl-4-hydroxybenzylmonoethylphosphonate) (Irganox 1425; **3.34**) and distearyl pentaerythrityl diphosphate (Weston 618; **3.113**) in the absence of other additives, at a total concentration of 0·2 wt% in polypropylene films.[51]

The shape of interaction between different stabilizers (and additives) can be determined only by experimental examination.

Synergistic effects in polymeric systems (plastics) are commonly employed in order to:[216,679]

(i) improve polymer stability;
(ii) reduce the cost of the stabilizing compositions, by using the cheaper additives in place of more expensive ones;
(iii) evade the patent regulations.

The mechanism of synergistic/antagonistic effects is still not well known; however, synergistic mixtures have been in practical use for polymer stabilization for several decades.

The synergistic effect that is observed depends very much on the type of effect being examined, e.g. induction period, oxygen uptake, concentration of hydroperoxide and/or carbonyl groups, change in mechanical properties, etc. One of the effects chosen to be examined can indicate a synergistic effect, whereas another effect examined may even show an antagonistic action in the same polymer–additive compositions. This gives rise to much confusion in published results.

Polymer ageing and stabilization processes are never uniform through the sample thickness due to non-uniform light absorption, oxygen diffusion, migration of additives in the polymer matrix and so on, and for that reason it is difficult to define experimental conditions in which synergistic/antagonistic effects should be measured.

Initial proposals about the influence on synergism effects of variation of light intensity and concentration of antioxidants/photostabilizers in a polymer matrix have been considered.[597,600,602,969] The nature of the effect is shown in Fig. 2.17 where the light and antioxidant (S_1) concentration profiles in a polymer film of thickness L undergoing photo-oxidation in the presence of a photostabilizer are

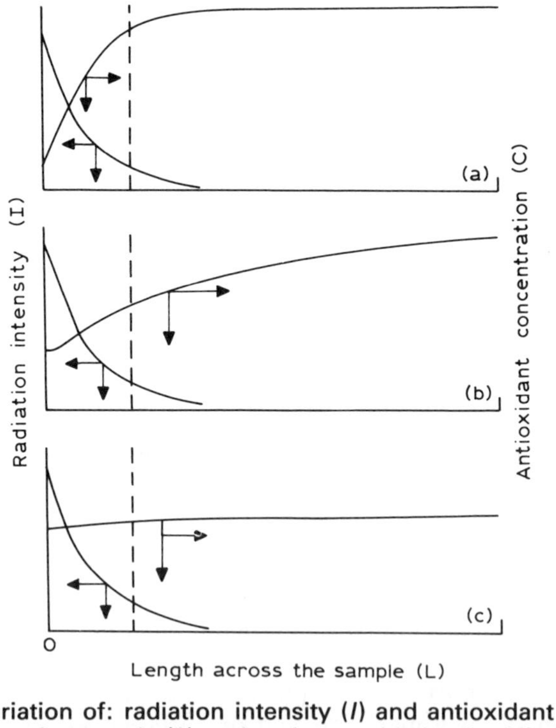

Fig. 2.17. Variation of: radiation intensity (I) and antioxidant concentration (C) across the sample (L) under photo-oxidation in different modes: (a) no diffusion; (b) slow diffusion; and (c) fast diffusion. The broken line at l marks the photoreaction zone boundary.[597,603]

shown. The photoreaction is mostly limited to the layer adjacent to the irradiated surface. During the reaction the antioxidant is consumed. The resultant concentration gradient gives rise to a diffusion flow of antioxidant (S_1) from the deeper layer (protected from the action of light) to the photoreaction zone, thereby increasing the induction period of photo-oxidation. A quantitative analysis of the mechanism of diffusion synergism was carried out. The synergism condition has been described by the equation:[603]

$$\frac{t_{AB} - t}{(t_A - t) + (t_B - t)} > 1 \qquad (2.13)$$

where t_A and t_B = times required for variation of a characteristic property under the action of the individual additives A and B, respectively,

t_{AB} = time required for variation of a characteristic property under the action of the mixture of A and B,

t = time in the absence of additives.

There are a number of different proposed mechanisms of synergistic or antagonistic behaviour of additives:[603]

(i) Mechanism based on fast and slow diffusion transfer, e.g. of antioxidant to the reaction zone.

(a) Fast diffusion: Antioxidant concentration rapidly becomes about the same across the polymer film. Almost all the antioxidant in the film is consumed for inhibition of the photoreaction. Criterion: $t_{AB} \gg t_l$.

(b) No diffusion: During time t_{AB} the diffusion is insignificant and each layer reacts independently. The reaction is inhibited by the antioxidant available in the reaction zone. Criterion: $t_{AB} \ll t_l$.

(c) Slow diffusion: Condition intermediate between fast diffusion and no diffusion. Antioxidant transfer has a strong effect, but diffusion is insufficient to equalize the concentration. Only a portion of antioxidant present in the film is used for inhibition of photoreaction. Criterion: $t_l < t_{AB} < t_L$.

Note: t_{AB} is the induction period in the presence of the mixture of additives A and B: t_l and t_L are diffusion times to distances l and L, respectively.

Table 2.7
Synergism in photostabilization of polymers[603]

Mechanism	*Effect under optimum conditions*	*Optimum concentration ratio*[a]	*Dependence on increasing parameters:*			*Essential properties of stabilizers*	*Essential properties of polymers*
			Film thickness	*Light intensity*	*Temperature*		
1. Antioxidant diffusion towards photoreaction zone	Strong	$c_1 = c_2$	Increases	Independent or decreases	Independent or increases	Light stabilizer is a strong absorber	$\tau_l \leq \tau_{1,2}$[b,c]
2. Protection of light stabilizer with antioxidant	Medium	$c_1 > c_2$	Increases	Independent or decreases	Independent or increases	Stabilizer is a strong absorber and has a low inhibition factor	—
3. Antioxidant S_1 protects polymer during processing stabilizer S_2 during service	Medium	$c_1 > c_2$	Increases	Independent or decreases	Independent or increases	Stabilizer: strong absorber, low inhibition factor; Antioxidant: low photochemical activity	—
4. Enhanced activity of UV absorber in the presence of antioxidant	Weak	—	—	—	—		—
5. Stabilizer S_1 protects polymer at the stabilizer S_2 activation stage	Weak	$c_1 > c_2$	—	—	—	Principal stabilizer conversion product is active	Readily thermo-oxidizable

6. Stabilizer S_1 is radical scavenger; stabilizer S_2 is peroxide decomposer	No critical effects during photo-oxidation. No synergism.						
7. Quenching of antioxidant S_1 excited states with quencher S_2	Medium	$c_1 > c_2$	—	—	Independent or increases	Antioxidant is photochemically active	Non-uniform oxidation $c_{loc} \gg c_{av}$ [d]
8. Increase of solubility and/or redistribution of stabilizer S_1 in the presence of an agent S_2	Weak	$c_1 \simeq c_2$	—	—	Decreases	Polar compounds; low solubility	Nonpolar
9. Formation of effective stabilizers from less active ones	Strong	$c_1 \simeq c_2$	—	—	Increases	Low effectiveness	—
10. Regeneration of stabilizer S_1 with agent S_2	Medium	$c_1 \ll c_2$	—	—	Increases	Light-stable S_1 and S_2 and their conversion products	Sufficiently high molecular mobility

[a] c_1, c_2 are concentrations of S_1 and S_2.
[b] τ_l is time of antioxidant diffusion to a distance $l = 1/\varepsilon_2 c_2$, where ε_2 is the absorption coefficient of UV absorber.
[c] $\tau_{1,2}$ is the induction period of oxidation in the presence of a mixture of stabilizers.
[d] c_{loc} and c_{av} are the local and average stabilizer concentrations.

(ii) Mechanism based on the assumption that antioxidant protects photostabilizer through reduction of the concentration of radicals formed during photoreactions of impurities, chromophoric groups in the polymer and photo-oxidation products.[262,603,1034] On the other hand, photostabilizer can protect the antioxidant.

(iii) Mechanism based on the complementary action of one stabilizer during processing (antioxidant) and of another during service (photostabilizer).[30,62,262,266]

(iv) Mechanism based on the enhancement of the effect of a photostabilizer in the presence of an antioxidant due to a change in the termination modes.[380,629,759] The rate of chain termination in the presence of antioxidant is proportional to light intensity, whereas in the absence of antioxidant the chain termination is a bimolecular process and its rate is proportional to $\sqrt{I}$. This means that antioxidant enhances the effectiveness of photostabilizer which reduces the initiation rate.

(v) Mechanism based on polymer protection using a subsidiary stabilizer which protects it during the activation of the primary stabilizer.[1060] In this proposed mechanism the added stabilizer in its original form does not participate in the stabilization mechanism; however, the products of its chemical transformation may participate effectively in such a mechanism.

(vi) Mechanism based on interaction of free-radical scavengers and hydroperoxide decomposers.[598,1061] It has been shown that mixtures of radical scavengers and hydroperoxide decomposers do not show any synergistic effect during photo-oxidation.

(vii) Mechanism based on deactivation of excited states of antioxidants (donors) and quenchers (acceptors).[724]

(viii) Mechanism based on improving compatibility of additives with the polymer matrix with the help of other additives (e.g. surfactants such as hydroxyethylated cetyl alcohol).[366,367]

(ix) Mechanism based on formation of effective stabilizers by transformation of stabilizing or non-stabilizing additives.[71]

(x) Mechanism based on regeneration of the effective stabilizer by some reducing agents. It has been observed in the thermo-oxidation of polymers but not in photo-oxidation.[603]

In Table 2.7 comparisons are presented of proposed mechanisms (i)–(x). Mechanisms (i)–(v) and (vii)–(ix) have been observed experimentally in polymers undergoing photo-oxidation. Mechanism (vi) probably does not occur, because no synergistic effect has been observed. Mechanism (x) also has not been realized yet. The highest synergism is obtained from the systems acting by more than one stabilization mechanism.

A number of important conclusions for selecting and testing synergistic systems have been made:[365,599–603,725,969]

(i) The light stability of a sample is proportional to its thickness (fast diffusion, samples not very thick).
(ii) The synergistic effect depends on light intensity, since with increasing intensity it becomes possible to pass from the fast diffusion condition where the effect is maximal to the slow diffusion mode.
(iii) The synergistic effect depends on the molecular weight of anitoxidant since the greater the molecular weight and, consequently, the greater the molecular size, the lower the diffusivity.
(iv) The effectiveness of stabilizer mixtures depends on the absorptivity of the UV photostabilizer. Therefore, for a given spectrum of active light the additive which has a higher absorption in this wavelength range will be more effective. The light stability of the sample containing a stabilizing system will be dependent on the incident light spectrum.
(v) For a given total concentration of stabilizer, the synergistic effect is maximal when the stabilizers are mixed in the ratio 1 : 1.

Synergistic and antagonistic effect observed in different types of mixtures of antioxidants with photostabilizers are discussed separately in sections devoted to these additives.

3

Antioxidants

3.1 HINDERED PHENOLS

Hindered phenols are kinetic chain-breaking antioxidants. In the primary reaction the phenoxy radical (**3.1**) is formed by one of the following mechanisms:[500,906,911,1171]

(i) In a solution (e.g. methanol), phenols may form charge-transfer (CT) complexes with dissolved oxygen, which under UV irradiation are further photolysed into phenoxy radicals and hydroperoxy radicals ($HO_2\cdot$):

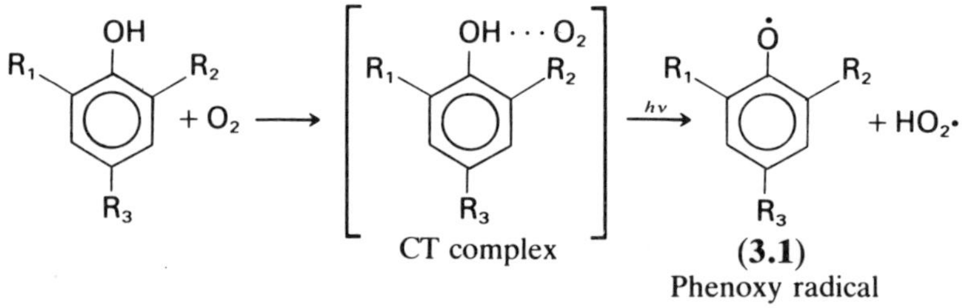

(**3.1**)
Phenoxy radical

(3.1)

Photolysis of hindered phenols is a complicated process and the final products formed from the photolysis are the result of different transformation reactions of free radicals.[907] The mechanism of these transformation reactions depends on the structure of the hindered phenols (*ortho* and *para* positions of substituents R_1, R_2 and R_3, their size, etc).

(ii) Phenoxy radicals may also form as a result of a hydrogen abstraction reaction from the phenols by different types of free radicals such as alkyl (R•), alkoxy (RO•) and/or alkylperoxy (ROO•) radicals which can be formed from internal and/or external impurities in the polymer:

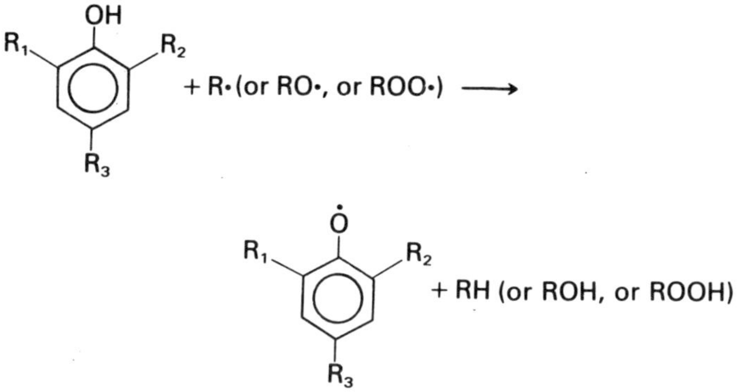

(3.2)

This radical chain-breaking reaction is, in general, considered to be an important first step in the stabilization process of hindered phenols.

The stability and reactivity of phenoxy radicals (**3.2–3.5**) is determined by steric effects of the substituents R_1, R_2 and R_3 and by the extent of delocalization of the unpaired electron.[132,228,321,355,709,751,950,951,1150] In a series of phenoxy radicals, the lifetime increases as steric hindrance in the *ortho* and *para* positions increases:

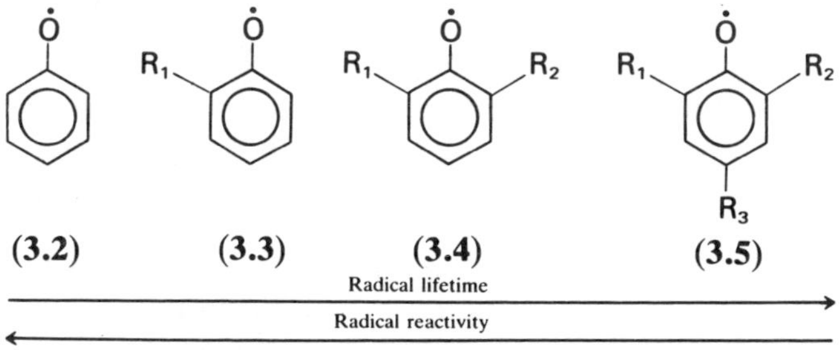

(**3.2**) (**3.3**) (**3.4**) (**3.5**)

Melting and boiling points of the 2,4,6-trisubstituted phenols are shown in Table 3.1. Phenoxy radicals with small substituents (R_1 and R_2) in the *ortho* positions are less stable then those with bulky groups because of the ease of dimerization. A bulky substituent such as a tertiary butyl group in the position *ortho* or *para* to the phenolic carbon has a significant steric hindrance effect, thus minimizing

Table 3.1
Melting and boiling points of the 2,4,6-trisubstituted phenols[767]

Substituent in the 4-position	*Melting point (°C)*	*Boiling point (°C)*
H	41	180
CH_3	70	—
C_2H_5	44	—
C_3H_7	36	—
tert.-Butyl	130	136
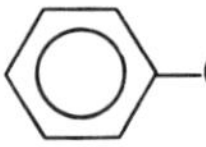 CH_2	59	—
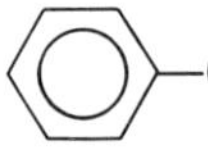 $(CH_3)_2$—	76	150
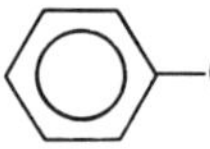 $(CH_3)CH$	41	109
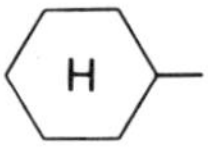	98	—
—$CH_2CH_2COOC_{18}H_{37}$	50	—

dimerization. The following order of steric hindrance is valid:

tert.-butyl > benzyl > alkyl > methyl > hydrogen

The formation and concentration of phenoxy radicals can be determined by ESR spectroscopy. The spin concentration measurements can provide a basis for comparing the radical scavenging activity of different antioxidant systems. The concentration of radicals formed from antioxidant cannot be measured directly within the bulk polymer, because of their low concentration in relation to the polymer radicals formed. With a model system composed of the antioxidant in solution with a known concentration of initiating radicals it is possible to measure ESR spectra of antioxidant radicals and their concentration (cf. Section 3.1.2).

The hindered phenoxy radicals (**3.5**), depending on substituents, can be more or less stable (even over a period of 400 h), because they can isomerize into more resonance-stabilized cyclohexadienonyl radicals

(**3.6** and **3.7**):

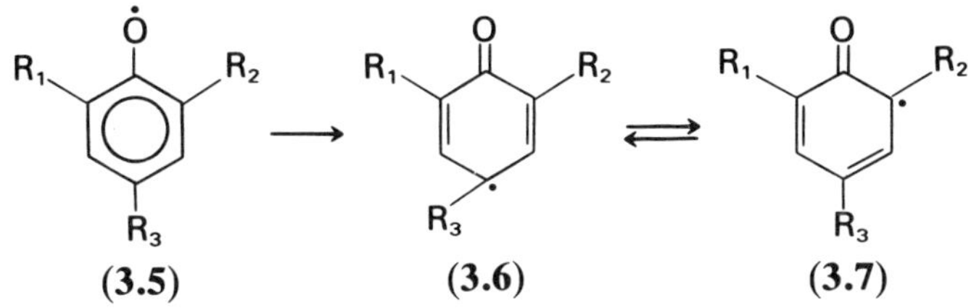

(3.3)

The rate of isomerization is of the order of 10^8 litre mol^{-1} s^{-1}.[286,355]

End-of-pulse transient absorption spectroscopy of 2,6-di-*tert.*-butyl-*p*-cresol (Topanol OC) (**8.3**) shows the formation under UV irradiation of phenoxy radicals, which can further decay by three bimolecular processes, namely:[98,696]

(i) Head-to-head to give a peroxy compound:

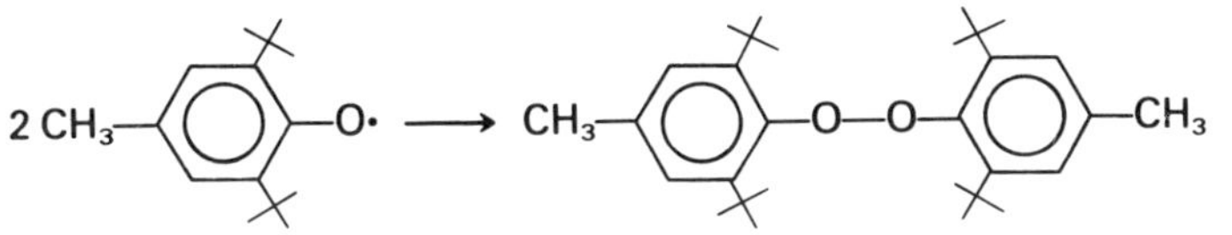

(3.4)

(ii) Tail-to-tail to give a bisphenol coupling product:

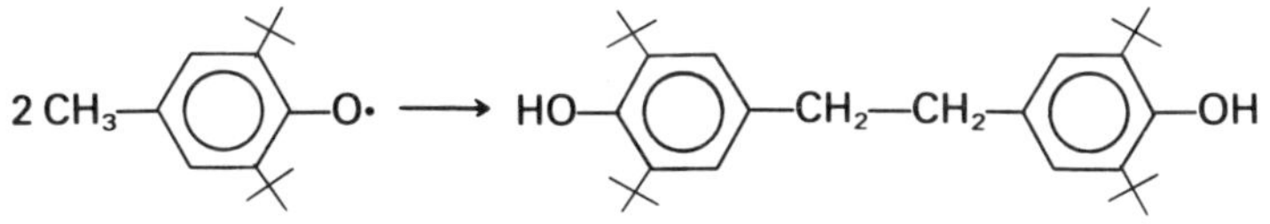

(3.5)

(iii) Head-to-tail to give a phenyl-ether-phenol:

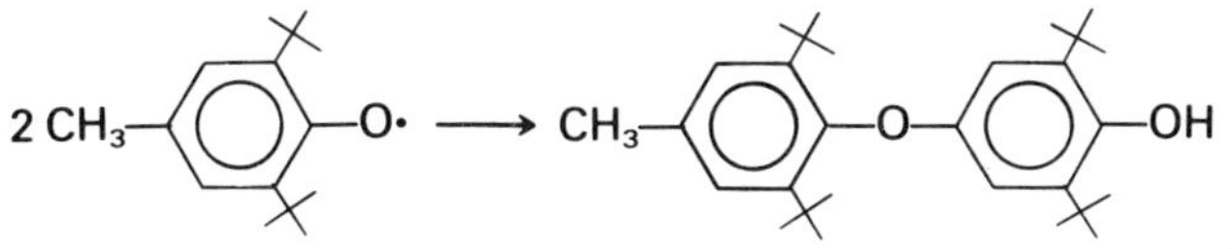

(3.6)

In secondary reactions cyclohexadienonyl radicals undergo, subject to the steric effects of the R_1, R_2 and R_3 groups a C—C coupling reaction to cyclohexanone dimers (**3.8**):

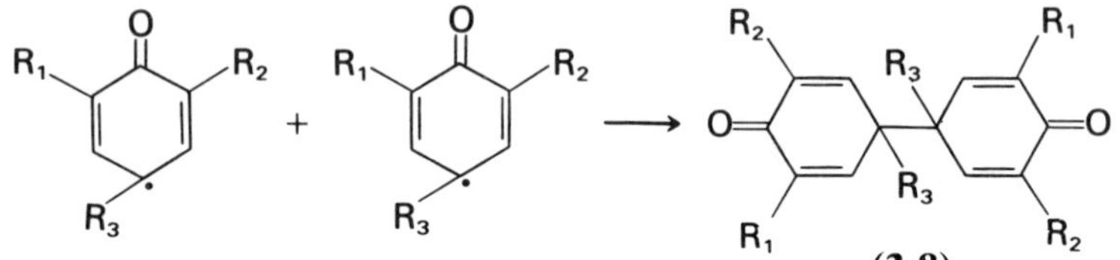

(3.7)

(**3.8**)

This C—C coupling reaction occurs for cresol ($R_1 = H$ and $R_3 = CH_3$),[621] for 2,4- or 2,6-dialkyl phenols[565] and for 2,4,6-trisubstituted phenols which have a non-bulky and readily cleavable substituent in the *ortho* or *para* position.[132]

The cyclohexadienones (**3.9**) formed from disubstituted phenols may, under UV irradiation, intermolecularly isomerize to 4,4′-bisphenyldiols (**3.10**):

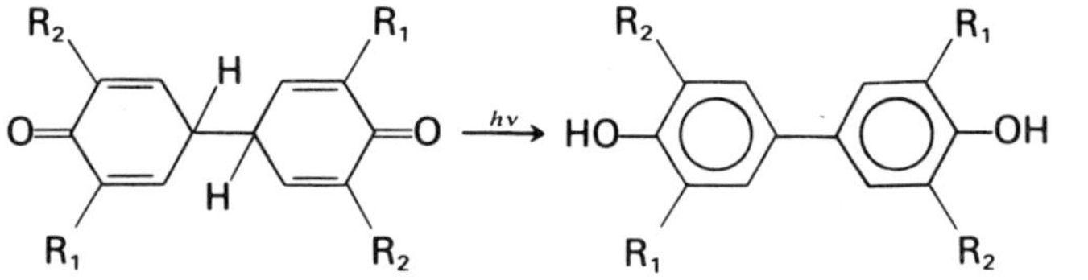

(3.8)

(**3.9**) (**3.10**)

Phenols which have a methyl group in the *para* position may, in the presence of alkyloxy (RO•) and/or alkylperoxy (ROO•) radicals, isomerize to 4,4′-ethylenebis(2,6-alkylphenol) (**3.11**) according to the following mechanism:[163,588]

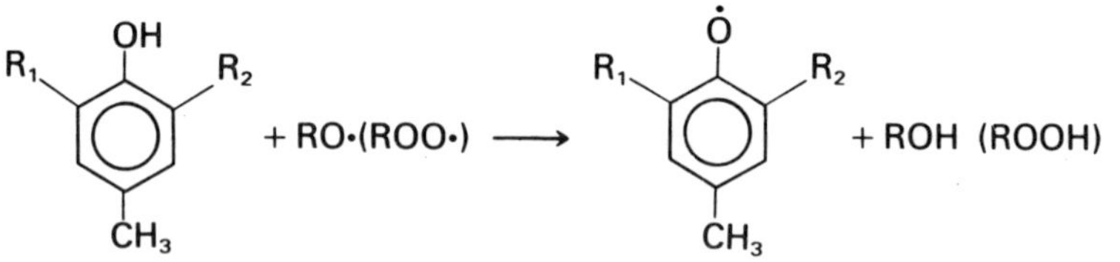

(3.9)

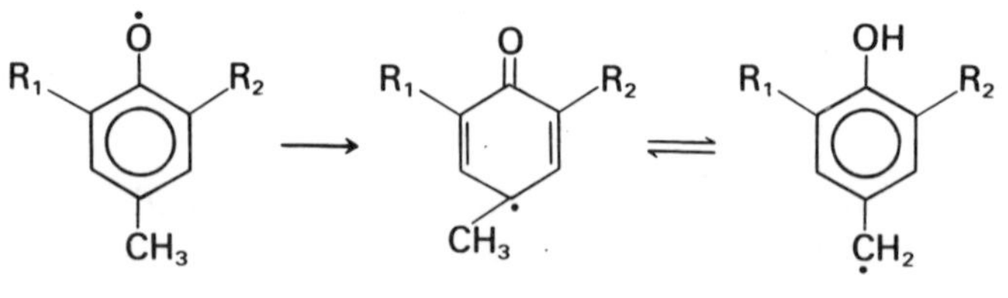

(3.10)

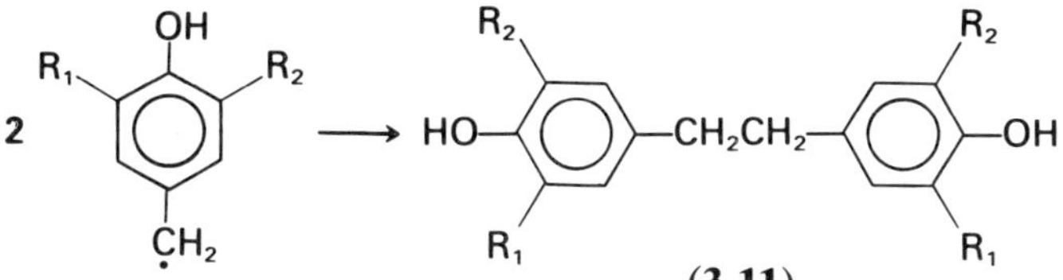

(3.11)

(3.11)

Substituted 4,4′-bisphenyldiols (**3.12**) and 4,4′-ethylenebis(2,6-alkylphenols) (**3.14**) also act as stabilizers in reactions similar to reactions (3.1) and (3.2), during which diphenylquinone (**3.13**) and diquinone methide (**3.15**) are formed, respectively:[906,1012]

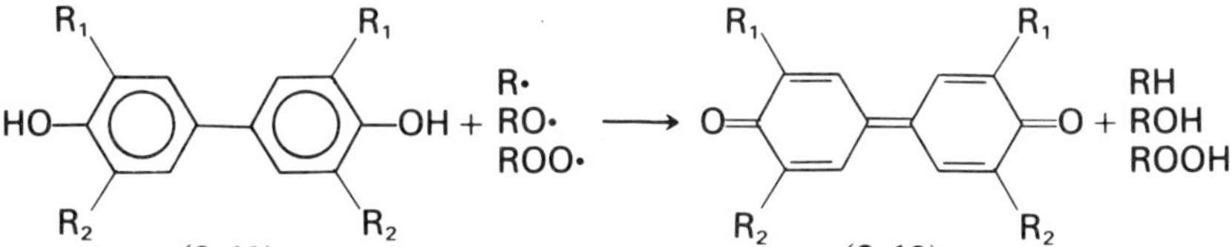

(**3.12**) (**3.13**)

(3.12)

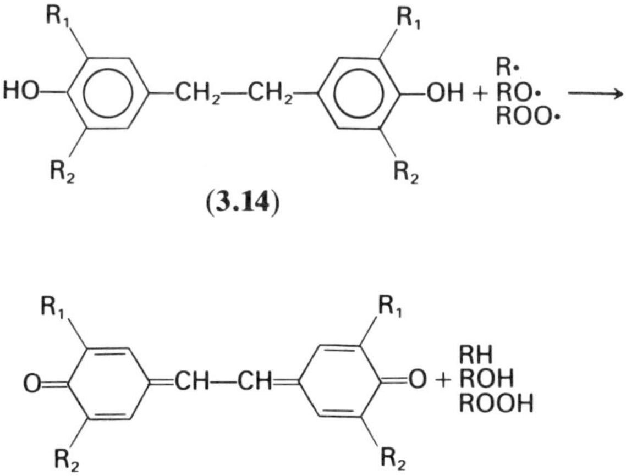

(**3.14**)

(**3.15**)

(3.13)

The conversion of substituted phenols to active oxidation products is facilitated by the presence of hydroperoxide groups in the polymer.

Phenoxy radicals (**3.5**) may also react with cyclohexadienonyl radicals (**3.6**) (C—O coupling) according to the reaction:[1033]

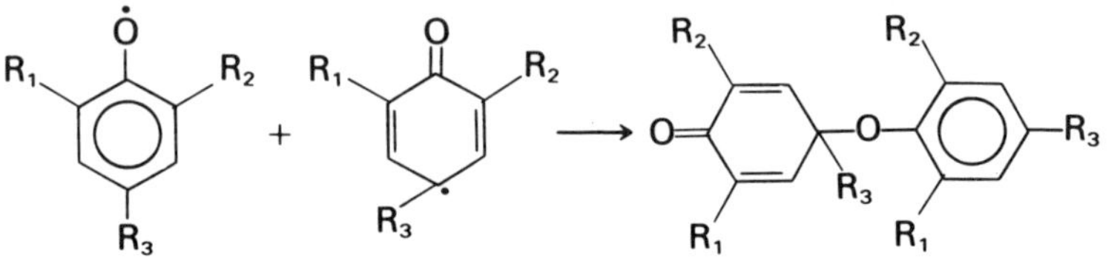

(**3.5**) (**3.6**)

(3.14)

In the case of disubstituted phenols (where $R_3 = H$, e.g. 2,6-di-*tert.*-butylphenol), under UV irradiation intermolecular isomerization may occur:[588,621]

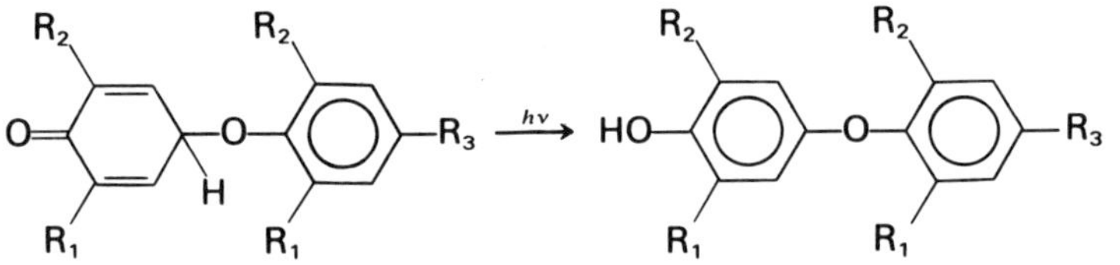

(3.15)

It is generally accepted that the chain-breaking reaction involves mainly cyclohexadienonyl radicals (**3.6** and **3.7**), which can react (scavenge) with different types of radicals formed during the thermal and/or photo-degradation of polymers, e.g. polymer alkyl (P•), polymer alkyloxy (PO•) and polymer alkylperoxy (POO•) radicals [reactions (3.17) and (3.18)].[906,907,923,1012]

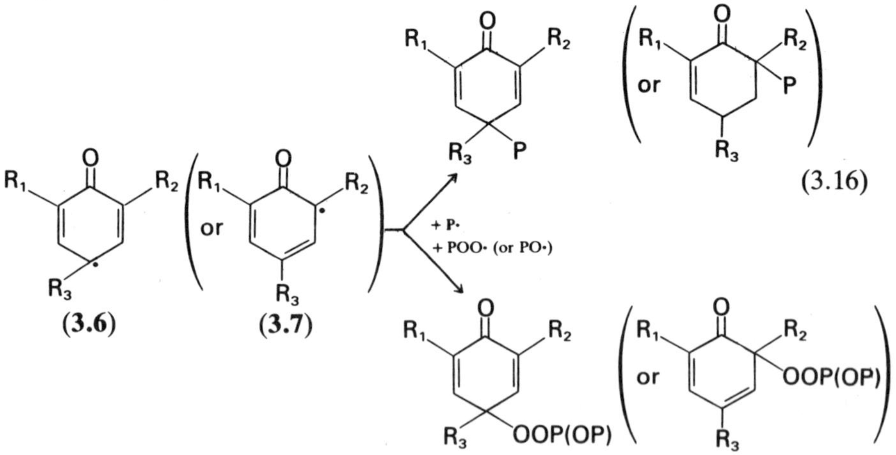

(**3.6**) (**3.7**)

(3.16)

(3.17)

It should be taken into consideration that the alkylperoxycyclohexadienones (**3.16**) formed are thermally and/or photochemically split into a new group of oxy radicals (**3.17**).

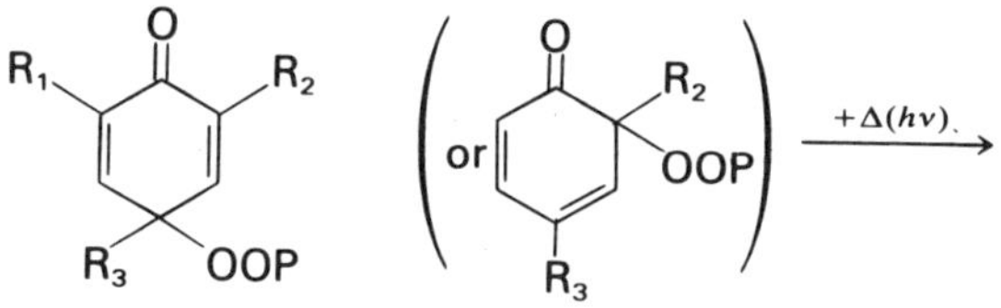

(**3.16**)

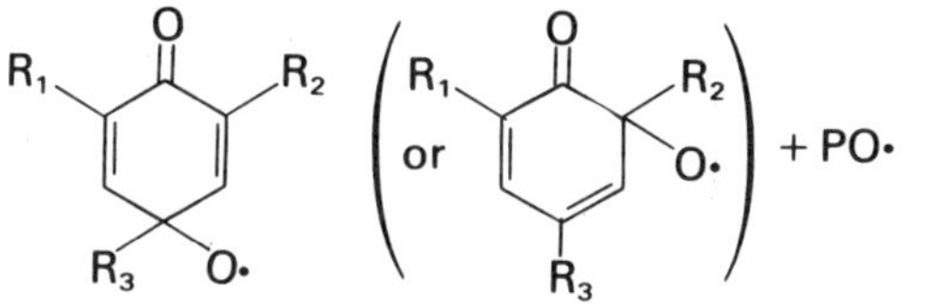

(**3.17**)

(3.18)

Cyclohexadienonyl radicals **3.6** and **3.7** may also react with oxygen and form cyclohexadienonyl peroxy radicals (**3.18**):

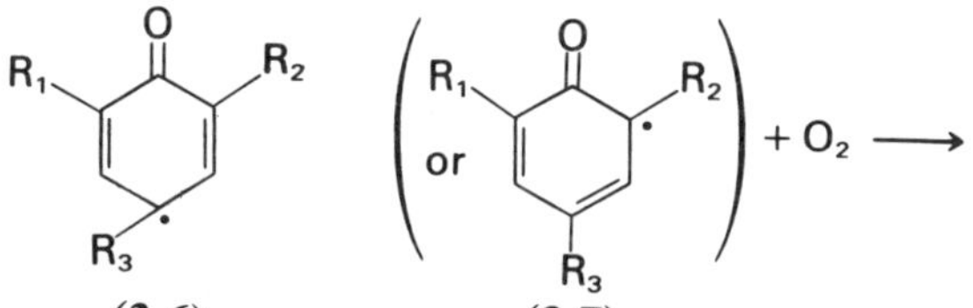

(**3.6**) (**3.7**)

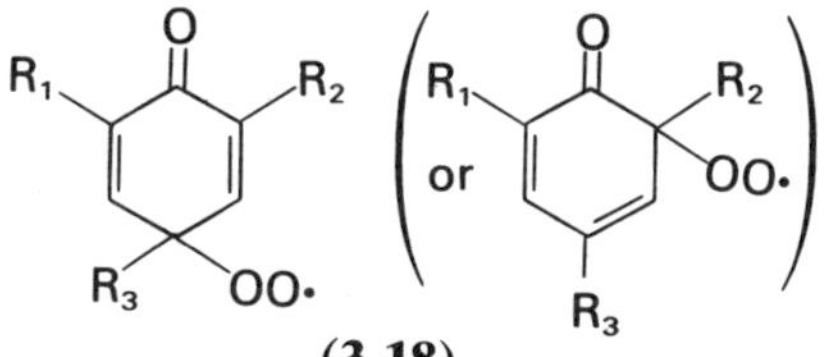

(**3.18**)

(3.19)

Both radicals (**3.18**) may further initiate free radical oxidation (they are pro-oxidants) and/or degradation and/or crosslinking of polymers.

Alkyloxycyclohexadienone radicals (**3.17**) can undergo α-cleavage reaction with formation of corresponding quinones (**3.19**) and alkyl radicals (R•):

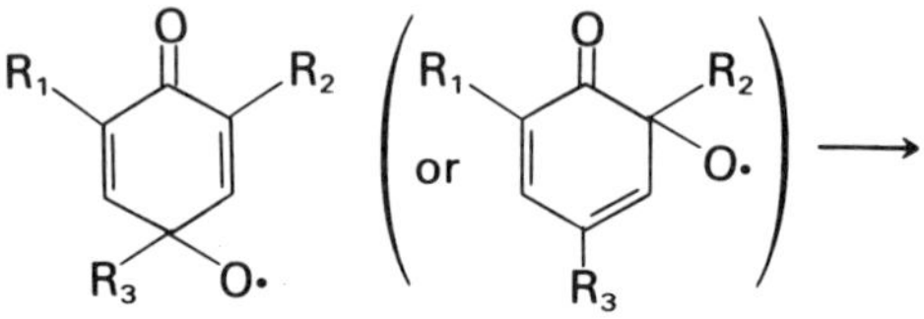

(3.17)

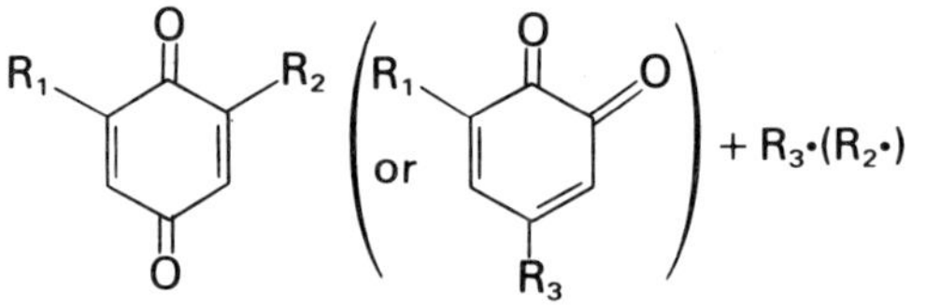

(3.20)

(3.19)

Alkyloxycyclohexadienone radicals (**3.17**) may also abstract hydrogen and produce corresponding hydroxyquinones (**3.20**):

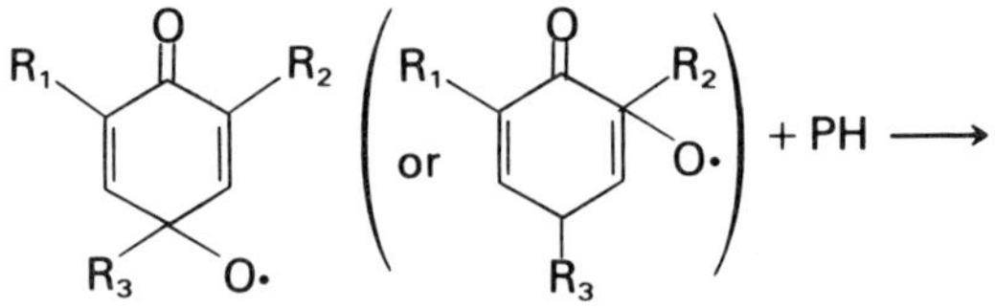

(3.17)

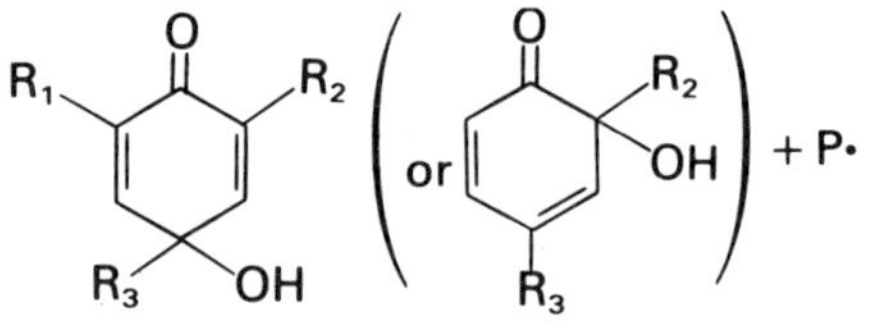

(3.21)

(3.20)

The above mechanisms involved in the chain-breaking processes do not include other possibilities such as the termination reaction between different types of radicals formed, transformation of different products, etc.

Hindered phenols have been carefully investigated, as antioxidants against free-radical thermal and/or photo- oxidation of polypropylene,[217,719,1176] polystyrene,[1127] and coatings.[550] A number of hindered phenols of different structures are produced commercially (Table 3.2).

The main disadvantage in practice is that most phenols are not very compatible with nonpolar solid polymers, and tend to form crystalline aggregates in the polymer matrix, leaving large volumes of unstabilized material. Since the centres for the antioxidant activity are, in general, polar groups (OH), there is a certain incompatibility between inert polymers and hindered phenols. In practice this effect can be overcome by incorporating long hydrophobic groups into the phenol molecules (cf. Table 3.2).

The stabilizing efficiency of hindered phenols can be enhanced by coupling them with other compounds, e.g. hindered amines (HALS) (Table 6.1).

3.1.1 Reaction of Hindered Phenols with Singlet Oxygen

Phenolic antioxidants are easily oxidized to quinones by singlet oxygen ($^1O_2$); however, they can also form phenoxy radicals by the following reactions:[772–774,854,1089,1090,1098]

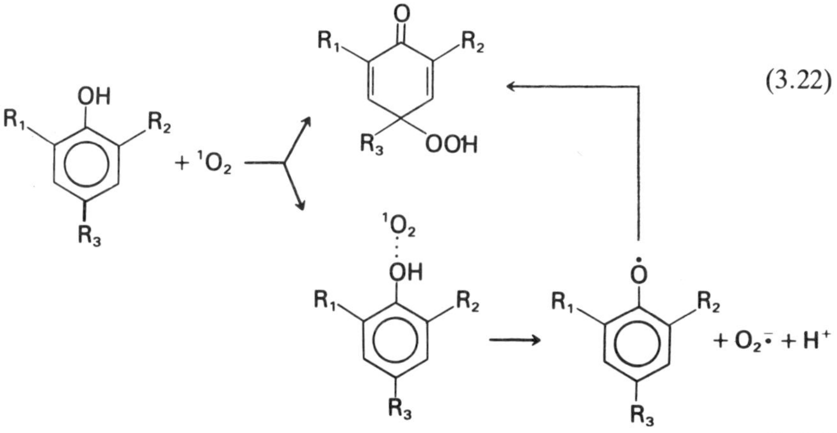

(3.22)

(3.23)

2,2′-Methylenebis(4-methyl-6-*tert.*-butylphenol) (**3.39**) is oxidized by the singlet oxygen (generated by methylene blue in methanol solution) according to reactions (3.24) and (3.25).[1091] Both phenyl rings are oxidized with formation of hydroperoxides of the 2,5-cyclohexadienone type (**3.40** and **3.41**) as main products, whereas other oxidized compounds are formed in an insignificant amount.

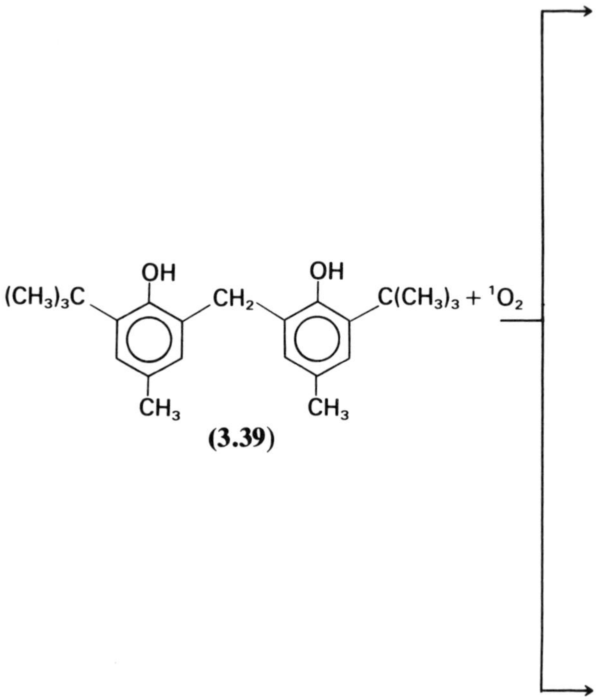

(3.39)

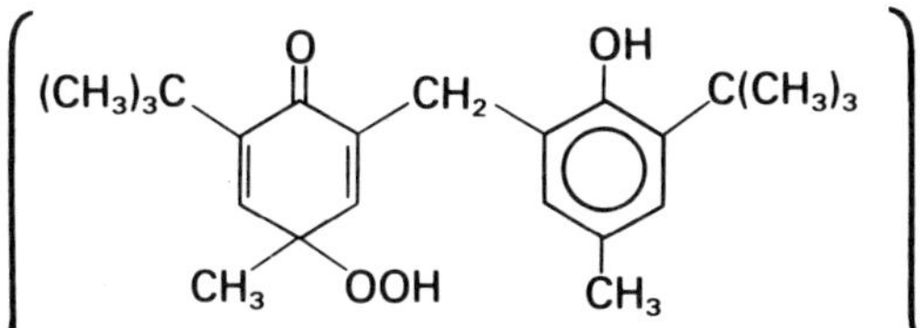

(3.40)

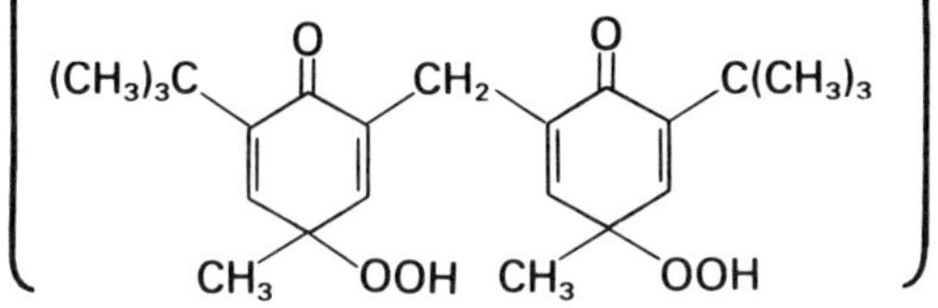

(3.41)

Main products (3.24)

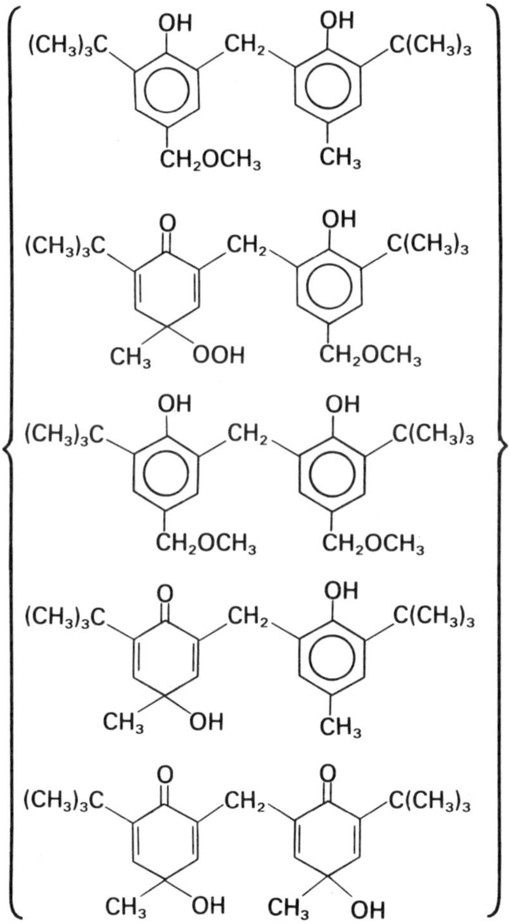

(3.25)

Table 3.2
Antioxidant and heat stabilizers based on hindered phenols: physical properties and solubility

Chemical name	*Commercial name*	*Chemical structure*
Octadecyl 3-(3,5-di-*tert*.-butyl-4-hydroxyphenyl)propionate *or*: Octadecyl 3,5-di-*tert*.-butyl-4-hydroxyhydrocinnamate *or CAS name*: Benzenepropanoic acid, 3,5-bis(1,1-dimethylethyl)-4-hydroxy, octadecyl ester; *CAS number* 2082-79-3	Irganox 1076 (Ciba–Geigy)	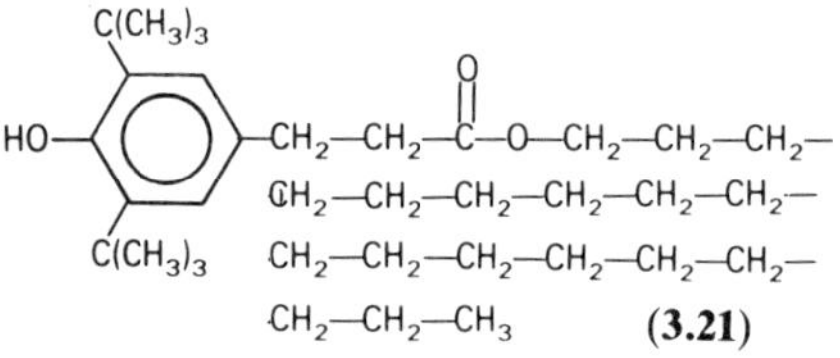 **(3.21)**
Triethyleneglycol bis[3(3-*tert*.-butyl-4-hydroxy-5-methylphenyl)propionate]	Irganox 245 (Ciba–Geigy)	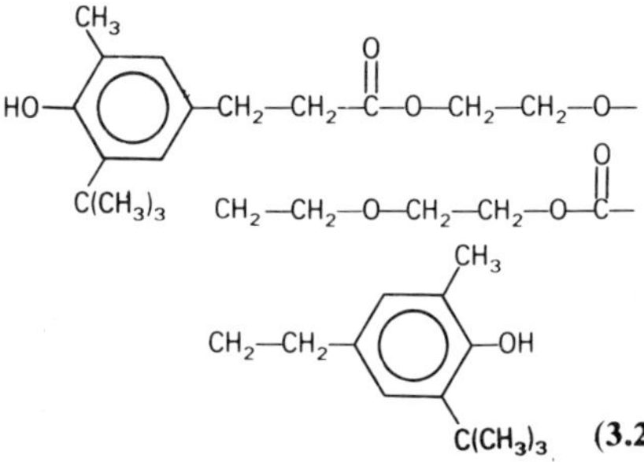 **(3.22)**
1,6-Hexanediol bis[3-(3,5-di-*tert*.-butyl-4-hydroxyphenylpropionate]	Irganox 259 (Ciba–Geigy)	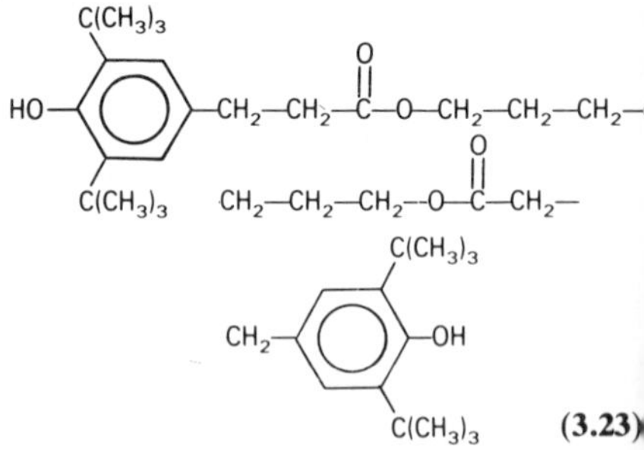 **(3.23)**

Molecular weight	*Appearance*	*Solubility (wt %)*							*Melting point (°C)*	*Recommended applications*
		Acetone	*Benzene*	*Chloroform*	*Ethyl acetate*	*Hexane*	*Methyl alcohol*	*Water*		
531	White to slightly yellowish powder	26	57	57	42	31	0·6	<0·01	49–54	Polyolefins, polystyrene, ABS, SBS, rubbers, poly(vinyl chloride), polyesters, polyamides, polyurethanes, adhesives
586·8	White to slightly yellowish powder	>50	18	>40	37	<0·1	12	<0·01	76–79	Polystyrene, ABS, SBS, rubbers, poly(vinyl chloride), polyurethanes, polyamides, thermoplastic polyesters
639	White crystalline powder	36	39	48	25	1·5	1·7	<0·01	103–108	Polyacetal and its copolymers

(continued)

Table 3.2—*contd.*

Chemical name	*Commercial name*	*Chemical structure*
CAS name: Benzenepropanoic acid 3,5-bis(1,1-dimethylethyl)-4-hydroxy-2[3-[3,5-bis(1,1-dimethylethyl)-4-hydroxyphenyl]-1-oxopropyl]hydrazide; *CAS number* 32687-78-8	Irganox MD1024 (Ciba–Geigy)	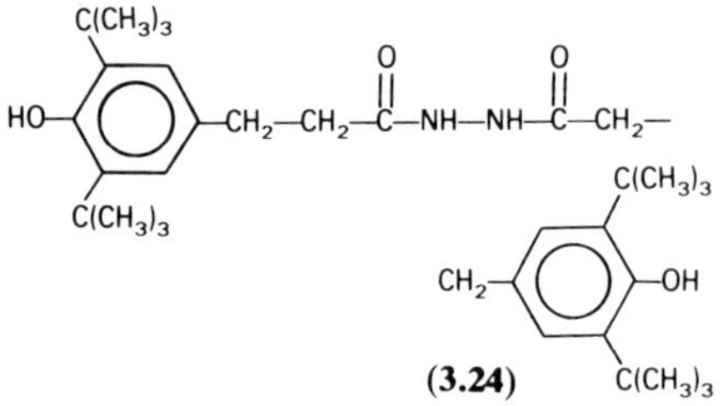 **(3.24)**
N,N′-Trimethylenebis(3,5-di-*tert.*-butyl-4-hydroxyhydrocinnamide)	Irganox 1019 (Ciba–Geigy)	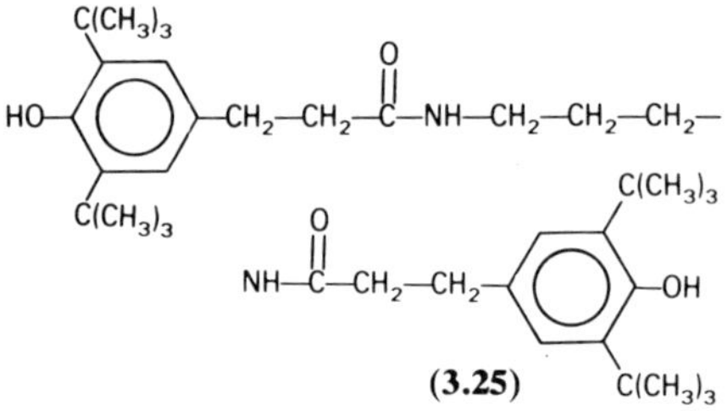 **(3.25)**
N,N-Hexamethylenebis(3,5-di-*tert.*-butyl-4-hydroxyhydrocinnamide)	Irganox 1098 (Ciba–Geigy)	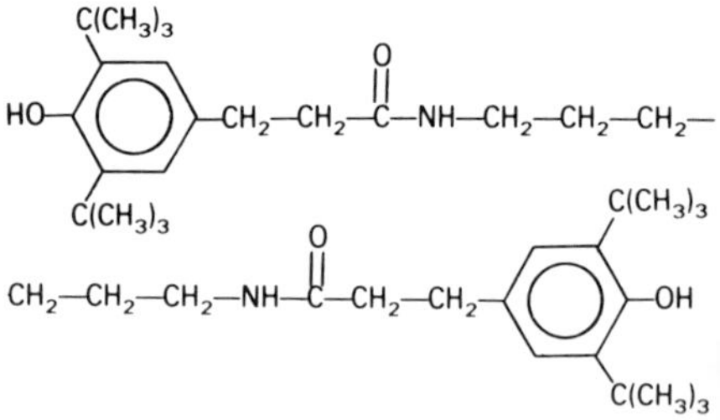 **(3.26)**
2,2′-Thiodiethylenebis[3-(3,5-di *tert.*-butyl-4-hydroxyphenyl)propionate] *or*: Thiodiethylenebis-(3,5-di-*tert.*-butyl-4-hydroxy)hydrocinnamate *or CAS name*: Benzenepropanoic acid, 3,5-bis(1,1-dimethylethyl)-4-hydroxythiodi-2,1-ethanediyl ester; *CAS number* 41484-35-9	Irganox 1035 (Ciba–Geigy)	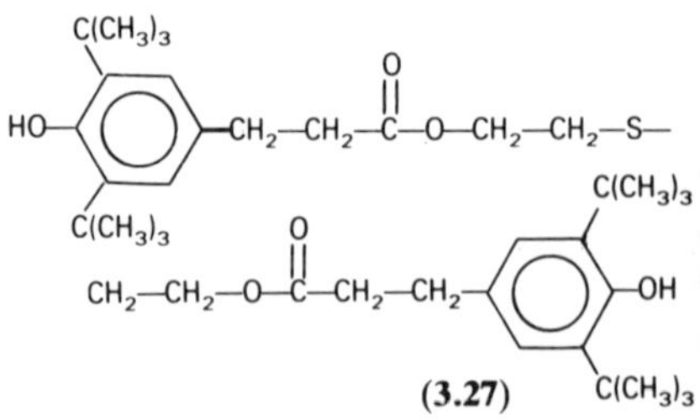 **(3.27)**

Molecular weight	Appearance	Solubility (wt %)							Melting point (°C)	Recommended applications
		Acetone	Benzene	Chloroform	Ethyl acetate	Hexane	Methyl alcohol	Water		
553	White powder	4			0·5	<0·01	4	<0·01	224–229	Polyolefins
590	White to off-white, free-flowing powder	2	0·1	14	0·5	0·1	30	0·01	175–179	Polyesters, polyamides, polyether-polyesters
637	White, free-flowing crystalline powder	2·8	0·01	6·6	1·2	0·01	6·0	0·01	156–161	Polyamides
643	White, free-flowing crystalline powder	>50	36	>50	50	5·0 at 50°C	6·7	<0·01	63–67	Polyolefins, EPDM, polybutadiene, SBS, polystyrene, ABS, poly(vinyl acetate), polyurethanes

(continued)

Table 3.2—*contd.*

Chemical name	*Commercial name*	*Chemical structure*
2,4-*Bis*(*n*-octylthio)-6-(4-hydroxy-3,5-di-*tert.*-butylanilino)-1,3,5-triazine	Irganox 565 (Ciba–Geigy)	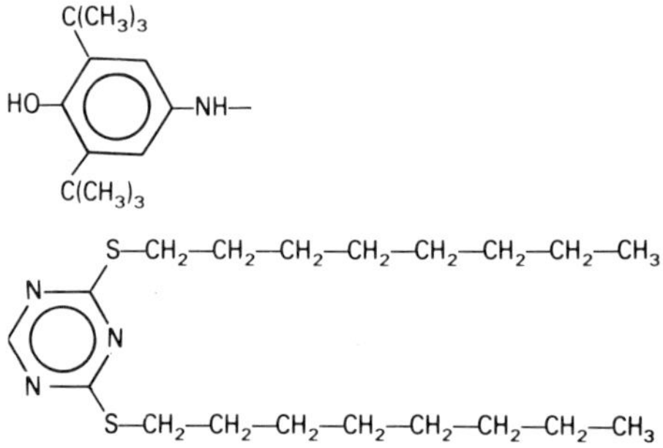 (3.28)
2,2′-Thiobis(4-methyl-6-*tert.*-butylphenol)	Irganox 1081 (Ciba–Geigy)	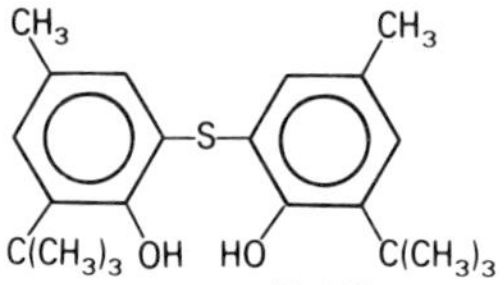 (3.29)
1,3,5-Trimethyl-2,4,6-tri(3,5-di-*tert.*-butyl-4-hydroxybenzyl)benzene *or CAS name*: Phenol, 4,4′,4″-[(2,4,6-trimethyl-1,3,5-benzenetriyl) tri(methylene)]tri[2,6-bis(1,1-dimethylethyl)]; *CAS number* 1709-70-2	Irganox 1330 (Ciba–Geigy) A0 330	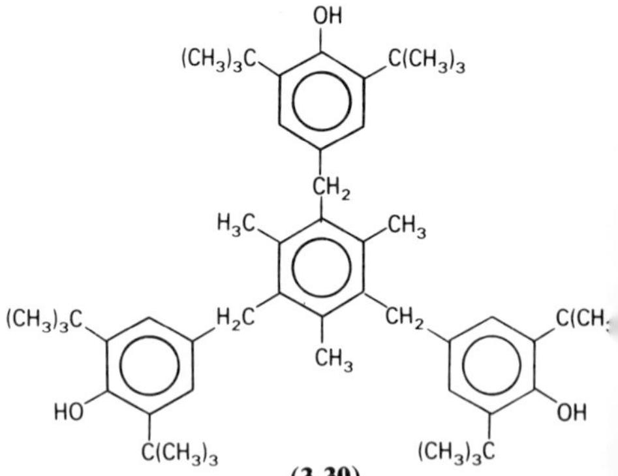 (3.30)

Molecular weight	*Appearance*	*Solubility (wt %)*							*Melting point (°C)*	*Recommended applications*
		Acetone	*Benzene*	*Chloroform*	*Ethyl acetate*	*Hexane*	*Methyl alcohol*	*Water*		
589	White crystalline powder	20·2	43·4			9·1	1·4	0	93–98	Rubbers
358·5	Off-white crystalline powder	>50	>50	>50	>50	28	35	<0·01	81–86	Polyolefins, EPDM, EPM
775	White to slightly yellowish powder	18	34	28	27	10	3	0·01	Min. 240	Polyolefins; cable insulations and films

(continued)

Table 3.2—*contd.*

Chemical name	*Commercial name*	*Chemical structure*
1,3,5-Tri(3,5-di-*tert.*-butyl-4-hydroxybenzyl)isocyanurate *or CAS name*: 1,3,5-Triazine-2,4,6(*1H,3H,5H*)trione,1,3,5-tris[[3,5-bis(1,1-dimethylethyl)-4-hydroxyphenyl]methyl]; *CAS number* 27676-62-6	Irganox 3114 (Ciba–Geigy) Goodrite 3114	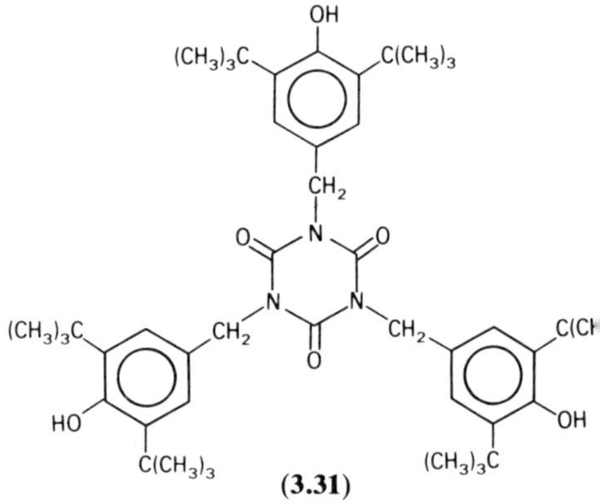 **(3.31)**
Pentaerythrityl tetrakis [3-(3,5-di-*tert.*-butyl-4-hydroxyphenyl)propionate] *or*: Tetrakis[methylene(3,5-di-*tert.*-butyl-4-hydroxyhydrocinnamate)]methane *or CAS name*: 2,2-Bis[[3-[3,5-bis(1,1-dimethylethyl)-4-hydroxyphenyl]-1-oxopropoxy]methyl]-1,3-propanediyl-3,5-bis(1,1-dimethylethyl)-4-hydroxybenzenepropanoate; *CAS number* 6683-19-8	Irganox 1010 (Ciba–Geigy)	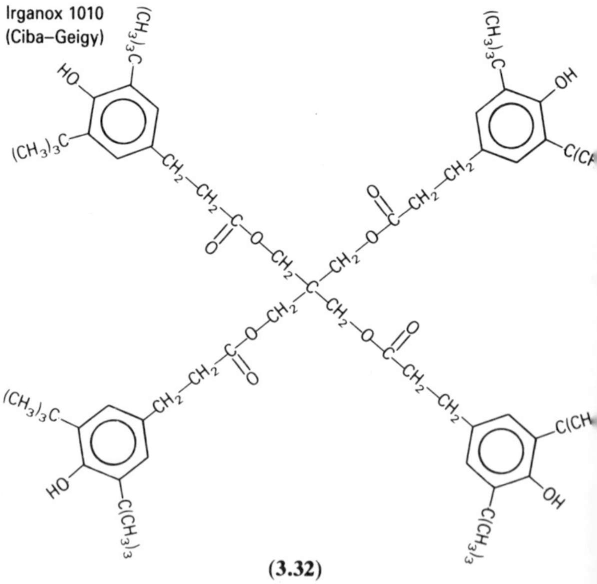 **(3.32)**

Molecular weight	*Appearance*	*Solubility (wt %)*							*Melting point (°C)*	*Recommended applications*
		Acetone	*Benzene*	*Chloroform*	*Ethyl acetate*	*Hexane*	*Methyl alcohol*	*Water*		
784	White to slightly yellowish powder	29	10	21		1·6	0·5	<0·01	218–223	Polyolefins, polystyrene, ABS, SBS, rubbers, poly(vinyl chloride), polyesters, polyamides, polyurethanes, additives
1178	White to slightly yellowish powder	47	56	71	46	0·3	1	<0·01	110–125	Polyolefins, polystyrene, ABS, SBR, rubbers, poly(vinyl chloride), polyesters, polyamides, polyurethanes, additives

(continued)

Table 3.2—*contd.*

Chemical name	*Commercial name*	*Chemical structure*
Diethyl ester of 3,5-di-*tert.*-butyl-4-hydroxybenzylphosphonic acid	Irganox 1222 (Ciba–Geigy)	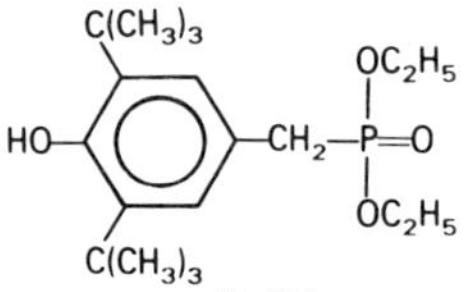 **(3.33)**
Calcium bis(3,5-di-*tert.*-butyl-4-hydroxybenzylmonoethylphosphonate) *or CAS name*: Phosphonic acid, [[3,5-bis(1,1-dimethylethyl)-4-hydroxyphenyl]methyl]-, monoethyl ester, calcium salt (2:1); *CAS number* 65140-91-2	Irganox 1425 (Ciba–Geigy)	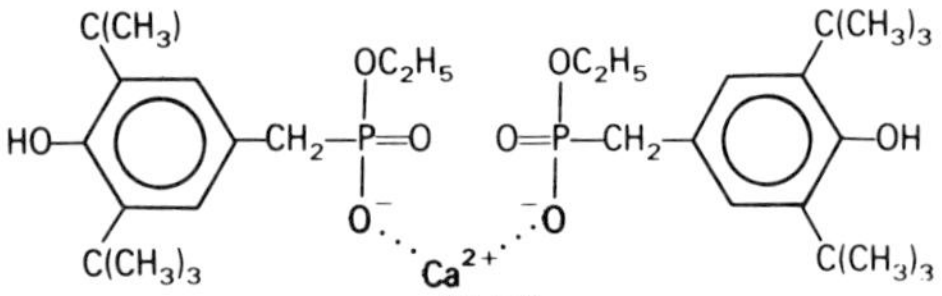 **(3.34)**
Tris(2,4-di-*tert.*-butylphenyl)phosphite; *CAS number* 31570-04-4	Irganox B (Ciba–Geigy); Hostanox VP PAR (Hoechst AG)	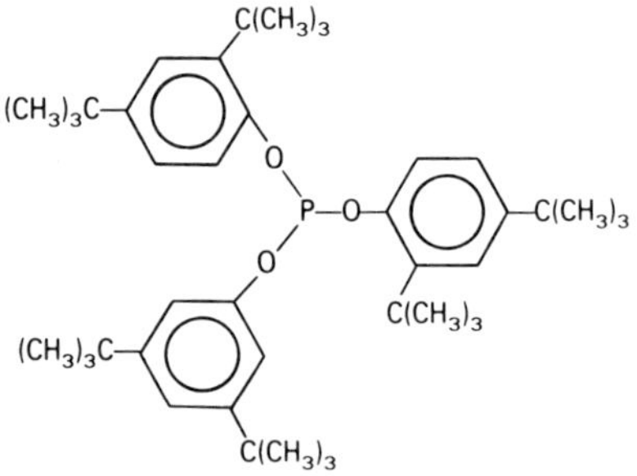 **(3.35)**
2,2′-Methylenebis(6-*tert.*-butyl-4-methylphenol) *or*: Bis(3-*tert.*-butyl-2-hydroxy-5-methylphenyl)methane	Cyanox 2246 (Cyanamid); Plastanox 2246 (Cyanamid)	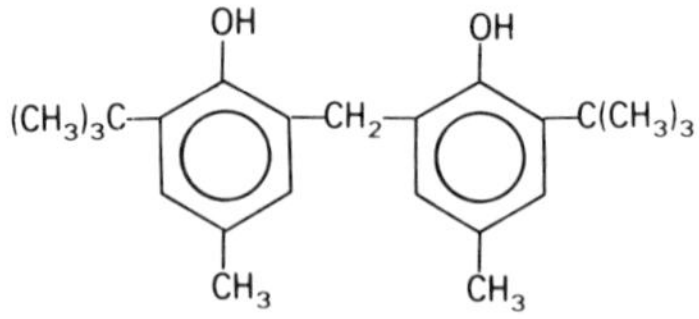 **(3.36)**

Molecular weight	Appearance	Solubility (wt %)							Melting point (°C)	Recommended applications
		Acetone	Benzene	Chloroform	Ethyl acetate	Hexane	Methyl alcohol	Water		
356	White to slightly yellowish, free-flowing powder	27	33	50	20	0·6	62	0·01	119–125	Polyester fibres
663·6	White to off-white powder	0·05	0·05	0·05	0·05	0·05	8	0·01	>260	Polyolefins, rubbers, adhesives
647	White, free-flowing powder	1	34	36	4	11	0·01	0·01	180–185	Polyolefins, polybutene, polystyrene, ABS, SAN, polycarbonates, polyamides, ethylene vinyl acetate (EVA) copolymers
340·5	Cream to white, free-flowing powder	>60	35·8	47·6	54·4	5·8	36	<0·1	121–128	Polyolefins, ABS, co(ethylene–vinyl acetate)

(continued)

Table 3.2—*contd.*

Chemical name	*Commercial name*	*Chemical structure*
2,2′-Methylenebis(6-*tert.*-butyl-4-ethylphenol)	Cyanox 425 (Cyanamid); Plastanox 425 (Cyanamid)	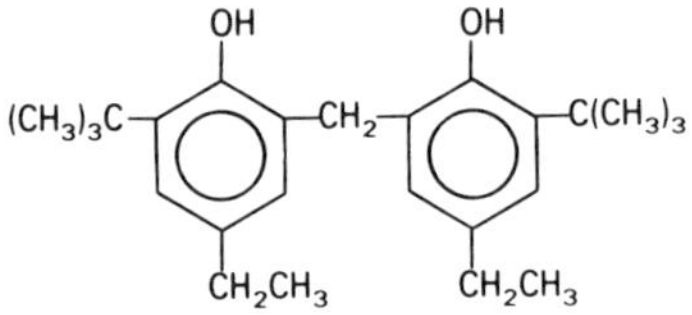 **(3.37)**
1,1,3-Tri(2′-methyl-4′-hydroxy-5′-*tert.*-butylphenyl)butane	Topanol CA (ICI USA)	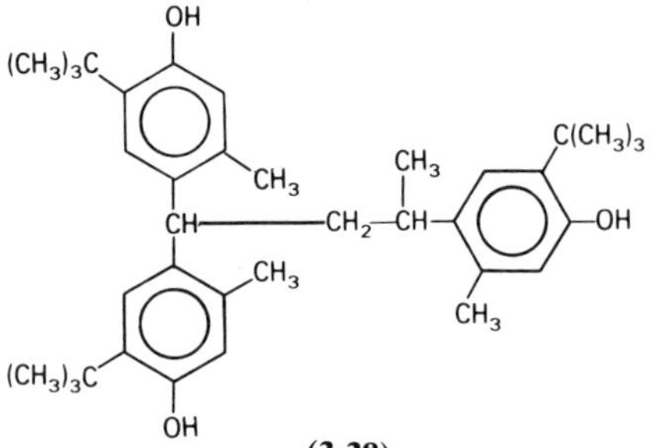 **(3.38)**

[a] NA, Data not available.

Hindered phenols have been investigated as quenchers of singlet oxygen in the photo-oxidation of polybutadienes.[923]

3.1.2 ESR Spectra of Phenoxy Radicals

ESR spectra of phenoxy radicals have been the subject of intensive investigation by a number of authors.[360,1105,1106,1149,1150]

According to ESR spectra of the phenoxy radicals generated, the differently substituted phenols may be divided into five main groups:[1106]

(i) Phenols which form stable primary free radicals, e.g. phenols with two *tert.*-butyl groups in the *ortho* and *para* positions substituted with alkyl groups such as —CH_3, —CH_2CH_3, or —$C(CH_3)_3$ (cf. Table 3.3).

Molecular weight	Appearance	Solubility (wt %)							Melting point (°C)	Recommended applications
		Acetone	Benzene	Chloroform	Ethyl acetate	Hexane	Methyl alcohol	Water		
368·5	Cream to white, free-flowing powder	60	33·3	50	56·5	9	40	0·1	117–123	Polyacrylics, ABS
545					NA[a]					Polyolefins, all thermoplastic resins

(ii) Phenols which form stable secondary free radicals but not in the *para* position (Table 3.3A). Primarily formed unstable radicals change rapidly into more stable secondary radicals, as registered by ESR spectroscopy. When two methyl groups are in the *ortho* positions, oxidative coupling leads to the formation of polyphenyl ether radicals (Table 3.3B). In the case of 2-*tert.*-butyl-4-methylphenol, in contrast with other phenols unsubstituted in the *ortho* position, the delocalization of the free electrons in this position leads to a rapid dimerization (Table 3.3C).

(iii) Phenols which give stable phenoxy radicals after coordination in the ligand field of the transition metal. Phenols, biphenols, naphthols, and dihydroxybenzophenones belong to this group (Table 3.4).

Table 3.3
ESR spectra of stable free phenoxy radicals generated with $Bu^tOO\cdot$ and $Bu^tO\cdot$ radicals from *ortho*-substituted phenols[1106]

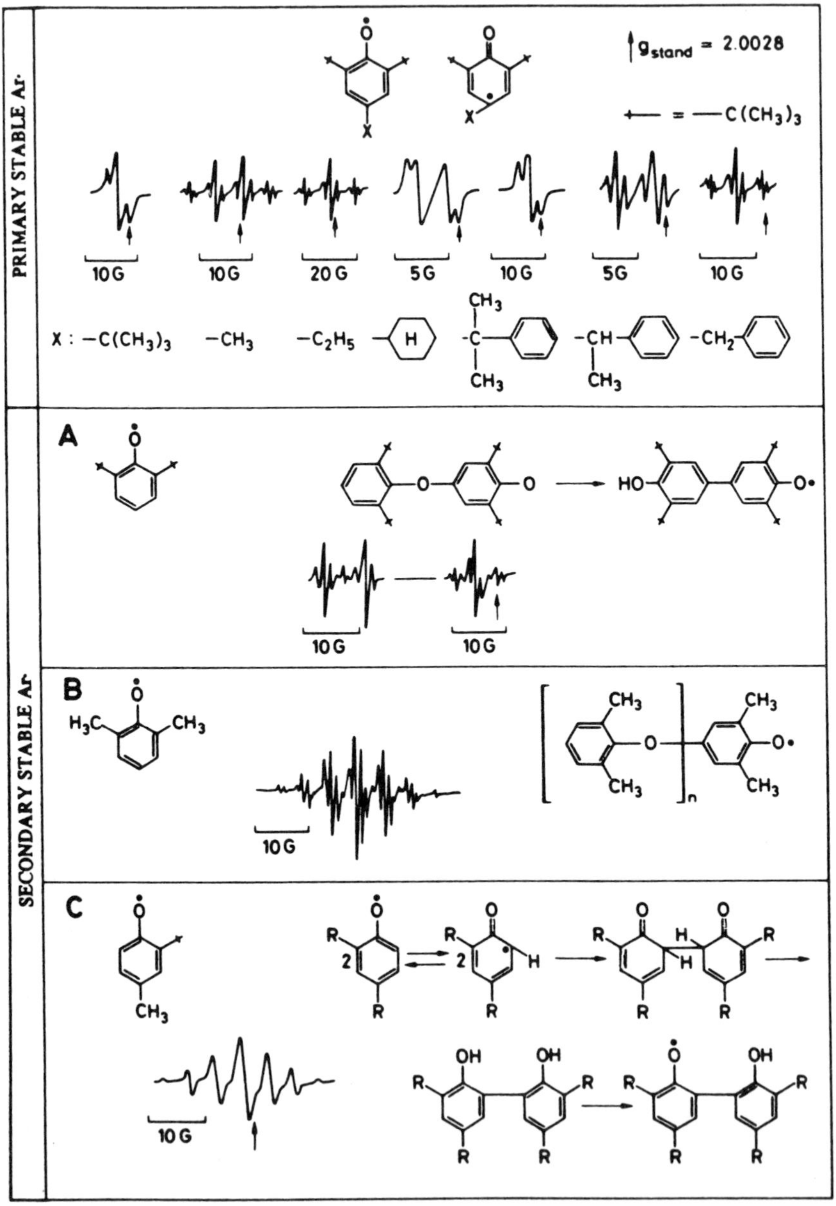

Table 3.4

ESR spectra of complex-bonded phenoxy radicals of Co(III)$(acac)_2$OH and generated with Bu^{t}OO· and Bu^{t}O· from phenols with a free *ortho* position[1106]

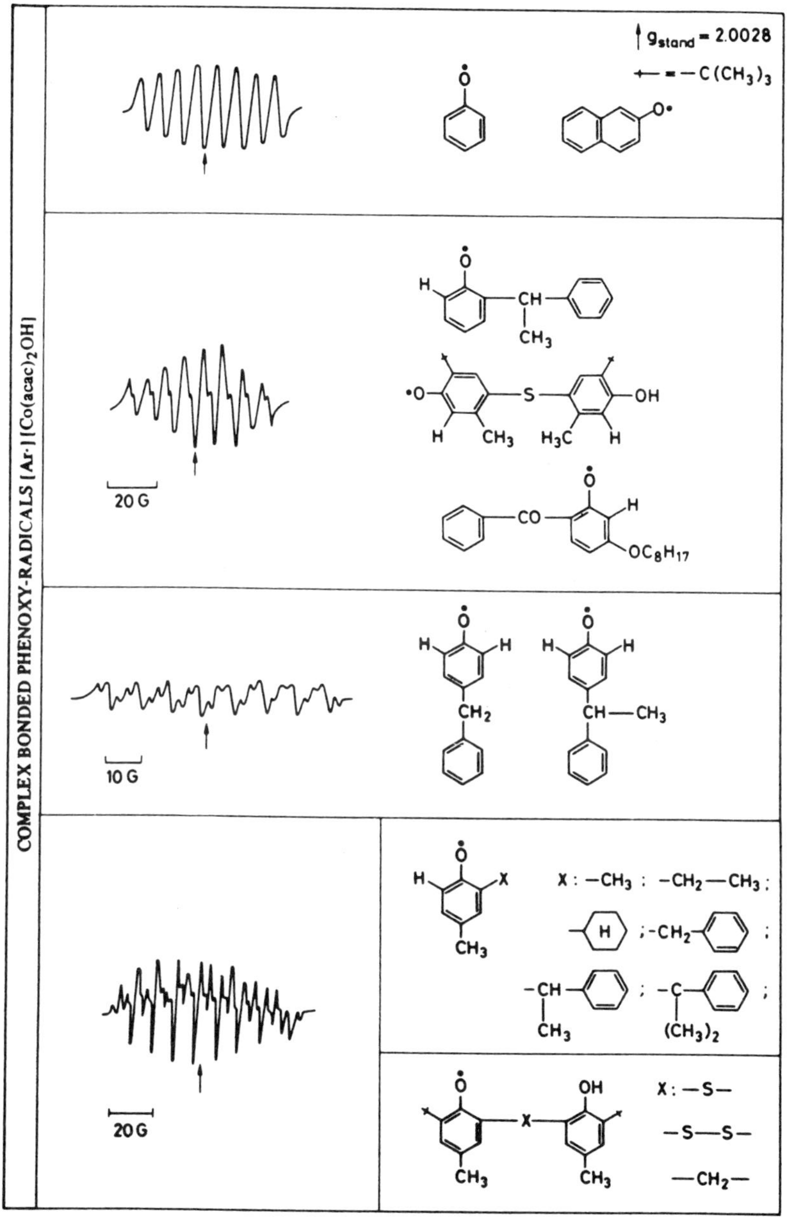

Table 3.5
ESR spectra obtained from phenolic antioxidants[360]

Antioxidant	Trade name	Chemical structure	ESR spectra $T = 25°C$ $T = 65°C$
2,6-Di-*tert*.-butyl-4-methylphenol	BHT 35	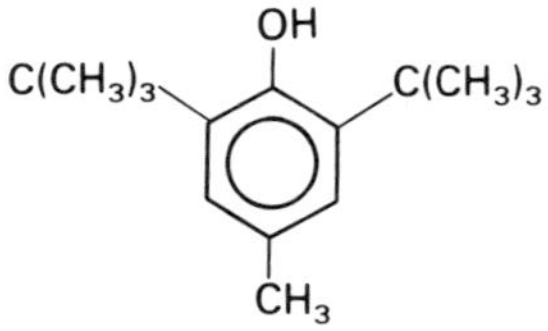	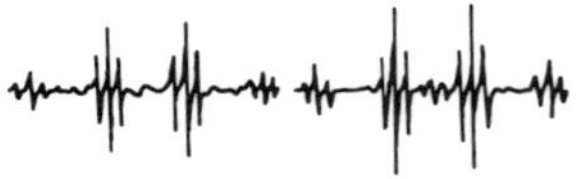
Octadecyl-3-(3,5-di-*tert*.-butyl-4-hydroxyphenyl)propionate (**3.21**)	Irganox 1076	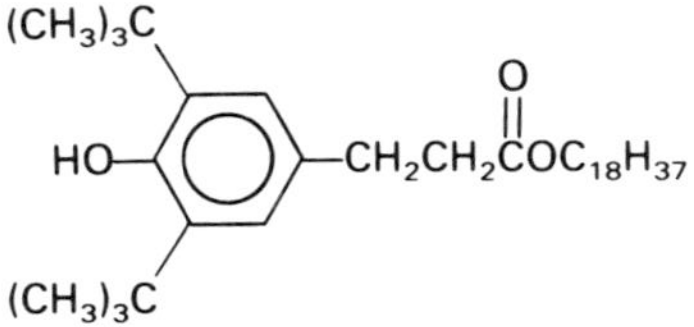	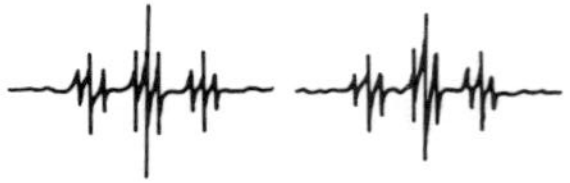
Pentaerithrityl tetrakis[3-(3,5-di-*tert*.-butyl-4-hydroxyphenyl)propionate (**3.32**)	Irganox 1010	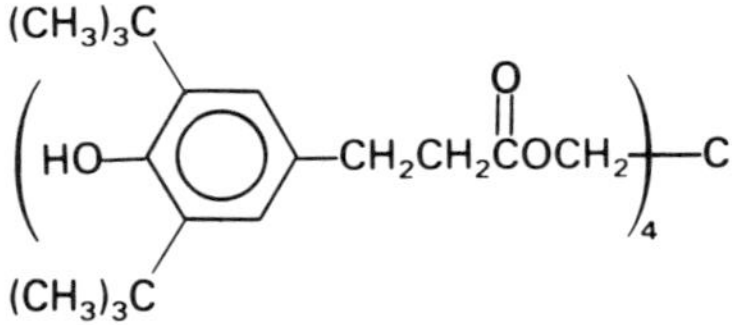	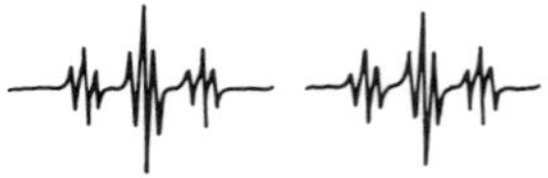

1,3,5-Trimethyl-2,4,6-tris(3,5-di-*tert*.-butyl-4-hydroxybenzyl)benzene (**3.30**)	Irganox 1330	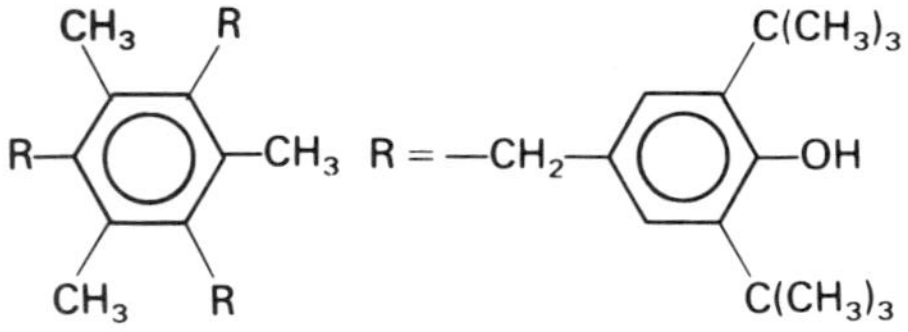	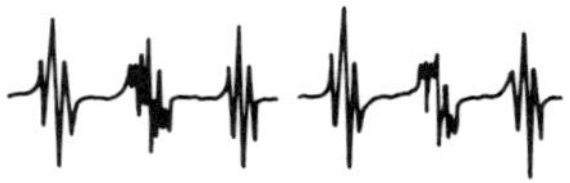
Tri(3,5-di-*tert*.-butyl-4-hydroxybenzyl)isocyanurate (**3.31**)	Irganox 3114	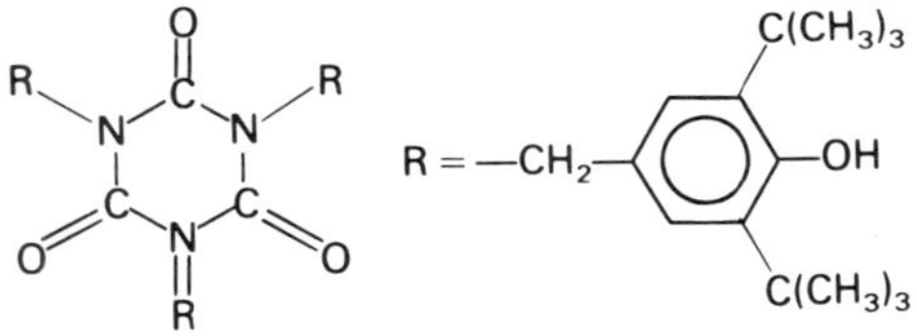	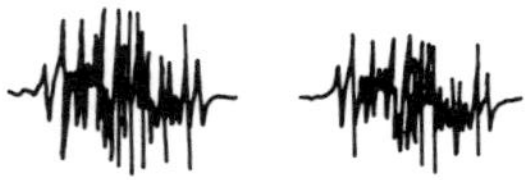
1,1,3-Tri(2′-methyl-4′-hydroxy-5′-*tert*.-butylphenyl)butane (**3.38**)	Topanol CA	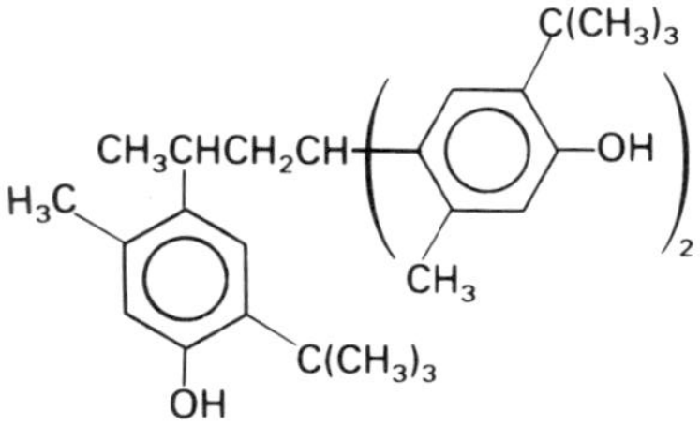	

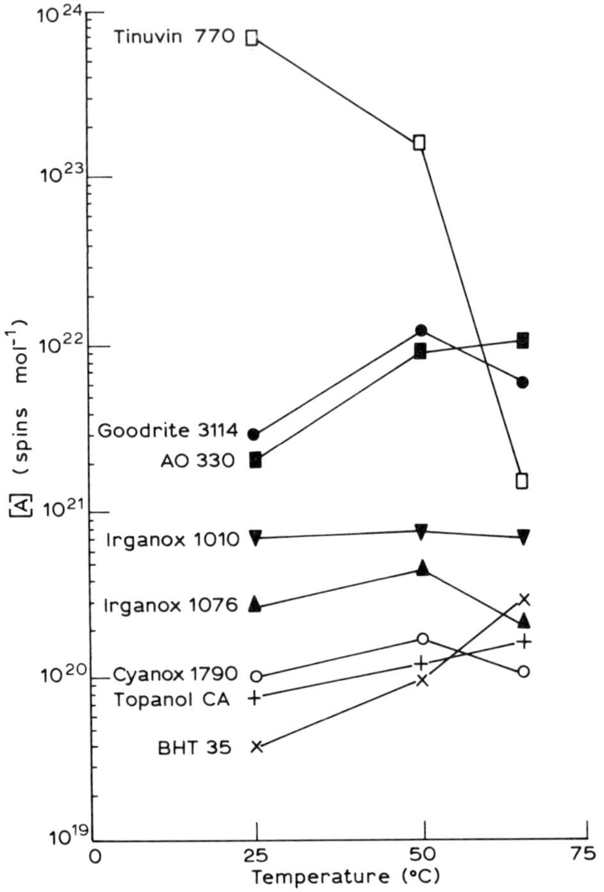

Fig. 3.1. Molar spin concentration as a function of temperature for different antioxidants such as bis(2,2,6,6-tetramethyl-4-piperidinyl) sebacate (Tinuvin 770; **6.3**), 1,3,5-tri(3,5-di-*tert.*-butyl-4-hydroxybenzyl)isocyanurate (Goodrite 3114; **3.31**), 1,3,5-trimethyl-2,4-6-tri(3,5-di-*tert.*-butyl-4-hydroxybenzyl)benzene (A0 330 or Irganox 1330; **3.30**), pentaerythrityl tetrakis[-3(3′,5′-di-*tert.*-butyl-4′-hydroxyphenyl)propionate (Irganox 1010; **3.32**), octadecyl-3-(3,5-di-*tert.*-butyl-4-hydroxyphenyl)propionate (Irganox 1076; **3.21**), 1,3,5-tri(4-*tert.*-butyl-3-hydroxy-2,6-dimethylbenzyl)isocyanurate (Cyanox 1070), 1,1,3-tri(2′-methyl-4′-hydroxy-5′-*tert.*-butylphenyl)butane (Topanol CA; **3.38**), 2,6-di-*tert.*-butyl-4-methylphenol (BHT 35) as a measure of radical scavenging activity in the model system.[360]

(iv) Phenols whose radicals undergo disproportionations, e.g. dihydroxythiophenols.
(v) Phenols substituted with NO_2 groups, e.g. nitrophenols which do not form radicals in nonpolar solvents.

ESR spectra of some of commercially available hindered phenolic antioxidants are shown in Table 3.5.

These phenoxy radicals are stable even at temperatures exceeding 50°C (Fig. 3.1). The influence of molecular structure on radical stability and, consequently, on radical scavenging activity is believed to be an important factor in the interpretation of the spin concentrations observed.[360]

3.2 HYDROXYGALVINOL/GALVINOXY RADICAL SYSTEM

Hydroxygalvinol (hydrogalvinoxyl) (**3.42**) is a bright yellow crystalline product ($T_m = 158$–159°C), which has absorption maximum $\lambda_{max} =$ 380 nm. The concentration of hydroxygalvinol in a polymer can easily be estimated using a calibration curve (Fig. 3.2).[145]

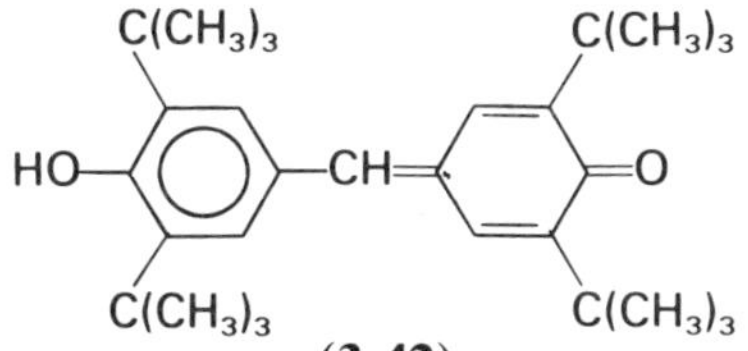

(**3.42**)

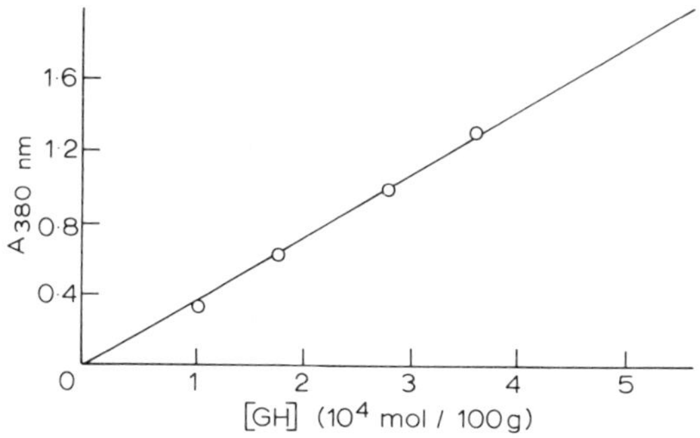

Fig. 3.2. Calibration curve for hydroxygalvinol (**3.42**) concentration [GH] using the absorbance at 380 nm.[145]

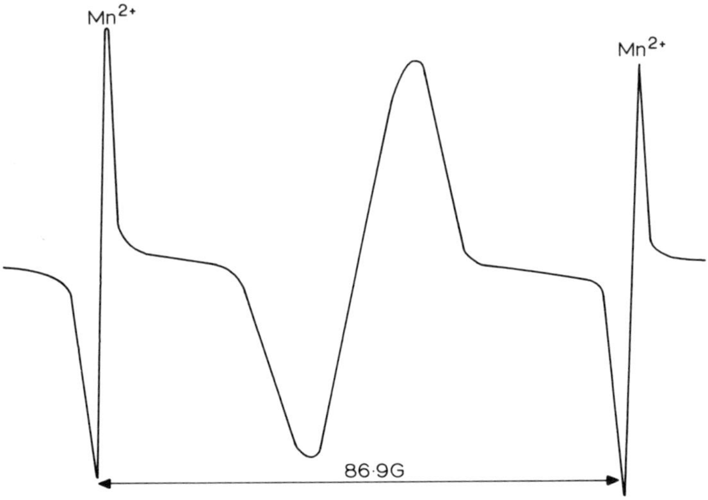

Fig. 3.3. Typical ESR spectrum of galvinoxy radical (**3.43**) [G·] in polypropylene ($g = 2{\cdot}018$).[145]

Galvinoxy radical [2,6-di-*tert.*-butyl-α-(3,5-di-*tert.*-butyl-4-oxo-2,5-cyclohexadien-1-ylidene)-4-tolyloxy] (**3.43**) is a crystalline violet product ($T_m = 158$–159°C) which gives a characteristic ESR spectrum (Fig. 3.3). The radical concentration in a polymer sample can be determined by comparing the ratio $Y_m(\Delta H_{pp})^2$ for the polymer with $Y_m(\Delta H_{pp})^2$ for the reference sample (Fig. 3.4):[145]

$$\frac{Y_m(\Delta H_{pp})^2_{\text{sample}}}{Y_m(\Delta H_{pp})^2_{\text{reference}}} = \frac{\text{Concentration of radicals in test sample}}{\text{Concentration of radicals in reference sample}} \tag{3.26}$$

where Y_m is in centimetres and ΔH_{pp} in gauss.

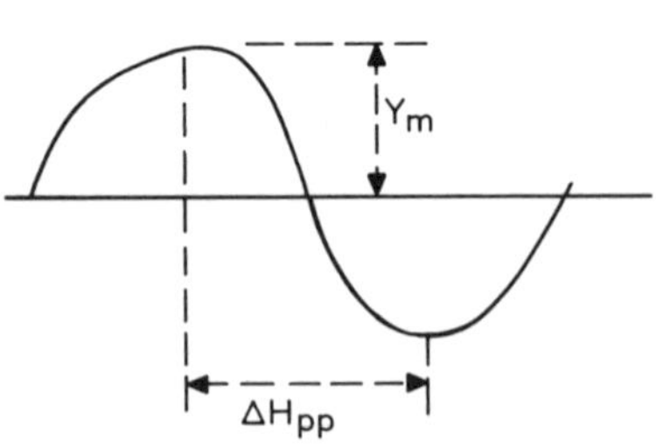

Fig. 3.4. Description of Y_m and ΔH_{pp} notation.[145]

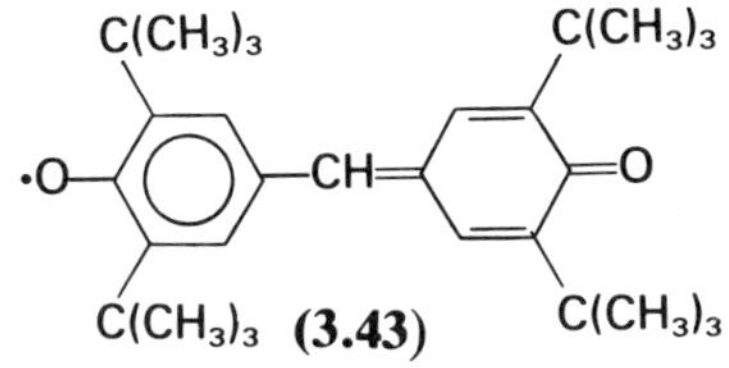

(**3.43**)

Hydroxygalvinol (**3.42**) (abbreviated here as GH) reacts with alkylperoxy radicals (ROO•) producing a stable galvinoxy radical (**3.43**) (abbreviated here as G•):[1016,1017]

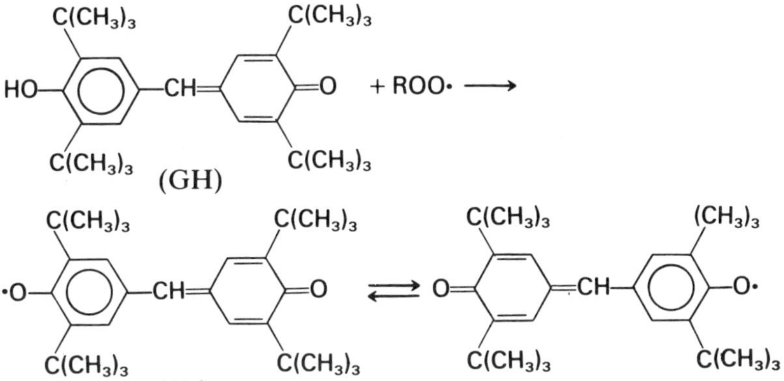

(3.27)

The hydroxygalvinol/galvinoxy radical system is an effective thermal antioxidant and melt processing stabilizer for polypropylene.[143,145,146,268,1016]

The galvinoxy radical (G•) mixed with polypropylene and processed for 5 min at 200°C in an oxygen-free atmosphere is almost completely converted into hydroxygalvinol (GH) (stage 1). During the next 5 min a rapid increase of galvinoxy radical concentration is observed (stage 2). After this time the concentration of these two compounds again changes in a correlated fashion (stage 3) (Fig. 3.5).[145]

Stage 1 of the reaction (decrease of G• and increase of GH) can be explained by the following reactions:

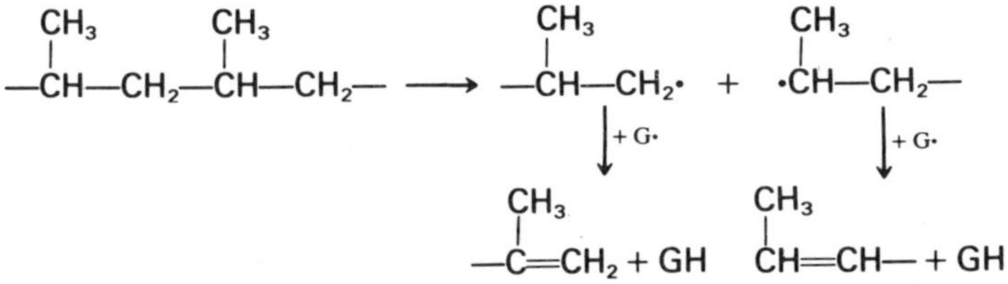

(3.28)

(3.29) (3.30)

Stage 2 of the reaction (decrease of GH and increase of G•) is a result of the following reactions:

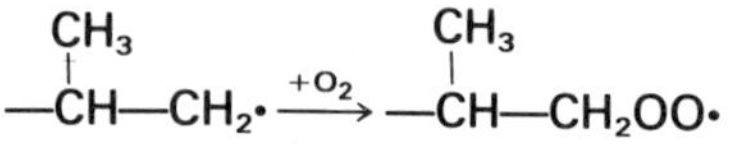

(3.31)

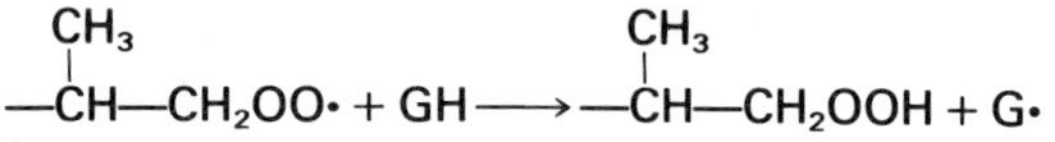

(3.32)

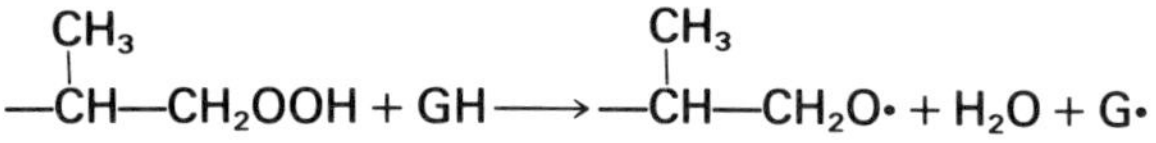

(3.33)

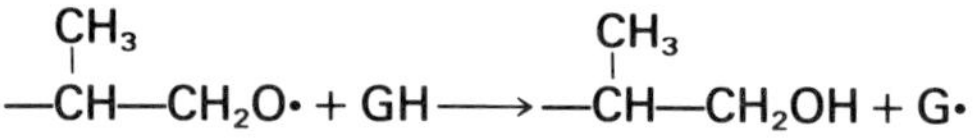

(3.34)

Stage 3 of the reaction (decrease of G• and increase of GH) occurs by the mechanism:

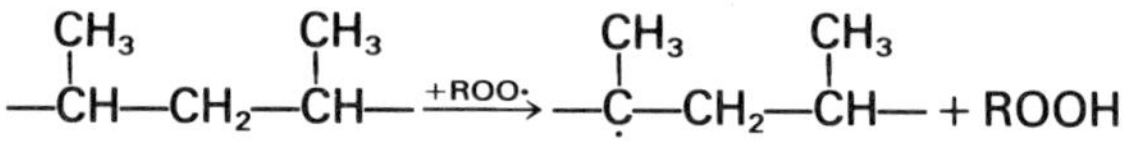

(3.35)

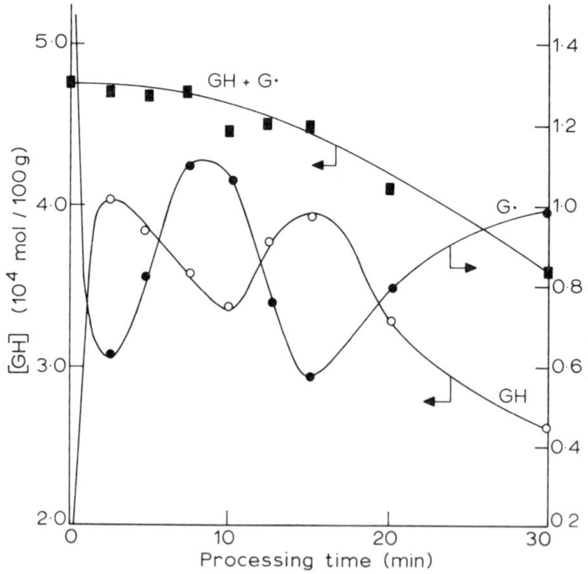

Fig. 3.5. Changes in hydroxygalvinol (**3.42**) [GH] and galvinoxy radical (**3.43**) [G·] concentrations in polypropylene during processing at 200°C (closed mixer). Initial [G·] = $4{\cdot}74 \times 10^{-4}$ mol/100 g.[145]

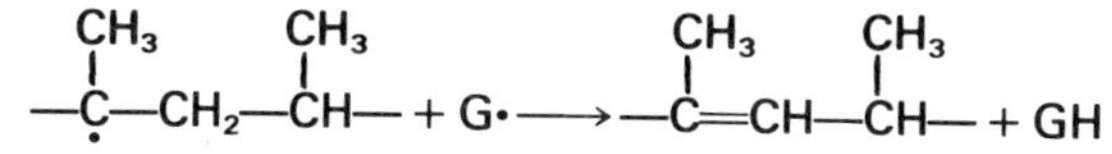

(3.36)

This mechanism is based on the assumption of the depletion of oxygen initially present in the mixer or dissolved in a polymer.

In an open mixer, the galvinoxy radical (G•) is a very poor melt stabilizer for polypropylene. Oxygen leads to the destruction of the G•/GH redox couple:

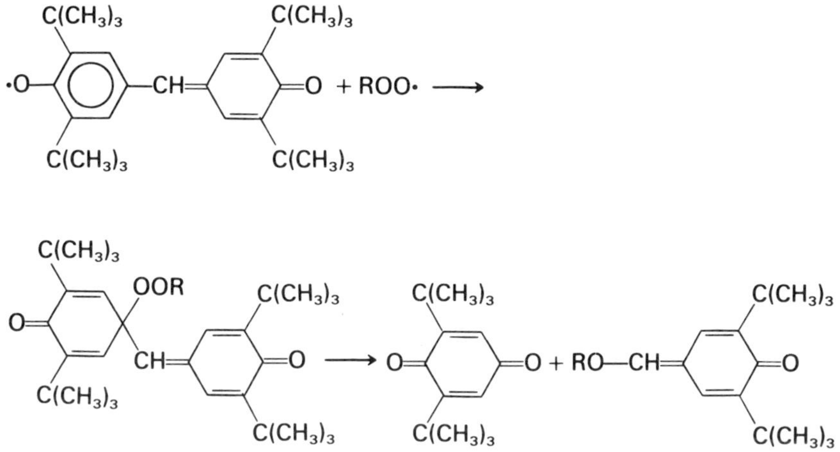

(3.37)

No evidence has been found for the reaction of polymer radicals with galvinoxy radicals.

Photoantioxidant activity of hydroxygalvinol in polypropylene processed at 200°C is shown in Table 3.6.[268]

Table 3.6
Photoantioxidant activity of hydroxygalvinol (GH) in polypropylene (processed 200°C/10 min in closed mixer)[268]

Initial [*GH*] (10^4 *mol/100 g*)	*Embrittlement time* (*h*)
2·37	171
4·75	220
9·50	286
Control	85

3.3 PHENOLIC SULPHIDES

Phenolic sulphides such as 3,5-di-*tert.*-butyl-4-hydroxyphenylmethane thiol (**3.44**) and its analogues (**3.45–3.48**) are effective kinetic chain-breaking antioxidants for many polymers such as polyolefins, poly(vinyl chloride), ABS and rubbers.[658,858,1027,1028,1034]

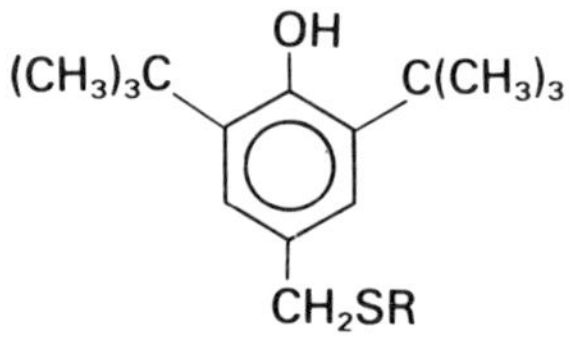

(**3.44**)

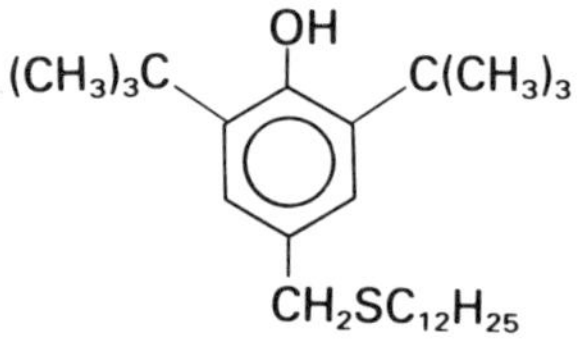

(**3.45**)

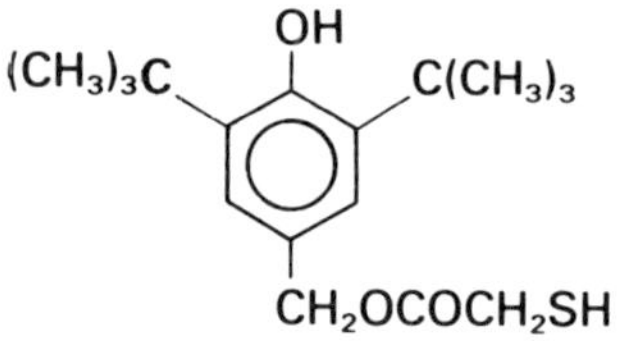

(**3.46**)

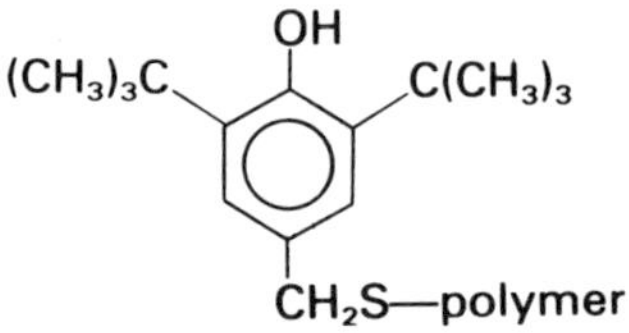

(**3.47**)

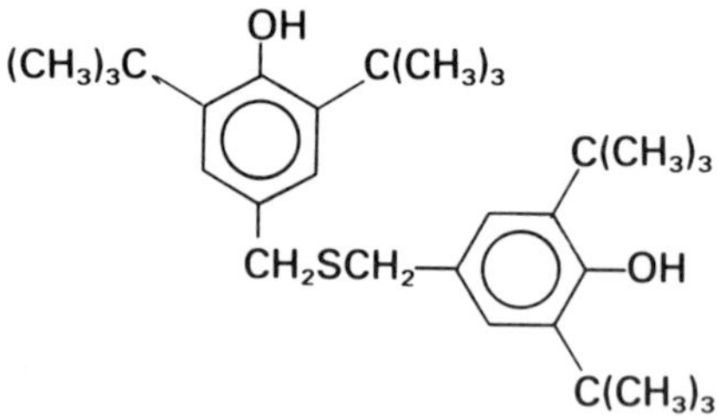

(**3.48**)

Addition of phenolic sulphides to olefinic double bonds suggests that the chain-breaking reactions occur predominantly through the thiol radical (**3.50**) rather than the phenoxy radical (**3.49**):[1027]

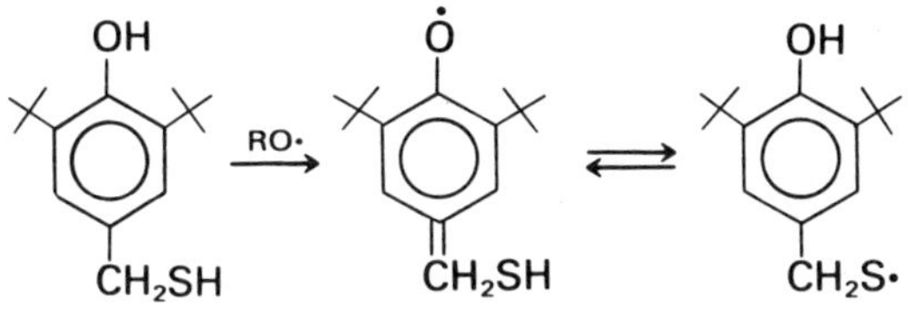

(3.44) **(3.49)** **(3.50)** (3.38)

The phenoxy radical (**3.49**) can undergo a reaction during which quinone methide (**3.51**) and •SH radical are formed:

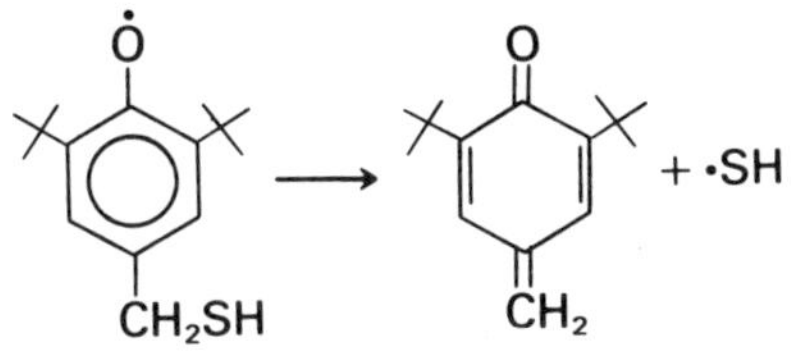

(3.49) **(3.51)** (3.39)

The thiol radical (**3.50**) can react with hydroperoxy groups (OOH) by the following mechanism:[1027]

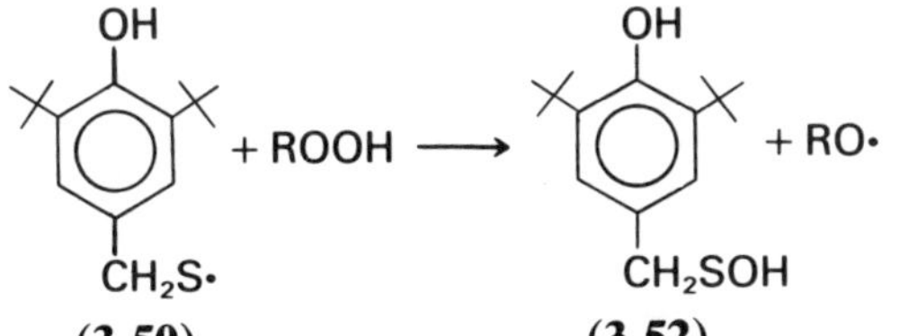

(3.50) **(3.52)** (3.40)

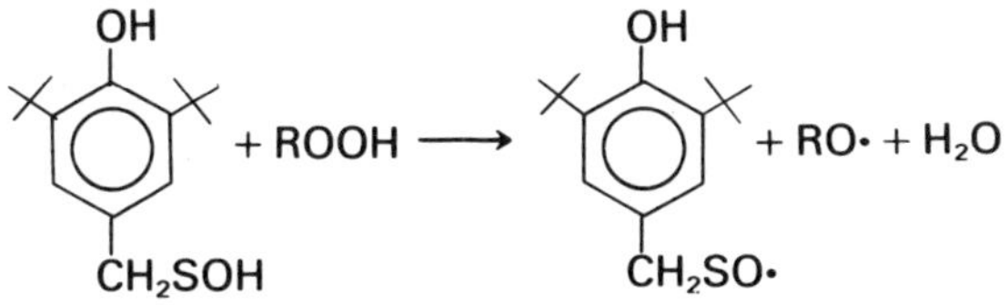

(3.41)

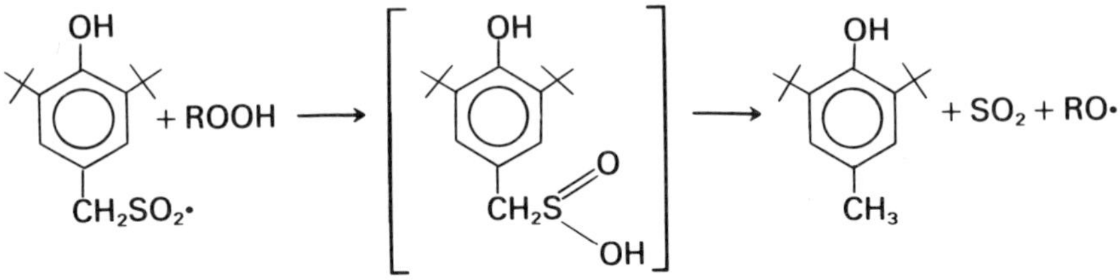

(3.53)

(3.42)

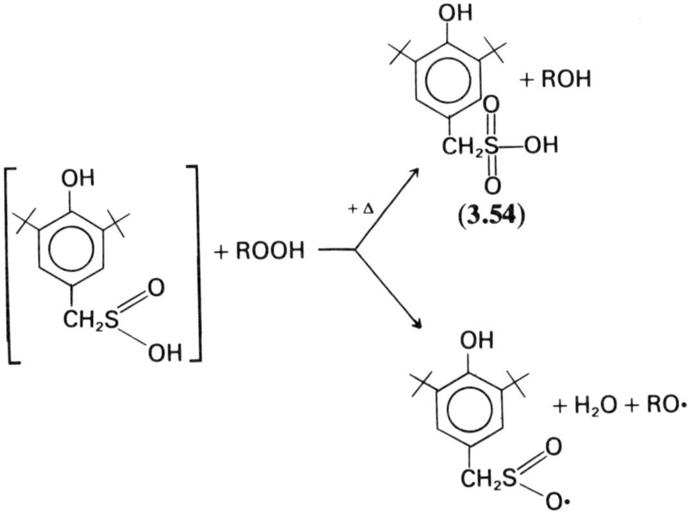

(3.54)

(3.43)

(3.44)

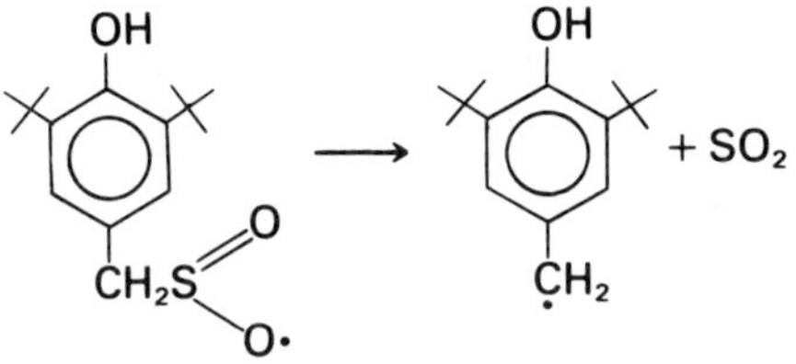

(3.45)

In this reaction a number of products are formed, such as sulphenic acids (**3.52**), sulphinic acids (**3.53**), sulphonic acids (**3.54**) and sulphur dioxide. These are effective radical scavengers and participate in the antioxidation mechanism. The other important antioxidant function of the acidic sulphur species is the ability to destroy hydroperoxides in a catalytic process.

3.4 MONOSULPHIDES

Monosulphides (**3.55**) are effective hydroperoxy-group (ROOH) decomposers, according to the following reactions:[104,1012]

$$(R_1CH_2CH_2)_2S \; (\mathbf{3.55}) + ROOH \longrightarrow (R_1CH_2CH_2)_2S{=}O + ROH \qquad (3.46)$$

$$(R_1CH_2CH_2)_2S{=}O \longrightarrow R_1CH_2CH_2SOH + R_1CH{=}CH_2 \qquad (3.47)$$

$$R_1CH_2CH_2SOH + ROOH \longrightarrow R_1CH_2CH_2SO\cdot + RO\cdot + H_2O \qquad (3.48)$$

$$R_1CH_2CH_2SOH + ROOH \longrightarrow R_1CH_2CH_2S({=}O)OH \qquad (3.49)$$

$$2\,R_1CH_2CH_2SOH \longrightarrow R_1CH_2CH_2S({=}O){-}SCH_2CH_2R_1 + H_2O \qquad (3.50)$$

$$R_1CH_2CH_2S({=}O){-}SCH_2CH_2R_1 \longrightarrow R_1CH_2CH_2S({=}O)\cdot + \cdot SCH_2CH_2R_1 \qquad (3.51)$$

$$2\,R_1CH_2CH_2S({=}O)\cdot \longrightarrow R_1CH_2CH_2S({=}O){-}S({=}O){-}CH_2CH_2R_1 \qquad (3.52)$$

$$2\,R_1CH_2CH_2S\cdot \longrightarrow R_1CH_2CH_2S{-}SCH_2CH_2R_1 \qquad (3.53)$$

A number of these antioxidants are commercially produced (Table 3.7).

Monosulphides show synergistic effects with phenolic antioxidants.

3.5 THIOPHENOLS AND DITHIOBISPHENOLS

Thiophenols (**3.60** and **3.61**)[277,617,633,671,728,913,1028] and dithiophenols (**3.62** and **3.63**)[670,1071] are efficient antioxidants. 2,2′-Thiobis(4-methyl-

Table 3.7
Examples of commercially produced dialkylthiodipropionate type antioxidants: physical properties and solubility

Chemical name	*Commerical name*	*Chemical structure*
Dilauryl thiodipropionate *or*: Dilauryl ester of β,β′-thiodipropionic acid *or CAS name*: Propanoic acid, 3,3′-thiobis-, didodecyl ester; *CAS number* 123-28-4	Irganox PS 800 (Ciba–Geigy) Cyanox LTDP (Cyanamid)	$S(CH_2CH_2{-}C({=}O){-}O{-}C_{12}H_{25})_2$ **(3.56)**
Ditridecyl ester of β,β′-thiodipropionic acid	Cyanox 711 (Cyanamid)	$S(CH_2CH_2{-}C({=}O){-}O{-}C_{13}H_{27})_2$ **(3.57)**
Dimyristyl ester of β,β′-thiodipropionic acid	Irganox 801 (Ciba–Geigy) Cyanox MTDP (Cyanamid)	$S(CH_2CH_2{-}C({=}O){-}O{-}C_{14}H_{28})_2$ **(3.58)**
Distearyl thiodipropionate *or*: Distearyl ester of β,β′-thiodipropionic acid *or CAS name*: Propanoic acid, 3,3′-thiobis-, dioctadecyl ester; *CAS number* 693-36-7	Irganox PS 802 (Ciba–Geigy); Cyanox STDP (Cyanamid); Plastanox STDP	$S(CH_2CH_2{-}C({=}O){-}O{-}C_{18}H_{37})_2$ **(3.59)**
Mixed lauryl–stearyl ester of β,β′-thiodipropionic acid	Cyanox 1212 (Cyanamid)	

Molecular weight	*Appearance*	*Solubility (wt %)*[a]							*Melting point (°C)*	*Recommended applications*
		Acetone	*Ethanol*	*Benzene*	*Toluene*	*Dichloromethane*	*n-Heptane*	*Water*		
515	White waxy crystalline flakes	37	2	40	40	40		0·05	38–40	Polyolefins
543	Colourless to slightly yellow clear liquid	cm	13·8	cm	cm		cm			Polyolefins, ABS, SBR
571	White waxy crystalline flakes	9	2	40	40	40		0·05	50	Polyolefins
683	White waxy crystalline flakes	1	1	11	10	13		0·05	53–65	Polyolefins
	White waxy crystalline flakes	1·8	0·05		30		4·2	0·05	49	Polyolefins

[a] cm, Completely miscible.

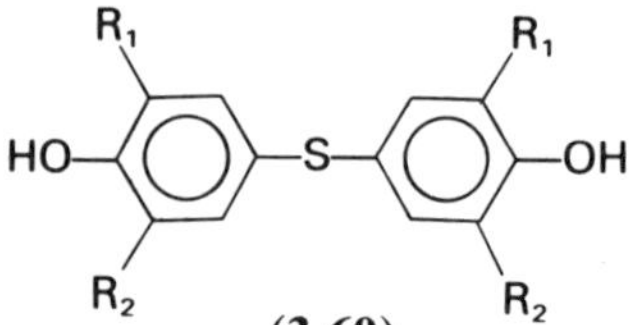

(3.60)

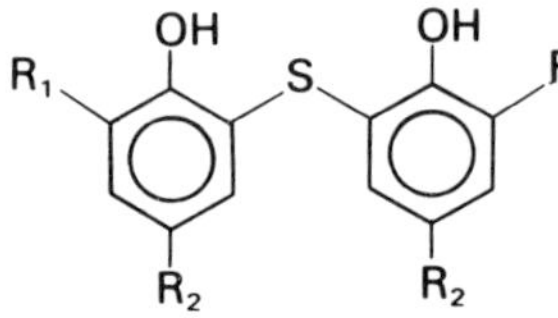

$R_1 = C(CH_3)_3$
$R_2 = CH_3, C(CH_3)_3$

(3.61)

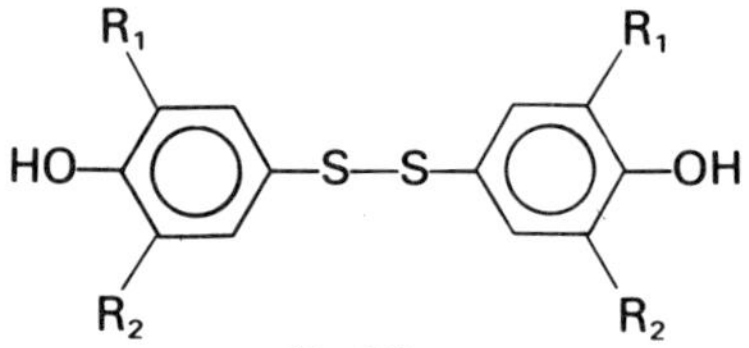

(3.62)

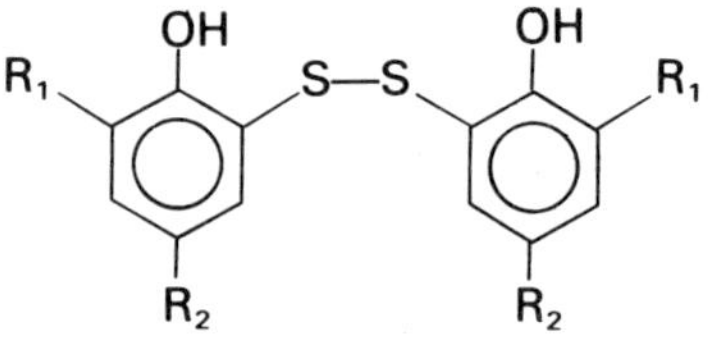

$R_1 = C(CH_3)_3$
$R_2 = CH_3, C(CH_3)_3$

(3.63)

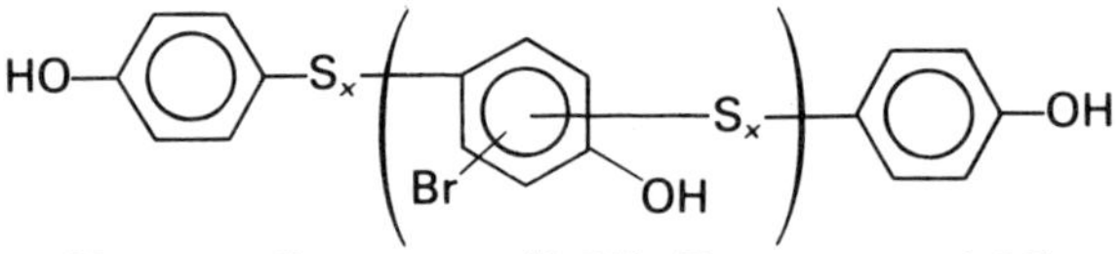

$\bar{M}_n = 1100$; $n = 5$; S content = 42·3%; Br content = 1·8%, $x = 1$–8.

(3.64)

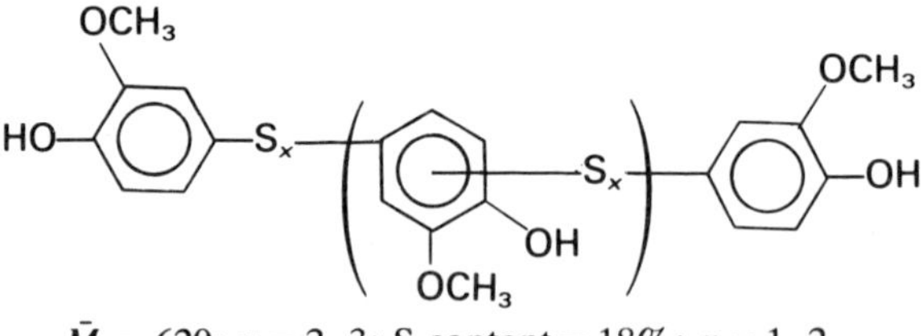

$\bar{M}_n = 620$; $n = 2$–3; S content = 18%; $x = 1$–2.

(3.65)

6-*tert.*-butylphenol) is commercially produced as Irganox 1081 (Ciba–Geigy) (**3.29**, Table 3.2). In addition the thio-oligomers hydroxyphenylene sulphide (**3.64**) and hydroxymethoxyphenylene sulphide (**3.65**) have been tested as antioxidants.[728]

The contribution of the sulphidic function to the antioxidant activity of phenolic sulphides is enhanced according to the number of sulphur atoms in the bridge connecting the two phenolic moieties. Disulphides are more efficient than monosulphides.[670,671] The increase of the stabilization efficiency of dithiobisphenols in comparison with the corresponding thiobisphenols is probably due to active participation of the disulphide bridge in the chain-breaking [reaction (3.54)] and hydroperoxide-decomposing [reaction (3.57)] stabilizing mechanisms:[615,617,670]

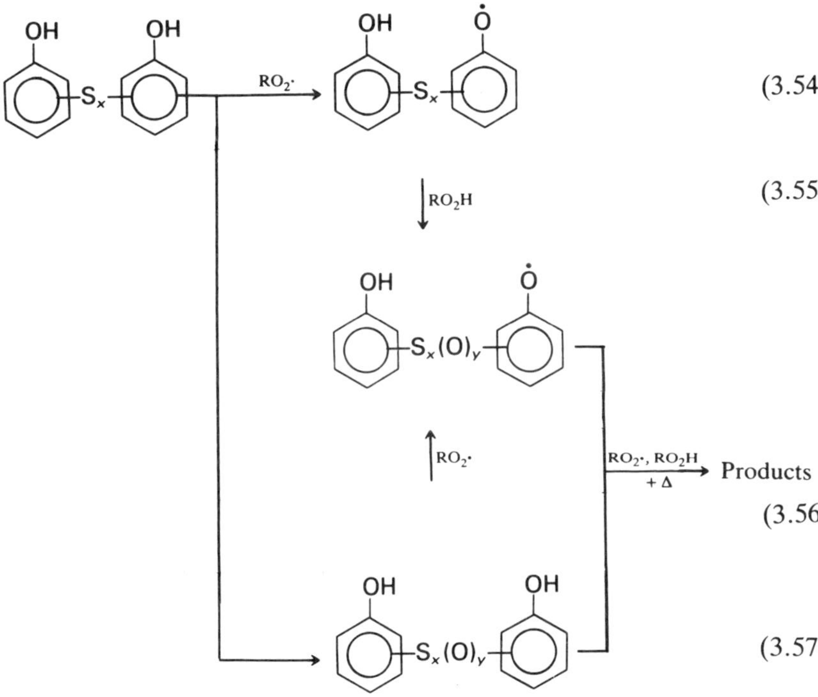

(3.54)

(3.55)

(3.56)

(3.57)

The sulphide group may react directly with hydroperoxide (OOH) groups and/or with free radicals formed from the decomposition of this group (HO• and PO•):[728]

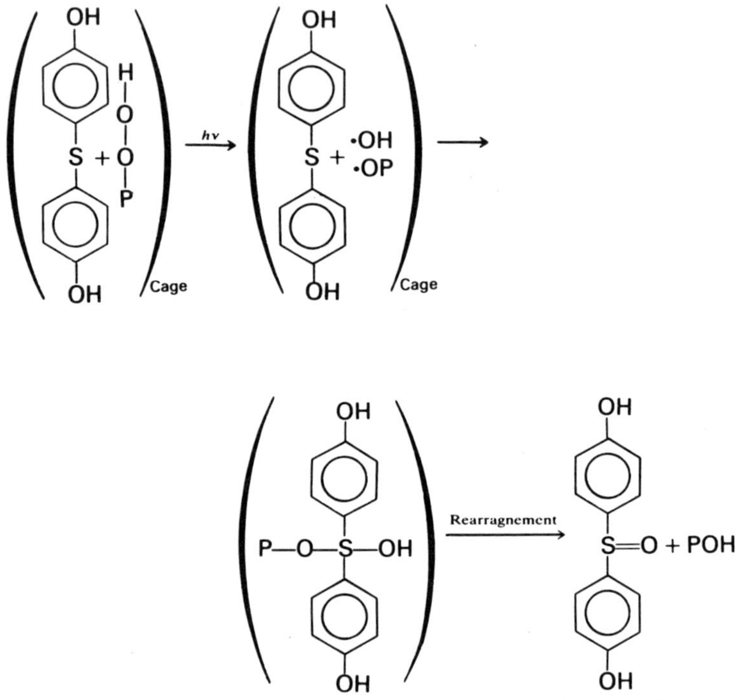

(3.58)

Sulphoxide

3.6 SULPHOXIDES AND SULPHONES

Sulphoxides (**3.66**)[215,277,468,469,614,616] and sulphones (**3.67**)[277,616] are effective antioxidants for polypropylene. During UV irradiation, sulphoxide (**3.66**) easily cleaves into sulphoxy (sulphinyl) (**3.68**) and

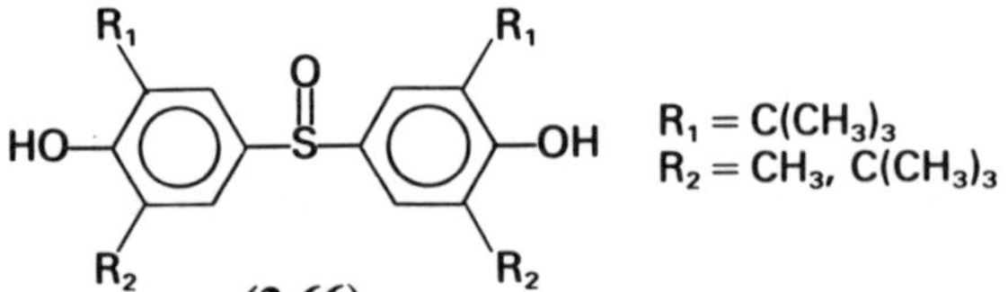

(**3.66**)

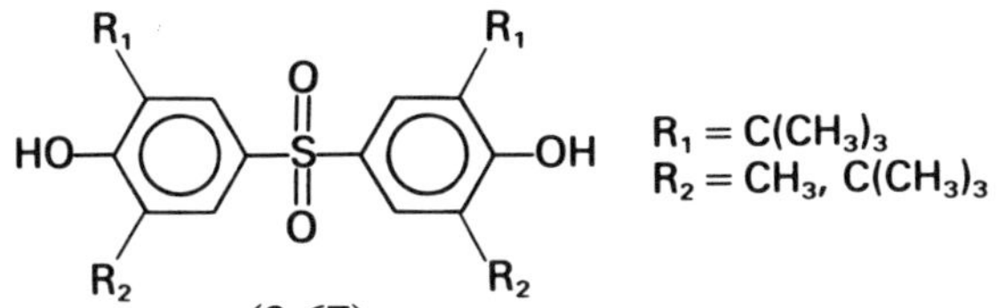

(3.67)

hydroxybenzyl (**3.69**) free radicals:

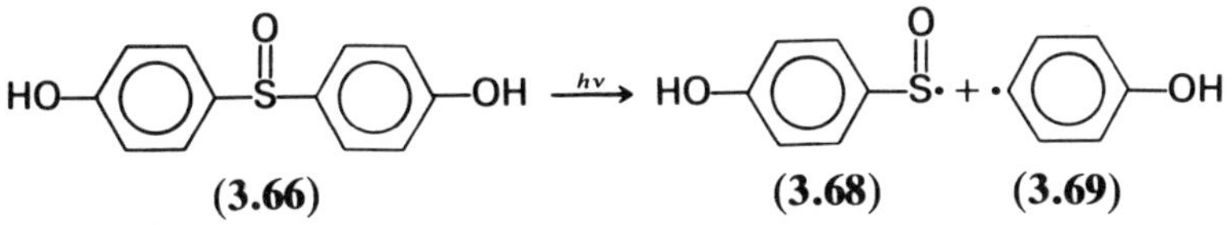

(3.59)

Sulphoxy free radicals (**3.68**) can easily abstract a hydrogen atom from polymer and/or from a hydroxyl group of other stabilizer molecules:

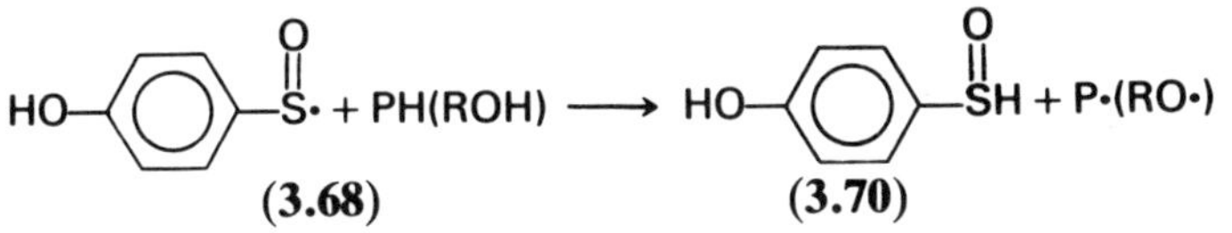

(3.60)

The sulphenic acids (RSOH) are unstable with respect to self-oxidation and reduction and are known as 'intermediate' products in various organic reactions.

Sulphoxy free radical (**3.68**) can also recombine with various oxygen-containing free radicals such as HO•, PO• and POO•:

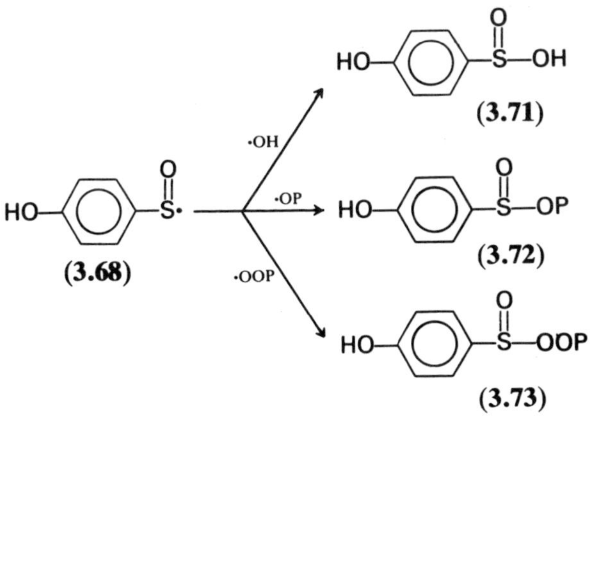

(3.61)

(3.62)

(3.63)

Sulphinic acid ester (**3.72**) is a stable product, but persulphinate (**3.73**) is not: it can easily be cleaved by light excitation into free radicals with simultaneous liberation of sulphur dioxide:

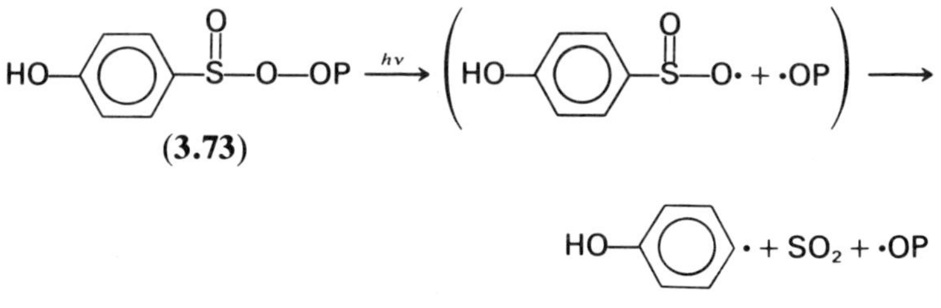

(3.64)

Hydrogen abstraction from hydroxyphenylene sulphinic acid (**3.71**) by various polymeric free radicals also results in liberation of sulphur dioxide:

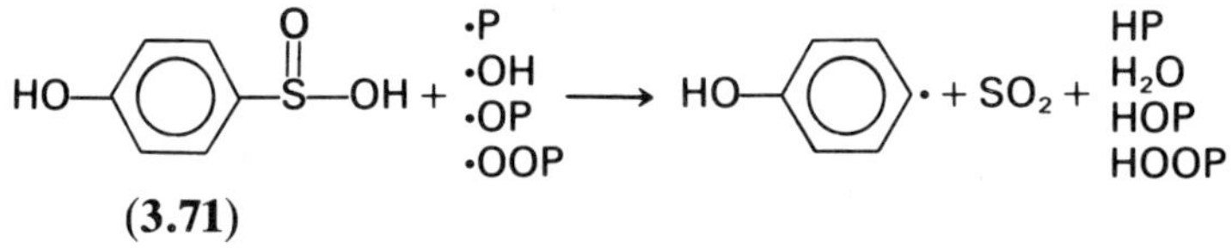

(3.65)

During UV irradiation, hydroxyphenylene sulphinic acid (**3.71**) may also react with hydroxyperoxides to give a stronger hydroxyphenylene sulphonic acid (**3.74**) or its ester (**3.75**) according to the following 'cage' mechanism:

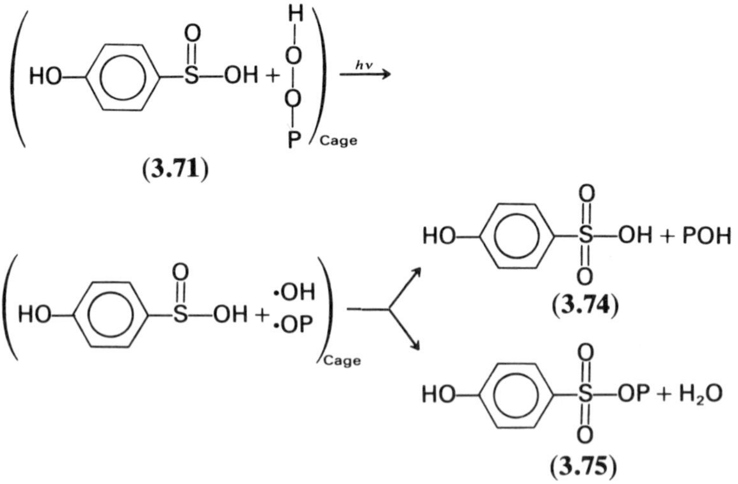

(3.66)

(3.67)

Sulphenic acid (**3.70**), sulphinic acid (**3.71**), and sulphonic acid (**3.74**) and sulphur dioxide (SO_2), which can be formed during the 'pro-oxidative' period [reactions (3.64) and (3.65)] act as catalysts for further catalytical decomposition of polymer hydroperoxides to non-radical products by non-radical processes.

It has been demonstrated that sulphur dioxide (SO_2) is the catalytic species responsible for the decomposition of hydroperoxides.[210,287]

3.7 SYNERGISTIC EFFECTS BETWEEN HINDERED PHENOLS AND SULPHUR-CONTAINING COMPOUNDS

Combinations of kinetic chain-breaking antioxidants (chain terminators) with peroxide decomposers may exhibit a synergistic effect.[290,637,1034] In the presence of a chain terminator, oxidative chains are shortened and the concentration of hydroperoxides is reduced so that the peroxide decomposer is, in effect, present at a higher concentration relative to the hydroperoxide formed. The effectiveness of peroxide decomposers increases markedly with concentration of the stabilizer, until at high concentrations oxidation becomes retarded with no evidence of autocatalysis.[531,535] Decomposition of hydroperoxides into non-radical products increases the effective concentration of the chain terminator since there are fewer oxidative chains to be interrupted. In this way each component of the synergistic system functions to increase the effective concentration of the other component relative to an intermediate which would promote oxidation of a polymer.

An example of such a synergistic effect is shown in the case of a mixture of 4,4′-bis(2-methyl-6-*tert.*-butylphenol) (**3.76**) and 2,2′-naphthyldisulphide (**3.77**) (Fig. 3.6).[531]

Marked synergistic actions have been observed between 4-alkoxy-2,6-diphenylphenols (**3.78**) and β,β'-disubstituted diethyl sulphoxides

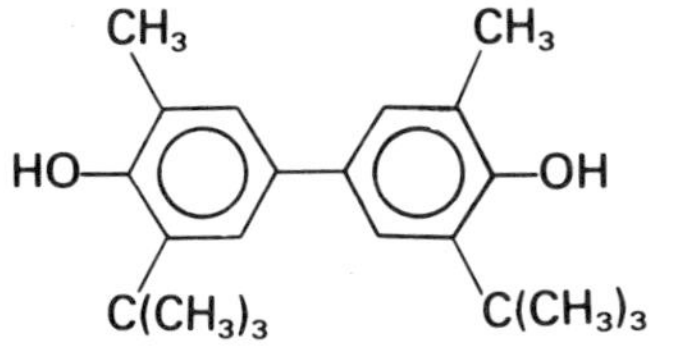

(**3.76**)

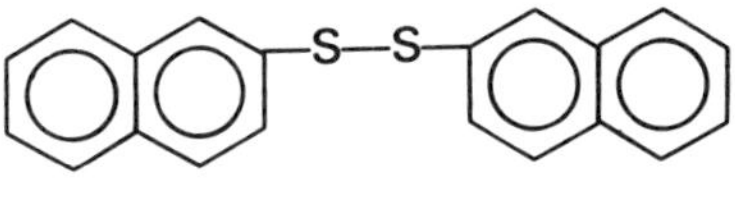

(**3.77**)

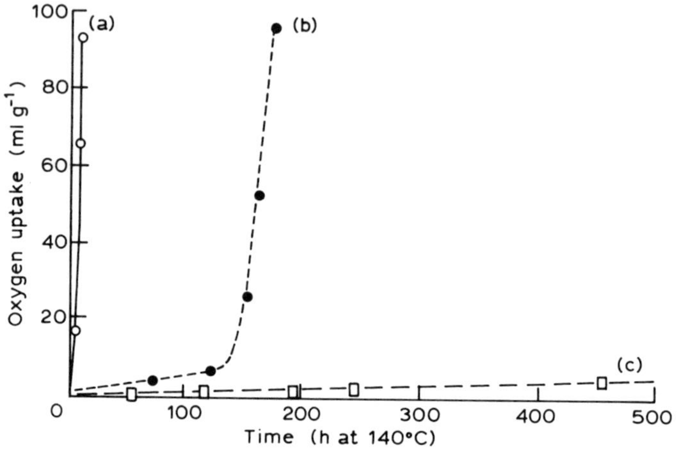

Fig. 3.6. Synergism between: (a) 2,2′-naphthyldisulphide (**3.77**) (0·1 wt%), (b) 4,4′-bis(2-methyl-6-*tert.*-butylphenol)(**3.76**)(0·1 wt%) and (c) a combination of (a) (0·1 wt%) and (b) (0·1 wt%).[531]

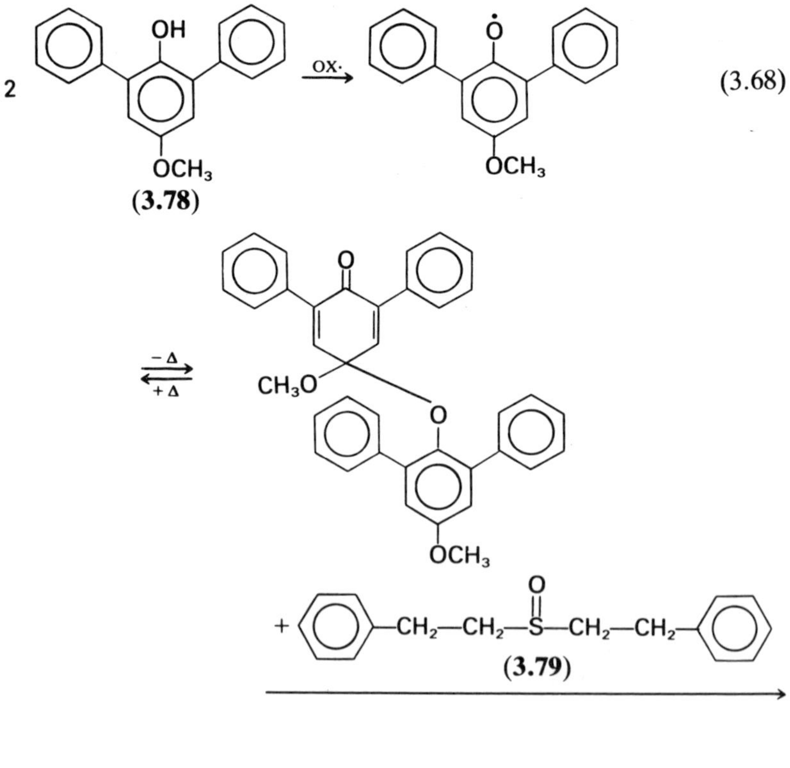

(3.68)

(3.78)

(3.79)

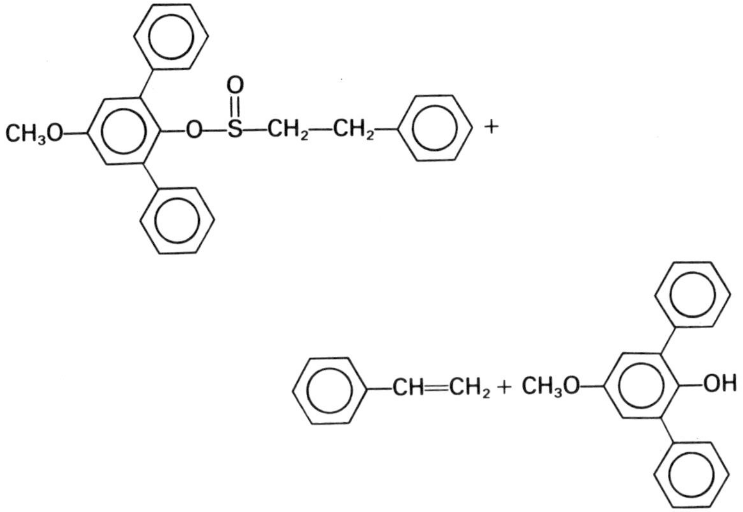

CH3O
O
S
CH2
CH2
+
CH=CH2 + CH3O
OH

(3.69)

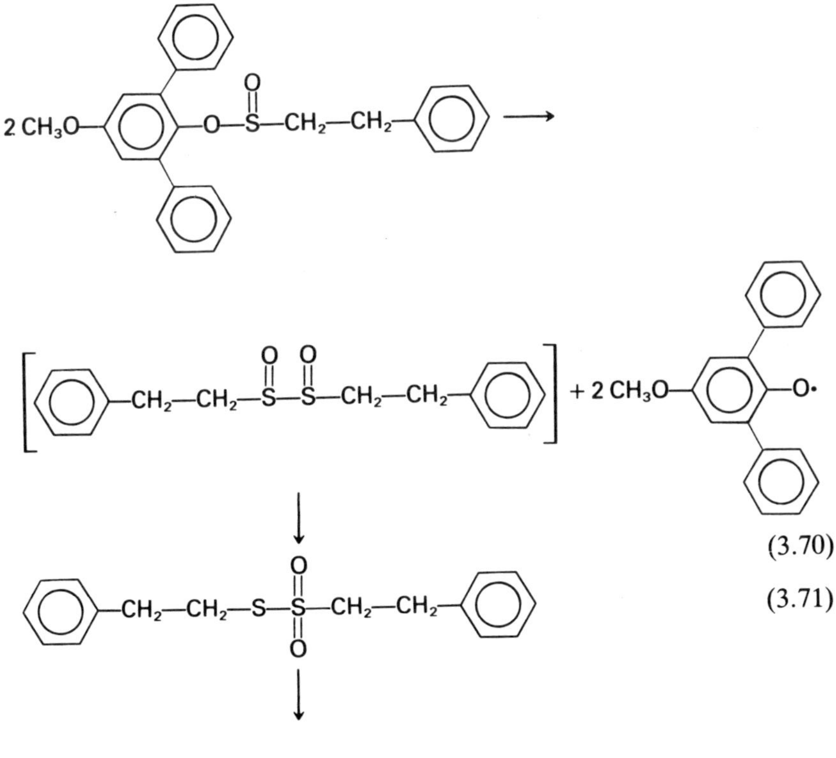

2 CH3O
O
S
CH2
CH2
O O
S S
CH2
CH2
+ 2 CH3O
O•
S
O
O

(3.70)

(3.71)

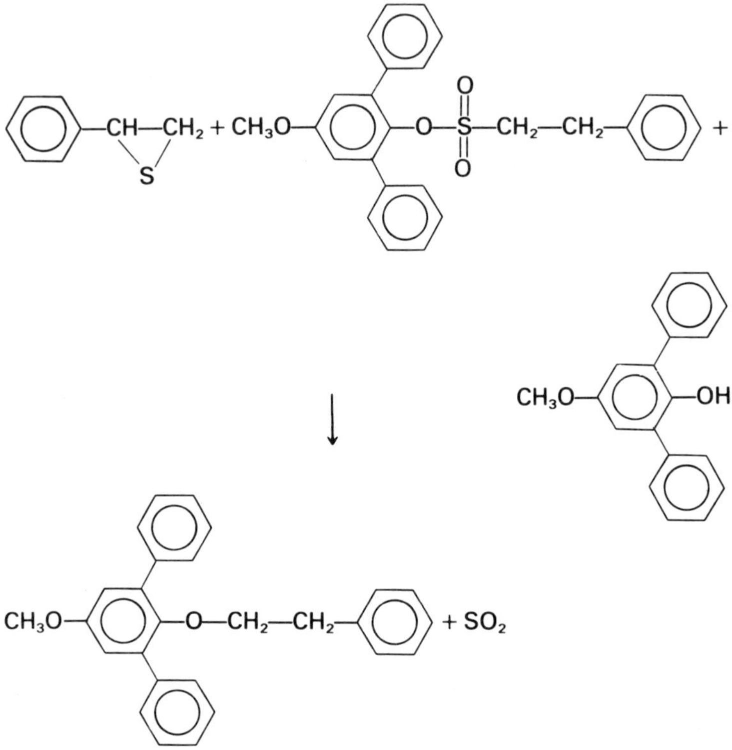

(3.72)

(3.73)

(**3.79**) during which a variety of sulphur compounds are formed which are capable of acting as hydroperoxide decomposers.[345]

3.8 DIXANTHOGENS AND THIOPHOSPHORYL DISULPHIDES

Dixanthogens (**3.80**) and thiophosphoryl disulphides (**3.81**) are effective processing antioxidants for polyolefins[101,108,110,130] and polydienes:[111]

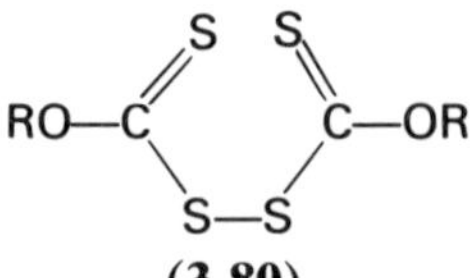

(**3.80**)

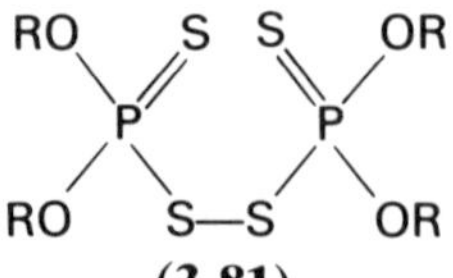

(**3.81**)

All of these compounds exhibit similar behaviour and they are effective hydroperoxide decomposers and free-radical scavengers (chain terminators). The antioxidant mechanism of these compounds can be presented here for thiophosphoryl disulphides (**3.81**) by the following reactions:

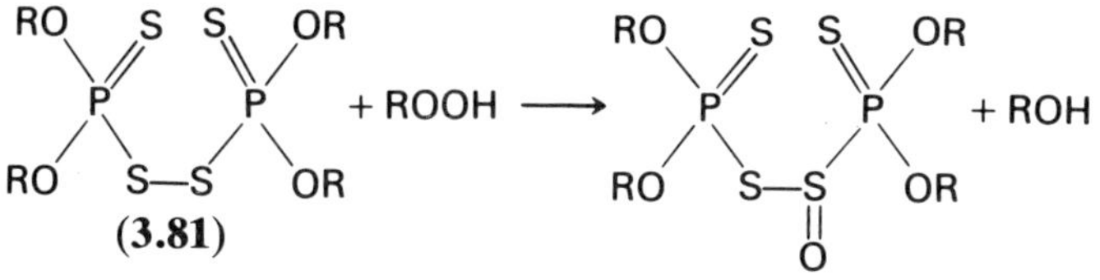

(**3.81**) (3.74)

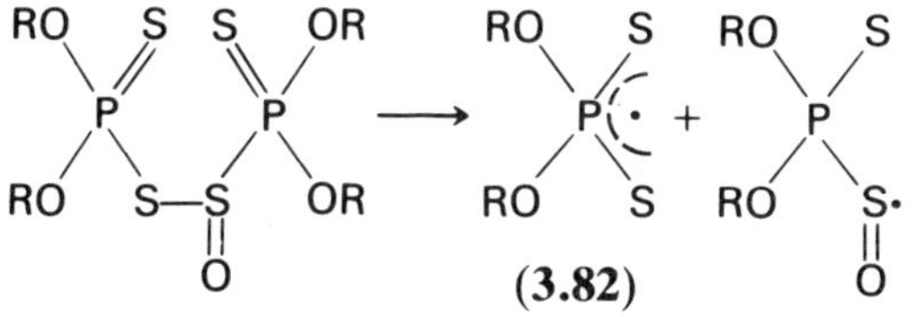

(**3.82**) (3.75)

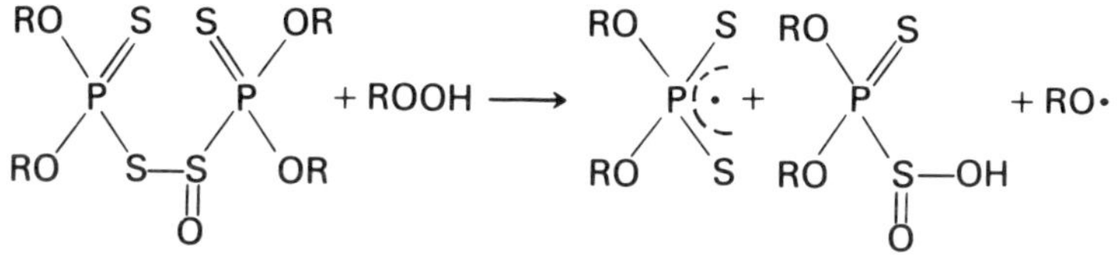

(3.76)

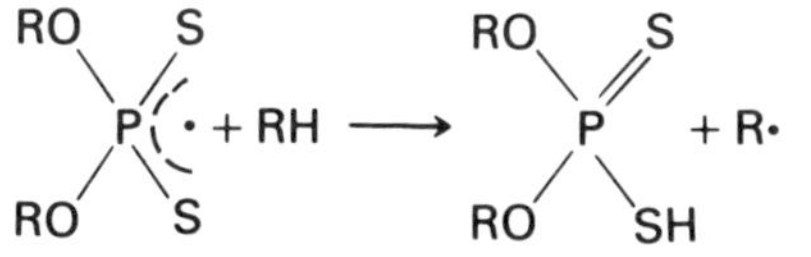

(3.77)

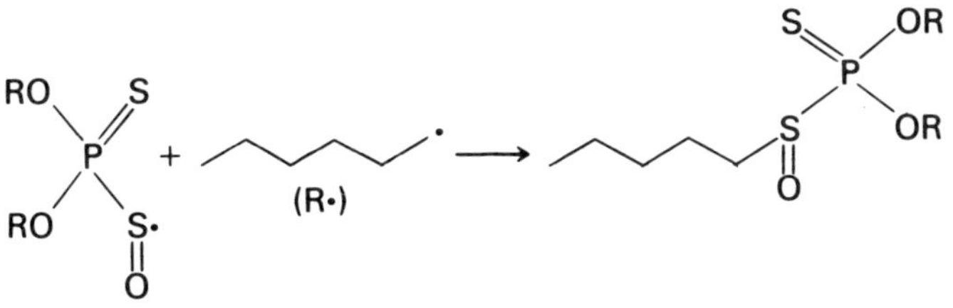

(3.78)

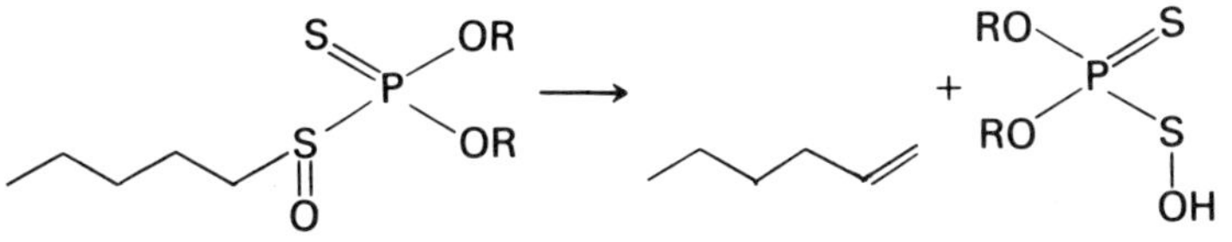

(3.79)

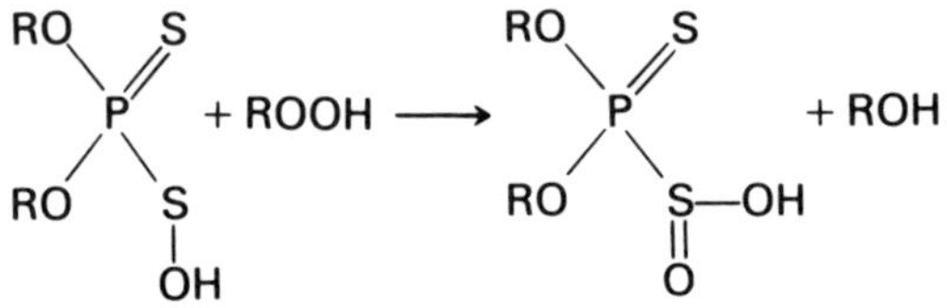

(3.80)

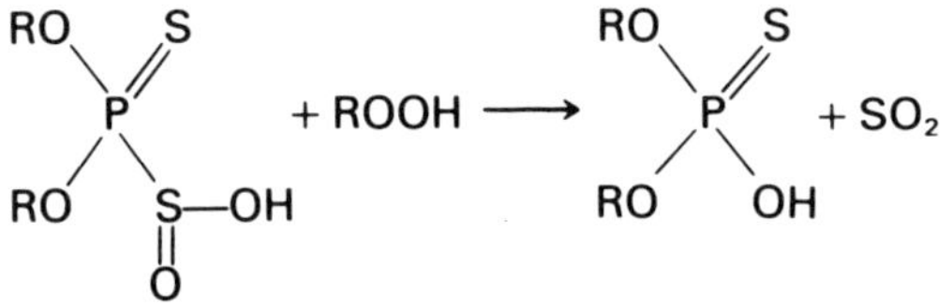

(3.81)

$$SO_2 + ROOH \longrightarrow SO_3(H_2SO_4) + ROH \qquad (3.82)$$

Thiophosphoryl sulphide radicals (**3.82**) react readily with polyisoprene (natural rubber) to give adducts (**3.83**) containing up to 50% of the compound **3.82** added, according to the mechanisms:[111]

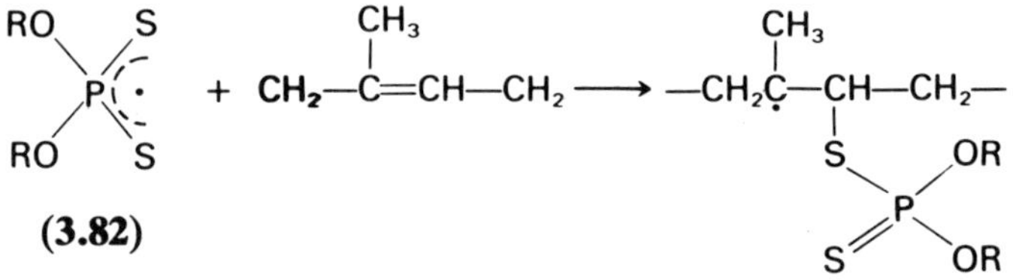

(3.83)

(3.82)

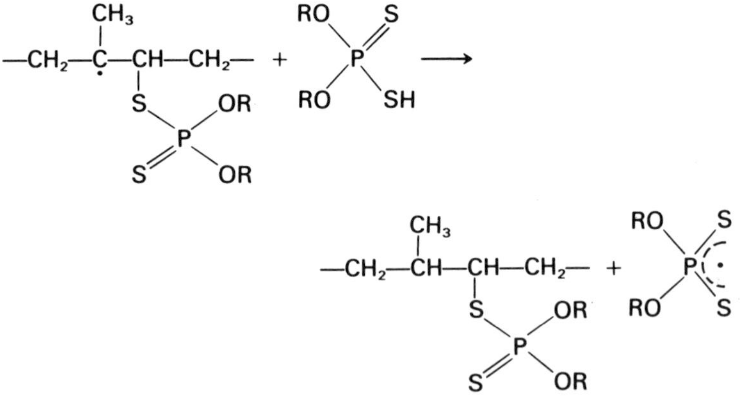

(3.84)

The adduct (**3.83**) is not stable in vulcanization temperatures of 140°C and is thermally decomposed [reaction (3.85)].

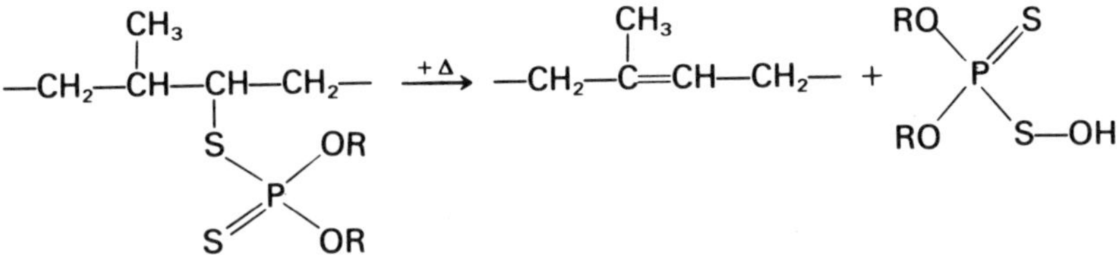

(3.83)

(3.85)

3.9 METAL DITHIOLATES AS ANTIOXIDANTS

Metal dithiolates are represented by three similar groups of compounds:

(i) Metal xanthates (**3.84**).[112,123,131,276,559]
(ii) Metal dithiocarbamates (**3.85**).[2,115–118,151,262,266,271,287,509,878,931–933,1008,1026,1117]
(iii) Metal dithiophosphates (**3.86**).[108,122–131,275,594]

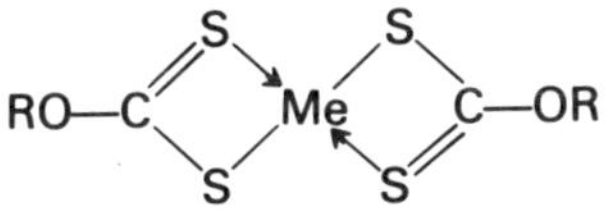

Me = Ni(II), Co(III), Fe(II), Cu(I)

(**3.84**)

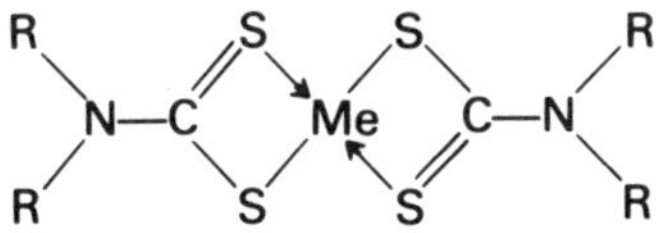

Me = Ni(II), Fe(IV), Fe(III), Zn(II)

(**3.85**)

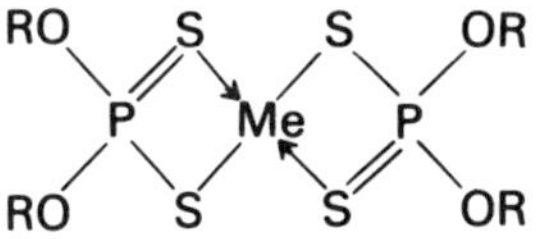

Me = Ni(II)

(**3.86**)

All of these dithiolates exhibit similar behaviour and they are effective hydroperoxide decomposers and free-radical scavengers (chain terminators).[104,115–118,124]

Two different processes are involved during the decomposition of hydroperoxide groups (ROOH):

(i) Initial catalytic homolysis of hydroperoxides (rapid reaction) which is favoured by a low molar ratio of hydroperoxide to metal dithiolate.
(ii) Catalytic ionic decomposition of hydroperoxide (slow reaction, which occurs after an induction period), which depends on a higher molar ratio of hydroperoxide to metal dithiolate.

These processes depend on:

(i) the type of metal dithiolate;
(ii) the ratio of metal dithiolate to hydroperoxide groups;
(iii) the nature of the metal ion in the dithiolate.

The mechanism of decomposition of hydroperoxides in a polymer is almost the same for all dithiolates and is presented below for nickel dialkyldithiophosphates (**3.87**) as a representative example for the whole family of dithiolates:[104,109,124,1066]

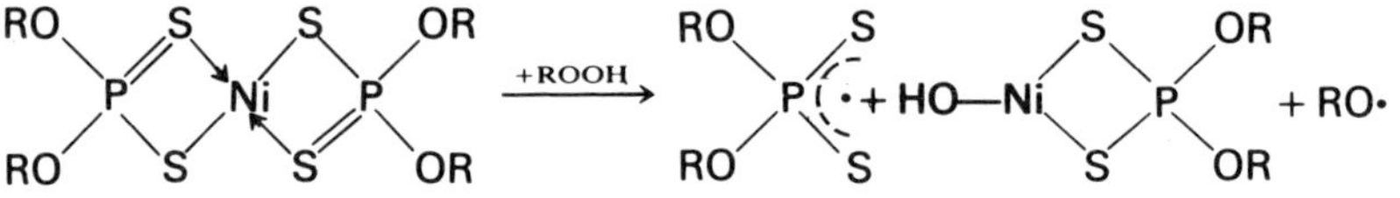

(**3.87**)

(3.86)

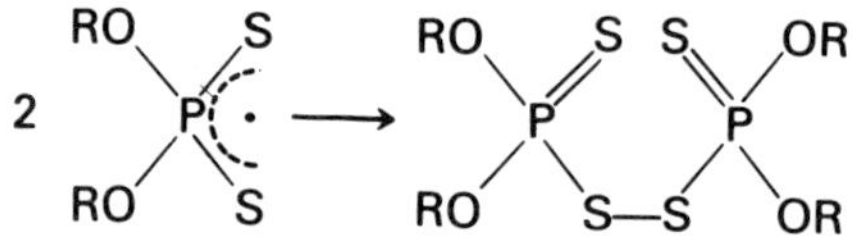

(3.87)

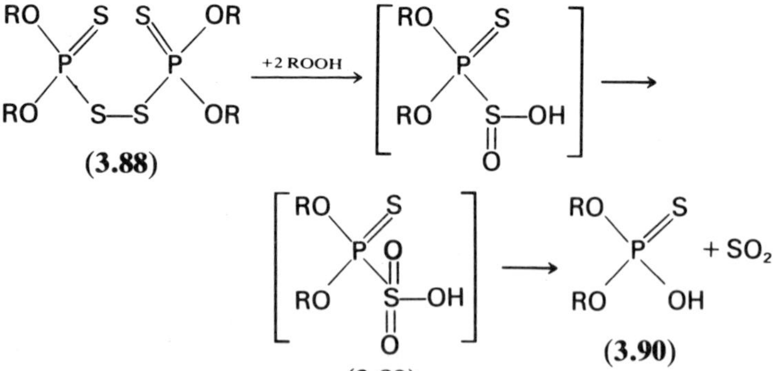

(**3.88**) (**3.89**) (**3.90**)

(3.88)

$$SO_2 + ROOH \longrightarrow SO_3(H_2SO_4) + ROH \tag{3.89}$$

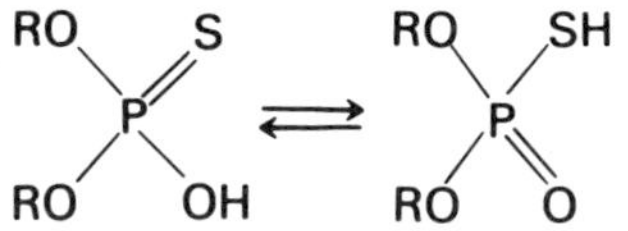

(3.90)

Metal dithiolates are not directly responsible for the decomposition of hydroperoxide groups, but rather their products such as disulphide (**3.88**) [reaction (3.87)] and sulphur dioxide (SO_2) [reaction (3.88)].[36,118,124] Decomposition of the sulphinic acid (**3.89**) formed gives rise to SO_2 and thionophosphoric acid (**3.90**).

The disulphide radical (**3.91**) may also participate in the decomposition of hydroperoxide groups [reaction (3.91)] or scavenging of free radicals [reaction (3.93)]:

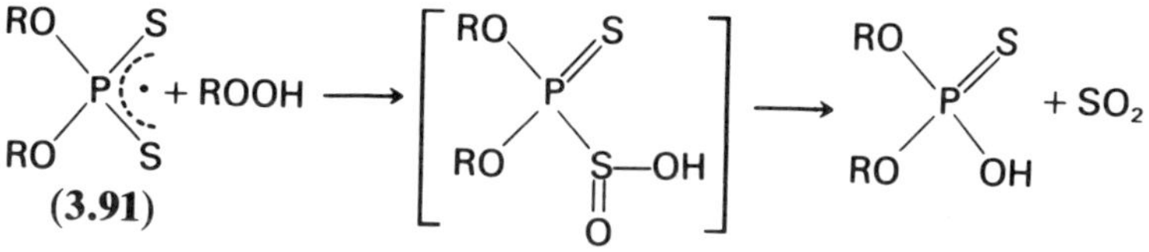

(**3.91**) (3.91)

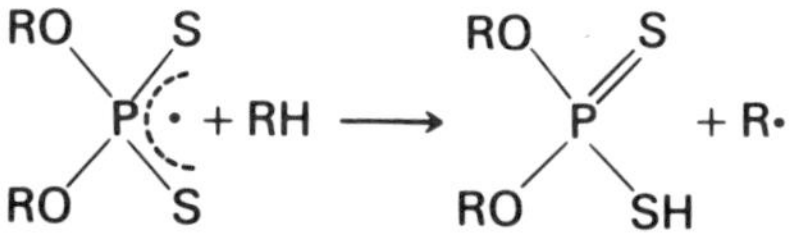

(3.92)

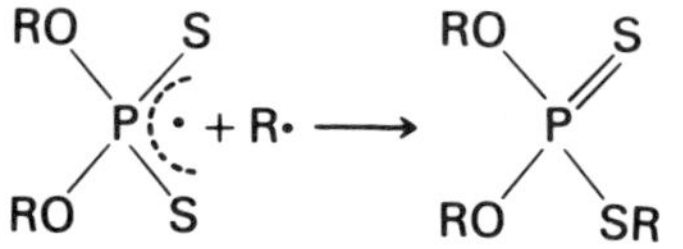

(3.93)

Among the final products formed are sulphonic acid and metal sulphate. The following reactions complete the antioxidant activity of metal dithiolates:

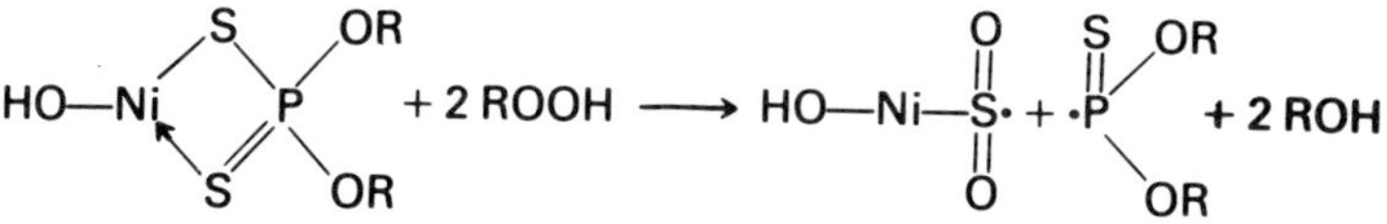

(3.94)

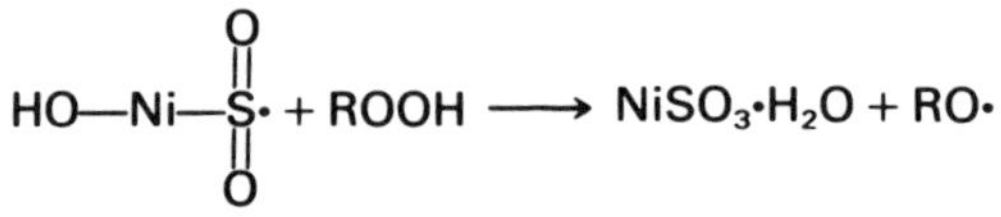

(3.95)

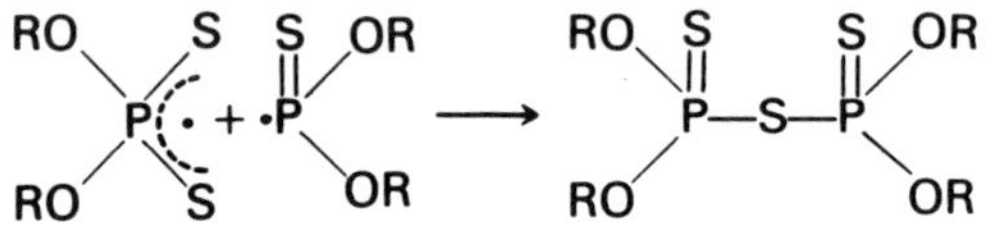

(3.96)

Metal dithiolates can also be used as photostabilizers (cf. Section 5.21).

3.10 4-ALKYL-2-MERCAPTOTHIAZOLINES

It has been shown that 4-alkyl-2-mercaptothiazolines (**3.92**) are effective antioxidants for polypropylene.[106,107]

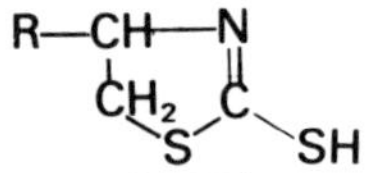

(3.92)

Their stabilizing mechanism is complex and probably occurs by the following mechanism:

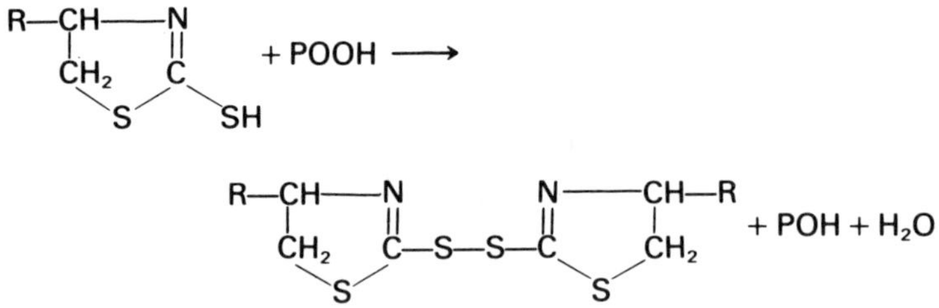

(3.97)

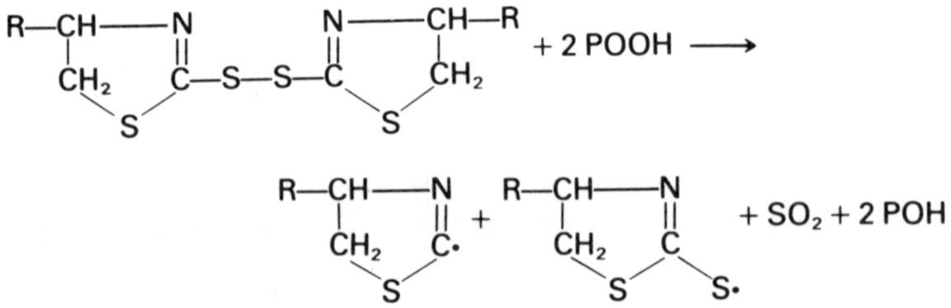

(3.98)

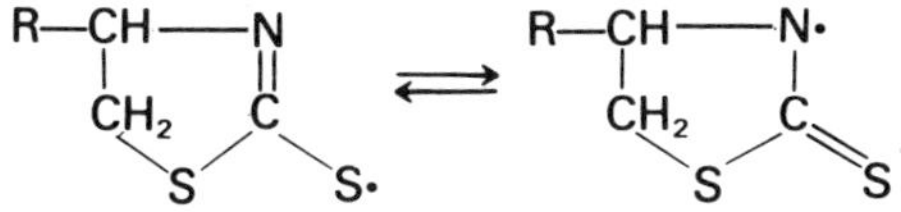

(3.99)

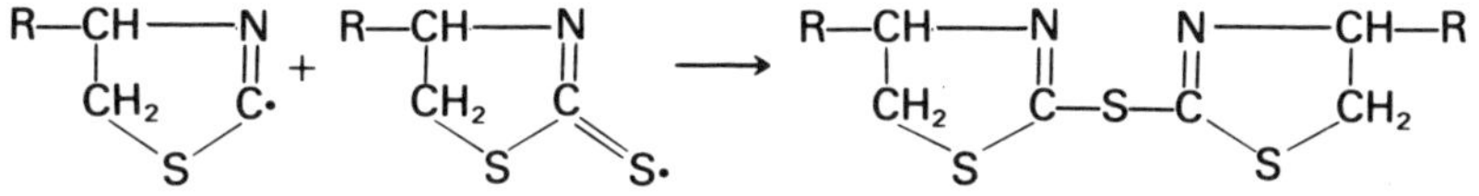

(3.100)

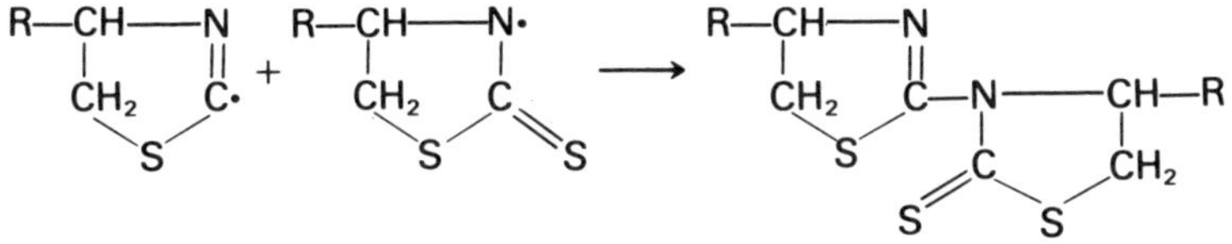

(3.101)

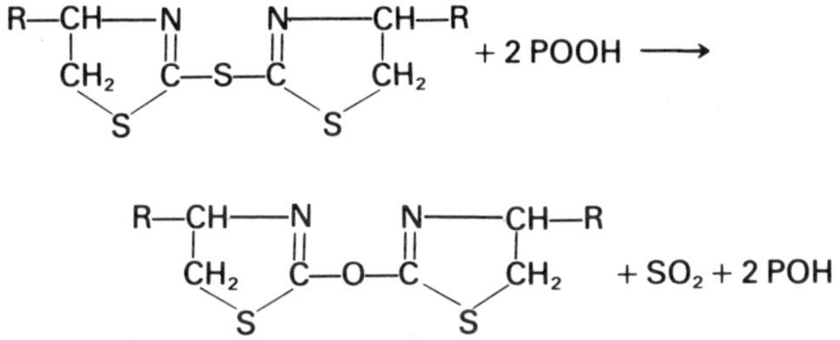

(3.102)

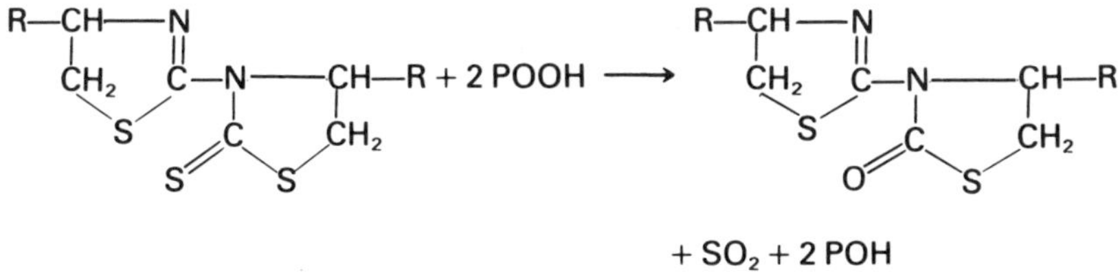

(3.103)

A number of free radicals are formed, which give different products which have been detected by gas chromatography and mass spectroscopy. All of these reactions are accompanied by evolution of SO_2.

The 4-alkyl-2-mercaptothiazolines are pro-oxidant in the initial stages of photo-oxidation of polypropylene; however, their oxidative transformation products contribute to their photo-antioxidant activity.[106]

3.11 SULPHUR DIOXIDE ROLE IN ANTIOXIDIZING SYSTEMS

Sulphur dioxide (SO_2) can be formed during antioxidant activity of various sulphur-containing antioxidants, e.g.

—phenolic sulphides (cf. Section 3.3),
—thiophenols and dithiobisphenols (cf. Section 3.5),
—sulphoxides and sulphones (cf. Section 3.6),
—thiophosphoryl disulphides (cf. Section 3.8),
—transition metal dithiolates (cf. Section 3.9),
—mercaptothiazolines (cf. Section 3.10).

Sulphur dioxide shows both chain-breaking and peroxide-decomposing functions in autoxidation systems, according to the following reactions:[580]

$$SO_2 + ROOH \longrightarrow SO_2RO\cdot + \cdot OH \qquad (3.104)$$

$$SO_2 + ROOH \longrightarrow SO_2HO\cdot + \cdot OR \qquad (3.105)$$

$$SO_2 + RO\cdot \longrightarrow SO_2RO\cdot \qquad (3.106)$$

$$SO_2 + ROO\cdot \longrightarrow SO_2ROO\cdot \qquad (3.107)$$

$$SO_2ROO\cdot \longrightarrow SO_3 + RO\cdot \qquad (3.108)$$

Formation of persulphonyl radicals ($SO_2RO\cdot$) has been observed by ESR spectroscopy.[410] However, another mechanism of interaction of SO_2 with hydroperoxy groups (—OOH) in polymers has been proposed:[233]

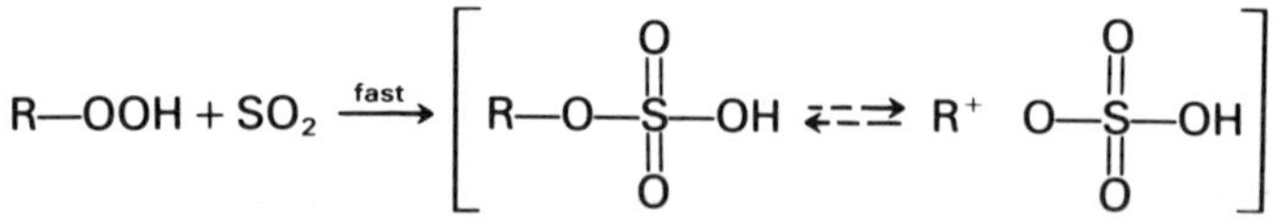

(3.109)

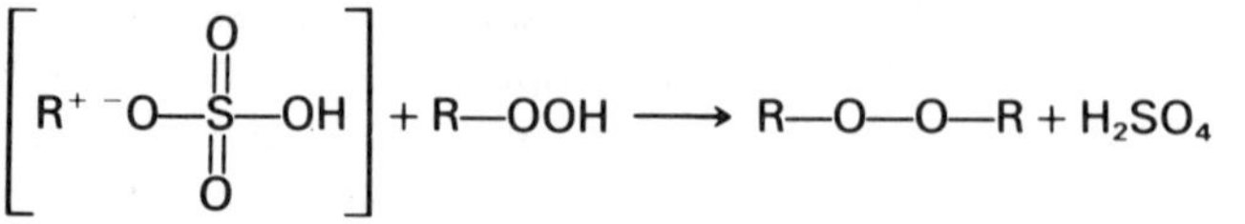

(3.110)

The reaction of SO_2 with hydroperoxy groups in oxidized polymers has been investigated by several authors[211,233,287,545,580,795] and used to quantify hydroperoxide content (cf. Section 9.6.1.8).[795] However, this method does not improve hydroperoxide content determination in comparison with the direct measurement of hydroperoxides by infrared spectroscopy[233] or titration methods.[431]

It has been reported that SO_2 treatment of polypropylene films substantially improves their photostability.[58,59]

3.12 ROLE OF THE ANTIOXIDANT STRUCTURE IN STABILIZATION PROCESSES

The chemical structure of antioxidants plays an important role in their compatibility with the host polymer, and in their reactivity. The *para* position in hindered phenols should be substituted with methyl or methylene groups, which participate in chemical reactions (cf. Section 3.1).

In designing antioxidants, a number of different structure elements have been taken into consideration:

(i) Spacers connecting the phenol ring with the rest of molecule: —CH_2—; —CH_2CH_2—.
(ii) Bridges connecting two or more phenolic groups.
(iii) Groups which increase compatibility with the polymer matrix: —$(CH_2)_{17}CH_3$; CO—O—CH_2CH_2—O—CH_2—CH_2—O—CH_2CH_2—O—CO—; —CO—NH—$(CH_2)_6$—NH—.
(iv) Functional groups with heteroatoms O, N, P or S.
(v) Substituents in the *ortho* position in the phenol rings (cf. Section 3.1).

For example, a phenolic antioxidant having the structure *N,N'*-hexamethylene-bis(3,5-di-*tert.*-butyl-4-hydroxy-hydrocinnamide) (**3.26**) (Irganox 1098, Ciba–Geigy) is a very efficient stabilizer against thermal oxidation and discoloration of Nylons 6 and 66, whereas pentaerythrityl tetrakis [3-(3',5'-di-*tert.*-butyl-4'-hydroxyphenyl)propionate] (**3.32**) (Irganox 1010, Ciba–Geigy) is more suitable for polyamides with long alkylene chains like Nylon 12.

A phenolic antioxidant benzenepropanoic acid, 3,5-bis(1,1-dimethylethyl)-4-hydroxy-2[3-[3,5-bis(1,1-dimethylethyl)-4-hydroxyphenyl]-1-oxopropyl]hydrazide (Irganox MD 1024, Ciba–Geigy) (**3.24**) is an efficient stabilizer for polyolefins and has been used in combination with Irganox 1010 for the stabilization of polyethylene communication cables in contact with copper.[810]

A number of other antioxidants with specially designed structures are presented in Table 3.2.

New phenolic antioxidants of 4-alkylcalix *n*-arenes (**3.93**) with strong hydrogen bonds formed between hydroxyl groups belonging to different phenyl rings of high-molecular-weight phenol–formaldehyde are condensates.[884] At a concentration of 0·2–0·4 wt% they are very effective as antioxidants and photo-antioxidants for polyolefins. Calixiranes are partially oxidized in reactions with hydroperoxide groups to hydroxybenzophenone and arylcalicylate structures.

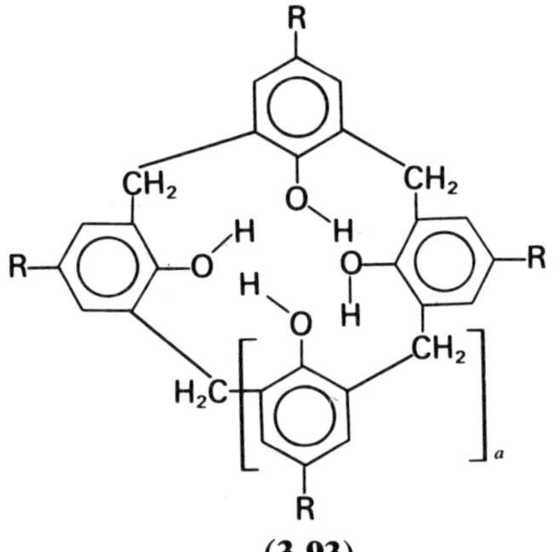

(3.93)

3.13 OVEN AGEING ANTIOXIDANT PERFORMANCE

During the thermal oxidation (oven ageing) of antioxidants, the stabilizing effects on the polymer film (polypropylene) are related to the thermal stability of antioxidant (Table 3.8).[68]

The oven ageing performance follows the order:

Irganox 1010 > Cyanox 1790 > Irganox 1076
> Cyanox 2246 > Topanol CA > Goodrite 3114

The thermal stability of the antioxidants follows the order:

Irganox 1010 > Cyanox 1790 > Irganox 1076
> Topanol CA > Goodrite 3114 > Cyanox 2246

Table 3.8
Stabilizing effectiveness of different antioxidants in polypropylene during oven ageing (130°C)[68]

Antioxidant	*Commercial name*	*Structure*	*Stabilizing effectiveness measured as embrittlement time (h)*
1,1,3-Tri(2′-methyl-4′-hydroxy-5′-*tert.*-butylphenyl)butane (**3.38**)	Topanol CA	See Table 3.2	75
1,3,5-Tri(4-*tert.*-butyl-3-hydroxy-2,6-dimethylbenzyl)isocyanurate	Cyanox 1790		240
2,2′-Methylenebis(6-*tert.*-butyl-4-methylphenol) (**3.36**)	Cyanox 2246	See Table 3.2	90
Stearyl 3-(3,5-di-*tert.*-butyl-4-hydroxyphenyl)propionate (**3.21**)	Irganox 1076	See Table 3.2	215
Pentaerythrityl tetrakis(3,5-di-*tert.*-butyl-4-hydroxyphenyl)propionate (**3.32**)	Irganox 1010	See Table 3.2	210
1,3,5-Tri(3,5-di-*tert.*-butyl-4-hydroxylbenzyl)isocyanurate (**3.31**)	Goodrite 3114	See Table 3.2	55
Distearyl pentaerythrityl diphosphite (**3.113**)	Weston 618	See Section 3.17	5
Distearyl thiodipropionate (**3.59**)	Plastanox STDP	See Table 3.7	2

Antioxidant performance and antioxidant stability are closely connected; the anomalous order of stability is probably influenced by volatilization during heating.

3.14 POLYQUINONES AS ANTIOXIDANTS

Polyquinones (at different quinone/hydroquinone ratios, depending on the nature of the catalyst used: (for **3.94**) H_2SO_4 or BF_3 and (for **3.95** and **3.96**) sodium ethoxide have been used as effective antioxidants for styrene–butadiene rubber.[977]

Polyquinones prepared using either H_2SO_4 or BF_3–acetic acid complex:

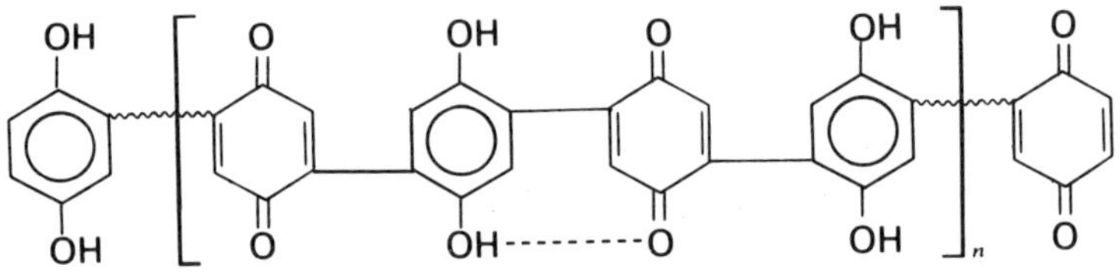

(3.94)

Polyquinones prepared using sodium ethoxide catalyst:

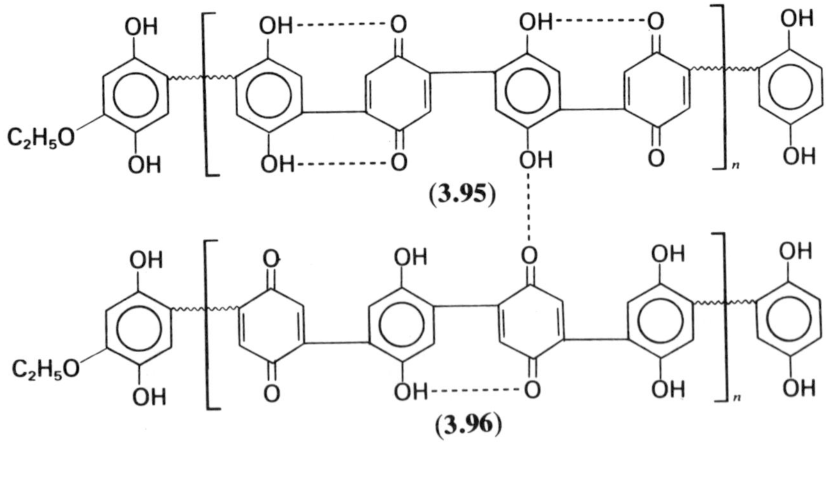

(3.95)

(3.96)

The antioxidant efficiency of polyquinone products seems to be due to the ability of both the hydroquinone and benzoquinone nuclei to react with the radical intermediates of the degradation process as well as their ability to block the odd electron sites created on the rubber chains, thus disrupting the chain degradation process. This may be represented by the following reactions:[978]

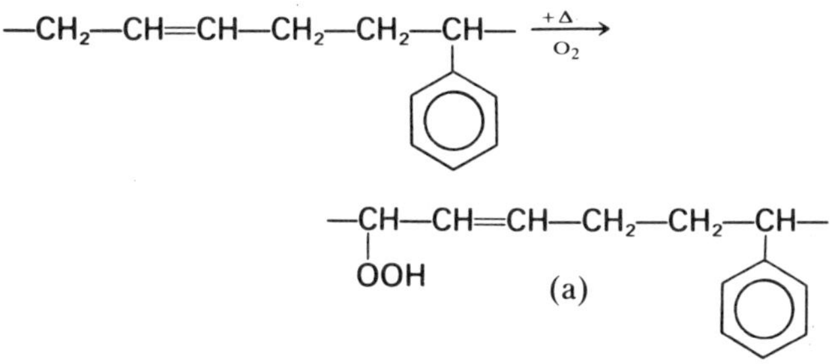

(3.111)

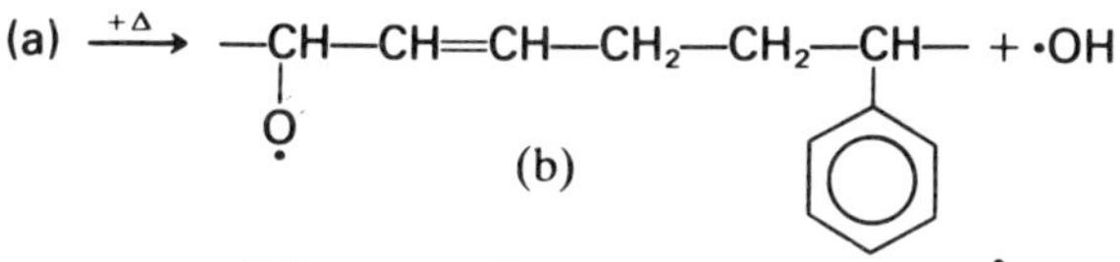

(3.112)

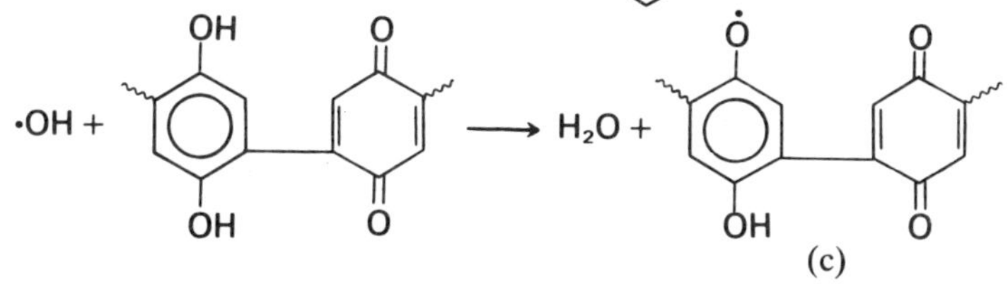

(3.113)

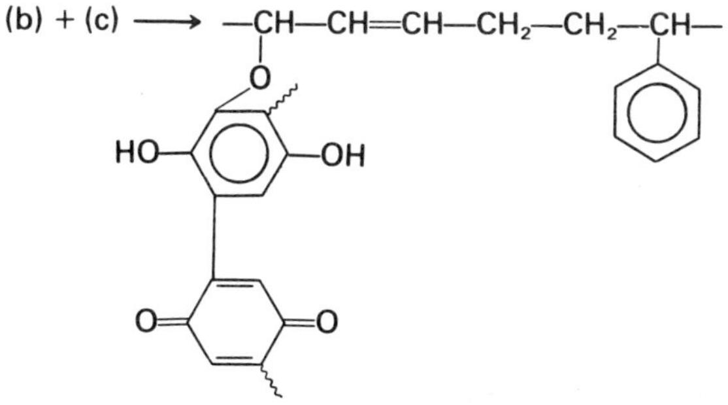

(3.114)

According to this mechanism, the hydroxy radical (HO•) formed abstracts hydrogen atoms from the hydroquinone nuclei of the stabilizer, thus leading to the formation of highly stabilized semiquinone radicals. The latter, due to their high stability, can probably recombine with the alkoxy radicals (PO•) derived from the rubber molecule leading not only to the blocking of the odd electron site formed on the rubber chain, but also to the bonding of the antioxidant molecule with its effective hydroquinone and benzoquinone nuclei.

p-Benzoquinone–tin derivatives (**3.97**), obtained by polycondensation of the quinone and $SnCl_4$ in the absence of solvent, have been investigated as photostabilizers for polybutadiene rubber.[1170] The stabilizing efficiency increases as the tin content of the stabilizer increases. The mechanism is based in the involvement of both the quinone part of the stabilizer and the Sn–Sn bonds in scavenging the radical species formed, as well as blocking the radical sites on the polymeric chain. The reactions involved may be represented by the following equations.

$$\text{R—H} + O_2 \longrightarrow \text{ROOH} \qquad (3.115)$$

$$\text{ROOH} \xrightarrow{+h\nu} \text{RO}\cdot + \cdot\text{OH} \qquad (3.116)$$

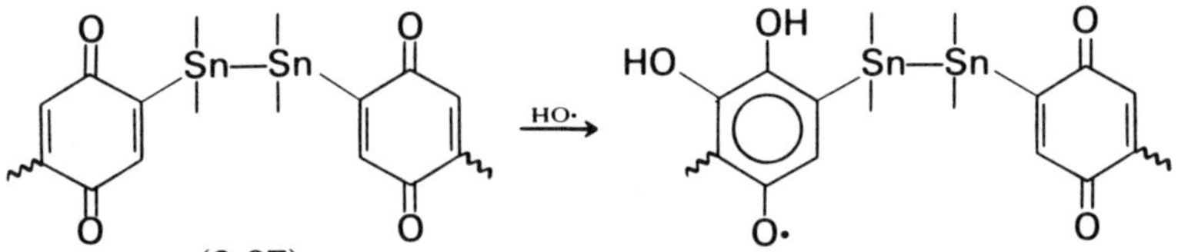

(**3.97**)

(3.117)

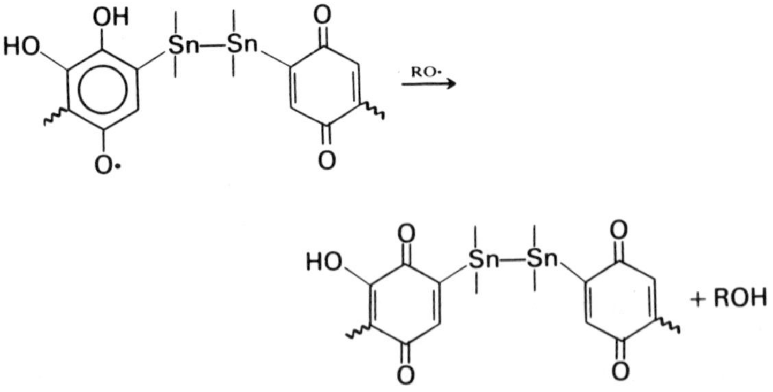

(3.118)

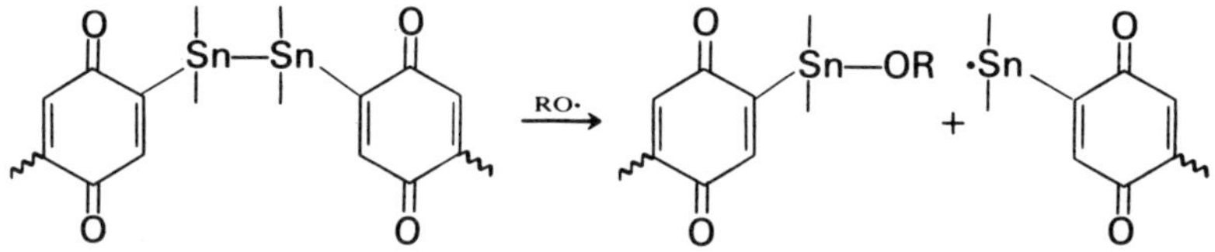

(3.119)

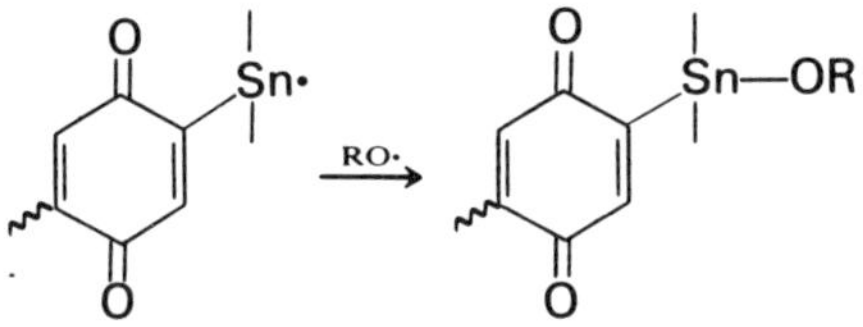

(3.120)

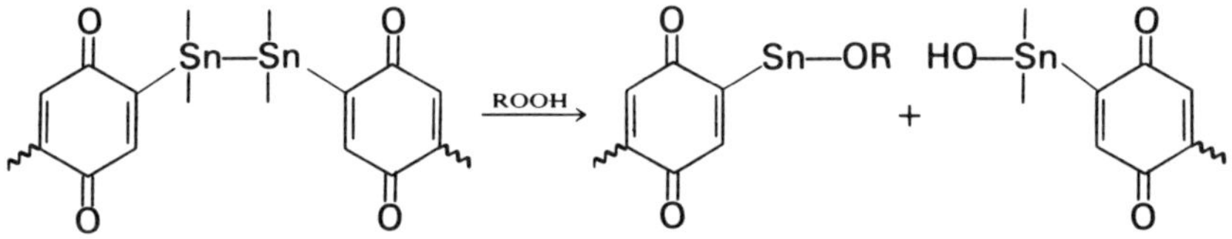

(3.121)

Because of the high sensitivity of Sn–Sn bonds towards oxidizing agents, it has been suggested that they react either with the hydroperoxides or with the radicals derived from them.

At a subsequent stage of oxidation, the quinone–tin stabilizers are probably transformed into the quinone–tin oxides which seem to be very effective UV absorbers, so that they can afford additional protection for the rubber against subsequent degradation by light:

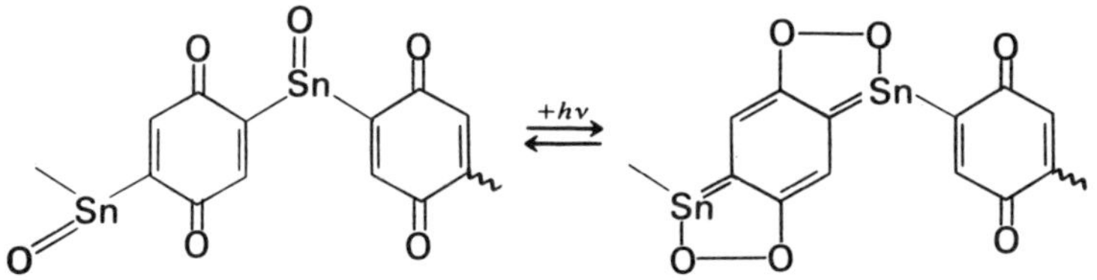

(3.122)

3.15 LIGNIN DERIVATIVES AS STABILIZERS

Lignin (**3.98**) has a complicated structure that contains 12 coniferyl units:

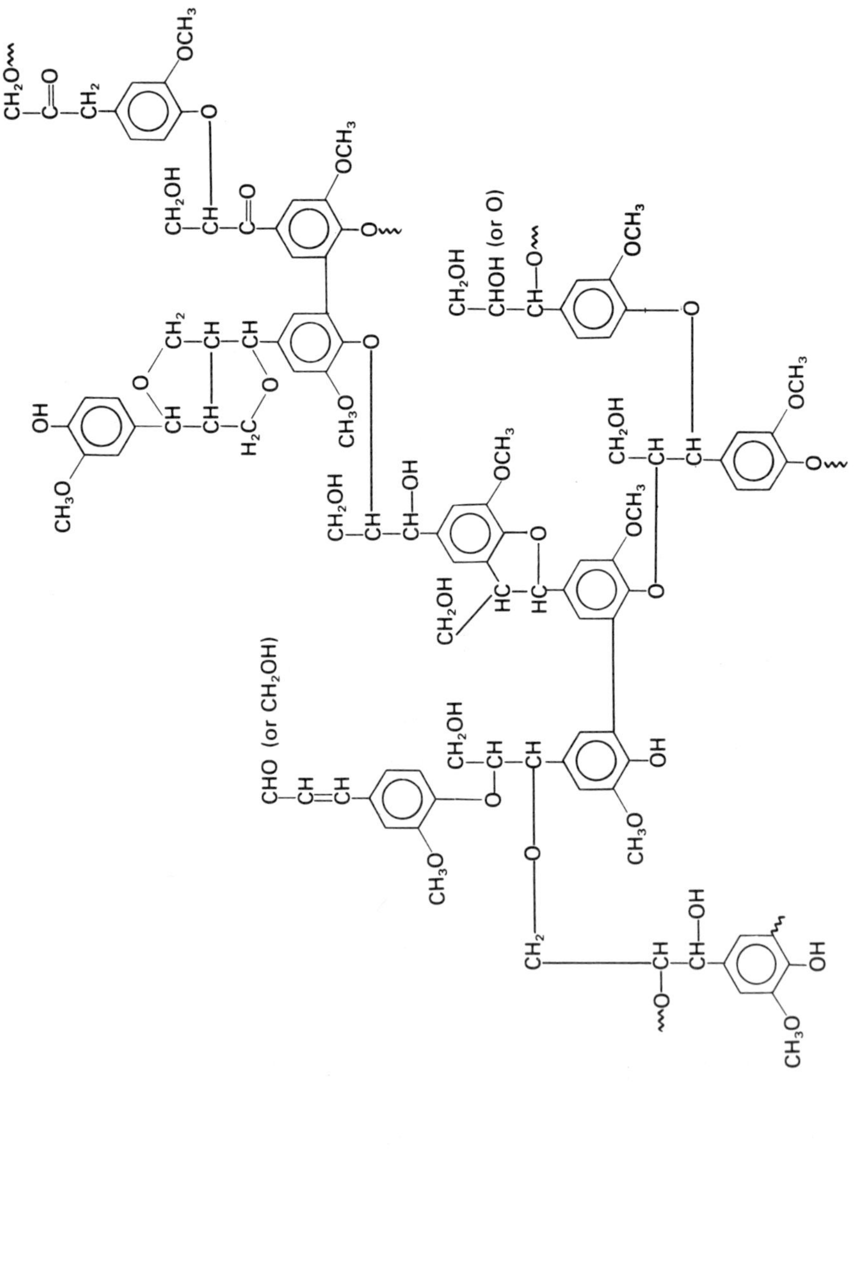

(3.98)

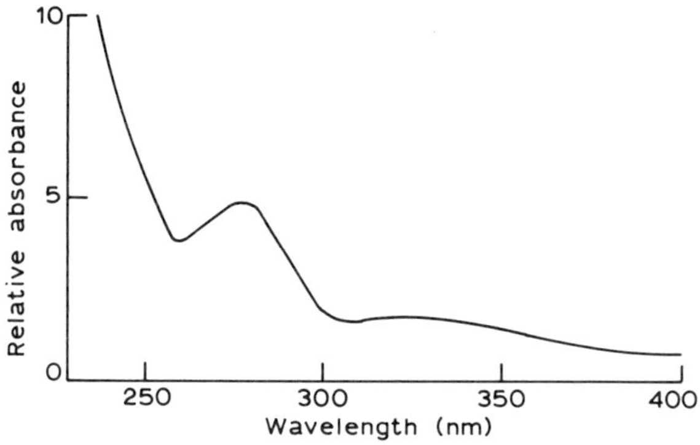

Fig. 3.7. Absorption spectrum of lignin.[562]

The structure of lignin is very different for each variety of plant and also according to the environment in which the plant was grown. Lignin is an excellent light absorber; its absorption spectrum (Fig. 3.7) extends with a tail to over 400 nm.[562]

Lignin always contains a small amount of stable phenoxy radicals which may be generated during sample preparation, handling or storage in the presence of light. During UV irradiation of lignin several other free radicals are formed, which have been detected by ESR spectroscopy. Because of the very complicated lignin structure (**3.98**) it is extremely difficult to identify free-radical sites formed.

It has been shown that lignins exhibit an antioxidant effect in the presence of oil and other fat-soluble substances.[258] Heat-treated lignin from softwood black liquor has been shown to stabilize rubbers against atmospheric oxidation.[203,391] Sugar cane bagasse lignin has been found to act as an antioxidant[348,425] and photostabilizer[347] for rubbers. The most probable explanation of a mechanism for these lignin properties is that the lignin can effectively trap alkyl and peroxy radicals.[361] Thus, the chemical stabilizing effect of lignin should be similar to that of hindered phenols. However, its high molecular weight considerably decreases its mobility in the polymer matrix, reducing or eliminating losses by migration and volatilization.

3.16 ORGANOPHOSPHORUS ANTIOXIDANTS

Organic alkyl (**3.99**) and aryl (**3.100–3.102**) phosphites are effective antioxidants and photostabilizers for polyolefins, poly(vinyl chloride) and rubbers.[407,578,579,664,876,877,995,996,1135]

(3.99)

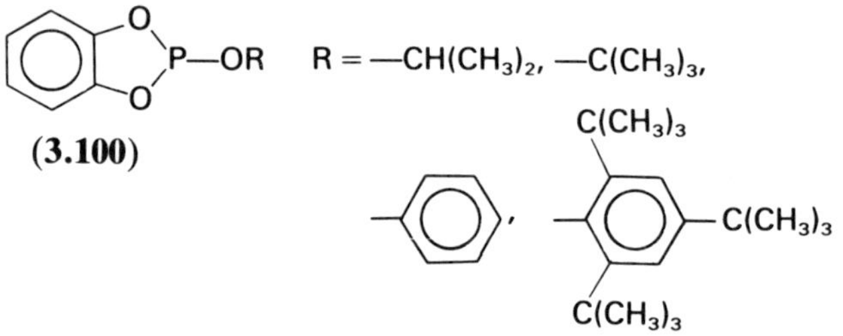

(3.100)

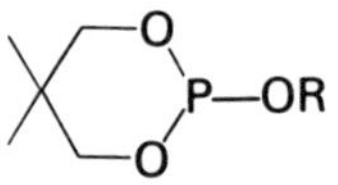

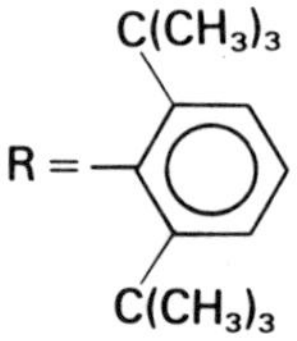

(3.101)

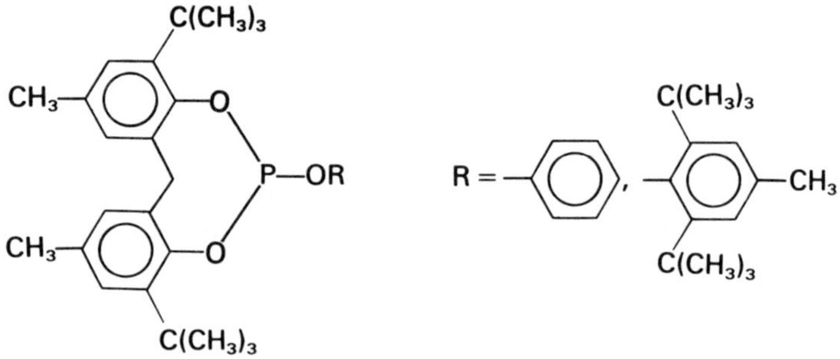

(3.102)

Phosphites (Table 3.9) are very compatible with almost all polymers. They have very low volatility and low reactivity towards oxygen. In addition, phosphites exhibit synergistic effects with other antioxidants and photostabilizers. They also preserve the original colour of the polymer during processing and use.

Phosphites may participate in reactions with:

(i) hydroperoxides (peroxide decomposers);
(ii) free radicals (kinetic chain-breaking antioxidants);
(iii) proton donors, in particular HCl;

(iv) ketoallyl chloride groups;
(v) metal salts.

Phosphites mainly act as hydroperoxide decomposers [reactions (3.123) and (3.124)] however, aryl phosphites are also capable of acting as kinetic chain-breaking antioxidants (chain terminators) reactions (3.125) and (3.126)]:[424,898,899,993–997]

$$(RO)_3P + R_1OOH \longrightarrow (RO)_2\overset{\displaystyle OR_1}{\overset{|}{P}}{=}O + ROH \qquad (3.123)$$

$$(ArO)_3P + R_1OOH \longrightarrow (ArO)_3P{=}O + ROH \qquad (3.124)$$

$$(ArO)_3P + R_1O\cdot \longrightarrow (ArO)_2P{-}OR_1 + ArO\cdot \qquad (3.125)$$

$$(ArO)_3P + R_1O_2\cdot \longrightarrow (ArO)_3P{=}O + R_1O\cdot \qquad (3.126)$$

Alkyl phosphites are not able to react in that way, because they are oxidized by alkoxyl radicals giving alkyl radicals which propagate the chain oxidation:

$$(RO)_3P + R_1O\cdot \longrightarrow (RO)_3P{=}O + R_1\cdot \qquad (3.127)$$

At higher temperatures, as in the processing of polypropylene (180°C), hindered aryl phosphites exhibit an antioxidative efficiency which resembles, or even surpasses, that of hindered phenols.[702,995,996]

ESR spectra of trialkylphosphites in the presence of oxygen at low temperatures indicate that alkoxyphosphoranyl (**3.103**) and alkylperoxyphosphoranyl (**3.104**) radicals were present:[334]

$$(RO)_3P + RO\cdot \longrightarrow \underset{\textbf{(3.103)}}{(RO)_3\dot{P}{-}OR} \qquad (3.128)$$

$$(RO_3)\dot{P}{-}OR \xrightarrow{+O_2} \underset{\textbf{(3.104)}}{(RO)_3\overset{\displaystyle \dot{O}}{\overset{\displaystyle O}{\overset{|}{P}}}{-}OR} \qquad (3.129)$$

$$(RO)_4PO_2\cdot + (RO)_3P \longrightarrow (RO_4)P{-}O\cdot + (RO)_3P{=}O \qquad (3.130)$$

$$(RO)_4P{-}O\cdot + (RO)_3P \longrightarrow (RO)_4P\cdot + (RO)_3P{=}O \qquad (3.131)$$

$$(RO)_4P{-}O\cdot \longrightarrow (RO)_3P{=}O + RO\cdot \qquad (3.132)$$

Phosphites have the ability to react and remove HCl formed, e.g.

Table 3.9
Examples of commercially produced phosphites

Chemical name	*Commercial name*	*Chemical structure*
CAS name: Phenol, 2,4-bis(1,1-dimethylethyl), phosphite (3:1); *CAS number* 31570-04-4	Irgafos 168 (Ciba–Geigy)	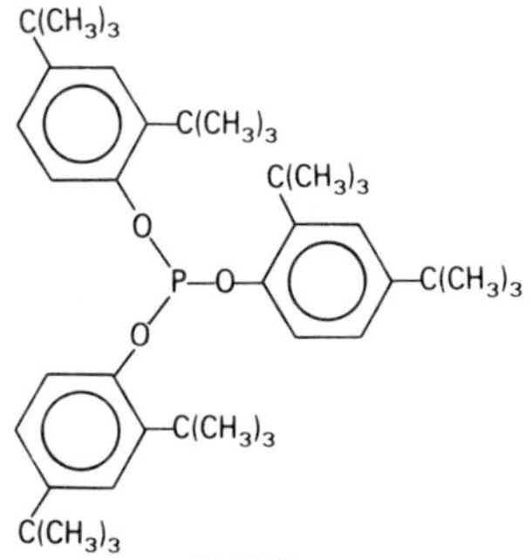 **(3.105)**
CAS name: Phosphonous acid, [1,1′-biphenyl]-4,4′-diylbis-, tetrakis[2,4,-bis(1,1-dimethylethyl)phenyl] ester; *CAS number* 38613-77-3	Irgafos P-EPQ (Ciba–Geigy)	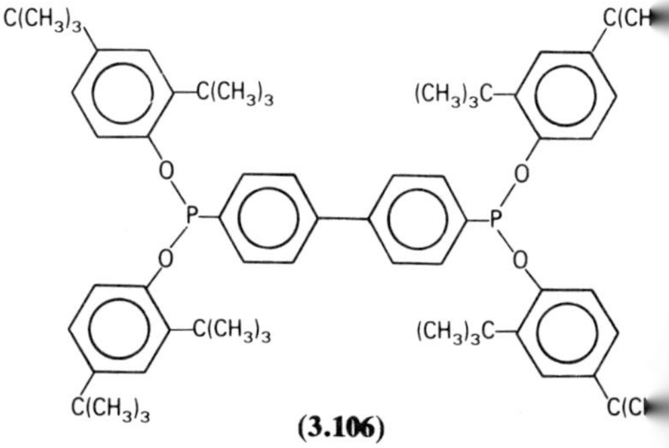 **(3.106)**

during the photo- and/or thermal degradation of poly(vinyl chloride), according to reactions (3.133)–(3.135).[572,1123]

$$(RO)_3P + HCl \longrightarrow (RO)_2PH{=}O + RCl \qquad (3.133)$$

$$(RO)_2PH{=}O + HCl \longrightarrow RO{-}\overset{\overset{\large OH}{|}}{P}H{=}O + RCl \qquad (3.134)$$

Molecular weight	*Appearance*	*Solubility (wt %)*					*Melting point (°C)*	*Recommended applications*
		n-Hexane	*Methanol*	*Acetone*	*Ethyl acetate*	*Water*		
647	White powder	11	<0·01	1	4	<0·01	180–186	Polyolefins
991	Off-white powder	>50	3·5	>50	>50	<0·01	85–110	Polyolefins

$$RO-\overset{\displaystyle OH}{\overset{|}{P}}H{=}O + HCl \longrightarrow H_3PO_3 + RCl \qquad (3.135)$$

Triarylphosphites are stable towards reaction with HCl.

Phosphites may also react with ketoallyl chloride groups, which can be formed during the dehydrochlorination of poly(vinyl chloride) according to the following mechanism:[794]

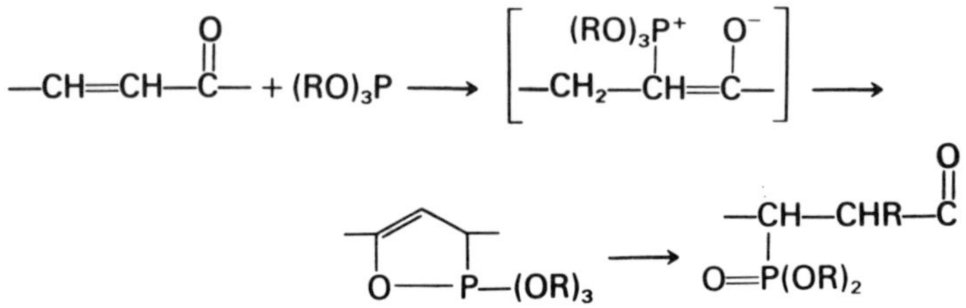

(3.136)

Organic phosphites have the ability to form complexes with various transition metal compounds; however, they are very sensitive to oxygen, moisture and temperature.[18]

Dialkyldithiophosphoric acids **(3.107)** and their ammonium salts undergo addition to the double bond in rubber in the presence of free radicals formed by redox reaction with hydroperoxides or by mechanical shear of the macromolecule:[110,111]

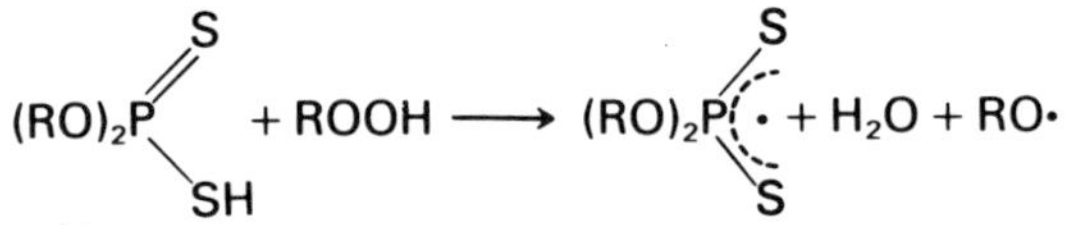

(3.107)

(3.137)

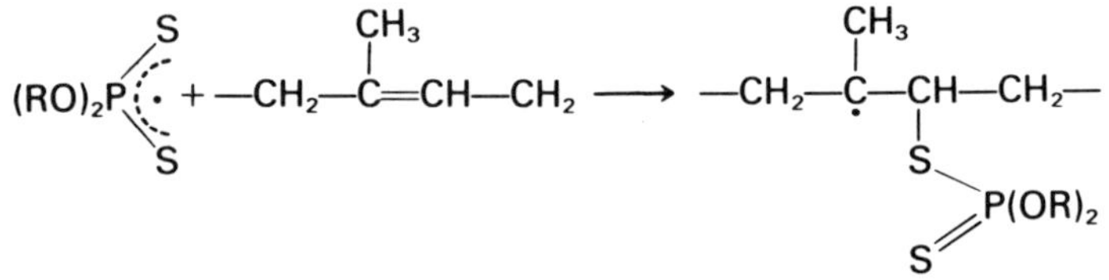

(3.138)

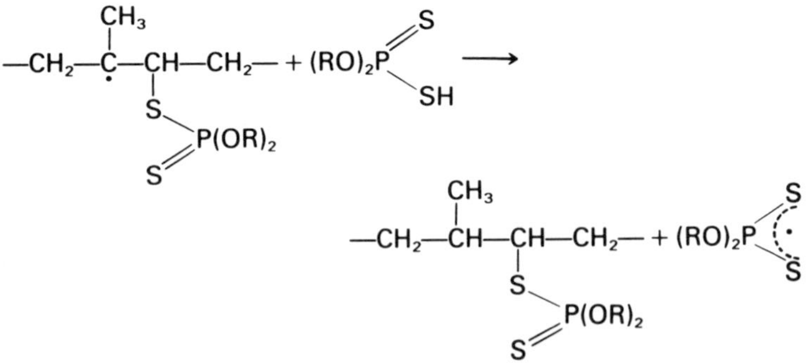

(3.139)

Transition metal ions facilitate reaction (3.137).

3.17 PENTAERYTHRITYL DIPHOSPHITES

Pentaerythrityl diphosphites (**3.108–3.112**) are used exclusively for the stabilization of polyolefins, mostly polypropylene.[558] In spite of relatively low antioxidant effectiveness (at higher temperatures) and sensitivity to hydrolysis, they have the desirable characteristic of not imparting colour to the polymer during degradation. In combination with other antioxidants (phenolic and sulphur types) they achieve very good antioxidant characteristics. They exhibit synergistic effects with 3,5-di-*tert.*-butyl-4-hydroxybenzoic acid,*n*-hexadecyl ester (Cyasorb UV 2908) (**5.1**).[99,100]

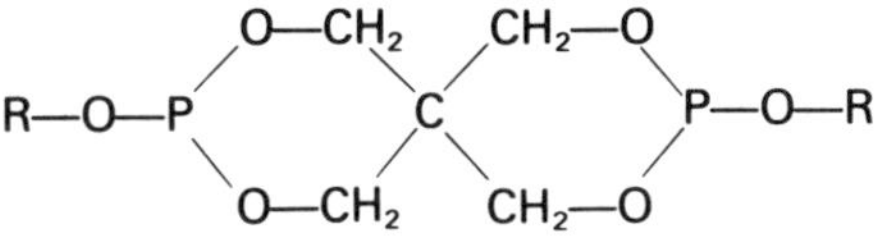

(**3.108**)

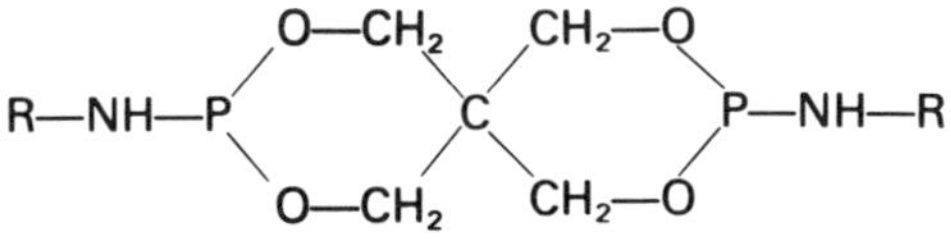

(**3.109**)

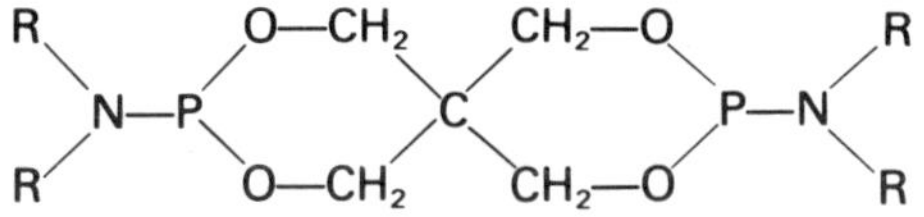

(**3.110**)

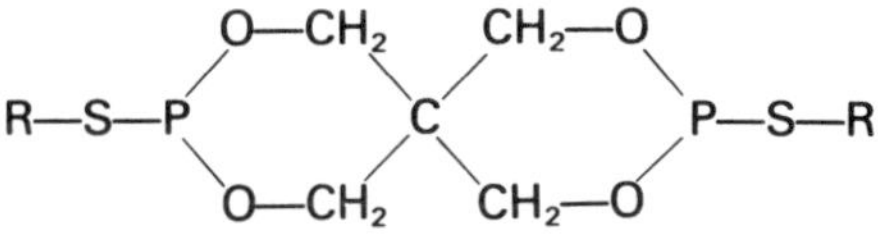

(**3.111**)

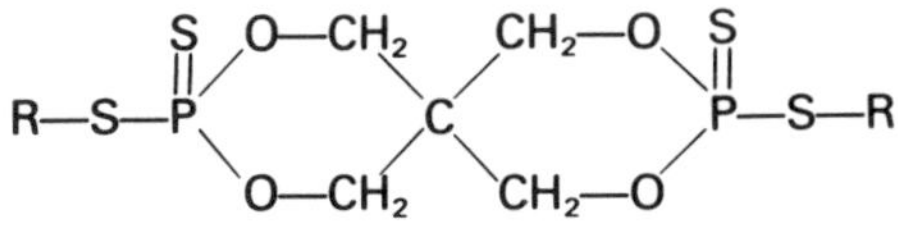

(**3.112**)

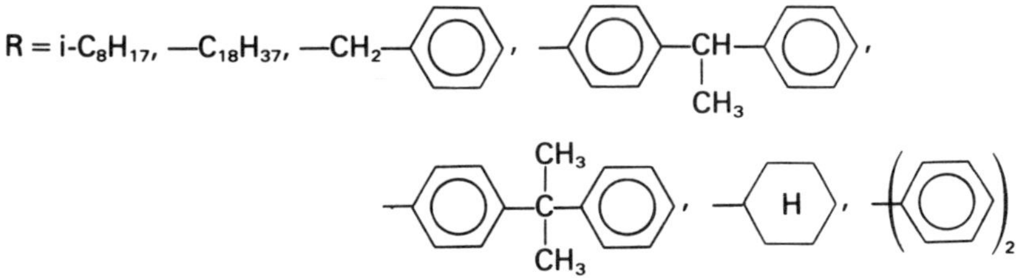

Distearyl pentaerythrityl diphosphate (Weston 618) (**3.113**) is commercially produced by the Borg Warner Corporation (The Netherlands).

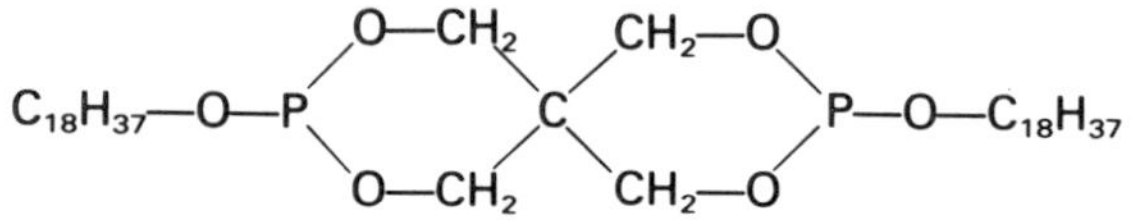

(**3.113**)

3.18 ORGANOBORON ANTIOXIDANTS

Organoboron compounds, particularly boric acid esters and arylboronic acid derivatives, are good antioxidants for polypropylene.[663]

3.19 MACROCYCLIC LIGANDS

The macrocyclic ligands, such as:

—1,4,8,11-tetra-azacyclotetradecane (**3.114**),
—1,4,8,11-tetramethyl-1,4,8,11-tetra-azacyclotetradecane (**3.115**),
—1,4,8,11-tetrathiacyclotetradecane (**3.116**),
—1,4,7,10-tetraoxa-acyclododecane (**3.117**),
—5,14-dihydrodibenzo[*b*,*i*]-1,4,8,11-tetra-azacyclotetradecene (**3.118**),
—1,7,10,16-tetraoxa-4,13-diazacyclo-octadecane (**3.119**)

are effective antioxidants to protect natural rubber against oxidation catalysed by metal ions with which they form inactive complexes.[474,476,477]

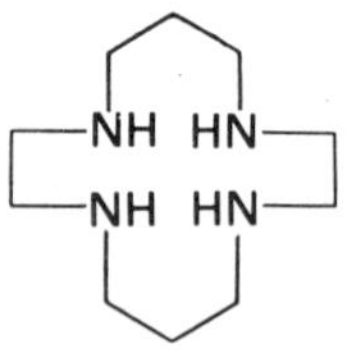

(**3.114**)

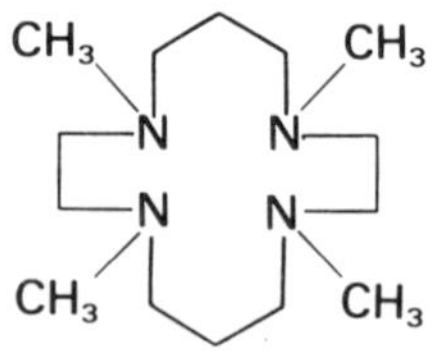

(**3.115**)

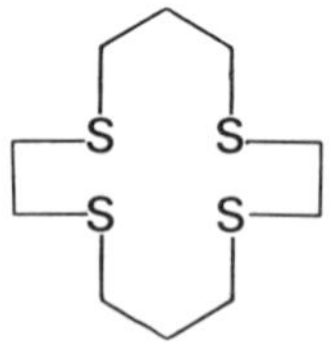

(**3.116**)

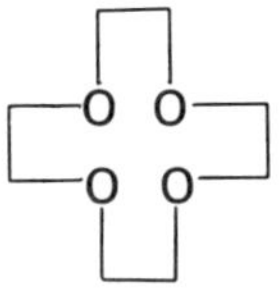

(3.117)

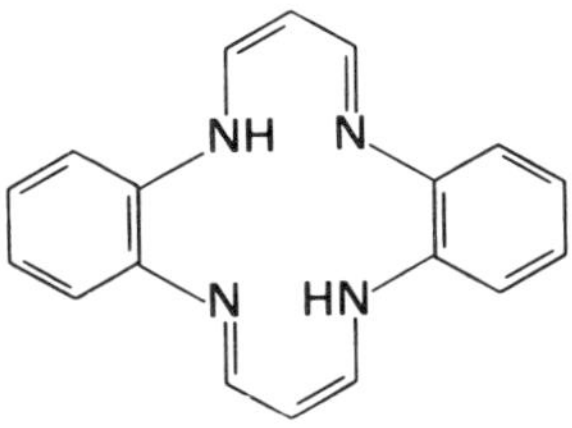

(3.118)

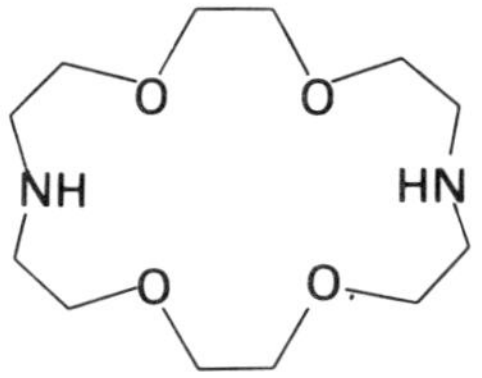

(3.119)

3.20 AROMATIC AMINE ANTIOXIDANTS

Properties of aromatic amines as chain-breaking antioxidants in polymeric systems have been reviewed in a number of publications.[909,910,1016]

Low-molecular hydroperoxides are decomposed by amines.[1107] For example, tertiary hydroperoxides are stable with pyridine for several hours at 60°C, whereas primary and secondary hydroperoxides are rapidly decomposed at room temperature.[750] Pyridine vapour partially decomposes hydroperoxides in photo-oxidized polystyrene.[727]

Reactivity of aromatic amines with hydroperoxides is generally lower than that of aliphatic amines.[587] The reaction rate decreases in the series:

primary amines < secondary amines < tertiary amines

and it has been suggested that amine antioxidant efficiency may be lowered in the presence of ROOH.[10]

Primary amines such as *N,N'*-disubstituted 1,4-phenylenediamines (**3.120**) are generally good antioxidants. They react with peroxy radicals (ROO•) by the following mechanisms:[1115]

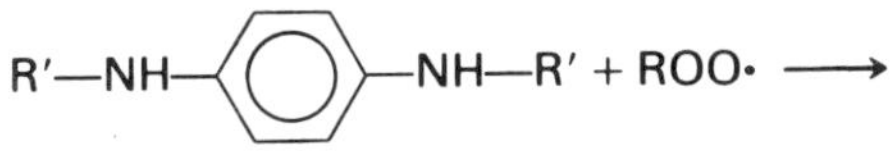

(3.120)

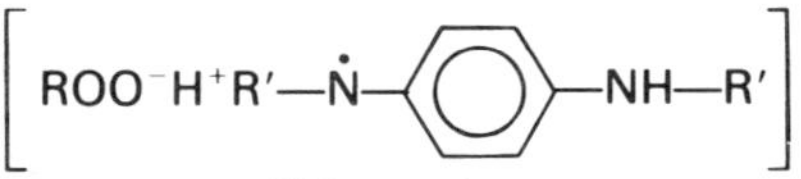

(3.140)

CT complex

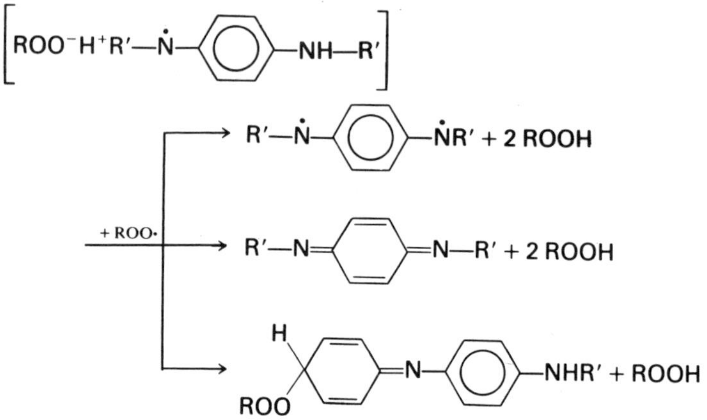

(3.141)

(3.142)

(3.143)

Secondary aromatic amines (diarylamines) (**3.121**) are effective antioxidants in the protection of polyolefins and rubbers. They react with peroxy radicals (ROO•) to give diaryl nitroxy radicals (**3.123**):[10]

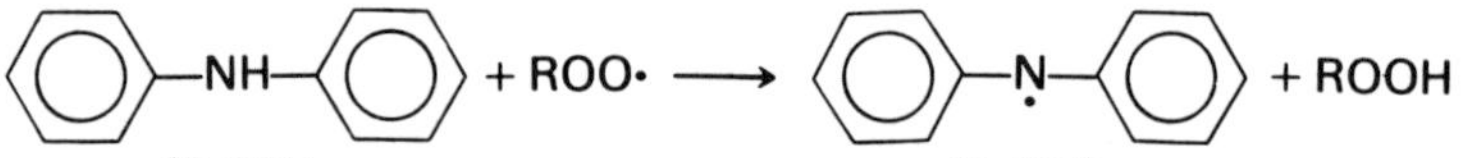

(3.144)

(**3.121**) (**3.122**)

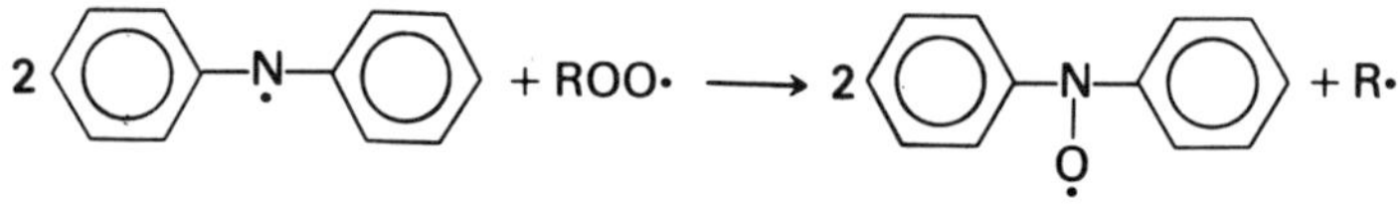

(3.145)

(**3.123**)

Diaryl nitroxy radicals (**3.123**) are more reactive than nitroxy radicals formed from hindered piperidines. They react easily with alkyl radicals (R•) to give diarylhydroxylamine ethers (**3.124**):

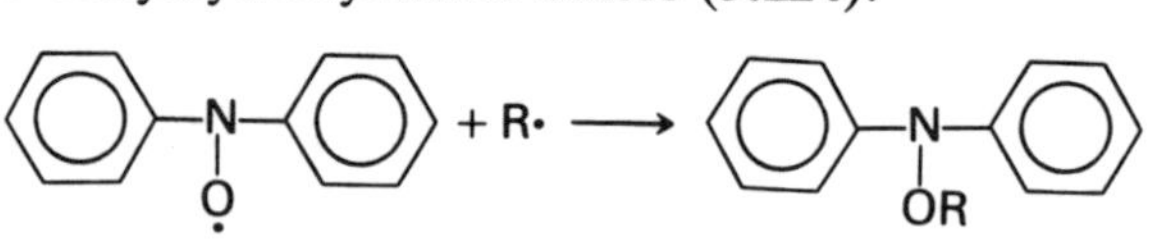

(3.146)

(**3.123**) (**3.124**)

or abstract hydrogen and form diarylhydroxylamine (**3.125**):

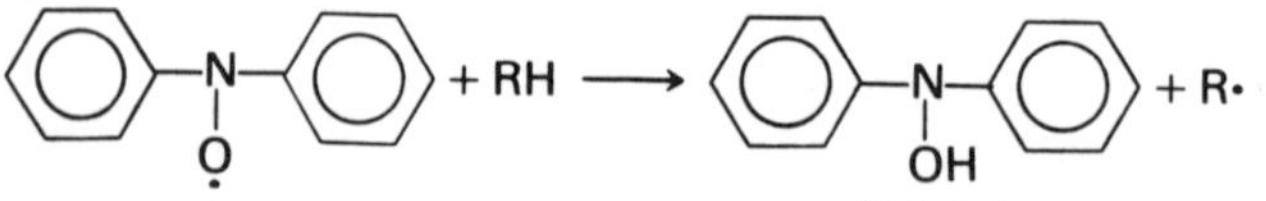

(3.147)

(**3.125**)

Diarylhydroxylamines may react further with peroxy radicals (ROO•) or hydroperoxides (ROOH) to regenerate diaryl nitroxy radicals **(3.123)**:[297]

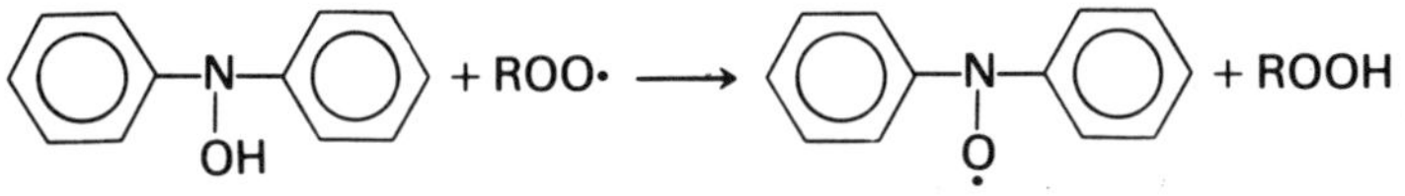

(3.148)

(3.123)

However, they also can react with peroxy radicals to yield ineffective species such as *N*-aryl-1,4-benzoquinone-1,4-phenylenediamine **(3.126)**[191] and further to benzoquinone and nitrobenzene:

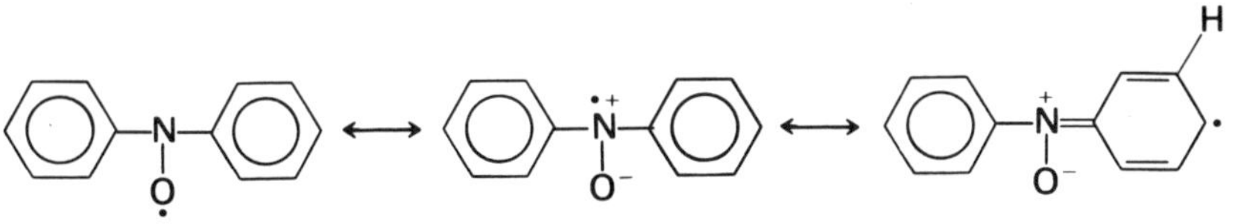

(3.149)

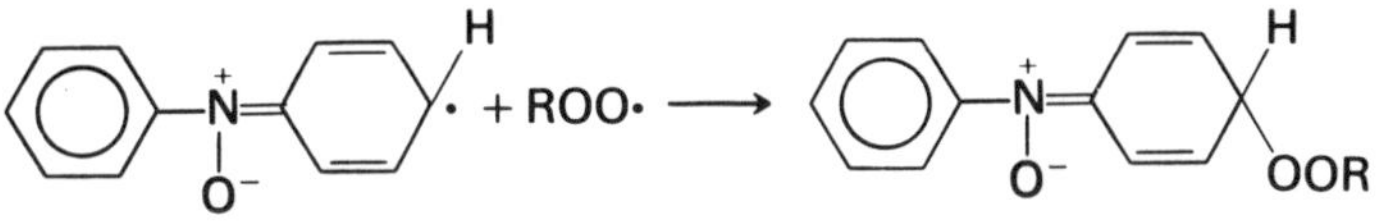

(3.150)

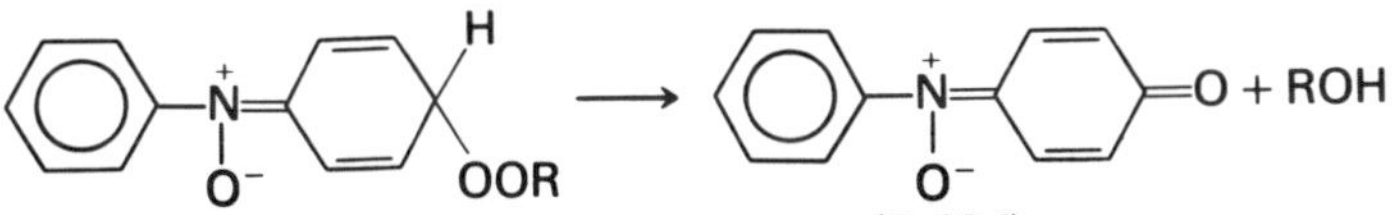

(3.151)

(3.126)

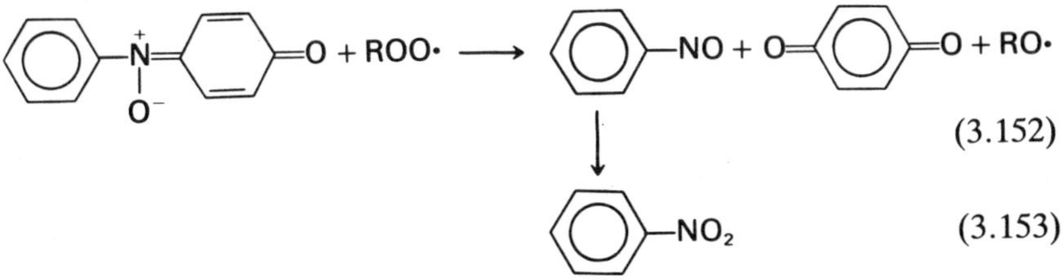

(3.152)

(3.153)

Secondary aromatic amines are very effective antiozonant additives.[698,711,910,944]

The aminyl radical (**3.122**) formed in reaction (3.144) can form polycyclic antioxidants (**3.127**) by dimerization,[272,906]

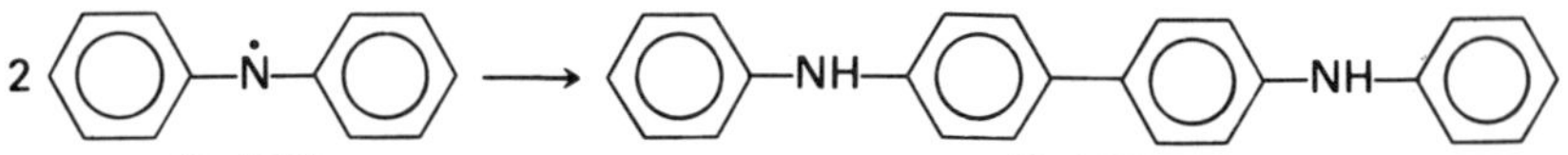

(**3.122**) (**3.127**)

(3.154)

which further react with peroxy radicals (ROO•):

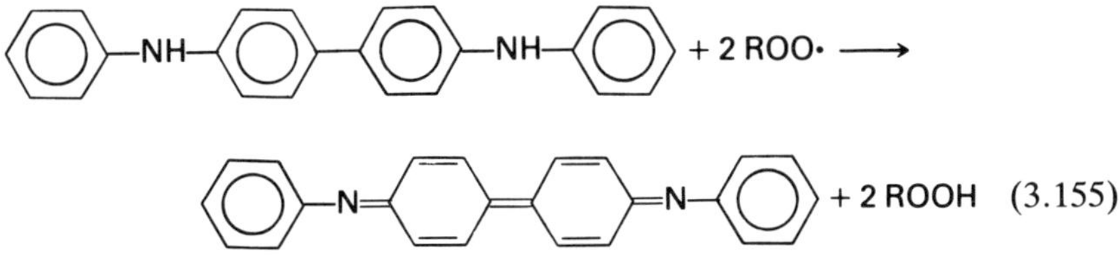

(3.155)

In the study of synergism between phenolic and amine antioxidants, 2,6-di-*tert.*-butyl-4-(4-phenylaminophenylimino)-2,5-cyclohexadiene-1-one (**3.128**) has been isolated.[1093]

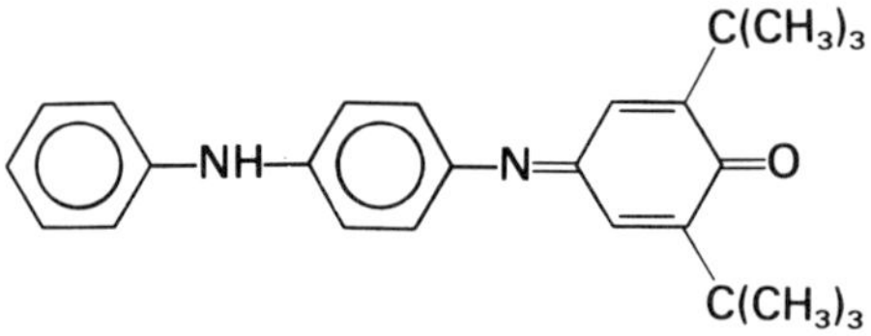

(**3.128**)

This compound acts as a kinetic chain-breaking antioxidant (chain terminator).

As ESR study of the photoinduced degradation of polypropylene in the presence of *N,N,N′,N′*-tetramethyl-*p*-phenylenediamine has also been published.[1088]

Antioxidants with heterocyclic rings such as substituted 3-anilino-1,5-diphenylpyrazoles (**3.129**),[768,989] *N*-phenyl-*N*-benzimidazolyl-2-amines (**3.130**)[769] and *N*-benzothiazolyl-2-*N*-benzimidazolyl-2-amines (**3.131**)[769] have been studied for their antioxidant efficiency in the thermal oxidation of polypropylene.

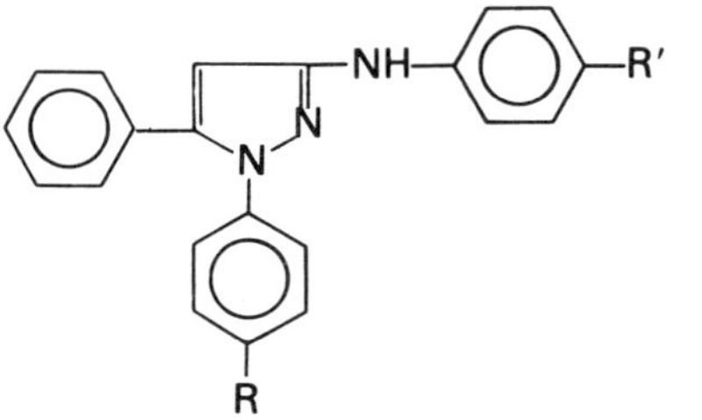

(3.129)

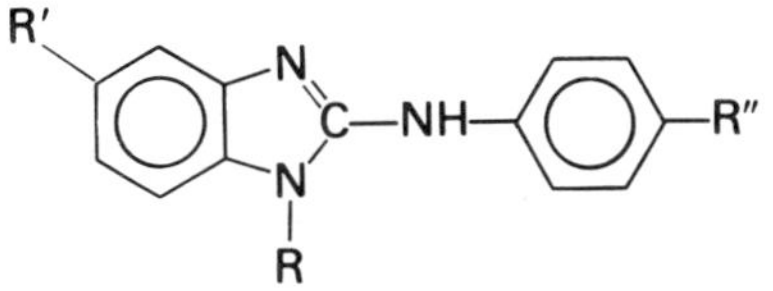

(3.130)

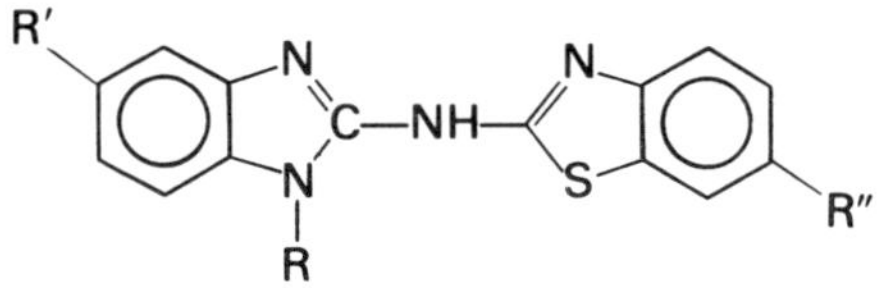

(3.131)

3.20.1 Photochemical Reactions of Aromatic Amines

Aromatic amines are almost non-effective in the stabilization of photo-oxidative degradation of polymers, because they are easily photo-oxidized.[665]

p-Phenylenediamine **3.132** is readily photo-oxidized to benzoquinone di-imine (**3.133**) and 4,4′-diaminoazobenzene (**3.134**):[749]

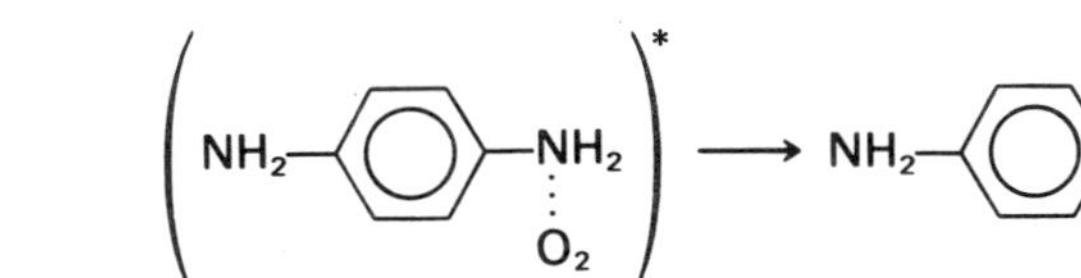

(3.132)

CT complex

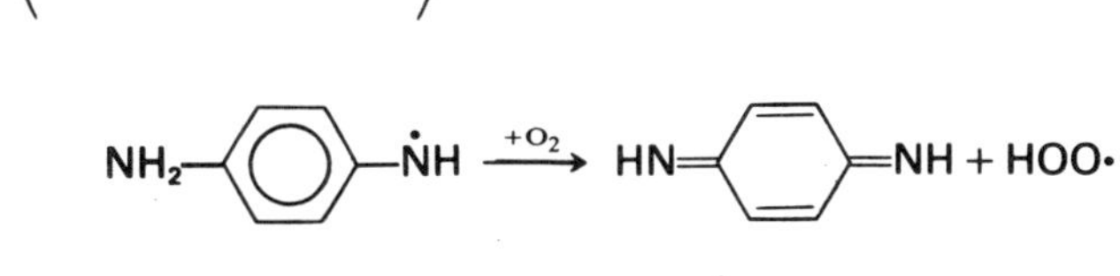

(3.156)

(3.157)

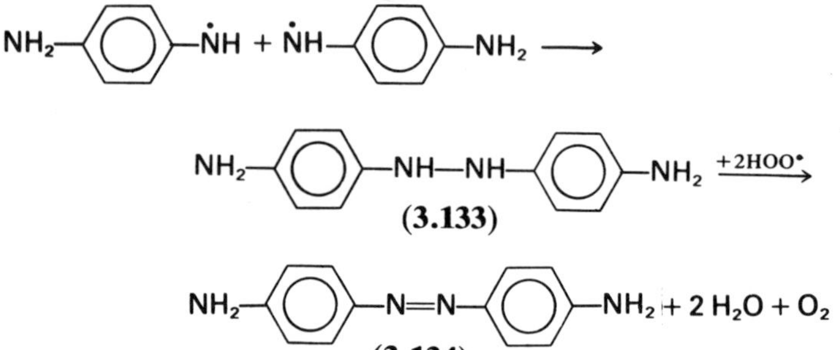

(3.133)

(3.134)

(3.158)

Much more complicated is the photo-oxidation mechanism of *N*-isopropyl-*N'*-phenyl-1,4-phenylenediamine (**3.135**), during which are formed *N*-isopropyl-*N'*-phenyl-1,4-benzoquinone di-imine (**3.136**) and 3-isopropylaminocarbazole (**3.137**):[708]

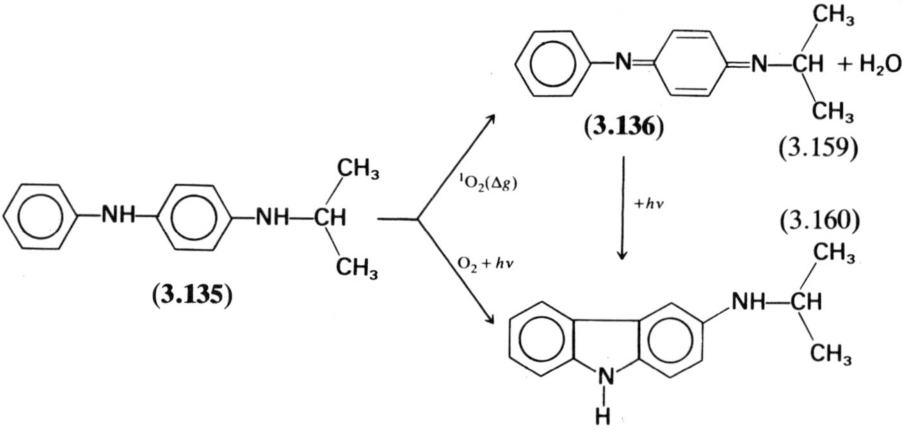

(**3.135**)

(**3.136**) (3.159)

(3.160)

(**3.137**) (3.161)

N-Isopropyl-*N'*-phenyl-1,4-phenylenediamine in admixture with 2,6-di-*tert.*-butylphenol shows an synergistic effect.[1092]

3.21 NITROSO COMPOUNDS AND NITRONES

Nitroso compounds and nitrones are very effective spin traps (radical scavengers) and are commercially available (Table 3.10). They react

Table 3.10
Spin traps commercially available[1197]

Chemical name	*Structure*
2-Methyl-2-nitrosopropane dimer (N-*t*-B)	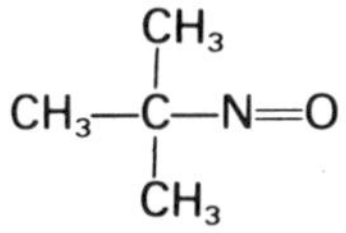 **(3.138)**
Nitrosobenzene, 97%	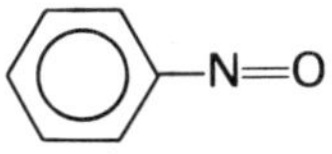 **(3.139)**
2,4,6-Tri-*tert.*-butylnitrosobenzene, 99% (BNB)	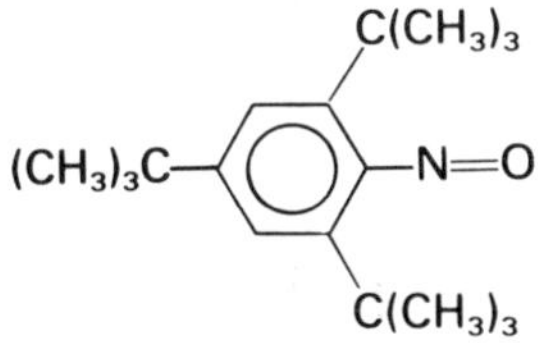 **(3.140)**
3,5-Dibromo-4-nitrosobenzenesulphonic acid, sodium salt hydrate, 98%	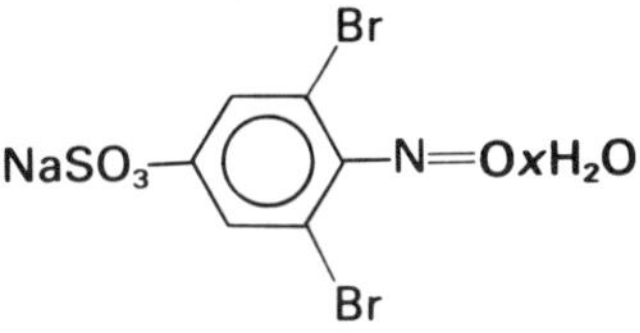 **(3.141)**
N-tert-Butyl-α-phenylnitrone (PBN)	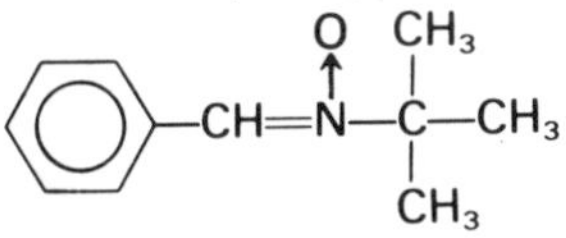 **(3.142)**
α-(4-Pyridyl 1-oxide)-*N-tert.*-butylnitrone, 99% (4-POBN)	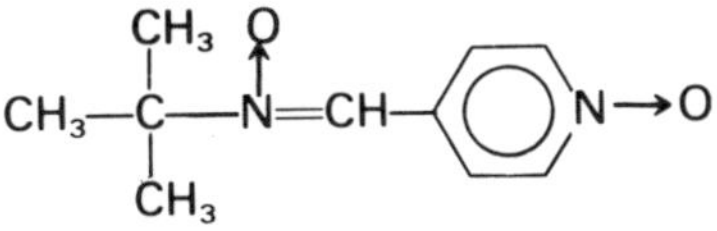 **(3.143)**
N-tert.-Butyl-α-(2-sulphophenyl)nitrone, sodium salt, 97%	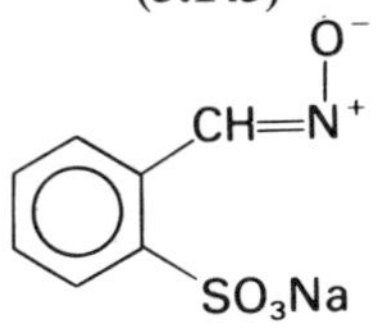 **(3.144)**

(*continued*)

Table 3.10—*contd.*

Chemical name	*Structure*
5,5-Dimethyl-1-pyrroline *N*-oxide, 97% (DMPO)	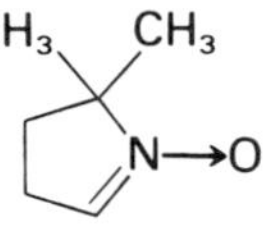 **(3.145)**
3,3,5,5-Tetramethyl-1-pyrroline *N*-oxide, 95%	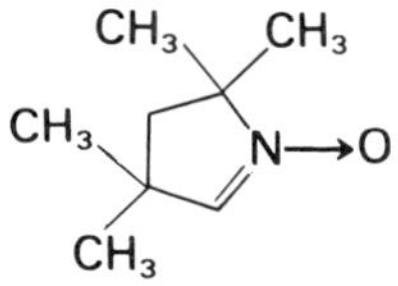 **(3.146)**

immediately with free radicals giving nitroxy radicals (spin adducts):[381,417]

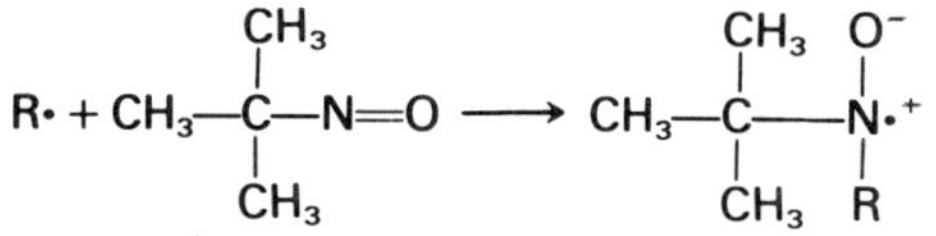

(3.138) (3.162)

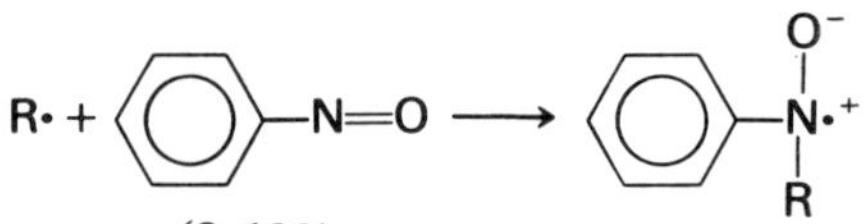

(3.139) (3.163)

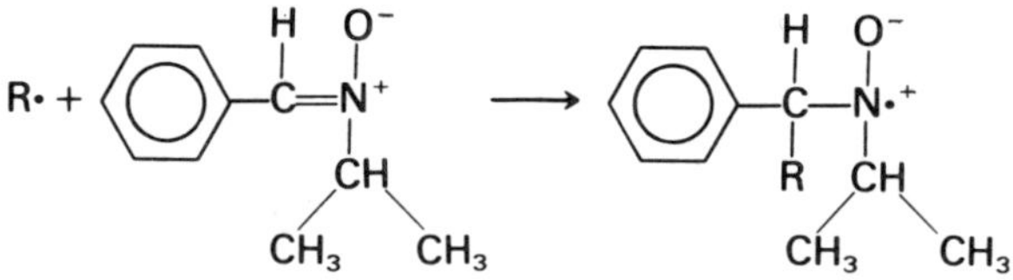

(3.164)

Nitrogen and proton hyperfine splittings in the ESR spectra of the spin adducts are diagnostic parameters for the identification of the free-radical spin adduct formed.[686,935]

2-Methyl-2-nitroso propane (**3.138**) is a very effective processing antioxidant (melt stabilizer) but it is ineffective as a thermal antioxidant. The antioxidant and photostabilizing activities of different nitroso compounds are shown in Table 3.11.[13,14,268]

Table 3.11
Photostabilizing and antioxidant activity of nitroso and nitrone compounds in polypropylene (10^{-3} mol/100 g, processed 10 min at 180°C)[268]

				Time to embrittlement of polypropylene films (h)	
				In accelerated weathering test	*In air oven at 140°C*
Control (no additive)				85	0·5
Nitroso compound 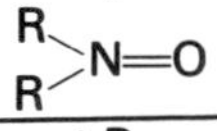					
t-Bu				770	1·0
t-Oct				690	1·0
Ph				330	7·0
2,4,6-Cl_3Ph				200	4·5
$4Me_2NPh$				260	11·0
2,3,4,5,6-Me_5Ph				200	11·5
Nitrone compound 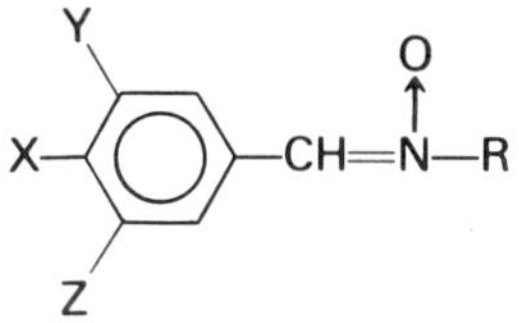					
R	*X*	*Y*	*Z*		
Ph	H	H	H	245	1·5
Ph	OH	H	H	210	10·0
Ph	OH	Me	Me	140	11·0
Ph	NO_2	H	H	210	—
Ph	OMe	H	H	150	—
t-Bu	H	H	H	90	1·0
t-Bu	OH	H	H	140	4·0
t-Bu	OH	Me	Me	130	7·5

3.22 ANTIOXIDANT PROPERTIES OF NITROSOAMINES

Aliphatic *N*-nitrosoamines (**3.147**) decompose with evolution of nitrogen:[948]

$$\underset{(\mathbf{3.147})}{R_2N{-}N{=}O} \rightleftharpoons [RN{=}N{-}OR] \longrightarrow R\cdot + N_2 + RO\cdot \qquad (3.165)$$

Table 3.12
Thermal antioxidant (oven ageing) activity at 140°C (concentration 10^{-3} mol/100 g, processed at 180°C) and effect of nitrosoamine concentration on the UV stability of polypropylene (concentration 10^{4} g/100 g, all processed for 10 min at 180°C)[272]

Processing time (min):	*7·5*	*10*	*15*	*20*	*1*	*5*	*10*
Nitrosoamine	*Thermal activity (h)*				*Embrittlement time (h)*		
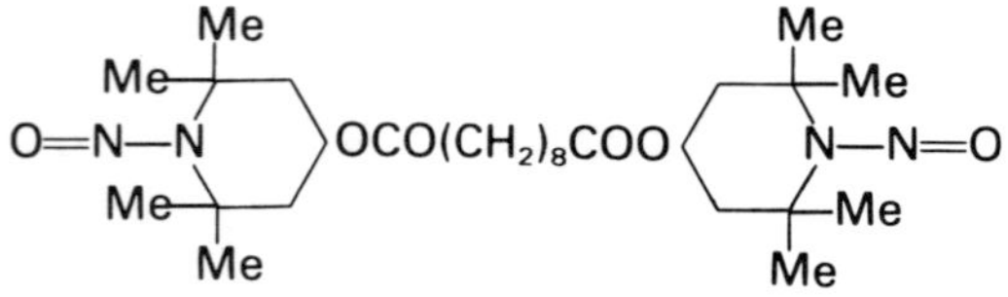	1·5	1·5	1·0	1·0	440	1340	1580
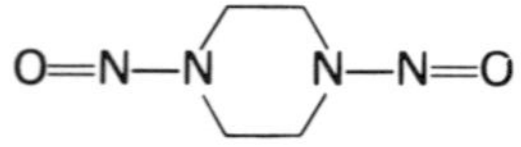	1·0	1·0	0·5	0·5	155	300	570

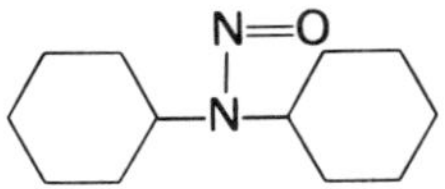	1·0	1·0	1·0	0·5	135	190	230
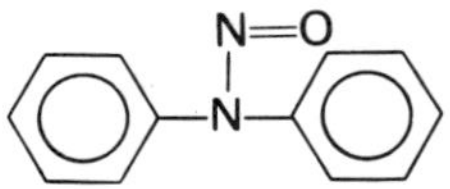	37·0	39·0	36·0	31·0	170	200	220
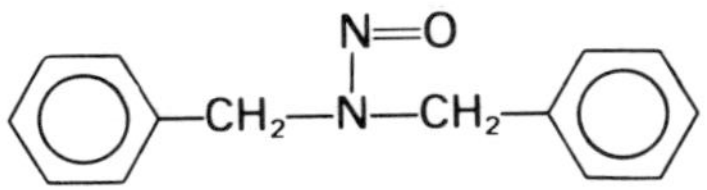					80	80	80
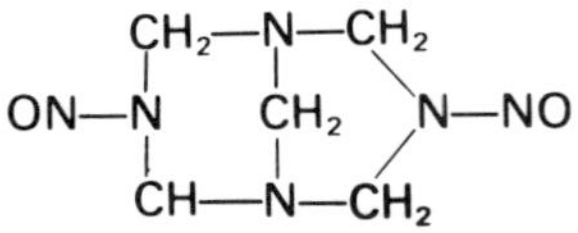					80	80	80
Control (no additive)	0·5	0·5	0·5				

On the other hand, aromatic *N*-nitrosoamines (**3.148**), known as effective anitfatigue agents in rubber products, dissociate into diaryl nitroxy (**3.150**) and di-*tert.*-alkyl nitroxy (**3.151**) radicals:[272]

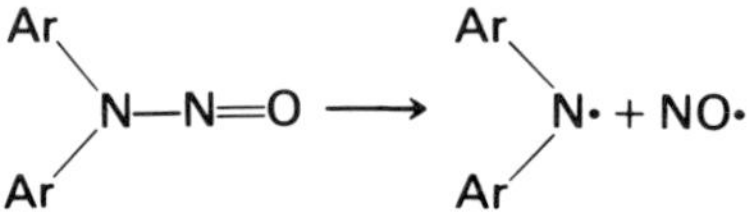

(**3.148**) (**3.149**) (3.166)

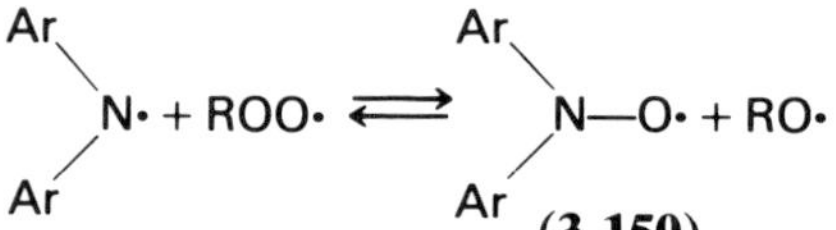

(**3.150**) (3.167)

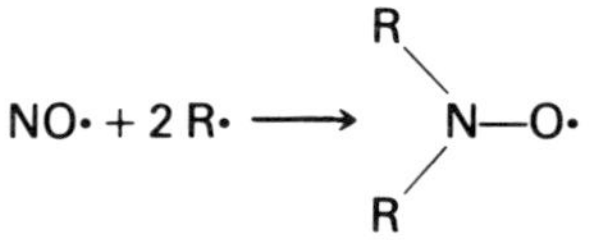

(**3.151**) (3.168)

Diaryl nitroxy radicals (**3.150**) may abstract hydrogen and produce hydroxylamine (**3.152**), whereas di-*tert.*-alkyl nitroxy radicals (**3.151**) may react with alkyl radicals (R·) giving alkoxyamines (**3.153**):

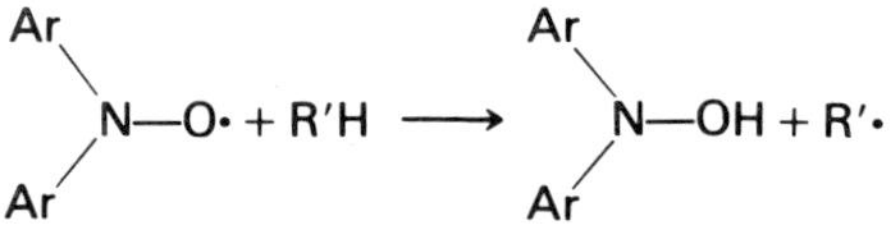

(**3.150**) (**3.152**) (3.169)

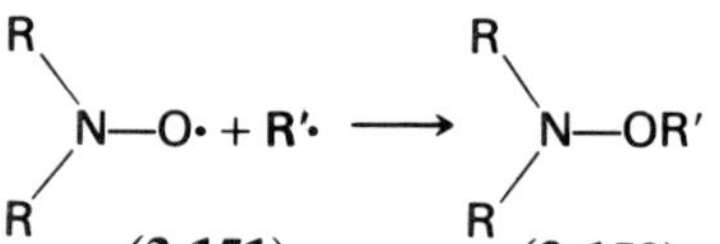

(**3.151**) (**3.153**) (3.170)

Aminyl radicals (**3.149**) may also form polycyclic antioxidants.

A number of structurally different *N*-nitrosoamines (Table 3.12) have been found to be effective melt stabilizers and photostabilizers.[272] The *N*-nitrosoamine obtained from bis(2,2,6,6-tetramethyl-4-piperidinyl) sebacate (Tinuvin 770) (**6.3**) is almost twice as effective as Tinuvin 770 as a photostabilizer in polypropylene under the same conditions (Table 3.12) and is also considerably more effective than the corresponding nitroxy radical.

3.23 ANTIOXIDANT PROPERTIES OF ANTISTATIC AGENTS (TENSIDES)

Antistatic agents (tensides) commerically available such as *N,N'*-bis(2-hydroxyethyl)octadecylamine (**3.154**) (Ethomeen 18/12 from Akzo Chemie) and *N,N'*-bis(2-hydroxyethyl)amide of lauric acid (**3.155**) (Dehydat 10 from Henkel) are effective antioxidant hydroperoxide decomposers in polyethylene:[901–904]

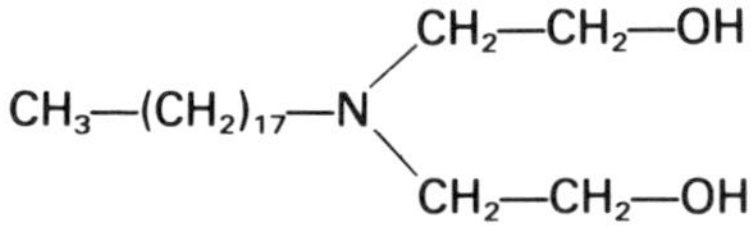

(**3.154**)

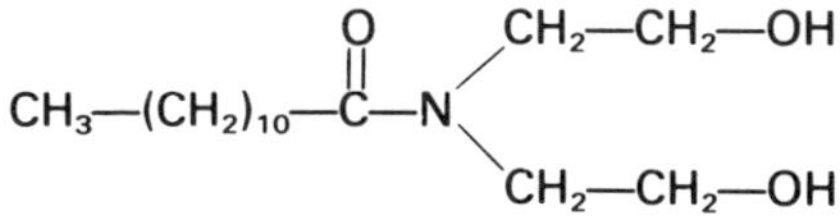

(**3.155**)

The amine tenside **3.154** forms with hydroperoxides the complex **3.156** which is further decomposed into the amino-oxide **3.157** and alkanol:

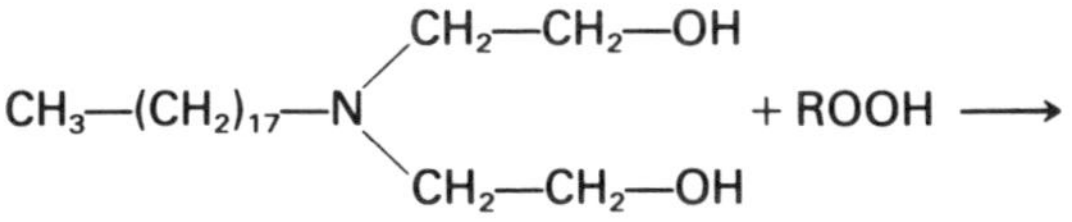

(**3.154**)

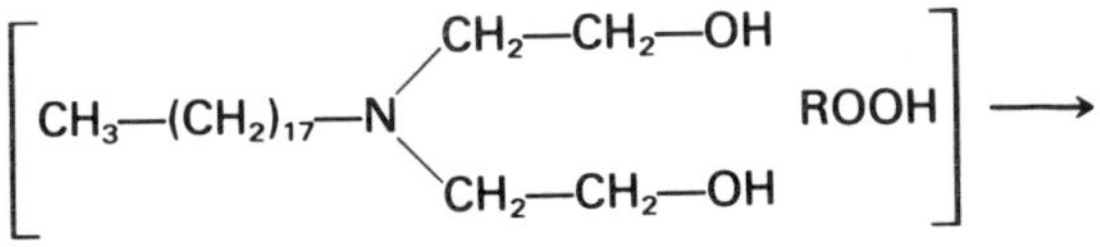

(**3.156**)

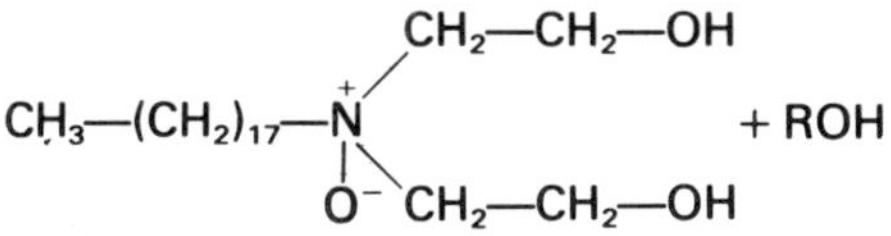

(3.171)

(**3.157**)

Both amine (**3.154**) and amide (**3.155**) tensides are easily photolysed. Amide tensides accelerate the photo-oxidative degradation of polyethylene films (in weathering conditions), whereas amine tensides in the presence of hydroxybenzotriazole (**3.158**) have a photostabilizing effect on the weathering of polyethylene. The amine tenside may form intramolecularly bound complexes (**3.159** and/or **3.160**):[902]

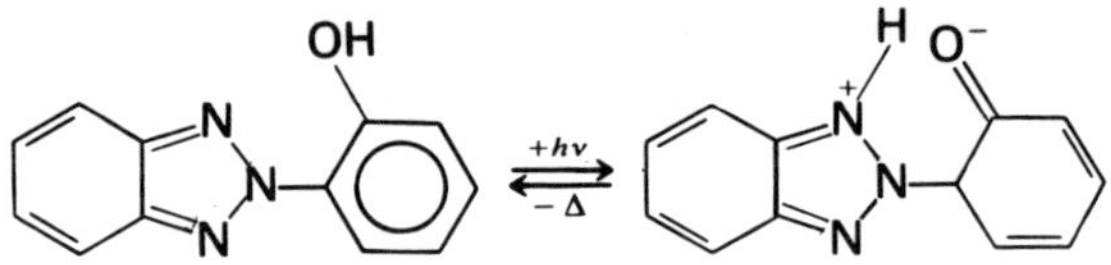

(3.172)

(**3.158**)

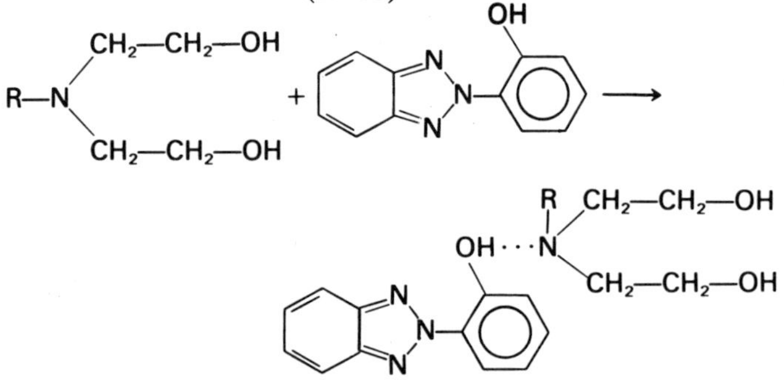

(3.173)

(**3.159**)

or

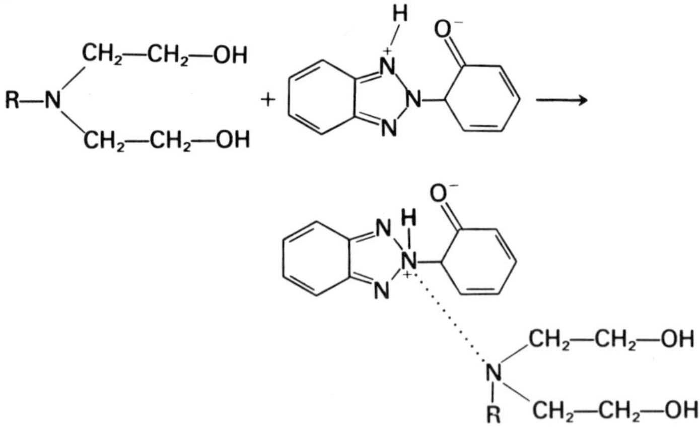

(3.174)

(**3.160**)

Polyethylene films stabilized with tensides also show antistatic properties, which disappear under UV-irradiation.

4
Pigments

4.1 PHOTOSTABILIZING PROPERTIES OF PIGMENTS

Pigments are insoluble inorganic (mineral) compounds, e.g. titanium dioxide, iron oxides, cadmium yellow, cadmium red, ultramarine blue, chromium oxide and molybdate yellow (Table 4.1) and organic compounds of complex structure (Table 4.2) which, used as additives (fillers and/or photostabilizers), are incorporated into plastics, coatings and inks, for:[83,799,887,1153]

(i) light screening,
(ii) cost reduction,
(iii) reinforcement,
(iv) hardening,
(v) improving slip and skid resistance,
(vi) improving flow properties,
(vii) colour effects;
(viii) storage stability; and
(ix) gloss and flatting efficiency.

White pigments such as titanium oxide (TiO_2), zinc oxide (ZnO), magnesium oxide (MgO) and silicon oxide (SiO_2) are commonly used in the plastics and coating industries. TiO_2 (cf. Section 4.2) and ZnO (cf. Section 4.3) are photochemically reactive and exhibit photostabilization or photosensitization effects on polymers.

Calcium carbonate ($CaCO_3$) and zinc oxide (ZnO), because of the basic character of their metal ions, may react in acidic systems (e.g. in poly(vinyl chloride)), but they are stable in neutral or slightly alkaline environments.

Table 4.1
Examples of inorganic pigments[865]

Pigment			*Metal oxide*
Colour	*Type*	*Formula*	
Red	Cadmium red	CdS + CdSe	PbO
Orange	Cadmium orange	CdS + CdSe	Pb_3O_4
Yellow	Cadmium yellow	CdS	PbO_2
Yellow	Chromium yellow	$PbSO_4 \cdot PbCrO_4$	CdO
Green	Copper phthalocyanine green	Chlorinated copper	CuO
		phthalocyanine	Cu_2O
Green	Chromium oxide green	Cr_2O_3	Sb_2O_3
Blue	Cobalt blue	$CoO \cdot nAl_2O_3$	SnO
Blue	Ultramarine blue	$3NaAl \cdot SiO_4 \cdot Na_2S_2$	Cr_2O_3
			Al_2O_3
Blue	Milori blue	$Fe_4[Fe(CN)_6]_3 \cdot nH_2O$	TiO_2
			Fe_2O_3
Violet	Cobalt violet	$Co_3(PO_4)_2$	ZnO
Violet	Mineral violet	$(NH_4)Mn(P_2O_7)$	MnO_2
Brown	Red iron oxide	Fe_2O_3	Co_2O_3
White	Titanium dioxide	TiO_2 (rutile)	

Barium sulphate ($BaSO_4$), being an extremely insoluble salt of a strong acid and a strong base, is usually quite inert.

Cadmium sulphide (CdS) (cadmium yellow) is one of the most photochemically studied coloured pigments.[33,47,93,141,206,384,408,522,590,648,1119] Some controversy exists regarding the photocatalytic activity of this pigment.[47,93,141,206] The results have been attributed to the purity of the pigment used.[141] The addition of antioxidant 2,2′-methylenebis(6-*tert.*-butyl-1-cresol) markedly improve the photostability of polyethylene pigmented with cadmium yellow.[890]

Copper acetate hydrate (verdigris, a copper blue–green pigment) is an effective photostabilizer for linseed oil (**4.13**)-based paint media; however, it is discoloured after a long period. A catalytic antioxidant mechanism involves redox reactions of the copper ions:[942]

$$\underset{\textbf{(4.13)}}{ROCO(CH_2)_6COCH{=}CHCH{=}CH(CH_2)_4CH_3} \xrightarrow{+h\nu} ROCO(CH_2)_6CH_2\cdot + \cdot COCH{=}CHCH{=}CH(CH_2)_4CH_3 \quad (4.1)$$

$$2\,ROCO(CH_2)_6CH_2\cdot \xrightarrow[Cu^+/Cu^{2+}]{O_2} ROCO(CH_2)_5CH{=}CH_2 + ROCO(CH_2)_6CH_2OOH \quad (4.2)$$

Table 4.2
Examples of organic pigments[865]

Name	Structure	Commercial name
Phthalocyanine blue	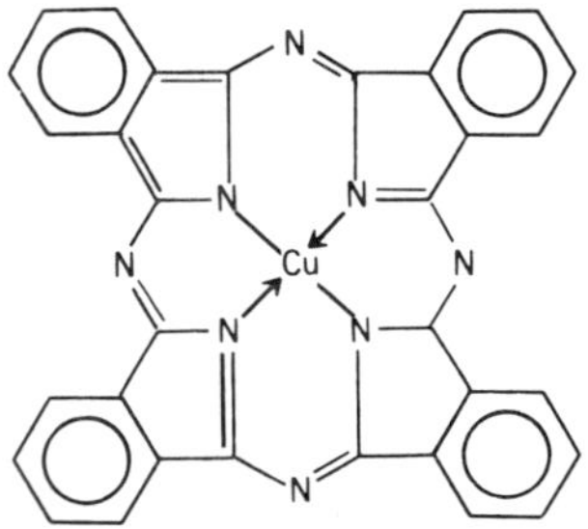 (4.1)	Pigment Blue 15:3
Dioxazine	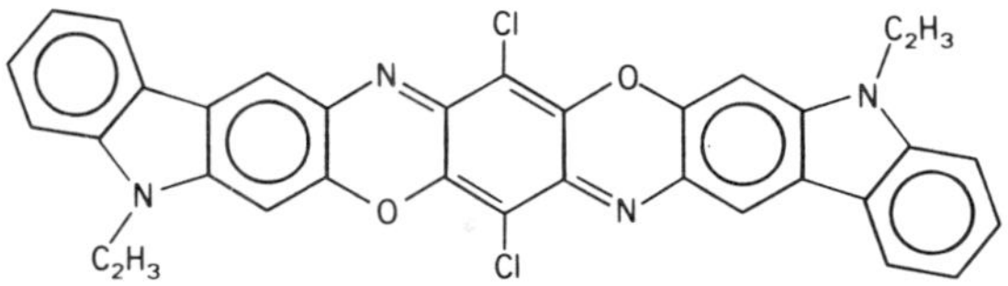 (4.2)	
Tetrachloroisoindolines	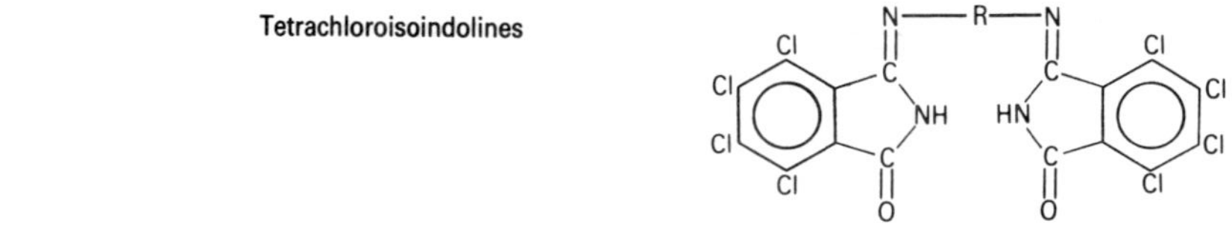 (4.3)	Pigment Yellow 109, 110

(continued)

Table 4.2—*contd.*

Name	*Structure*	*Commercial name*
Perylene	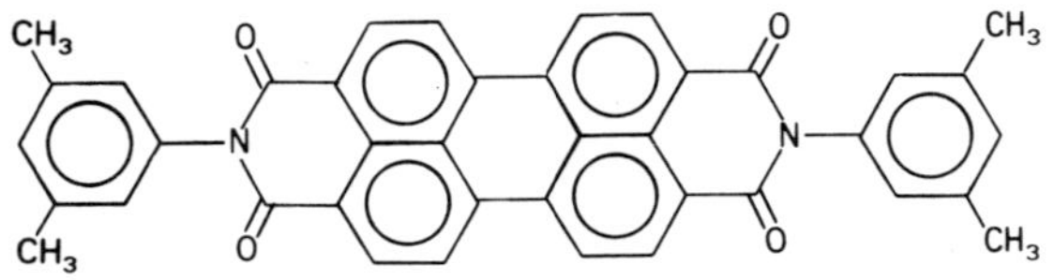 (4.4)	Pigment Red 149
Indanthrene blue	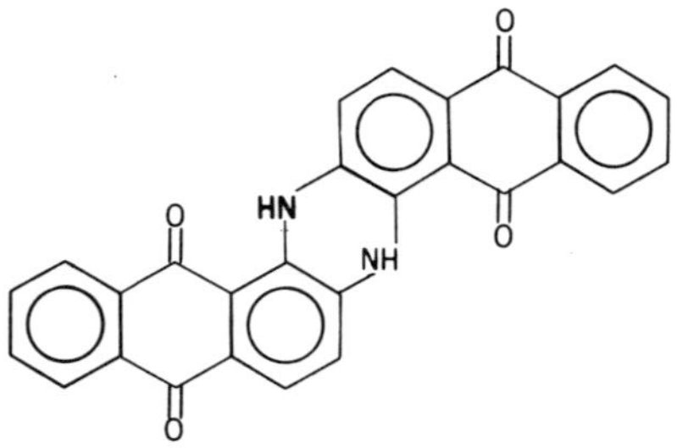 (4.5)	
Flavanthron	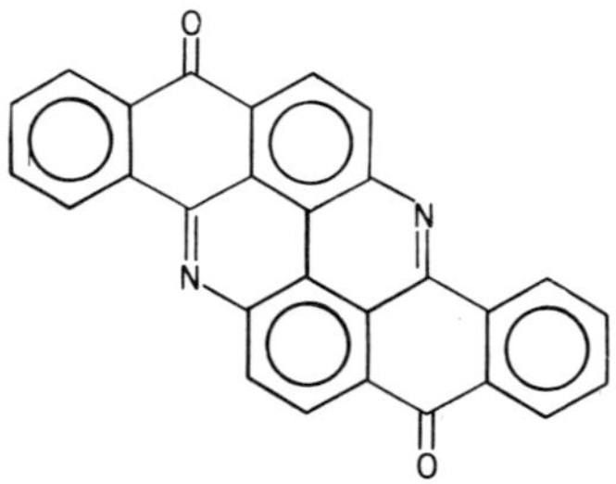 (4.6)	

Thioindigo	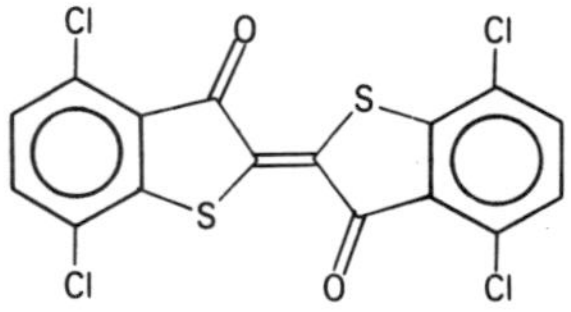 (4.7)	Pigment Red 88
Azo condensation	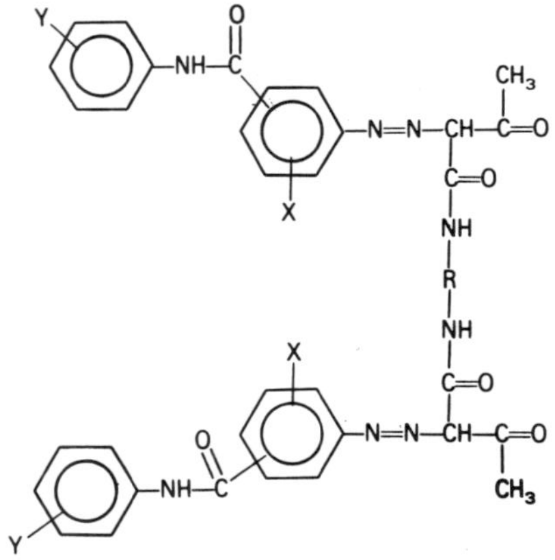 (4.8)	Pigment Yellow 93, 94, 95

(continued)

Table 4.2—*contd.*

Name	*Structure*	*Commercial name*
	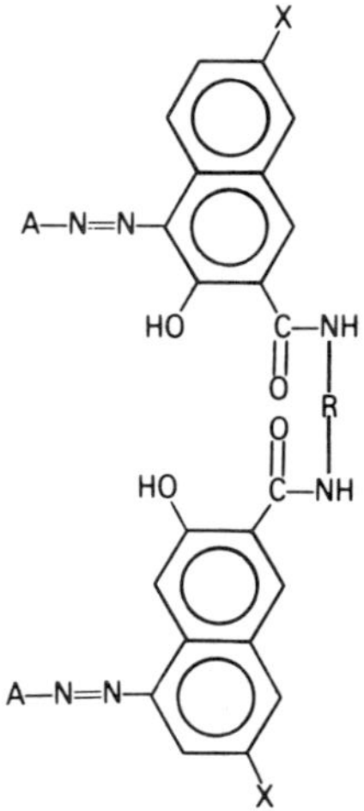 (4.9)	Pigment Orange 31 Pigment Red 144, 166
Diarylide	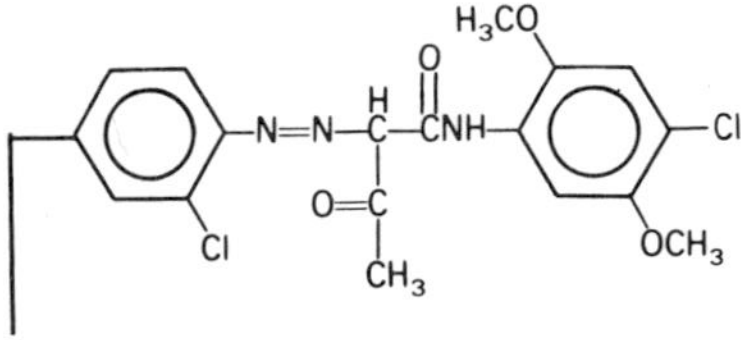	Pigment Yellow 83

	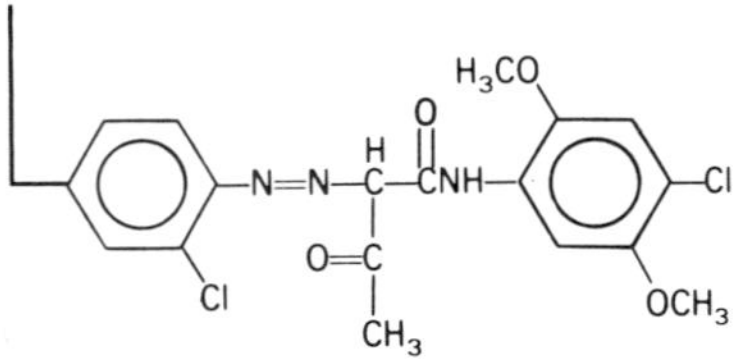 (4.10)
Quinacridone	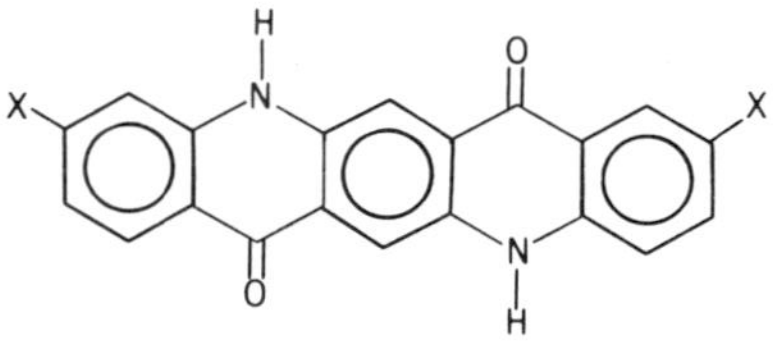 (4.11)
Quinophthalone	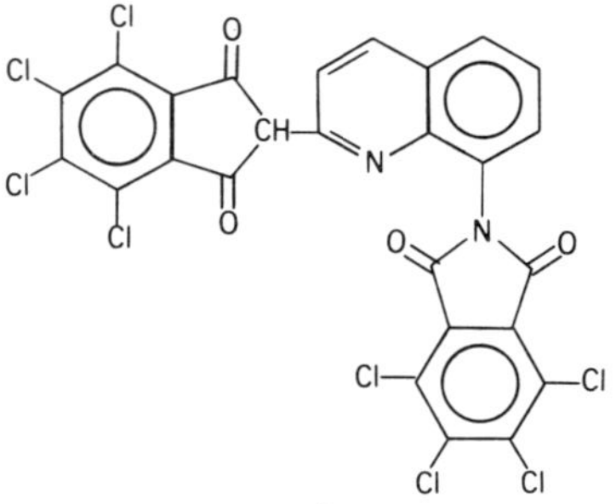 (4.12)
Channel carbon black	C (pH 3·0)
Channel carbon black	C (pH 5·8)
Furnace carbon black	C (pH 8·1)
Furnace carbon black	C (pH 9·5)

Comparison of stabilizing properties of different pigments in polypropylene gives the following order:[279]

pure polymer < cadmium yellow
< copper phthalocyanine < azo yellow pigment

Cadmium yellow and chrome oxide green exhibit synergistic effects in polyethylene (HDPE) in a mixture with cetyl-3,5-di-*tert.*-butyl-4-hydroxybenzoate (**5.1**) (Cyasorb UV 2908).[75]

White pigments scatter back all of incident light and the observer perceives the colour, typically white, of the light that permits his vision. Low additions will provide translucence and larger ones, opacity. Because the light is not absorbed, coloured pigments or dyes can be used simultaneously to create a wide range of colour in the product.

To be perfectly white, a pigment should not absorb any of the visible light that falls upon it. The absorption bands in the near-ultraviolet that extend slightly into the visible produce a yellowish colour by absorbing some blue–violet.

The reflecting power of some colourless pigments is listed in Table 4.3. It is much lower in a polymer matrix in comparison with air. This occurs because whitening and opacification are produced by scattering or diffuse reflection which depend upon the refractive index of the polymer matrix. The spectral characteristics of some pigments are shown in Fig. 4.1.

Table 4.3
Reflecting power of colourless pigments as a function of reflectance index in various media[257]

	Average n_D of powder in air	*Fresnel reflecting power (%)*		
		In air ($n_D = 1{\cdot}00$)	*In plastic* ($n_D = 1{\cdot}50$)	*In plastic* ($n_D = 1{\cdot}65$)
Rutile TiO_2	2·76	21·9	8·7	6·3
Anatase TiO_2	2·52	18·6	6·4	4·4
Sb_2O_3 (cubic)	2·09	12·4	2·7	1·4
ZnO	1·99	11·0	2·0	0·9
MgO	1·74	7·3	0·5	0·1
$BaSO_4$	1·64	5·9	0·2	0·0
$CaCO_3$	1·57	4·9	0·1	(0·1)
SiO_2	1·55	4·7	0·0	(0·1)

n_D, Refractive index.

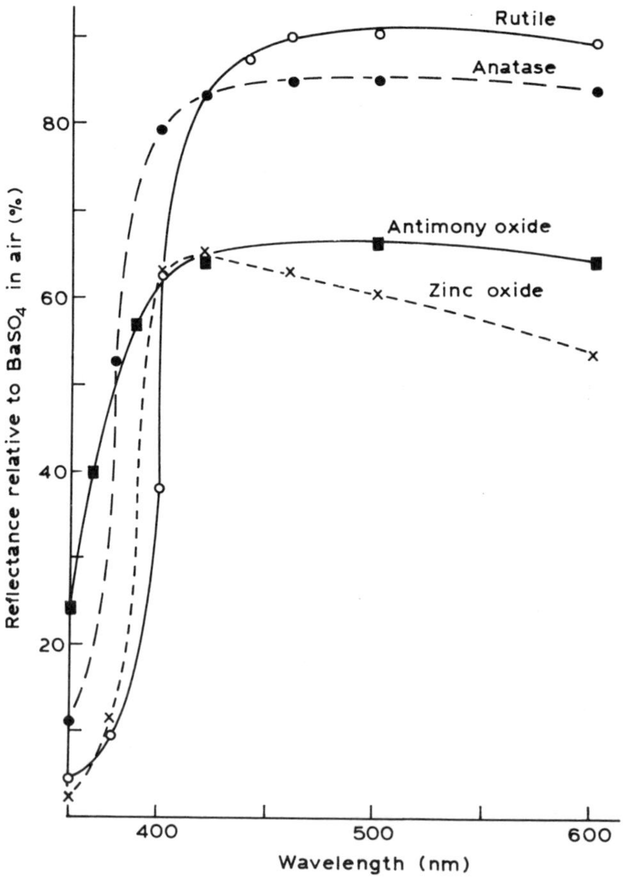

Fig. 4.1. Reflectance of white-pigmented poly(vinyl chloride).[257]

The efficiency with which white pigments reflect or scatter light depends upon their degree of subdivision, i.e. the number of potentially useful interfaces in a given volume. Particles smaller than about 0·15 μm scatter blue light much more strongly than red and particles larger than about 0·25 μm scatter all the colours more or less equally. At low concentrations, the scattering efficiency of white pigments is essentially independent of concentration. At higher concentrations the diffuse scattering coefficient begins to fall as the particles get closer and closer together.

The hiding power of a pigment is determined by its ability to scatter and absorb visible light. In the case of white pigments, hiding is accomplished by light-scattering, which depends primarily on the difference in refractive index between the pigment and the medium in which it is dispersed (binder). The greater this difference, the more effectively light is scattered. Since the refractive index of binder substances is relatively low (~1·5), the higher the refractive index of the pigment, the greater is the potential for hiding.

The degree of scattering is dependent on:

(i) The difference in refractive index between pigment and the vehicle.
(ii) The particle size of the pigment. Pigments of smaller average particle size impart superior stability against photodegradation of polymers.
(iii) The pigment concentration (volume fraction loading).
(iv) The wavelength of the radiation.

Calculation of the absorption by pigmented formulations is very complex due to scattering and cannot be handled by simple equations.[1152,1153]

The light available to surface layers of polymer in the presence of a given fraction (V) of pigment is given by the equation:[136]

$$I_l = I_0(1 - 3V/2)^{l/2r} \tag{4.3}$$

where I_l = light intensity at depth l,
I_0 = incident light intensity,
V = volume fraction of pigment,
l = depth to which light passes,
r = radius of a pigment particle,
$2r$ = laminar thickness.

The residual light at a given depth is a decreasing nonlinear function of the volume fraction of pigment in the polymer matrix.

The light attenuation by pigment particles is relatively rapid (Fig. 4.2), implying that the degradation of filled polymer will occur mostly in the surface layers (this is also true in the case of polymers containing UV absorbers).[136]

Pigments used in polymer (plastic) or coating industries should be:

(i) Light-stable for long-term performance without fading.
(ii) Heat-stable to withstand polymer processing conditions.
(iii) Migration-resistant.
(iv) Cheap, nontoxic.

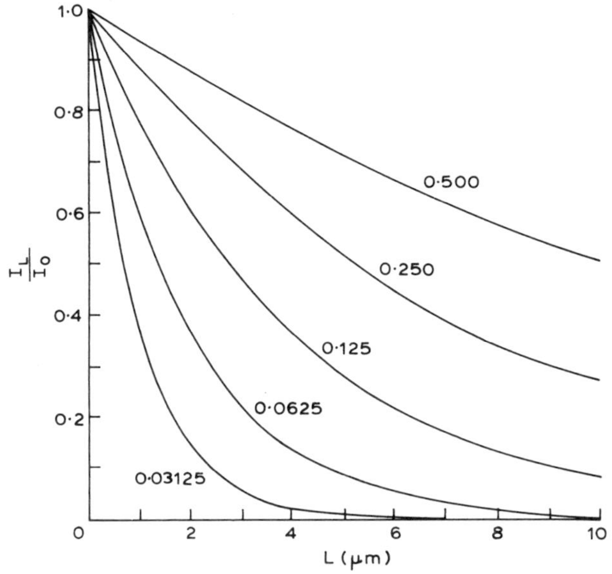

Fig. 4.2. The dependence of light intensity at a depth L (μm) from the polymer surface upon the particle radii (μm) indicated for each curve.[136]

Pigments may influence the light stability of the stabilized polymer in many ways:

(i) They can significantly increase UV stability (Table 4.4).[648] Generally green, blue and black pigments have favourable effects on polymer light stability.[84,252,341,351,626,797,1119]

(ii) They can sensitize photodegradation and thus drastically reduce the light stability.[17,141,369,479,648,797,865,890,1096] A number of white, yellow, orange and red pigments have such sensitizing properties in the presence of UV irradiation.

(iii) They can take over the role of the photostabilizer and prevent harmful UV radiation from reaching the deeper layers of the plastic material.

(iv) They can exhibit synergistic (Table 4.5) or antagonistic effects with other additives.[47,93,292]

(v) They increase a sample's surface temperature by strong absorption of the whole spectrum of solar radiation. Surface temperatures of exposed pigmented samples may increase in moderate climatic zones as follows: white 30–40°C, yellow and red

Table 4.4
Influence of pigments on photostability of polypropylene films (10 μm) during irradiation in the carbon-arc Weatherometer[1119]

	Stabilization factor	*UV absorbance/ml at 375 nm*
Control	1·0	—
Pigment red 177	1·0	0·0
Quinacridone magenta	1·0	0·18
Ultramarine blue	1·75	0·01
Cadmium yellow	2·0	0·36
Rutile titanium dioxide	2·25	1·09
Mercadium red	2·75	0·26
Chromium oxide	2·75	0·08
Iron oxide tan	3·25	0·33
Phthalocyanine blue	3·5	0·49
Phthalocyanine green	3·5	0·54
Channel carbon black	12	1·56

Table 4.5
Influence of pigments and 2-(2-hydroxy-3,5-di-*tert*.-amylphenyl)-2*H*-benzotriazole (Tinuvin 328, **5.47**)[a] on the photostability of low-density polyethylene (LDPE) films (80 μm) during irradiation in the carbon-arc Weatherometer[797]

Pigment at 3 wt % concn.	*Tinuvin 328 (%)*	*Time to 50% retention elongation (h)*	*Time to 0·50 carbonyl (h)*
Control	Control	140	290
Control	0·15	375	390
Control	0·20	430	400
Cadmium yellow	Control	175	200
Ultramarine blue	Control	250	350
Phthalocyanine green	Control	285	400
Chrome green	Control	335	500
Cadmium yellow	0·20	310	460
Ultramarine blue	0·20	435	>500
Phthalocyanine green	0·20	460	>500
Chrome green	0·20	530	>500

[a] Structure in Table 5.7.

40–42°C, green 44–46°C and black 50–60°C. In tropical climates dark pigmented samples with insufficient air ventilation can be heated 80–90°C above ambient temperature. Surface temperatures do not depend exclusively on pigmentation, but the nature of the polymer (plastic) has a pronounced effect. Increasing the surface temperature will accelerate oxidative degradation.[473,784]

The following properties of pigments may influence performance in polymers:

(i) Dispersibility.
(ii) Light-absorbing properties,[764] and/or scattering of radiation.[1153]
(iii) Adsorbing qualities.
(iv) Semiconductor properties.[881,1153]
(v) Metal contents (cf. Table 4.6).[865]
(vi) Polymer matrix influences.
(vii) Surface properties.
(viii) Nature and activity of fading products.
(ix) Photochemical behaviour, e.g. formation of free radicals from photolysis, sensitization of singlet oxygen ($^1O_2$) formation, energy transfer reactions from photoexcited pigment chromophores, etc.

During their manufacture, plastics are coloured in the molten state, in kneaders or extruders. They are then discharged, cooled and granulated. The coloration of plastic films, fibres and film tapes take place almost exclusively with the pigments in granule form, but it can also be carried out with powdered pigments.

Dispersion of a pigment in the polymer matrix is an important factor which may in consequence have an influence on:

(i) Photostabilization of polymer (plastic) material.
(ii) Fluctuation in colour intensity and deviation in shade.
(iii) Printing problems due to inhomogeneities at the surface.
(iv) Deterioration of mechanical properties.
(v) Processing (e.g. tearing of films due to pigment agglomerates).

The coloration level of pigment preparations and dispersion of a pigment in the polymer matrix should be very carefully controlled.

Table 4.6
Order of effect of metals in pigments and metal oxides in the degradation of various polymers[865]

Effect	*Polycarbonate*				*Polyacetal*		*Polypropylene*		*Poly(vinyl chloride)*	
	Pigment			*Metal oxide*	*Pigment*	*Metal oxide*	*Pigment*	*Metal oxide*	*Pigment*	*Metal oxide*
	Type	*pH*	*Contained metal*							
↑	Chromium yellow	5·4	Pb	Pb	Mineral violet	Pb	Milori blue	Cu	Red iron oxide	Zn
	Copper phthalocyanine blue	5·6	Cu	Cu	Channel carbon black-2	Cu	Copper phthalocyanine green	Co	Titanium dioxide	Fe
				Cd					Cobalt blue	Hg
	Channel carbon black-1	8·1	—	Sb	Chromium yellow	Cd	Cobalt blue	Fe	Cobalt violet	Co
					Copper phthalocyanine blue	Ti	Cobalt violet	Mn	Milori blue	Cu, Cr
	Cadmium yellow	6·0	Cd	Sn			Mineral violet	Cr	Copper phthalocyanine blue	Ti
	Cadmium red	6·0	Cd	Al	Copper phthalocyanine green	Fe	Red iron oxide	Al		
	Copper phthalocyanine green	5·2	Cu	Ti			Ultramarine blue	Ti	Copper phthalocyanine green	Sn
					Red iron oxide	Zn	Chromium oxide	Zn		

	Mineral violet	2·9	Mn	Mn	Cadmium red	Al	green		Cadmium red	Al
	Titanium dioxide	5·8	Ti	Fe	Cadmium yellow	Mn	Titanium dioxide	Pb	Chromium yellow	Mn
	Red iron oxide	5·0	Fe	Cr	Titanium dioxide	Cr	Chromium yellow	K	Chromium oxide green	Cd
	Ultramarine blue	5·6	Al	Zn	Cobalt violet	Co	Indanthrene blue	Cd	Ultramarine blue	Pb
	Cobalt violet	5·7	Co	Co	Chromium oxide green	K	Cadmium red	Na	Channel carbon black-2	Ba
↑	Chromium oxide green	5·5	Cr	K	Cobalt blue	Na	Channel carbon black-1	Ba	Channel carbon black-1	Ca
The order of increasing effect	Channel carbon black-2	3·7	—	Na	Channel carbon black-1	Ba	Channel carbon black-2			

Effect	*Polystyrene*		*Poly(methyl methacrylate)*		*Acrylonitrile–styrene copolymer*	
	Pigment	*Metal oxide*	*Pigment*	*Metal oxide*	*Pigment*	*Metal oxide*
↑	Red iron oxide	Fe		Fe	Red iron oxide	Fe
	Cadmium red	Sn	Cadmium yellow	Sn	Cadmium yellow	Sn
	Cadmium yellow	Cd	Red iron oxide	Ti	Cadmium red	Cd
The order of increasing effect	Chromium oxide green	Cr	Cadmium red	Cr	Titanium dioxide	Ti
	Chromium yellow	Co	Chromium yellow	Cd	Chromium yellow	Pb
	Cobalt violet	Pb	Cobalt violet	Co	Cobalt violet	Cr, Co
	Copper phthalocyanine blue	Cu	Copper phthalocyanine blue	Cu, Pb	Copper phthalocyanine blue	Cu

4.2 TITANIUM DIOXIDE

Titanium dioxide (TiO_2) exists in two morphological crystalline forms, which exhibit different photoactivities when incorporated into a polymer matrix:

(i) Anatase (TiO_2)—which is photochemically active and sensitizes polymer photodegradation,
(ii) Rutile (TiO_2)—which is photochemically relatively inactive.

These two TiO_2 forms, anatase and rutile, have been the subject of intensive studies.[39,63,76,77,79,80,84–86,89,408,409,416,590,648,706,1083]

Both morphological forms, anatase and rutile, show 6:3 coordination, i.e. each titanium atom is surrounded by six oxygen atoms and each oxygen atom is surrounded by three titanium atoms. The way in which the groups are linked together is different for the two morphological forms:[157,204,844]

Anatase (TiO_2)—forms near-regular octahedral crystals (Fig. 4.3).
Rutile (TiO_2) —forms slender, prismatic crystals which are frequently twinned (Fig. 4.4).

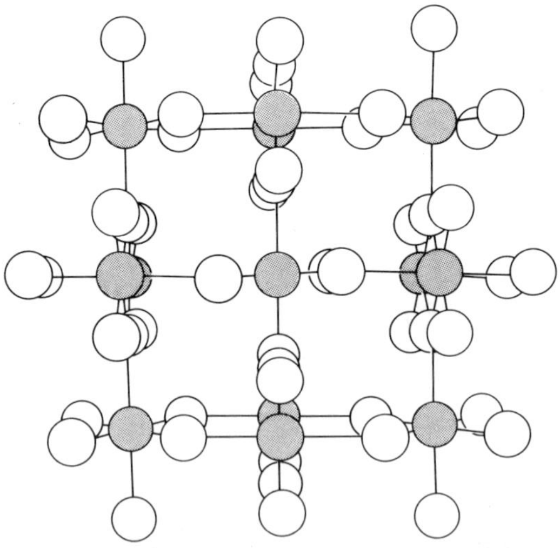

Fig. 4.3. Structure of anatase TiO_2 (titanium dark spheres, oxygen light).[844]

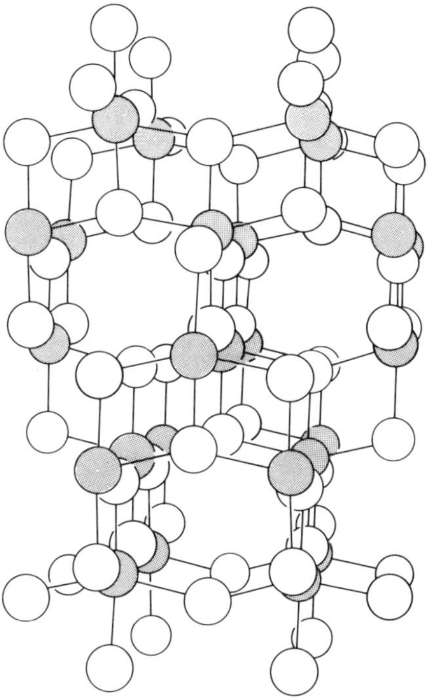

Fig. 4.4. Structure of rutile TiO_2 (titanium dark spheres, oxygen light).[844]

A summary of the crystallographic properties of these two forms is given in Table 4.7.

The three-dimensional periodic atomic structure is interrupted at the surface of crystalline particles. This results, in the case of an ionic crystal lattice, in formation of unsaturated (ionic) bonds. These

Table 4.7
Crystallographic properties of anatase and rutile (TiO_2)[844]

Property	*Anatase*	*Rutile*
Crystal habit	Tetragonal	Tetragonal
Optical	Uniaxial, negative	Uniaxial, positive
Density (g/ml)	3·9	4·23
Hardness, Mohs' scale	$5\frac{1}{2}$–6	6–$6\frac{1}{2}$
Unit cell dimensions (Å)		
a,b	3·758	4·584
c	9·514	2·953

unsaturated bonds are capable of binding all kinds of molecules, atoms or ions tightly (this is called *Chemisorption*). Besides chemisorption, physical absorption can also occur at a surface due to weak van der Waals forces. Most solid particles are normally covered with surface oxides or hydroxides, which strongly determine the surface properties of these particles.[654,1122]

Titanium dioxide (TiO_2) surfaces are covered with surface hydroxyls of an amphoteric character.[188–190] How the adsorption of water leads to the formation of hydroxyls on the anatase single crystal surface is shown in Fig. 4.5.[188]

The most important technique used to identify the nature of the surface hydroxyls on TiO_2 is FTIR spectroscopy.[807–809]

The IR spectra show two distinct groups:

(i) The group at 3000–3700 cm^{-1} contains distinct peaks at 3420, 3632, 3647, 3659 and 3676 cm^{-1} (Fig. 4.6) and shoulders at 3614, 3639 and 3728 cm^{-1} (Fig. 4.7), which can be attributed to OH-stretching vibrations of free and hydrogen-bonded surface OH groups (chemisorbed water).

(ii) The group at 1200–1700 cm^{-1} containing peaks at 1361, 1454, 1624, 1585 and 1622 cm^{-1}, which can be assigned to physically adsorbed water (H—O—H bending) and carbonates.[806]

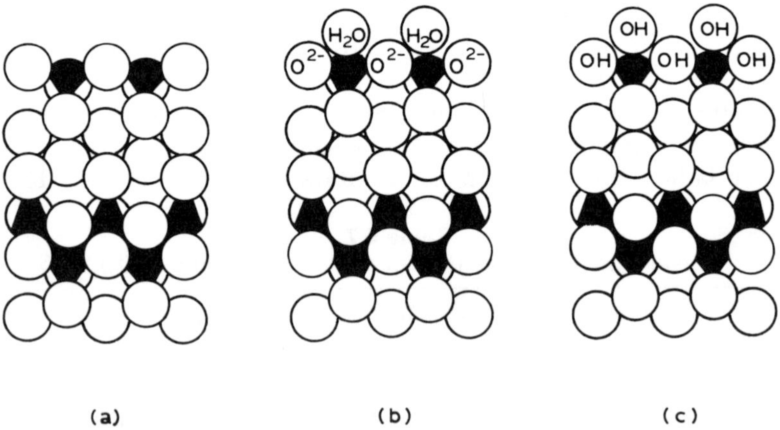

Fig. 4.5. Formation of surface hydroxyls (OH) on the anatase TiO_2 surface: (a) an uncovered surface; (b) coordination of Ti^{4+} ions by water molecules; and (c) the formation of surface hydroxyl ions by proton transfer from water to O_2^- ions.[188]

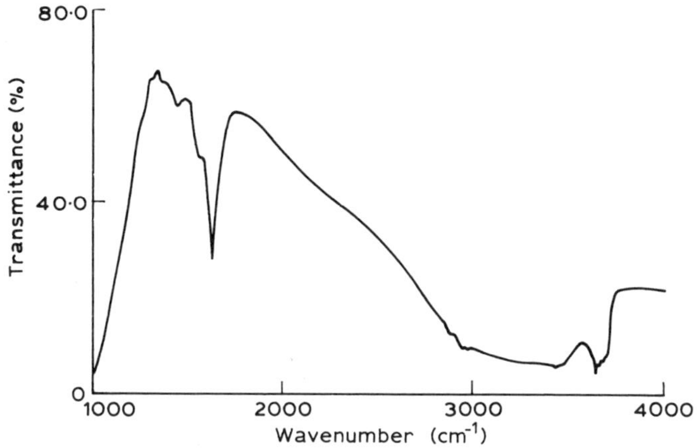

Fig. 4.6. IR spectrum of TiO_2.

Surface hydroxyls can be removed from a TiO_2 surface by heating the sample above 600°C in oxygen atmosphere.

Anatase and rutile have different absorption spectra (Fig. 4.8).[883]

TiO_2 is an extrinsic semiconductor of the *n*-type and the distances between the valence and the conduction bands are: anatase 3·29 eV (385 nm) and rutile 3·05 eV (385 nm). With the absorption of a quantum of energy higher than 3·3 eV or 3·05 eV, an electron is lifted

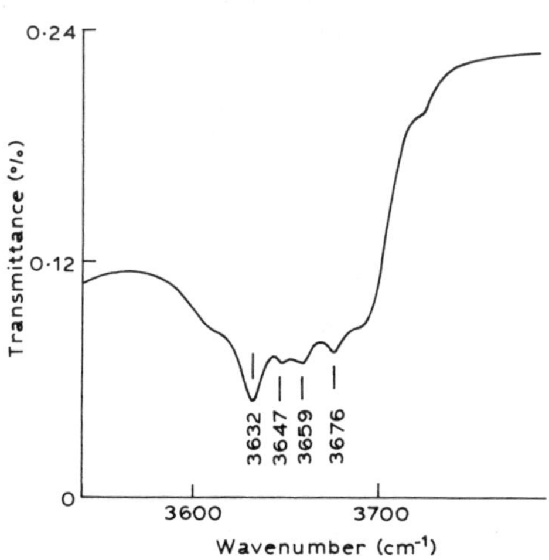

Fig. 4.7. Enlargement of spectrum shown in Fig. 4.6.

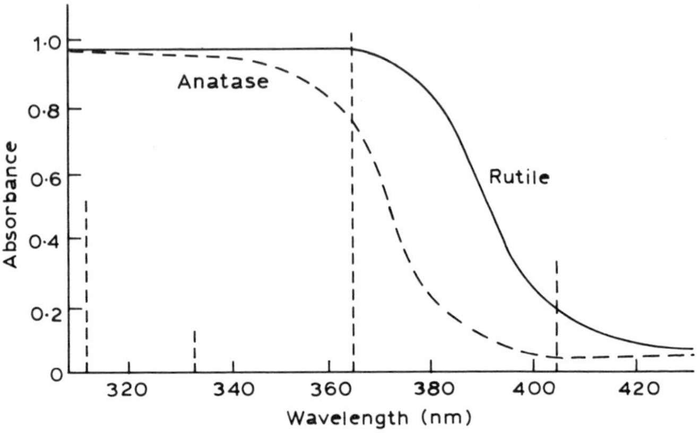

Fig. 4.8. Absorption spectra of: – –, anatase TiO_2; ——, rutile TiO_2.[883]

out of the valence band and into the conduction band, thereby forming an *exciton* (Fig. 4.9).[325,1138] The fact that rutile is more stable is explained by the more rigid bonding of its surface hydroxyl groups. Although rutile more readily forms excitons, the positive hole reacts more slowly with the hydroxyl anions (OH^-).

A measure of the energy of the electron/hole pair may be derived from emission studies, since the wavelength of emission corresponds to the energy released on electron/hole recombination. Emission studies have been reported on TiO_2 molecules in a neon matrix at 4K,[776] single-crystal rutile at 77K, and polycrystalline anatase and rutile powders at 77K (Table 4.8).[12,86] These results indicate that the available energy of the electron/hole pairs of photoexcited anatase is substantially greater than that of rutile.[1153]

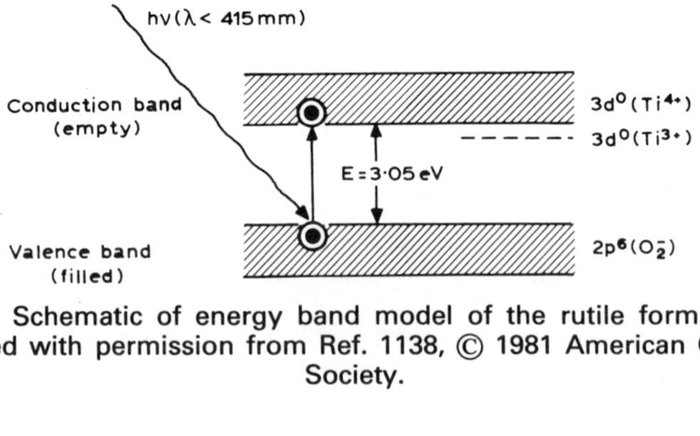

Fig. 4.9. Schematic of energy band model of the rutile form of TiO_2. Reprinted with permission from Ref. 1138, © 1981 American Chemical Society.

Table 4.8
Emission data from TiO_2[1153]

TiO_2	*Modification temperature (K)*	*Emission max. (nm)*	*$E_{emission}$ (kcal mol^{-1})*
Anatase	77	530	54
Rutile	77	850	34

At low temperatures, TiO_2 pigments emit infrared radiation, which can be related to their photoactivity in the polymers:[85,86,89]

(i) Anatase (TiO_2) has strong green emission with a wavelength maximum at 540 nm (on exciting with light of 340 nm).
(ii) Rutile (TiO_2) has a weak infrared emission at 815 nm and 1015 nm (on exciting with light of 375 nm).

Luminescence properties of rutile pigments depend also on their manufacture:[38,39]

(i) rutile produced by the 'sulphate' process has an emission band at 815 nm (related to its photoactivity in polyethylene);
(ii) rutile produced by the 'chloride' process has a broad emission with a band maximum at 1015 nm.

Anatase (but not rutile) has been found to quench the long-lived phosphorescence emission from polyolefins.[86,89,1153] The following electron transfer mechanism between a triplet excited (T_1) carbonyl group and TiO_2 has been proposed:[1153]

$$>C{=}O \xrightarrow{+h\nu} (>C{=}O)^*(T_1) \tag{4.4}$$

$$(>C{=}O)^*(T_1) + TiO_2\,(\text{anatase}) \longrightarrow >\dot{C}{-}O^{\bar{\cdot}} + TiO_2^{+} \tag{4.5}$$

This mechanism can be depicted graphically as shown in Fig. 4.10. In

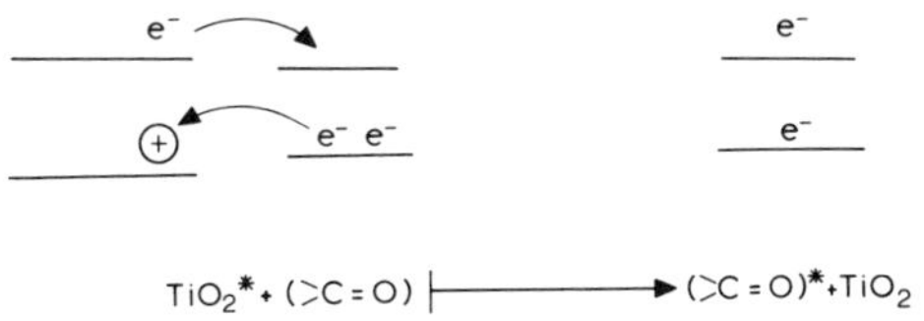

Fig. 4.10. A semiconductor model for energy transfer from an excited TiO_2^* molecule to a carbonyl group.[881]

this figure, photoexcited anatase (TiO_2) is represented by a valence band which possesses an electron vacancy or hole (p^+) and an upper energy level which possesses the promoted electron (e^-). Overlap of these levels with occupied and unoccupied levels of the acceptor (carbonyl group) may allow electron transfer processes which raise the acceptor to an excited state.[880,881,1153]

During UV irradiation of TiO_2 in the presence of oxygen, the oxygen radical anion ($O_2^{\overline{\bullet}}$) is formed:[76,77,881,1062,1136,1138]

$$TiO_2 + O_2 \xrightarrow{+h\nu} TiO_2^{\overset{+}{\bullet}} + O_2^{\overline{\bullet}} \tag{4.6}$$

In a reversible annihilation reaction different excited forms of oxygen such as singlet oxygen ($^1O_2$)[880,881] and atomic oxygen radicals (O•),[547] are produced:

$$TiO_2^{\overset{+}{\bullet}} + O_2^{\overline{\bullet}} \longrightarrow TiO_2 + {}^1O_2 \tag{4.7}$$

On the TiO_2 surface, in the presence of water, hydroxyl groups $Ti^{4+}OH^-$ are formed (Fig. 4.11).[1138] Under light irradiation with energy higher than 3·0 eV the following reactions may occur.

(i) An exciton [electron (e^-)/hole (p^+) pair] is formed:

$$Ti^{4+}OH^- \xrightarrow{+h\nu} e^- + p^+ \tag{4.8}$$

(ii) The exciton reacts further instantaneously with a surface hydroxide ion, and a Ti^{4+} ion of the lattice forms a hydroxyl

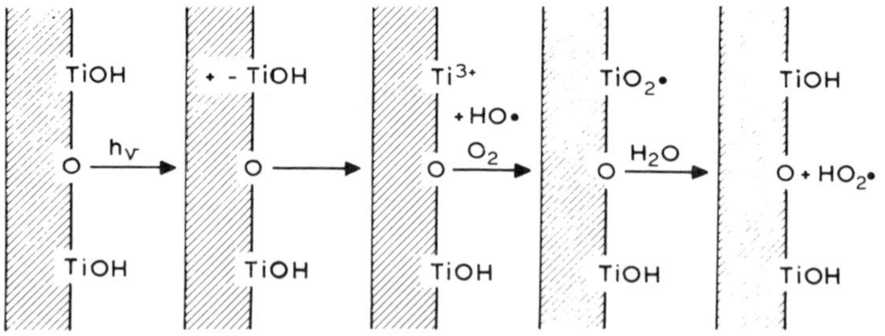

Fig. 4.11. Surface reactions on TiO_2 generate deleterious free radicals. Reprinted with permission from Ref. 1138, © 1981 American Chemical Society.

radical (HO•) and Ti^{3+} ion:[672,881,1138,1139]

$$p^+ + Ti^{4+}OH^- \longrightarrow Ti^{4+} + HO\bullet \quad (4.9)$$

$$e^- + Ti^{4+} \longrightarrow Ti^{3+} \quad (4.10)$$

$$HO\bullet + \bullet OH \longrightarrow H_2O_2 \quad (4.11)$$

$$TiO_2 + H_2O_2 \longrightarrow TiO_3 + H_2O \quad (4.12)$$

$$Ti^{4+}OH^- + H_2O_2 \longrightarrow Ti^{3+}OOH + H_2O \quad (4.13)$$

ESR spectroscopy studies have demonstrated formation of hydroxyl radicals on the surface of TiO_2 (Fig. 4.12).[1138] Irradiation of a suspension of anatase in normal primary alcohols (methanol, ethanol, or *n*-butanol), in a nitrogen atmosphere, led to the formation of Ti^{3+} ions and corresponding aldehyde in solution.[174,222,326,754] The hydroxyl radicals formed are responsible for the reaction:

$$2\,HO\bullet + RCH_2OH \longrightarrow RCHO + 2\,H_2O \quad (4.14)$$

During irradiation of isopropanol in the presence of TiO_2, acetone is formed by a reaction analogous to that above.[589] This reaction has been proposed to predict the behaviour of TiO_2 in polymeric media.

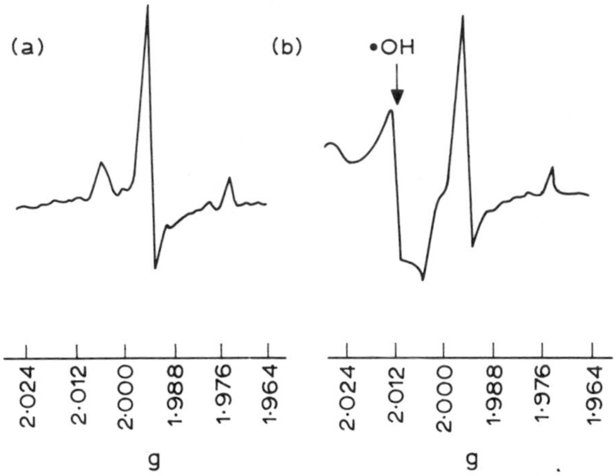

Fig. 4.12. ESR signal of hydroxy radical (HO·) observed during UV irradiation of anatase TiO_2/water: (a) before irradiation and (b) after irradiation. Reprinted with permission from Ref. 1138, © 1981 American Chemical Society.

(iii) The next step is the addition of an atmospheric oxygen molecule, which takes over the electron from Ti^{3+} and turns into the superoxide radical ion ($O_2^{\overline{\cdot}}$):

$$Ti^{3+} + O_2 \longrightarrow Ti^{4+} + O_2^{\overline{\cdot}} \tag{4.15}$$

(iv) This is followed by the important reaction in which water is consumed. The water reacts in the form of its dissociation products:

$$O_2^{\overline{\cdot}} + H^+ \longrightarrow HO_2\cdot \tag{4.16}$$

$$Ti^{4+} + OH^- \longrightarrow Ti^{4+}OH^- \tag{4.17}$$

As a result of this reaction hydroperoxy radicals ($HO_2\cdot$) are formed. Hydroxyl ions (HO^-) remain on the TiO_2 surface to participate in the cycle.

(v) It is also well known that water can be photolysed at an n-TiO_2 electrode illuminated with UV radiation.[422,1083]

$$H_2O \xrightarrow[TiO_2]{+h\nu} H^+ + HO\cdot + e_{aq}^- \tag{4.18}$$

(vi) Reactive hydroxy radicals (HO•) formed on the surface of the TiO_2 are completely separated from the titanium atom after the absorption of UV radiation.[193,1139]

The reactive hydroxyl (HO•) and hydroperoxy ($HO_2\cdot$) radicals may further react with polymer and initiate its degradation. For example, it has been reported that the effect of moisture on photodegradation of Nylon 66 is much greater in the presence than in the absence of TiO_2.[722] TiO_2 photocatalyses decomposition of *tert.*-butyl hydroperoxide and atactic polypropylene hydroperoxides.[687]

This effect is particularly important in destruction of coatings pigmented with TiO_2 and it is known as *chalking,* during which the polymer surface gradually erodes away.[998,1136,1138,1139] Similar photocatalytic oxidation can be caused by zinc oxide (ZnO).

According to the ASTM D658 definition, chalking is the phenomenon manifested in paint films by the presence of loose removable powder, evolved from the film itself, at or just beneath the surface.

With anatase pigments, destruction by the photocatalytic oxidation cycle takes place more rapidly than by UV degradation, and therefore governs the overall picture of chalking (Fig. 4.13). After weathering, the anatase particles remain on the paint surface in what might be

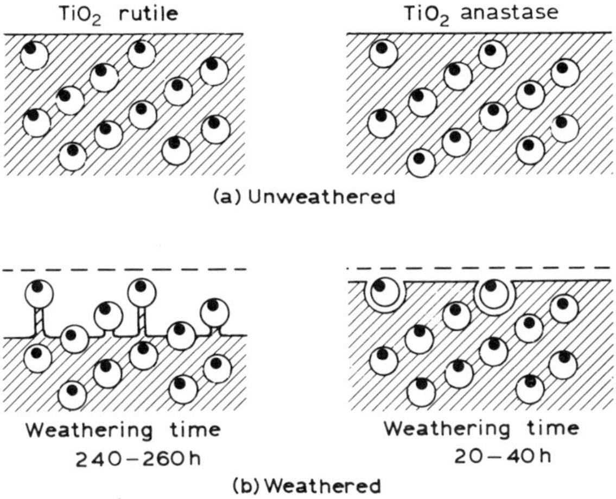

Fig. 4.13. Schematic of degradation processes during the weathering of binders pigmented with TiO_2. Reprinted with permission from Ref. 1138, © 1981 American Chemical Society.

Table 4.9
Photoactivity of reagent anatase TiO_2 coated with metal acetates[a]

Metal acetate coating, 3 wt%	Φ_{rel} *in isopropyl alcohol*	*Metal acetate coating, 3 wt%*	Φ_{rel} *in isopropyl alcohol*
None	1·0	Neodymium	0·3
Silver	0·9	Cupric	0·3
Thalium	0·9	Magnesium	0·3
Gallium	0·9	Barium	0·3
Iron	0·6	Yttrium	0·3
Lead	0·6	Sodium	0·3
Rubidium	0·5	Lithium	0·2
Strontium	0·5	Chromic	0·2
Aluminium	0·5	Stannous	0·2
Lanthanum	0·5	Didymium	0·2
Zirconium	0·4	Nickelous	0·2
Uranyl	0·4	Calcium	0·2
Potassium	0·4	Cerous	0·02
Samarium	0·4	Zinc	0·01
Praseodymium	0·3	Cobaltous	0·01
Niobium	0·3	Manganous	<0·001

[a] Reprinted with permission from Ref. 590, © 1980 American Chemical Society.

described as ‘pits’. Rutile pigment particles therefore stand on ‘pedestals’ which have remained behind because the pigment particles cast a shadow preventing the UV radiation from reaching the paint in these regions.[1138]

The difference in photoactivity between anatase and rutile has been attributed to the difference in the energies of their excited states. The anatase form of TiO_2 has a photocatalytic effect on the degradation of polymers, such as polyolefins and their copolymers,[47,86,589,619,1067] polystyrene,[619] ABS,[619] polyamides,[85,805,958,1078] and poly(vinyl chloride),[257,619,1068] whereas the rutile form of TiO_2 has a stabilizing effect on photodegradation.

To retard the photocatalytic effect many commercial TiO_2 pigments are usually coated with one or several layers of inorganic barrier coatings such as alumina, silica, zinc oxide or manganese acetate (Table 4.9). Alumina and silica, which are strong electron-acceptors, probably attach to the surface of hydroxyl groups:[573,590,591,1137]

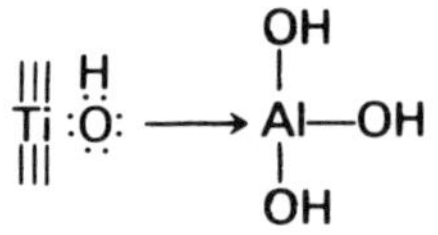

(4.19)

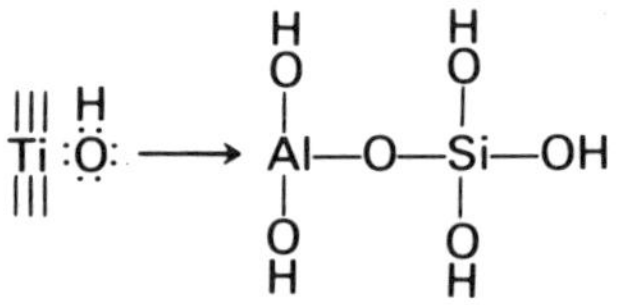

(4.20)

This can reduce the activity to a few per cent of that of uncoated anatase pigment, but it does not completely eliminate it.

It is well established in the literature that manganese(II)-compounds-coated TiO_2 pigments are efficient light stabilizers for Nylon polymers.[298,364,609,1190]

Anatase with a surface coating of manganese(II) acetate does not exhibit phosphorescence at 540 nm.[85] The level of manganese ions added has an important effect on the photochemical properties of anatase.

The role of the acetates of manganese, cobalt, zinc and cerium can be attributed to their effectiveness in bonding metal acetates on the surface of the pigments, which serve as sites for the rapid annihilation of excitons by the preferential participation of the metal acetate, thereby avoiding reactions involving titanium ions:[590,591]

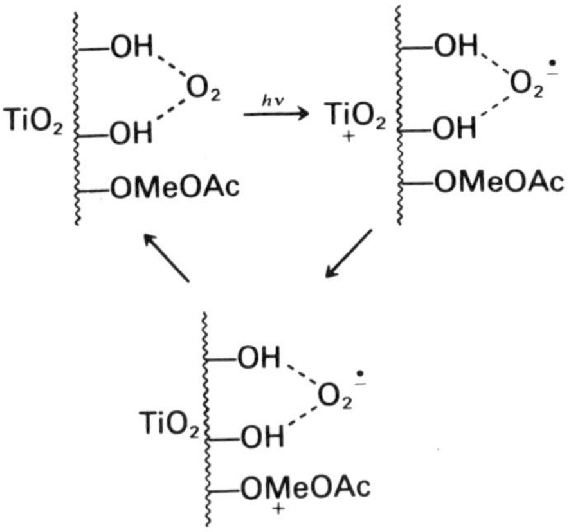

(4.21)

(4.22)

(4.23)

Titanium oxide (TiO_2) can be used simultaneously with zinc oxide (ZnO) in formulations for the stabilization of polyolefins and polystyrene,[257,619] and poly(vinyl chloride).[257,333,359,619]

Combinations of zinc oxide, titanium oxide and zinc dialkyldithiocarbamate improve the weather resistance of polyethylene and polypropylene.[165,252,253,340,957,1191] Synergistic stabilization of polyolefins by titanium oxide and zinc dialkyldithiocarbamates has also been reported.[846]

The International Lead Zinc Research Organization has sponsored work on the influence of combinations of zinc oxide and organic additives upon the weatherability of polyolefins. The following results have been reported:[165,252,253,340,1191]

—After four years outdoors in Connecticut, the mechanical strength of polypropylene containing zinc oxide and either tri(mixed mono- and di-nonylphenyl) phosphite or zinc diethyldithiocarbamate (**3.85**) was found to be better than that of a similar compound containing 2-hydroxy-4-*n*-octoxybenzophenone (**5.30**) (Cyasorb UV 531).

—In high-density polyethylene (HDPE) and polypropylene, zinc oxide combined with zinc diethyldithiocarbamate or zinc dimethyldithiocarbamate imparts better resistance to weathering than a substituted benzophenone.

—No advantage was found over the benzophenone in low-density polyethylene (LDPE).

—Neither zinc oxide nor zinc diethyldithiocarbamate was effective in HDPE when used alone.[886]

The results observed from the addition of titanium dioxide or zinc oxide to polymers containing zinc dialkyldithiocarbamates appear to be a consequence of the combined effect of the scattering and absorbing properties of the oxides and the hydroperoxide decomposition promoted by the dithiocarbamates.[757]

Studies of synergistic and antagonistic effects between TiO_2 and other additives as photostabilizers and/or antioxidants have been reported by a few authors.

There is some synergistic effect between rutile (TiO_2) and photostabilizer 2-hydroxy-4-*n*-octoxybenzophenone (Cyasorb UV 531, **5.30**).[89]

Titanium oxide shows antagonistic effects with hindered phenolic antioxidants.[561]

The hindered amine light stabilizers (HALS) were found to be ineffective in protecting polypropylene against the photocatalytic action of anatase. In the case of rutile, good photoprotection of the polymer was observed, the effect being greater for a coated pigment. However, the presence of antioxidant (Irganox 1010, **3.32**) antagonized the stabilizing action of the hindered piperidine compounds when in the presence of an uncoated pigment. These results can be explained in terms of additive absorption onto the pigment particle surface versus the photocatalytic activity of the pigment.[63]

Both crystalline forms of TiO_2 show antagonistic behaviour with metal chelate photostabilizers such as[89,93]

—nickel(II) 2,2′-thiobis(4-*tert.*-octylphenolato)-*n*-butylamine (Cyasorb UV 1084) (**5.63**);
—nickel(II) *o*-butyl-3,5-di-*tert.*-butyl-4-hydroxybenzyl)phosphonate (Irgastab 2002) (**5.67**);
—nickel(II) bis(1-phenyl-3-methyl-4-decanoyl-5-pyrazolate) (Sanduvor NPU) (**5.68**).

The antagonism is also present when the antioxidant is included in the system.

The calcium bis(3,5-di-*tert.*-butyl-4-hydroxybenzyl)monoethyl phosphonate (Irganox 1425, **3.34**) also displays antagonism with anatase TiO_2 in the absence and presence of the antioxidant (pentaerythrityl tetrakis[3(3′,5′-di-*tert.*-butyl-4′-hydroxyphenyl)propionate] (Irganox 1010, **3.32**)). However, the rutile form shows an additive effect with this chelate on its own and only slight antagonism in the presence of the antioxidant. In fact, the very effective Cyasorb UV 1084 (**5.63**) is only marginally better when used with rutile TiO_2 in the absence and presence of the antioxidant than Irganox 1425 under the same conditions.[47]

4.3 ZINC OXIDE

Zince oxide (ZnO) is a very widely used white pigment in the plastics and coating industries, and has the greatest UV absorption of any pigment commercially available.[177]

The photochemistry of solid ZnO is complex. Under UV irradiation an exciton (electron (e^-)/hole (p^+) pair) is formed, which reacts further with ZnO by the following reactions:[672]

$$ZnO \xrightarrow{+h\nu} e^- + p^+ \tag{4.24}$$

$$p^+ + ZnO \longrightarrow Zn^+ + O \tag{4.25}$$

$$p^+ + Zn^+ \longrightarrow Zn^{2+} \tag{4.26}$$

$$e^- + Zn^+ \longrightarrow Zn^0 \tag{4.27}$$

$$e^- + O_2 \longrightarrow O_2^{\overline{\cdot}} \tag{4.28}$$

$$e^- + O \longrightarrow O^{\overline{\cdot}} \tag{4.29}$$

$$O_2 + O^{\overline{\cdot}} \longrightarrow O_3^{\overline{\cdot}} \tag{4.30}$$

$$2\,O_2^{\overline{\cdot}} + H_2O \rightleftharpoons 2\,HO_2\cdot + 2\,OH^- \tag{4.31}$$

$$2\,HO_2\cdot \longrightarrow H_2O_2 + O_2 \tag{4.32}$$

The species $O_3^{\overline{\cdot}}$ has been detected by ESR spectroscopy, with several oxide systems: ZnO, MgO and TiO_2.[737] It is obvious that all of these oxygen species may participate in complex reactions in a polymer matrix, analogous to TiO_2 reactions (cf. Section 4.2). ZnO accelerates the photo-oxidation of polyolefins,[687,688] ethylene–propylene copolymers[1067] and polyamides.[958] In spite of ZnO's photocatalytic properties, it has been used for stabilization against UV radiation of polyolefins,[340,886] poly(vinyl chloride),[359] and paints and rubber.[177]

4.4 METAL SALTS COATINGS

A cerous chloride ($CeCl_3$)/poly(vinyl alcohol) complex has been applied to protect polycarbonate surfaces by coating.[647] This complex is effective as a UV screener (Fig. 4.14). In order to prevent ion leaching by water, the cerous ion coated films were further overcoated with poly(vinyl acetate). This inexpensive solar glazing material can be used for protection of solar collectors.

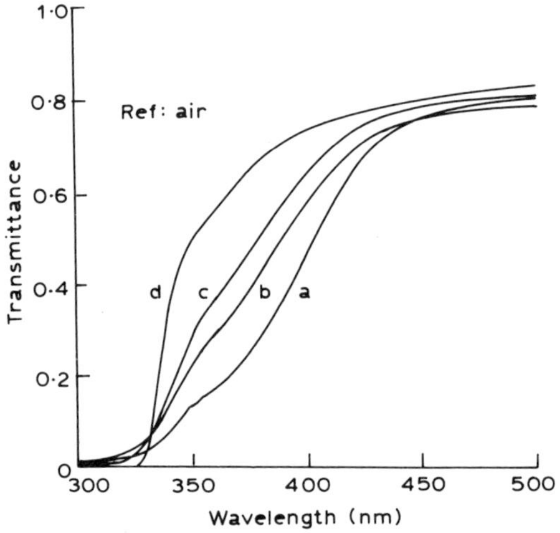

Fig. 4.14. Transmittance spectra of $CeCl_3$/poly(vinyl alcohol)-coated Lexan sheet sealed with a poly(vinyl acetate) overcoat after 20 h of UV irradiation: (a) 0·1 g poly(vinyl alcohol); (b) 0·005 g $CeCl_3$ in 0·1 g poly(vinyl alcohol); (c) 0·01 g $CeCl_3$ in 0·1 g poly(vinyl alcohol); (d) 0·05 g $CeCl_3$ in 0·1 g poly(vinyl alcohol).[647]

4.5 CARBON BLACK

Carbon black is formed by incomplete combustion of many organic substances: solid, liquid, or gaseous. The classification of carbon black of various types has been based traditionally on the manufacturing processes.[354] The following classes of carbon black are at present distinguished according to the range of average particle size expressed in Å:

(i) furnace black (170–700 Å);
(ii) impingement black (100–270 Å);
(iii) thermal black (1500–5000 Å);
(iv) lamp black (500–900 Å);
(v) acetylene black (350–500 Å).

The particles of a given carbon black do not all have the same diameters, but show a size distribution curve characteristic of the particular sample.

Carbon blacks of smaller average particle size impart superior stability against photodegradation of polymers. The effectiveness of carbon black is dependent upon its type, the particle size and the degree of dispersion in the polymer matrix.

Carbon blacks interact with polymer matrices by means of their surface. Since the particles are very small, their specific surface area must be very large. Generally, carbon blacks have been considered to be polycrystalline solids. The chemical and physical properties of carbon blacks are dependent upon the nature of the oxygenated structure on their surface. Oxygen, which is present in the flame during particle formation, plays a vital role in carbon black characteristics. All commercial carbon blacks contain oxygen and oxidized groups such as quinone, phenol, carboxyphenol, lactone, etc. (Fig. 4.15) and Table 4.10.[354]

The high efficiency of carbon blacks as light stabilizers for polymers

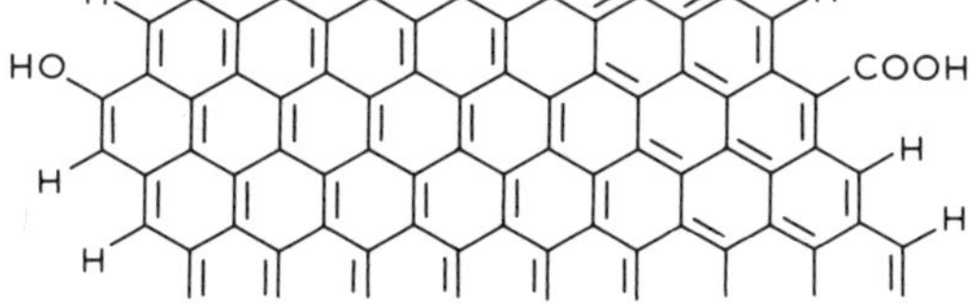

Fig. 4.15. Carbon black model.

Table 4.10
Chemical and physical properties of carbon blacks[666]

Carbon black	*Type*	*Specific surface area, BET* ($m^2\ g^{-1}$)	*pH*	*Oxygen content* (%)	*Active groups* (μ *equiv.* g^{-1})			
					COOH	*Lactone*	*Phenol*	*CO*
Farbruss FW_2	HCC(0)	460	3	9·9	626	104	580	640
Spezialschwarz 4	LFI(0)	180	3	8·4	196	157	143	108
Farbruss S 160	MCC	150	5	4·0	57	42	36	4
Farbruss S 170	MCC	200	5	2·5	—	—	—	—
Printex U	RCC	110	5	2·4	46	58	16	64
Printex 140 V	CC	96	5	2·4	—	—	—	—
Flammruss 700	Lamp	48	5	6·8	—	—	—	—
Flammruss 101	Lamp	21	7	0·4	—	—	—	—
Corax 9	Furnace	200	9	0·6	—	—	—	—

is probably due to their ability to act as:

(i) An inner filter for UV radiation. Carbon blacks absorb essentially all of the incident UV and visible radiation and a sample made opaque in this way limits the range of colour attainable.
(ii) A free-radical scavenger, thus reducing the propagation of the initial step. Many carbon blacks contain stable free radicals.
(iii) A quencher of excited singlet and/or triplet states of polymers.
(iv) An antioxidant, due to their surface phenolic groups.
(v) A peroxide decomposing agent.

The stabilization of polymers increases with increasing concentration of carbon black. A comparison of the photostabilizing efficiency of carbon black with that of other inorganic pigments is shown in Table 4.11. Carbon black is an excellent photostabilizer and wide application is limited only by its black colour.

Carbon black has been used extensively to stabilize polyolefins against photo-oxidation.[531–533,795,1140,1160] The efficiency of carbon black is shown by the fact that polyethylene samples containing only 1 wt% of this pigment have been found to be stable even after outdoor exposure (in Florida) for more than 30 years.[530] Carbon black also helps to maintain the elongation at break of polyethylene during natural weathering.[955] The efficiency of protection of polyethylene (against photoinduced loss of flexibility) by carbon black increases markedly with decreasing particle size of the carbon black.[1140] In the case of polyethylene containing 2 wt% of carbon black, the photo-oxidation was found to be localized mostly in a 'skin' layer 10–15 μm thick.[532]

Carbon blacks form synergistic or antagonistic combinations with the conventional amine-, phenolic- or sulphur-containing varieties of antioxidants.[529,533,534,536,666,1094,1119,1157]

In carbon black–antioxidant–polyethylene systems several reactions occur simultaneously.[666] Carbon blacks adsorb any kind of antioxidant on their surface. The strength of these adsorptive secondary forces depends on the chemical nature of the carbon blacks. By these bonds the antioxidants may be withdrawn from the system by the reduction of their mobility.

In the case of highly acidic carbon blacks, the number of the phenolic and quinone active sites on the carbon surface exceeds that of the antioxidants, if they are used in the usual commercial ratio. For neutral carbons the number of active sites on the carbon is at least one

Table 4.11
Stabilization of polypropylene by inorganic pigments[340]

Pigment	*Initial condition*		*Time to initial change (h)*				*Final aged appearance*		
	Clarity	*Colour*	*Colour*	*Gloss*	*Surface cracking*	*Flex. break*	*Time (h)*	*Colour*	*Surface*
Control	Clear	White	89	303	423	159	423	Yellow	All cracked
Carbon black	Opaque	Black	None	1240	2184	None	2184	Black	Fine cracks
TiO_2	Opaque	White	808	472	472	1992	1992	Yellow	Cracks and pits
Al silicate	Cloudy	Off-white	161	161	425	161[a]	425	Yellow	All cracked
Silica	Clear	White	273	237	457	93	457	Yellow	All cracked

[a] Partial break.

order of magnitude less than the number of active groups on the antioxidant.

The loss of stabilizer effectiveness with high oxygen content 'acidic' carbon blacks can be attributed to the fact that the quinone-type oxygen of the carbon is a stronger hydrogen acceptor than is the propagating radical; hence, the antioxidant (both for amine or phenolic type) is consumed by the former:

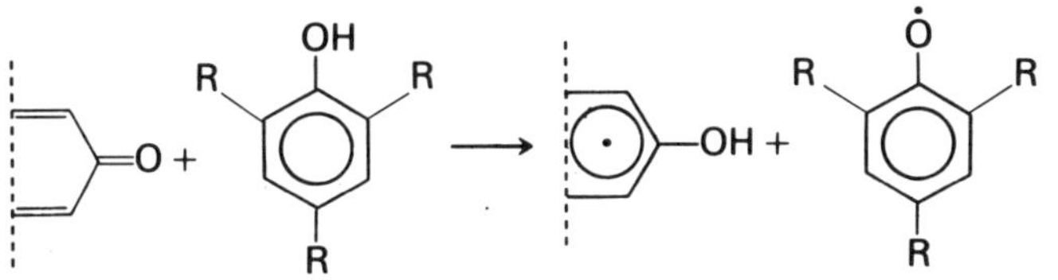

(4.33)

In addition, the phenolic groups of the carbon surface and the amine-type antioxidants may form hydrogen bonds with each other, resulting in sterically hindered proton transfer reactions:

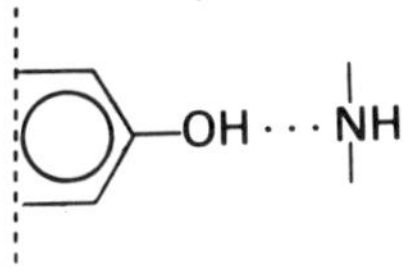

Synergistic interaction of carbon black and an antioxidant is relatively rare. It may happen, however, than the common effect is better than the effect of the antioxidant or carbon black separately, but does not exceed the sum of the two.[666] The reason for the synergism, e.g. between *N,N*-di-2-naphthylphenylenediamine (Nonox CI) and carbon black can be the peroxide-decomposing feature of the latter. The synergism shown between carbon black having a high phenol group content, and the phenolic antioxidant 2,2-methylenebis[4-methyl-6(α'-methylcyclohexyl)phenol] (Nonox WSP) causes the carbon black to donate hydrogen to phenoxy radicals:

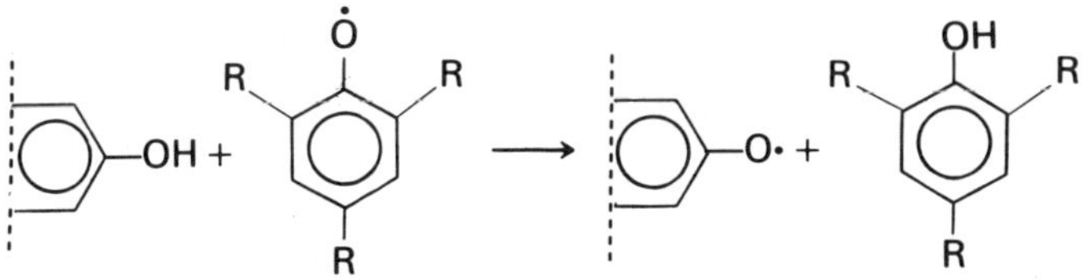

(4.34)

Interaction of carbon black with other types of antioxidants and/or photostabilizers, e.g. hindered amine light stabilizers (HALS), is unknown yet.

5
Photostabilizers

5.1 ALKYL-SUBSTITUTED *p*-HYDROXYBENOZATES

Alkyl-substituted *p*-hydroxybenzoates, e.g. cetyl-3,5-di-*tert.*-butyl-4-hydroxybenzoate (**5.1**), stabilize polymer oxidative and/or photodegradation by the following mechanism.[75, 95–100]

(i) They suppress hydroperoxide formation during processing (i.e. they are effective thermal chain-breaking macroradical terminators).
(ii) They terminate with polymer alkoxy (PO•) and hydroxy (HO•) radicals.
(iii) They are resistant to photodimerization in the polymer because of the presence of the carboxylic acid group directly attached to the *para* position of the phenyl ring. This prevents them from taking part in further photoreactions in which harmful and undesirable yellow stilbenequinones are formed (cf. Section 3.1).
(iv) They are synergistic with other antioxidants which act as hydroperoxide decomposers.

Cetyl 3,5-di-*tert.*-butyl-4-hydroxybenzoate (or 3,5-di-*tert.*-butyl-4-hydroxybenzoic acid, *n*-hexadecyl ester), **5.1**, which is commercially produced as Cyasorb UV 2908 by the American Cyanamid Company, USA, is an excellent antioxidant and photostabilizer for polyolefins, particularly pigmented opaque formulations.[67,75,95] Its physical properties are listed in Table 5.1, and its absorption spectrum is shown in Fig. 5.1. It is well compatible with polymers and recommended for

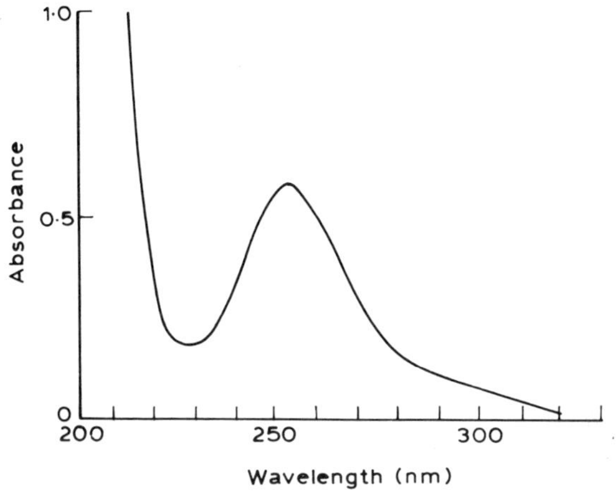

Fig. 5.1. Absorption spectrum of 3,5-di-*tert.*-butyl-4-hydroxybenzoic acid, *n*-hexadecyl ester (**5.1**; Cyasorb UV 2908) (2×10^{-5} M in *n*-hexane).[67]

applications such as piping, crates, drums, seating, automotive, marine, garden and recreational products. Cyasorb UV 2908 can also be used in highly loaded polyolefin concentrates for subsequent let-down, eliminating cracking and bridging problems normally associated with concentrates containing 10% or more of photostabilizer.

Cyasorb UV 2908 exhibits synergistic effects with different inorganic pigments such as cadmium yellow and chrome green.[75]

5.2 PHENYL ESTERS OF BENZOIC ACID

Phenyl esters (*ortho, meta* and *para*) of benzoic acid (known as salicylates) were the first technically used photostabilizers and they are still available because of their low price. The main disadvantage of salicylates is that they turn yellow when they are exposed in a polymer matrix to UV irradiation (Table 5.2). The yellowing of these compounds is due to the *photo-Fries rearrangement* in which hydroxybenzophenones are formed [cf. reactions (5.1) and (5.2)].

The phenyl ester of 2-hydroxybenzoic acid (Salol) (**5.2**) (absorption spectrum in Fig. 5.2) under UV irradiation is photo-rearranged into 2,2′-dihydroxybenzophenone (**5.3**) and 2,4′-dihydroxybenzophenone

Table 5.1
Physical properties and solubility of cetyl-3,5-di-*tert.*-butyl-4-hydroxybenzoate

Chemical name	*Commercial name*	*Chemical structure*	*Absorption spectrum*
Cetyl 3,5-di-*tert.*-butyl-4-hydroxybenzoate *or*: 3,5-Di-*tert.*-butyl-4-hydroxybenzoic acid, *n*-hexadecyl ester	Cyasorb UV 2908 (Cyanamid)	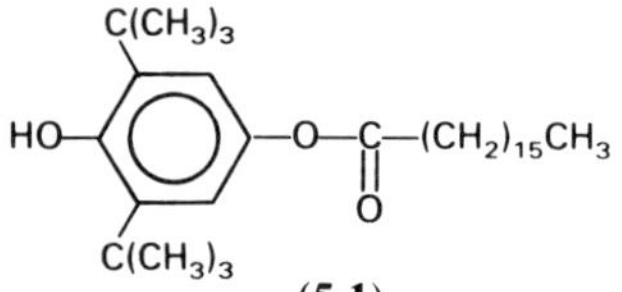 (**5.1**)	See Fig. 5.1

(**5.4**):[845]

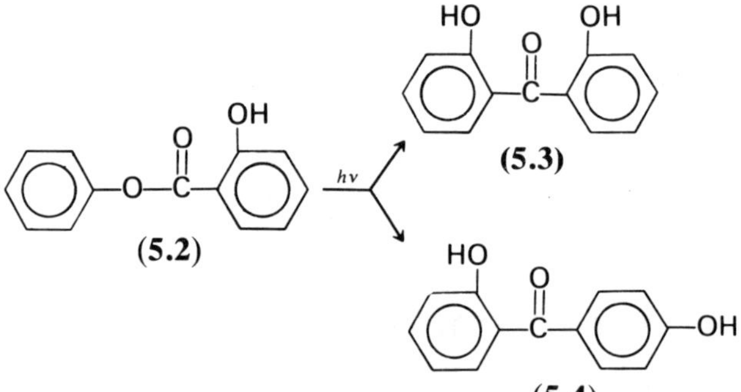

(**5.3**) (5.1)

(**5.2**) (**5.4**) (5.2)

The 3-hydroxyphenyl ester of benzoic acid (resorcinol monobenzoate) (**5.5**) (absorption spectrum, Fig. 5.2) under UV irradiation also undergoes the photo-Fries rearrangement, to 2,4-dihydroxybenzophenone (**5.6**):[845]

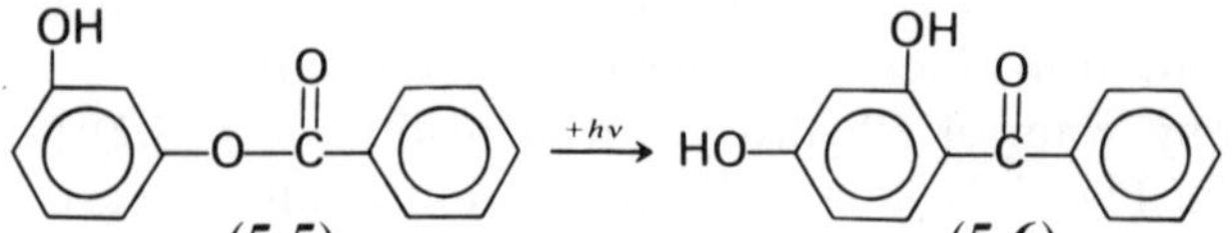

(**5.5**) (**5.6**) (5.3)

Molecular weight	*Appearance*	*Solubility (wt%)*					*Melting point (°C)*	*Recommended applications*
		Methanol	*Acetone*	*Chloroform*	*Hexane*	*Toluene*		
475	Near-white free-flowing powder	1·3	8·3	61·1	35·9	53·5	60	Polyolefins, particularly pigmented opaque formulations

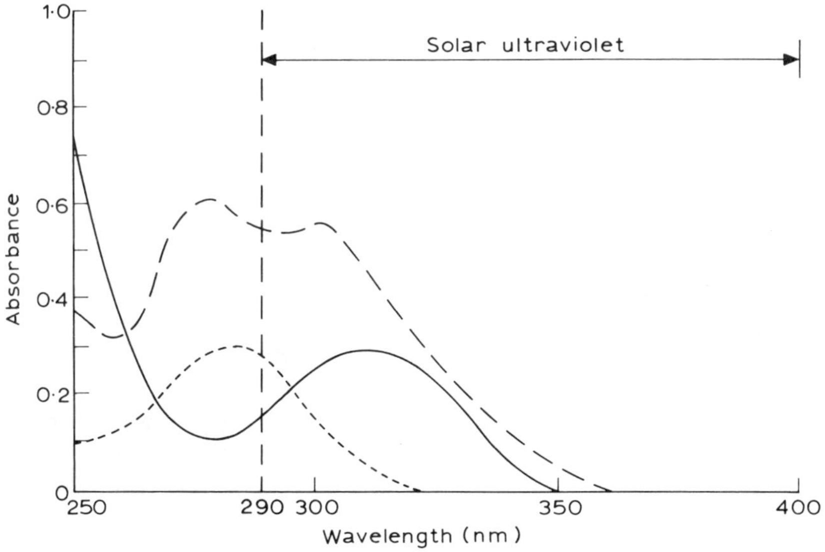

Fig. 5.2. Absorption spectra of: ——, phenyl salicylate (**5.2**); · · ·, resorcinol monobenzoate (**5.5**); — — —, resorcinol monobenzoate, after UV irradiation. Concentration of phenyl esters in CH_2Cl_2, 6×10^{-4} M; 0·1 cm pathlength.[697]

Table 5.2
Stabilizing effectiveness of the rearranging salicylates in polypropylene during irradiation in the Weatherometer[845]

Additive[a]	*Major hypothetical rearranged product*	*Stabilizing effectiveness measured as stress-crack life* (*h*)	*Yellowing*
Control		90	None
Phenyl salicylate (Salol)	2,2′-Dihydroxybenzophenone	130	Very slight
Phenyl 3-phenylsalicylate	2,2′-Dihydroxy-3-phenylbenzophenone	140	Some
2-Dodecyl-4-methylphenyl salicylate	2,2′-Dihydroxy-3-dodecyl-5-methyl-benzophenone	280	Much
o-Octylphenyl salicylate	2,2′-Dihydroxy-3-octylbenzophenone	525	Some
2,6-Dimethylphenyl 5-*tert.*-butylsalicylate	None	420	None
2,6-Dimethylphenyl 4-dodecyl-oxysalicylate	None	460	None
p-*tert.*-Butylphenyl 4-dodecyl-oxysalicylate	2,2′-Dihydroxy-4-dodecyloxy-5′-*tert.*-butylbenzophenone	470	Some
2,6-Dimethylphenyl salicylate	None	500	None
2,6-Dimethyl-4-octadecylphenyl salicylate	None	550	None
p-Octylphenyl salicylate (OPS)	2,2′-Dihydroxy-5-octylbenzophenone	1200	Much
p-Octylphenyl 5-octylsalicylate	2,2′-Dihydroxy-5,5′-dioctyl-benzophenone	1660	Very much

[a] Concentration 1 wt %.

2,4-Di-*tert.*-butylphenyl 4′-hydroxy-3′,5′-di-*tert.*-butylbenzoate (**5.7**) (absorption spectrum in Fig. 5.3) is an effective photostabilizer for polyolefins. Under UV irradiation it undergoes the photo-Fries rearrangement to a 4-hydroxybenzophenone (**5.8**):[93,242,540,541]

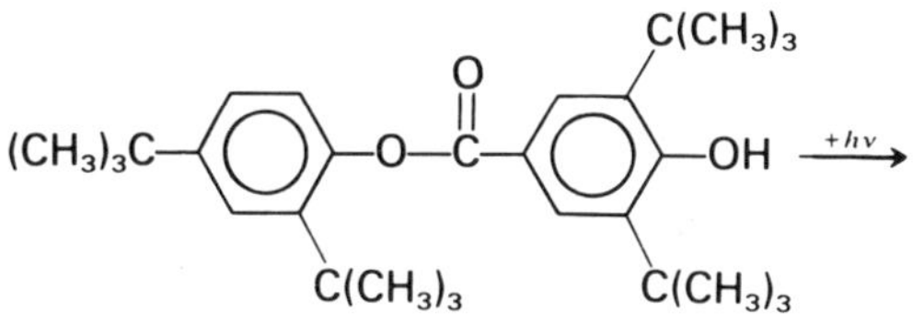

(**5.7**)

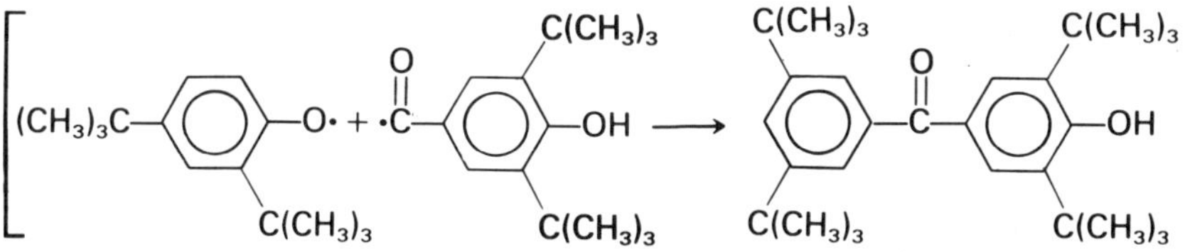

(**5.8**)

(5.4)

The Fries rearrangement occurs, however, at a slow rate (Fig. 5.3) and cannot be considered as the main photostabilizing mechanism of *p*-hydroxybenzoates. A protective mechanism is not known yet.

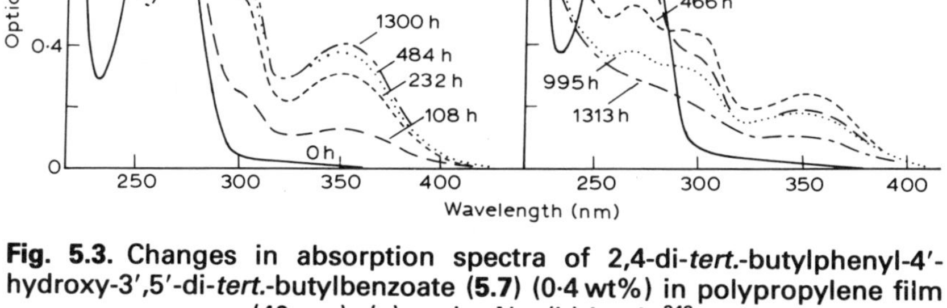

Fig. 5.3. Changes in absorption spectra of 2,4-di-*tert.*-butylphenyl-4′-hydroxy-3′,5′-di-*tert.*-butylbenzoate (**5.7**) (0·4 wt%) in polypropylene film (40 μm): (a) under N_2; (b) in air.[242]

Table 5.3
Stabilizing effectiveness of the salicylates in polyethylene in outdoor weathering[845]

Additive[a]	*Yellowing*	*Stabilizing effectiveness measured as stress-crack life (months)*
Control	None	12
Phenyl salicylate (Salol)	Very slight	22
p-tert.-Butylphenyl salicylate (TBS)	Slight	40
p-Biphenylyl 5-phenylsalicylate	Much	49
p-Biphenylyl salicylate	Much	72
Phenyl 3-phenylsalicylate	Some	>85
o-Octylphenyl salicylate	Some	>85

[a] Concentration 1 wt %.

Photo-rearrangement with resultant yellowing is obviously not a prerequisite for stabilizing efficiency in the aromatic salicylates (Table 5.3). For example, 2,6-dimethylphenyl salicylate fairly successfully inhibited Weatherometer stress cracking without yellowing or rearranging. However, none of the non-yellowing salicylates has equalled the effectiveness of the better yellowing varieties, such as *p*-octylphenyl salicylate.

Stabilizing effectiveness in the non-rearranging salicylates in polypropylene is connected with the NMR chemical shifts observed for the *o*-hydroxy protons (Table 5.4). This dependence has been pro-

Table 5.4
Stabilizing effectiveness of the non-rearranging salicylates in polypropylene during irradiation in the Weatherometer[845]

Salicylate[a]	*Stabilizing effectiveness measured as stress-crack life (h)*
2,6-Dimethylphenyl-5-*tert.*-butylsalicylate	415
2,6-Dimethylphenyl 4-dodecyloxysalicylate	460
2,6-Dimethylphenyl salicylate	500
2,6-Dimethyl-4-octadecylphenyl salicylate	550
2,2′-Dihydroxybenzophenone	330

[a] Concentration 1 wt%.

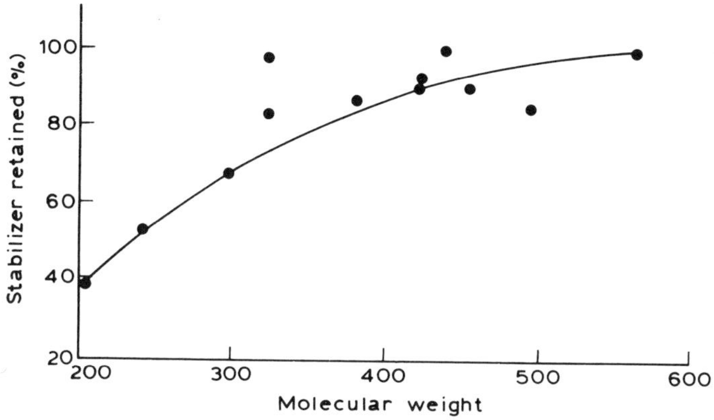

Fig. 5.4. Loss of aromatic salicylates from polypropylene by volatilization during compounding.[845]

posed as a measure of the photostabilizing efficiency in the non-yellowing salicylates.[845]

Loss of aromatic salicylates from plastics during compounding depends on their molecular weight (Fig. 5.4).

Salicylic acid derivatives have been widely investigated as photostabilizers for polyolefins,[279,540,845] poly(vinyl chloride),[463,464,928] poly(methyl methacrylate)s (in contact lenses),[946] polyurethanes,[1082] and acetal- and butyral-cellulose.[279,845]

5.3 SALICYLANILIDES AND OXANILIDES

o-Hydroxysalicylanilides can form three different intramolecular hydrogen bonds, depending on solvent polarity:[370]

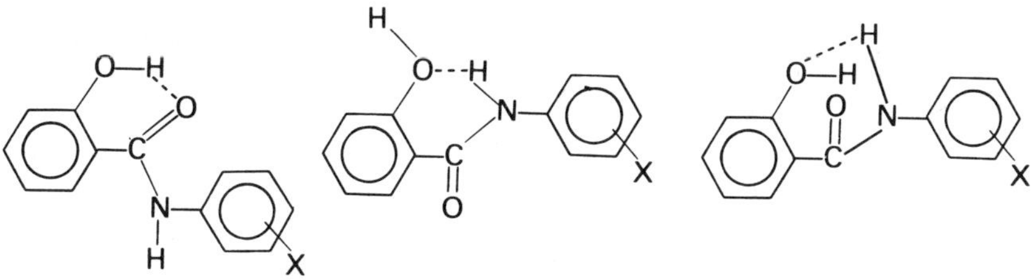

α-form β-form γ-form

The α-form of o-hydroxysalicylanilide formed by intramolecular hydrogen bonding between the hydroxyl group and the carbonyl group dominates and can in addition exist in two resonance structures according to the intramolecular charge transfer:

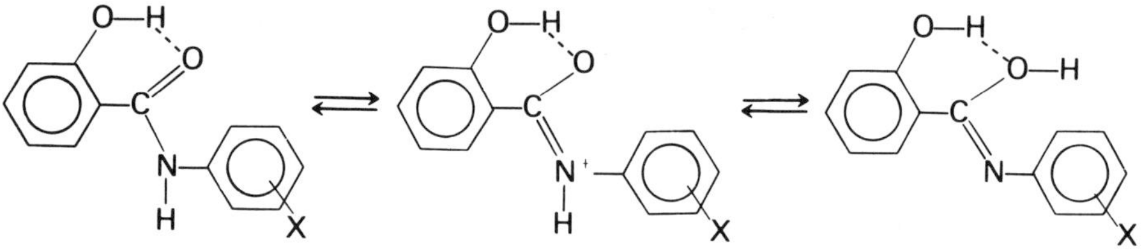

(5.5)

Salicylanilide (**5.9**) under UV irradiation decomposes to the benzyl radical (**5.10**) and aniline radicals (**5.11**) according to the reaction:[680]

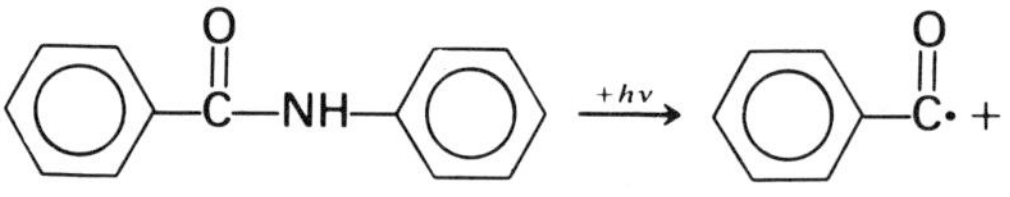

(**5.9**) (**5.10**)

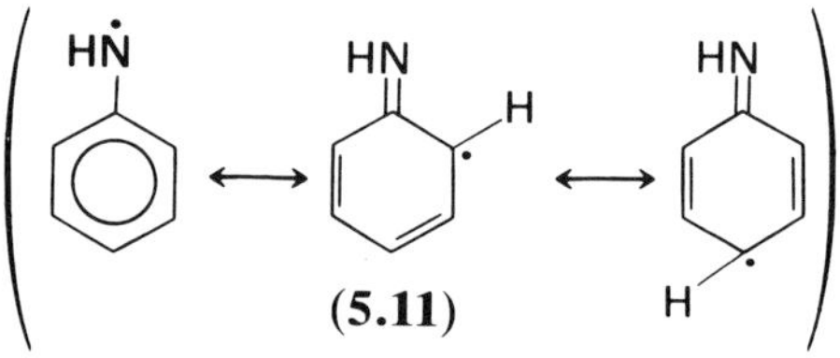

(**5.11**) (5.6)

As a result of *photo-Fries rearrangement*, o-aminobenzophenone and p-aminobenzophenone are formed:

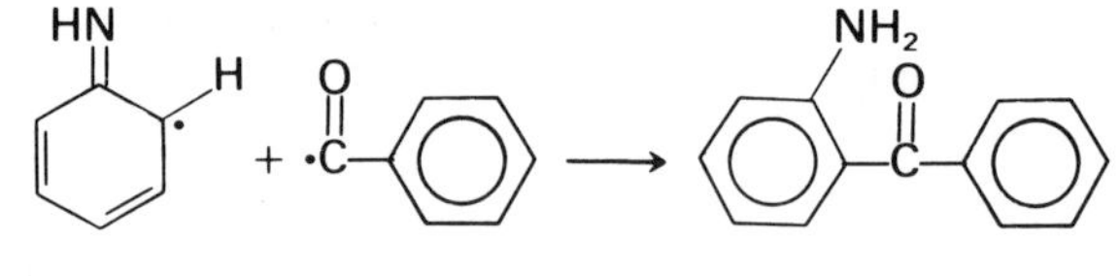

(5.7)

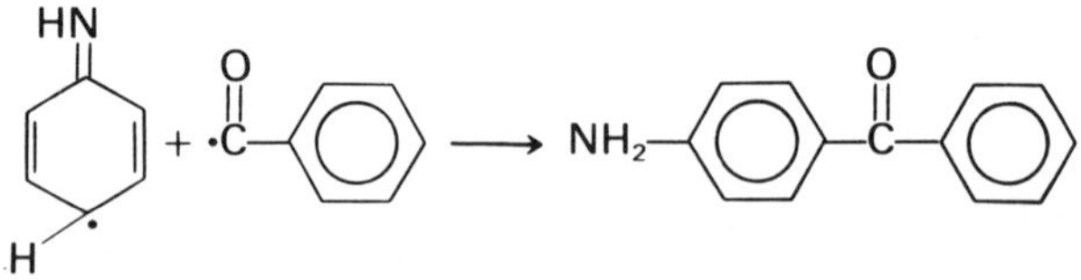

(5.8)

Derivatives of salicylanilides (**5.12**) are effective photostabilizers for natural rubber.[370,640]

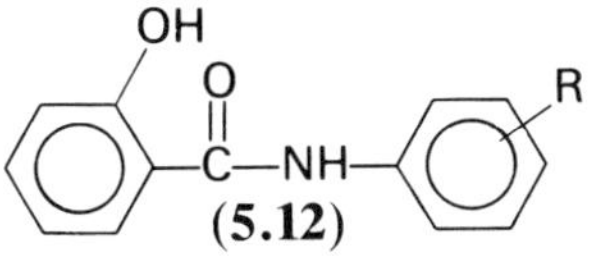

(5.12)

R = H, Cl, CH_3, OCH_3

Aromatic oxanilides (**5.13**) used as photostabilizers show an extremely high photostability, related to a very short lifetime of the excited state. The deactivation probably results from an intramolecular proton transfer. They are also very stable against attack by free radicals and they are not free-radical scavengers.[22]

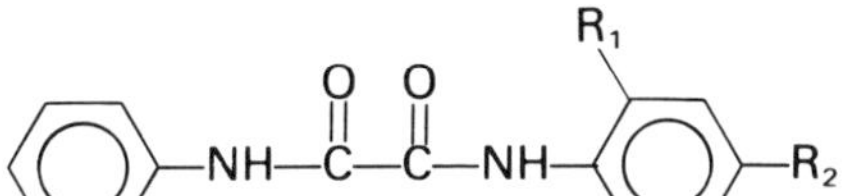

(5.13)

R_1 = H, C_2H_5, OC_8H_{17}-*n*
R_2 = H, $C_{12}H_{25}$-*i*, $C_{12}H_{25}$-*n*

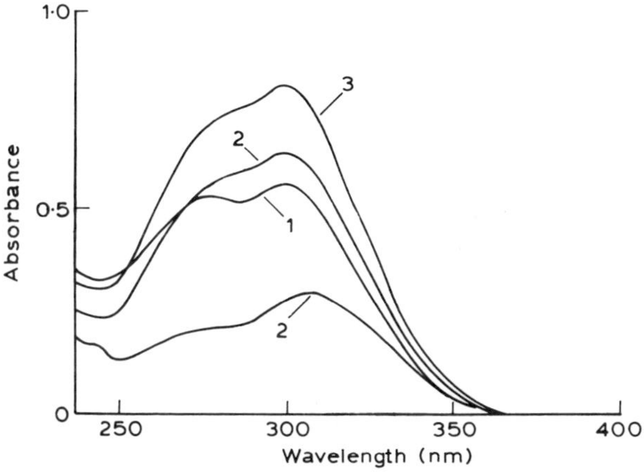

Fig. 5.5. Absorption spectra of differentially substituted oxanilides in ethanol.[22]

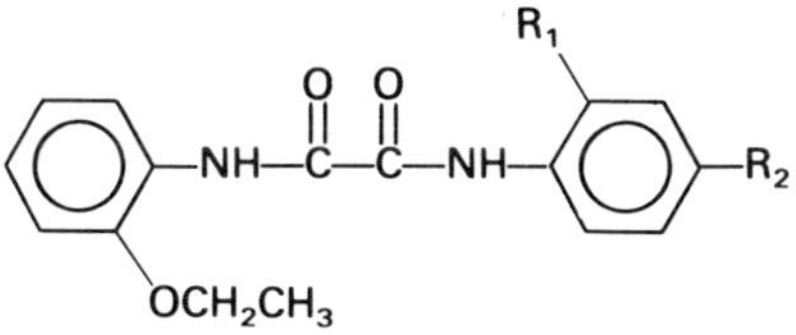

Curve no.	R_1	R_2
1	C_2H_5	H
2	H	$C_{12}H_{25}(i)$
3	H	$C_{12}H_{25}(n)$
4	$OC_8H_{17}(n)$	H

The photostabilization action is due essentially to the inner filter effect of the oxanilides which absorb strongly in the near- and far-UV regions (Fig. 5.5).

The quenching of oxanilides excited states formed by absorption by the polymer is relatively unimportant.

5.4 DIKETONES

It has been found that β-diketones such as dibenzoylmethane (**5.14**), benzoyl acetone (**5.15**) and diethyl terephthaloyldiacetate (**5.16**) have photostabilizing properties in polybutadiene.[1164] The photo-

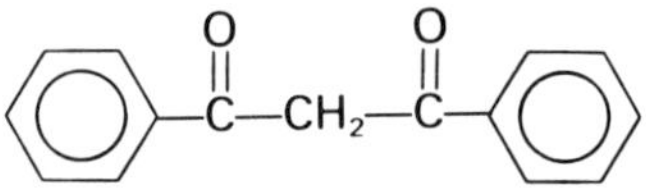

(**5.14**)

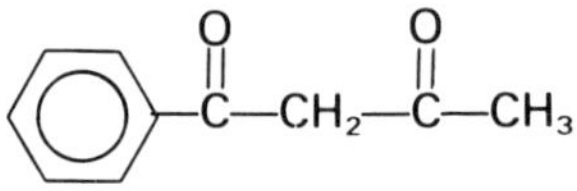

(**5.15**)

(**5.16**)

stabilization mechanism for this group of compounds can be explained by the possibility of the formation of tautomeric forms (**5.17** and **5.18**) by transfer of hydrogen to the oxygen atom in the carbonyl group and formation of a hydroxy group:

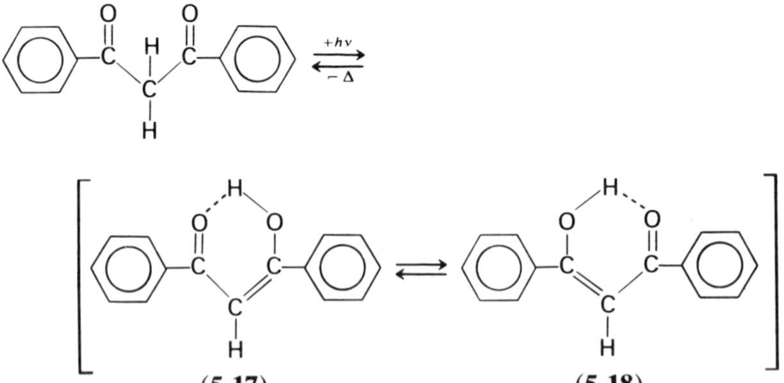

(**5.17**) (**5.18**)

(5.9)

5.5 BENZYLIDENE MALONATE

Benzylidene malonate (*p*-methoxybenzilidenemalonicdimethylester) (**5.19**) (absorption spectrum in Fig. 5.6) is commercially produced as Cyasorb UV 1988 (American Cyanamid).

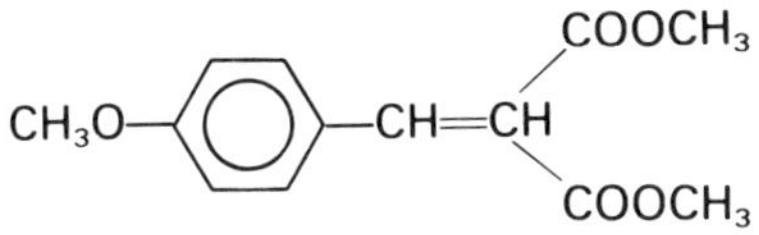

(**5.19**)

It has been found to stabilize poly(vinyl chloride) effectively; however it is photochemically destroyed under UV irradiation. The mechanism of these reactions is not yet well known.[518]

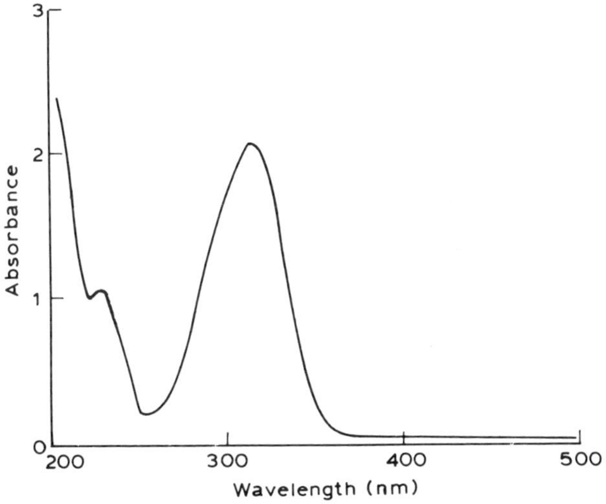

Fig. 5.6. Absorption spectrum of benzylidene malonate (*p*-methoxybenzilidenemalonicdimethylester) (**5.19**). Reprinted with permission from Ref. 518, © 1988 Pergamon Press PLC.

5.6 ESTERS OF α-CYANO-β-PHENYLCINNAMIC ACID

Esters of α-cyano-β-phenylcinnamic acid e.g. 2-cyano-3,3-diphenylacrylate (**5.20**) (absorption spectrum in Fig. 5.7) have been used for

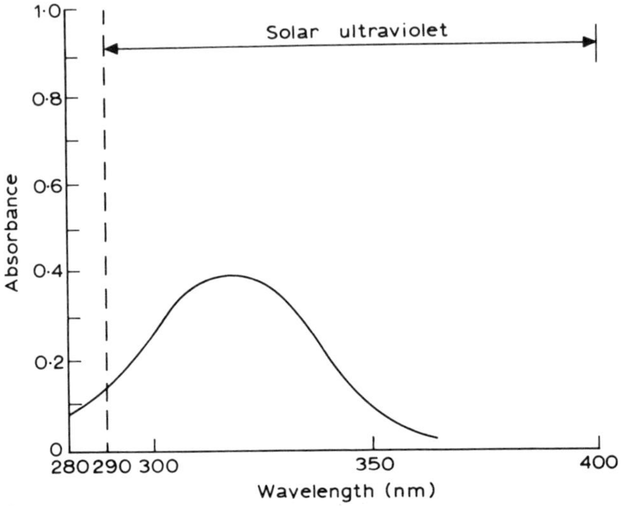

Fig. 5.7. Absorption spectrum of ethyl 2-cyano-3,3-diphenylacrylate (**5.20**) (6×10^{-4} M in CH_2Cl_2, 0·1 cm pathlength).[697]

photostabilization of polymers, which cannot be stabilized by compounds containing phenolic groups.[229,1081]

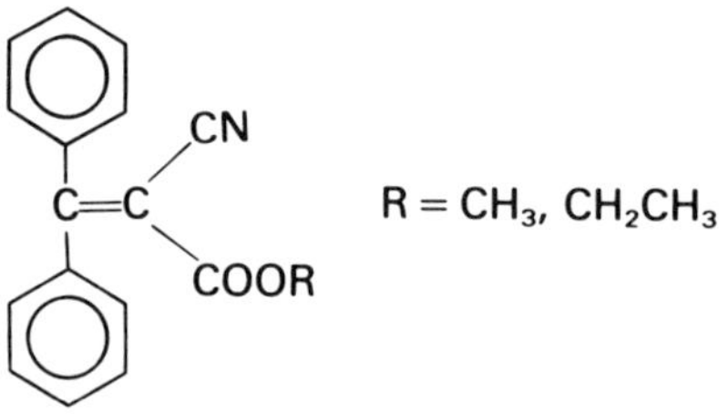

(5.20)

Methyl and ethyl esters of α-cyano-β-phenylcinnamic acid have been found to be very effective photostabilizers for certain polymers, especially for polyacetals, which are sensitive to high concentrations of phenol-containing antioxidants and photostabilizers. However, at 200°C, the temperature at which polyacetals are fabricated, they show significant volatility and losses in stabilizer concentration. Consequently, a number of attempts have been made to decrease the volatility of these photostabilizers by increasing their molecular weight, e.g. by preparing diesters[230] and polymers[1084] containing derivatives of α-cyano-β-phenylcinnamic acid (cf. Section 7.3.1).

5.7 HYDROXYBENZOPHENONES

The photochemical difference between the two isomers, *ortho*- and *para*-hydroxybenzophenone, is very distinct:

(a) *ortho*-hydroxybenzophenones are in most cases photochemically stable, and are employed for the photostabilization of a number of polymers:[66,91,430]
(b) *para*-hydroxybenzophenones are photoreduced by a mechanism similar to that of photoreduction of benzophenone and are used as photoinitiators.[919]

The *o*-hydroxybenzophenones are divided into two classes: 2-hydroxybenzophenones (**5.21**) and 2,2′-dihydroxybenzophenones (**5.22**):

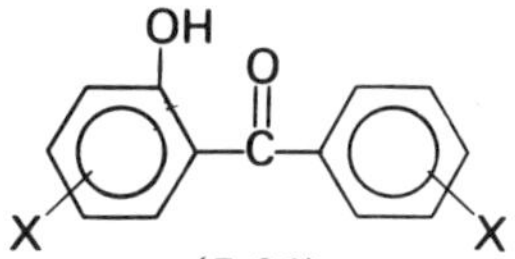

(**5.21**)

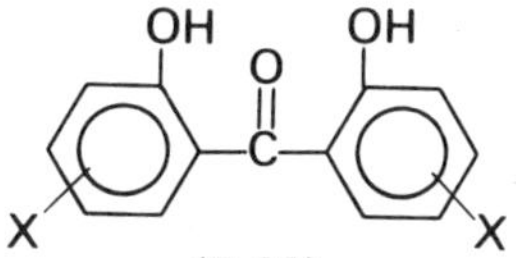

(**5.22**)

The 2-hydroxybenzophenones usually have a faint yellow colour and an ability to stabilize polymers to light below 370 nm. The 2,2′-dihydroxybenzophenones absorb light effectively up to 400 nm and also have a faint yellow colour. They photostabilize polymers to light below 400 nm.

Figure 5.8 shows the spectra of some hydroxybenzophenones. The *p*-hydroxybenzophenones exhibit only one strong band with the normal red shift in polar solvents. The *o*-hydroxybenzophenones, on the other hand, exhibit a double-headed band, of which the peak at the longer wavelength shows a blue shift in polar solvents.

The visual hypsochromic effect of the methoxy group in the *para* position to the carbonyl group is due to a reduction of the bandwidth of the very flat long-wavelength peak of the simple *o*-hydroxybenzophenone. The short-wavelength peak exhibits the normal bathochromic effect expected from the introduction of an auxochrome. These results indicate the formation of a strong hydrogen bond. Substitution in the *para* position by a methoxy group increases the strength of the hydrogen bond. The stable position of the proton is at the phenolic oxygen atom in the ground state, and at the carbonyl group in the first excited triplet state. This behaviour is very pronounced in the case of *o*-hydroxy *p*-methoxy-benzophenone.

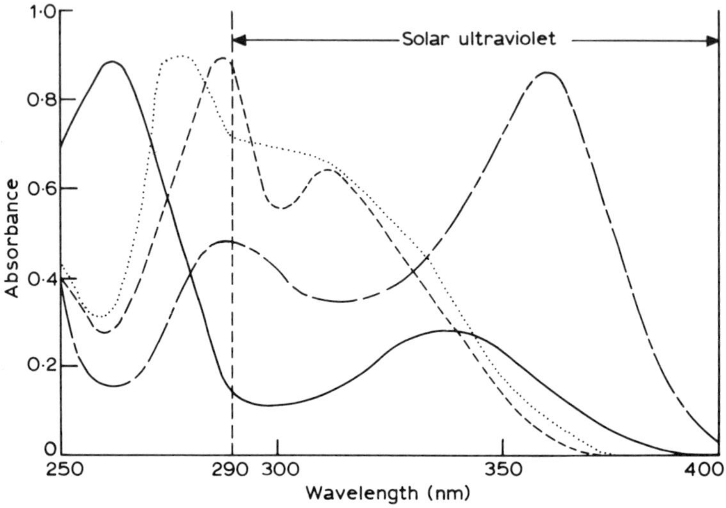

Fig. 5.8. Absorption spectra of: ——, 2-hydroxybenzophenone (**5.21**); ···, 2,4-dihydroxybenzophenone; — — —, 2-hydroxy-4-methoxybenzophenone (**5.28**); – – –, 2,2′-dihydroxy-4,4′-dimethoxybenzophenone, all at concentration of 6×10^{-4} M in CH_2Cl_2 (0·1 cm pathlength).[697]

The photostabilizing mechanism of *o*-hydroxybenzophenones has been the subject of many investigations.[66,167,515,541,542,581,653,695,785,855]

The most probable mechanism involved in the photostabilization is the change of energy of an absorbed photon to intramolecular proton transfer during which a 'quinone structure' is formed. This reaction may occur by one of four proposed cycles:

(i) The *o*-hydroxybenzophenone molecule in the ground state (S_0) absorbs a photon and is excited to the singlet (S_1) charge-transfer state:[539,653,785]

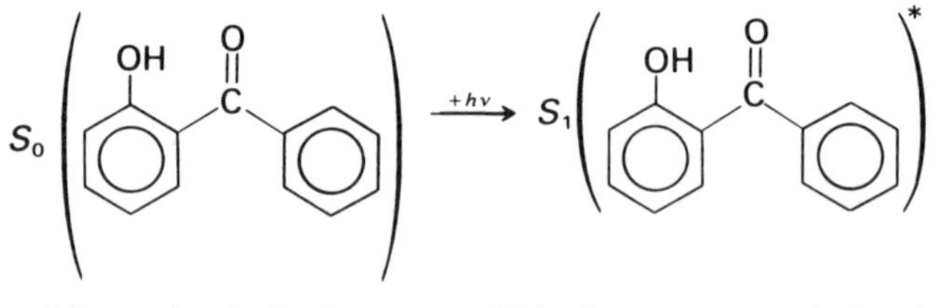

(5.10)

In this excited singlet state (S_1) the proton of the hydroxyl group is transferred to the carbonyl group, during which the 'quinone structure' (possibly of n,π^* character) is formed:

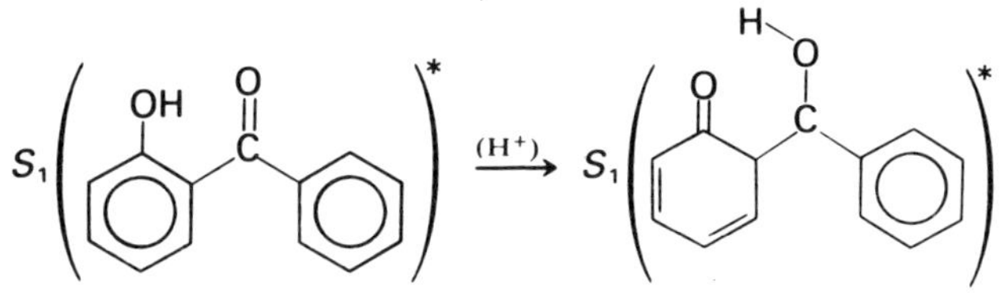

(5.11)

The excited singlet state (S_1) of 'quinone structure' passes by the intersystem crossing (ISC) process to the triplet state (T_1):

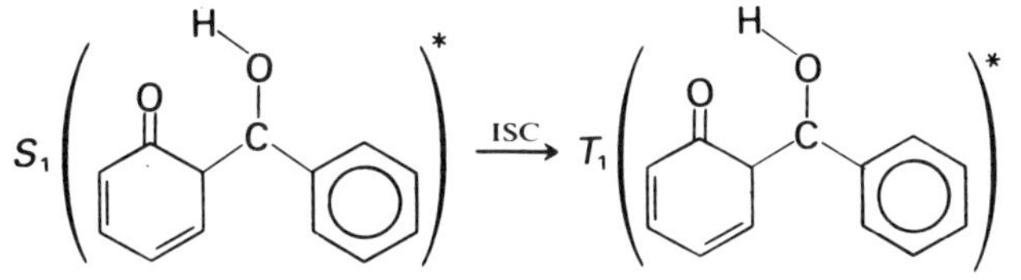

(5.12)

Since in the excited triplet state (T_1) the stable structure is the 'ketone structure', the proton tunnels back:

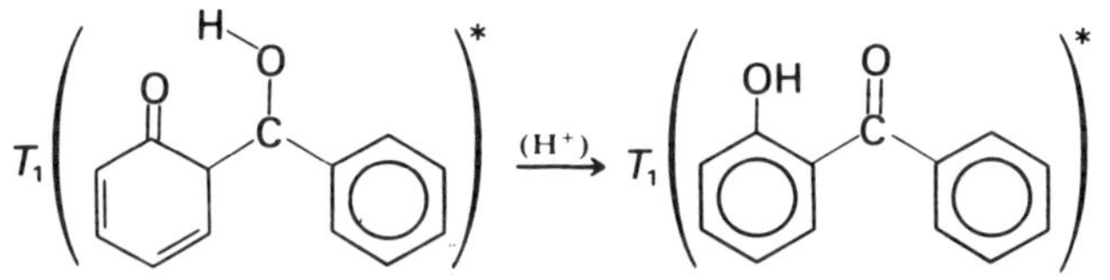

(5.13)

The ISC process leads the excited triplet state (T_1) of the 'ketone structure' directly to the ground state (S_0):

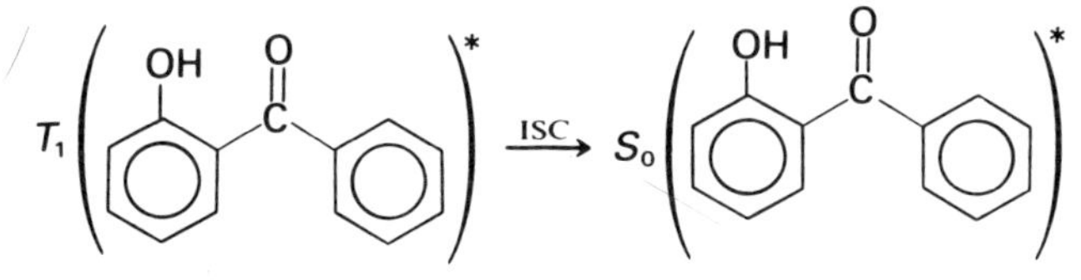

(5.14)

(ii) Another explanation postulates that the rotation of the hydroxyphenyl group in the excited singlet state (S_1) of the 'quinone structure' enhances the Franck–Condon factor between the excited singlet state (S_1) and the ground state (S_0), thus leading to a high rate for the internal conversion (IC) process to the ground state:

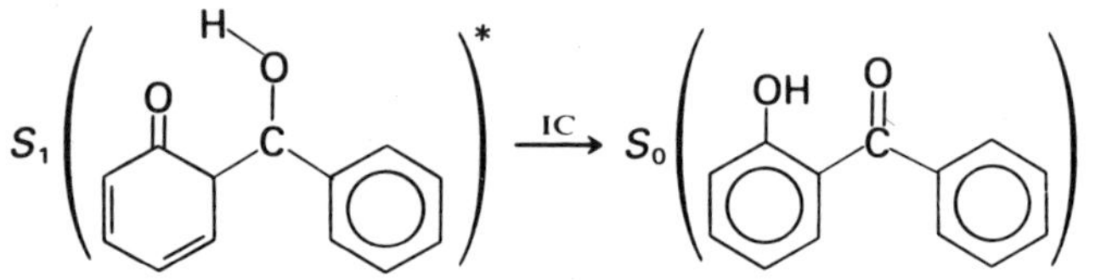

(5.15)

This scheme does not consider formation of the excited triplet state (T_1).

Both cycles (i) and (ii) are shown in the energy diagram in Fig. 5.9.

(iii) This cycle has been proposed on the basis of nanosecond flash spectroscopy:[515,785]

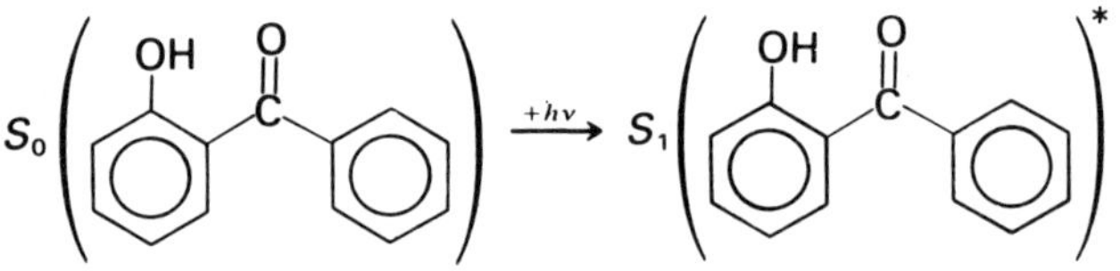

(5.16)

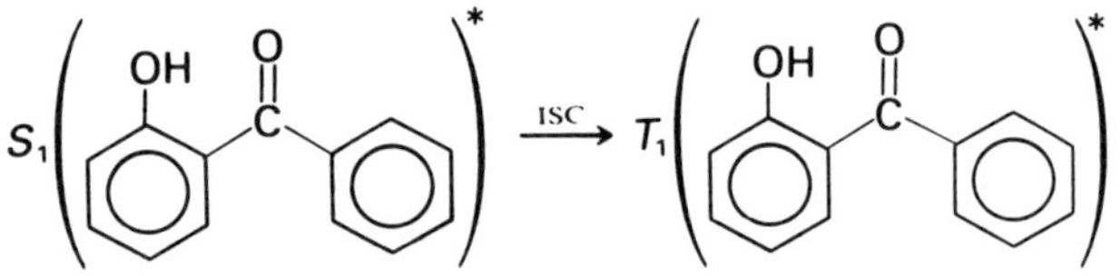

(5.17)

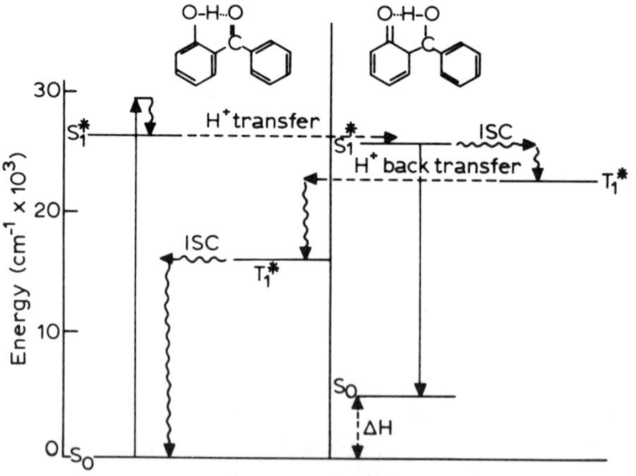

Fig. 5.9. Energy diagram of photophysical processes involved in the photostabilizing mechanism of *o*-hydroxybenzophenones.[653]

The lifetime of the excited triplet state (T_1) of *o*-hydroxybenzophenone in solution is very short (10^{-8} s):[695]

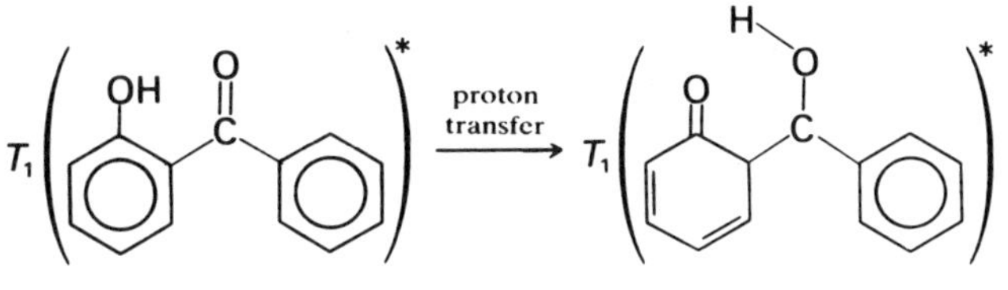

(5.18)

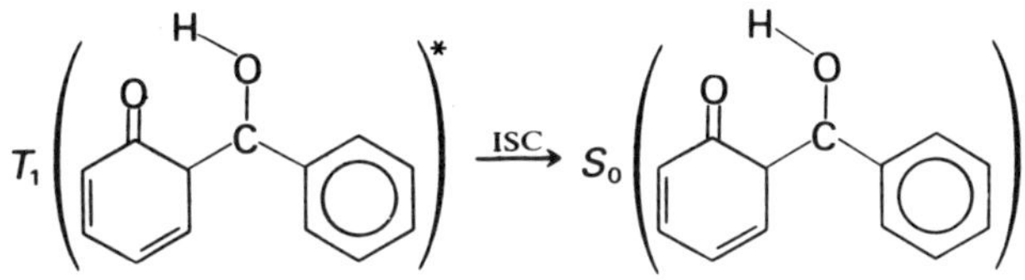

(5.19)

The lifetime of the 'quinone structure' in the ground state (S_0) is shorter than 10^{-9} s:[515]

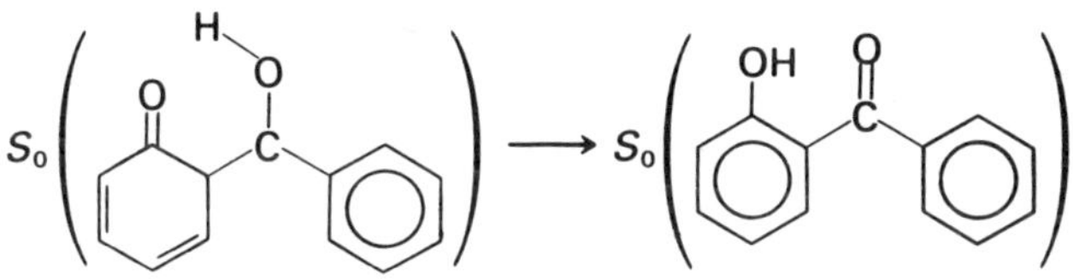

(5.20)

(iv) This cycle considers intramolecular proton transfer through excited singlet states, with the exception of the triplet state of benzophenone:[581,653]

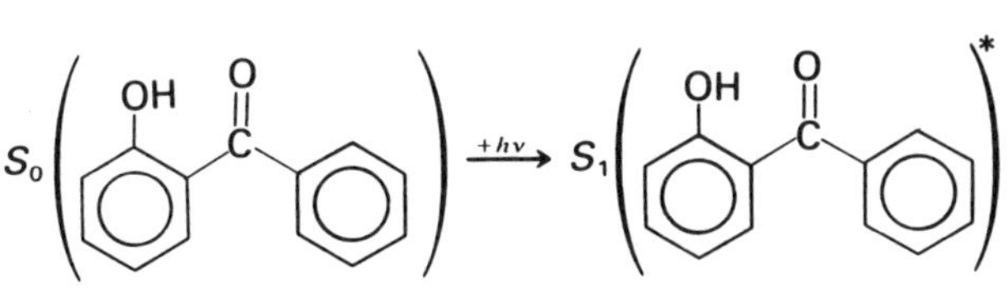

(5.21)

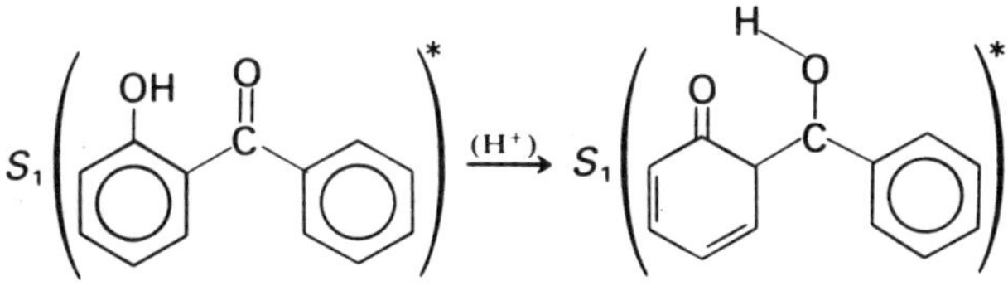

(5.22)

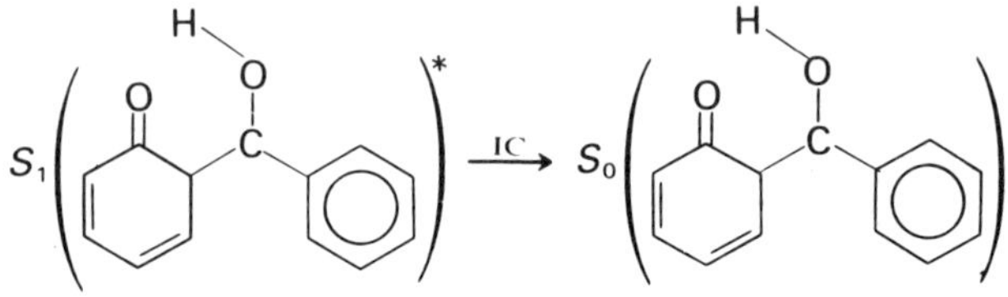

(5.23)

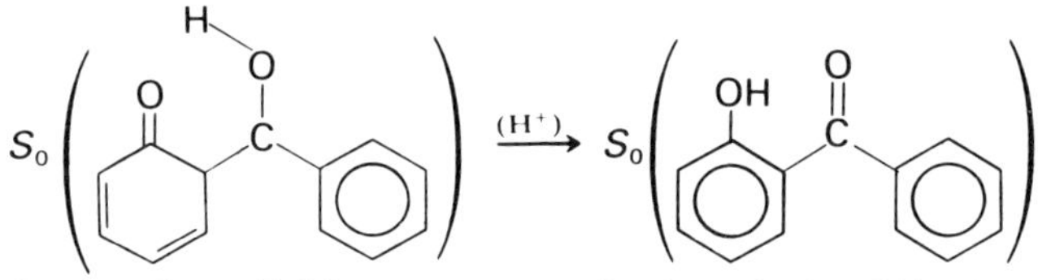

(5.24)

On the basis of available spectroscopic data it is difficult to decide which of these proposed cycles (i)–(iv) operate in intramolecular proton transfer.

It is doubtful whether intramolecular proton tranfer cycles alone can be responsible for the photostabilization mechanism. Substitution of the *ortho*-hydroxyl group by a urethane **(5.23)** or trimethylsilane **(5.24)** group does not decrease the photostabilizing activity of such substituted *o*-hydroxybenzophenones.[863] These compounds cannot

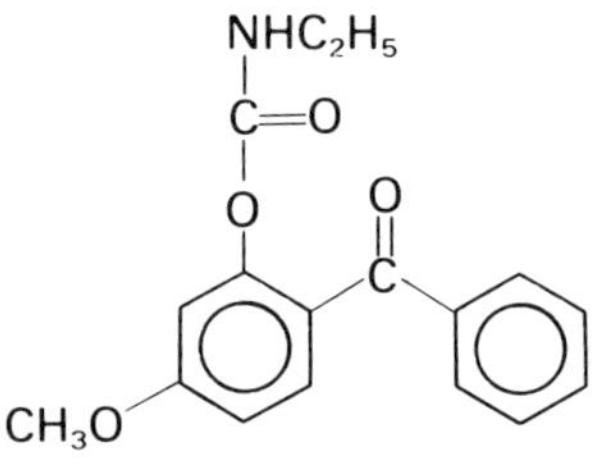

(5.23)

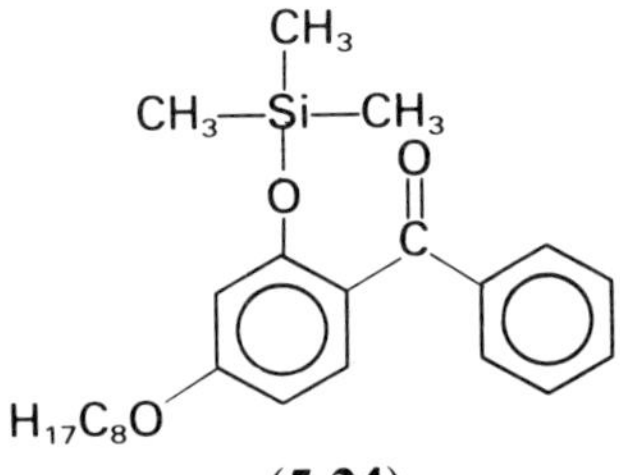

(5.24)

photostabilize by the proposed mechanism of intramolecular proton transfer.

Several other propositions have been made in order to explain the photostabilization properties of *o*-hydroxybenzophenones. These compounds can also act in the following ways.[91,571,1177]

(i) As light screeners.
(ii) To quench excited states of chromophoric groups (e.g. carbonyl groups) present in polymers by an energy-transfer mechanism.[72,209,441,511,570] Much of the evidence has originated from the observation of a reduction in the fluorescence and phosphorescence emission intensities and lifetimes of carbonyl groups in polymer systems.[72,895,1147]

Hydroxybenzophenones have very little quenching properties of singlet oxygen ($^1O_2$).[570]

(iii) As free-radical scavengers.[91,280,1126]

(iv) As hydroperoxy group (OOH) decomposers. It has been found that *o*-hydroxybenzophenones are photodecomposed more rapidly in processed polypropylene compared with unprocessed polymer (Fig. 5.10).[61,66] The rate of photodecomposition corre-

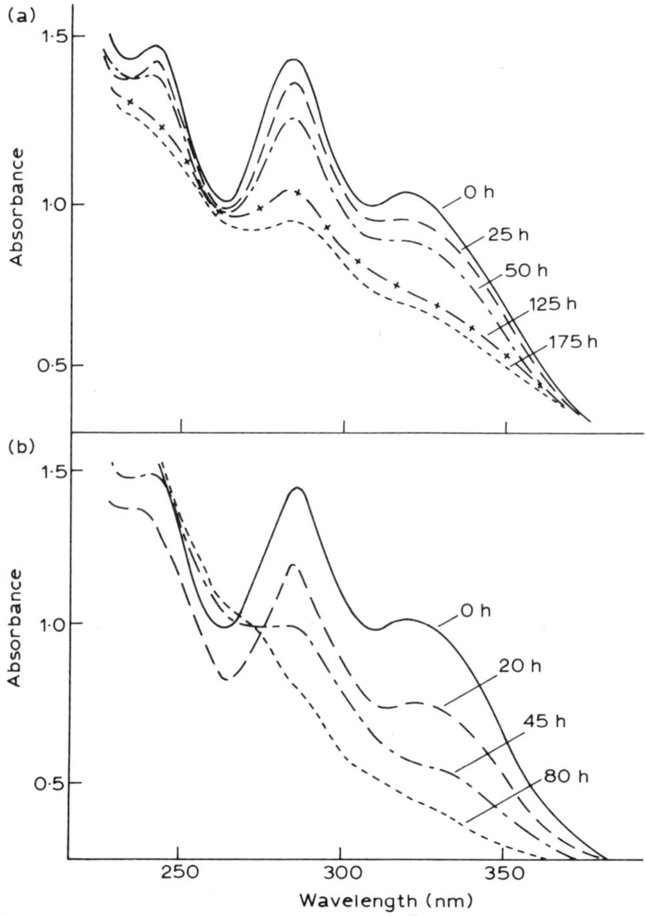

Fig. 5.10. Change in the absorption spectrum of 2-hydroxy-4-*n*-octoxybenzophenone (**5.30**; Cyasorb UV 531) in polypropylene film (300 μm thick) (a) before, and (b) after 10 min of processing and during irradiation in the Microscal unit.[61] ——, 100 mg litre^{-1}; — — —, 1000 mg litre^{-1}.

lates exactly with initial hydroperoxide concentrations in the polymer.[66] These results can be explained by the mechanism in which polymer oxy (PO•) and hydroxy (HO•) radicals produced in the photolysis of hydroperoxides abstract a hydrogen atom of the hydroxyl group in *o*-hydroxybenzophenone:[265]

$$\text{POOH} \xrightarrow{+h\nu} \text{PO}\bullet + \bullet\text{OH} \qquad (5.25)$$

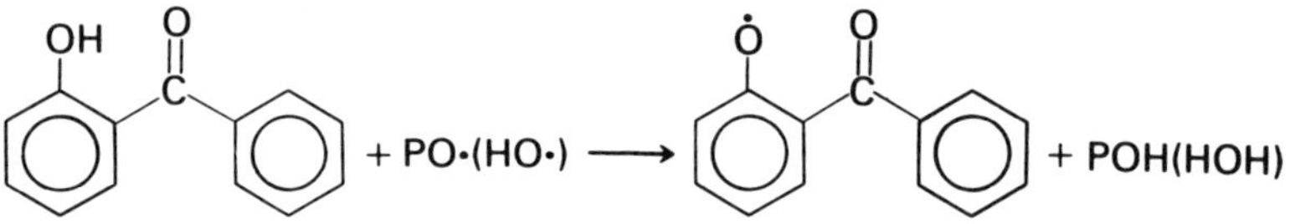

(5.25)

(5.26)

The oxy radical (**5.25**) is no longer capable of intramolecular hydrogen bonding.

It has been reported that 4-methoxy (**5.26**) and 4-*n*-octoxy (**5.27**) derivatives of 2-hydroxybenzophenone exhibit higher stabilization effects than pure 2-hydroxybenzophenones.[265] This effect has been explained by the fact that during UV radiation these groups may photolyse to give the corresponding alkyl radicals:

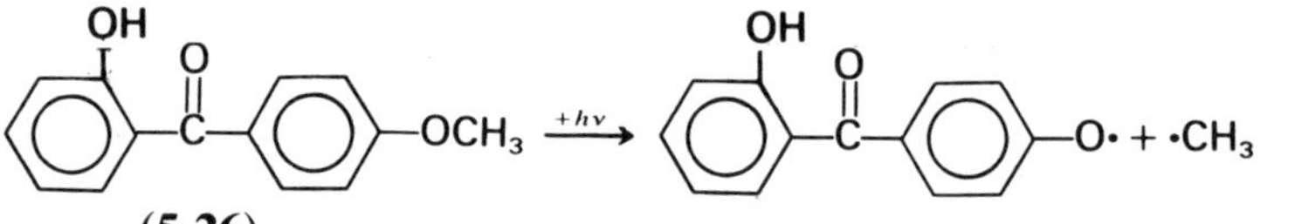

(5.27)

(5.26)

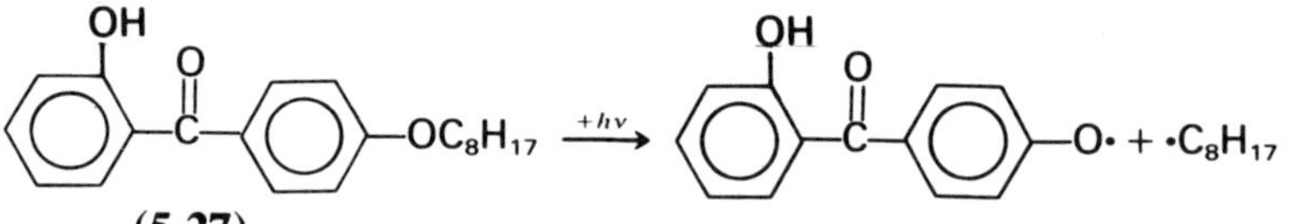

(5.28)

(5.27)

The smaller methyl radicals ($\bullet CH_3$) may move easily out of the polymer cage, whereas the larger octyl radicals ($\bullet C_8H_{17}$) may not. In the latter case radical recombination could therefore increase the light stabilities of the 4-methoxy and 5-*n*-octoxy derivatives of *o*-hydroxybenzophenones.

The list of commercially available *o*-hydroxybenzophenones is very extensive. The most important photostabilizers from this group have been collected in Table 5.5. Their transmission spectra are given in Figs. 5.11–5.15.

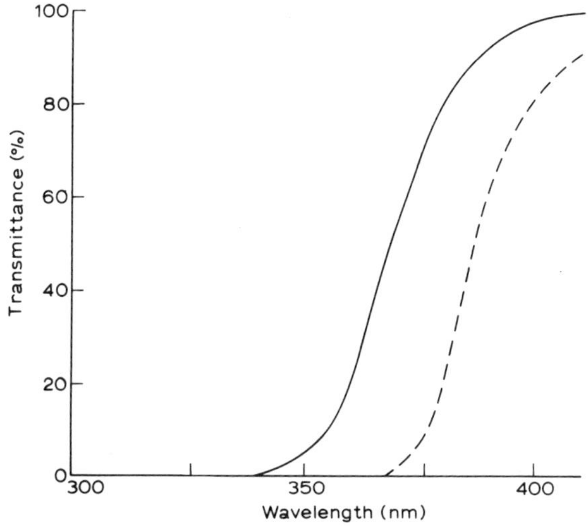

Fig. 5.11. Transmittance spectra of 2-hydroxy-4-*n*-dodecyloxybenzophenone (**5.31**; Chimassorb 125) in methanol.[1198]

In common industrial applications they are added at concentrations of 0·5–1·0 wt% (monohydroxybenzophenones) or 0·1–0·2 wt% (dihydroxybenzophenones).

Hydroxybenzophenones have been widely investigated as photostabilizers for polyethylene,[91,878] polypropylene,[66,90,151,510,1126]

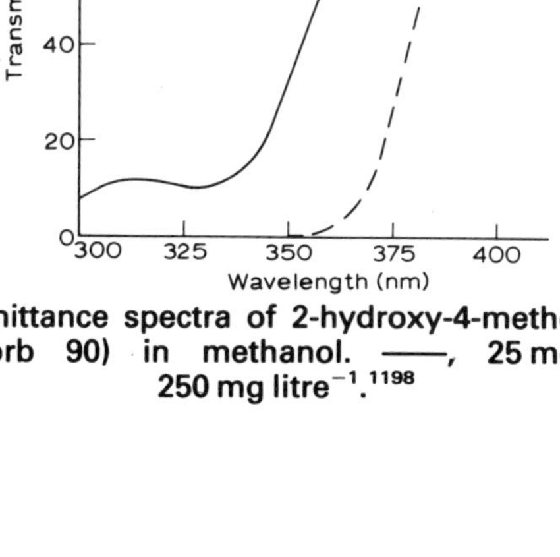

Fig. 5.12. Transmittance spectra of 2-hydroxy-4-methoxybenzophenone (**5.28**; Chimassorb 90) in methanol. ——, 25 mg litre^{-1}; – – –, 250 mg litre^{-1}.[1198]

Table 5.5
Examples of commercially produced hydroxybenzophenone photostabilizers: physical properties and solubility

Chemical name	*Commercial name*	*Chemical structure*	*Molecular weight*
2-Hydroxy-4-methoxybenzophenone	Cyasorb UV 9 (Cyanamid); Chimassorb 90 (Ciba–Geigy)	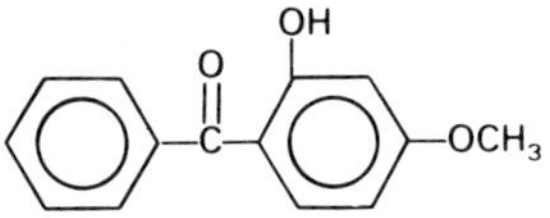 **(5.28)**	228
2,2′-Dihydroxy-4-methoxybenzophenone	Cyasorb UV 24 (Cyanamid)	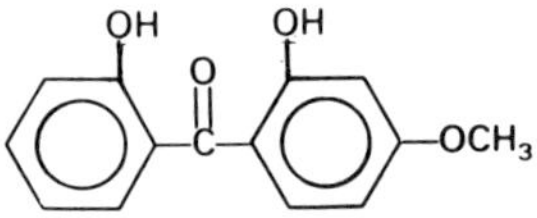 **(5.29)**	244·2
2-Hydroxy-4-*n*-octoxybenzophenone	Cyasorb UV 531 (Cyanamid); Chimassorb 81 (Ciba–Geigy)	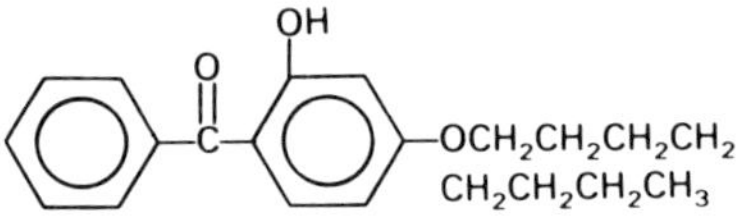 **(5.30)**	326·1
2-Hydroxy-4-*n*-dodecyloxybenzophenone	Chimassorb 125 (Ciba–Geigy)	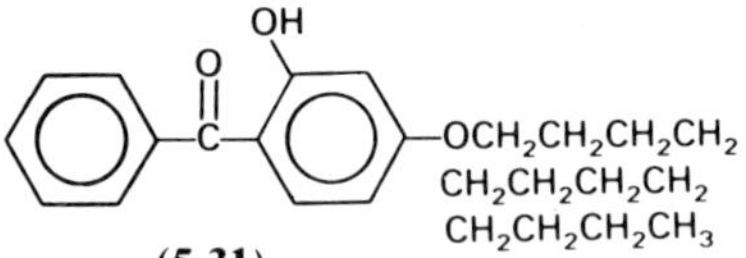 **(5.31)**	367·3
Poly[4-(2-acryloxyethoxy)-2-hydroxybenzophenone]	Cyasorb UV 2126 (Cyanamid)	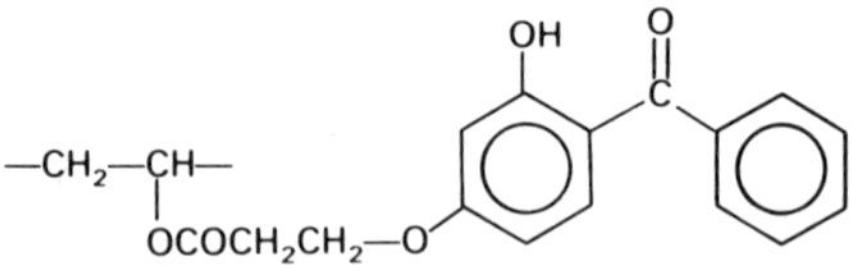 **(5.32)**	50 000 (approx.)

Appearance	*Solubility (wt %)*								*Melting point (°C)*	*Recommended applications*
	Benzene	*n-Hexane*	*95% Ethanol*	*Methyl ethyl ketone*	*Methylene chloride*	*Carbon tetrachloride*	*Di(2-ethylhexyl) phthalate*	*Water*		
Pale cream to white powder	56·2	4·3	5·8	—	59·5	34·5	18·7	<0·01	62	Flexible and rigid poly(vinyl chloride) Unsaturated polyesters Polystyrene Polyacrylics
Yellow powder	46·6	2·3	21·4	55·3	—	22·2	31·1	<0·01	68	Polyurethane coatings Alkyd coatings Phenolic and other oil-based systems
Light yellow powder	72·7	40·1	2·6	74·3	69·8	—	20·5	<0·01	47	Polyolefins Poly(vinyl chloride) Co(ethylene-vinyl acetate)
Pale yellow powder	80	70	5	80	—	—	—	<0·01	50–51	Polyolefins
Light yellow powder	30–40	0·1	0·1	50	—	—	—			Polymers where low extraction or migration is desired In automotive, greenhouse, home construction and solar applications

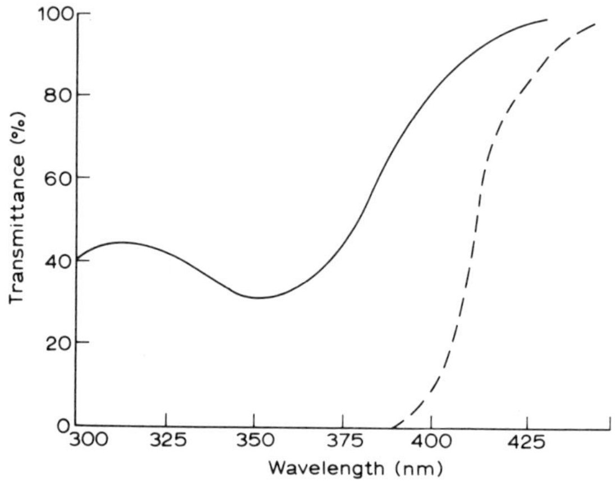

Fig. 5.13. Transmittance spectra of 2,2′-dihydroxy-4-methoxybenzophenone (**5.29**; Cyasorb UV 24) in methanol. ——, 10 mg litre^{-1}; – – –, 100 mg litre^{-1}.[1198]

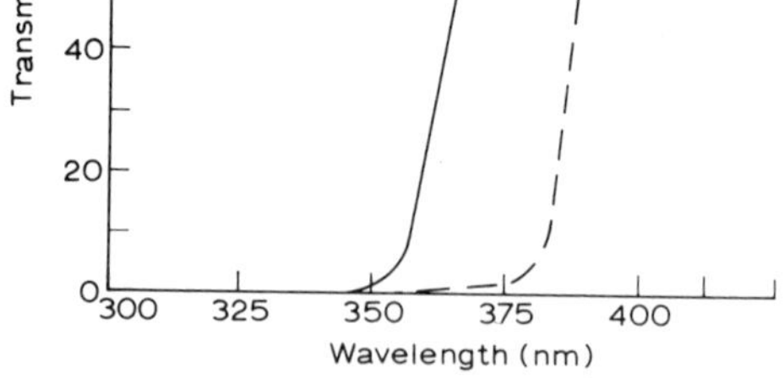

Fig. 5.14. Transmittance spectra of 2-hydroxy-4-*n*-octoxybenzophenone (**5.30**; Chimassorb 81) in methanol. ——, 100 mg litre^{-1}; – – –, 1000 mg litre^{-1}.[1198]

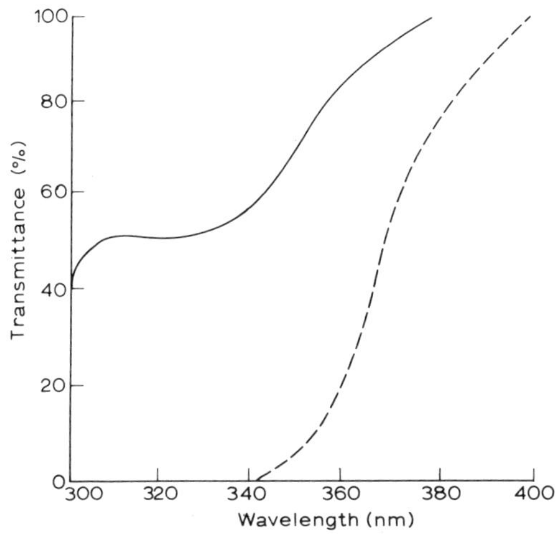

Fig. 5.15. Transmittance spectra of poly[4-(2-acryloxyethoxy)-2-hydroxybenzophenone] (**5.32**; Cyasorb UV 2126) in dimethylacetamide. ——, 10 mg litre^{-1}; — — —, 100 mg litre^{-1}.[1198]

polystyrene,[441,896,1048] poly(methyl methacrylate),[1058] poly(vinyl chloride),[464,763] polyesters,[313] unsaturated crosslinked polyesters,[613] polyurethanes,[1082] melamine–formaldehyde resins,[985] cellulose[985] and wool.[259,962] Specially sulphonated 2-hydroxybenzophenones (**5.33**) and 2,2′-dihydroxybenzophenones are useful photostabilizers for retarding wool photodegradation.[791,792]

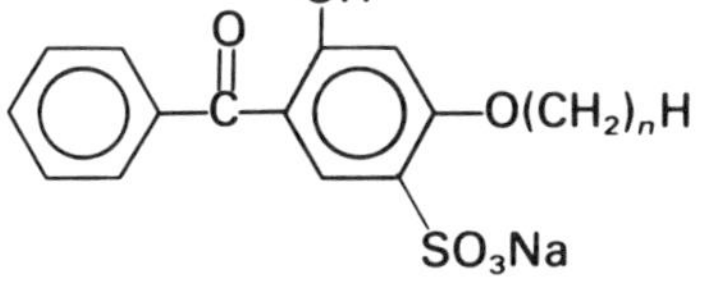

(**5.33**)

In Table 5.6 examples of times to embrittlement are shown for polypropylene films stabilized by various hydroxybenzophenones.[66]

In low-density polyethylene, *o*-hydroxybenzophenones operate by migration to the surface of the polymer, where both screening and radical scavenging become the dominant modes of photoprotection.[91]

o-Hydroxybenzophenones exhibit synergistic effects with hindered amine light stabilizers (HALS; cf. Section 6.1) and phenols.[597]

Table 5.6
Time to embrittlement (0·06 carbonyl units) for polypropylene films containing various hydroxybenzophenones exposed in a Microscal Unit [(Weatherometer Microscal Ltd, UK), employing a 500 W high-pressure fluorescent lamp (50°C, relative humidity, ambient)][66]

Additive	*Embrittlement time (h)*
2,2′-Dihydroxybenzophenone	240
2-Hydroxybenzophenone	175
2-Hydroxy-4′-*tert.*-butyl-4-methoxybenzophenone	330
2,2′-Dihydroxy-4-methylbenzophenone	550
2,2′-Dihydroxy-4,4′-dimethoxybenzophenone	840
2,2′-Dihydroxy-4-*n*-octoxybenzophenone	2 000
2-Hydroxy-4,4′-di-*n*-octoxybenzophenone	1 800

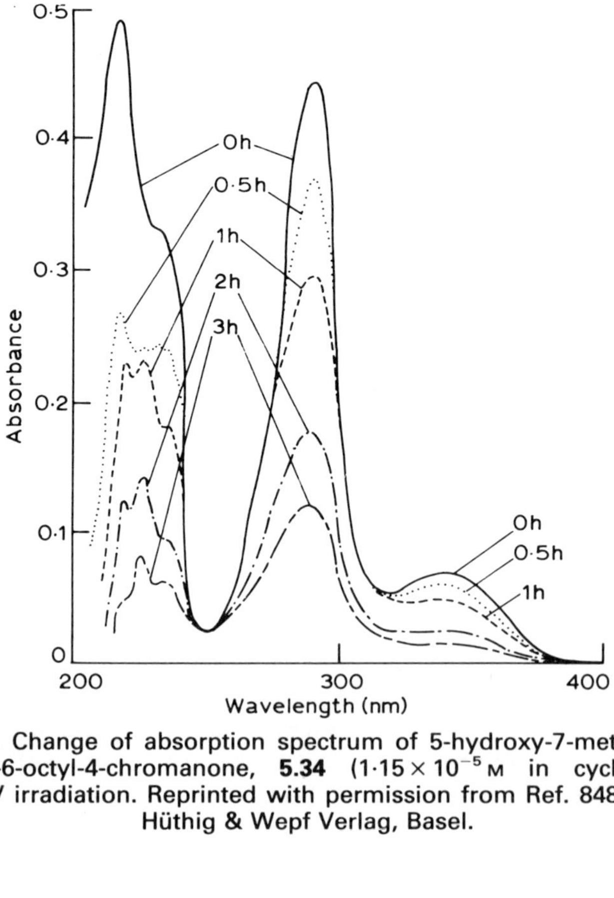

Fig. 5.16. Change of absorption spectrum of 5-hydroxy-7-methoxy-2,2-dimethyl-6-octyl-4-chromanone, **5.34** ($1{\cdot}15 \times 10^{-5}$ M in cyclohexane) under UV irradiation. Reprinted with permission from Ref. 848, © 1983 Hüthig & Wepf Verlag, Basel.

5.8 CHROMANONE AND TETRALONE PHOTOSTABILIZERS

Compounds with 4-chromanone (**5.34**) and 1-tetralone (**5.35**) structures have been found to be useful as photostabilizers for poly(vinyl chloride)[848–850] The UV absorption spectrum of 5-hydroxy-7-methoxy-2,2-dimethyl-6-octyl-4-chromanone is shown in Fig. 5.16. This photostabilizer is photolysed after prolonged UV irradiation.

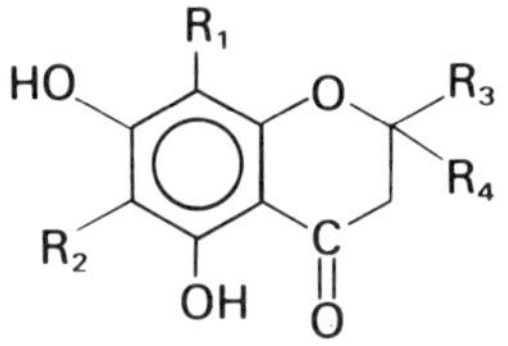

(**5.34**)

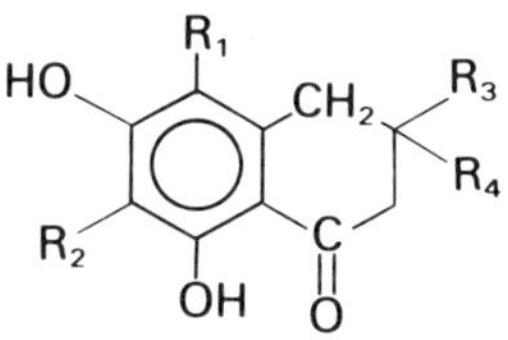

(**5.35**)

5.9 HYDROXYPHENYLBENZOTRIAZOLES

Absorption spectra (Fig. 5.17) measurements indicate that *o*-hydroxybenzotriazoles (**5.36**) are in equilibrium between two or more conformers in the ground state (S_0). Two conformers absorbing at 302 and 340 nm may be stabilized by a combination of intra- and intermolecular hydrogen bonding:[329]

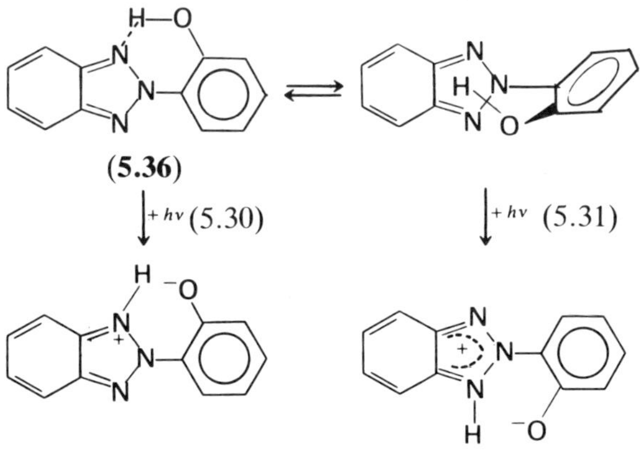

(5.29)

(**5.36**)

$+h\nu$ (5.30)

$+h\nu$ (5.31)

The photostabilizing mechanism of *o*-hydroxyphenylbenzotriazoles has been the subject of many investigations.[34,187,411,516,519,541,582,583,680,875,1146,1162] The most probable mechanism involved in photostabilization is the exchange of energy of an absorbed photon to

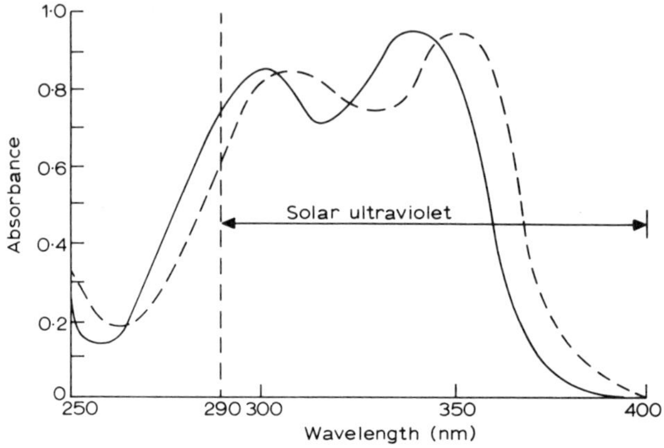

Fig. 5.17. Absorption spectra of: ——, 2-(2-hydroxy-5-methylphenyl)-2*H*-benzotriazole (Tinuvin P; **5.43**); — —, 2-(2-hydroxy-3,5-di-*tert.*-butylphenyl)-5-chloro-2*H*-benzotriazole (Tinuvin 327) (**5.46**); both at a concentration of 5×10^{-4} M in CH_2Cl_2; 0·1 cm pathlength.[697]

intramolecular proton transfer, during which a 'quinone structure' is formed (a similar mechanism to that described for *o*-hydroxybenzophenones, cf. Section 5.7).

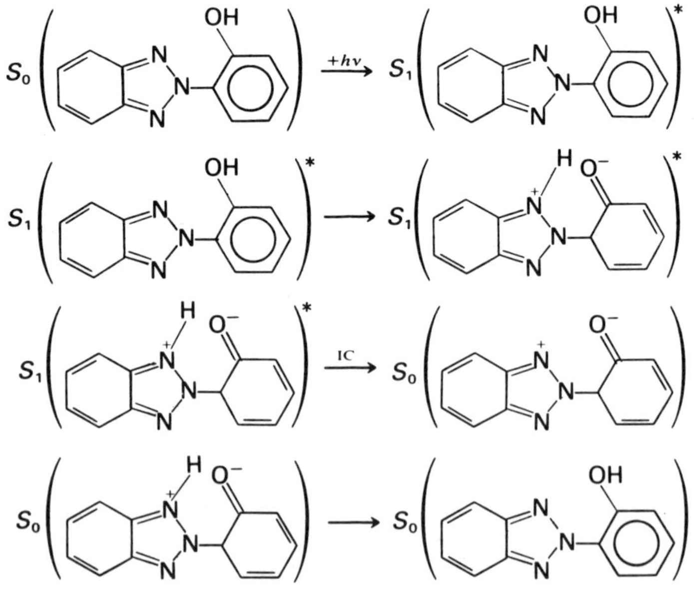

(5.32)

(5.33)

(5.34)

(5.35)

The lifetime of the 'quinone structure' in the ground state (S_0) is less than 1·0 ns.[515] Rotation of the hydroxyphenyl group around the central C–N bond may contribute to a rapid radiationless deactivation of the singlet excited state.[541]

The 2-(2′-hydroxy-5′-methylphenyl)-2H-benzotriazole (Tinuvin P; **5.43**) also shows phosphoresence in ethanol at 77 K and in ethanol/ether/pyridine at 90 K (with a phosphorescence decay time in the range of seconds).[1148]

It has also been proposed that *o*-hydroxyphenylbenzotriazoles act by the screening effect (light-absorbing effect in a spectral region where the absorption spectra of polymer and photostabilizer overlap) and by an energy transfer from the excited polymer triplet state (T_1) to the stabilizer triplet state (T_1).[1148]

It is doubtful whether the intramolecular proton-transfer cycle can be responsible for the photostabilization mechanism of polymers by *o*-hydroxyphenylbenzotriazoles. Substitution of the *ortho*-hydroxyl group with a urethane (**5.37**) or trimethylsilane (**5.38**) group does not decrease the photostabilizing activity of such substituted *o*-hydroxyphenylbenzotriazoles.[863]

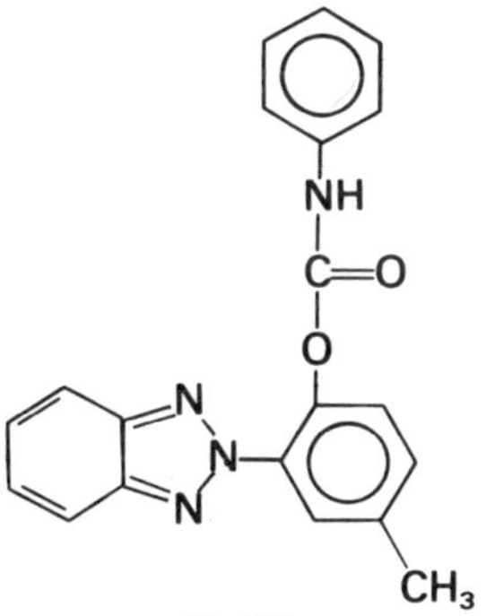

(5.37)

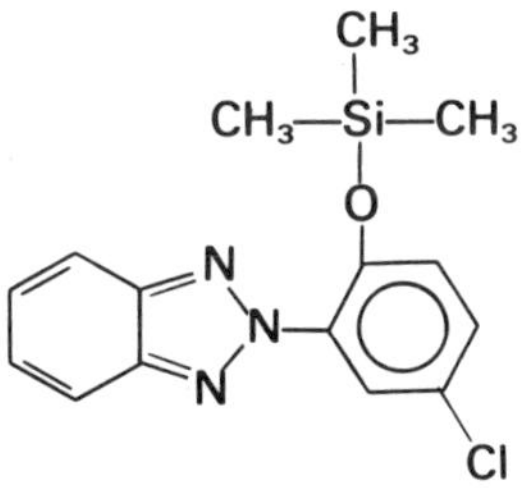

(5.38)

Several other propositions have been made in order to explain photostabilization properties of *o*-hydroxyphenylbenzotriazoles. These compounds can also[441,511,1147]

(i) act as light screeners, and
(ii) quench excited states of chromophoric groups (e.g. carbonyl groups) present in polymers by an energy-transfer mechanism.

It has been reported that processing has a marked effect on the

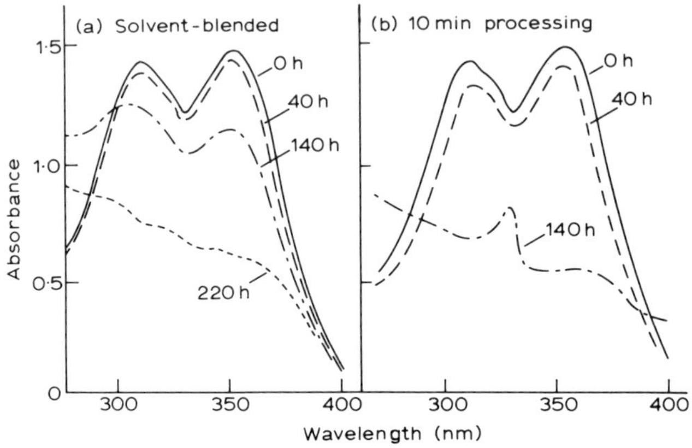

Fig. 5.18. Change in the absorption spectrum of 2-(2-hydroxy-3,5-di-*tert.*-butylphenyl)-5-chloro-2*H*-benzotriazole (Tinuvin 327; **5.46**) in polypropylene film (300 μm thick), (a) before, and (b) after 10 min of processing and during irradiation in the Microscal unit.[26] The numbers on the curves are the times (h) of irradiation.

photodecomposition of *o*-hydroxyphenylbenzotriazoles (Fig. 5.18) which can be due to the reaction of polymer oxy (PO•) and hydroxy (HO•) radicals formed from the thermal decomposition of polymer hydroperoxide groups (P—OOH). Both radicals may abstract hydrogen atoms from the hydroxyl group in *o*-hydroxyphenylbenzotriazoles:[61]

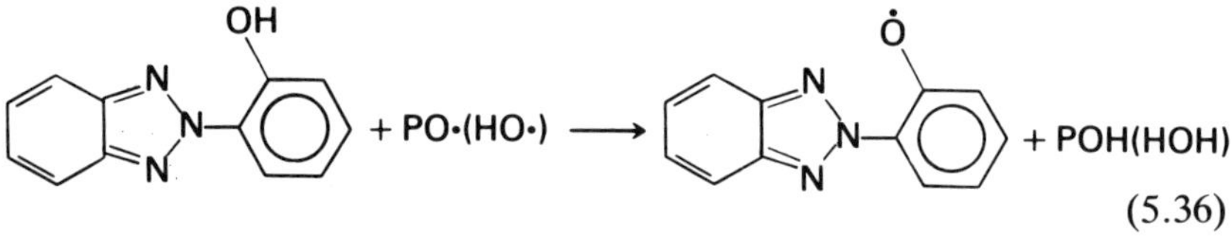

(5.36)

This reaction has also been confirmed by reaction of *o*-hydroxyphenylbenzotriazole with low-molecular-weight peroxy radicals in the dark.

The acetate ester of 2-(2-hydroxy-5-methylphenyl)-2H-benzotriazole (**5.39**), upon UV irradiation in methylene chloride or in poly(methyl methacrylate) films, undergoes photocleavage in which 2-(2-hydroxy-5-methylphenyl)benzotriazole (**5.40**) is regenerated. No photo-Fries rearrangement was observed:[515,715]

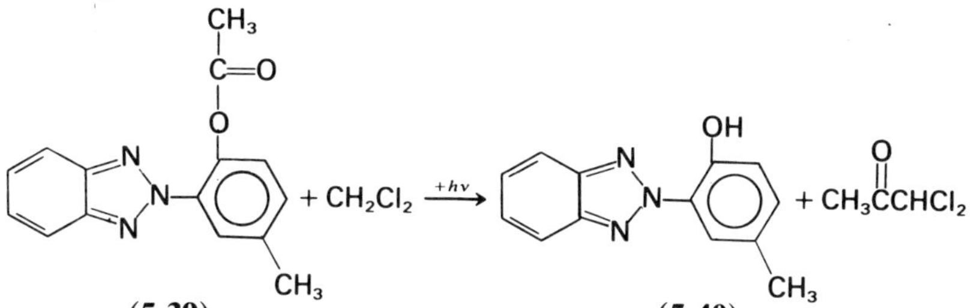

(5.39) **(5.40)** (5.37)

The 2-(2-hydroxy-5-*tert.*-butylphenyl)benzotriazole-5-carbonic acid anilide (**5.41**) has been found to be an effective photostabilizer for polyamides.[680]

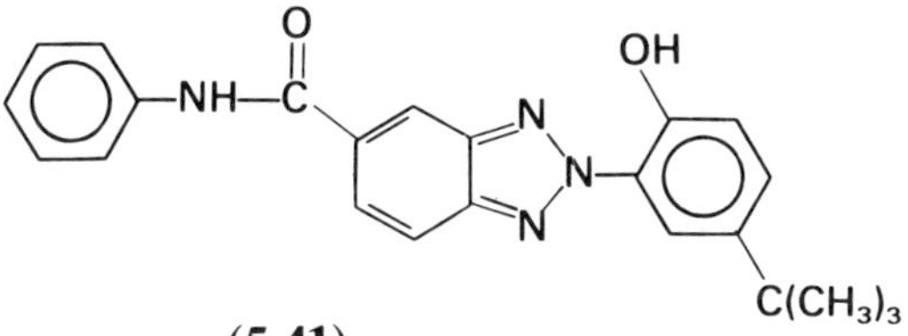

(**5.41**)

Benzenesulphonate esters of hydroxyphenylbenzotriazole (**5.42**) were used in UV-curable coating formulations.[861,862] During UV exposure, they undergo a photo-Fries rearrangement to form the corresponding *o*-benzenesulphonyl phenol derivatives.

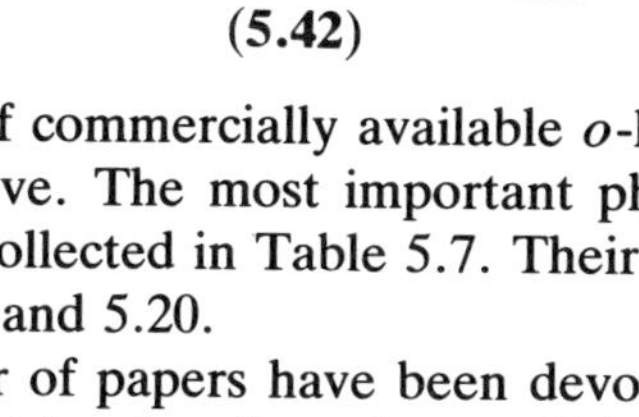

$R = CH_3, C_8H_{17}$

(**5.42**)

The list of commercially available *o*-hydroxyphenylbenzotriazoles is very extensive. The most important photostabilizers from this group have been collected in Table 5.7. Their transmission spectra are given in Figs 5.19 and 5.20.

A number of papers have been devoted to the study of *o*-hydroxyphenylbenzotriazoles for photostabilization of polypropylene,[151,510]

Table 5.7
Examples of commercially produced hydroxy-benzotriazole type photostabilizers physical properties and solubility

Chemical name	*Commercial name*	*Chemical structure*	*Molecular weight*
2-(2-Hydroxy-5-methylphenyl)-2*H*-benzotriazole	Tinuvin P (Ciba–Geigy)	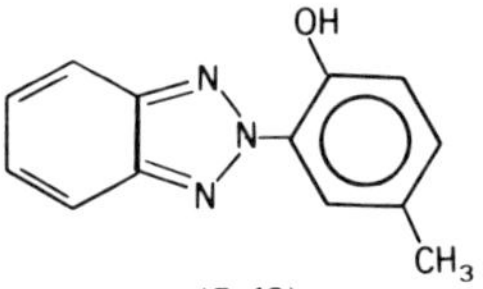 **(5.43)**	225·2
2-(2-Hydroxy-3,5-di-*tert.*-butylphenyl)-2*H*-benzotriazole	Tinuvin 320 (Ciba–Geigy)	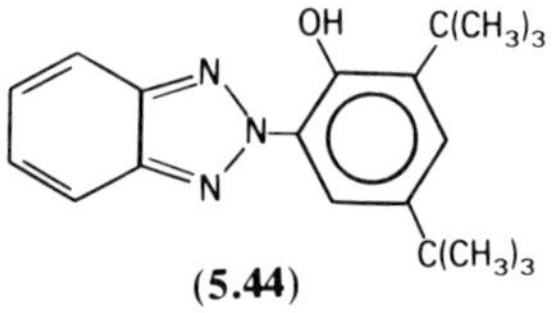 **(5.44)**	323·4
2-(2-Hydroxy-3-*tert.*-butyl-5-methylphenyl)-2*H*-benzotriazole	Tinuvin 326 (Ciba–Geigy)	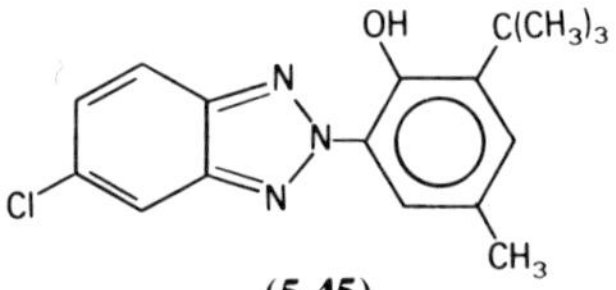 **(5.45)**	301·6
2-(2-Hydroxy-3,5-di-*tert.*-butylphenyl)-5-chloro-2*H*-benzotriazole	Tinuvin 327 (Ciba–Geigy)	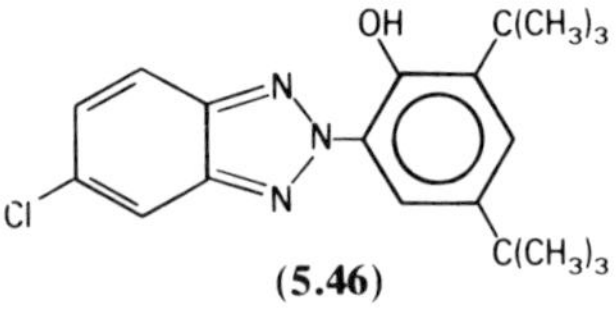 **(5.46)**	358
2-(2-Hydroxy-3,5-di-*tert.*-amylphenyl)-2*H*-benzotriazole	Tinuvin 328 (Ciba–Geigy)	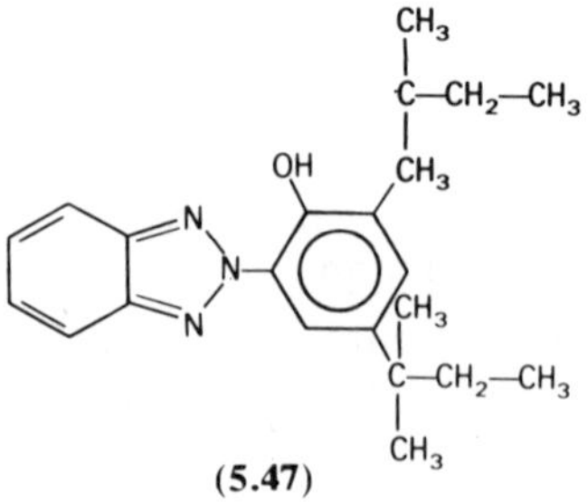 **(5.47)**	351

Appearance	Solubility (wt %)														Melting point (°C)	Recommended applications
	Benzene	*Toluene*	*Styrene*	*Xylene*	*Chloroform*	*Ethyl acetate*	*Acetone*	*Methyl ethyl ketone*	*Methanol*	*Ethanol*	*Methylene chloride*	*n-Hexane*	*Cyclohexane*	*Water*		
Slightly yellowish crystalline powder	6·9	6·0	7·2	—	—	3·5	2·5	4	0·2	0·3	—	—	0·7	<0·01	128–132	Polystyrenes Poly(vinyl chloride) Polyesters Polyacrylics Polyurethanes Polycarbonates Cellulose esters
Yellowish crystalline powder	19·5	18·3	17·4	—	—	6·5	2·9	6·0	0·2	0·3	—	—	11·3	<0·01	152–154	Rigid PVC Polycarbonates Polyurethanes Polyamides Synthetic fibres Lacquers
Slightly yellow crystalline powder	9·2	9·5	12·6	—	—	2·5	1·1	2·9	0·1	0·1	—	—	4·1	<0·01	140–141	Polyolefins Polyesters lacquers and coatings
Light yellow crystalline powder	16	16	—	—	19	5	1	—	0·1	0·1	17	4	7	<0·01	154–158	Polyolefins Poly(vinyl chloride) Unsaturated polyesters ABS Polyamides Chlorinated rubbers
Yellowish crystalline powder	39	33	29	—	—	16	6	14	0·4	0·7	—	—	26	0·01	79–87	Polyolefins Polyurethanes Coatings

(continued)

Table 5.7—*contd.*

Chemical name	*Commercial name*	*Chemical structure*	*Molecular weight*
2-[2-Hydroxy-3,5-di(1,1-dimethylbenzyl)-phenyl]-2*H*-benzotriazole	Tinuvin 900 (Ciba–Geigy)	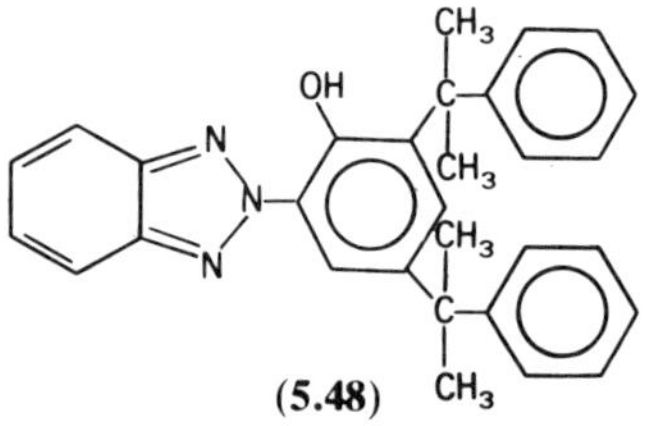 (5.48)	455
2-(2-Hydroxy-5-*tert.*-octylphenyl)-2*H*-benzotriazole	Cyasorb UV 5411 (Cyanamid)	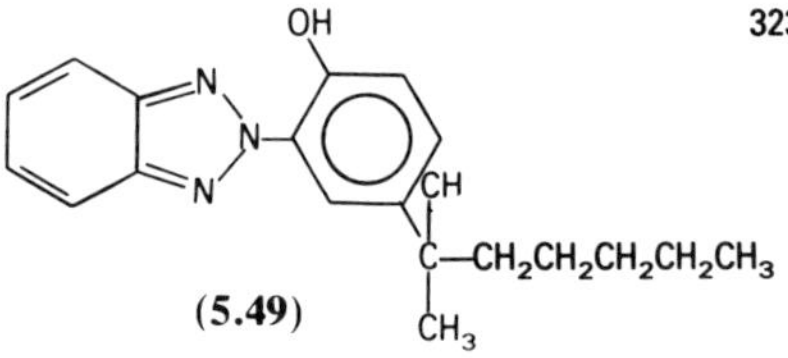 (5.49)	323
CAS name: Poly(oxy-1,2-ethanediyl), -(3-(3-(2*H*-benzotriazol-2-yl)-5(1,1-dimethylethyl)-4-oxopropyl)-α-hydroxy); *CAS number* 104810-48-2	Tinuvin 1130 Component 1 (Ciba–Geigy)	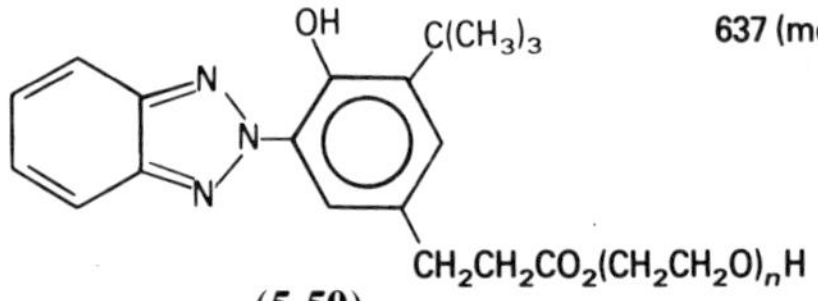 (5.50)	637 (mol.)
CAS name: Poly(oxy-1,2-ethanediyl), -(3-(3-(2*H*-benzotriazol-2-yl)-5-(1,1-dimethylethyl)-4-hydroxyphenyl)-1-oxopropyl)-(3-(3-(2*H*-benzotriazol-2-yl)-5-(1,1-dimethylethyl)-4-hydroxyphenyl)-1-oxopropoxy); *CAS number* 104810-47-1	Tinuvin 1130 Component 2 (Ciba–Geigy)	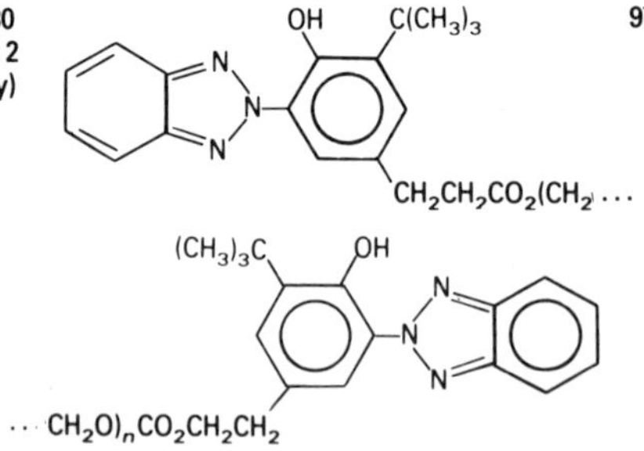 (5.51)	975 (mol.)

Appearance	*Solubility (wt %)*														*Melting point (°C)*	*Recommended applications*
	Benzene	*Toluene*	*Styrene*	*Xylene*	*Chloroform*	*Ethyl acetate*	*Acetone*	*Methyl ethyl ketone*	*Methanol*	*Ethanol*	*Methylene chloride*	*n-Hexane*	*Cyclohexane*	*Water*		
Light yellow to yellow powder	—	—	—	12	—	7	—	1	—	—	—	—	—	—	135–143	Thermoplastic coatings Extrusion coatings
lear-white free-flowing powder	39·4	—	33·2	—	—	21·4	—	—	—	—	—	—	—	<0·01	101–105	Polypropylene Poly(vinyl chloride) Polycarbonates Polyacrylates Polystyrenes Polyurethanes Coatings
lear yellow to light amber liquid	100	—	—	—	—	>50	—	100	—	—	—	—	—	<0·01	—	Coatings, especially in water-borne resin systems

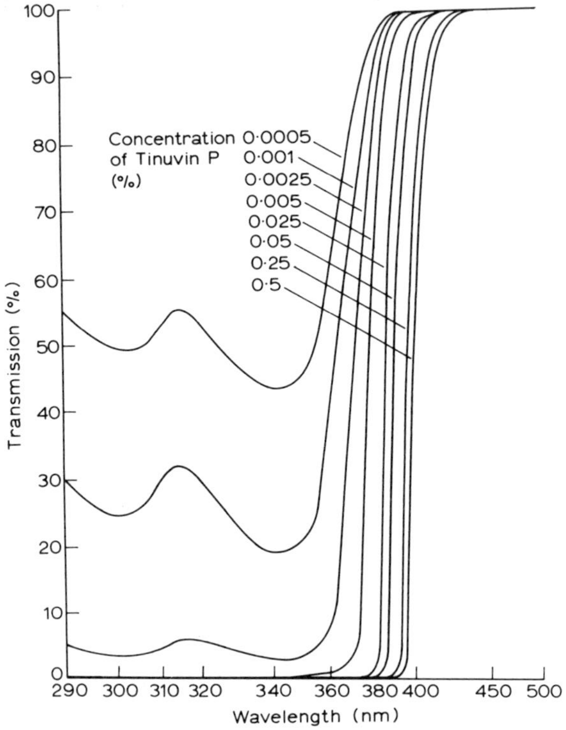

Fig. 5.19. Transmittance spectra of 2-(2-hydroxy-5-methylphenyl)-2*H*-benzotriazole (Tinuvin P; **5.43**).

polystyrene,[441,896] polyacrylates and poly(methyl metacrylates),[1058,1130] poly(vinyl chloride),[138,464] polyamides,[681,1148] unsaturated crosslinked polyesters,[613] polybutadiene,[329] ABS[658] and wool.[383,701,791,1141,1142] Specially sulphonated hydroxybenzotriazoles (**5.52** and **5.53**) are useful photostabilizers for retarding wool photodegradation.[383,791]

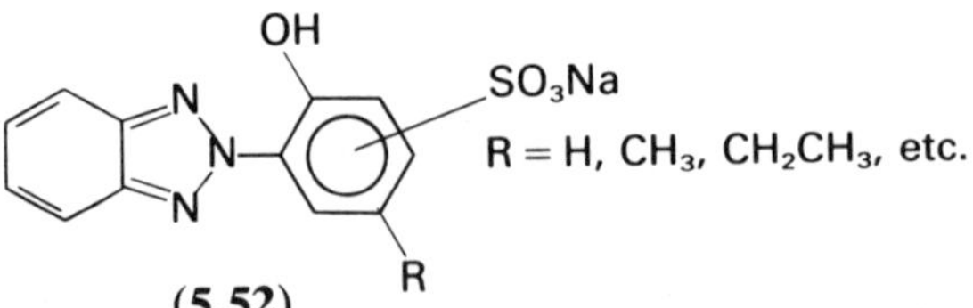

(5.52)

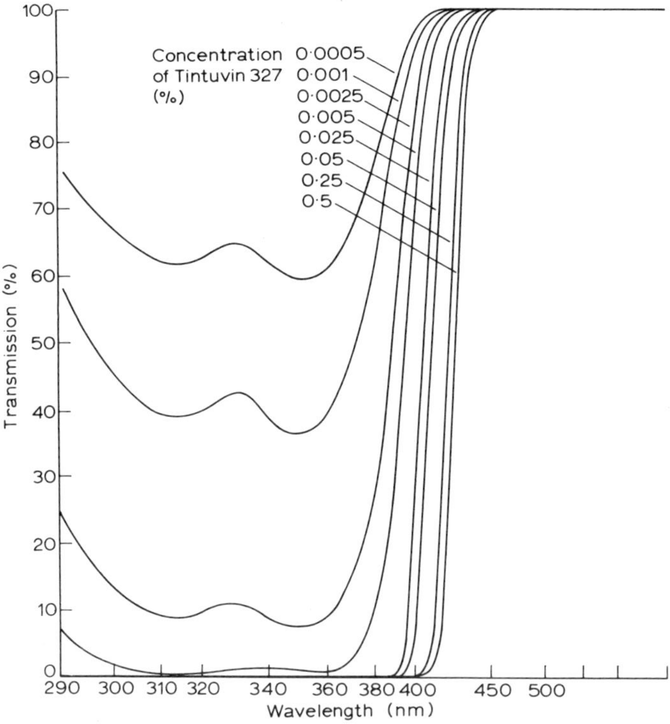

Fig. 5.20. Transmittance spectra of 2-(2-hydroxy-3,5-di-*tert.*-butylphenyl)-5-chloro-2*H*-benzotriazole (Tinuvin 327; **5.46**).

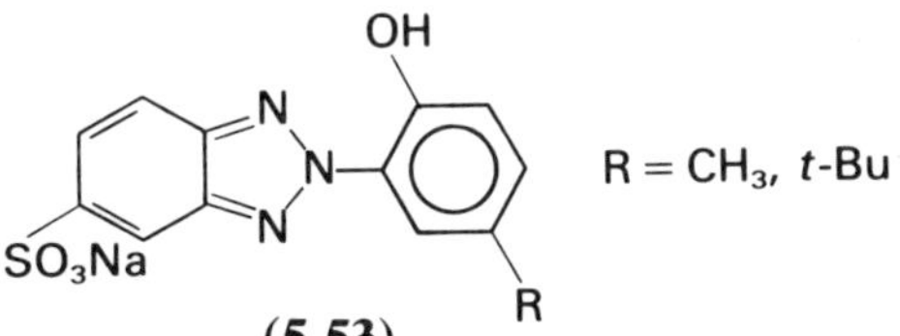

(**5.53**)

5.9.1 New Di- and Tri-benzotriazole Photostabilizers

Reaction of 2-nitrobenzenediazonium salts (**5.54**) with an excess of resorcinol (1,3-dihydroxybenzene) (**5.55**) or phloroglucinol (1,3,5-trihydroxybenzene) (**5.57**) give (after reductive cyclization) not only monosubstituted benzotriazole derivatives but also di- (**5.56**) and in the case of phloroglucinol, even the tri-benzotriazole-substituted compounds (**5.58**).[1131]

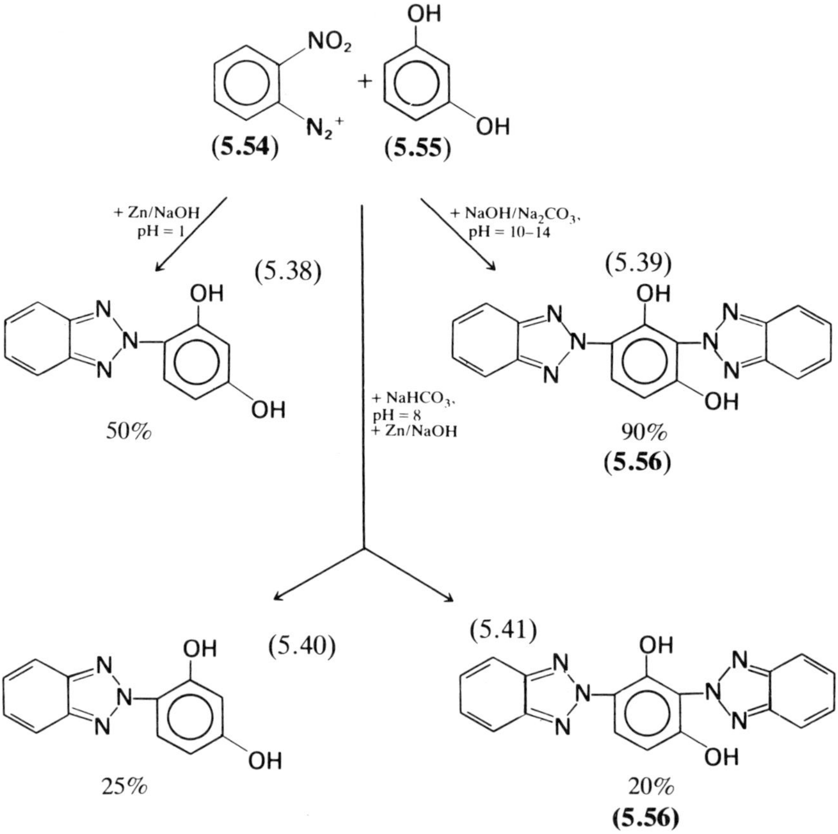

Absorption spectra (Fig. 5.21) show that the disubstituted compounds have a very high extinction, more than twice that of monosubstituted 2(2-hydroxyphenyl)-2*H*-benzotriazole, and an ultraviolet spectrum broadened to shorter wavelengths.[582,1131] Addition of a third benzotriazole ring does not significantly change the extinction coefficient, which is increased by only 20%.

These disubstituted compounds are starting materials for a family of polymerizable monomers (cf. Section 7.3.4).[713]

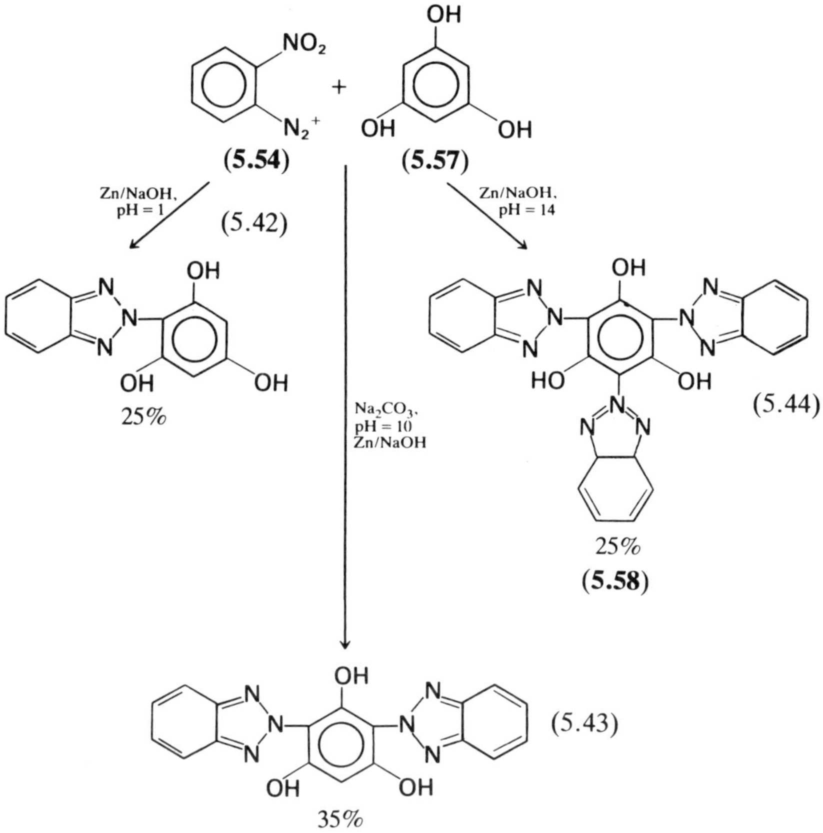

5.10 ORGANIC METAL PHOTOSTABILIZERS

Organic metal photostabilizers include a wide group of very different compounds such as metal chelates, metal stearates, metal salts of organic acids, etc. The stabilization mechanism is different and characteristic for each separate group.

Processing history plays a dominant role in controlling the photostabilizing performance of organic metal photostabilizers. Each stabilizer operates differently, being dependent on its relative thermal stability during processing and photostability.[40]

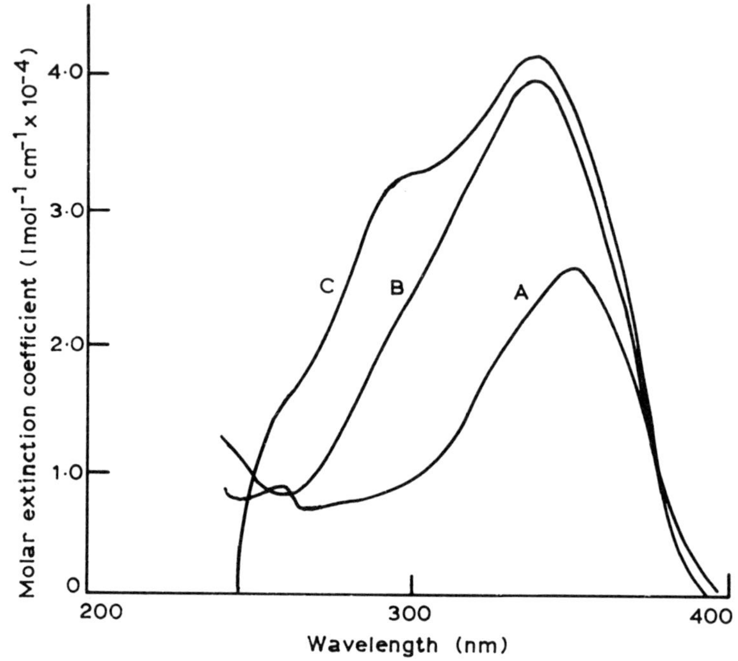

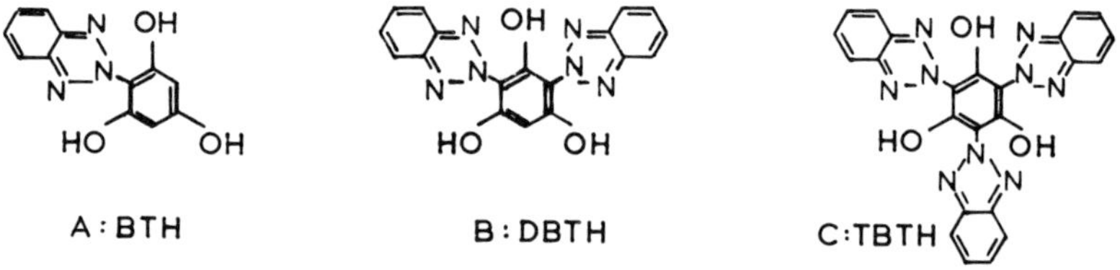

Fig. 5.21. Absorption spectra of di- and tri-benzotriazole-substituted compounds. Reprinted with permission from Ref. 1131, © 1985 Butterworth & Co. (Publishers) Ltd.

Metal chelates operate mainly as powerful radical traps (radical scavengers) and/or hydroperoxide-group decomposers. They inhibit polymer hydroperoxide (POOH) formation during processing and act as UV-stable chain terminators or by giving products during the early stages of photo-oxidation which are themselves effective stabilizers.[699]

The following reactions present the general mechanism of photostabilization by nickel chelates (abbreviated here as NiL_2):[170,699]

$$NiL_2 + ROOH \longrightarrow {}^{+}NiL + L^- + RO_2\cdot + H\cdot \quad (5.45)$$

$$^{+}NiL \rightleftharpoons Ni^{2+} + L^- \quad (5.46)$$

$$L^- + RO_2\cdot \longrightarrow L\cdot + RO_2^- \quad (5.47)$$

$$L\cdot + RO_2\cdot \longrightarrow L\text{—}RO_2 \quad (5.48)$$

where L^- is the ligand.

Nickel chelates decompose *tert.*-butyl hydroperoxide to di-(*tert.*-butyl) peroxide and undergo a reversible two-electron reduction:[287,987,988]

$$NiL_2 + RO_2H \rightleftharpoons (RO_2NiL_2)^-H^+ \quad (5.49)$$

$$(RO_2NiL_2)^-H^+ + RO_2H \rightleftharpoons (NiL_2)^{2-}2H^+ + RO_2R + O_2 \quad (5.50)$$

$$(NiL_2)^{2-}2H^+ + RO_2H \rightleftharpoons (RONiL_2)^-H^+ + H_2O \quad (5.51)$$

$$(RONiL_2)^-H^+ + RO_2H \rightleftharpoons NiL_2 + H_2O + RO_2R \quad (5.52)$$

The dianion is virtually colourless and the monoanion is only faintly coloured.

Nickel chelates were found to quench singlet and triplet states of carbonyl groups in ketones[9,40,72,244,249,441,479,509,511,524,570] and the excited triplet state of anthracene.[212,852,853]

The excited carbonyl group quenching rate constants of metal chelates can be roughly correlated with stabilization effectiveness in thin polypropylene film.[511]

Nickel chelates are also effective quenchers of singlet oxygen ($^1O_2$) by two kinds of reaction:[243,246,249,427,511,512,570,690,699,1187,1188]

$$NiL_2 + {}^1O_2 \longrightarrow NiL_2{}^* + {}^3O_2 \quad \text{(physical quenching)} \quad (5.53)$$

$$NiL_2 + {}^1O_2 \longrightarrow NiL_2 \text{ oxidation products} \quad \text{(chemical quenching)} \quad (5.54)$$

The rate constants for $^1O_2$ quenching (k_q) by some nickel chelates are listed in Table 5.8. Diamagnetic nickel chelates are excellent quenchers with rate constants which vary from the diffusion-controlled rate limit to about one-tenth of this value. Paramagnetic nickel chelates quench $^1O_2$ at 1/100 to 1/1000 of this value, or less. Diamagnetic nickel complexes have the coordinating atom arranged in square-planar geometry. Singlet oxygen has a relatively unhindered approach to the nickel atom from above or below the plane formed by the co-ordinating atoms, and can easily interact with an orbital perpendicular to the plane. In paramagnetic complexes the approach to the

Table 5.8
Rate constants of quenching (k_q) of singlet oxygen ($^1O_2$) by different metal chelates[804]

Chelate	*Magnetic properties*	k_q (*litre* mol^{-1} s^{-1})
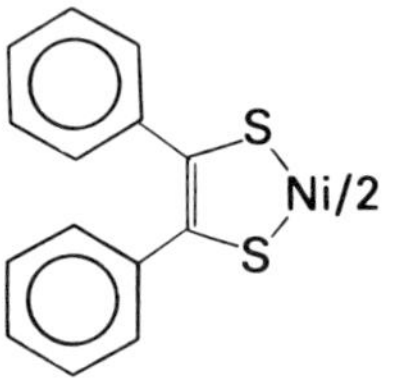	Diamagnetic	$1 \cdot 1 \times 10^{10}$
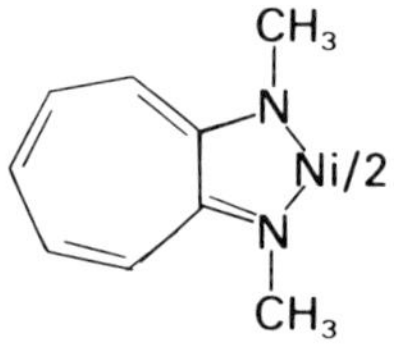	Diamagnetic	$6 \cdot 1 \times 10^{9}$
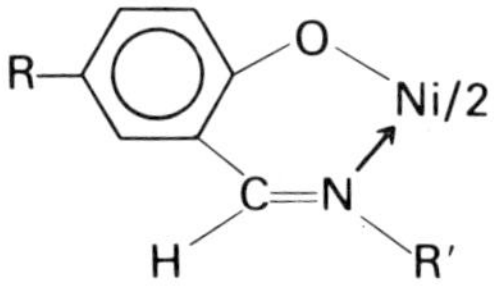 R = Br, n-C_4H_9	Diamagnetic	$3 \cdot 7 \times 10^{9}$
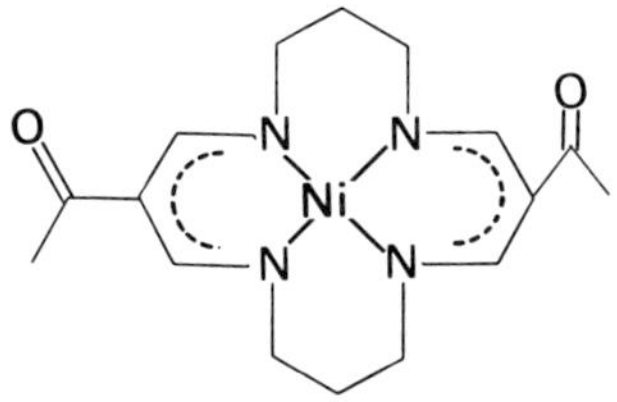	Diamagnetic	$1 \cdot 6 \times 10^{9}$
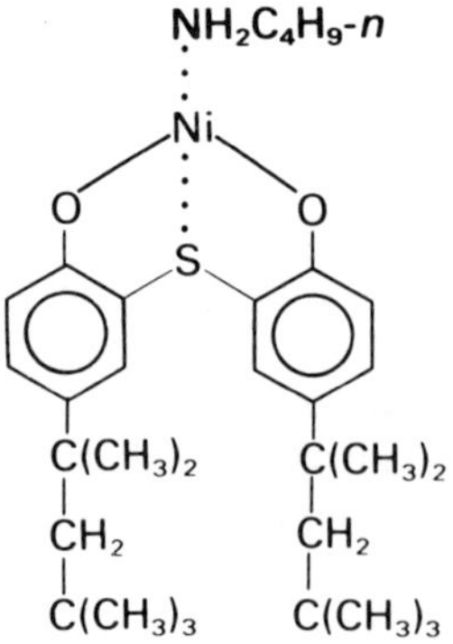	Paramagnetic	$1 \cdot 7 \times 10^{8}$

Table 5.8—*contd.*

Chelate	*Magnetic properties*	k_q (*litre* mol^{-1} s^{-1})
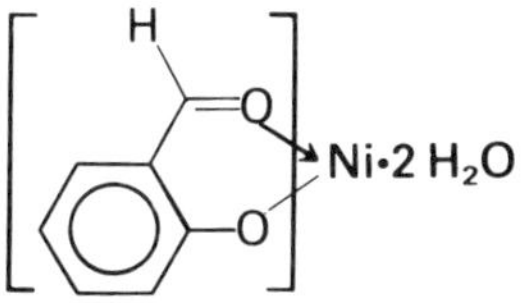	Paramagnetic	$4{\cdot}6 \times 10^7$
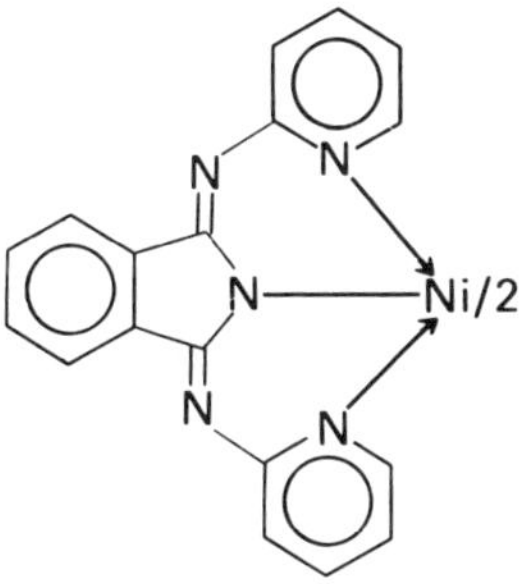	Paramagnetic	$1{\cdot}6 \times 10^8$
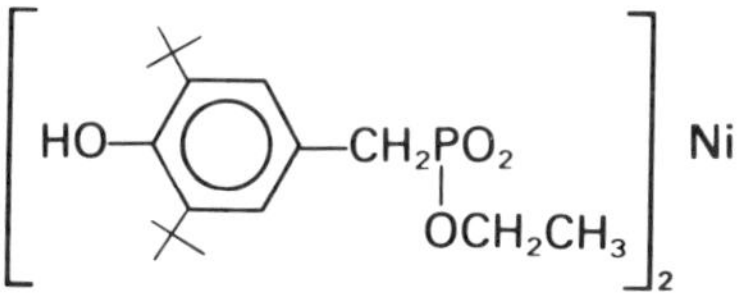	Paramagnetic	$2{\cdot}2 \times 10^7$
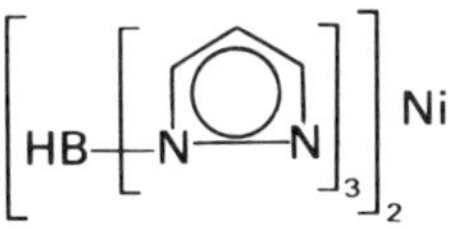	Paramagnetic	$2{\cdot}1 \times 10^6$

nickel atoms is blocked by the tetrahedral or octahedral arrangements of the ligands.[804]

It has been shown that in the following series of nickel chelates, the stabilizing efficiency was dependent on their photolysing stability:[690] bis(*tert.*-butylphosphine)nickel(II) dichloride (**5.59**); cyclopentadienyl(triphenylphosphine)nickel(II) chloride (**5.60**); bis(dithiobenzil)nickel(II) (**5.61**); bis(4-dimethylaminodithiobenzil)nickel(II) (**5.62**); and 2,2′-thiobis(4-*tert.*-octylphenolato)-*n*-butylaminenickel(II) (**5.63**) (absorption spectra shown in Fig. 5.22).

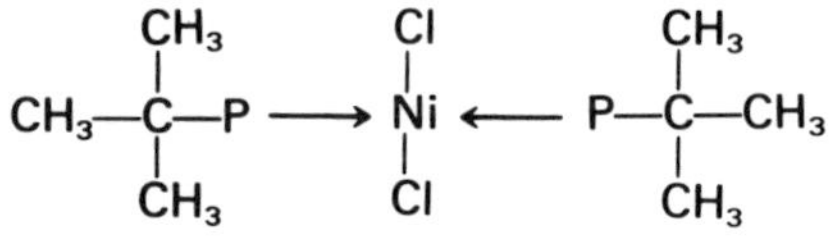

(5.59)

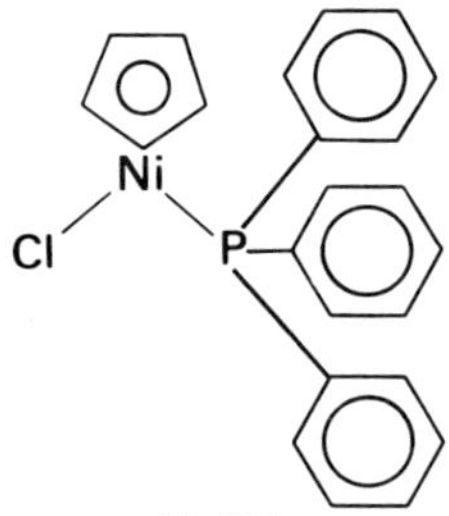

(5.60)

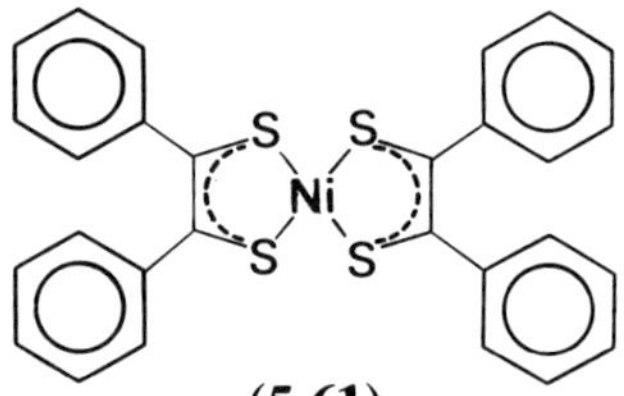

(5.61)

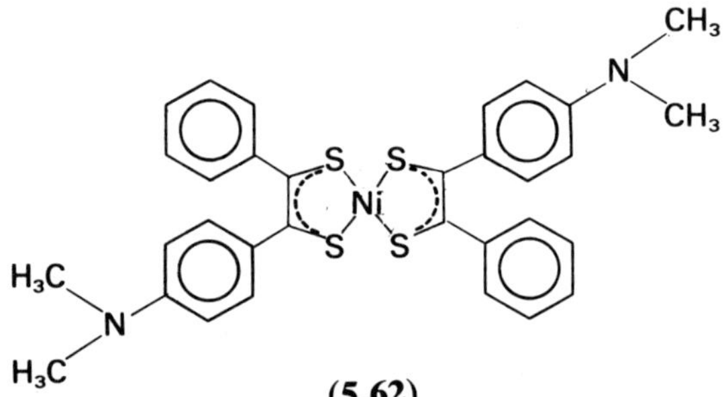

(5.62)

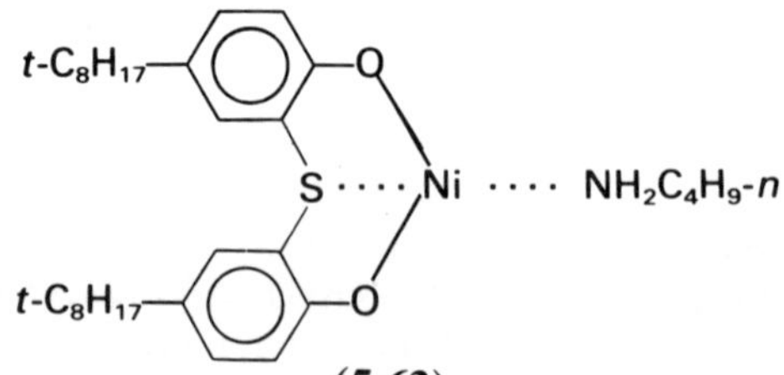

(5.63)

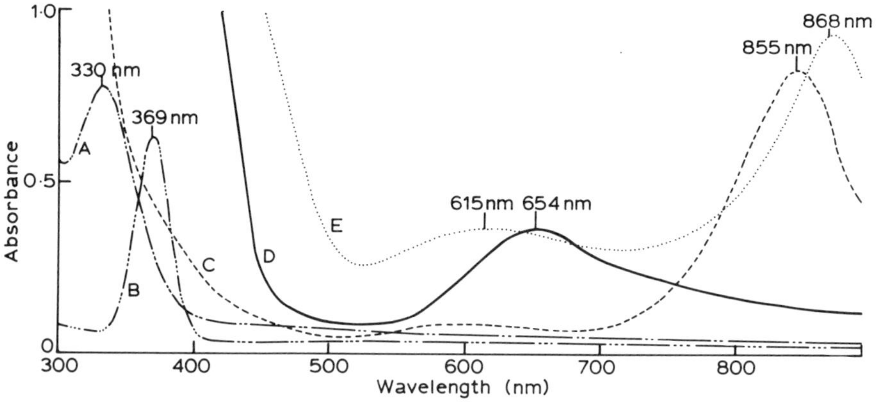

Fig. 5.22. Absorption spectra of nickel chelates: A, bis(*tert.*-butylphosphine)nickel(II) dichloride (**5.59**); B, cyclopentadienyl-(triphenylphosphine)nickel(II) chloride (**5.60**); C, bis(dithiobenzil)nickel(II) (**5.61**); D, bis(4-dimethylaminodithiobenzil)nickel(II) (**5.62**); E, 2,2′-thiobis(4-*tert.*-octylphenolato)-*n*-butylaminenickel(II) (**5.63**).[690]

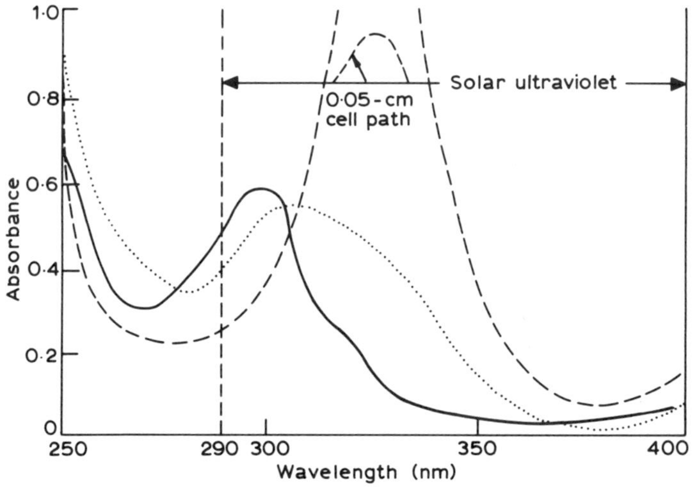

Fig. 5.23. Absorption spectra of nickel chelates: ——, bis[2,2-thiobis-4-(1,1,3,3-tetramethylbutyl)phenolato]nickel(II) (Ferro AM 101; **5.64**); · · ·, 2,2′-thiobis(4-*tert.*-octylphenolato)-*n*-butylamine nickel(II) (Cyasorb UV 1084; **5.63**); — —, dibutyldithiocarbamate nickel(II) (**5.66**); all at concentration 6×10^{-4} M in CH_2Cl_2, 0·1 cm pathlength.[697]

Table 5.9
Examples of commercially available nickel chelate photostabilizers for polyolefins

Name	*Structure*	*Commercial name*
2,2′-Thiobis(4-*tert*.-octylphenolato)-*n*-butylamine nickel(II)	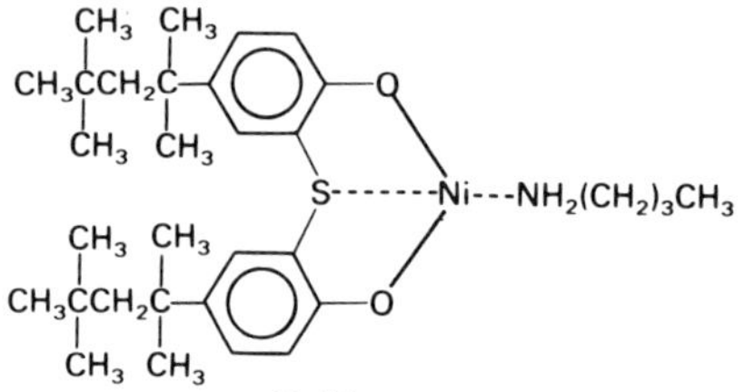 (5.63)	Cyasorb UV 1084
Bis[2,2-thiobis-4-(1,1,3,3-tetramethylbutyl)-phenolato]nickel(II)	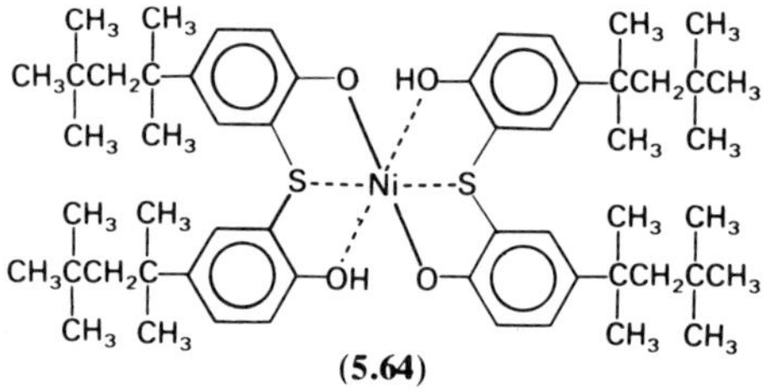 (5.64)	Ferro AM 101
2-Hydroxy-4-methyl-acetophenone oxime nickel(II)	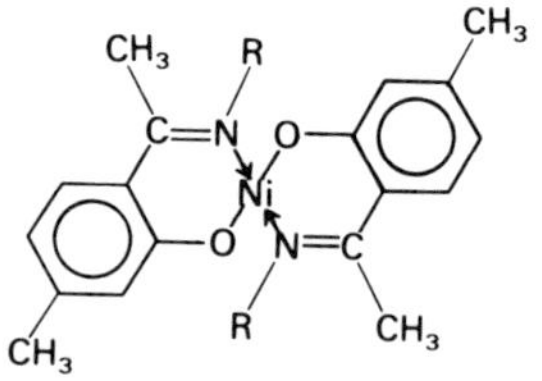 R = OH, n-C_4H_9, C_6H_5 (5.65)	
Dibutyldithiocarbamate nickel(II)	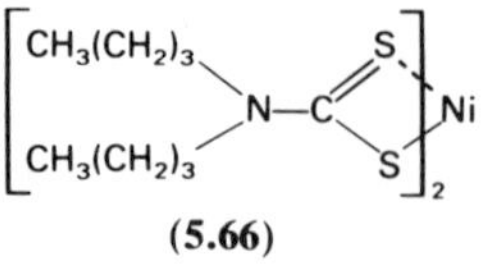 (5.66)	NBC

Table 5.9—*contd.*

Name	*Structure*	*Commercial name*
Bis(*o*-butyl-3,5-di-*tert.*-butyl-4-hydroxybenzyl phosphonate) nickel(II)	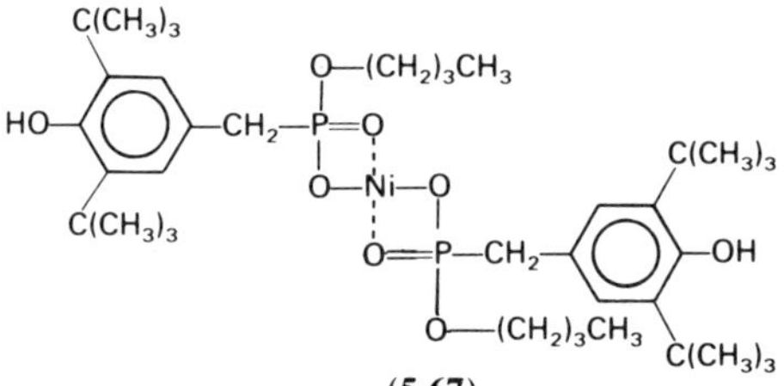 (5.67)	Irgastab 2002
Bis[(1-phenyl-3-methyl)-4-decanolyl-5-pyrazolate] nickel(II)	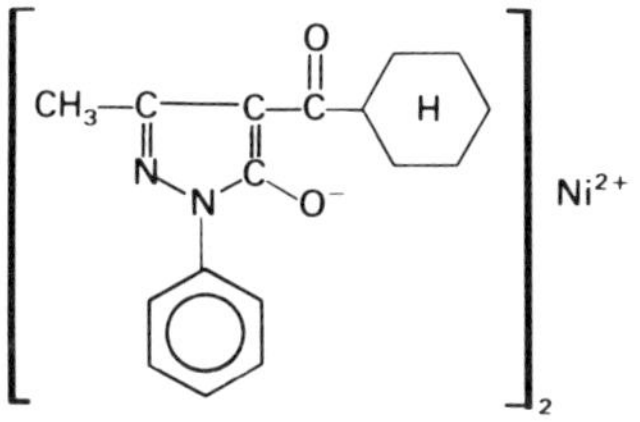 (5.68)	Sanduvor NPU

The list of commercially available metal chelates is very extensive. The most important photostabilizers from this group have been collected in Table 5.9. Some of their absorption spectra are given in Fig. 5.23.

A number of papers have been devoted to the study of metal chelates for stabilization of polyolefins,[9,40–43,47,71,72,213,242,245,248,249,412,878,932] polystyrene,[441,524] and polydienes.[208,690,1187,1188]

Metal chelates show both synergistic and antagonistic effects when they interact with other additives.[46,47,50,51] These effects are dependent on the chemical structure of a given metal chelate, and additives and formulations. The mechanisms of these interactions are not well understood.

5.11 METAL STEARATES

Metal stearates are commonly used additives in the processing of polymers. It has been reported that cupric stearate, for example, is an effective antioxidant and photostabilizer for polyolefins.[147,520,864,865,870] The rate of photo-oxidative degradation is much reduced, indicating that Cu^{2+} is able to compete effectively with oxygen for polymer alkyl radicals (P•):[867]

$$P\bullet + Cu^{2+} \longrightarrow PCu^{+} \tag{5.55}$$

$$P\bullet + O_2 \longrightarrow POO\bullet \tag{5.56}$$

Reaction (5.56) also competes with the redox decomposition of hydroperoxides:

$$POOH + Cu^{2+} \longrightarrow POO\bullet + Cu^{+} + H^{+} \tag{5.57}$$

$$POOH + Cu^{+} \longrightarrow PO\bullet + Cu^{2+} + OH^{-} \tag{5.58}$$

The antioxidant mechanism of cupric stearate during processing closely resembles that of stable aryloxy and nitroxy radicals (cf. Section 6.2).

In mixtures with other antioxidants and/or photostabilizers, metal stearates show antagonistic or synergistic effects depending on the formulation of the plastic material, e.g.[43,50,51]

—Calcium stearate antagonizes the photostabilizing action of nickel chelate (Irgastab 2002; **5.67**), whereas it strongly synergizes with the other nickel chelate, Cyasorb 1084 (**5.63**).
—Nickel stearate synergizes with the calcium chelate Irganox 1425 (**3.34**) in unprocessed polymer, whereas in processed polymer it exhibits antagonism.

Rates of photodecomposition of the metal chelates are also affected by metal stearates. There can be some degree of metal exchange with certain metal chelates.

5.12 METAL SALTS OF ORGANIC ACIDS

Metal salts (Ni^{2+}, Zn^{2+}) of 3,5-di-*tert.*-butyl-4-hydroxybenzoic acid (Shell acid; **5.69**) has been found to be an effective photostabilizer for polypropylene.[53] Their behaviour is explained in terms of macroalkyl radical scavenging during processing, which in turn leads to a lower concentration of hydroperoxide groups prior to film irradiation. Addition of calcium stearate gives rise to metal exchange between the calcium and the corresponding metal in the salt.[43,50]

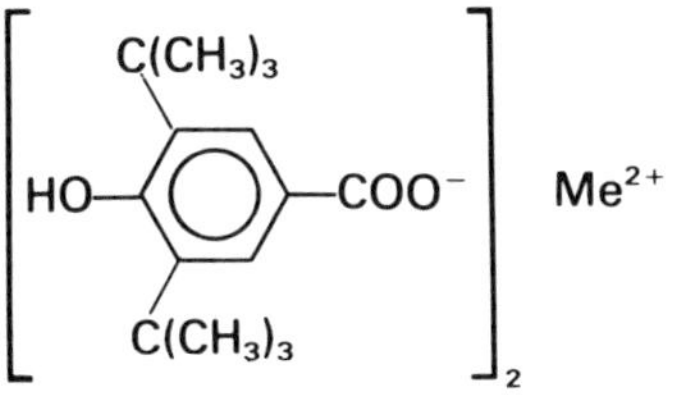

(5.69)

Metal salts (Co^{2+}, Mn^{2+}, Ni^{2+}) of organic acids such as 2-(2,4-dihydroxybenzoyl)benzoic acid (**5.70**) (absorption spectra in Fig. 5.24)[720] and 2-[(2-hydroxy-1-naphthalenyl)carbonyl]benzoic acid (**5.71**) (absorption spectra in Fig. 5.25)[924] are effective stabilizers against

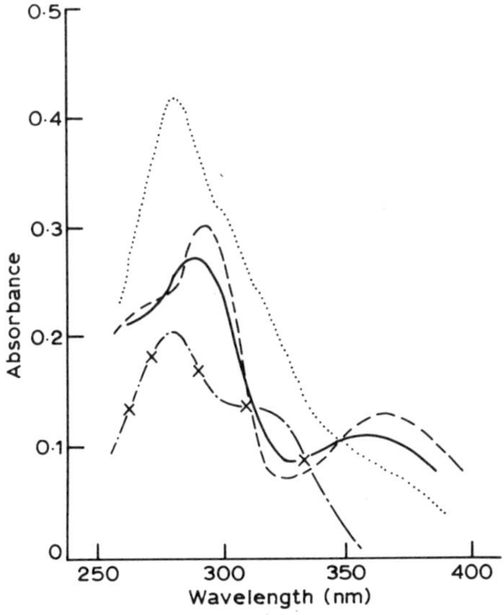

Fig. 5.24. Absorption spectra of: ×, 2-(2,4-dihydroxybenzoyl)benzoic acid (**5.70**) and its salts; ——, Co(II); — —, Ni(II); · · ·, Mn(II).[720]

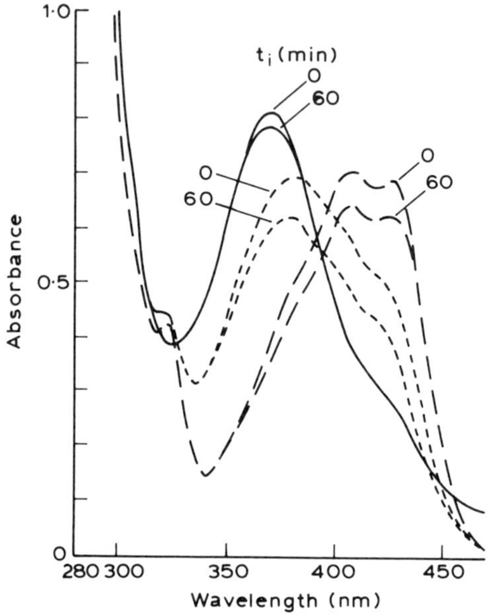

Fig. 5.25. Absorption spectra of metal salts of 2-[(1-hydroxy-2-naphthalenyl)carbonyl]benzoic acid (**5.71**): ···, Co(II); ——, Mn(II); – – –, Ni(II). Reprinted with permission from Ref. 924, © 1982 Pergamon Press PLC.

free-radical photo-oxidation and singlet oxygen ($^1O_2$) oxidation of polybutadiene.

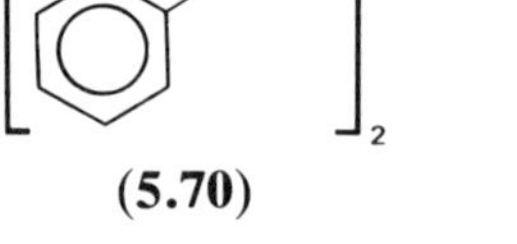

(5.70)

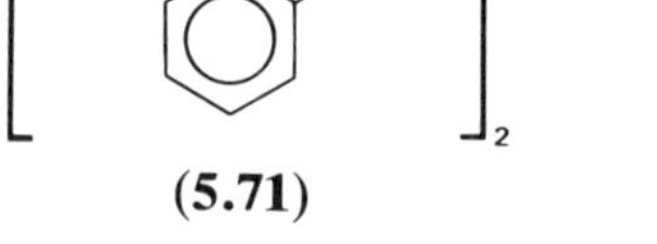

(5.71)

These metal salts are active by a combination of several important effects such as reducing the number of photons absorbed by polymer chromophores, deactivation of excited carbonyl groups, scavenging of free radicals and quenching of singlet oxygen (Fig. 5.26).

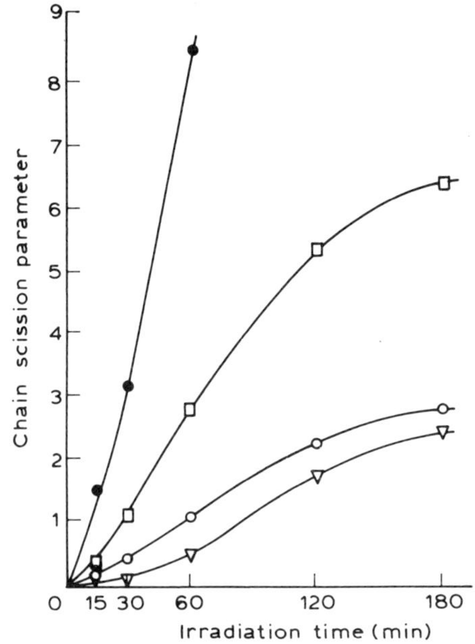

Fig. 5.26. Change of chain scission parameter after UV irradiation of polybutadiene in benzene solution (1·0 wt%) in the presence of metal salts of 2-[(1-hydroxy-2-naphthalenyl)carbonyl]benzoic acid (**5.71**): ●, control; ○, Co(II); ▽, Mn(II); □, Ni(II). Reprinted with permission from Ref. 924, © 1982 Pergamon Press PLC.

5.13 METAL OXIME CHELATES

Antioxidant and photostabilizing effects are displayed by cobalt(III) and nickel(II) oxime chelates (**5.72**).[71,133,524,932,933]

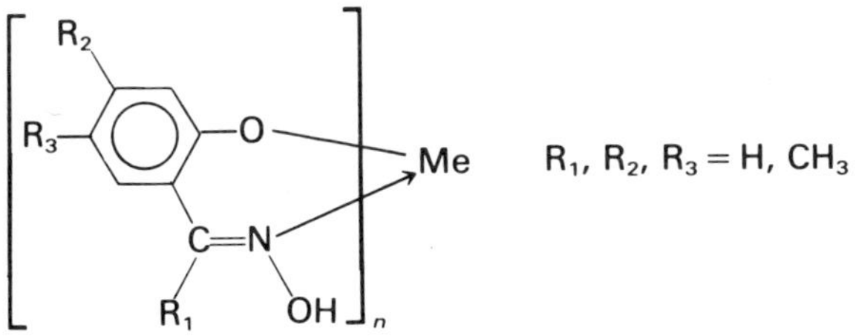

(**5.72**)

The antioxidant effect of nickel oxime chelate, e.g. nickel(II) 2-hydroxy-4-methylacetophenone oxime (**5.73**) is due to the ability of these compounds to reduce alkylperoxy radicals with the simultaneous formation of stabilized phenoxy radicals:[699,930]

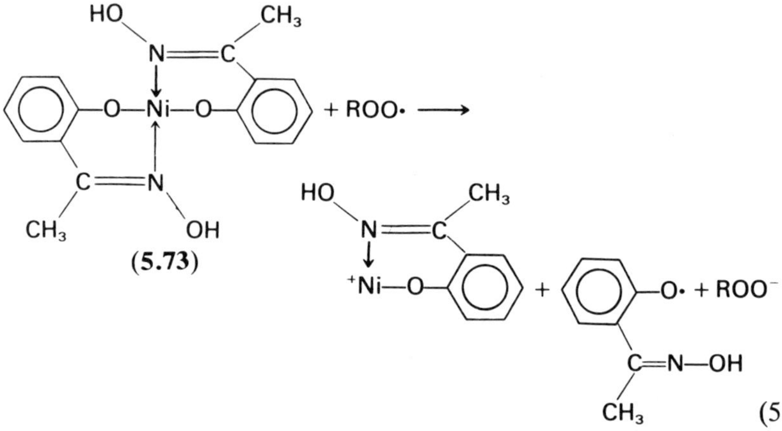

(5.73)

(5.59)

The behaviour of Schiff base chelates as oxidation inhibitors can be explained by a mechanism similar to reaction (5.59):

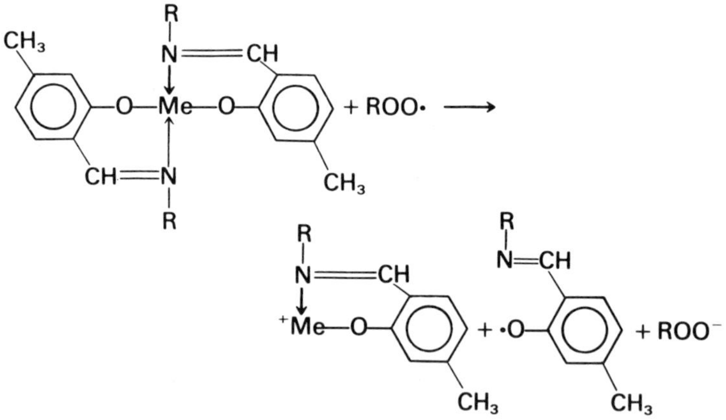

(5.60)

Furthermore, if reaction (5.60) does occur, then the further reaction (5.61) may take place:[170]

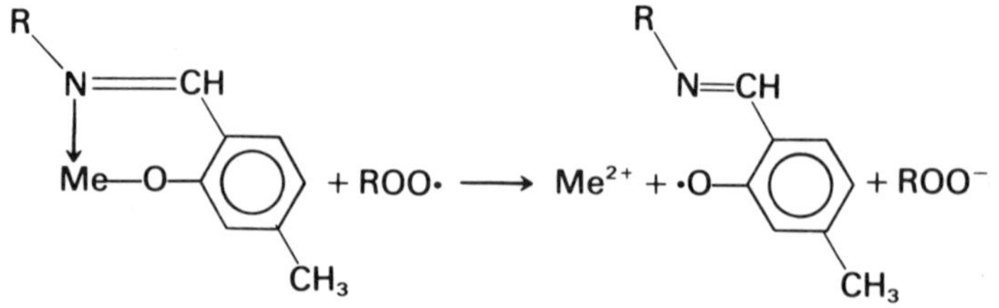

(5.61)

One consequence of reactions (5.59) and (5.60) would be the formation of free metal ions in the system. Thus, as soon as the metal chelate is decomposed by alkylperoxy radicals, the metal ions catalyse oxidation by induced decomposition of hydroperoxides (cf. Section 1.16.2). This mechanism explains the dual behaviour of the Schiff base chelates as both inhibitors and catalysts. In the same way acetophenone oxime chelates (**5.74**) which have a somewhat similar structure to Schiff base chelates, strongly catalyse decomposition of the hydroperoxides:[932,933]

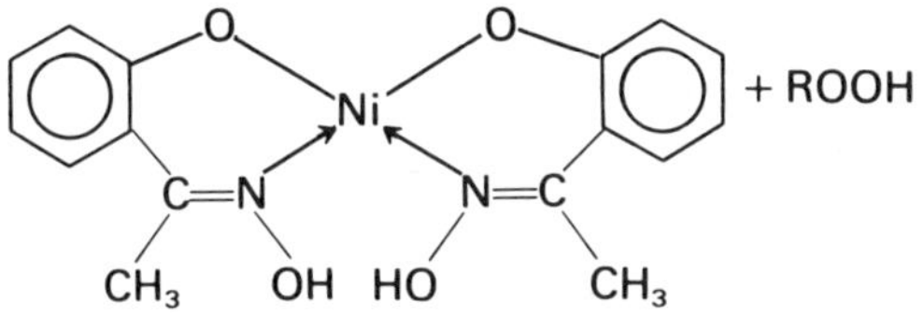

(**5.74**)

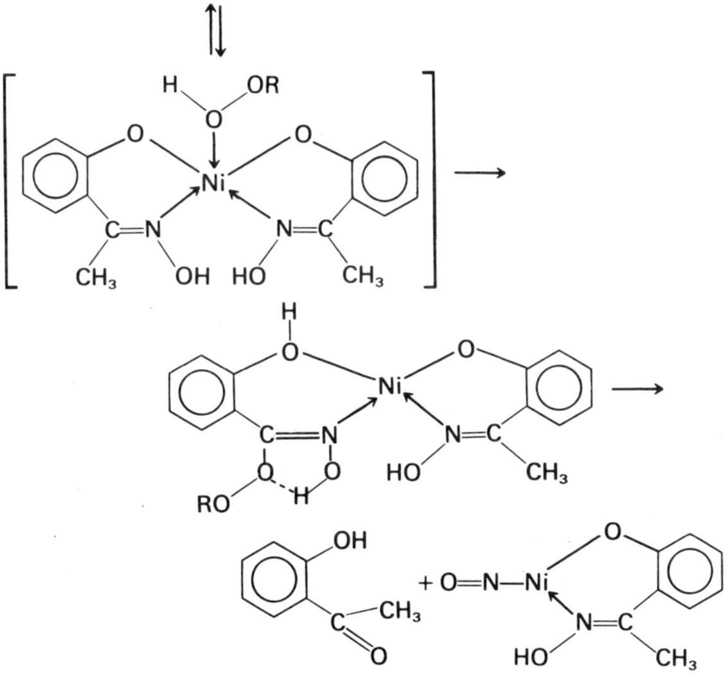

(5.62)

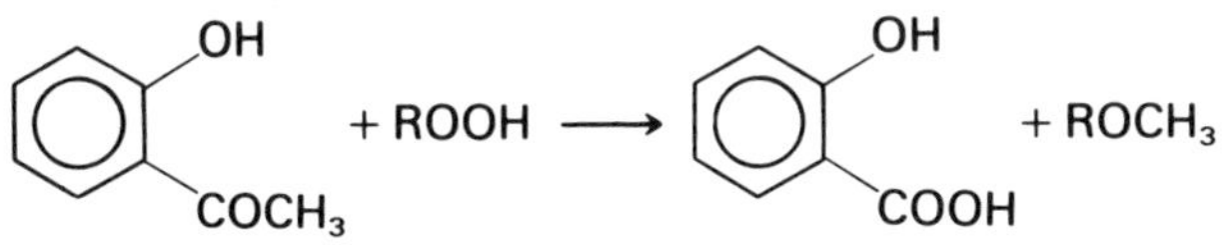

(5.63)

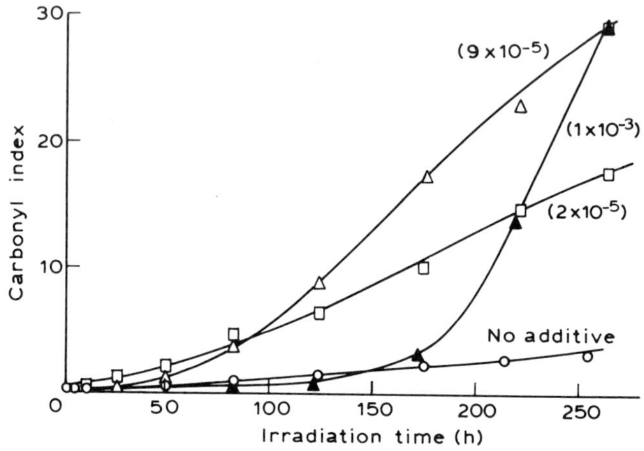

Fig. 5.27. UV degradation of low-density polyethylene (LDPE) containing iron(III) 2-hydroxyacetophenone oxime. Numbers in parentheses on the curves are concentrations in mol/100 g of polymer. Reprinted with permission from Ref. 133, © 1974 Pergamon Press PLC.

Iron(III) 2-hydroxyacetophenone oxime acts as a photoinitiator of the degradation of polyethylene (LDPE) (Fig. 5.27).[133]

Nickel(II) Schiff base chelates are efficient quenchers both of carbonyl triplet excited states[9,72] and singlet oxygen ($^1O_2$).[1187,1188]

Nickel oxime chelates have been investigated as photostabilizers for polypropylene[213] and polystyrene.[524]

5.14 NICKEL(II) MONO-OXIME (HYDRAZONE) CHELATES

Nickel(II) mono-oxime (hydrazone) chelates such as diacetyl mono-oxime(2-methylbenzoyl)hydrazone nickel(II) (**5.75**), diacetyl mono-oxime(2–chlorobenzoyl)hydrazone nickel(II) (**5.76**), and diacetyl mono-oxime(4-nitrobenzoyl)hydrazone nickel(II) (**5.77**) are effective photostabilizers for *cis*-1,4-polybutadiene in solutions and films. They act as UV-absorbers and also as singlet oxygen ($^1O_2$) quenchers. It has also been suggested that they trap polymer oxy (PO•) and polymer peroxy (POO•) radicals. They also show a synergistic effect with β-carotene.[1]

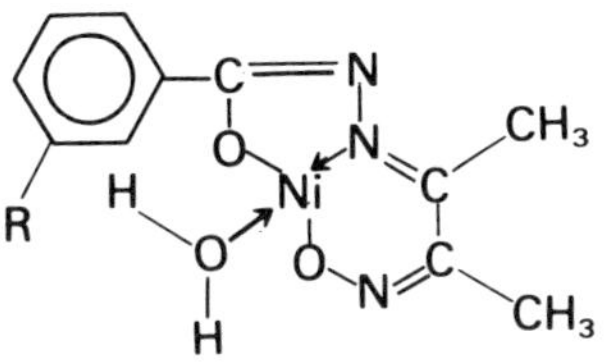

$R = CH_3$ **(5.75)**
$R = Cl$ **(5.76)**

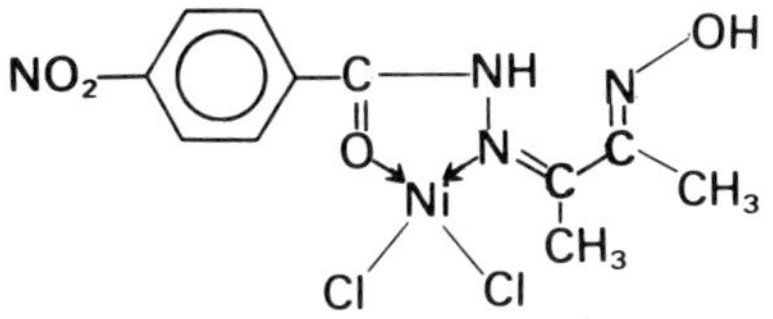

(5.77)

5.15 METAL SALICYLATE CHELATES

Antioxidant and photostabilizing effects are displayed by bis(3,5-diisopropyl salicylate) chelates **(5.78)**.

The antioxidant effect of metal salicylate chelates is analogous to that of the metal oxime chelates (cf. Section 5.14) and is due to the ability of these compounds to reduce alkylperoxy radicals with the simultaneous formation of stabilized phenoxy radicals:[170,699]

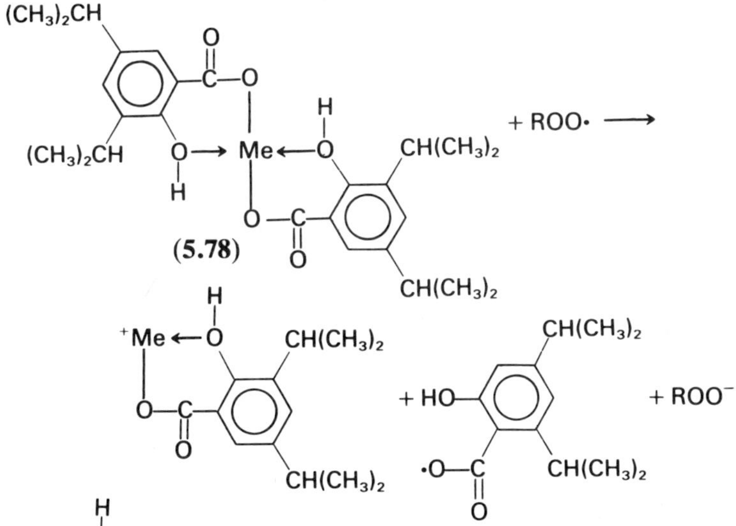

(5.78)

(5.64)

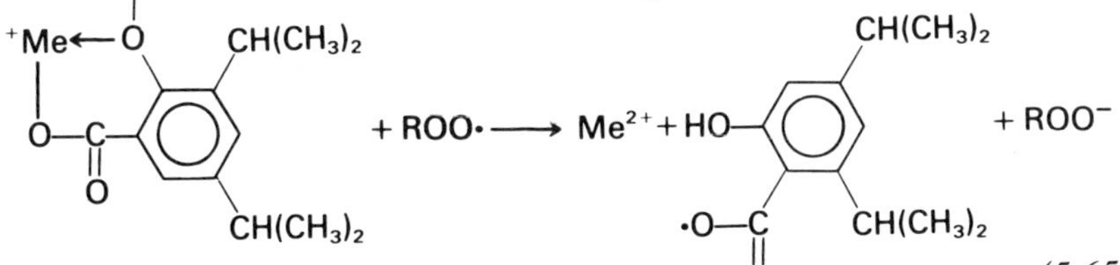

(5.65)

The cobalt(II), nickel(II), copper(II) and chromium(III) bis(3,5-di-isopropyl salicylate) chelates are effective antioxidants for polypropylene and carboxy-terminated polybutadiene.[170]

5.16 NICKEL(II) BIS(STILBENE DITHIOLATE) CHELATES

The nickel(II) bis(stilbene dithiolate) [or nickel(II) dithiobisbenzil] (**5.79**) and nickel(II) bis(4-dimethylaminostilbene dithiolate) [or nickel(II) bis(4-dimethylaminothiobenzil)] (**5.80**) show antioxidant and photostabilizing effects in the oxidation of polypropylene[170] and polydienes.[170,690]

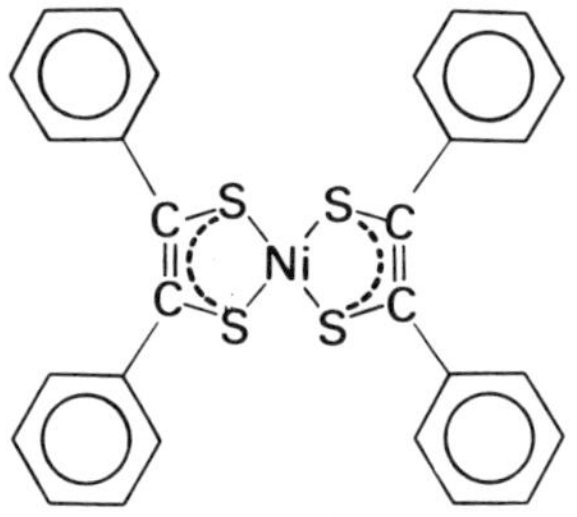

(**5.79**)

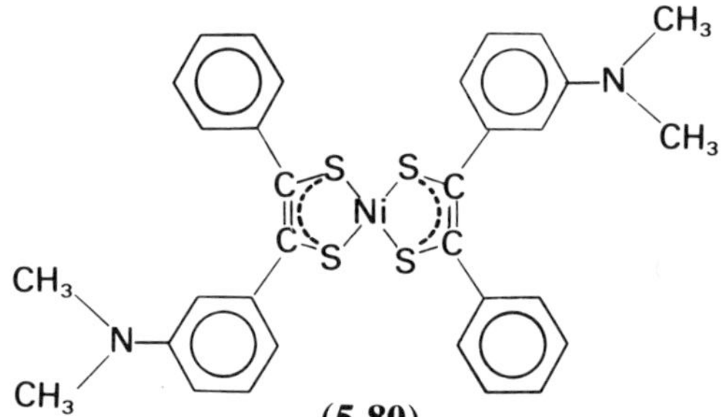

(**5.80**)

5.17 PYRAZOLE CHELATES

Aminopyrazoles (**5.81**) act as chain-breaking antioxidants by forming pyrazolyl radicals:[949,990]

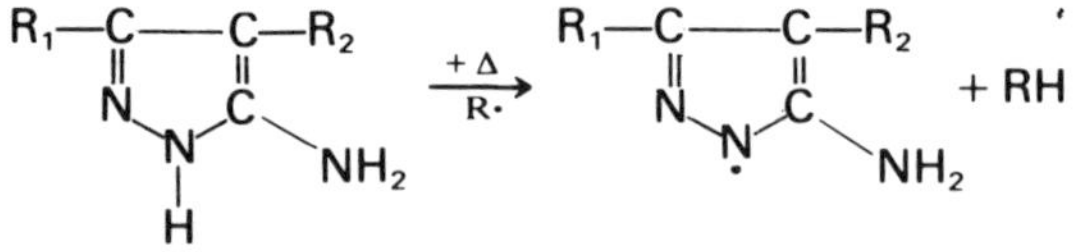

(**5.81**)

(5.66)

Commercially produced nickel(II) bis[(1-phenyl-3-methyl)-4-decanolyl-5-pyrazolate] [Sanduvor NPU (Sandoz UK); **5.82**] (absorption spectrum in Fig. 5.28) has been found to be an effective photostabilizer in polypropylene.[44,48]

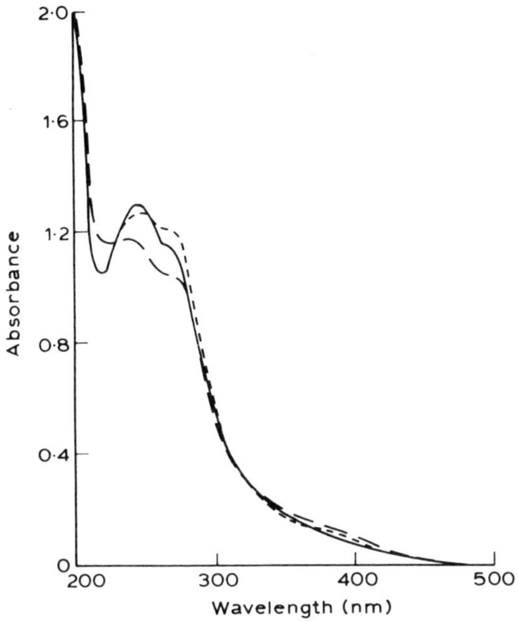

Fig. 5.28. Absorption spectrum of bis[(1-phenyl-3-methyl)-4-decanolyl-5-pyrazolate] nickel(II) (Sanduvor NPU; **5.82**), in polypropylene film after: ——, no heating; - - - -, 3 h pre-heating; — — —, 6 h pre-heating in an air oven. Reprinted with permission from Ref. 44, © 1984 Pergamon Press PLC.

Processing history plays a major role in controlling the photostabilizing performance of this chelate (Fig. 5.29). Upon heating, the chelate is partially decomposed according to the reactions:

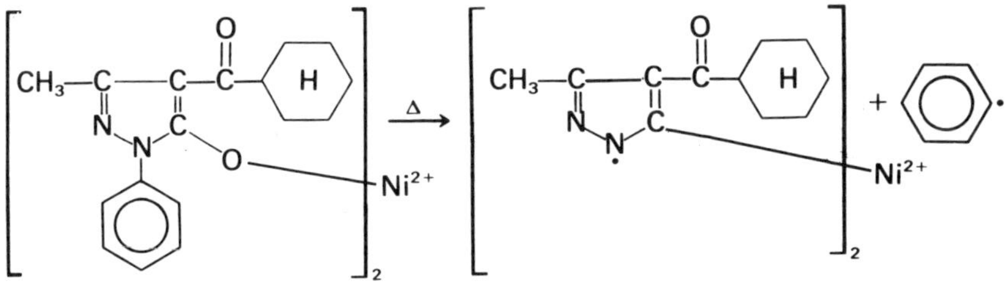

(5.82) (5.83)

(5.67)

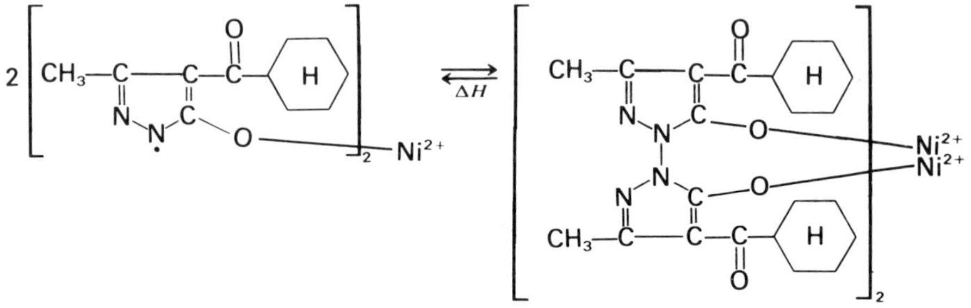

(5.84)

(5.68)

The bispyrazole (**5.84**) has a thermally unstable N–N bond which, on heating, decomposes to form pyrazolyl radicals (**5.83**). Hence the original compound (**5.82**) can be regenerated by the reverse process. Excessive heating or long processing may cause side reactions and so regeneration is either not observed or only a small effect is produced.

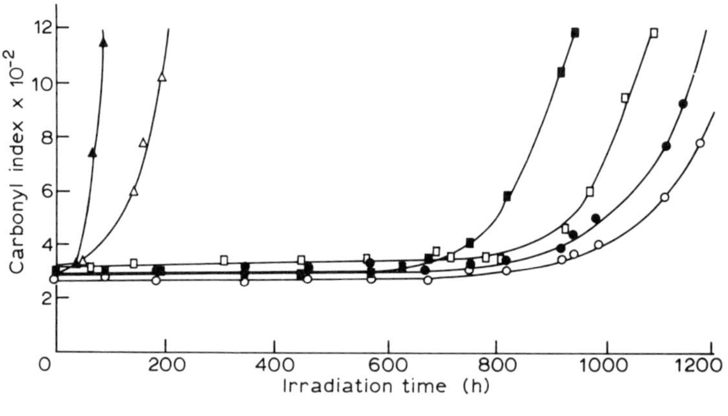

Fig. 5.29. Rate of photo-oxidation, in a Weatherometer (Microscal Ltd, UK) employing a 500 W high-pressure mercury fluorescent lamp (50°C, relative humidity ambient), of polypropylene film 200–300 μm thick containing: △, no additives; ▲, no additives after 1 h heating; and 0·1 wt% of nickel(II) bis[(1-phenyl-3-methyl)-4-docanolyl-5-pyrazolate] (Sanduvor NPU; **5.82**) after: ○, no heating; ●, 1 h heating; □, 3 h heating; ■, 6 h heating (at 135°C) in an air oven. Reprinted with permission from Ref. 44, © 1984 Pergamon Press PLC.

The inhibition process is assigned to polymer alkyl radical (P•) scavenging both during processing and photo-oxidation:

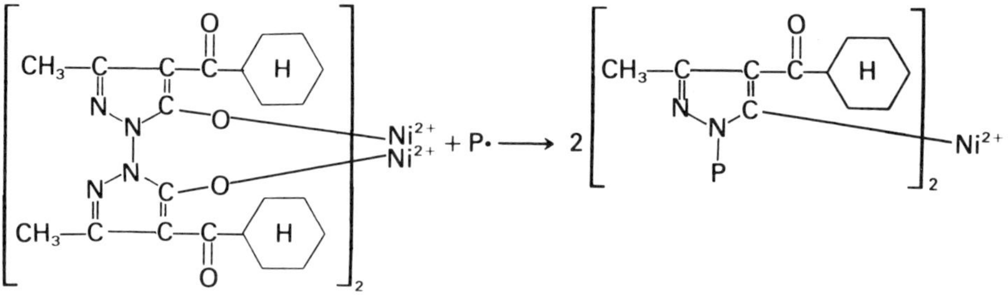

(5.69)

Another possible reaction is the ionic deactivation of polymer peroxy radicals (POO•) by the ligand anion:

$$L^{-} + POO\bullet \longrightarrow L\bullet + POO^{-} \qquad (5.70)$$

Pyrazole chelates do not decompose hydroperoxides. In the later stages of decomposition of the chelate by heat and/or UV radiation, the photoreactive products from the stabilizer promote the degradation of the polymer.

5.18 NICKEL(II) THIOPICOLINE ANILIDE CHELATE

The nickel(II) thiopicoline anilide (**5.85**) has been studied as an effective photo-antioxidant for photostabilization of poly(2,6-dimethyl-1,4-phenylene oxide).[274]

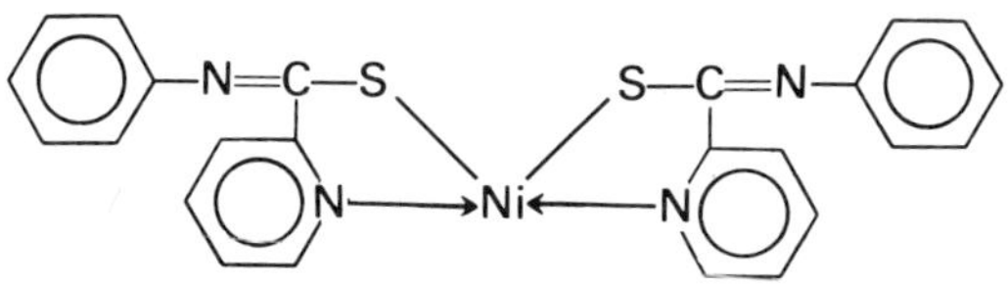

(5.85)

It decomposes hydroperoxy groups by the following mechanism:

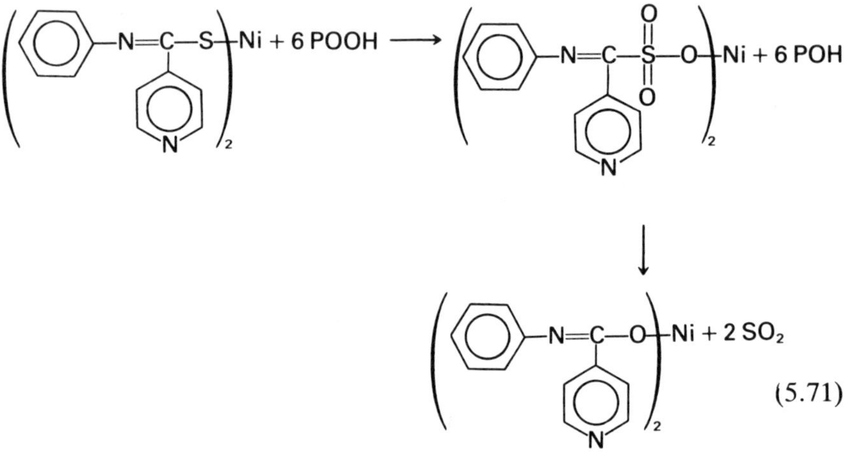

(5.71)

The nickel(II) thiopicoline anilide can also react with polymer peroxy radicals (POO•) in a two-stage electron-transfer mechanism:

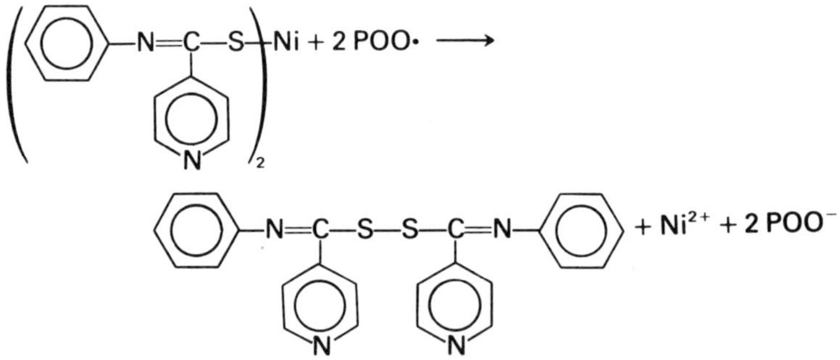

(5.72)

5.19 NICKEL(II) 2,2′-THIOBIS(4-*tert.*-OCTYLPHENOLATO)-*n*-BUTYLAMINE

Nickel(II) 2,2′-thiobis(4-*tert.*-octylphenolato)-*n*-butylamine (**5.86**) (Cyasorb UV 1084) (physical properties in Table 5.10) is a very effective thermal and photo- stabilizer for polyolefins through its ability to destroy hydroperoxides. The other possible photostabilizing route is radical scavenging, which has been shown to occur in solution.[41,49,878] Excited-state quenching and UV screening have been found to be unimportant in the overall mechanism of photostabilization.[40,43,45] The detailed mechanisms of photostabilization in which this photostabilizer participates are still unknown.

Cyasorb UV 1084 has found application in photostabilization of low-density polyethylene for agricultural applications (cf. Section 8.5).

It shows synergistic effects with calcium and strontium stearates, whereas with zinc, sodium, magnesium and cobalt stearates it exhibits antagonistic effects.[43,50,51] In admixture with hindered amine light stabilizers (HALS) it shows antagonistic or synergistic effects depending on the particular formulation.[46,51]

5.20 NICKEL(II) BIS(*o*-BUTYL-3,5-DI-*tert.*-BUTYL-4-HYDROXYBENZYL)PHOSPHONATE

Nickel(II) bis(*o*-butyl-3,5-di-*tert.*-butylhydroxybenzyl)phosphonate (**5.87**; Irgastab 2002) is a well-known nickel chelate photostabilizer. However, the photochemistry of this stabilizer is very little known. It quenches anthracene and reacts with singlet oxygen ($^1O_2$).[852,853]

Under UV irradiation this nickel chelate is slowly photolysed, according to reaction (5.73).[852]

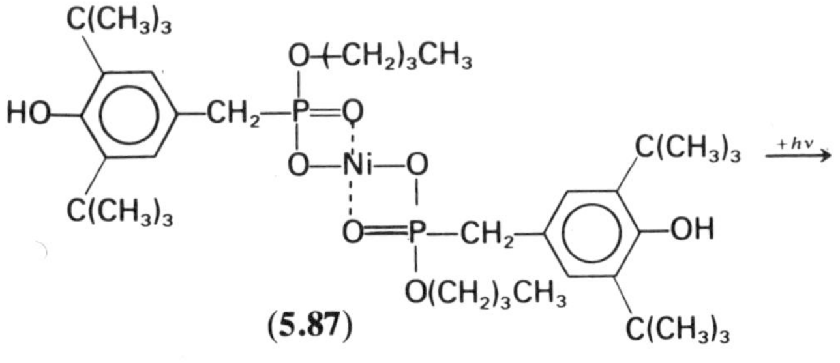

(**5.87**)

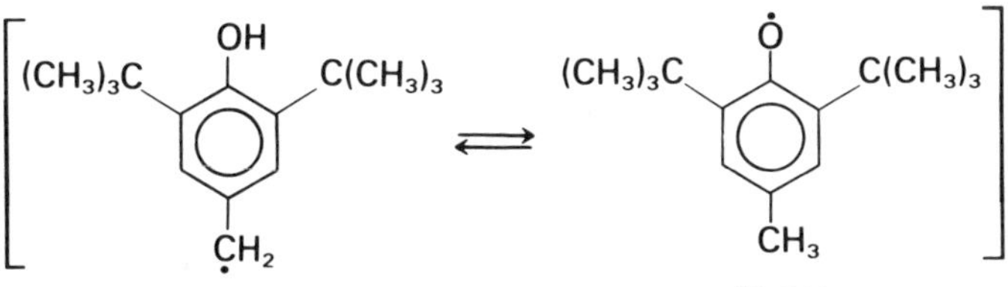

(**5.88**) (**5.89**) (5.73)

Table 5.10
Physical properties and solubility of 2,2′-thiobis(4-*tert.*-octylphenolato)-*n*-butylamine nickel(II) chelate

Chemical name	*Commercial name*	*Chemical structure*
2,2′-Thiobis(4-*tert.*-octyl-phenolato)-*n*-butylamine nickel(II)	Chimassorb N705 (Ciba–Geigy); Cyasorb UV 1084 (Cyanamid)	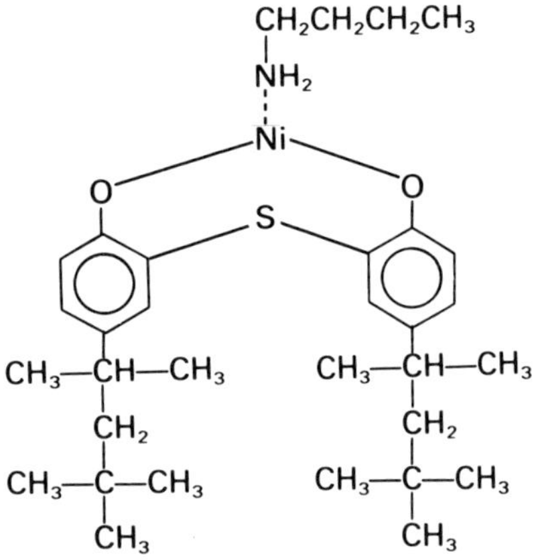 (5.86)

The hindered phenoxy radicals (**5.88** and **5.89**) which are formed stabilize polymers according to the mechanism presented in Section 3.1.

5.21 METAL DITHIOLATES AS PHOTOSTABILIZERS

The metal dithiolates which are represented by three similar groups of compounds:

(i) metal xanthates (**5.90**),
(ii) metal dithiocarbamates (**5.91**), and
(iii) metal dithiophosphates (**5.92**)

Molecular weight	*Appearance*	*Solubility (wt%)*					*Melting point (°C)*	*Recommended applications*
		Ethanol	*n-Heptane*	*Methyl ethyl ketone*	*Tetrahydrofuran*	*Toluene*		
571	Pale green powder	1·0	51·2	1·2	48·8	42·8	258–261	Polyolefins

are not only effective as antioxidants (cf. Section 3.9), but can also be used as photostabilizers.

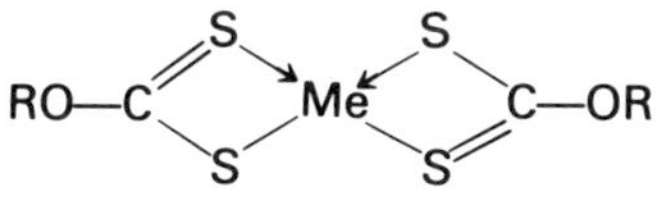

(5.90)

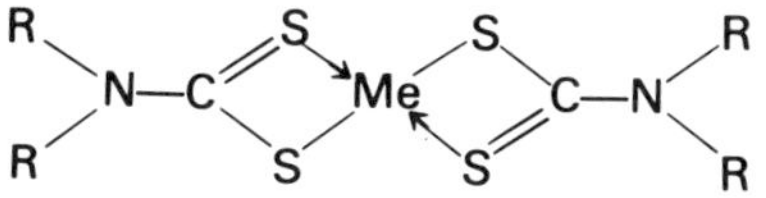

(5.91)

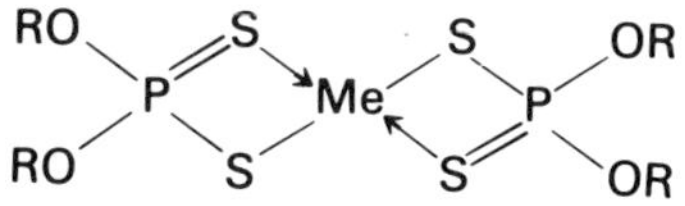

(5.92)

Stabilization of polypropylene by benzophenone-substituted xanthates show a three-step behaviour: initial rapid photo-oxidation, followed by a secondary inhibition period and a third autoacceleration step.[131] The first step is mainly due to the benzophenone substituent, whilst the length of the inhibition period is directly related to the sulphur content. During this period xanthates therefore lead to complete inhibition of photo-oxidation with regard to both chemical oxidation and physical change.

Nickel(II) dibutyldithiocarbamate (absorption spectrum in Fig. 5.30) and zinc(II) dibutyldithiocarbamate show the best performance as photostabilizers and each functions as a combined UV screen,

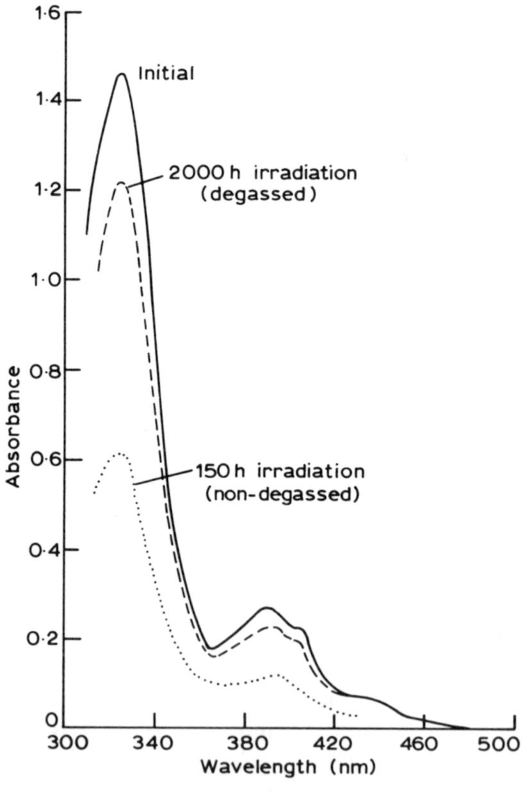

Fig. 5.30. Absorption spectra of dibutyldithiocarbamate nickel(II) (**5.66**) in polypropylene films after UV irradiation (· · ·) in the presence of air and (———) without air.[509]

hydroperoxide decomposer and radical trap. Both these photostabilizers are able to quench the excited triplet state of carbonyl groups.[509,511] Nickel dibutyldithiocarbamate is an effective singlet oxygen ($^1O_2$) quencher whereas zinc dibutyldithiocarbamate does not quench at all.[509,512]

Nickel dibutyldithiocarbamate under prolonged UV irradiation is slowly photolysed (Fig. 5.30).

Iron(III) dialkyldithiocarbamate, instead of stabilizing, accelerates the photo-oxidative degradation of polyolefins (cf. Section 5.22).

Metal diethyldithiocarbamates show synergistic effects in the presence of 2-hydroxy-4-*n*-octoxybenzophenone (Cyasorb UV 531; **5.30**).[119,262,266,270]

The piperidine and 1,4-diazabicyclo[2,2,2]octane form complexes with zinc(II) diethyldithiocarbamate which are more effective photostabilizers in polypropylene at the same molar concentration than zinc dithiolate itself.[269]

Nickel dithiophosphates (**5.92**) and their corresponding thiophosphoryl disulphides (**3.81**) are more effective photostabilizers than the closely related nickel xanthates (**5.90**) and dixanthogens (**3.80**) (cf. Section 3.8). In general, the nickel complexes in both series are more effective photostabilizers than the disulphides.[130]

5.22 METAL CHELATES AS PRO-OXIDANTS AND PHOTOINITIATORS

A number of metal chelates are very effective initiators (pro-oxidants) of thermal and/or photo- oxidative degradation of polymers.[865,1008] The main group of such pro-oxidants are acetylacetonates and some iron chelates such as iron(III) 2-hydroxyacetophenone oxime and iron(III) dialkyldithiocarbamate.

The iron chelates are commercially used as effective photoinitiators (pro-oxidants) for polymers with a time-controlled decay of mechanical properties.[1008] A comparison of the behaviour of these compounds is shown in Fig. 5.31.

An essential technical requirement for time-controlled degradable materials is that they should be capable of being processed normally, i.e. the additives should not be pro-oxidants during processing or in storage.[104,105] Iron dithiocarbamate is unique in behaving as an effective thermal antioxidant, while at the same time giving rise to polymer soluble carboxylates during UV irradiation (cf. Section 5.23).

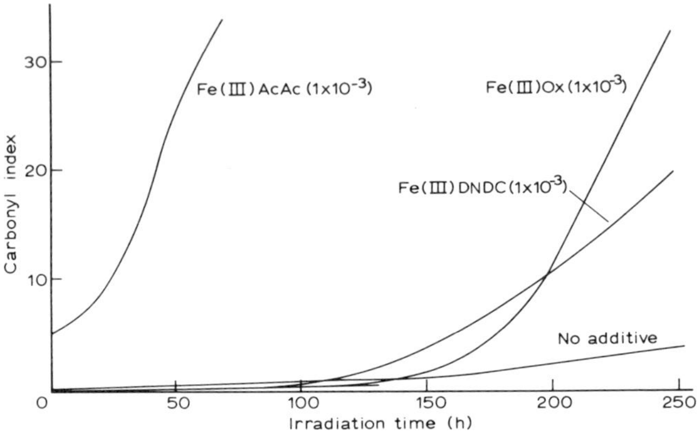

Fig. 5.31. Comparison of the effects of different kinds of photoinitiators at similar concentrations (1×10^{-3} M) on carbonyl group formation in low-density polyethylene (LDPE).[1008] Fe(III)AcAc, iron acetylacetonate; Fe(III)Ox, iron(III) 2-hydroxyacetophenone oxime; Fe(III)DNDC, iron dialkyldithiocarbamate.

The iron dithiocarbamates, when used alone, impart a range of outdoor lifetimes to polyethylene film in the mid-lifetime range. They are not, however, as effective photoinitiators as oxygen-chelated iron compounds. For example, iron acetylacetonates are known to be highly active photoinitiators for polyethylene, but they suffer from the serious practical deficiency that they are also powerful prodegradants for polyolefins during processing.[133,1004,1005]

Stabilizers which initially protect polymers completely from photo-oxidative degradation for a predetermined period depending on the use of the material[119,470] i.e. 'delayed action' photo-activators (e.g. for polyolefins) have to satisfy a number of conflicting requirements:

(i) They must not catalyse the degradation of the polymer during processing.
(ii) They should not sensitize the polymer to degradation during storage.
(iii) They must give a reproducible induction period to property changes during outdoor exposure.
(iv) The rate of photo-oxidation at the end of the induction period should be rapid so that physical disintegration occurs in as short a time as is physically possible.

As an example, photodegradable plastic mulching film for agricultural purposes should:

(i) retain its strength throughout its entire service lifetime;
(ii) rapidly lose its mechanical strength shortly before cropping so that it is easily disintegrated under the mechanical action of harvesting machines.

5.23 DITHIOCARBAMATE IRON COMPLEX ACTING AS PHOTOINITIATOR OF POLYMER DEGRADATION

Iron(III) dialkyldithiocarbamate, [tri(*N,N*-di-*n*-butyldithiocarbamate)-iron(III)] (**5.93**; absorption spectrum in Fig. 5.32) is photounstable and acts as a photoinitiator of oxidative degradation (photopro-oxidant) (Fig. 5.33) according to the following mechanism:[104,105,117,118,470,1008,1026]

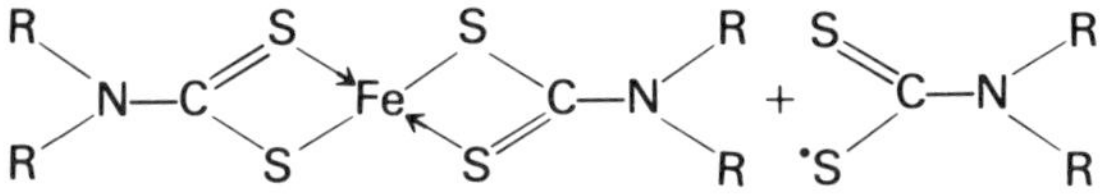

(5.74)

(**5.94**)

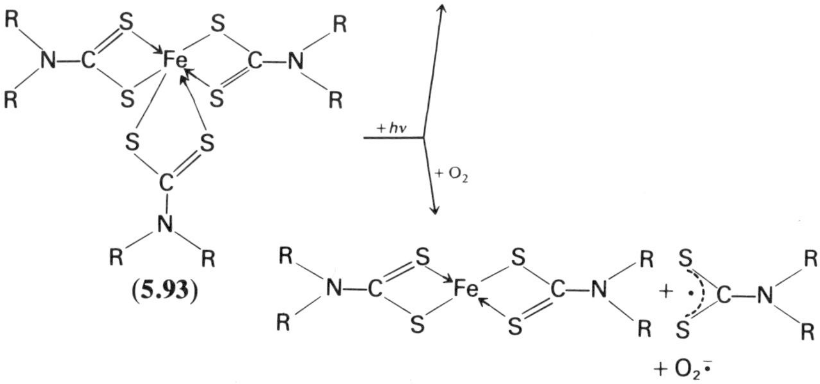

(**5.93**)

(5.75)

$R = CH_3, CH_3CH_2, (CH_3)_3C—$

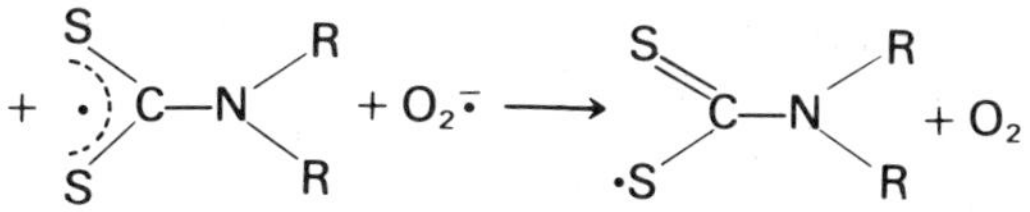

(5.76)

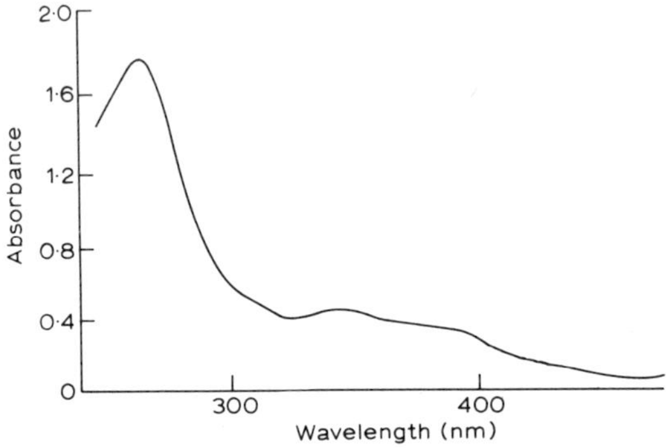

Fig. 5.32. Absorption spectrum of tri(*N,N*-di-*n*-butyldithiocarbamate) iron(III); (**5.93**) ($1{\cdot}2 \times 10^{-4}$ M in *n*-hexane). Reprinted with permission from Ref. 1026, © 1978 Pergamon Press PLC.

The formation of the highly reactive superoxide radical ion ($O_2^{\bar{\cdot}}$) explains the exceptional photoactivating effect of the iron dithiocarbamate in polyethylene when degraded in air.[133]

The disulphide radical (thiyl radical, xanthogen radical) (**5.94**) can abstract hydrogen from a polymer molecule (PH) and in this way produce a polymer alkyl radical (P•), which is susceptible to oxidation

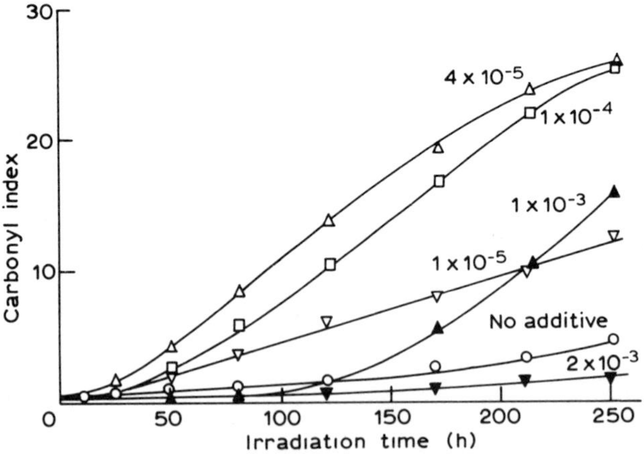

Fig. 5.33. UV degradation of low-density polyethylene (LDPE) containing dinonyldithiocarbamate iron(III) (**5.93**). The numbers on the curves are concentrations in mol/100 g of polymer. Reprinted with permission from Ref. 133, © 1974 Pergamon Press PLC.

to polymer alkylperoxy radical (POO•) and further to the formation of polymer hydroperoxide (POOH):

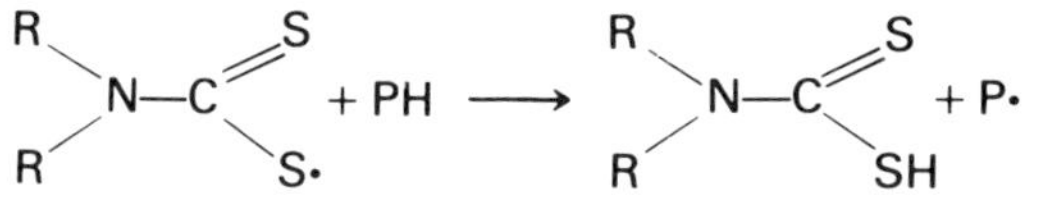

(5.77)

(5.94)

$$P\cdot + O_2 \longrightarrow POO\cdot \xrightarrow{+PH} POOH + P\cdot \qquad (5.78)$$

Iron dithiocarbamate is an important additive for producing photodegradable polyolefins with time-controlled decay of their mechanical properties.[133,1004–1006,1008] Their potential uses range from very short-lived materials such as packaging which is required to degrade in the environment immediately it is discarded, through protective films for annual crops where a lifetime of several months is required, to a protective mulch for young trees which may be required to stand up to 3–4 years in exposed environments before embrittlement.

Outdoor exposure tests performed on commercial photodegradable polyethylene (LDPE) packaging based on iron dithiocarbamate have

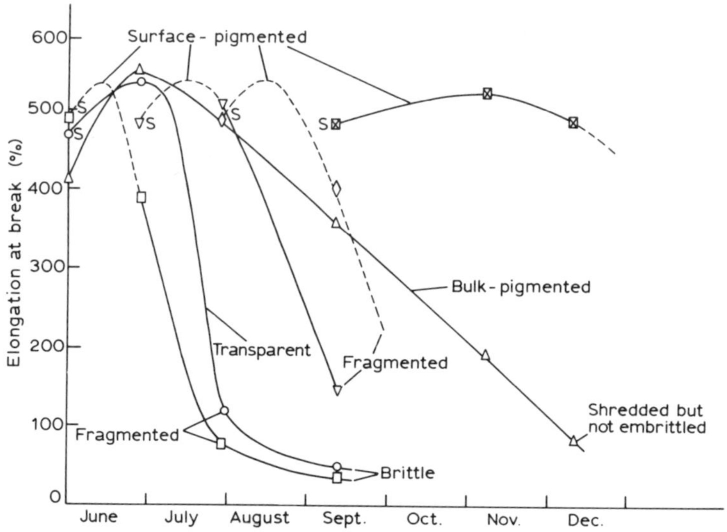

Fig. 5.34. Effect of seasonal sunlight variation on the degradation of carrier bags photosensitized with iron dithiocarbamate (**5.93**). S indicates start of outdoor exposure in Birmingham.[1008]

shown that fragmentation occurs under the unfluence of wind and weather when the elongation at break is reduced to about 20% of the original (Fig. 5.34).[1008] Titanium oxide (TiO_2-rutile) pigments retard the process but do not inhibit photodegradation of polyethylene. In polyethylene (HDPE), particularly in thick samples, iron dithiocarbamate in combination with titanium oxide produces a rapid surface degradation of the polymer which causes the development of a chalky layer on its surface which protects it from the further effects of sunlight. This powdered layer is readily removed by abrasion and if this is done deliberately or incidentally by mechanical erosion the degradation continues through the polymer bulk.[1008]

Mixtures of iron dithiocarbamate (photopro-oxidant) with nickel dithiocarbamate (photoantioxidant) allow a very wide range of embrittlement times for polyethylene to be obtained, in which the induction period is controlled by the nickel complex and the post-induction period rate by the iron complex.[117,119]

5.24 METAL ACETYLACETONATES

Metal acetylacetonates (**5.95** and **5.96**) show increasing pro-oxidant activity in the sequence

$$\mathrm{Ni(II) = Zn(II)(both\ inert) < Fe(III) < Co(III)}$$

(Table 5.11). They photoinitiate degradation of polyethylene (LDPE) (Figs 5.35 and 5.36),[133,857,1008,1117] polystyrene,[644] poly(vinyl chloride)[866] and polyurethanes.[282]

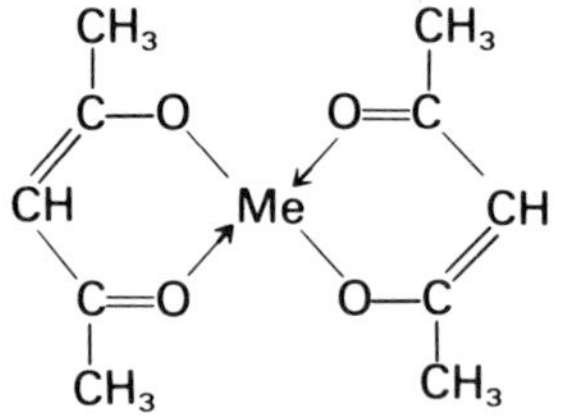

Me = Ni, Zn

(**5.95**)

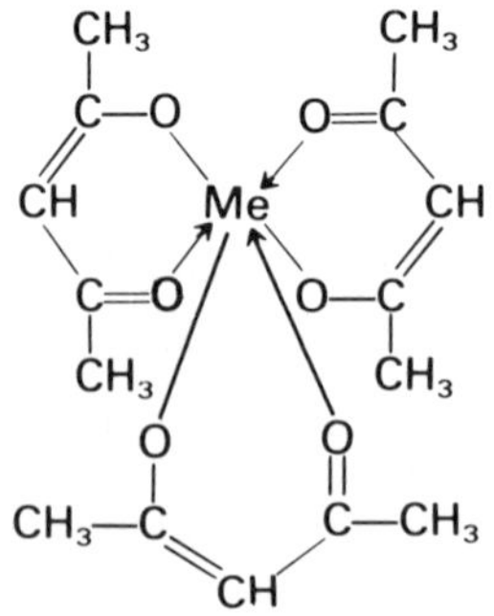

Me = Fe, Co

(**5.96**)

Table 5.11
Effect of transition metal acetylacetonates on the oxidation of low-density polyethylene (LDPE) during processing and on UV irradiation[1008]

Additive[a]	*Concentration (mol/100 g)*	*Carbonyl index before exposure*	*Time to embrittlement (h)*
$Fe(acac)_3$	2×10^{-3}	10·22	120
	1×10^{-3}	4·64	216
	$0{\cdot}7\times10^{-5}$	1·10	260
$Co(acac)_3$	2×10^{-3}	43·57	0
	$1{\cdot}4\times10^{-4}$	22·82	50
$Ni(acac)_2$	2×10^{-3}	0·51	1 968
$Mn(acac)_2{\cdot}2\,H_2O$	2×10^{-3}	0·64	600
	$1{\cdot}7\times10^{-4}$	2·70	510
$Ce(acac)_3$	2×10^{-3}	0·48	700
$Zn(acac)_2$	1×10^{-3}	0·38	1 370
None	—	0·15	2 100

[a] Abbreviation: acac, acetylacetonate.

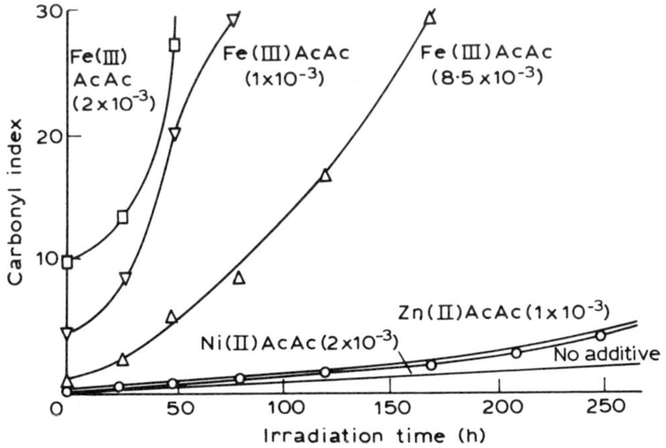

Fig. 5.35. UV degradation of low-density polyethylene (LDPE) containing different metal acetylacetonates. The numbers in parentheses are mol/100 g of polymer. Reprinted with permission from Ref. 133, © 1974 Pergamon Press PLC.

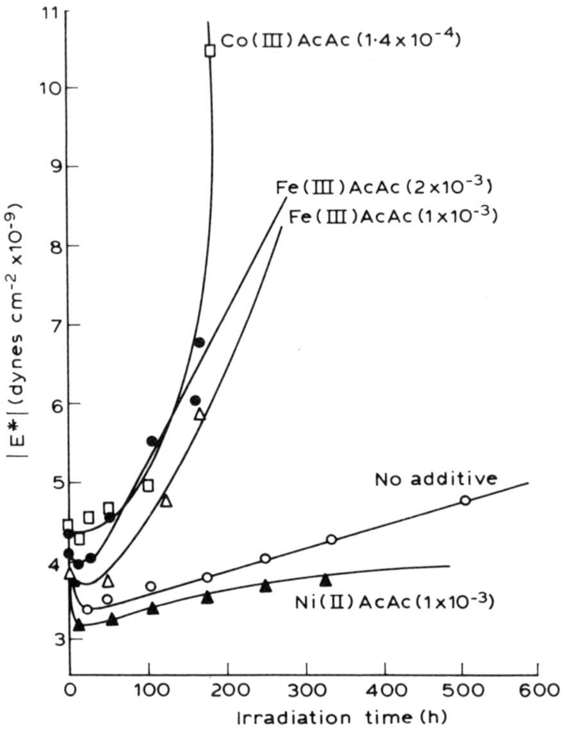

Fig. 5.36. Effect of different metal acetylacetonates on the complex modulus of low-density polyethylene (LDPE) during UV irradiation. The numbers in parentheses are mol/100 g of polymer. Reprinted with permission from Ref. 133, © 1974 Pergamon Press PLC.

Cobalt(III) acetylacetonate is a very powerful pro-oxidant during processing; at a concentration of 2×10^{-3} mol/100 g of polymer, it causes rapid embrittlement even without UV exposure. This correlates with a very high carbonyl group concentration after processing. Similar behaviour was observed with iron(III) acetylacetonate but to a lesser degree. Both complexes were destroyed rapidly during the processing operation. The rapid photodegradation occurring with iron(III) or cobalt(III) acetylacetonate is primarily a result of the oxidation of the polymer which occurs during processing.[779]

The metal acetylacetonates differ from the metal dithiocarbamates in two respects:[133,1004,1005]

(i) Except for the nickel complex, they are much less oxidatively stable and are destroyed rapidly during processing.
(ii) Acetylacetonates have no antioxidant activity.

The metal acetylacetonates are effective quenchers of the excited triplet state of the carbonyl group as well as singlet oxygen ($^1O_2$).[570]

5.25 METAL DEACTIVATORS

Metal deactivators are stabilizers which remove or partially deactivate the effect of metal pro-oxidants and photoinitiators. Trace of pure metals, metal salts, metal oxides (pigments) and organic metal compounds (e.g. catalysts) may catalyse thermal and/or photo-oxidative degradation of polymers (cf. Section 1.16).

The mechanism of the reactions that are involved in the deactivation processes of metal deactivators are specific for a given metal deactivator–metal pro-oxidant system, and are not very well known.

Of all the metal-catalysed degradation processes which occur in

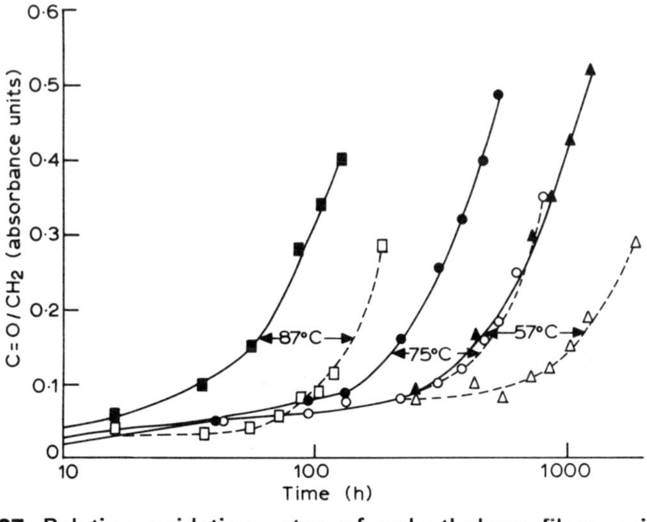

Fig. 5.37. Relative oxidation rates of polyethylene films oxidized as polymer–metal laminates at various temperatures.[531] ———, copper surface; – – –, outer surface.

Table 5.12

Commerical copper deactivators and their melting points[865]

Structure	*Producer*	*Melting point (°C)*
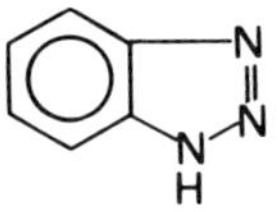	Sherwin–William	97–100
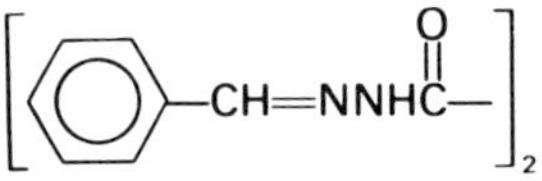	Eastman Chemical Products	333–336
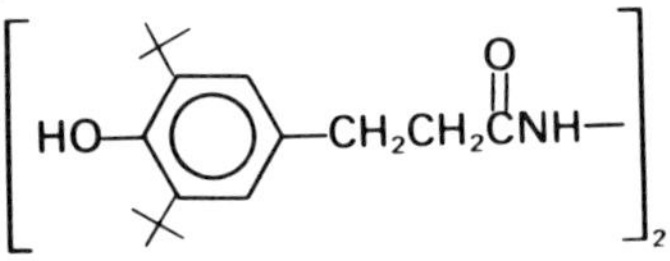	Ciba–Geigy	231
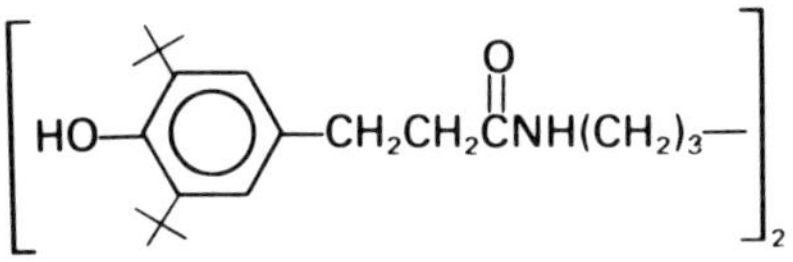	Ciba–Geigy	161–162
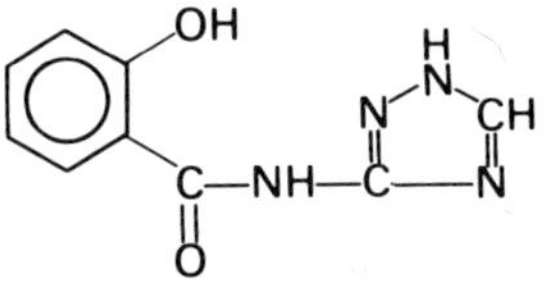	Argus Chemical	307–310
Proprietary	Eastman Chemical Products	276–286
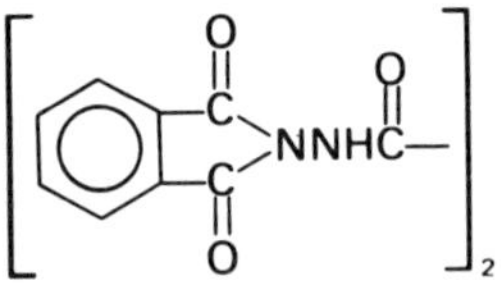	R. T. Vanderbilt	295
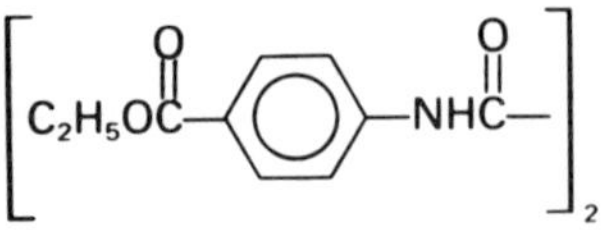	Monsanto Industrial Chemicals	286

Table 5.12—*contd.*

Structure	*Producer*	*Melting point (°C)*
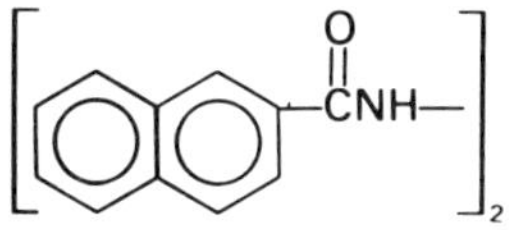 (BNH)	American Cyanamid	249
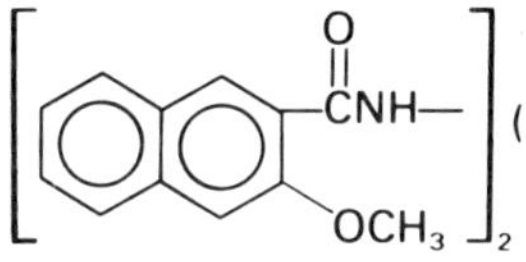 (MBNH)	American Cyanamid	252–257
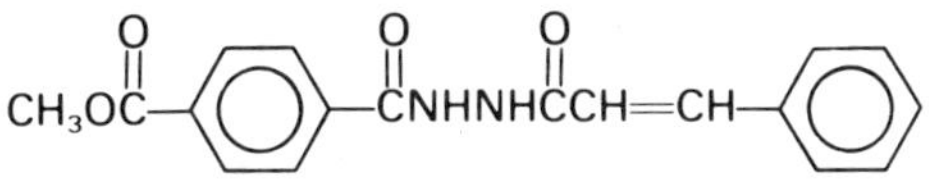	Ube Industries	206–208
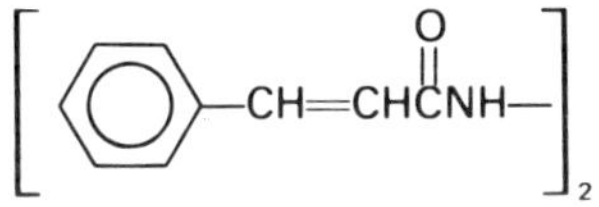	Ube Industries	279–281
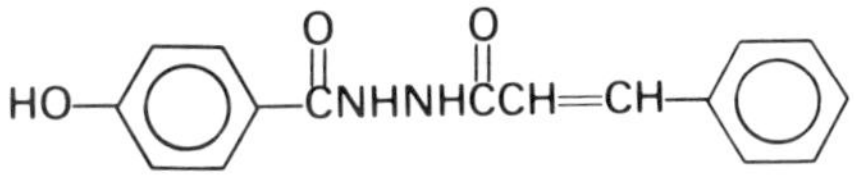	Ube Industries	302–305
Proprietary	Ube Industries	303–307
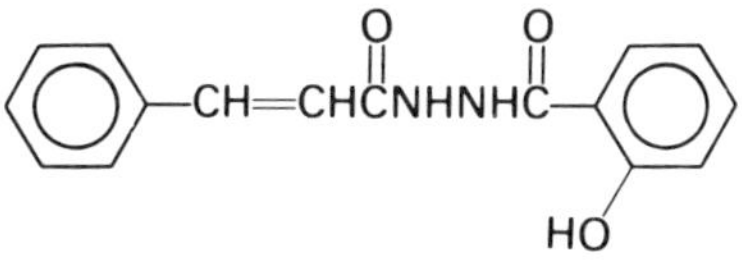	Ube Industries	281–283

polymers, most attention has been given to the copper-catalysed thermo-oxidative degradation of polyolefins (Fig. 5.37), since polyethylene is of great interest as an insulation material for electrical cables.[531,810,865,869] Examples of commercial copper deactivators are shown in Table 5.12.

5.26 ORGANOTELLURIUM COMPOUNDS AS PHOTOSTABILIZERS

Organotellurium compounds behave similarly to organotin compounds as thermal and UV stabilizers for poly(vinyl chloride).[3] A further potential advantage of tellurium over tin is that the former is capable of reacting with hydroperoxide,[704,1051] a reaction of potential importance in the process of poly(vinyl chloride) stabilization.

6

Hindered Amine Light Stabilizers (HALS)

6.1 MECHANISMS OF HALS REACTIONS

The most common group of hindered amine light stabilizers (abbreviated to HALS) originates from 2,2,6,6-tetramethylpiperidine (**6.1**). The name HALS is generally accepted to be used for hindered piperidine analogues. However, HALS include a number of other structurally different hindered amines (Table 6.1). This group of stabilizers has been commercially developed by the Sankyo Co. Ltd (Japan) and Ciba–Geigy Ltd (Switzerland).

Since the discovery of the effectiveness of hindered amines as photostabilizers in the 1960s, more than 800 patent applications involving different HALS compounds were filed.[811] The history of the discovery, chemistry, progress and application of HALS has been the subject of a number of review articles.[241,401,498,556,649,650,766,811,885,966,1038,1059,1110,1155,1156]

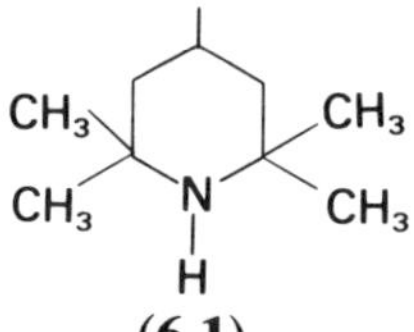

(**6.1**)

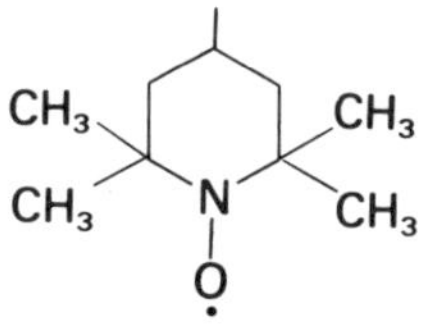

(**6.2**)

Commercially available HALS stabilizers are based on 2,2,6,6-tetramethylpiperidine (**6.1**; Table 6.1). Further substitution of the piperidine ring is employed to improve compatibility of the additive with the polymer and to enhance long-term retention of the additive during ageing of the polymer.

Only amines capable of forming stable 2,2,6,6-tetramethyl-

Table 6.1
Examples of commercially produced hindered amine light stabilizers (HALS): physical properties

Chemical name	*Commercial name*	*Chemical structure*
Bis(2,2,6,6-tetramethyl-4-piperidinyl) sebacate *or CAS name:* Bis(2,2,6,6-tetramethyl-4-piperidinyl) decanedioate	Tinuvin 770 (Ciba–Geigy); Sanol LS-770 (Sankyo); Mark LA-77 (Adeka Argus)	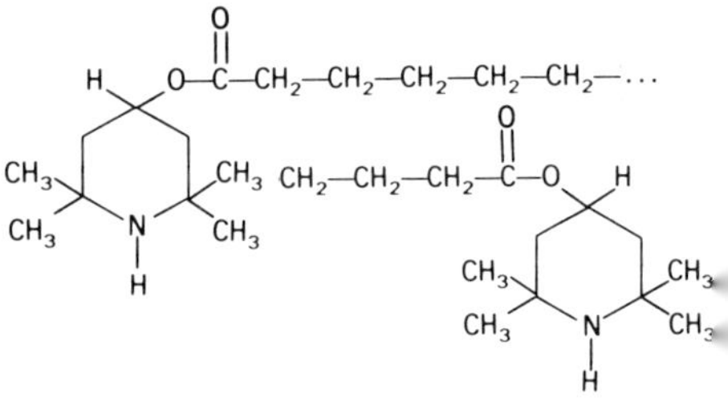 (6.3)
Bis(1,2,2,6,6-pentamethyl-4-piperidinyl) sebacate *or CAS name:* Bis(1,2,2,6,6-pentamethyl-4-piperidinyl) decanedioate	Tinuvin 292 (Ciba–Geigy); Sanol LS-292 (Sankyo)	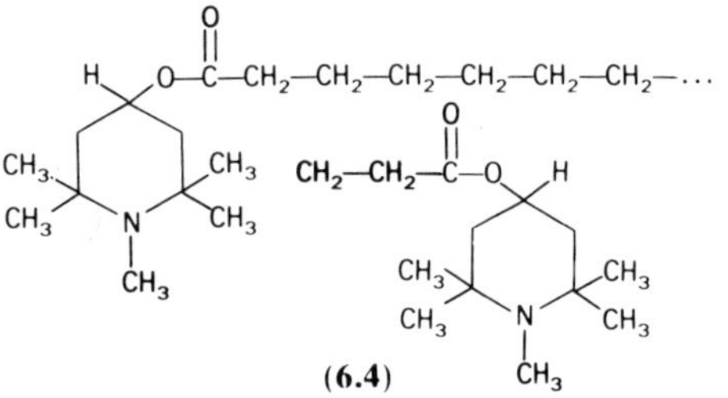 (6.4)
Polyester of succinic acid with *N-β*-hydroxyethyl-2,2,6,6-tetramethyl-4-hydroxypiperidine *CAS name:* Butenedioic acid, polymer with 4-hydroxy-2,2,6,6-tetramethyl-1-piperidine-ethanol; *CAS number* 70198-29-7	Tinuvin 622LD (Ciba–Geigy)	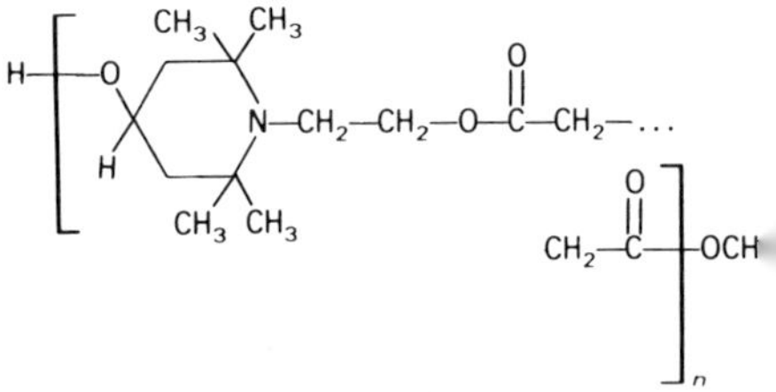 (6.5)
Poly[2-*N,N'*-di(2,2,6,6-tetramethyl-4-piperidinyl)-hexane-diamine-4-(1-amino-1,1,3,3-tetramethyl-butane)sym-triarine] *CAS name:* Poly{[6-[(1,1,3,3-tetramethylbutyl)-amino]-1,3,5-triazine-2,4-diyl]-[(2,2,6,6-tetramethyl-4-piperidinyl)-imino]-1,6-hexanediyl[(2,2,6,6-tetramethyl-4-piperidinyl)imino]}; *CAS number* 71878-19-8	Chimassorb 944 LD/FL (Ciba–Geigy)	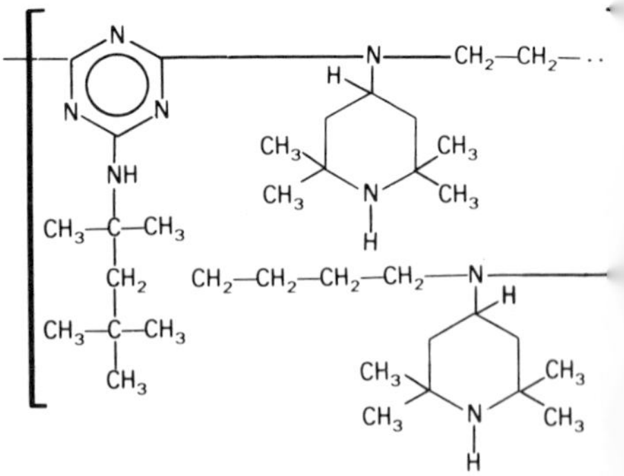 (6.6)

Molecular weight	Appearance	Solubility (wt %)									Melting point (°C)	Recommended applications
		Acetone	Benzene	Toluene	Chloroform	Ethyl acetate	Hexane	Methanol	Methylene chloride	Water		
481	White to slightly yellowish crystalline granules	19	46	—	45	24	5	38	56	<0·01	81–85	Polyolefins, EPDM, polystyrene, impact polystyrene, ABS, SAN, polyurethanes, polyamides, polyacetals
509	Slightly yellow viscous liquid	>50	—	—	—	>50	>50	>50	—	<0·01	Liquid	Polyolefins
283 (mol.) $\bar{M}_n > 2500$	White to slightly yellowish powder	4	—	15	>40	3	<0·01	0·05	>40	<0·01	55–70[a]	Polyolefins, olefin copolymers, polypropylene–elastomer blends, polyurethanes, polyamides, polyesters
599 (mol.) $\bar{M}_n > 2500$	White to slightly yellowish powder	>50	>50	—	>50	>50	40	3	>50	<0·01	103–135[a]	Polyolefins, olefin copolymers, polypropylene–elastomer blends, cross-linked polyethylene

(*continued*)

Table 6.1—*contd.*

Chemical name	*Commercial name*	*Chemical structure*
Bis(1,2,2,6,6-pentamethyl-4-piperidinyl) 2-*n*-butyl-(3,5-di-*tert*.-butyl-4-hydroxybenzyl)malonate *or CAS name:* Bis(1,2,2,6,6-pentamethyl-4-piperidinyl) [3,5-bis(1,1-dimethylethyl-4-hydroxyphenyl) methyl]butylpropanedioate; *CAS number* 63843-89-0	Tinuvin 144 (Ciba–Geigy)	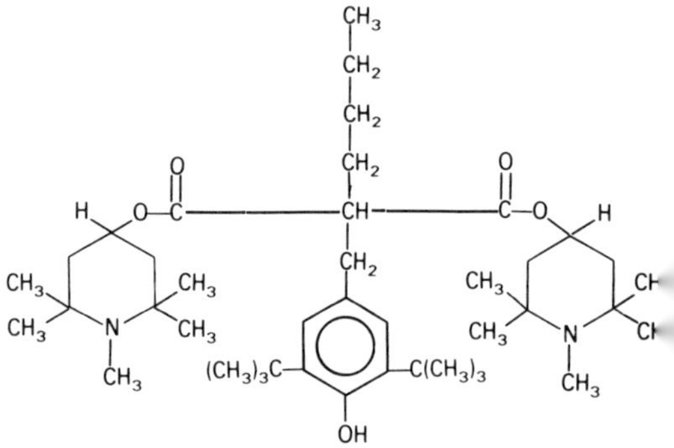 **(6.7)**
CAS name: 8-Acetyl-3-dodecyl-7,7,9,9-tetramethyl-1,3,8-triazaspiro[4.5]decane-2,4-dione; *CAS number* 82537-67-5	Tinuvin 440 (Ciba–Geigy)	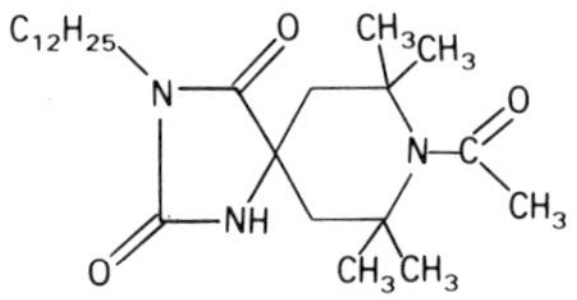 **(6.8)**
4-Benzoyloxy-2,2,6,6-tetramethylpiperidine	Sanol LS-744 (Sankyo)	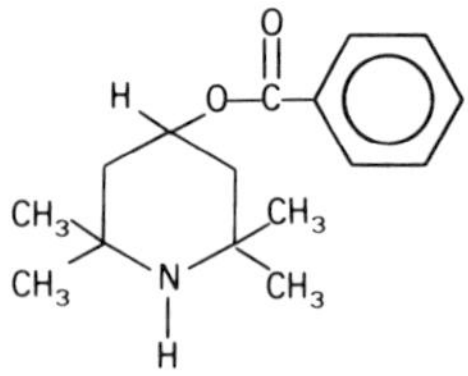 **(6.9)**

Molecular weight	*Appearance*	*Solubility (wt %)*									*Melting point (°C)*	*Recommended applications*
		Acetone	*Benzene*	*Toluene*	*Chloroform*	*Ethyl acetate*	*Hexane*	*Methanol*	*Methylene chloride*	*Water*		
685	Off-white crystalline powder	4	29	—	50	8	2	1	45	<0·01	146–150	Polyolefins, polystyrene, polyurethanes
436	White to yellow crystalline powder	—	—	70	—	—	—	37	—	0·1	74–78	Coating resins, polyurethanes, polyesters, polyacrylics
261·5							NA[b]				98	Polyolefins

(continued)

Table 6.1—*contd.*

Chemical name	*Commercial name*	*Chemical structure*
Tetra(2,2,6,6-tetramethyl-4-piperidinyl) 1,2,3,4-butanetetracarboxylate	Mark LA-57 (Adeka Argus)	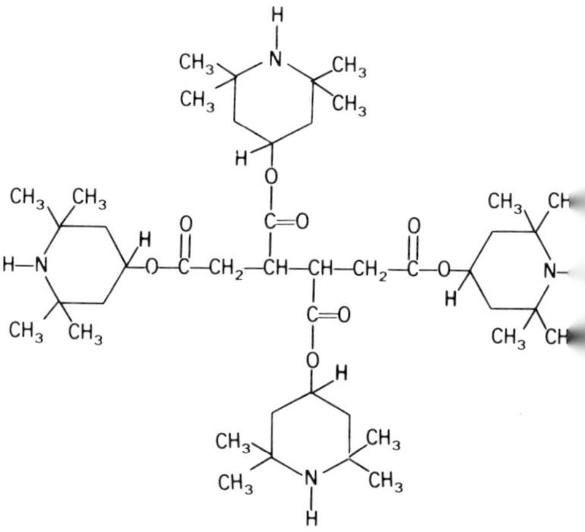 **(6.10)**
1-[2-(3,5-di-*tert*.-butyl-4-hydroxyphenyl-propionyloxy) ethyl]-4-(3,5-di-*tert*.-butyl-4-hydroxyphenylpropionyloxy)-2,2,6,6-tetramethylpiperidine	Sanol LS-2626 (Sankyo)	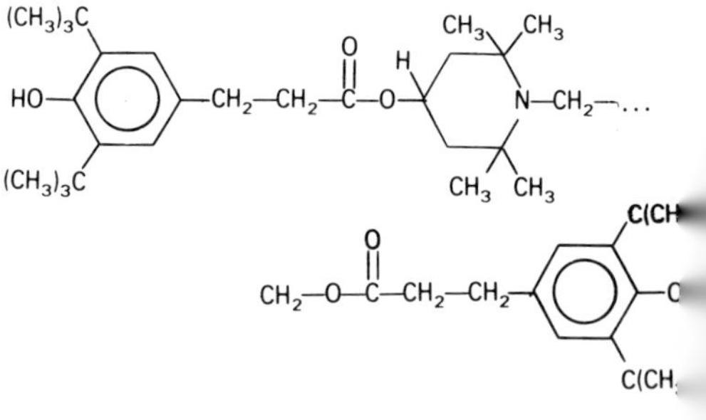 **(6.11)**
1,1′-(1,2-ethenadiyl)bis(3,3,5,5-tetramethyl-2-piperazinone)	Goodrite 3034 (Goodyear)	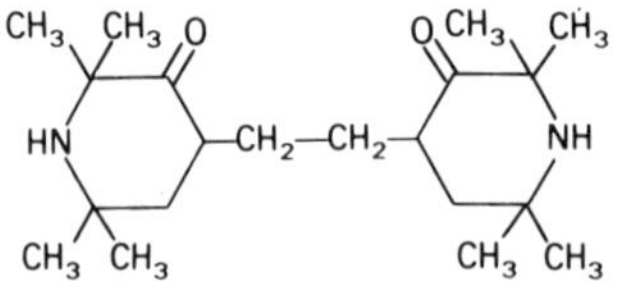 **(6.12)**

[a] Softening range.
[b] NA, Data not available.

Molecular weight	Appearance	Solubility (wt %)									Melting point (°C)	Recommended applications
		Acetone	Benzene	Toluene	Chloroform	Ethyl acetate	Hexane	Methanol	Methylene chloride	Water		
791							NA					
720							NA					
328												

piperidinoxy radicals (**6.2**) (abbreviated here for simplicity to nitroxy radicals (cf. Section 6.2) are effective photostabilizers. Tetra-substituted amines (i.e. at the 2,2,6,6-positions in the piperidine ring) which produce stable nitroxy radicals are more efficient photo-stabilizers than the tri-substituted hindered amines, the nitroxy radicals of which undergo self-reactions.[1070]

HALS stabilizers, depending on their chemical structure, can exist as liquids or solid powders (Table 6.1). They are colourless compounds, and except in the case of specific interactions with some other additives (synergistic or antagonistic reactions), do not affect the colour of the final product article. Most of the HALS stabilizers do not absorb in the UV wavelength region of terrestrial solar radiation (Fig. 6.1).

IR spectra of HALS based on 2,2,6,6-tetramethylpiperidine (**6.1**) exhibit the following absorption peaks: NH vibration at 3400 cm^{-1}, N—CH_3 vibration (doublet) at 2810–2790 cm^{-1}, NO vibration at 1345 cm^{-1}, and CO vibration at 1720 cm^{-1}.

HALS stabilizers (**6.1**) during UV irradiation in the presence of

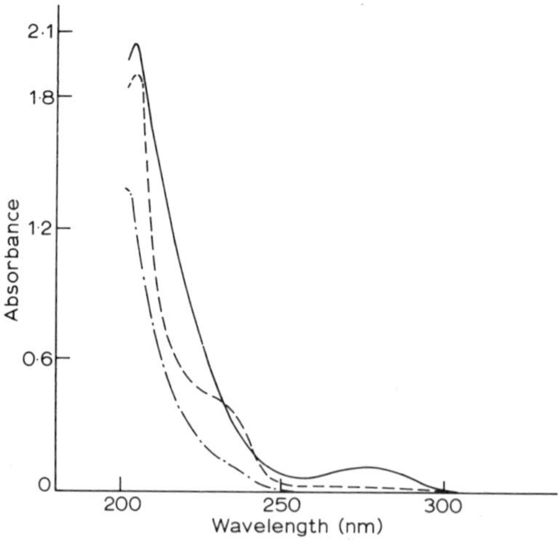

Fig. 6.1. Absorption spectra of bis(hindered piperidines) in methanol: ——, bis(1,2,2,6,6-pentamethyl-4-piperidinyl)-2-*n*-butyl-(3,5-di-*tert.*-butyl-4-hydroxybenzyl)malonate (Tinuvin 144; **6.7**) (5×10^{-5} M); – –, bis(1,2,2,6,6-pentamethyl-4-piperidinyl) sebacate (Tinuvin 292; **6.4**) (1×10^{-3} M); —·—, bis(2,2,6,6-tetramethyl-4-piperidinyl) sebacate (Tinuvin 770; **6.3**) (1×10^{-3} M).[1168]

oxygen (air) and free radicals (R•) produce nitroxy (piperidinoxy) radicals, **6.2** (cf. Section 6.2) according to the reaction:[31,82,554,1041,1060]

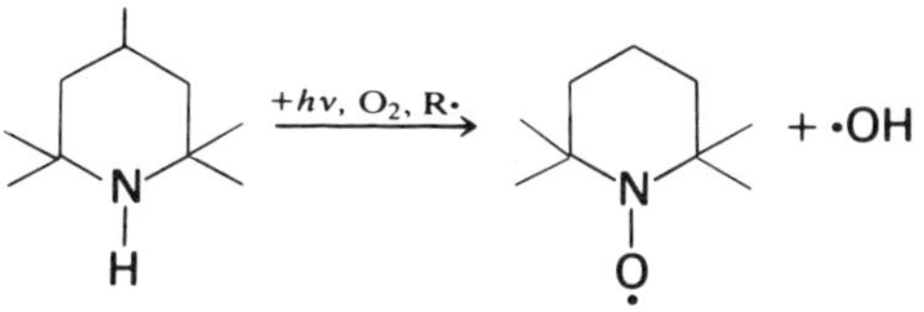

(6.1) **(6.2)** (6.1)

The mechanism of formation of nitroxy radicals (**6.2**) is not completely clear:

—What is the role of the free radicals (R•)? It has been proven that the tri-*tert.*-butylphenoxy radical[1040] or diphenylpicrylhydrazyl[1070] cannot abstract hydrogen from hindered amines.

—Hydroxyl radicals (HO•) do not react with nitroxy radicals, but probably react with free radicals (R•) to R—OH.

—What is the role of oxygen? It has been proposed that a charge-transfer (CT) complex (**6.13**) formed between HALS and oxygen can be an intermediate step in the formation of nitroxy radicals:[729]

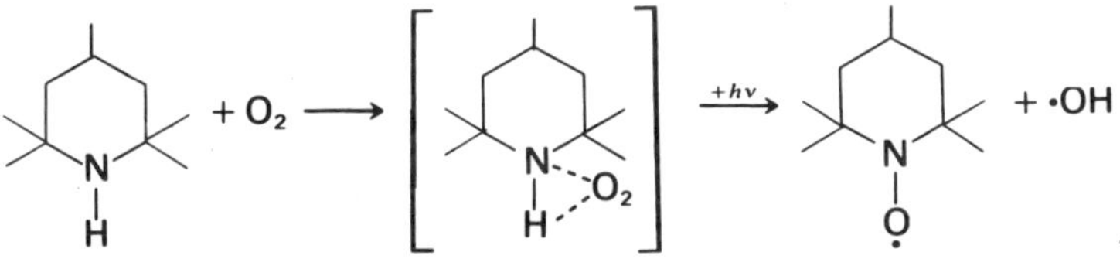

(6.13) (6.2)

Hindered amines are also capable of reacting with singlet oxygen ($^1O_2$). During the reaction nitroxy radicals are formed according to the mechanism:[155,397,734,803,925,1070,1169]

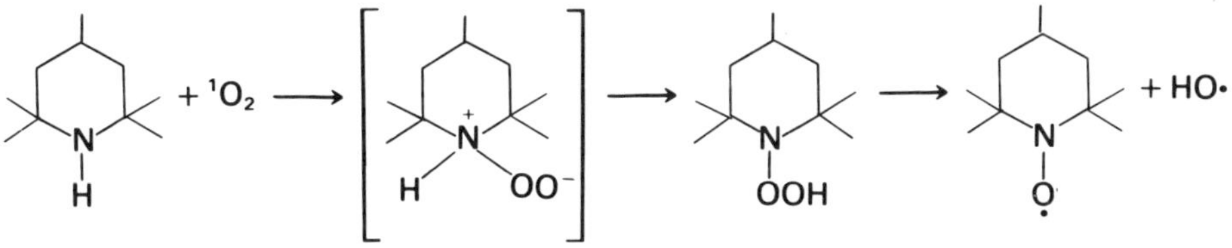

(6.3)

Table 6.2
Rate constants of quenching (k_q) of singlet oxygen ($^1O_2$) by different piperidines[803]

Piperidine	k_q *(litre mol^{-1} s^{-1})*
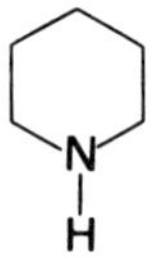	$5{\cdot}8 \times 10^6$
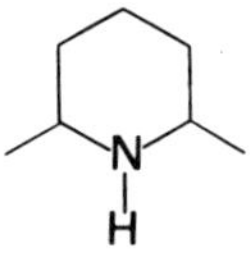	$3{\cdot}0 \times 10^6$
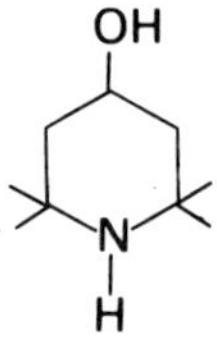	$<2 \times 10^5$
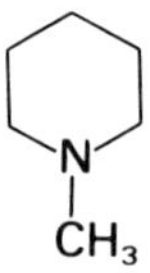	$5{\cdot}3 \times 10^7$
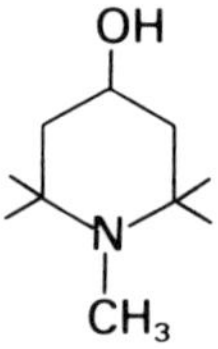	$9{\cdot}2 \times 10^7$
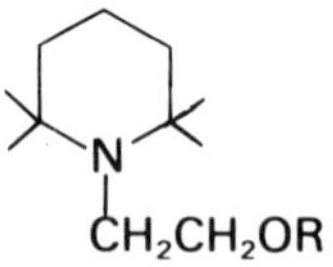 (R = OH, OCH_3)	$<2 \times 10^5$

In Table 6.2 the rate constants for singlet oxygen quenching (k_q) by some piperidines are given. The rate of $^1O_2$ quenching by piperidines depends on two factors:

(i) the ionization potential of the amine;
(ii) the amount of substitution of the carbon atoms α to the nitrogen atom. Close approach of the nitrogen to the $^1O_2$ molecule must be necessary, since highly hindered piperidines show no measurable quenching.[803]

Ozone reacts with hindered amines and nitroxy radicals with opening of the heterocyclic ring and production of 2,6-dimethyl-2-hydroxy-6-nitroheptane (**6.14**) according to the following mechanism:
Step I:[734]

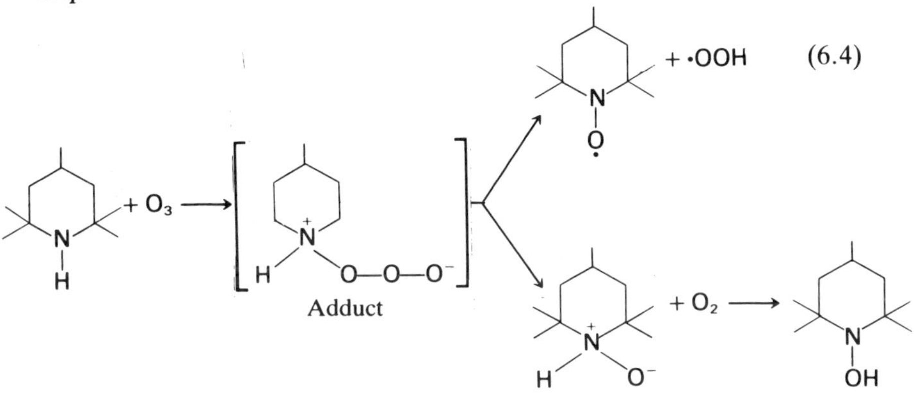

(6.4)

(6.5)

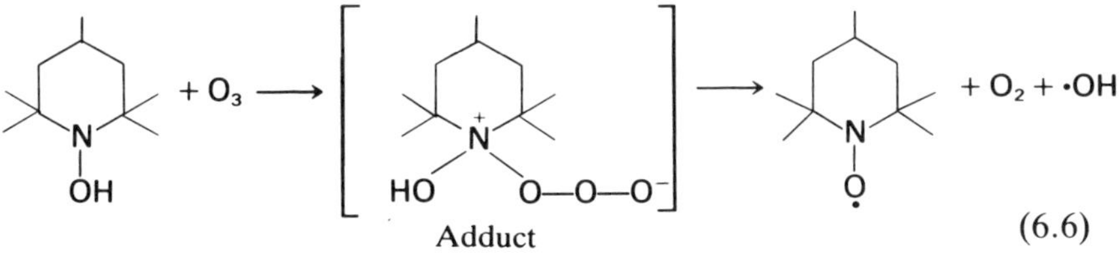

(6.6)

Step II:

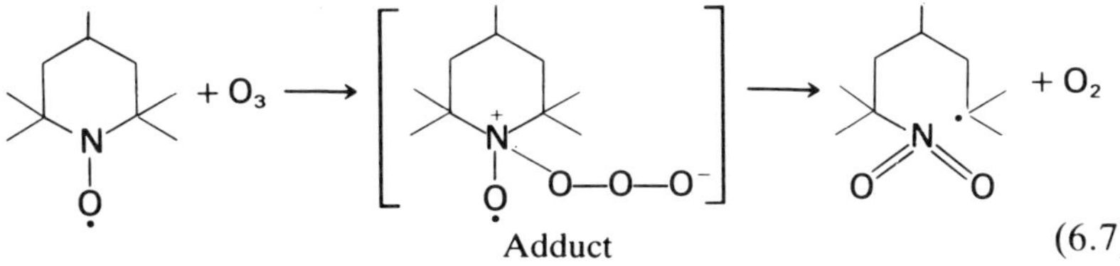

(6.7)

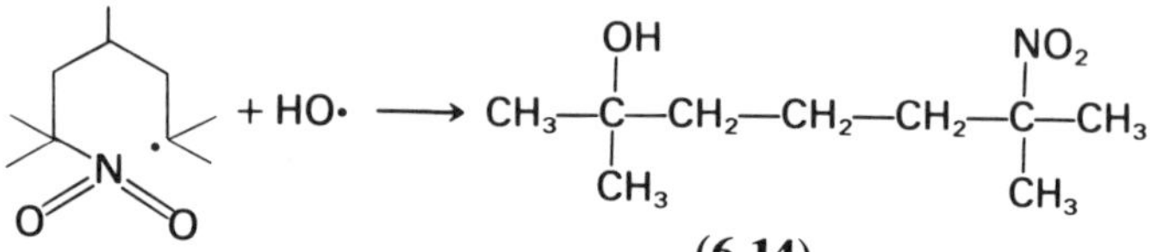

(6.8)

(6.14)

Ozonization of the hindered amine appears to occur via formation of piperidinoxy radicals (Step I) which are further ozonized to aliphatic 2,6-dimethyl-2-hydroxy-6-nitroheptane (Step II). It is well known that amines in general are very effective antiozonants.[698]

The CO_2 dissolved in H_2O and formic acid do not show any effect on HALS photostabilization properties. Other acids show an increasing effect in the order $SO_2/H_2O < HCl < NO_x/H_2O < HBr$.[1180]

Hindered amines are complexed with transition metal ions, such as titanium ions from the catalyst residue, and in this way prevent the formation of a metal ion–hydroperoxide complex which is the prime initiating species in a metal-catalysed auto-oxidation (e.g. of polyolefins).[388,389]

A number of mechanistic studies on reactivity of hindered amines in the photo-oxidation of low-molecular compounds such as iso-octane,[396] *tert.*-butyl hydroperoxide,[650,1167] tetralin,[650] benzaldehyde,[650] cyclohexane[650] and propylene tetramer[1167] have been carried out.

Hindered amines[400,1039,1040] and nitroxy radicals[273,491,678,729] form with hydroperoxides (OOH) associates **(6.15)** which have hydrogen-bond character

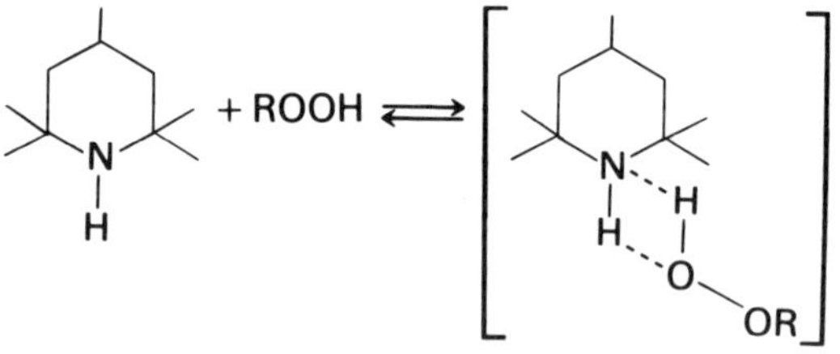

(6.9)

(6.15)

The existence of HALS–OOH associates has been proved by IR spectroscopy for 2,2,6,6-tetramethylpiperidine (and other HALS) and *tert.*-butyl hydroperoxide or cumene hydroperoxide (Fig. 6.2). Such associates exhibit a broad band absorption extending from 3700 to 2100 cm^{-1}. New absorptions at 3100 and 2800 cm^{-1} superimposed on this band are characteristic also of the associate formation

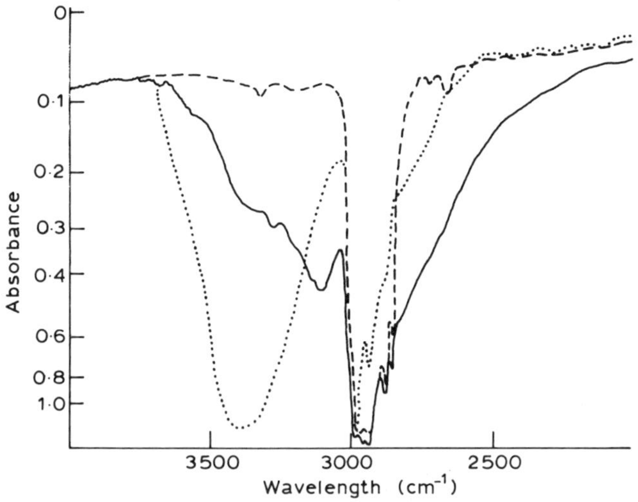

Fig. 6.2. IR spectra of: – –, 2,2,6,6-tetramethylpiperidine (**6.1**); · · ·, *tert*.-butyl hydroperoxide; ——, associate between 2,2,6,6-tetramethyl-piperidine and *tert*.-butyl hydroperoxide (1 : 1).[1039]

whereas the strong hydroperoxide dimer absorption at 3400 cm^{-1} is suppressed.[1038–1040]

HALS–polymeric hydroperoxide associates (complexes) are formed probably only for an isolated (monomeric) hydroperoxide group. In polypropylene hydroperoxide, groups may form a local dimeric 1,3-dihydroperoxide structure (**6.16**), which cannot form such associates.[289]

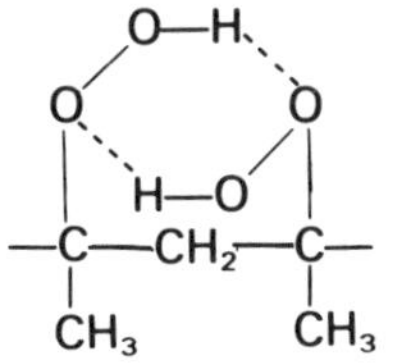

(6.16)

Thus the HALS molecules in the proximity of such a hydroperoxide will have to compete with hydroperoxide monomer–hydroperoxide dimer equilibrium. For that reason it is probably difficult to observe an IR spectrum of HALS–polymeric hydroperoxides in polypropylene film similar to that presented in Fig. 6.2.

Hindered amines react very slowly with simple low-molecular-weight hydroperoxides at moderate temperatures and probably this reaction is not significant in the polymer stabilization.[650] Studies of the mechanism of reaction of hindered amines with hydroperoxide groups have been reported in a number of publications;[25,67,154,234,236,251,401,457,490,556,678,1038,1059,1156,1175] it can be presented as a non-radical process:

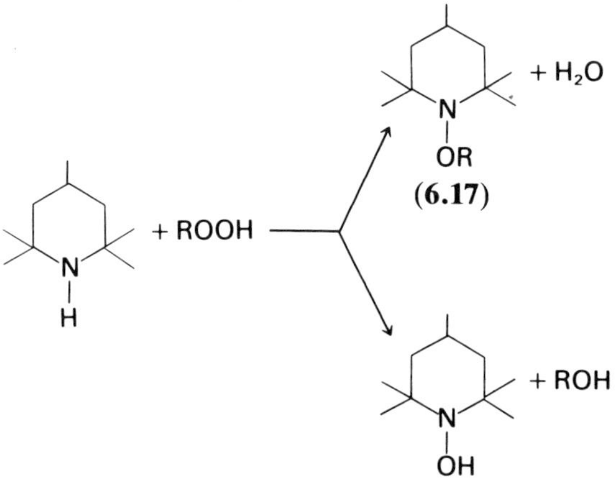

(6.10)

(6.11)

The alkoxyamine **6.17** formed in reaction (6.10) reacts with another hydroperoxide to produce a nitroxy radical **(6.2)**:[191,490,556,826,1055]

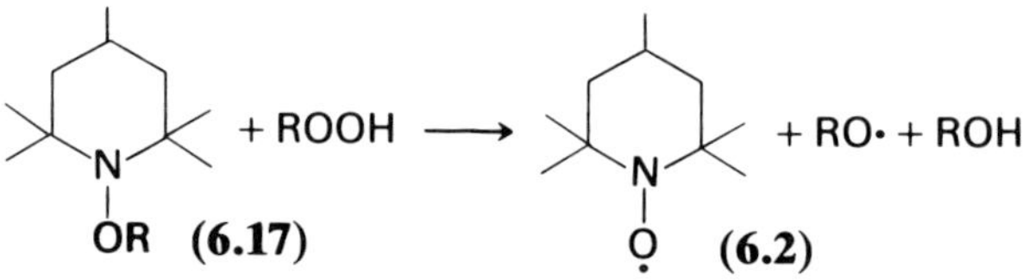

(6.12)

The alkoxyamine **6.17** also reacts with alkylperoxy readicals (ROO•) and forms nitroxy radicals **(6.2)**:[191,556,567,1041,1055]

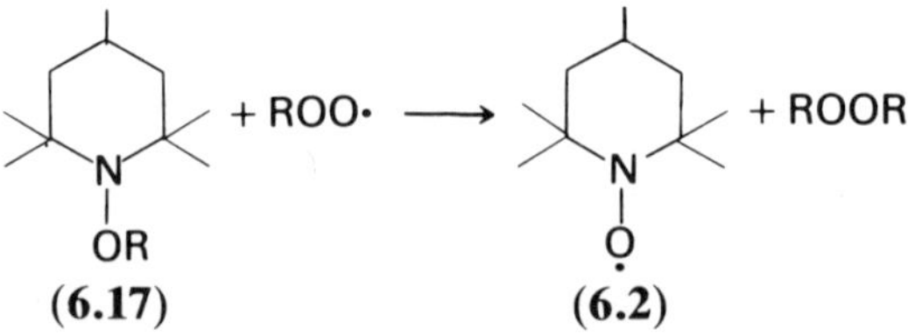

(6.13)

The rate constants for this reaction are ~100 times faster than for peroxy radical attack on *tert.*-C—H sites.[490] Formation of nitroxy radicals can be followed by ESR spectroscopy.[555]

The effect of different amines on the decomposition of *tert.*-butyl hydroperoxide decreases in the order as the number of α-H to the N:[678]

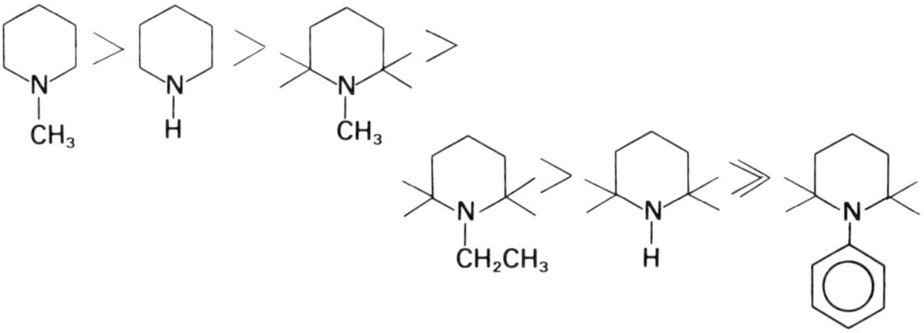

whereas the reaction rate of amines after 8 h of reaction ranks in a different order:

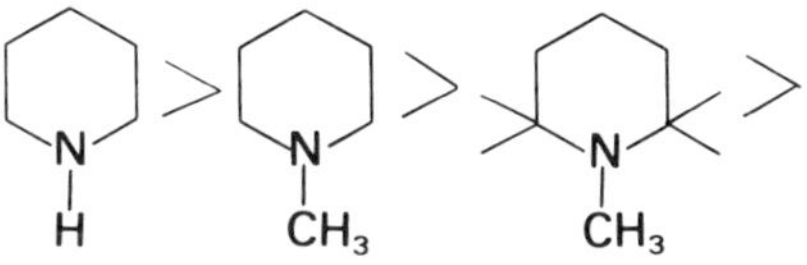

Reaction rate 100% 39·8% 33·4%

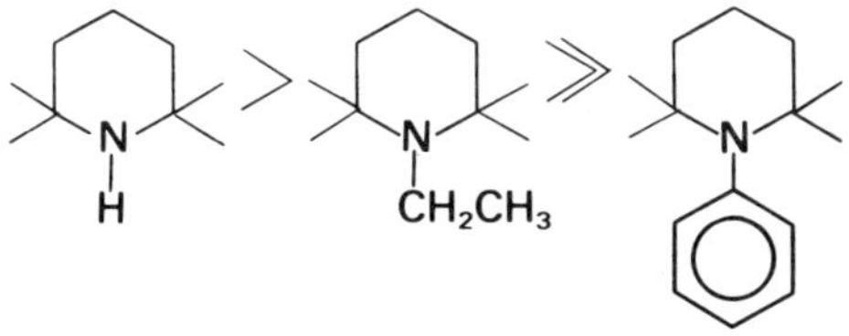

21·0% 17·0% 0%

These differences are probably because the amines do not react stoichiometrically with the hydroperoxide, and because the first reaction product, hydroxylamine, also decomposes *tert.*-butyl hydroperoxide.

Hindered amines and nitroxy radicals are capable of interrupting several hundred hydrocarbon auto-oxidation chains in the liquid phase at elevated temperature.[191]

Hindered amines act as free-radical scavengers by two mechanisms:

(i) By slow reactions with propagating polymer peroxy radicals (POO•), in this way reducing the kinetic chain length of the propagating oxidation step.[236,490,650]

(ii) Reducing the number of initiated chain reactions by fast scavenging of peracid radicals (acylperoxy radicals, R—CO—OO•) produced by non-terminating recombination of primary polymer peroxy radicals (POO•) within initiated radical pairs:[396]

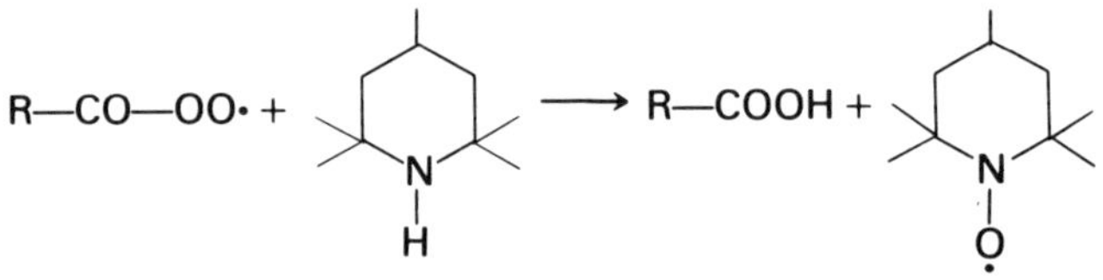

(6.14)

Hindered amines react rapidly with peracids (RCOOOH), resulting in the formation of nitroxy radicals:[1111]

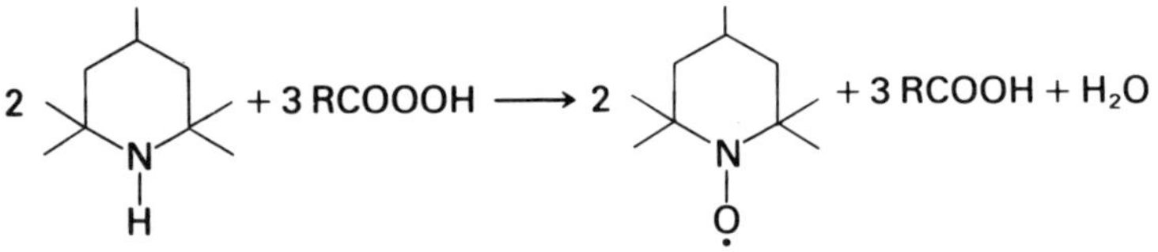

(6.15)

It has been reported that 1,2,2,6,6-pentamethylpiperidine **(6.18)** reacts with *p*-nitroperbenzoic acid **(6.19)** in methylene chloride yielding two types of nitroxy radicals **(6.20** and **6.21)**:[452]

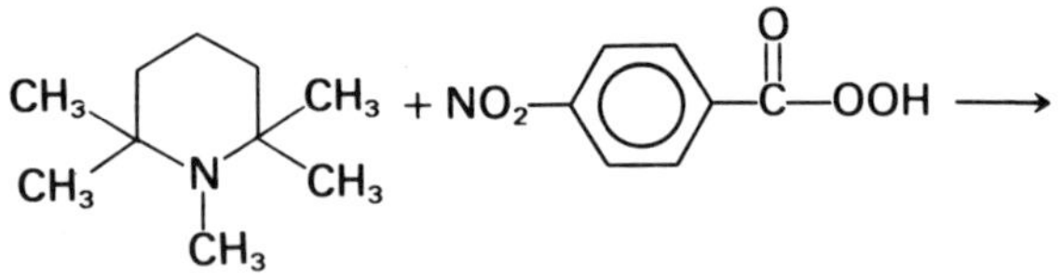

(6.18) **(6.19)**

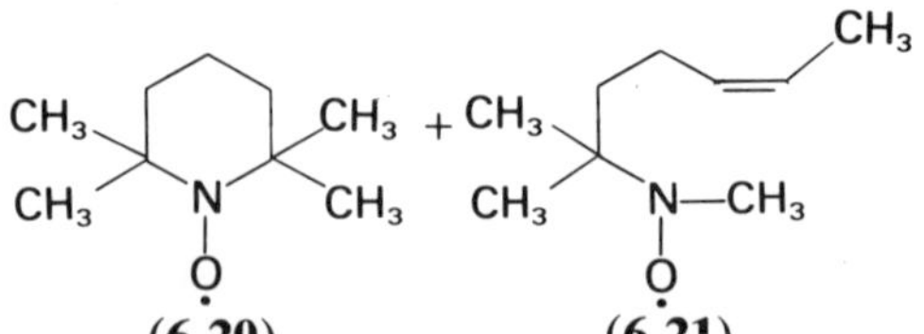

(6.20) **(6.21)**

(6.16)

Hindered amines are capable of quenching physically excited states of carbonyl groups.[399] They can also react with ketones in benzene solution under UV irradiation as follows:[729,733]

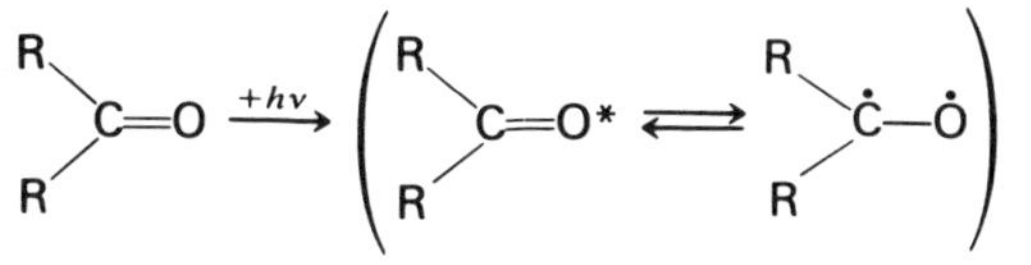

(6.17)

Excited (singlet or triplet) state

Biradical formed in excited triplet state

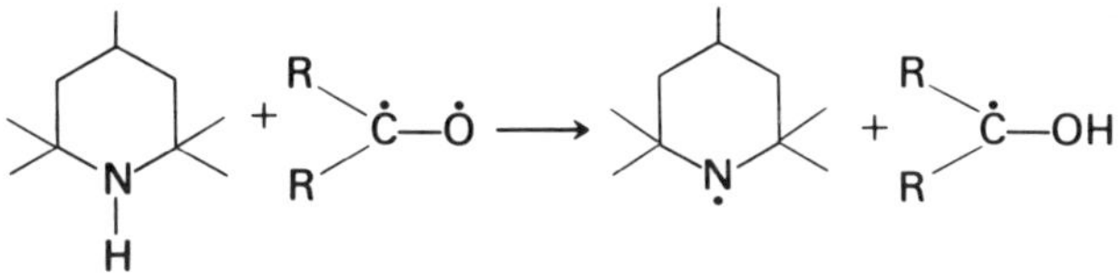

(6.18)

(6.22)

The hindered nitrogen radical (**6.22**) may further react with oxygen to form a nitrogenperoxy radical (**6.23**) which abstracts hydrogen and forms unstable hydroperoxide (**6.24**), which further decomposes to a nitroxy radical (**6.2**):

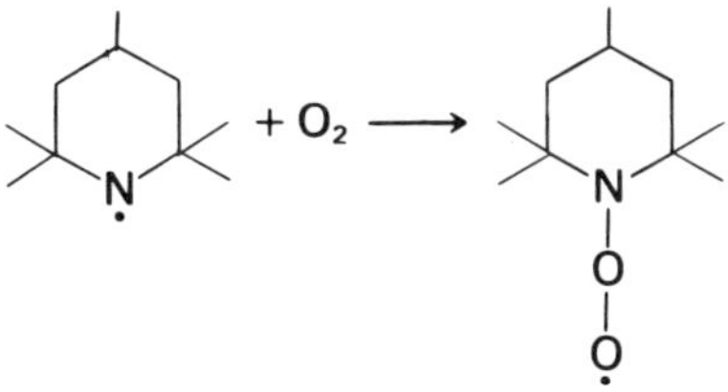

(6.19)

(6.22) **(6.23)**

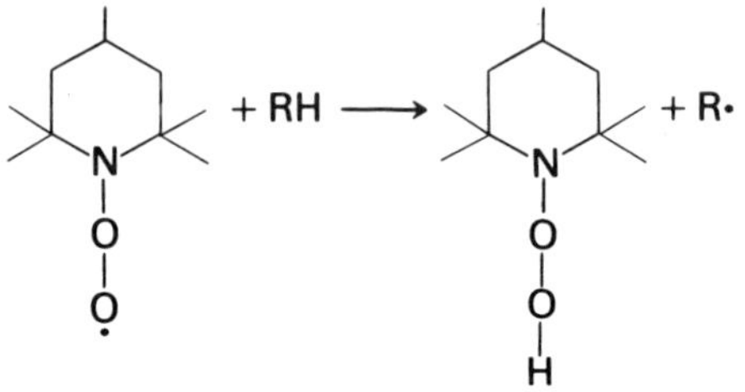

(6.20)

(6.24)

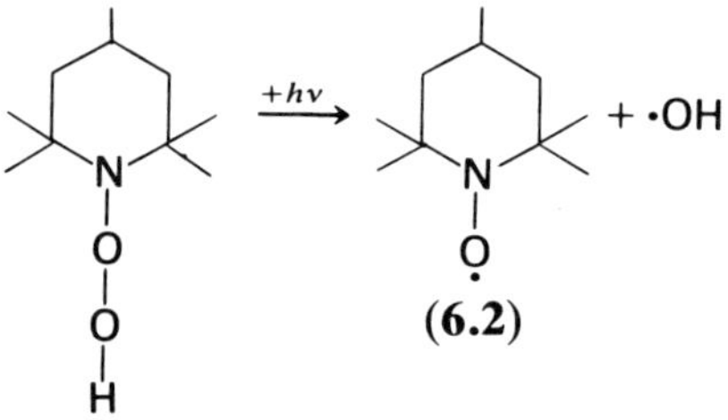

(6.2)

(6.21)

In polypropylene films containing bis(2,2,6,6-tetramethyl-4-piperidinyl) sebacate (Tinuvin 770; **6.3**) and benzophenone such reactions have not been observed.[28,55,64] However, nitroxy radicals react very efficiently with the excited triplet states of benzophenone.[61]

The excited triplet state of anthraquinone (**6.25**) reacts effectively with hindered amines (Tinuvin 770; **6.3**), abstracting hydrogen and giving the semiquinone radical **6.26**:[25,29,31,55]

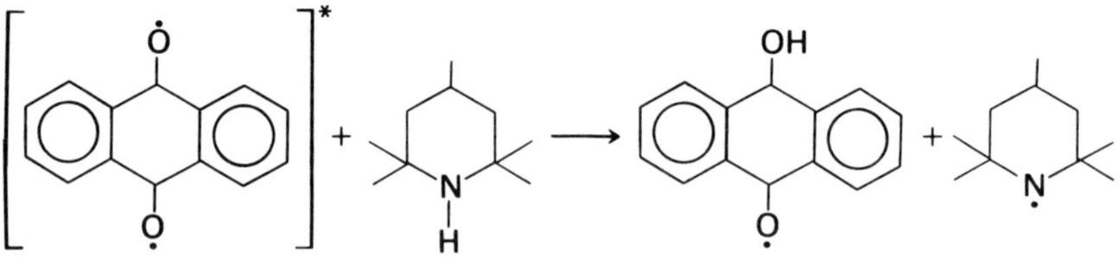

(6.25) **(6.26)**

(6.22)

Addition of the nitroxy radical (**6.2**) inhibits hydroquinone formation and has no effect on the intensity or lifetime of the phosphorescence emission from anthraquinone:[25,31]

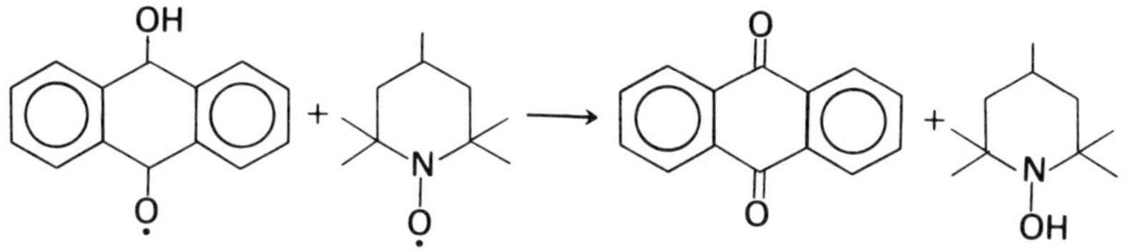

(6.2) (6.23)

Bis(2,2,6,6-tetramethyl-4-piperidinyl) sebacate (Tinuvin 770; **6.3**) reacts with 2-hydroxyanthraquinone (**6.27**) in dichloromethane giving an orange–red crystalline bis[*N*-(2-anthraquinonyl)-2,2,6,6-tetramethyl-4-piperidinyl] sebacate (**6.28**):[64]

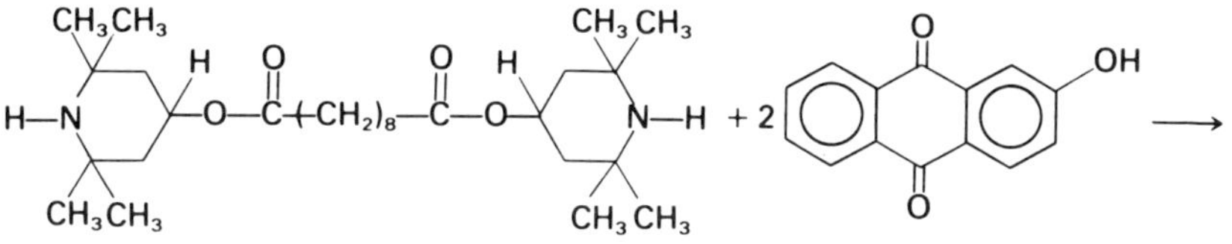

(**6.3**) (**6.27**)

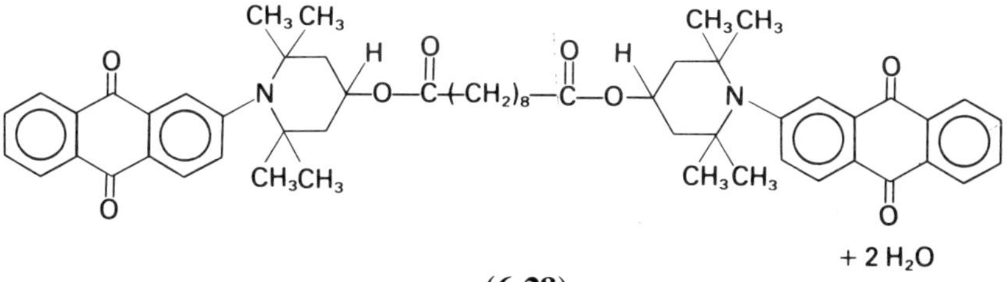

(**6.28**)

(6.24)

Tinuvin 770 (**6.3**) and its nitroxy radical inhibit the fading of the dye 1-piperidinoanthraquinone dispersed in epoxy resin films.[37]

It has been proposed that hindered amines and/or nitroxy radicals may cause a conversion of α,β- to β,γ-unsaturated ketone groups in photolysed polyolefins. In this mechanism the first-formed radical (in the α position) is trapped either by hydrogen donation from the hindered amine or by reaction with a nitroxy radical at the γ-position to form a polymer-bonded alkoxyamine.[28,31,78,81,82,88]

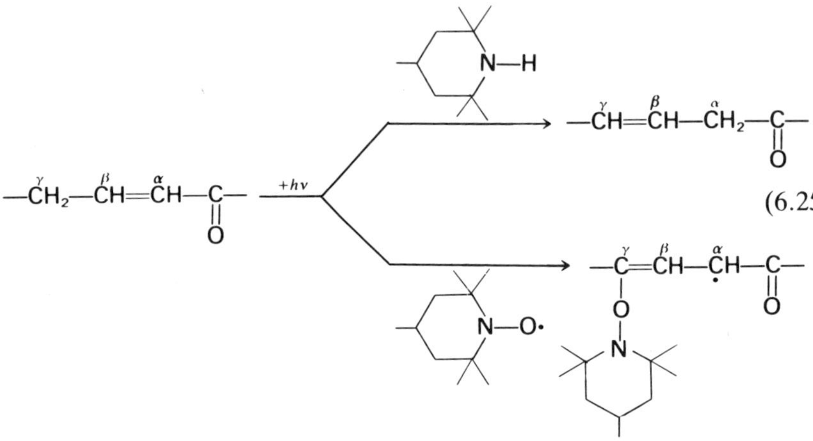

(6.25)

(6.26)

The photostabilizing mechanisms and phostabilizing effects of hindered amines have been widely investigated in many polymers such as polyethylene,[67,73,74] polypropylene,[26,28–31,55,61,63,64,67,68,73,74,78,82,88,92,94,152,234,236,251,267,296,362,396,490,499,554,556,676–678,1060,1110,1114,1116,1117] polystyrene,[295] poly(vinyl chloride),[13] polyamides,[69] rubbers,[917,1060,1168,1169] EPDM,[349] acrylic/melamine and acrylic/urethane coatings,[162,448,449,451,452,859] and epoxy/diacrylate resins.[550]

6.2 MECHANISMS OF NITROXY RADICAL REACTIONS

The nomenclature of nitroxy radicals (instead of nitroxyl radicals) is used here, according to Rule C-205 recommended by the IUPAC Commission of Nomenclature of Organic Chemistry.[952] However, nitroxy radicals originating from hindered amine light stabilizers (HALS) (based on hindered piperidine) should be called piperidinoxy radicals. For simplicity, nitroxy radical nomenclature is accepted in most papers devoted to HALS stabilizers.

The physicochemical properties, reactivity and reaction mechanisms of nitroxy radicals have been reviewed in a number of publications[328,636,966,1109] and books.[219,417,964,965,968,1181]

Nitroxy radicals exist as solid coloured powders (Table 6.3). The absorption spectrum of the 4-hydroxy-2,2,6,6-tetramethylpiperidine-

Table 6.3
Physical properties of some nitroxy radicals[294,675]

Chemical name	*Structure*	*M.p. (°C)*	*Colour*	*Crystallized from*	*Hyperfine splitting constant, $a_N(G)$*
4-Oxo-2,2,6,6-tetramethyl-piperidine-1-oxy (**6.29**)	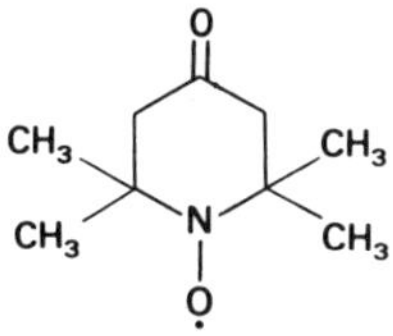 **(6.29)**	35	Orange crystals	*n*-Hexane Benzene	14·5
4-Hydroxy-2,2,6,6-tetra-methylpiperidine-1-oxy (**6.30**)	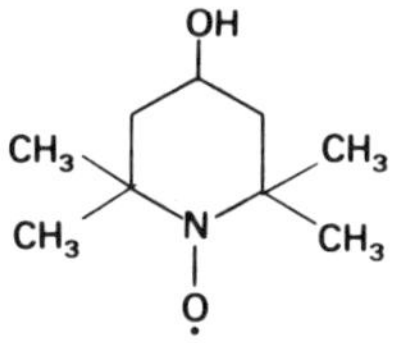 **(6.30)**	71–72	Orange crystals	Cyclohexane Benzene	17·0
-Acetoxy-2,2,6,6-tetra-methylpiperidine-1-oxy (**6.31**)	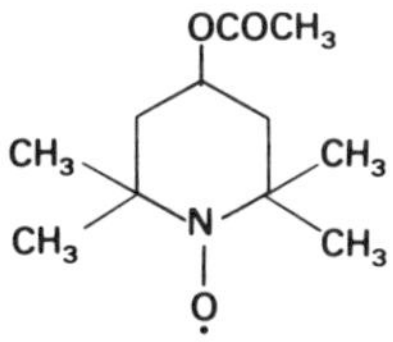 **(6.31)**	47–51	Red crystals	Cyclohexane Benzene	15·6
-Methacryloxy-2,2,6,6-tetramethylpiperidine-1-oxy (**6.32**)	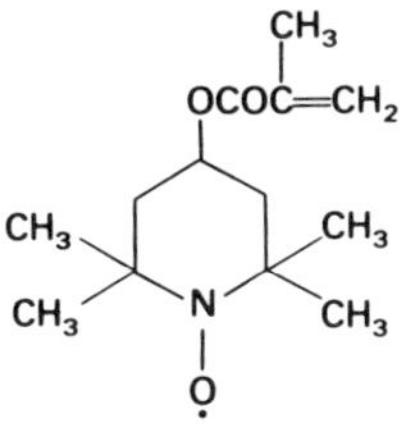 **(6.32)**	87–88	Red crystals	Cyclohexane Benzene	17·0

(continued)

Table 6.3—*contd.*

Chemical name	*Structure*	*M.p. (°C)*	*Colour*	*Crystallized from*	*Hyperfine splitting constant, $a_N(G)$*
4-Acetylamino-2,2,6,6,-tetramethylpiperidine-1-oxy (**6.33**)	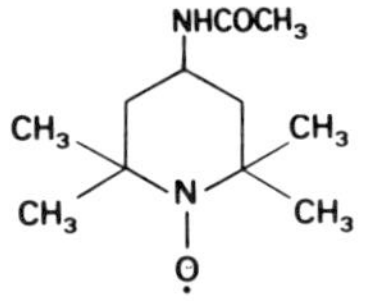 (**6.33**)	142–146	Red crystals	*n*-Hexane Methanol	16·3
4-Methacrylamino-2,2,6,6-tetramethylpiperidine-1-oxy (**6.34**)	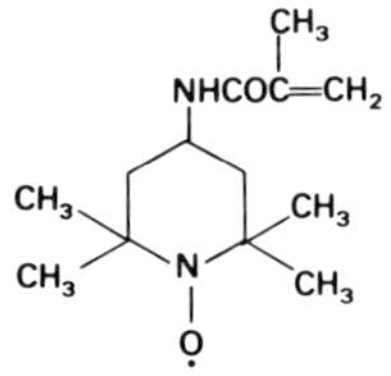 (**6.34**)	146–148	Brick red crystals	*n*-Hexane Methanol	16·4

N-oxy radical (**6.30**) is shown in Fig. 6.3. Nitroxy radicals have extraordinarily high stability. Their ESR spectra show the symmetric triplet with equal components of equal intensities (Fig. 6.4). In a restrictive environment, e.g. a polymer matrix, the ESR spectra of nitroxy radicals appear as asymmetric signals (Fig. 6.5).[153,449,557,637] Increasing the temperature of the sample or dissolving it in benzene causes transformation of the asymmetric component to the symmetric triplet (1 : 1 : 1) (as shown in Fig. 6.4).

Decreasing the temperature of a solution containing nitroxy radicals causes a change of the symmetric signal to asymmetric.[153] These changes are the consequences of the mobility change of the nitroxy radicals.

Nitroxy radicals are localized only in the amorphous phase in a polymer matrix.[494] The rotational and translational motions of a 2,2,6,6-tetramethylpiperidine-1-oxy radical (**6.2**) in a polypropylene matrix depend on the stretch ratio of the polymer film.[156]

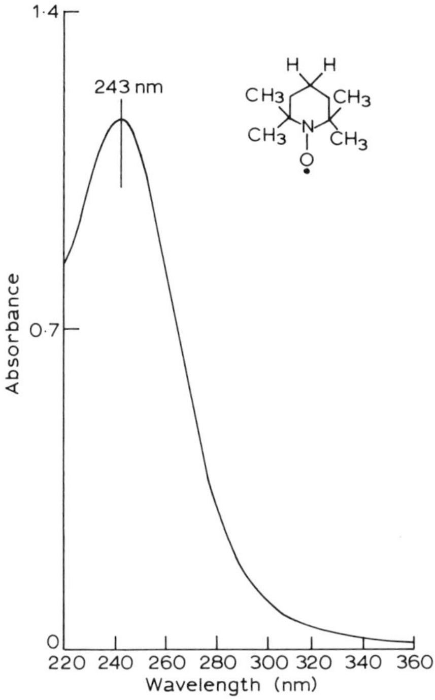

Fig. 6.3. Absorption spectrum of 4-hydroxy-2,2,6,6-tetramethylpiperidine-*N*-oxy radical, **6.30** (0·2 wt% in polypropylene film 300 μm thick).[734]

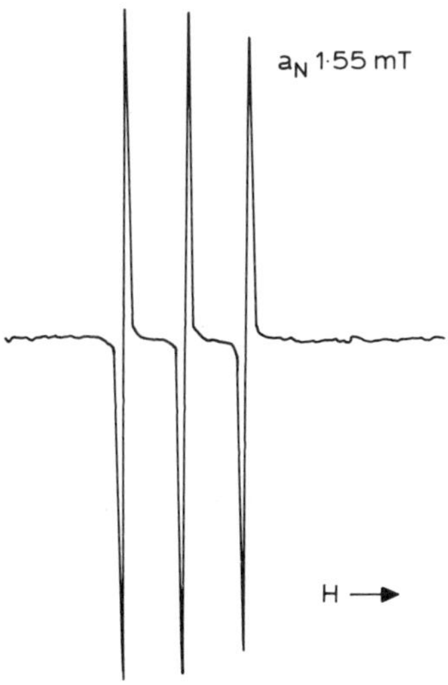

Fig. 6.4. ESR spectrum of a typical piperidinoxy radical (**6.2**) formed in photo-oxidized polypropylene films containing 4-hydroxy-2,2,6,6-tetramethylpiperidine.[554]

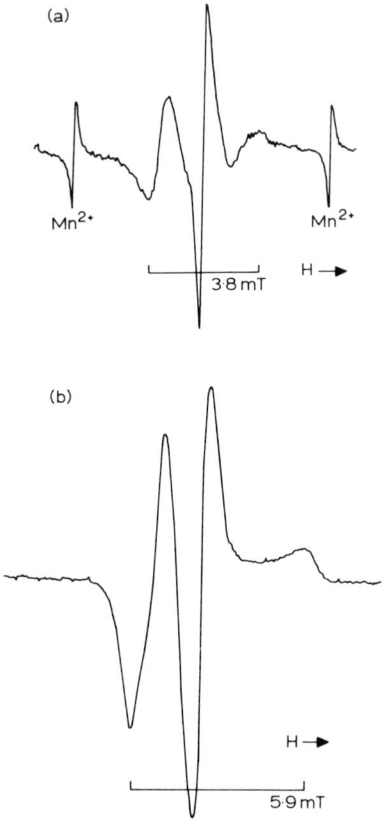

Fig. 6.5. ESR spectra of nitroxy radicals formed in polypropylene film after UV irradiation from: (a) 4-hydroxy-2,2,6,6-tetramethylpiperidine (**6.30**) and (b) bis-(2,2,6,6-tetramethyl-4-piperidinyl) sebacate (**6.50**).[554]

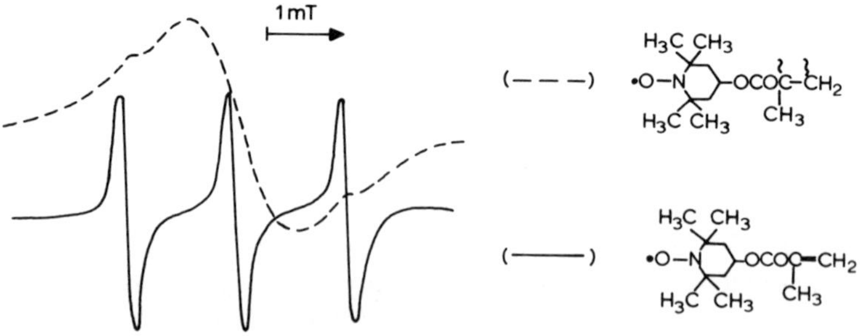

Fig. 6.6. Comparison of ESR spectra of monomeric and polymeric nitroxy radicals in benzene solution.[294]

The ESR spectrum of polymeric nitroxy radicals (**6.34**) considerably differs from that of monomer (**6.33**) (Fig. 6.6). The spin–spin interaction of the vicinal *N*-oxy groups is responsible for the change, e.g. the broadening of the bands.[31,74,218,294,555,800]

ESR spectroscopy has been widely used to monitor the kinetics of formation of nitroxy radicals and their decay during photo-oxidative degradation,[82,153,251,368,394,445–449,554,637,731,733,1117] thermal oxidation in processing,[120] and gamma-irradiation[158] of polymers stabilized with either HALS and/or hindered amine based nitroxy radicals. Low levels ($\sim 10^{-4}$ M) of nitroxy radicals can be detected by ESR spectroscopy during photo-oxidation of polypropylene containing HALS.[242,401,1060]

A number of useful ESR spectral data of nitroxy and polymeric nitroxy radicals can be obtained from publications devoted to spin labelling and spin probe techniques in ESR spectroscopy.[173,227,684,717]

Nitroxy radicals are known to be efficient quenchers of the excited singlet[278,397,493,682,1143] and triplet[224,467,569,992,1065] states of organic molecules and excimers.[492]

The quenching of excited singlet states (^1S) by stable radicals (R•) may occur via the following mechanisms:

(i) Exchange interaction in a collision complex:[332]

$$^1S + R\cdot \rightleftharpoons (S \cdots R\cdot)^* \rightarrow {^1T} + R\cdot \quad (6.27)$$

$$(S \cdots R\cdot)^* \rightarrow S_0 + R\cdot \quad (6.28)$$

where ^1T is the excited triplet state and S_0 is the ground state of the molecule.

(ii) Partial or complete reversible formation of a charge-transfer (CT) complex:

$$^1S + R\cdot \rightleftharpoons (S^{\pm} \cdots R\cdot^{\mp}) \rightarrow {^1T} + R\cdot \quad (6.29)$$

$$(S^{\pm} \cdots R\cdot^{\mp}) \rightarrow S_0 + R\cdot \quad (6.30)$$

(iii) Energy transfer

$$^1S + R\cdot \rightleftharpoons (S \cdots R\cdot)^* \longrightarrow S_0 + R\cdot^* \quad (6.31)$$

Details of the quenching mechanisms are still not clear.

Monomeric and polymeric nitroxy radicals quench the triplet state of triphenylene effectively. The analysis of the decay curves of phosphorescence of triphenylene in poly(methyl methacrylate) allowed calculation of the radii of the quenching spheres of some stable nitroxy radicals (Table 6.4).[294]

Table 6.4
Radii of quenching spheres of some nitroxy radicals for the quenching of triplet triphenylene in poly(methyl methacrylate) film at 77 K[294]

Chemical name	*Structure*	*Radius of quenching sphere (nm)*
4-Oxo-2,2,6,6-tetramethylpiperidine-1-oxy (**6.29**)	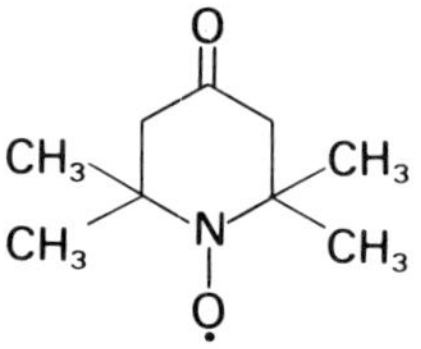 **(6.29)**	1·72
4-Hydroxy-2,2,6,6-tetramethylpiperidine-1-oxy (**6.30**)	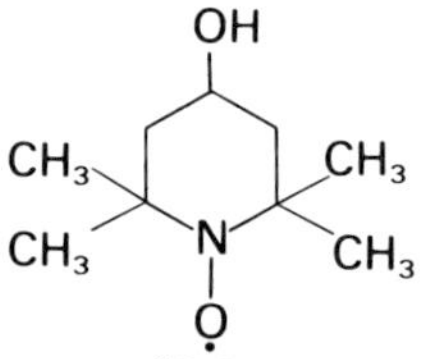 **(6.30)**	2·00
4-Acetoxy-2,2,6,6-tetramethylpiperidine-1-oxy (**6.31**)	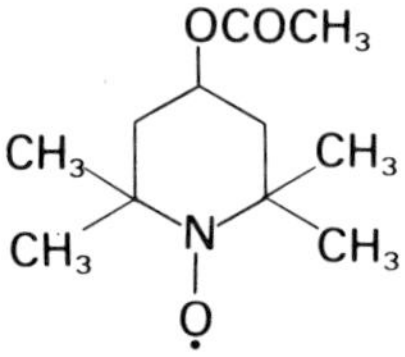 **(6.31)**	1·78
4-Methacryloxy-2,2,6,6-tetramethylpiperidine-1-oxy (**6.32**)	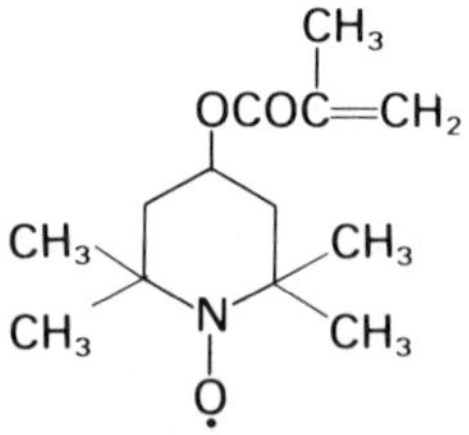 **(6.32)**	1·91

Table 6.4—*contd.*

Chemical name	*Structure*	*Radius of quenching sphere (nm)*
4-Acetylamino-2,2,6,6-tetramethylpiperidine-1-oxy (**6.33**)	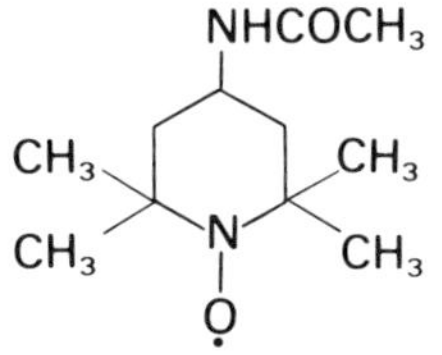 (**6.33**)	1·81
4-Methacrylamino-2,2,6,6-tetramethylpiperidine-1-oxy (**6.34**)	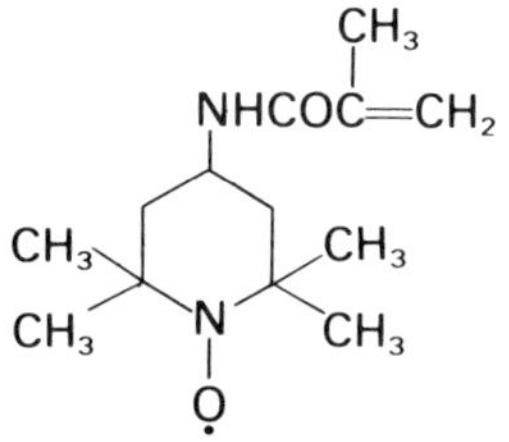 (**6.34**)	2·04
Poly(4-methacryloxy-2,2,6,6-tetramethylpiperidine-1-oxy) (**6.35**)	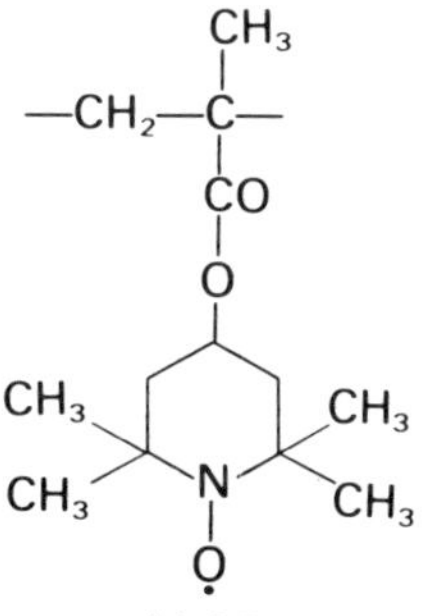 (**6.35**)	3·20

Nitroxy radicals quench singlet oxygen ($^1O_2$),[155,169,332,397,541] probably via electron-exchange interaction in collision complexes:[332]

$$^1O_2 + \rangle N{-}O\cdot \rightleftharpoons \underset{\text{Collision complex}}{(^1O_2 \cdots N{-}O\cdot)^*} \longrightarrow (O_2 \cdots N{-}O\cdot) \longrightarrow O_2 + NO\cdot \qquad (6.32)$$

(the star represents an excited state formed).

Table 6.5
Rate constants of singlet oxygen quenching by stable nitroxy radicals[332]

Compound	*Solvent*	k_q ($M^{-1}s^{-1}$)
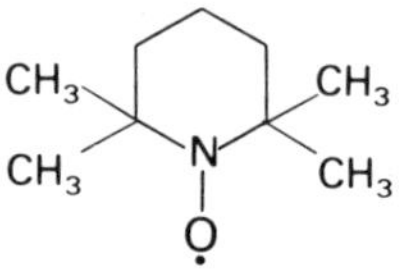	Toluene Acetonitrile Ethanol	$(5{\cdot}5 \pm 0{\cdot}17) \times 10^5$ $(4{\cdot}3 \pm 0{\cdot}27) \times 10^5$ $(2{\cdot}5 \pm 0{\cdot}10) \times 10^5$
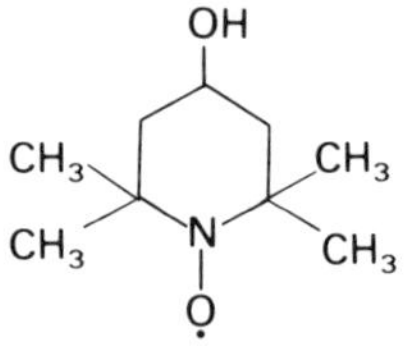	Toluene Chloroform Acetonitrile Ethanol	$(1{\cdot}56 \pm 0{\cdot}05) \times 10^5$ $(1{\cdot}14 \pm 0{\cdot}04) \times 10^5$ $(1{\cdot}40 \pm 0{\cdot}10) \times 10^5$ $(1{\cdot}24 \pm 0{\cdot}10) \times 10^5$
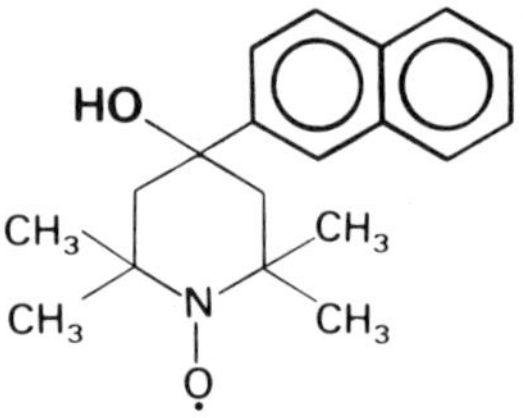	Chloroform	$(1{\cdot}00 \pm 0{\cdot}10) \times 10^5$
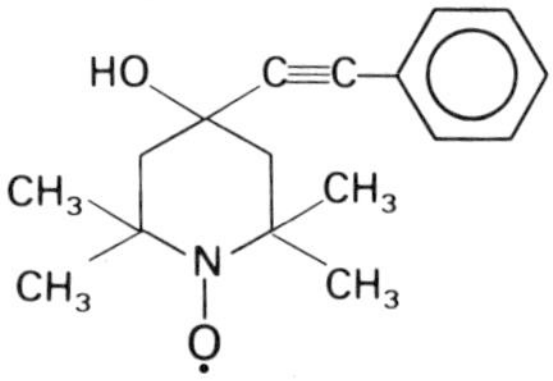	Chloroform Ethanol	$(7{\cdot}5 \pm 0{\cdot}8) \times 10^4$ $(1{\cdot}7 \pm 0{\cdot}15) \times 10^5$
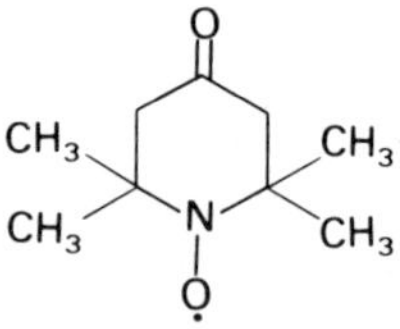	Toluene Acetonitrile Ethanol	$(1{\cdot}2 \pm 0{\cdot}1) \times 10^6$ $(1{\cdot}3 \pm 0{\cdot}15) \times 10^6$ $(1{\cdot}3 \pm 0{\cdot}1) \times 10^6$
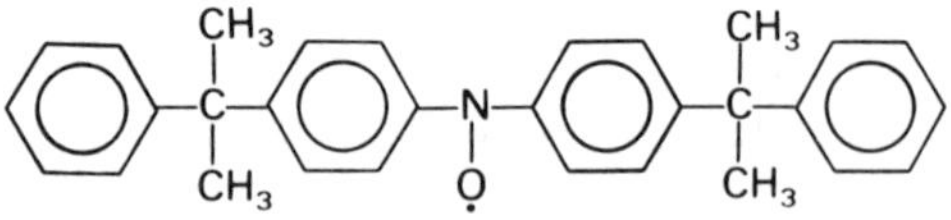	Toluene	$(2{\cdot}5 \pm 0{\cdot}2) \times 10^6$
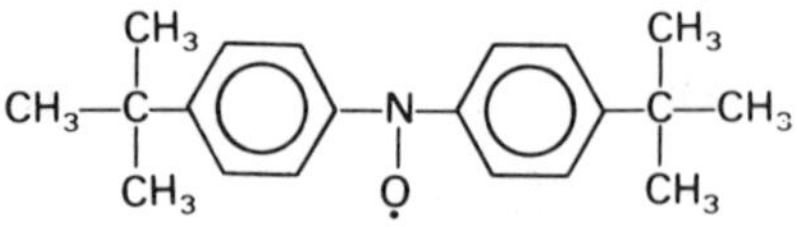	Toluene Acetonitrile	$(1{\cdot}6 \pm 0{\cdot}15) \times 10^7$ $(1{\cdot}7 \pm 0{\cdot}12) \times 10^7$

The rate constants of $^1O_2$ quenching (k_q) can be calculated from the equation:[332]

$$k_{lum} = \tau_0^{-1} + k_q[R\cdot] \quad (6.33)$$

where k_{lum} = luminescence decay rate (recorded at 1·27 μm),
τ_0 = lifetime of singlet oxygen in the pure solvent (ethanol 14 ± 1·5 μs, toluene 28 ± 2 μs, acetonitrile 65 ± 5 μs, chloroform 250 ± 20 μs).

Other methods for the determination of the quenching rate constant (k_q) are described in Section 9.13. Rate constants of $^1O_2$ quenching by nitroxy radicals are collected in Table 6.5. These data show that the nitroxy radicals are weak quenchers of singlet oxygen with $k_q \simeq 10^5$–10^6 M^{-1} s^{-1}: k_q values are lower than those of the triplet state (T_1) of anthracene, perylene and tetracene by different nitroxy radicals ($k_q \simeq 10^6$–10^8 M^{-1} s^{-1}).[467,683,1144] The quenching reaction of singlet oxygen ($^1O_2$) occurs mainly by the electron-exchange interactions:

$$^1O_2 + R\cdot \rightleftharpoons (^1O_2 \cdots R\cdot)^* \longrightarrow {}^3O_2 + R\cdot \quad (6.34)$$

Low-molecular-weight nitroxy radicals are unstable at elevated temperatures. For example, the 2,2,6,6-tetramethyl-4-oxopiperidine-1-oxy radical (**6.29**) decomposes at 105–110°C to 1-hydroxy-2,2,6,6-tetramethyl-4-oxopiperidine (**6.36**) and a trace of 2,6-dimethyl-2,5-heptadiene-4-one (**6.37**) according to the mechanism:[1110,1174]

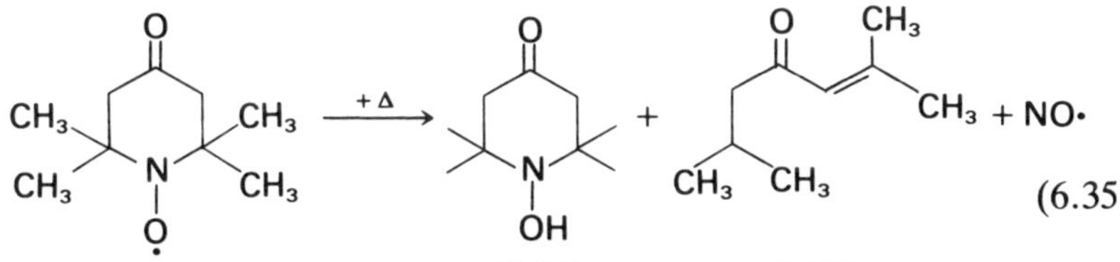

(6.35)

(**6.29**) (**6.36**) (**6.37**)

The crystals of (**6.29**), after six months' standing at room temperature, yield 1-hydroxy-2,2,6,6-tetramethyl-3-(2,2,6,6-tetramethyl-4-oxopiperidinoxy)-4-oxopiperidine (**6.38**):

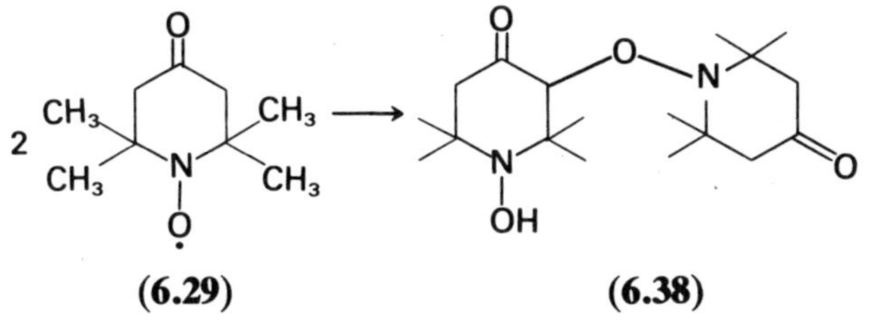

(6.36)

(**6.29**) (**6.38**)

Because of high polymer processing temperatures (150–280°C), low-molecular-weight nitroxy radicals are useless in the stabilization of polymers.

Nitroxy radicals are reactive species and recombine with alkyl radicals (R•) to alkoxyamines (hydroxylamine ethers) (**6.17**) at a rate of 10^6–10^8 litre mol^{-1} s^{-1}, which is a factor of ~20 slower than O_2 combination with alkyl radicals.[25,362,490,491,596,604,605,1155]

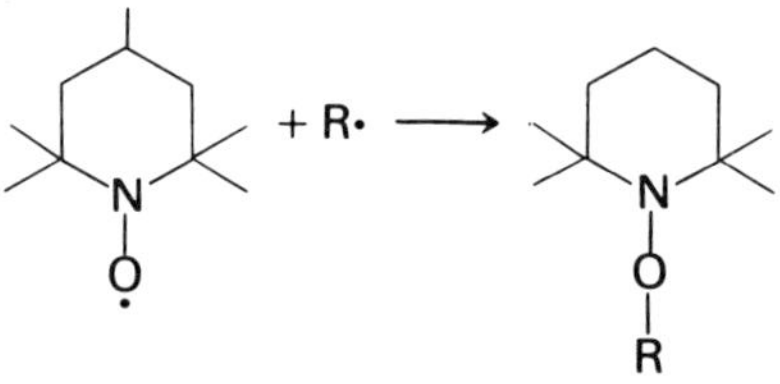

(6.37)

(6.17)

$$R\cdot + O_2 \longrightarrow ROO\cdot \qquad (6.38)$$

They also react vigorously with hydroxy radicals:

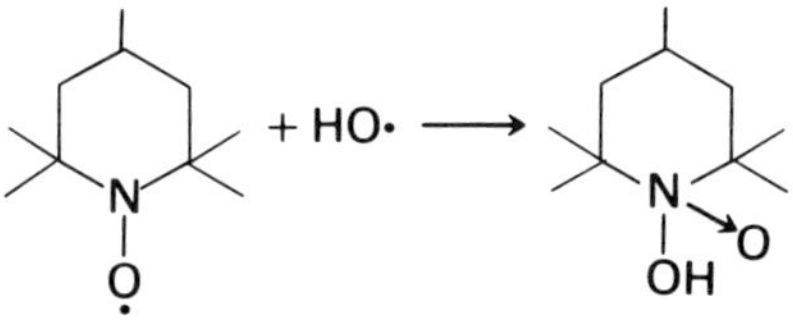

(6.39)

Nitroxy radicals do not react with alkylperoxy radicals (ROO•); however they can probably form charge-transfer (CT) complexes, **6.39**:[399,401]

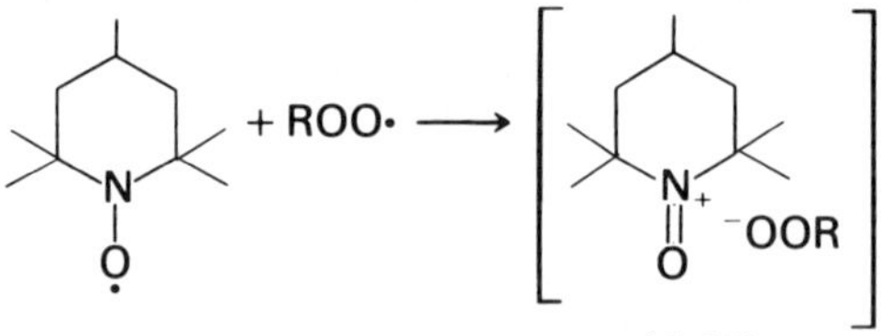

(6.40)

(6.39)

Reaction of nitroxy radicals with alkylperoxy radicals (ROO•) is still contradictory. It has been reported that bis(1,2,2,6,6-pentamethyl-4-piperidinyl) sebacate (Tinuvin 770; **6.3**) reacts very slowly with

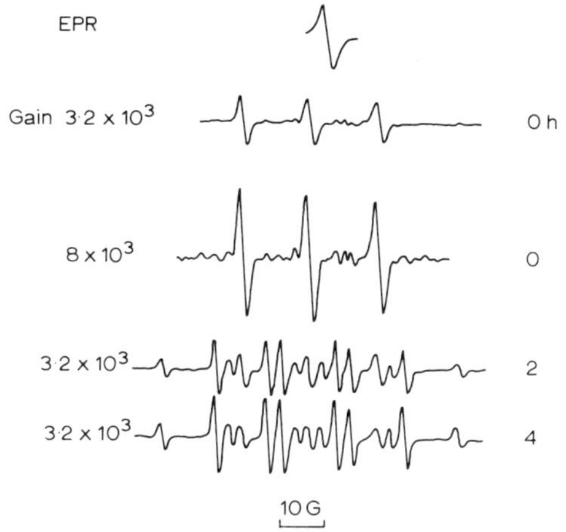

Fig. 6.7. Changes in the ESR spectra with time of reaction of bis(1,2,2,6,6-pentamethyl-4-piperidinyl) sebacate (Tinuvin 292; **6.4**) with *tert.*-butyl hydroperoxide at 40°C under nitrogen.[1167]

tert.-butyl hydroperoxide.[1167] The initial product is the nitroxy radical (Fig. 6.7), but a more complicated ESR spectrum can be observed after several hours (Fig. 6.8). This new ESR spectrum has been attributed to another nitroxy radical (**6.40**) formed by a ring-opening reaction via the *Cope rearrangement,* in which ROO• radicals

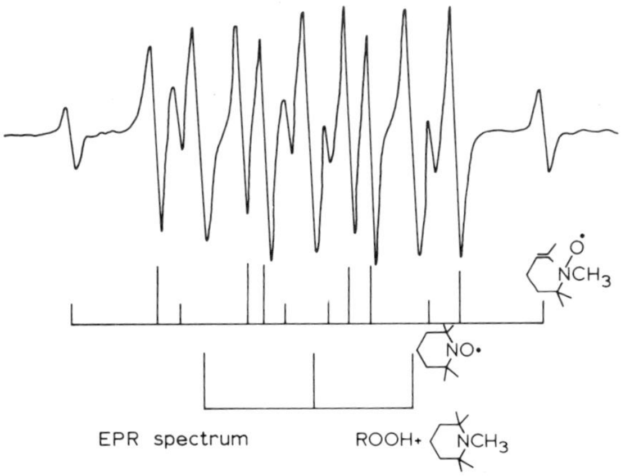

Fig. 6.8. ESR spectrum of nitroxy radical (**6.40**) formed by ring opening reaction via *Cope rearrangement.*[1167]

participate:

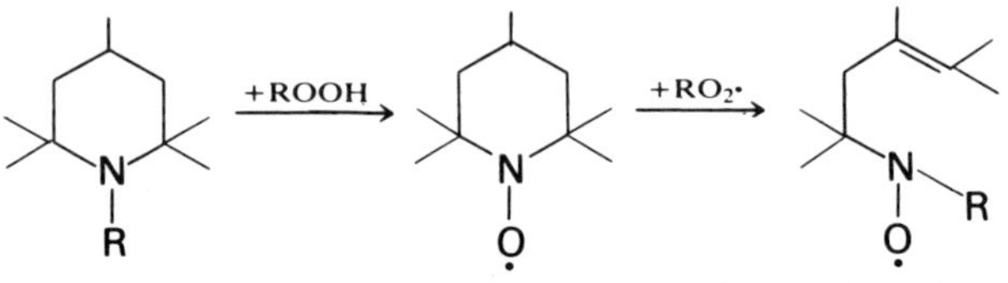

(6.41)

(6.40)

Nitroxy radicals are capable of oxidizing aliphatic alcohols to ketones by the following mechanism:[429]

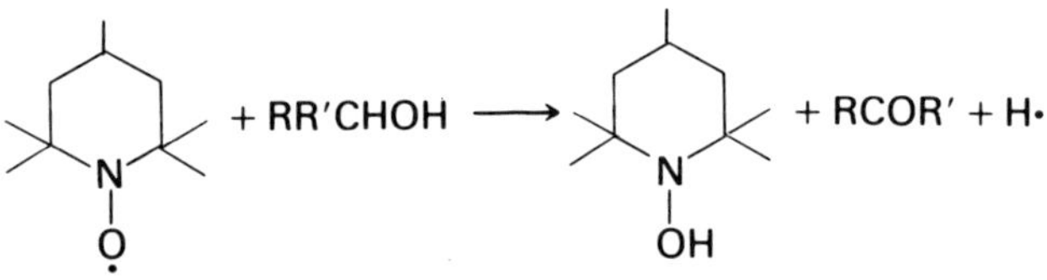

(6.42)

A similar reaction has been proposed between nitroxy radicals and a hindered phenolic antioxidant (Irganox 1010; **3.32**), during which a resonance-stabilized quinone radical (**6.41**) and hydroxylamine are formed:[24,30,637]

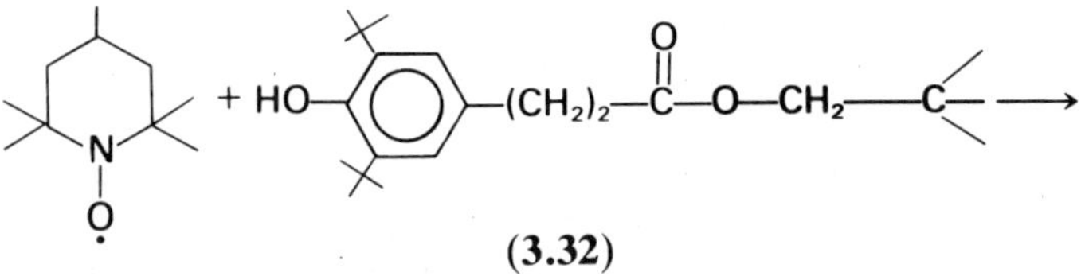

(3.32)

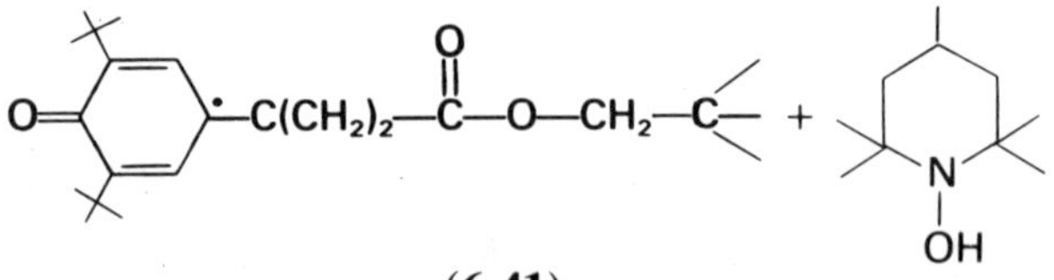

(6.43)

(6.41)

In acidic medium, nitroxy radicals disproportionate to produce hydroxylamines and oxoammonium salts (**6.42**):[357,480]

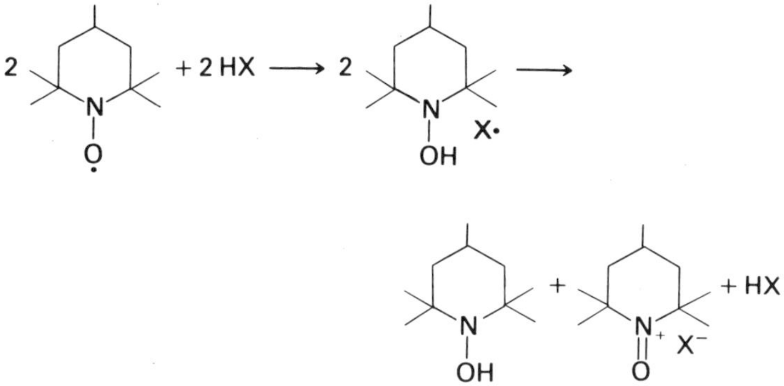

(**6.42**)

(6.44)

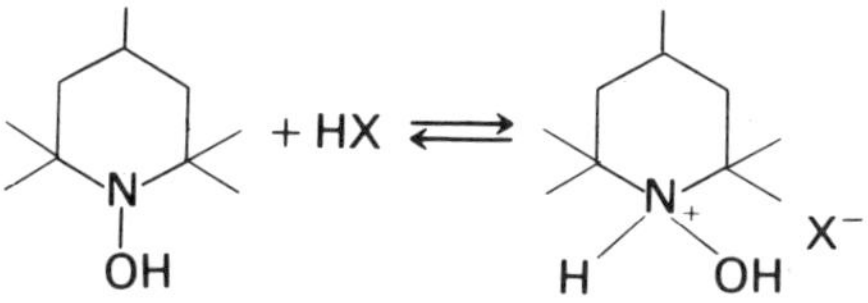

(6.45)

Oxoammonium salts are strong oxidizing agents. For example, nitroxy radicals in acid medium oxidize alcohols to the corresponding aldehydes.

Acylperoxides react with stable nitroxy radicals,[723,798,999] but there are inconsistent reports on the reaction of hydroperoxides with nitroxy radicals.[843]

According to some authors,[1040] 2,2,6,6-tetramethylpiperidine-*N*-oxy radicals (**6.2**) do not react with *tert.*-butyl hydroperoxide in benzene, whereas others report a fast decrease of the *tert.*-butyl hydroperoxide or cumene hydroperoxide content in chlorobenzene.[356,357] Hydrogen chloride evolved in photodecomposition of chlorobenzene can catalyse disproportionation of nitroxy radicals according to reaction (6.45). Each of the products formed may be further involved in the decomposition of hydroperoxide.

Irradiation of di-isopropyl ketone (**6.43**) in the presence of hindered amine (Sanol 744; **6.9**) results in formation of acetone, isobutyric acid and small amounts of isopropanol. At the same time the hindered amine is quantitatively oxidized to the corresponding nitroxy

radical:[398,399,401]

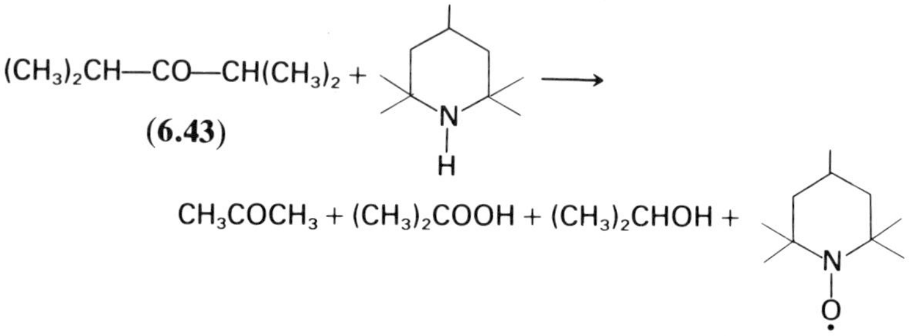

(6.43) (6.46)

It has been suggested that the isopropylacylperoxy radical **(6.44)** is responsible for the oxidation of hindered amines to nitroxy radicals:[396]

$$(CH_3)_2CH—CO—CH(CH_3)_2 \xrightarrow{+h\nu} (CH_3)_2CH—CO\cdot + \cdot CH(CH_3)_2 \quad (6.47)$$

$$(CH_3)_2CH—CO\cdot + O_2 \rightarrow (CH_3)_2CH—\overset{\overset{\displaystyle O}{\|}}{C}—OO\cdot \quad (6.48)$$

(6.44)

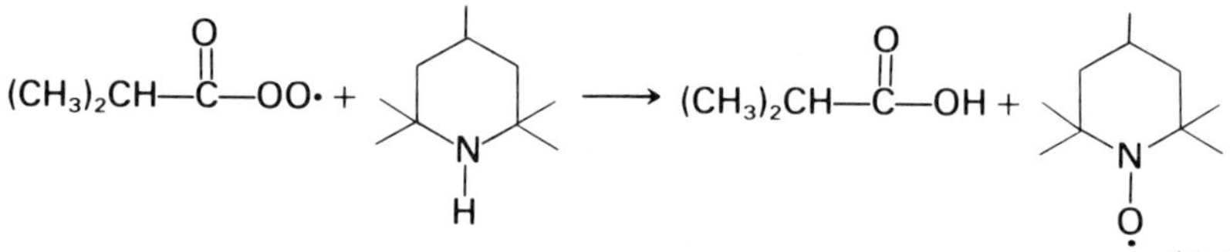

(6.49)

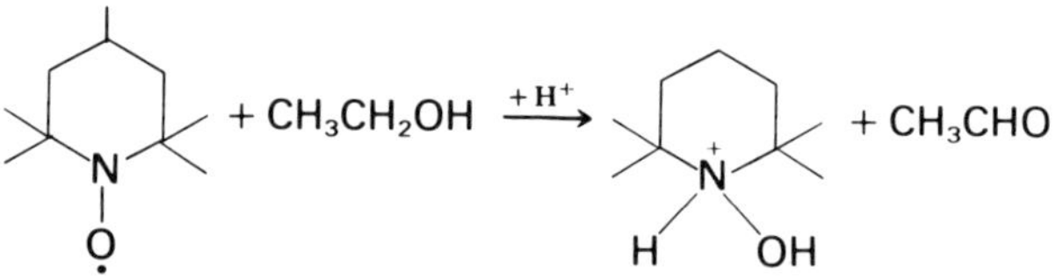

(6.50)

The interaction of acylperoxy radicals with hindered amines has been further proved by others.[650]

Strong oxidants such as chlorine and bromine oxidize nitroxy radicals to the corresponding oxoammonium salts **(6.45)**:[968]

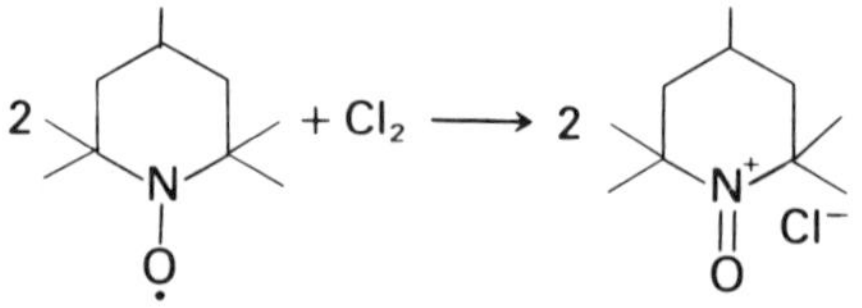

(6.51)

(6.45)

An extensive review on the role of nitroxy radicals in the oxidation of organic compounds has been published.[766]

Nitroxy radicals are capable of trapping radicals formed during the photolysis of acetone, and form *N*-acetyloxypiperidine (**6.46**):[1038]

$$CH_3COCH_3 \xrightarrow{+h\nu} CH_3CO\cdot + CH_3\cdot \qquad (6.52)$$

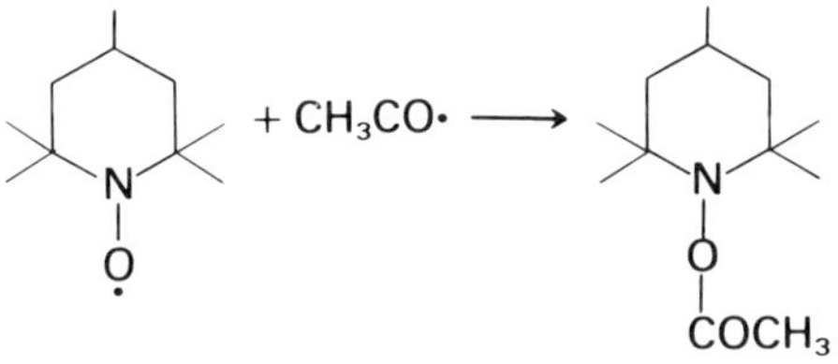

(6.53)

(**6.46**)

Bis(2,2,6,6-tetramethylpyridinyl) sebacate (Tinuvin 770; **6.3**) can complex effectively with various transition metals—iron(II), iron(III), cobalt(II), titanium(III), titanium(IV)—in the solution phase:[387,388,390]

—On adding aqueous iron(II) solution to acetone solutions of Tinuvin 770, a dark green colour (characteristic of iron(II) complexes) is formed. After a few seconds, the solution turns yellow/orange and, within 60 s, an orange product precipitates. This reaction occurs only in the presence of oxygen.

—Iron(III) and titanium(IV) are more stable oxidation states than their lower-charged analogues. The rapid oxidation of HALS complexes of iron(II) and titanium(III) occurs in the presence of oxygen. Cobalt(II) tends to be more stable towards oxidation.

The complexing of metal ions by HALS has a positive effect on stabilization of polymers. In this way metal ions are deprived of their catalyst activity for the production of hydroperoxide groups.[1118]

The 2,2,6,6-tetramethylpiperidine-1-oxy radical (**6.2**) forms coloured complexes with the perchlorates of Mn^{2+}, Fe^{2+}, Co^{2+}, Ni^{2+}, Zn^{2+} and Cu^{2+}.[967] Nitroxy radicals may also participate in redox reactions with metallic ions:[892]

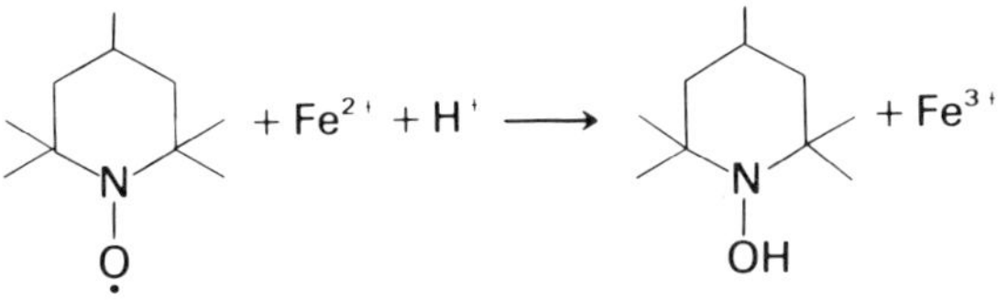

(6.54)

(**6.2**)

Nitroxy radicals show some complexing ability with certain ions, e.g. iron(II); however, the reaction is not as general or as efficient as that observed for the parent HALS. An explanation of this behaviour is not easy to formulate, but seems to be connected with the paramagnetic nature of the nitroxide radical and the differing electronic structures of the various transition metal ions encountered.[390] Other transition metal compounds such as titanium(II) and titanium(IV) salts react with hindered amines in acetone solutions yielding a white precipitate.[390]

Non-transition metal chlorides such as $AlCl_3$ and $MgCl_2$ in acetone solutions react with hindered amines to give white precipitates, which were not determined structurally.[390]

Nitroxy radicals cause yellowing of polymers, which may be due to weak absorption of visible light by these radicals themselves (Fig. 6.3) or to formation of coloured reaction products in the polymers.

The yellowing is especially intensive in the presence of phenolic antioxidants, e.g. 3,5-di-*tert.*-butyl-4-hydroxytoluene (**6.47**), which can give strongly coloured stilbenequinone compounds (**6.48**):[242,1120]

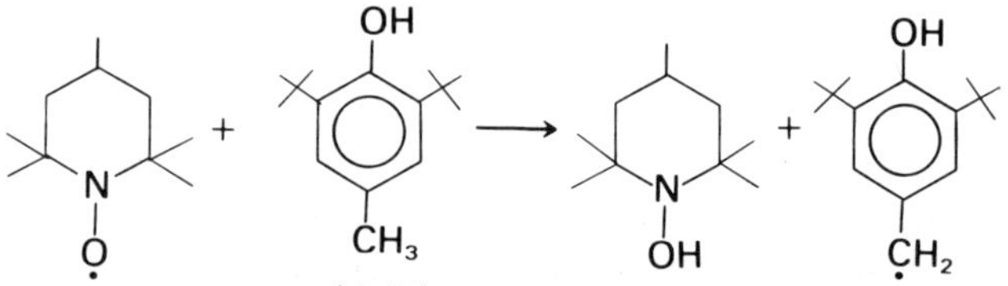

(**6.47**)

(6.55)

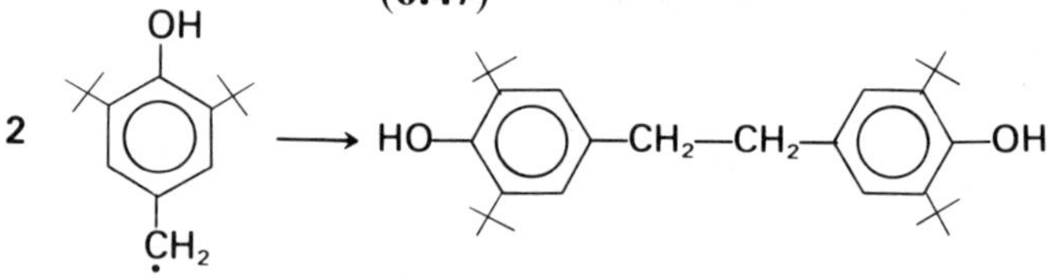

(6.56)

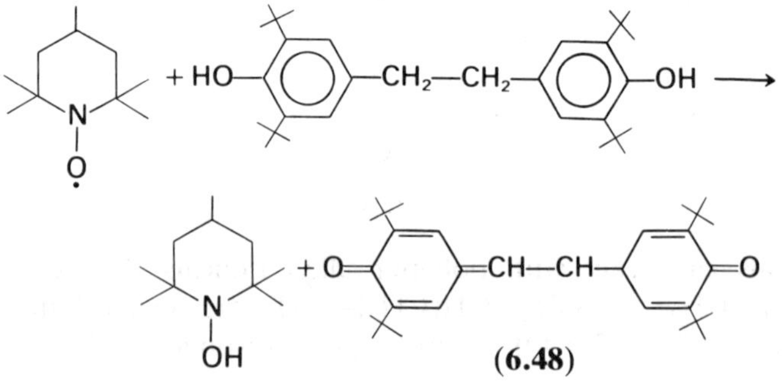

(**6.48**)

(6.57)

Nitroxy radicals react effectively with semiquinone radicals (QH•) which are a product of thermal reduction of the quinone (Q):[62,65,242]

$$Q + PH(polymer) \longrightarrow QH\bullet + P\bullet \quad (6.58)$$

$$2QH\bullet \longrightarrow Q + QH_2 \quad (6.59)$$

$$QH\bullet + PH \longrightarrow QH_2 + P\bullet \quad (6.60)$$

$$QH\bullet + \rangle N{-}O\bullet \longrightarrow Q + \rangle N{-}OH \quad (6.61)$$

$$QH_2 + \rangle N{-}O\bullet \longrightarrow QH\bullet + \rangle N{-}OH \quad (6.62)$$

The presence of the hindered piperidine compounds, particularly the nitroxy radical, inhibits the formation of the blue quinhydrone complex (**6.49**):[65]

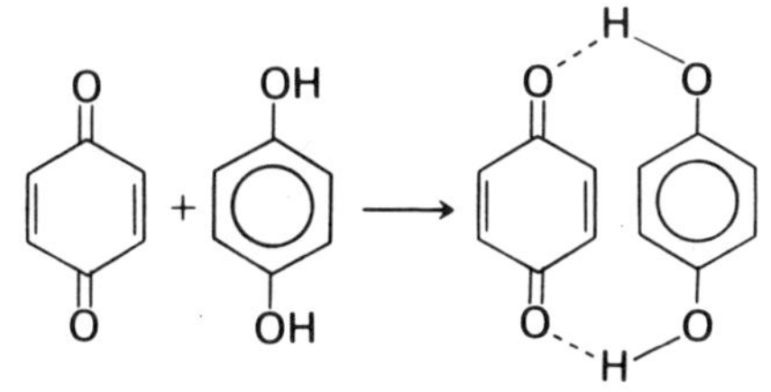

(6.63)

(**6.49**)

Nitroxy radicals are probably capable of abstracting hydrogen from polymer molecules and producing hydroxylamine and ethylenic unsaturation by the following mechanism:[67,73,94,144,267]

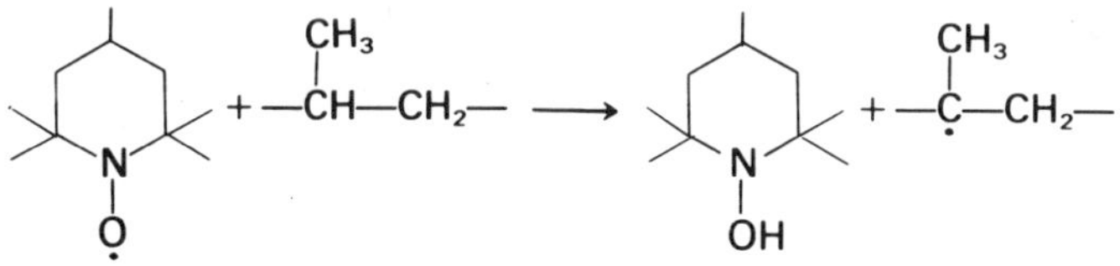

(6.64)

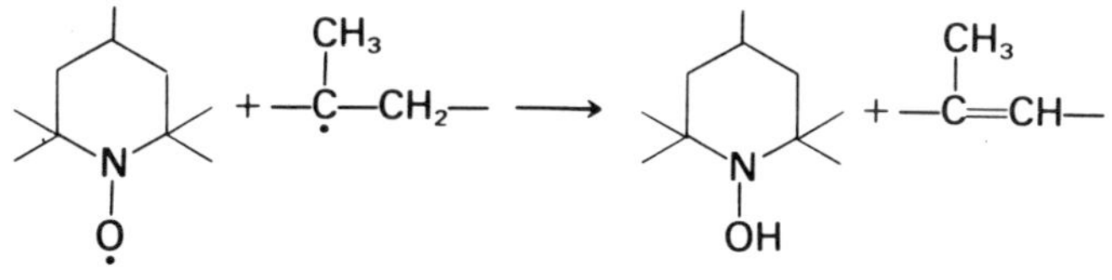

(6.65)

Nitroxy radicals can react with polymer alkyl radicals (P•) [e.g. $—CH_2-\dot{C}(CH_3)—CH_2—$, $—CH_2—CH(CH_3)—\dot{C}H_2$ and, in the latter stages of ageing, $—CH_2—\dot{C}{=}O$] to form polymer-bonded alkoxyamine derivatives of the stabilizer, most of which will retain these photostabilizing properties:[236,242,362,555,556]

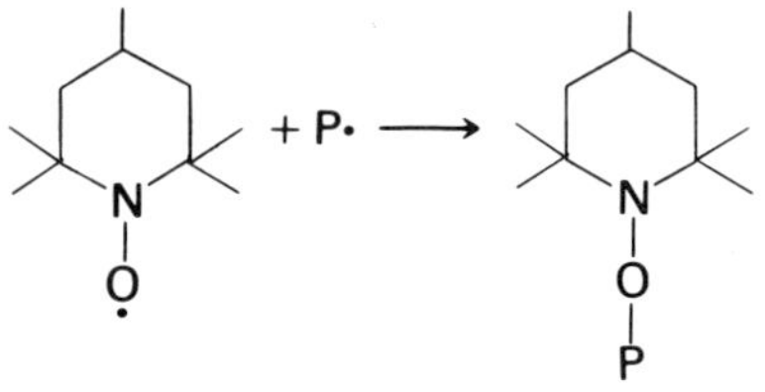

(6.66)

For example, bis(2,2,6,6-tetramethyl-4-piperidinyl) sebacate (Tinuvin 770; **6.3**) can form bispiperidinoxy radicals (**6.50**) which, after grafting to a polymer alkyl radical (P•) from one side (**6.51**), can be detected by ESR spectroscopy (Fig. 6.5).[554–556]

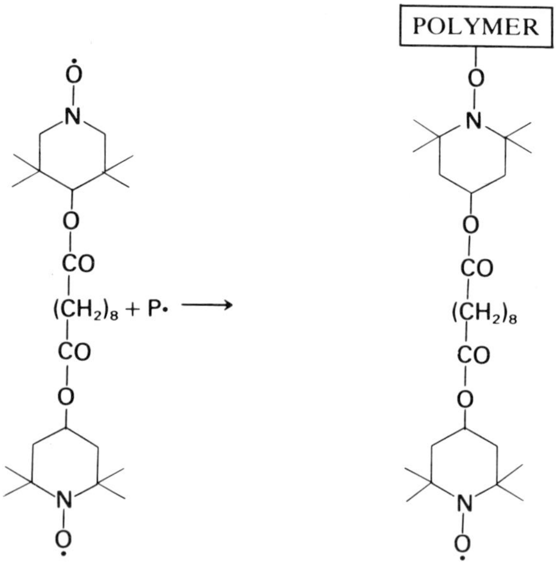

(6.67)

(6.50) **(6.51)**

Polymer-bonded alkoxyamine derivatives (**6.52**) can be decomposed (especially at elevated temperatures) into nitroxy radicals and polymer alkyl radicals (P•):[235,236,488,556,1055]

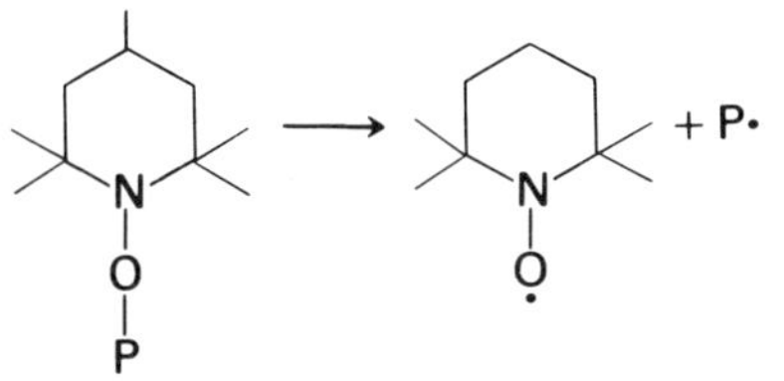

(6.52)

(6.68)

6.3 TRANSFORMATIONS OF HINDERED AMINES AND NITROXY RADICALS DURING PROCESSING OF POLYMERS

The thermal oxidation of low-molecular-weight hindered amines has been investigated in polypropylene film by oven ageing in air at 140°C.[94] The amine bis(2,2,6,6-tetramethyl-4-piperidinyl) sebacate (Tinuvin 770; **6.3**) shows a rapid growth in nitroxy radical formation due to reaction with hydroperoxides, whereas the bisnitroxy (**6.50**) derivative is relatively stable and decreases in concentration very slowly, giving hydroxylamine. The ageing process results in an improvement in photostability and this is associated with the reaction of the nitroxy radicals with a tertiary radical to give the hydroxylamine and ethylenic unsaturation.

The 4-hydroxy-2,2,6,6-tetramethylpiperidine-*N*-oxy radical (**6.30**) behaves differently during the early stages of ageing in that there is a rapid fall in photostability. Only during the latter stages of ageing do hydroxylamine products appear to control photostability.[94]

The concentration of this radical (**6.30**) in polypropylene shows a rhythmic variation during the early stages of processing.[120,121,126] A dynamic equilibrium exists between **6.30** and the hydroxylamine formed (Fig. 6.9). The corresponding >NOH curves show a striking mirror image to the changes in concentration of >NO•.

Quantitative measurements of piperidinoxy radical and hydroxylamine concentrations show that, under conditions of either restricted oxygen or oxygen excess, the concentration of both species changes reciprocally throughout the processing operation. In a closed mixer, the total overall concentration of piperidinoxy radicals and hydroxylamine remains at a high constant level, whereas in an open mixer their

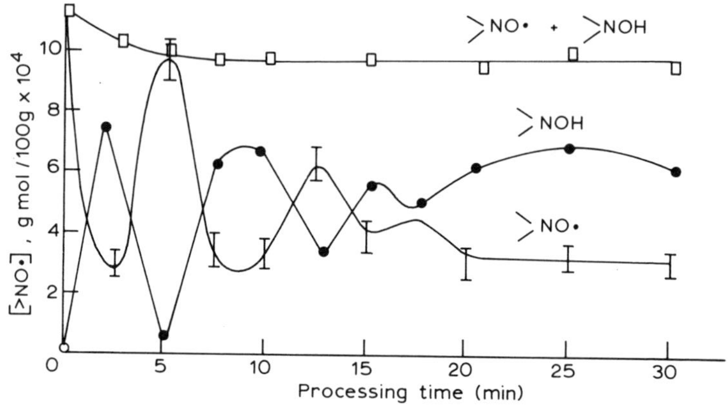

Fig. 6.9. Changes in concentration of nitroxy radicals (>N—O·) and hydroxylamine (>NOH) in polypropylene during processing at 180°C. Every point on the curves is the result of a separate processing experiment where the polymer is processed with $1{\cdot}16 \times 10^{-5}$ mol/100 g nitroxy radical for the time periods shown.[120]

total concentration decreases continuously to a low level, indicating the irreversible destruction of both compounds. The photo-oxidative stability of the polypropylene was found to depend on the total concentration of piperidinoxy radicals and hydroxylamine in the system after processing.[121]

Hindered amines behave as pro-oxidants during both melt-processing and thermal processing of low-density polyethylene.[293,352]

6.4 STABILIZING PROPERTIES OF HALS IN COMMERCIAL POLYMERS

In spite of a great effort by a number of investigators, the photo-stabilizing mechanism caused by hindered amine/nitroxy radical systems is still not understood. The general mechanism of photo-stabilization can be described by the reactions shown in Fig. 6.10. The kinetics of these reactions are complex because of different reaction rates.

On the basis of general knowledge of the chemistry hindered amines (cf. Section 6.1) and nitroxy radicals (cf. Section 6.2) all the possible

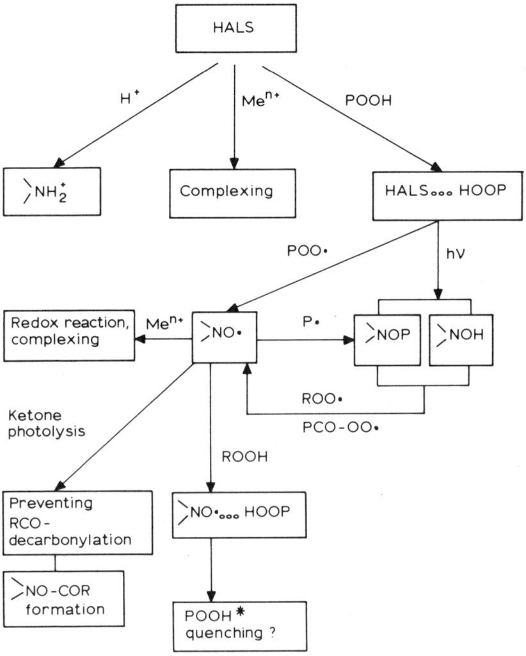

Fig. 6.10. Stabilizing pathways of hindered amine light stabilizers (HALS).[1038]

reactions can be divided into two groups:[162,396,448,449,451]

(i) Reactions which are too slow to play an important role in photostabilization:

$$>NH + POOH \longrightarrow >NOH + POH \qquad (6.69)$$

$$>NOR + POO\cdot \longrightarrow >NO\cdot + POOR \qquad (6.70)$$

It is important to stress that hindered amines react neither fast enough nor efficiently enough with peroxy radicals and simple peroxides at ambient conditions to be responsible for major stabilizing mechanisms.[650]

(ii) Reactions which are significant in photostabilization mechanisms:

$$>NH + POOH \xrightarrow{200°C} >NOH + POH \qquad (6.71)$$

$$>NH + POOH(activated) \longrightarrow >NO\cdot + POH \qquad (6.72)$$

$$>\text{NH} + \text{POO}\cdot \xrightarrow{\text{oxygen}} >\text{NO}\cdot + \text{POOH} \tag{6.73}$$

$$>\text{NOR} + \text{POO}\cdot \longrightarrow >\text{NO}\cdot + \text{R{=}O} + \text{POH} \tag{6.74}$$

$$>\text{NH} + \text{P—}\overset{\overset{\text{O}}{\|}}{\text{C}}\text{—OO}\cdot \longrightarrow >\text{NO}\cdot + \text{PCOOH} \tag{6.75}$$

$$>\text{NO}\cdot + \text{P}\cdot \longrightarrow >\text{NOP} \tag{6.76}$$

$$>\text{NO}\cdot + \text{P—}\overset{\overset{\text{O}}{\|}}{\text{C}}\text{—O}\cdot \longrightarrow >\text{N—O—O—}\overset{\overset{\text{O}}{\|}}{\text{C}}\text{—P} \tag{6.77}$$

$$>\text{NOR} + \text{P—}\overset{\overset{\text{O}}{\|}}{\text{C}}\text{—OO}\cdot \longrightarrow >\text{NO}\cdot + \text{PCOOH} + \text{R{=}O} \tag{6.78}$$

Hindered amines react rapidly with acylperoxy radicals ($\text{P—}\overset{\overset{\text{O}}{\|}}{\text{C}}\text{—OO}\cdot$). Hydroxylamines (or alkyloxyamines) also react efficiently with acylperoxy radicals, providing regeneration of nitroxy radicals. The reactivity of hindered amines, nitroxy radicals and alkoxyamines (and hydroxylamine) are compared in Table 6.6

During photo-oxidative degradation of polymers, added hindered amines and their reaction products, nitroxy radicals and alkoxyamines must compete with oxygen and tertiary hydrogen in the polymer to result in a significant reduction in the rate of formation of hydroperoxides, i.e.:

$$\text{P}\cdot \xrightarrow{+\text{O}_2} \text{POO}\cdot \tag{6.79}$$

$$\text{P}\cdot \xrightarrow[>\text{N—O}\cdot]{} \text{P—O—N}< \tag{6.80}$$

$$\text{POO}\cdot \xrightarrow{\text{PH}} \text{POOH} + \text{P}\cdot \tag{6.81}$$

$$\text{POO}\cdot \xrightarrow[>\text{N—O}\cdot]{} >\text{N—O}\cdot + \text{other products} \tag{6.82}$$

Table 6.6
Comparison of reactivity of different radicals with hindered amines, nitroxy radicals and hydroxy(alkoxy)amines

Type of free radical	*Abbreviation*	*Hindered amine*	*Nitroxy radical*	N-*Hydroxylamine* (N-*alkoxylamine*)
		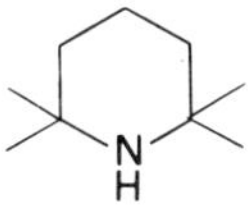	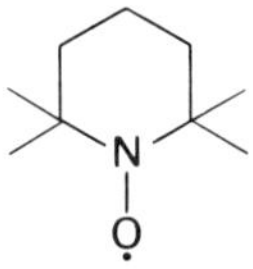	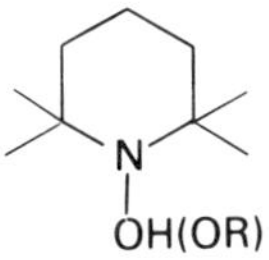
Allyl	R•	Reacts	Reacts	
Alkyloxy	RO•	Reacts very slowly		
Alkylperoxy	ROO•	Reacts very slowly		
Acyl	R—C(=O)•		Reacts	Reacts slowly
Acyloxy	R—C(=O)—O•	Reacts fast		Reacts fast
Acylperoxy	R—C(=O)—OO•	Reacts fast		Reacts fast

The effectiveness of the hindered amine stabilizers is a function of the kinetic chain length and of the lifetime of the stabilizer in the polymer matrix.

Experimental results show that the hindered amine, its nitroxy radical and model alkoxyamines are all effective photostabilizers.[267,556] Cyclohexyloxyamine (**6.55**) shows a slightly higher induction period compared with 4-hydroxy-2,2,6,6-tetramethylpiperidine (**6.53**) and the 4-hydroxy-2,2,6,6-tetramethylpiperidine-*N*-oxy radical (**6.54**) (Fig. 6.11).[556]

Derivatives of 1,2,2,6,6-pentamethylpiperidines exhibit higher photostabilizing efficiency in polypropylene than 2,2,6,6-tetramethylpiperidines.[296] Similar results were obtained in the stabilization of polybutadiene.[1168]

Various *N*-substituted derivatives of 4-benzoyloxy-2,2,6,6-tetramethylpiperidine (**6.56**) on the photostabilization of polypropylene have been investigated (Table 6.7).[676–678] The order of photostabilizing

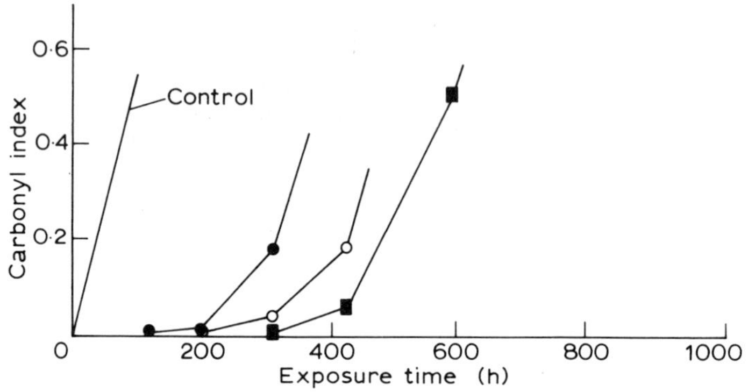

Fig. 6.11. Changes in carbonyl index during accelerated photoageing of polypropylene films containing hindered amine derivatives: ●, 4-hydroxy-2,2,6,6-tetramethylpiperidine (**6.53**); ○, 4-hydroxy-2,2,6,6-tetramethylpiperidine-*N*-oxy radical (**6.54**); ■, cyclohexyloxyamine (**6.55**).[556]

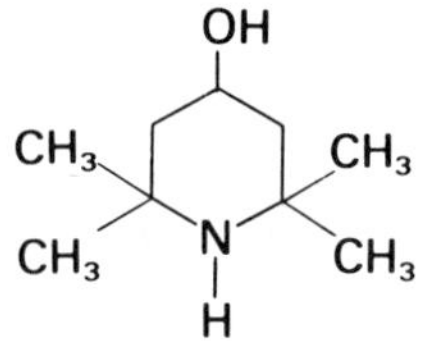

(6.53)

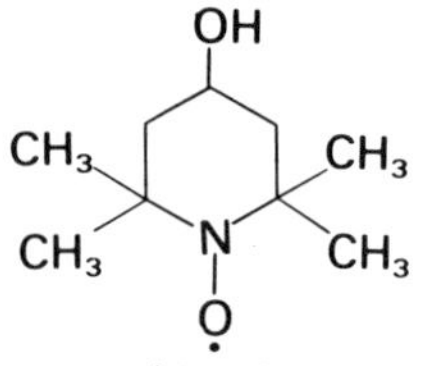

(6.54)

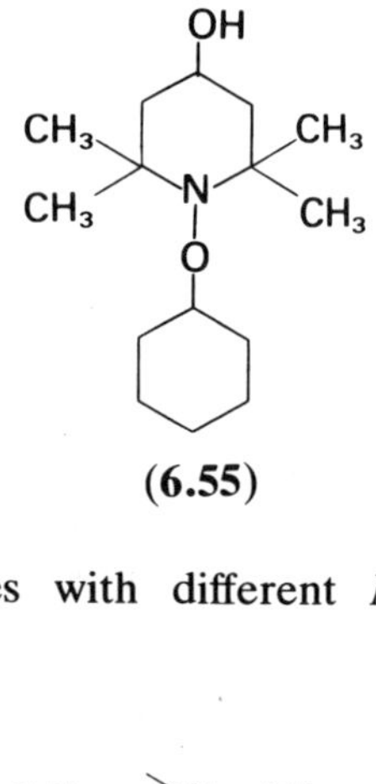

(6.55)

effectiveness of the tertiary amine derivatives with different *N*-substituents was as follows:

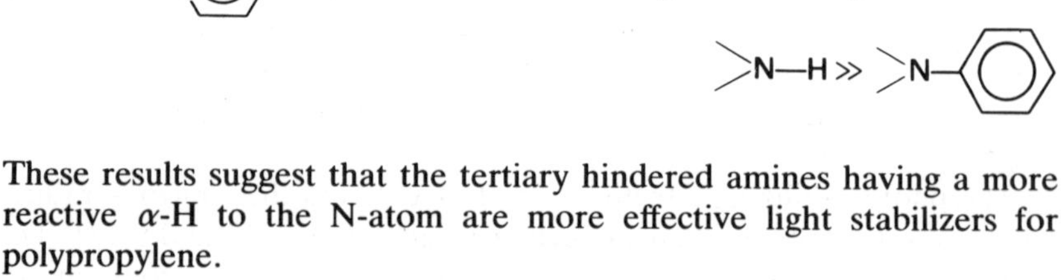

These results suggest that the tertiary hindered amines having a more reactive α-H to the N-atom are more effective light stabilizers for polypropylene.

Table 6.7
Photostabilizing effects of the different derivatives of 4-benzoyloxy-2,2,6,6-tetramethylpiperidine in polypropylene films (0·1 mm thick, concentration 5×10^{-3} mol kg^{-1}) upon irradiation in xenon Weatherometer (65WR, Atlas Electric Devices, black panel temperature 63 ± 3°C, without spray)[678]

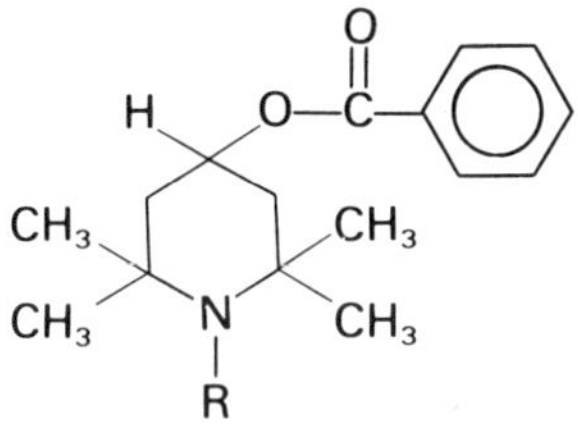

(6.56)

R	*Half-lifetime (h)*[a]
Control	75
H	1 100
O	1 730
CH_3	1 220
C_4H_9(*n*-)	1 350
$CH_2CH{=}CH_2$	1 700
CH_2Ph	2 200
Ph	650
$COCH_3$	850
$COCH{=}CH_2$	1 500
COPh	1 550

[a] Defined as time to 50% loss of initial elongation.

Hindered amine light stabilizers (HALS) are generally much more effective photostabilizers in the protection of polyolefins than a number of other commercially available photostabilizers and antioxidants (Fig. 6.12).

The photostabilizing effectiveness of HALS and their nitroxy radical analogues depends on their molecular weight (Figs 6.13 and 6.14). The reduced stabilization by the two lower-molecular-weight compounds,

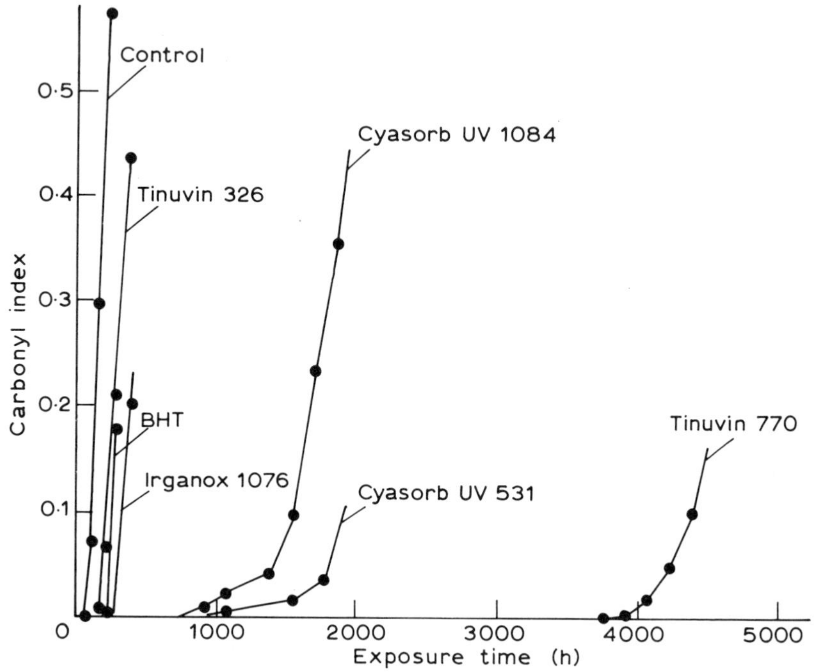

Fig. 6.12. Changes in carbonyl index during accelerated photoageing of polypropylene films containing photostabilizers and antioxidants (0·5 wt%): Tinuvin 770, bis(2,2,6,6-tetramethyl-4-piperidinyl) sebacate (**6.3**); Tinuvin 326, 2-(2-hydroxy-3-*tert.*-butyl-5-methylphenyl)-2*H*-benzotriazole (**5.45**); Cyasorb UV 531, 2-hydroxy-4-*n*-octoxybenzophenone (**5.30**); Cyasorb 1084, 2,2′-thiobis(4-*tert.*-octylphenolato)-*n*-butylamine nickel(II) (**5.63**); Irganox 1076, octadecyl-3-(3,5-di-*tert.*-butyl-4-hydroxyphenyl)propionate (**3.21**); BHT, 2,6-di-*tert.*-butyl-4-methylphenol.[556]

4-hydroxy-2,2,6,6-tetramethylpiperidine and its nitroxy radical analogue, is due to greater losses of the stabilizer by evaporation (Fig. 6.14) and emphasizes the importance of the molecular weight of the additive in the long-term retention and stabilization of a polymer.[1114] However, polymeric HALS (cf. Section 7.4) are much less effective than two other low-molecular-weight HALS photostabilizers (Fig. 6.13); this may be a result of segregation of the polymeric additive in the polymer matrix or to greatly reduced diffusional mobility of the hindered amine structural units in the polymer matrix.[556]

The extent of photostabilization of a polymer by HALS and their

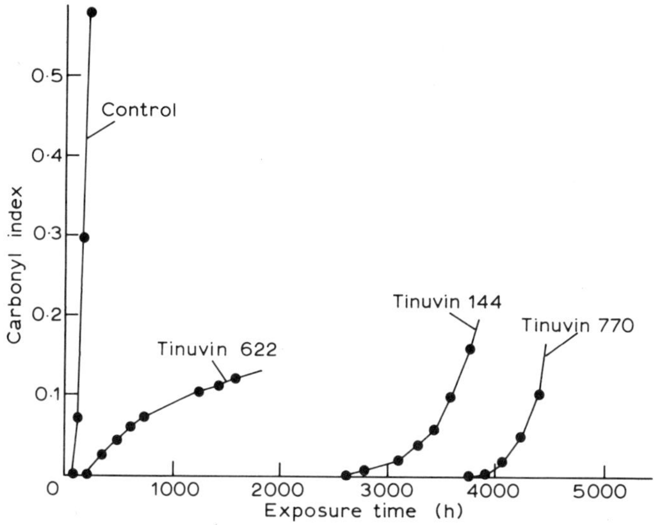

Fig. 6.13. Changes in carbonyl index during accelerated photoageing of polypropylene films containing commercial hindered amine light stabilizers (0·4 wt%): Tinuvin 770, bis(2,2,6,6-tetramethyl-4-piperidinyl) sebacate (**6.3**); Tinuvin 622, butanedioic acid polymer with 4-hydroxy-2,2,6,6-tetramethyl-1-piperidine-ethanol (**6.5**); Tinuvin 144, bis(1,2,2,6,6-pentamethyl-4-piperidinyl)-2-*n*-butyl-(3,5-*tert*.-butyl-4-hydroxybenzyl)-malonate (**6.7**).[556]

nitroxy radical analogues depends on the concentration of additive in the polymer (Figs 6.15 and 6.16). The induction period increases approximately linearly with increasing concentration of additive.[498,556]

The effectiveness of HALS is not dependent on the thickness of the polymer article. HALS are particularly useful for the protection of surface layers and are most effective in thin-section applications.

The stabilizing behaviour of HALS depends very much on the following factors (cf. Tables 6.8 and 6.9):[67]

(i) Stabilizer structure.
(ii) Polymeric environment (the pattern of behaviour in polypropylene is totally different from that in polyethylene).
(iii) Irradiation conditions.
(iv) Processing history (associated with oxidation of the amine groups to nitroxy radicals by reaction with hydroperoxides; cf. Section 6.3)[25,120,144,170,556]
(v) Presence of other additives (mainly antioxidants).

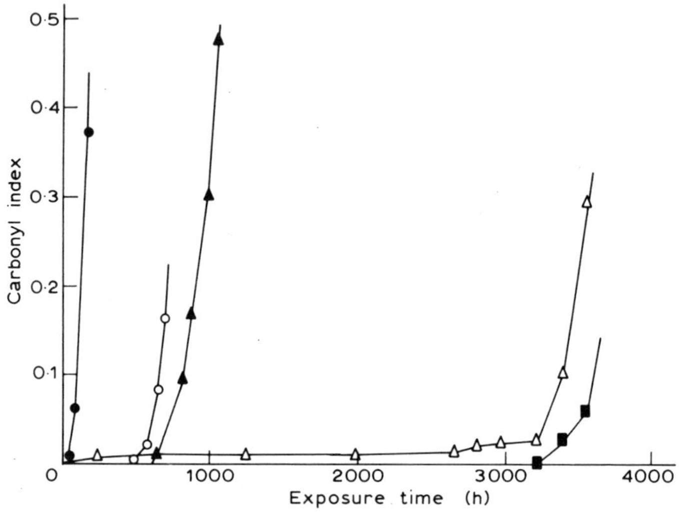

Fig. 6.14. Changes in carbonyl index during accelerated photoageing of polypropylene films containing hindered amines and nitroxides (concentrations of piperidine groups, 2×10^{-2} mmol g^{-1}): ●, control; ○, 4-hydroxy-2,2,6,6-tetramethylpiperidine; ▲, 4-hydroxy-2,2,6,6-tetramethylpiperidine-*N*-oxy radical (**6.30**); ■, bis(2,2,6,6-tetramethyl-4-piperidinyl) sebacate (Tinuvin 770; **6.3**); △, bis(2,2,6,6-tetramethyl-4-piperidinyl) sebacate *N*-oxy bi-radicals (**6.50**).[556]

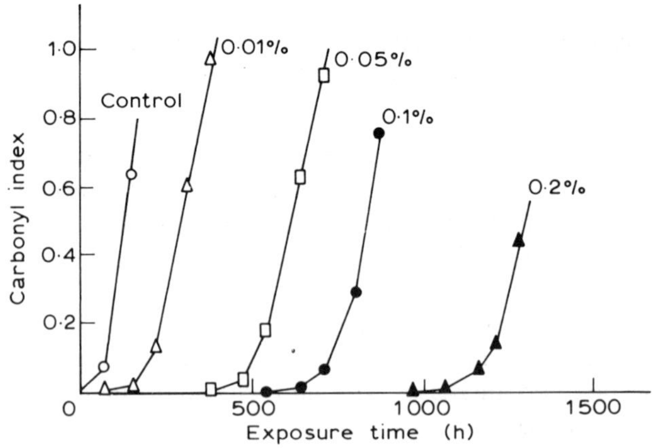

Fig. 6.15. Changes in carbonyl index during accelerated photoageing of polypropylene films containing increasing concentrations (wt%) of bis(2,2,6,6-tetramethyl-4-piperidinyl) sebacate (Tinuvin 770; **6.3**).[556]

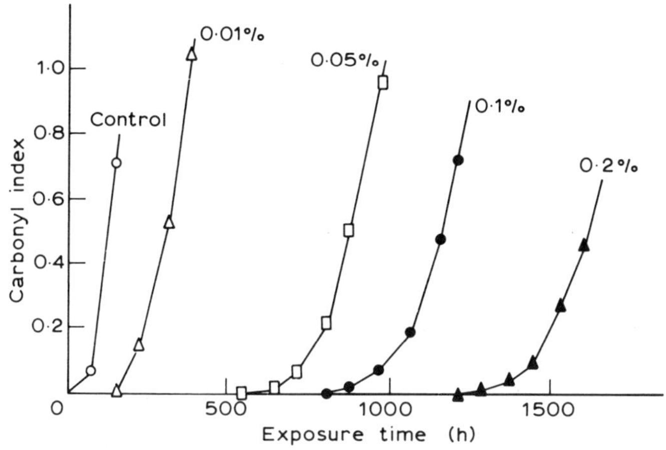

Fig. 6.16. Changes in carbonyl index during accelerated photoageing of polypropylene films containing increasing concentrations (wt%) of bis(2,2,6,6-tetramethyl-4-piperidinyl) sebacate *N*-oxy bi-radicals (**6.50**).[556]

The choice of the correct type of hindered amine and its concentration is essential for photostabilization (cf. Section 8.6).

6.5 SYNERGISTIC/ANTAGONISTIC EFFECTS OF HALS WITH OTHER ADDITIVES

Hindered amine light stabilizers (HALS) show very different synergistic or antagonistic effects with photostabilizers and antioxidants. Very contradictory results on this subject have been published.

Hindered amines have a synergistic effect with *ortho*-hydroxybenzophenones and *o*-hydroxybenzotriazoles. They inhibit the photolysis of both photostabilizers.[61,64,152,556,679,752] In the case of *o*-hydroxybenzophenones (Fig. 6.17) and its nitroxy radicals (Fig. 6.18) the synergistic effect increases with an increase of the concentration of added photostabilizers.

It has also been reported that nitroxy radicals formed from hindered amines are capable of abstracting hydrogen from the hydroxyl groups of 2-hydroxybenzophenone in benzene solution, causing in some

Table 6.8
UV embrittlement times (h to 0·06 carbonyl unit) for stabilized low-density polyethylene (LDPE) films in the Microscal[a], and SEPAP[b] units[67]

Stabilizer (0·1%w/w)	*Structure*	*No antioxidant*				*0·1% Goodrite 3125*[c]			
		Unprocessed		*Processed 5 min*		*Unprocessed*		*Processed 5 min*	
		Microscal	*SEPAP*	*Microscal*	*SEPAP*	*Microscal*	*SEPAP*	*Microscal*	*SEPAP*
Control		—	—	—	—	220	44	200	44
Chimassorb 944 (**6.6**)	See Table 6.1	2 440	840	2 430	840	2 480	800	1 330	750
Sanol 774 (**6.9**)	See Table 6.1	1 240	100	350	100	560	240	890	280
Tinuvin 770 (**6.3**)	See Table 6.1	1 910	840	1 350	800	2 130	1 000	3 400	1 050
HALS-1		2 060	650	1 680	610	2 830	840	2 600	1 200
Tinuvin 622 (**6.5**)	See Table 6.1	1 900	240	1 700	840	480	650	1 950	510
4-Hydroxy-2,2,6,6-tetramethylpiperidine		240	90	510	300	150	56	180	56

[a] Weatherometer Microscal Ltd, UK, employing a 500 W high-pressure mercury fluorescent lamp (50°C, relative humidity ambient).
[b] Apparatus employing a 400 W medium-pressure mercury lamp filtered by Pyrex envelope (50°C, 20% relative humidity).
[c] Goodrite 3125 is tri(3-(3′,5′-di-*tert.*-butyl-4′-hydroxybenzyl)-2″-acetoxyethyl) isocyanurate, whereas HALS-1 is bis[1,8(2,2,6,6-tetramethylpiperidinyl-4-carboxy)]octandioate (**6.57**).

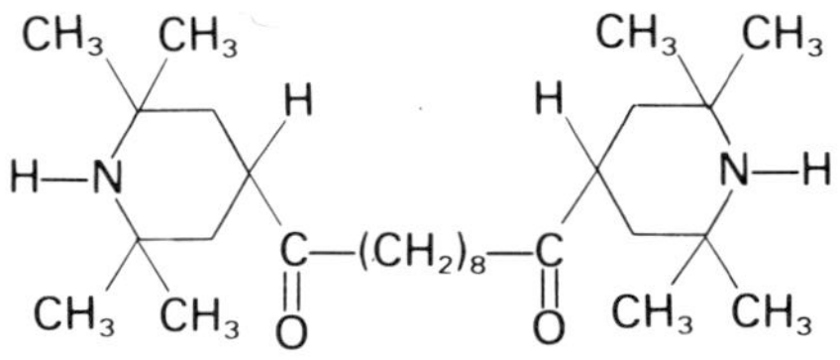

(6.57)

Table 6.9
UV embrittlement times (h to 0·06 carbonyl unit) for stabilized polypropylene films in the Microscal and SEPAP units[67,a]

Stabilizer (0·1%w/w)	*Structure*	*No antioxidant*				*0·1% Goodrite 3125*			
		Unprocessed		*Processed 5 min*		*Unprocessed*		*Processed 5 min*	
		Microscal	*SEPAP*	*Microscal*	*SEPAP*	*Microscal*	*SEPAP*	*Microscal*	*SEPAP*
Control		—	—	—	—	240	70	220	70
Chimassorb 944 (**6.6**)	See Table 6.1	660	300	790	300	900	360	500	200
Sanol 774 (**6.9**)	See Table 6.1	1 900	480	1 250	480	1 040	1 100	1 470	780
Tinuvin 770 (**6.3**)	See Table 6.1	2 300	1 400	1 650	1 010	2 170	1 400	2 370	950
HALS-1		2 560	1 400	2 940	1 450	3 180	1 600	2 630	1 400
Tinuvin 622 (**6.5**)	See Table 6.1	1 200	600	2 000	760	730	210	1 250	350
4-Hydroxy-2,2,6,6-tetramethylpiperidine		800	310	600	250	660	280	820	280

[a] Apparatus, Goodrite 3125 and HALS-1 as described in footnotes to Table 6.8.

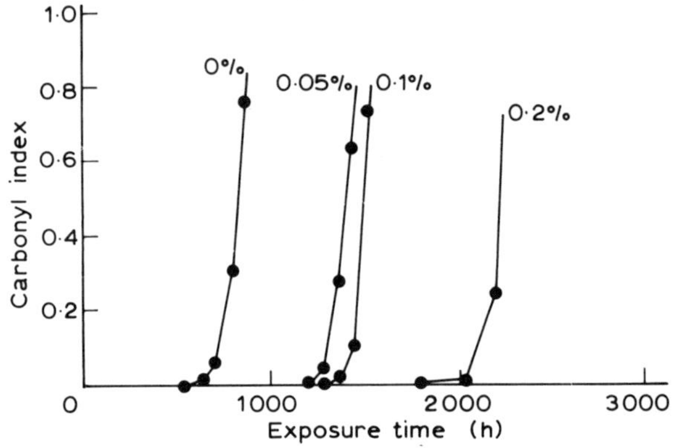

Fig. 6.17. Changes in carbonyl index during accelerated photoageing of polypropylene films containing bis(2,2,6,6-tetramethyl-4-piperidinyl) sebacate (Tinuvin 770; **6.3**) (0·1 wt%) and increasing concentrations of 2-hydroxy-4-*n*-octoxybenzophenone (Cyasorb UV 531; **5.30**).[556]

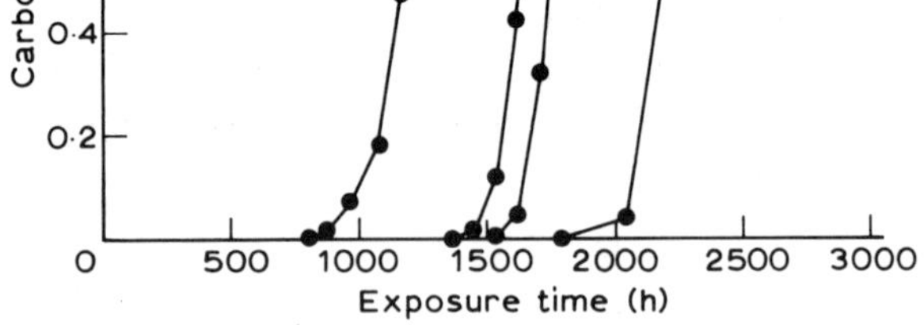

Fig. 6.18. Changes in carbonyl index during accelerated photoageing of polypropylene films containing bis(2,2,6,6-tetramethyl-4-piperidinyl) sebacate *N*-oxy bi-radicals (**6.50**) (0·1 wt%) and increasing concentrations of 2-hydroxy-4-*n*-octoxybenzophenone (Cyasorb UV 531; **5.30**).[556]

degree of antagonistic effect between these two compounds:[729]

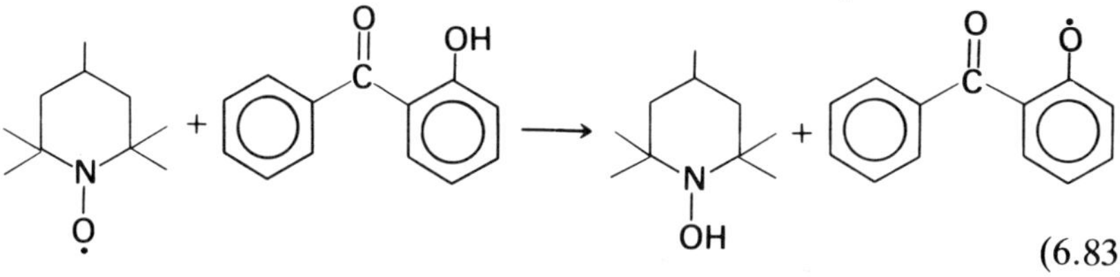

(6.83)

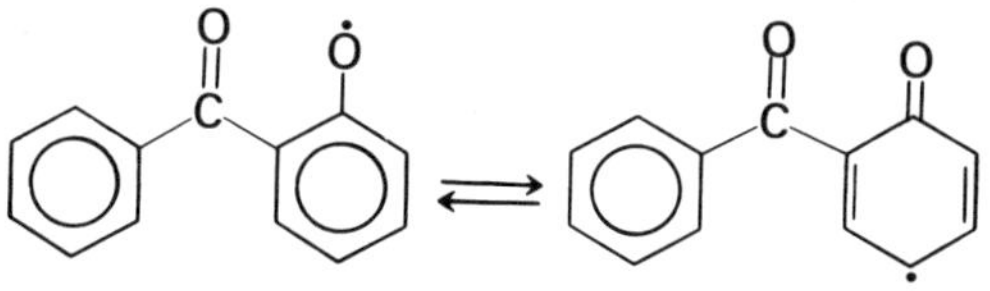

(6.84)

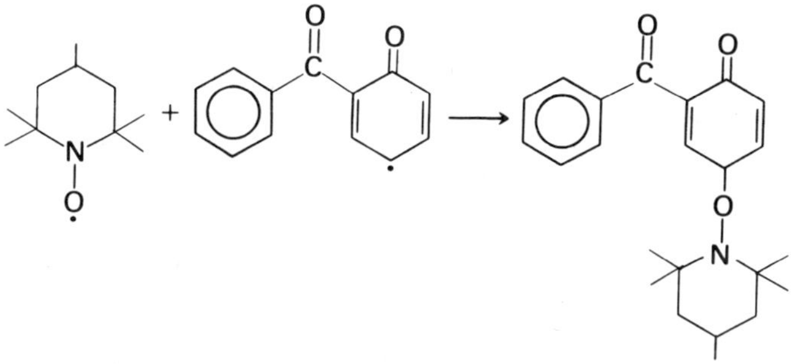

(6.85)

In order to improve the stabilizing efficiency of hindered amines, they were coupled with sterically hindered phenols (cf. Table 6.10). the idea behind such products was to combine the necessary increase in molecular weight with an additional stabilizing function. The phenolic moiety imparts thermal stability (antioxidant), whereas the amine part provides the light stability.

The nitrogen radical **6.22** formed from hindered amine in reaction (6.18) may recombine with phenoxy radical **6.64**, giving a substituted hydroxylamine, **6.65**:[729]

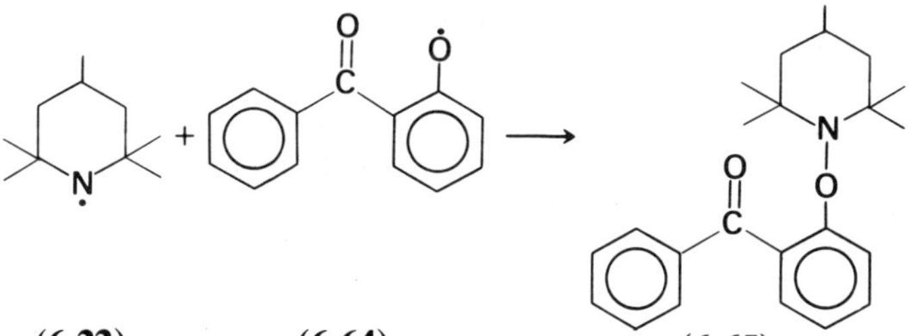

(6.86)

(6.22) **(6.64)** **(6.65)**

Table 6.10
Photostabilizing activity of different types of piperidine-phenols in polypropylene[1110]

Compound	*M.p. (°C)*	*Light stability measured as embrittlement time (h)*
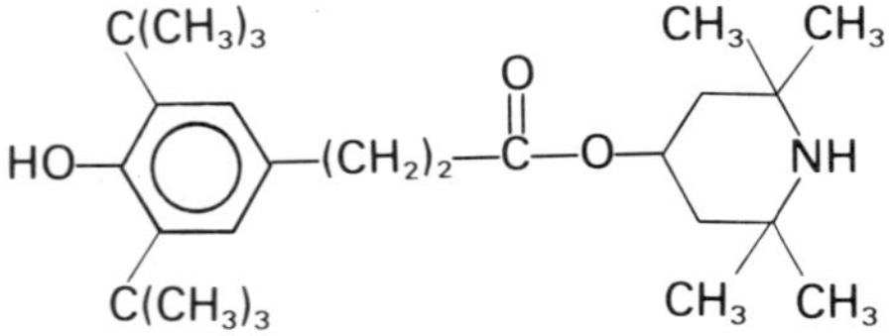 (**6.58**)	78	1170
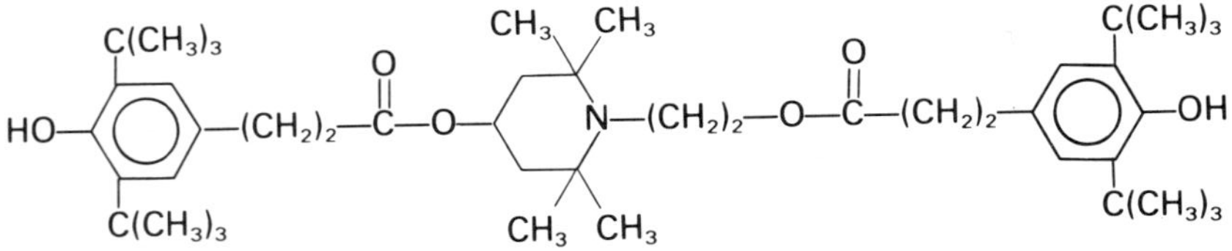 (Sanol LS-2626; **6.59**)	138	810
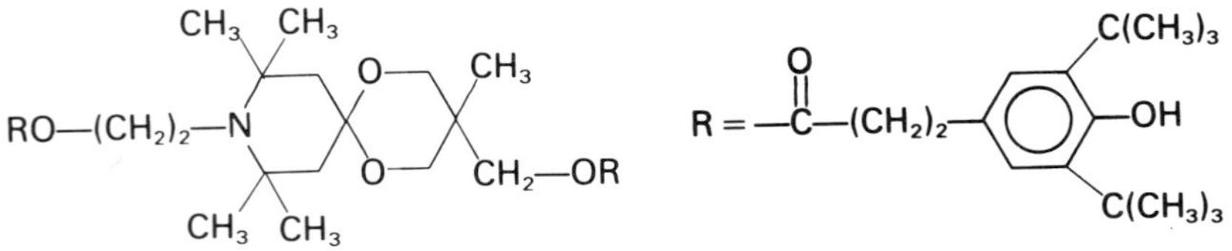 (**6.60**)	56	670

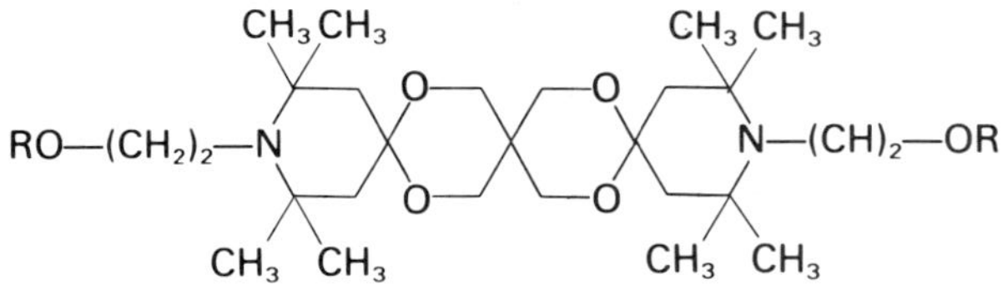 (6.61)	178	680
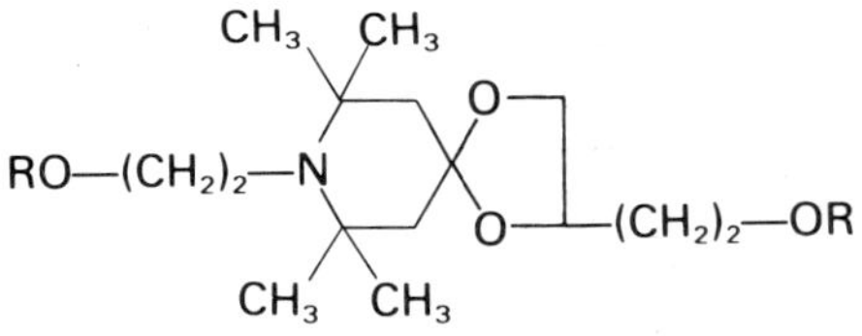 (6.62)	53	810
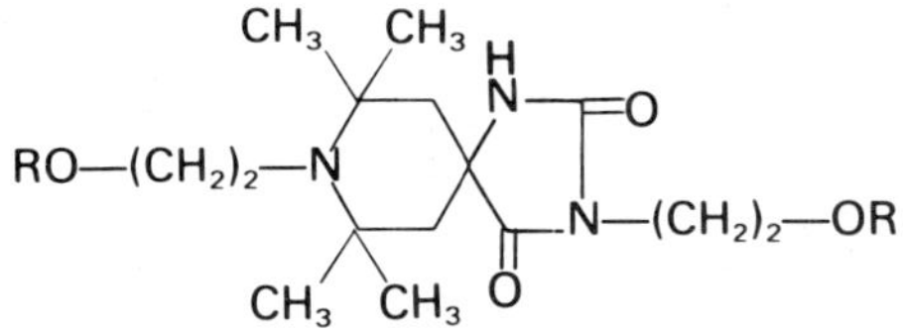 (6.63)	179	790
Control prepared with antioxidant		170

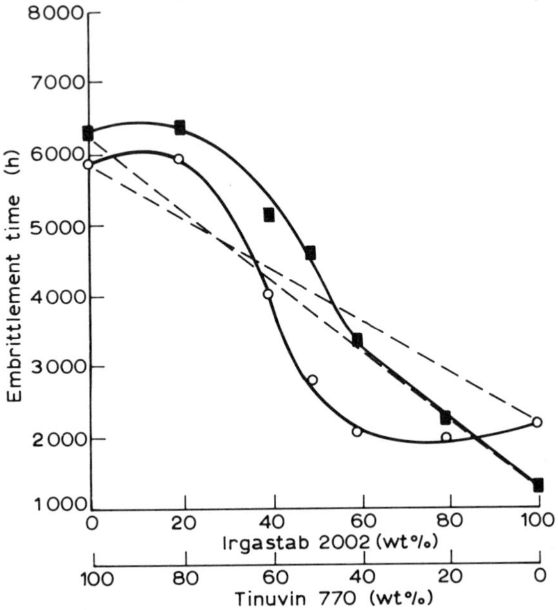

Fig. 6.19. Interaction between bis(*o*-butyl-3,5-di-*tert*.-butyl-4-hydroxybenzyl)phosphonate nickel (Irgastab 2002; **5.67**) and bis(2,2,6,6-tetramethyl-4-piperidinyl) sebacate (Tinuvin 770; **6.3**), at a total concentration of 0·1 wt% in polypropylene films, in the absence (○) and in the presence (■) of 0·1 wt% calcium stearate.[51]

A similar reaction may occur with 2-(2-hydroxy-5-methylphenyl)-2*H*-benzotriazole (Tinuvin P; **5.43**).[729]

Hindered amines show synergistic or antagonistic effects in mixture with metal chelates, depending on the formulation (Fig. 6.19)[26,31,51,152,153] The addition of a third component such as calcium stearate has an additional effect, which increases the synergism observed.[51]

Coupling of nitroxy radicals with metal xanthates (**6.66**) or with the disulphides (**6.67**) causes an antagonistic effect between these two stabilizers.[114]

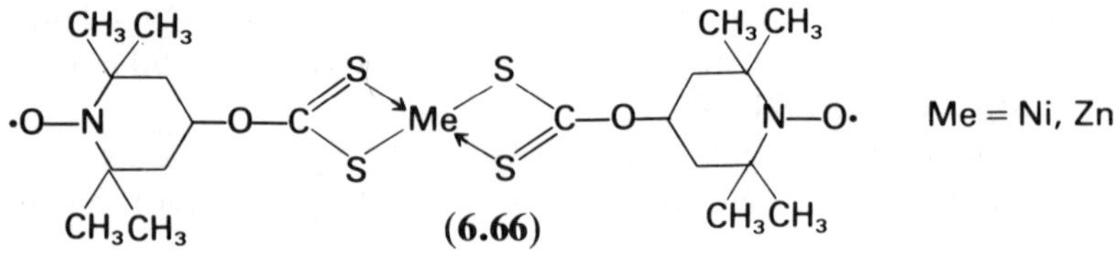

(6.66)

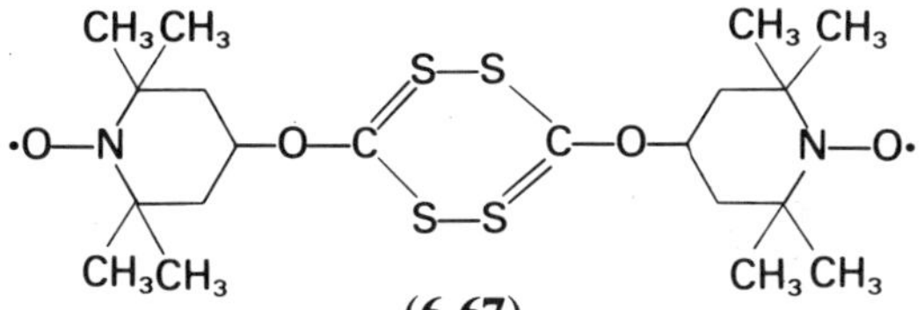

(6.67)

Interactions between hindered amines and antioxidants are complex. During thermal oxidation (oven ageing) the interactions are, with some exceptions, synergistic, whilst during photo-oxidation the interactions are generally antagonistic (Table 6.11).[24,26,30,31,62,68,81,87,92,732,925] In almost all cases the stabilizing effects on the polymer films are related to the stability of the antioxidant. The stabilizing performance of the phenolic antioxidants with each hindered piperidine stabilizer in thermal oxidation follows the order:

(i) *Tinuvin 770*: Irganox 1010 > Cyanox 1790 > Irganox 1076 > Goodrite 3114 > Topanol CA > Cyanox 2246.
(ii) *Chimassorb 994*: Topanol CA > Irganox 1076 > Cyanox 1790 > Irganox 1010 > Goodrite 3114 > Cyanox 2246.
(iii) *Cyasorb UV 3346*: Topanol CA > Goodrite 3114 > Irganox 1010 > Irganox 1076 > Cyanox 1790 > Cyanox 2246.
(iv) *Tinuvin 662*: Goodrite 3114 > Topanol CA > Irganox 1010 > Cyanox 1790 > Irganox 1076 > Cyanox 2246.

Antioxidants which are good thermal stabilizers tend to exhibit greater antagonism to the hindered piperidine compounds during photo-oxidation. The stabilizing performance of antioxidants with hindered piperidine compounds in photo-oxidation follows the order:

(i) *Tinuvin 770*: Goodrite 3114 > Topanol CA > Irganox 1076 > Weston 618 > Cyanox 1790 > Irganox 1010 > Cyanox 2246 > STDP.
(ii) *Chimassorb 994*: Irganox 1076 > Goodrite 3114 > Irganox 1010 > Cyanox 2246 > Topanol CA > Cyanox 1790 > Cyanox STDP > Weston 618.
(iii) *Cyasorb UV 3346*: Goodrite 3114 > Irganox 1076 > Topanol CA > Irganox 1010 > Cyanox 2246 = Cyanox 1790 > Weston 618 > Cyanox STDP.
(iv) *Tinuvin 622*: Irganox 1076 > Goodrite 3114 > Irganox 1010 > Cyanox 2246 > Weston 618 > Cyanox 1790 > Cyanox STDP > Topanol CA.

Table 6.11
Stabilizing effectiveness and additive effect of different stabilizers and their mixtures in polypropylene films (processed 10 min at 200°C)[68]

Additive (0·1% w/w) (commercial names)	*Stabilizing effectiveness measured as embrittlement time (h)*[a]	
	Oven ageing	*Photo-oxidation*
Tinuvin770	4·5	2 090
Chimassorb 944	2·0	1 090
Cyasorb UV 3346	2·5	810
Tinuvin 622	11	2 130
Topanol CA	75	385
Topanol CA + Tinuvin 770	175 (79·5)	2 750 (2 475)
Topanol CA + Chimassorb 944	520 (77)	950 (1 475)
Topanol CA + Cyasorb UV 3346	670 (77·5)	1 400 (1 195)
Topanol CA + Tinuvin 622	725 (86)	1 250 (2 515)
Irganox 1010	250	600
Irganox 1010 + Tinuvin 770	435 (254·5)	2 280 (2 690)
Irganox 1010 + Chimassorb 944	320 (252)	1 020 (1 690)
Irganox 1010 + Cyasorb UV 3346	290 (292·5)	1 300 (1 410)
Irganox 1010 + Tinuvin 622	580 (261)	1 790 (2 730)
Goodrite 3114	55	1 135
Goodrite 3114 + Tinuvin 770	180 (59·5)	>4 000 (2 225)
Goodrite 3114 + Chimassorb 944	55 (57)	1 550 (1 225)
Goodrite 3114 + Cyasorb UV 3346	400 (57·5)	2 300 (1 940)
Goodrite 3114 + Tinuvin 622	765 (66)	2 100 (3 265)
Irganox 1076	215	940
Irganox 1076 + Tinuvin 770	200 (219·5)	2 700 (3 030)
Irganox 1076 + Chimassorb 944	345 (217)	2 240 (2 030)
Irganox 1076 + Cyasorb UV 3346	110 (217·5)	2 240 (1 750)
Irganox 1076 + Tinuvin 622	375 (226)	>3 650 (3 070)
Cyanox 1790	240	245
Cyanox 1790 + Tinuvin 770	240 (244·5)	2 410 (2 335)
Cyanox 1790 + Chimassorb 944	320 (242)	880 (1 335)
Cyanox 1790 + Cyasorb UV 3346	75 (242·5)	950 (1 055)
Cyanox 1790 + Tinuvin 622	575 (251)	1 560 (2 375)
Cyanox 2246	90	370
Cyanox 2246 + Tinuvin 770	95 (94·5)	1 930 (2 460)
Cyanox 2246 + Chimassorb 944	40 (92)	1 030 (1 460)
Cyanox 2246 + Cyasorb UV 3346	30 (92·5)	950 (1 180)
Cyanox 2246 + Tinuvin 622	260 (101)	1 700 (2 500)
Weston 618	5	100
Weston 618 + Tinuvin 770	16 (9·5)	2 640 (2 190)
Weston 618 + Chimassorb 944	2 (7)	560 (1 190)

Table 6.11—*contd.*

Additive (0·1% w/w) (commercial names)	*Stabilizing effectiveness measured as embrittlement time (h)*[a]	
	Oven ageing	*Photo-oxidation*
Weston 618 + Cyasorb UV 3346	3 (7·5)	610 (910)
Weston 618 + Tinuvin 622	24 (16)	1 610 (2 230)
STDP	2	55
STDP + Tinuvin 770	4·5 (6·5)	1 500 (2 145)
STDP + Chimassorb 944	5 (4)	590 (1 145)
STDP + Cyasorb UV 3346	4 (4·5)	320 (856)
STDP + Tinuvin 622	47 (13)	1 320 (2 185)

Commercial name	*Chemical formula*
Tinuvin 770	Bis[2,2,6,6-tetramethyl-4-piperidinyl] sebacate (**6.3**)
Tinuvin 622	Polyester of succinic acid with *N*-β-hydroxyethyl-2,2,6,6-tetramethyl-4-hydroxypiperidine (**6.5**)
Chimassorb 944	Poly[2-*N,N*′-di(2,2,6,6-tetramethyl-4-piperidinyl)-hexanediamine-4-(1-amino-1,1,3,3-tetramethyl-butane)sym-triazine] (**6.6**)
Cyasorb UV 3346	Poly[2-*N,N*′-di(2,2,6,6-tetramethyl-4-piperidinyl)-hexanediamine-4-(*N*-morpholine)sym-triazine] (**8.1**)
Topanol CA	1,1,3-Tri(2′-methyl-4′-hydroxy-5′-*tert.*-butyl-phenyl)butane (**3.38**)
Cyanox 1790	1,3,5-Tri(4-*tert.*-butyl-3-hydroxy-2,6-dimethyl-benzyl)isocyanurate
Cyanox 2246	2,2′-Methylenebis(6-*tert.*-butyl-4-methylphenol) (**3.36**)
Irganox 1076	Stearyl 3-(3,5-di-*tert.*-butyl-4-hydroxyphenyl)-propionate (**3.21**)
Irganox 1010	Pentaerythrityl tetrakis(3,5-di-*tert.*-butyl-4-hydroxyphenyl) propionate (**3.32**)
Goodrite 3114	1,3,5-Tri(3,5-di-*tert.*-butyl-4-hydroxylbenzyl)-isocyanurate (**3.31**)
Weston 618	Distearyl pentaerythrityl diphosphite (**3.113**)
Plastanox STDP	Distearyl thiodipropionate (**3.59**)

[a] Additive effect is shown in parenthesis.

Table 6.12
Antagonistic (AN), synergistic (S) and additive (AD) effects for stabilized low-density polyethylene (LDPE) and polypropylene (PP) films containing 0·1 wt% Goodrite 3125 antioxidant and exposed in the Microscal and SEPAP units[67,a]

Stabilizer (0·1 wt%)	*Structure*	*LDPE*				*PP*			
		Unprocessed		*Processed*		*Unprocessed*		*Processed*	
		Microscal	*SEPAP*	*Microscal*	*SEPAP*	*Microscal*	*SEPAP*	*Microscal*	*SEPAP*
Chimassorb 944 (**6.6**)	See Table 6.1	AN	AN	AN	AN	S	AD	AN	AN
Sanol 774 (**6.9**)	See Table 6.1	AN	S	S	S	AN	S	AD	S
Tinuvin 770 (**6.3**)	See Table 6.1	AD	S	S	S	AN	AN	S	AN
HALS-1		S	S	S	S	S	S	AN	AN
Tinuvin 622 (**6.5**)	See Table 6.1	AN	S	S	AN	AN	AN	AN	AN
4-Hydroxy-2,2,6,6-tetramethylpiperidine		AN	AN	AN	AN	AN	AN	AD	AN

[a] Apparatus, Goodrite 3125 and HALS-1 as described in footnotes to Table 6.8.

Synergistic or antagonistic effects between hindered amines and phenol-type antioxidants (Goodrite 3125) depend also on the irradiation conditions (Table 6.12).[678]

The antagonistic effect between hindered piperidine compounds and antioxidants may also be due to the HALS ability to destroy hydroperoxides. This is true for the group of antioxidants which are peroxide-decomposers and react directly with hydroperoxide initially present in the polymer.

Nitroxy radicals generated during polymer processing oxidize phenolic antioxidants **(6.68)** to cyclohexadienonyl radicals **(6.69)** and finally to the corresponding quinones:

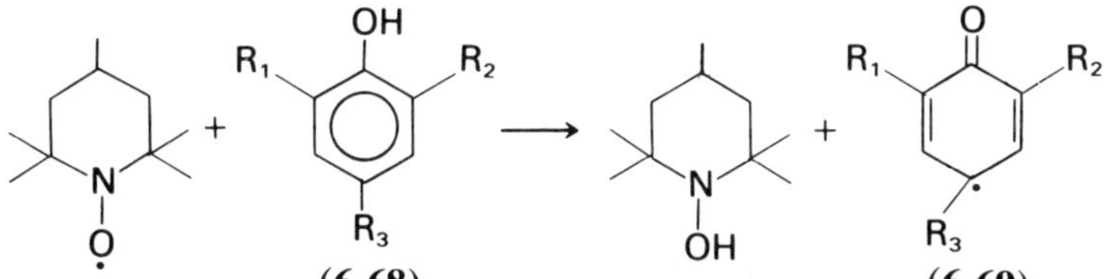

(6.68) **(6.69)** (6.87)

Cyclohexanedienonyl radicals **(6.69)** may also react with nitroxy radicals according to reaction (6.88):[732]

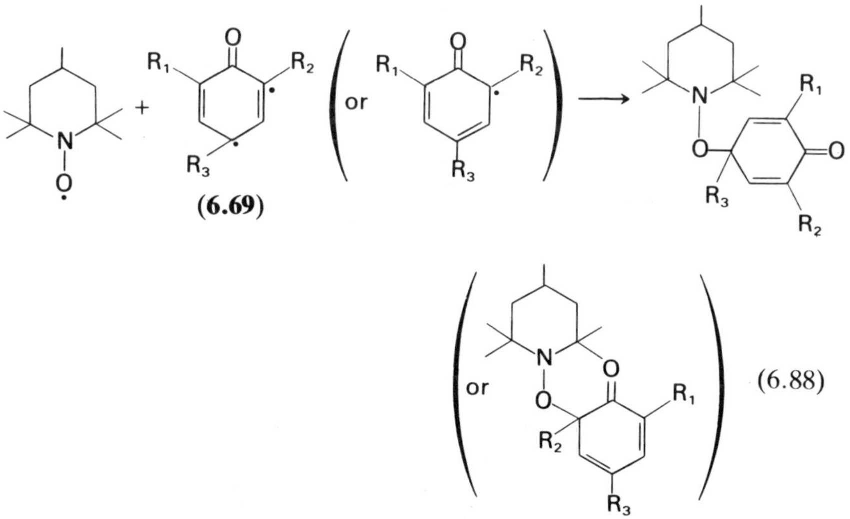

(6.69) (6.88)

The addition of calcium stearate, however, completely inhibits the formation of quinones and gives rise to an additive effect at low ratios of amine to complex and a synergistic effect at ratios greater than 50 : 50.[51]

The reactivity of nitroxy radicals with hindered phenols depends on the structure of the latter (the position and size of substituents in the phenyl ring). The following order of activity has been proposed:[732]

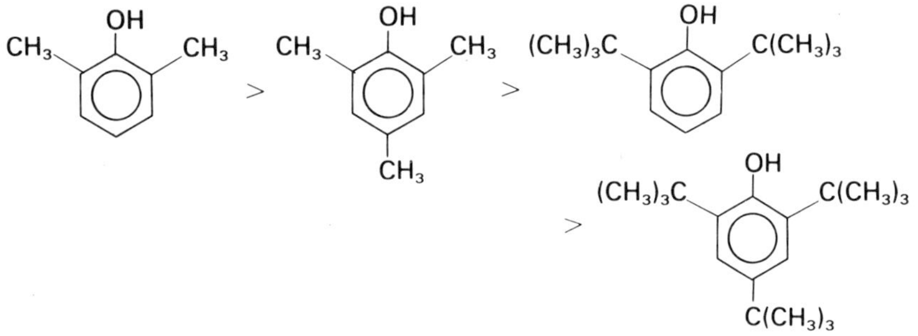

The presence of antioxidant free radicals produced from photolysis and/or reaction of the antioxidant with any polymer radicals/fragments can interfere with the nitroxy radicals and reduce the performance of the hindered stabilizers.

Sulphur-containing antioxidants, which are hydroperoxide group decomposers, antagonize the photostabilizing action of hindered amines.[26,31,264,641,670,671,730,733,827] The antagonistic effect can be explained by a mechanism in which sulphur-containing compounds decompose hydroperoxides in the polymer matrix, resulting in retardation or prevention of the formation of nitroxy radicals derived from HALS.[264]

Free radicals formed from photolysis of sulphur-containing antioxidants can also react with nitroxy radicals. For example, bis(4-hydroxyphenylene) sulphide, **6.70** (a model compound for other sulphide antioxidants), under UV irradiation decomposes into the hydroxyphenylthiyl radical (**6.71**) and the hydroxybenzyl radical (**6.72**):[730]

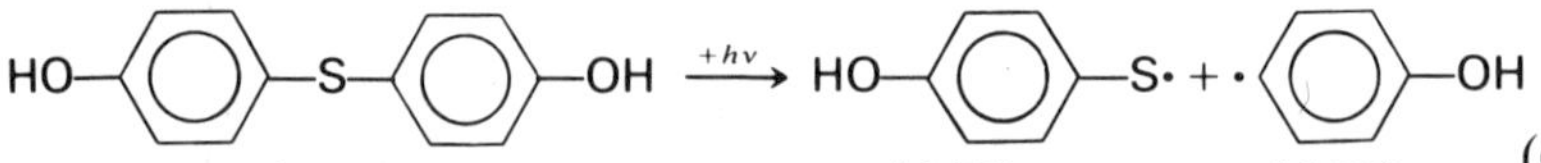

(**6.70**) (**6.71**) (**6.72**) (6.89)

These two radicals, **6.71** and **6.72**, may further react with nitroxy radicals:

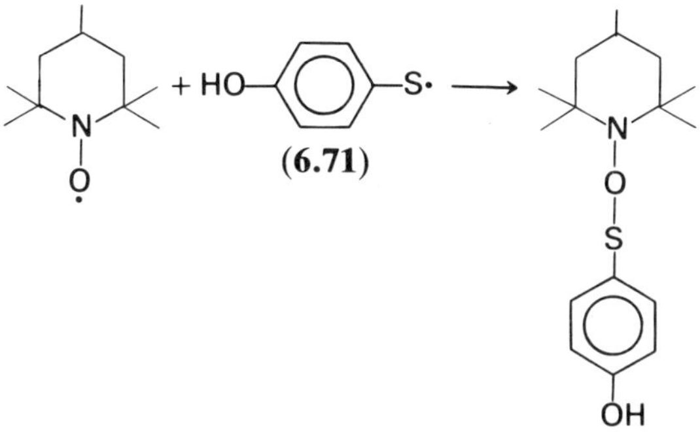

(6.71)

(6.90)

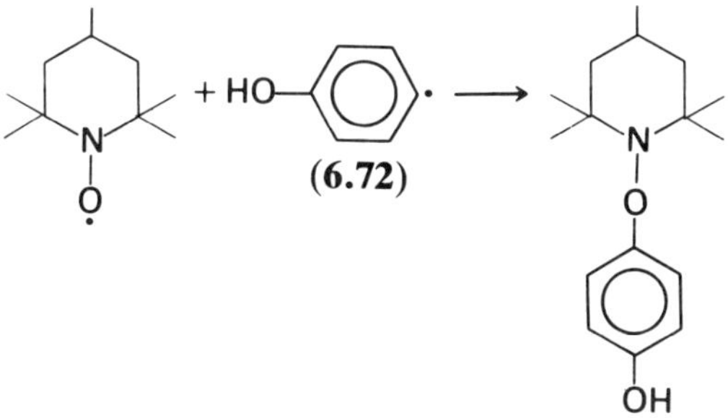

(6.72)

(6.91)

Bis-(4-hydroxyphenylene) sulphide (**6.70**) can also be photo-oxidized into bis-(4-hydroxyphenylene) sulphoxide (**6.73**), which can be further photolysed into the hydroxyphenylsulphoxy radical (**6.74**) and the hydroxybenzyl radical, **6.72**:[730]

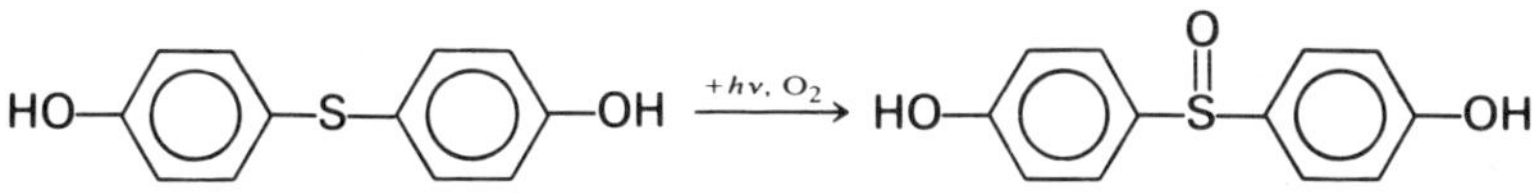

(**6.70**) (**6.73**)

(6.92)

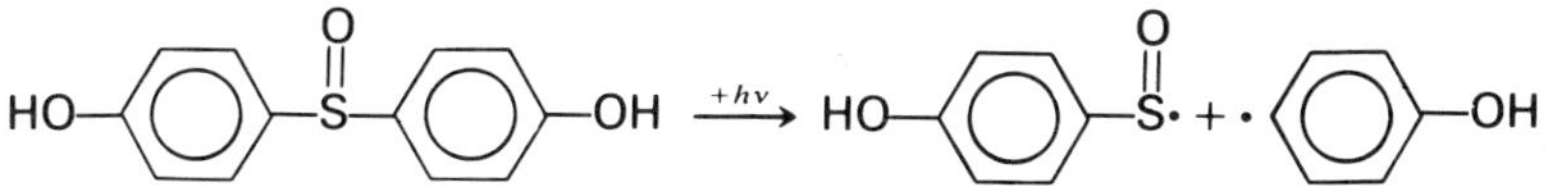

(**6.74**) (**6.72**)

(6.93)

These two radicals (**6.74** and **6.72**) may further react with nitroxy radicals:

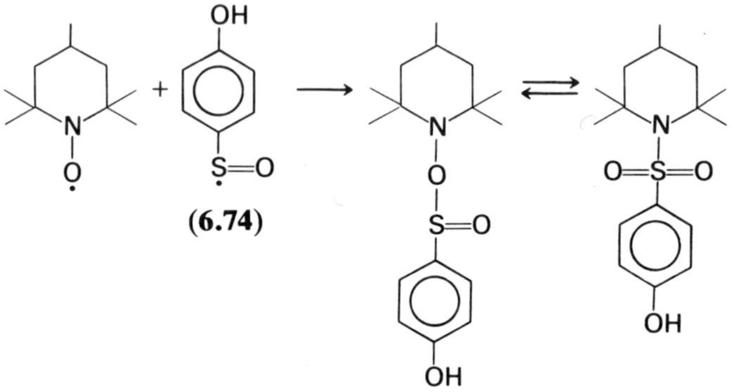

(6.74) (6.94)

Sulphur compounds readily reduce nitroxy radicals back to the corresponding amines.[827]

Formation of nitroxy radicals from HALS is strongly suppressed by the addition of *p*-toluenesulphonic acid, which forms a salt with HALS.[641] Deactivation of HALS cannot be due to the sulphur compound itself, but rather to its salts formed with the oxidized acidic products from the sulphur compounds. The antagonism between HALS and sulphur-containing compounds is the result of the formation of salts and/or reaction between a nitroxy radical ($>$N—O•), hydroxylamine ($>$N—OH) or hydroxylamine ether ($>$N—OR) and acidic products. The following reactions are involved in this mechanism:[641]

$$>N{-}O\cdot + RSO_3H \longrightarrow >N{-}OH + RSO_3\cdot \qquad (6.95)$$

$$>N{-}O\cdot + RSO_3H \longrightarrow >\overset{\cdot +}{N}{-}OH + RSO_3^- \qquad (6.96)$$

$$>N{-}OH + RSO_3H \longrightarrow >N{-}OH \cdots HO_3SR \qquad (6.97)$$

$$>\overset{+}{N}{=}O + RSO_3^- \longrightarrow >\text{ion pair (immonium oxide salt)} \qquad (6.98)$$

6.6 NOVEL HALS

A number of novel hindered amine light stabilizers (HALS), such as:

—piperidinyl esters (Table 6.13);
—oxazolidines and piperidinooxazolidines (Table 6.14);
—piperidinespiroacetals (Table 6.15).

Table 6.13
Photostabilizing activity of different types of piperidinyl esters[1110]

Compound	*M.p.* (°*C*)	*Light stability embrittlement measured as time* (*h*)
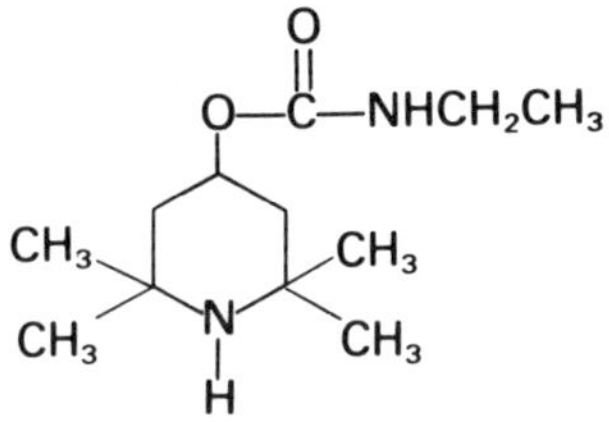 **(6.75)**	128	800
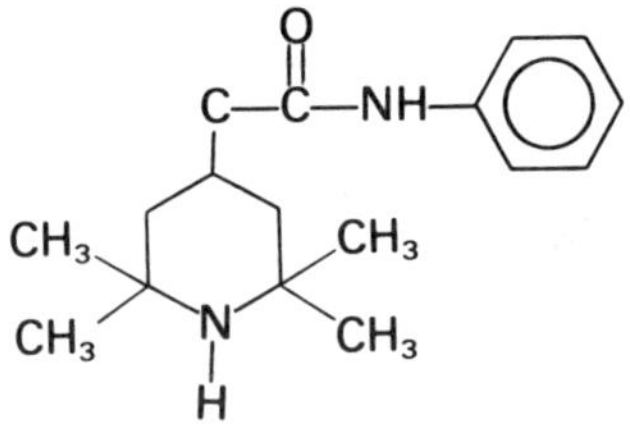 **(6.76)**	70	420
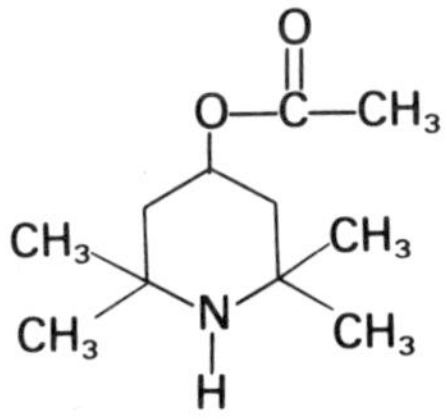 **(6.77)**	125	700
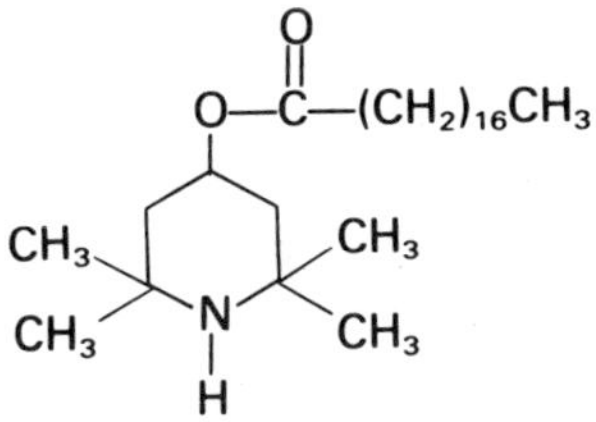 **(6.78)**	42	700

(continued)

Table 6.13—*contd.*

Compound	*M.p. (°C)*	*Light stability embrittlement measured as time (h)*
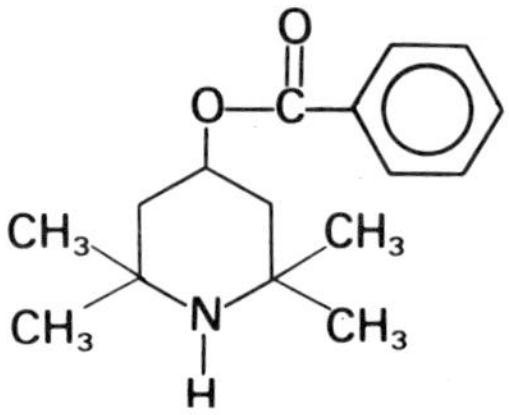 (Sanol LS-744; **6.79**)	98	500
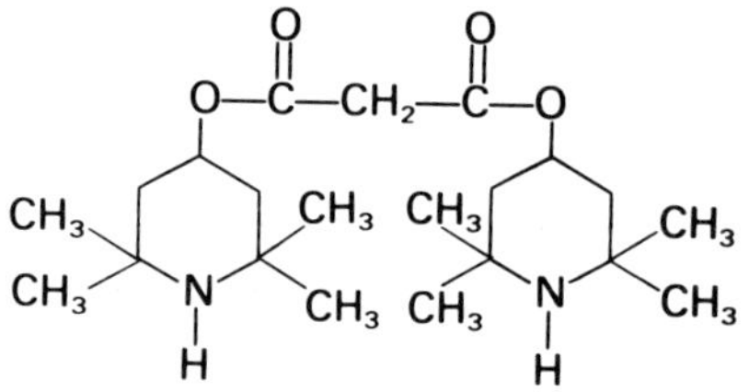 (**6.80**)	108	920
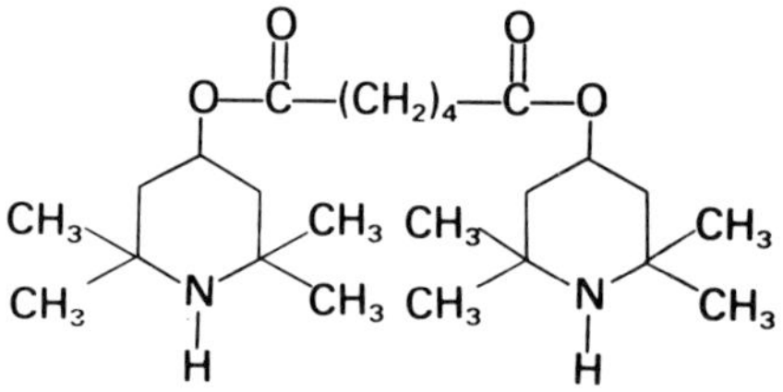 (**6.81**)	91	500
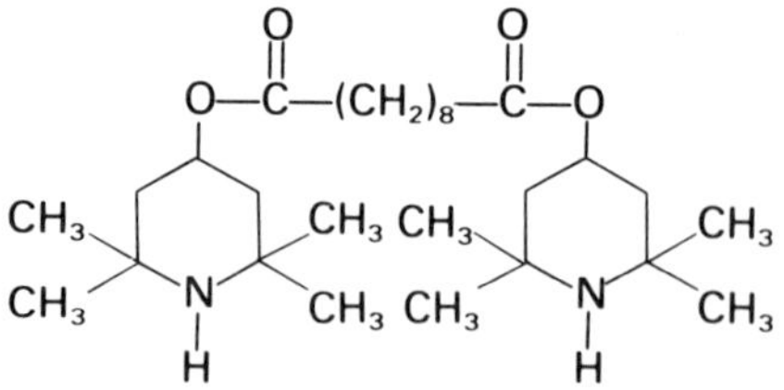 (Sanol LS-770; **6.82**)	86	720

Table 6.13—*contd.*

Compound	*M.p. (°C)*	*Light stability embrittlement measured as time (h)*
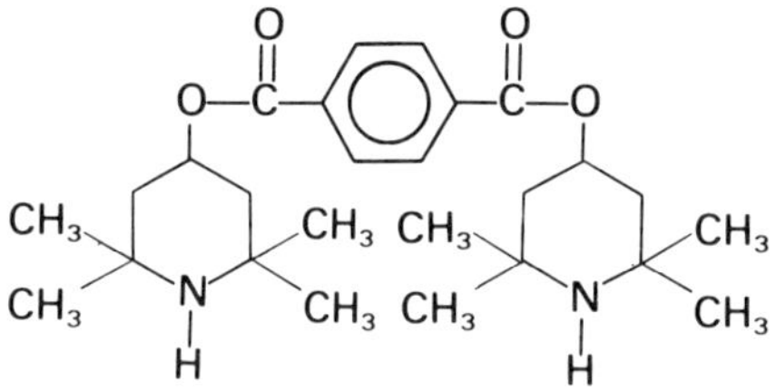 (6.83)	208	500
None	—	40

Table 6.14
Photostabilizing activity of different types of hindered amines[1110]

Compound	*M.p. (°C)*	*B.p. (°C)*	*Light stability measured as embrittlement time (h)*
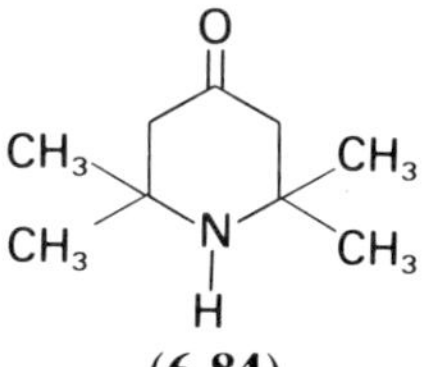 (6.84)	—	105/18 mm Hg	320
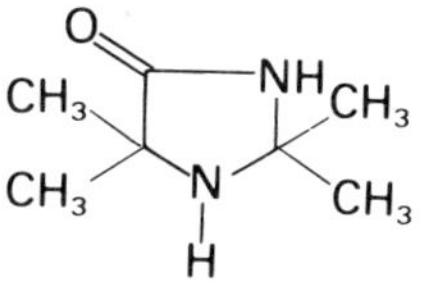 (6.85)	170	—	120

(continued)

Table 6.14—*contd.*

Compound	*M.p. (°C)*	*B.p. (°C)*	*Light stability measured as embrittlement time (h)*
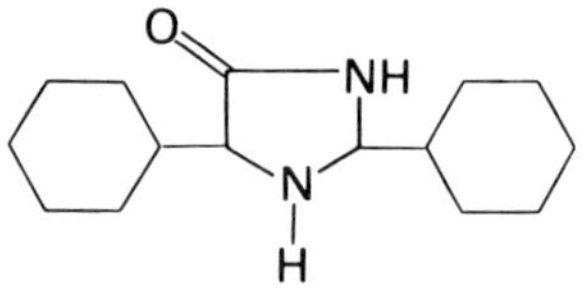 **(6.86)**	220	—	180
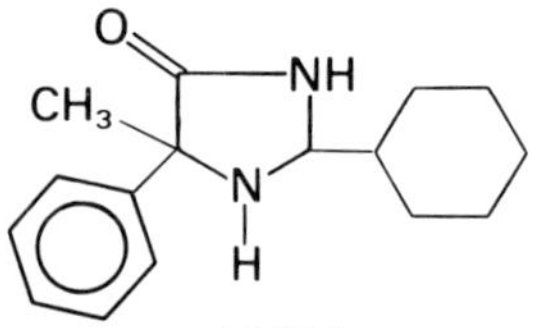 **(6.87)**	131	—	600
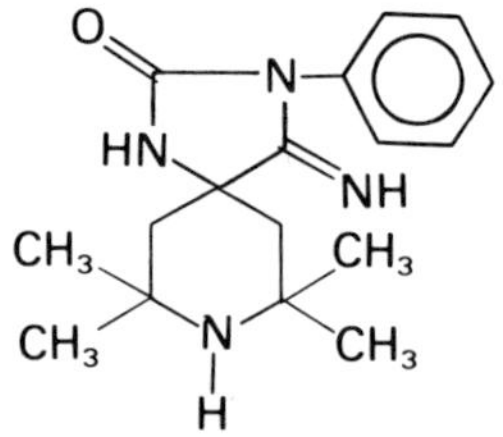 **(6.88)**	177	—	220
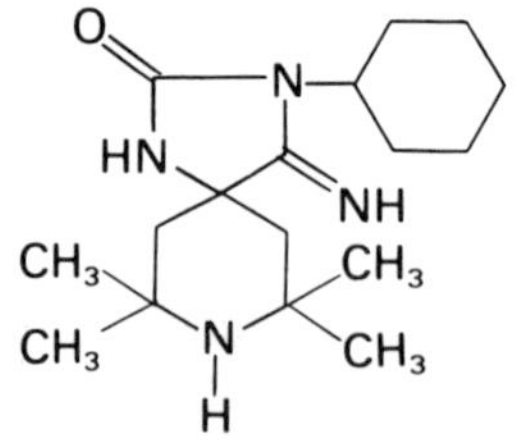 **(6.89)**	196	—	300

Table 6.14—*contd.*

Compound	*M.p. (°C)*	*B.p. (°C)*	*Light stability measured as embrittlement time (h)*
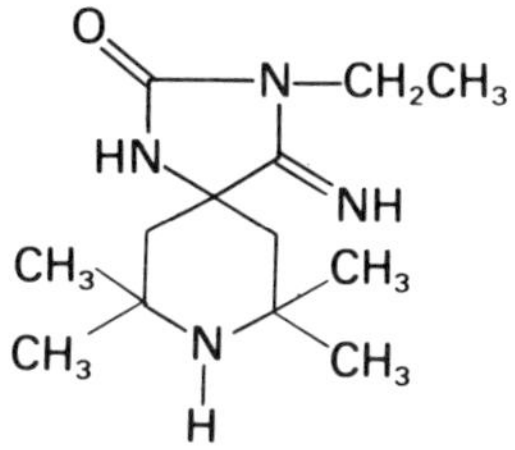 (6.90)	157	—	400
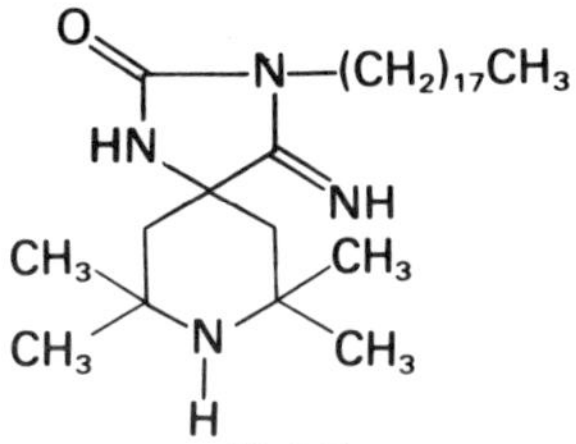 (6.91)	115	—	800
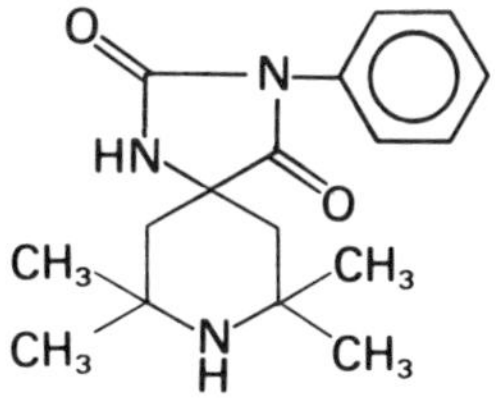 (6.92)	143	—	860
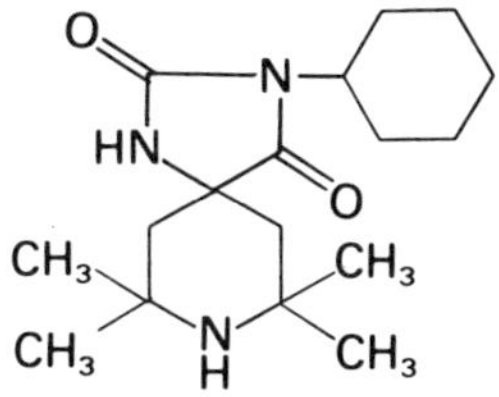 (6.93)	187	—	660

(continued)

Table 6.14—*contd.*

Compound	*M.p. (°C)*	*B.p. (°C)*	*Light stability measured as embrittlement time (h)*
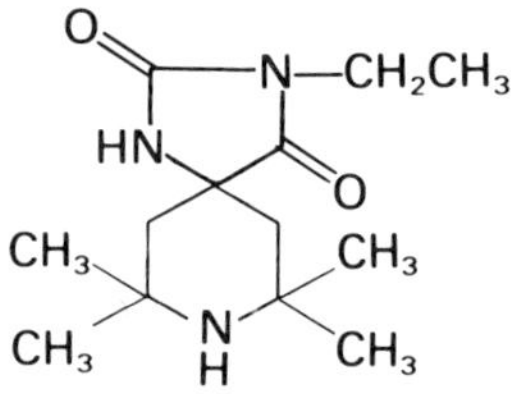 **(6.94)**	166	—	660
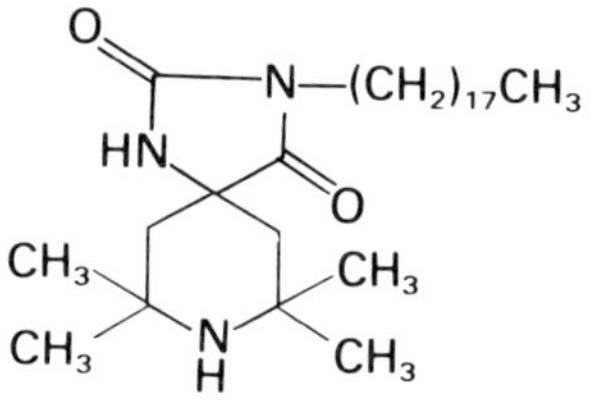 **(6.95)**	96	—	800
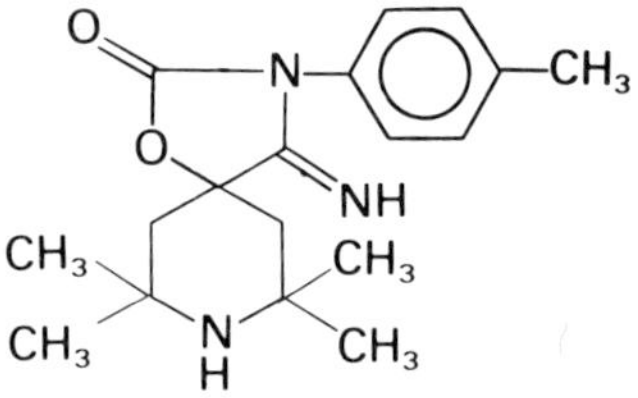 **(6.96)**	152	—	95
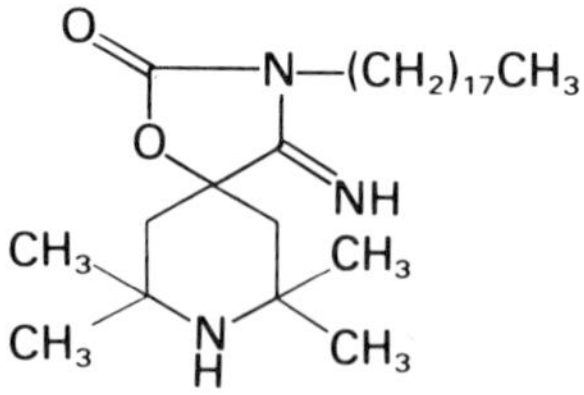 **(6.97)**	95	—	80

Table 6.14—*contd.*

Compound	*M.p. (°C)*	*B.p. (°C)*	*Light stability measured as embrittlement time (h)*
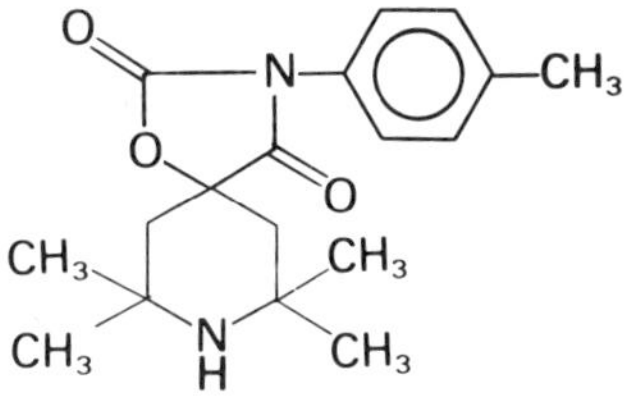 (6.98)	156	—	60
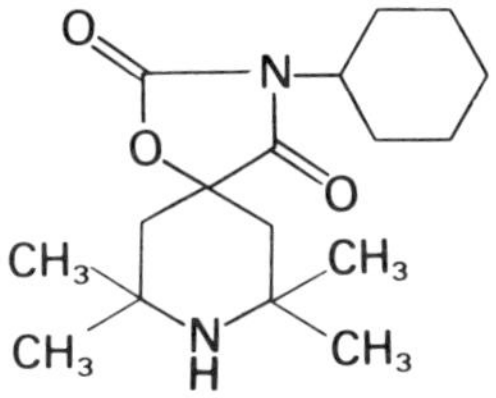 (6.99)	174	—	200
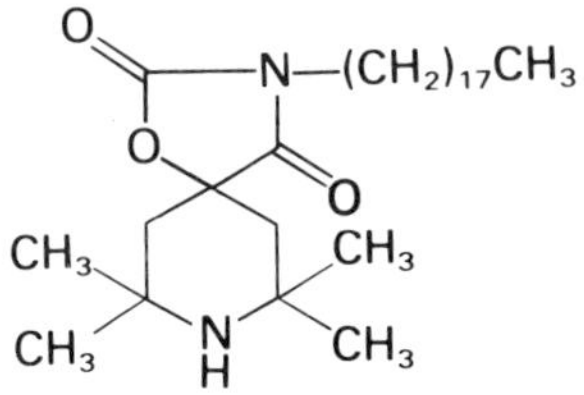 (6.100)	107	—	40
None	—	—	40

Table 6.15
Photostabilizing activity of different types of piperidinospiroacetals[1110]

Compound	*M.p. (°C)*	*B.p. (°C)*	*Light stability measured as embrittlement time (h)*
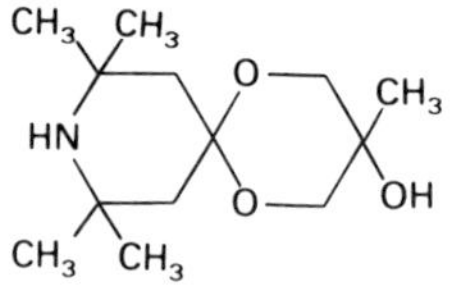 **(6.101)**	125	—	160
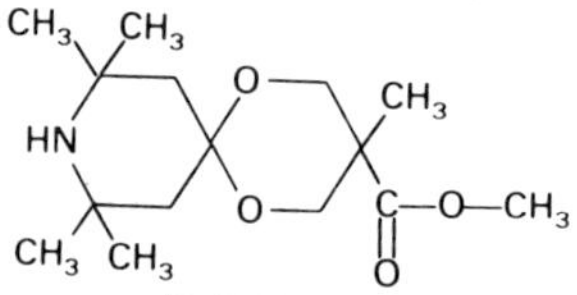 **(6.102)**	53	—	1 540
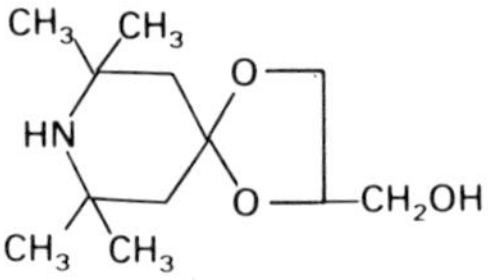 **(6.103)**	139	—	800
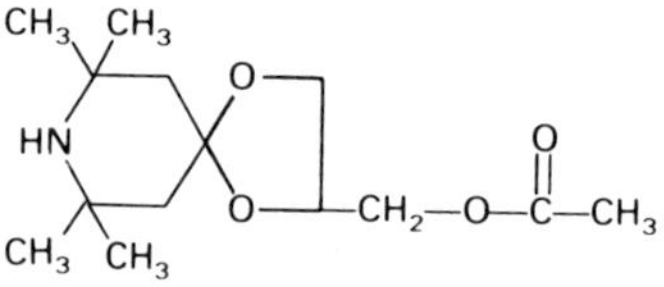 **(6.104)**	—	128/2 mmHg	460
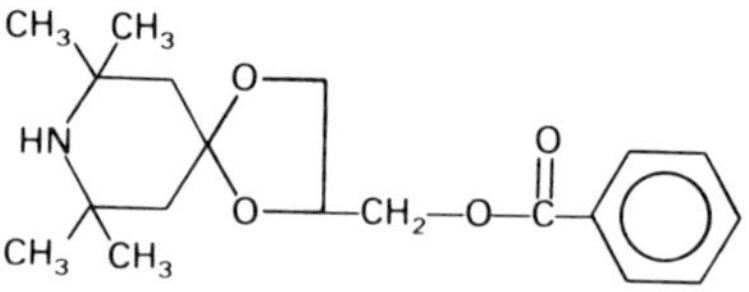 **(6.105)**	—	194/3 mmHg	600
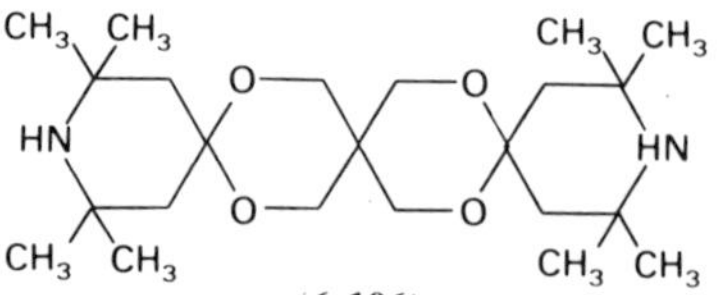 **(6.106)**	137	—	220

Table 6.15—*contd.*

Compound	*M.p. (°C)*	*B.p. (°C)*	*Light stability measured as embrittlement time (h)*
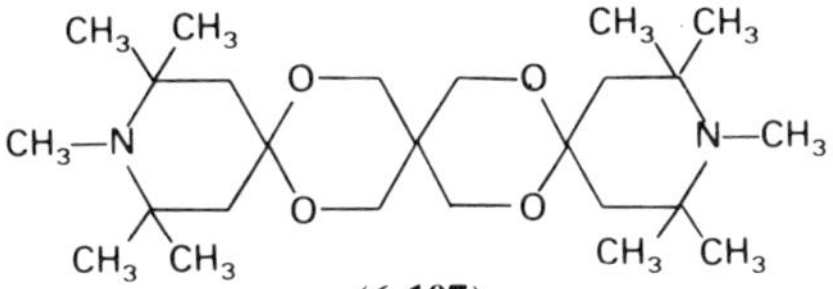 (**6.107**)	203	—	300
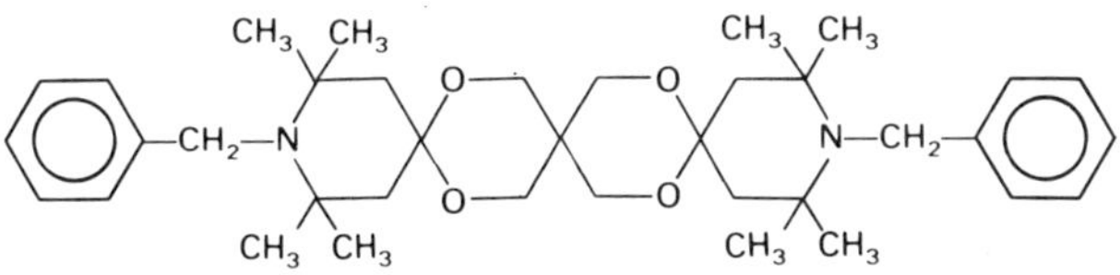 (**6.108**)	212	—	1 340
None	—	—	40

and other analogues have recently been developed and tested for their performance in polypropylene films.[929,1109,1110]

Especially interesting are hindered diazacycloalkanones, such as:

—1,1′-(1,2-ethanediyl)bis(3,3,5,5-tetramethyl-2-piperazinone) (Goodrite 3034; Table 6.1) (**6.12**);[689,700]

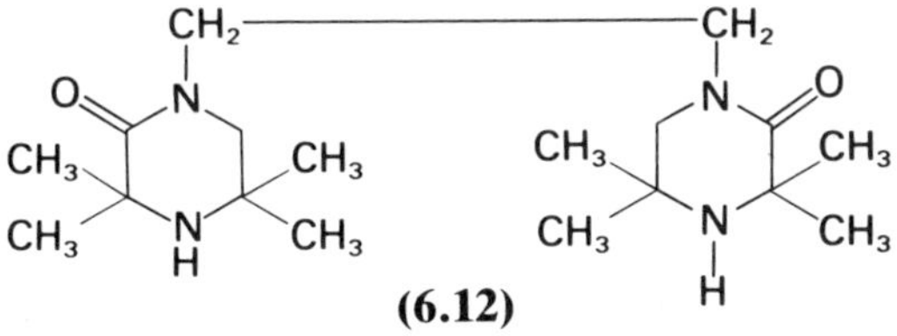

(6.12)

—1-tetradecyl-3,3-dimethylquinoxalin-2-one (**6.121**);[689,700]

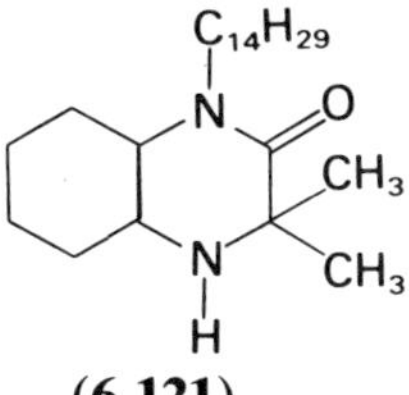

(6.121)

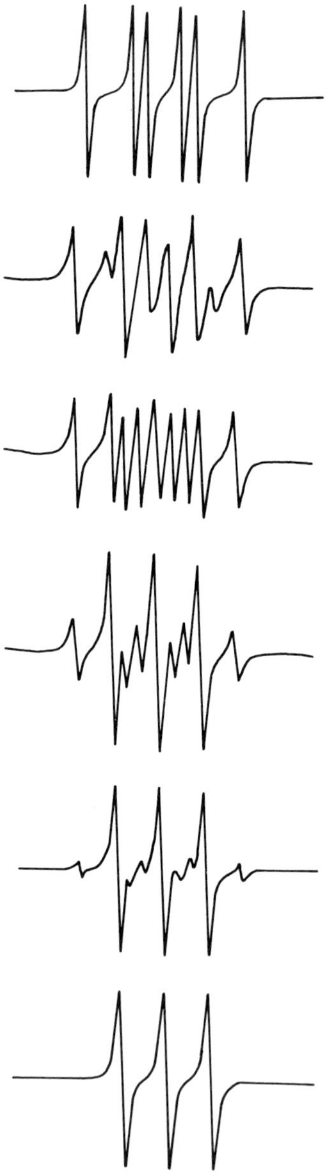

Fig. 6.20. ESR spectra obtained during the stepwise oxidation of 1-tetradecyl-3,3-dimethylquinoxalin-2-one (**6.121**). Reprinted with permission from Ref. 700, © 1985 American Chemical Society.

Table 6.16
Photostabilizing activity of different types of nitroxy radicals[1110]

Compound	*M.p. (°C)*	*B.p. (°C)*	*Light stability measured as embrittlement time (h)*
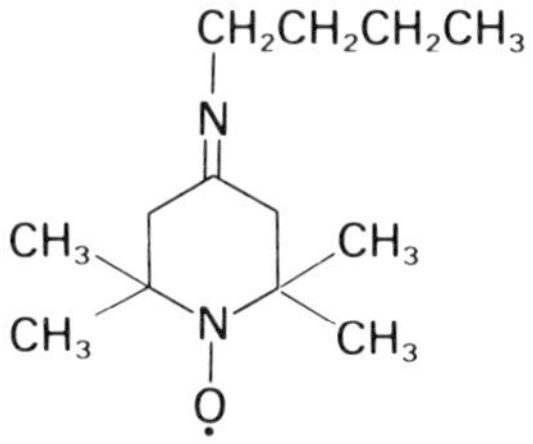 **(6.109)**	—	113/1 mmHg	300
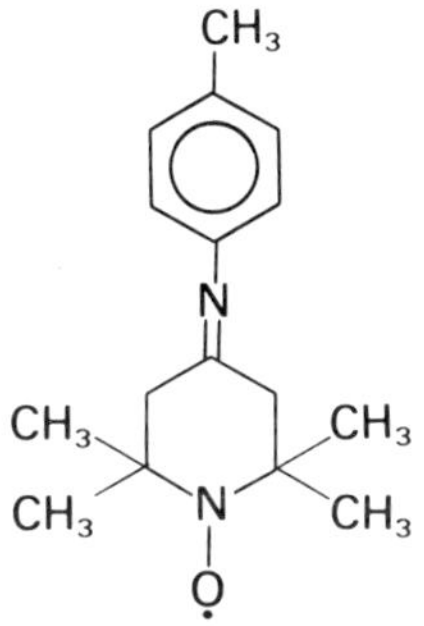 **(6.110)**	113	—	140
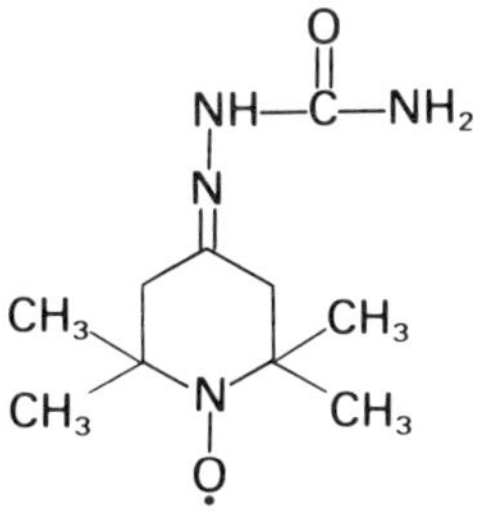 **(6.111)**	195	—	300
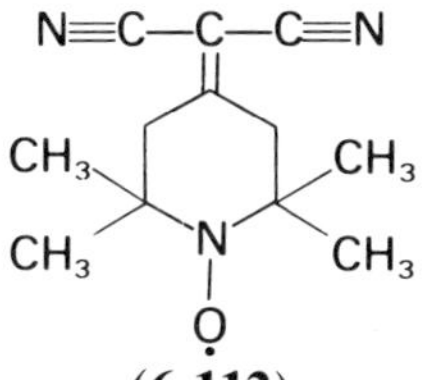 **(6.112)**	95	—	240

(continued)

Table 6.16—*contd.*

Compound	*M.p. (°C)*	*B.p. (°C)*	*Light stability measured as embrittlement time (h)*
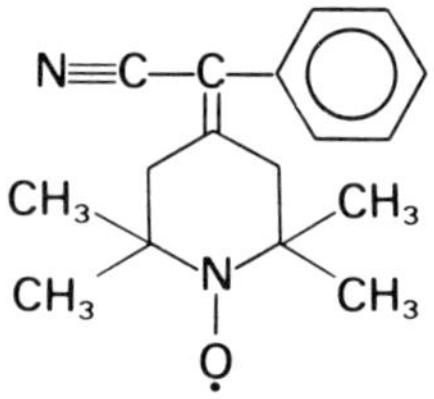 (6.113)	132	—	260
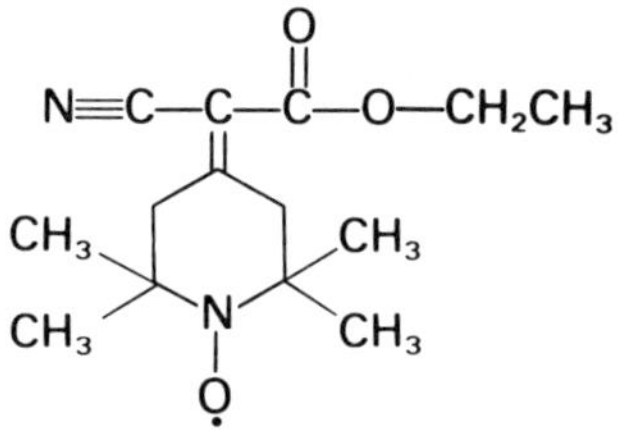 (6.114)	—	125/0·1 mmHg	380
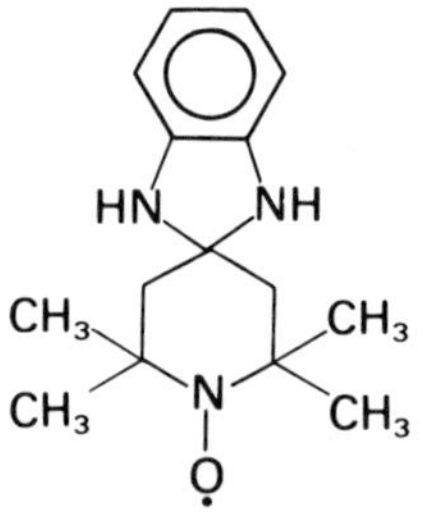 (6.115)	160	—	260
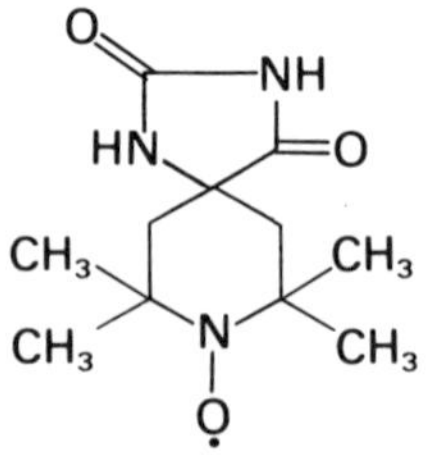 (6.116)	248	—	400

Table 6.16—*contd.*

Compound	*M.p. (°C)*	*B.p. (°C)*	*Light stability measured as embrittlement time (h)*
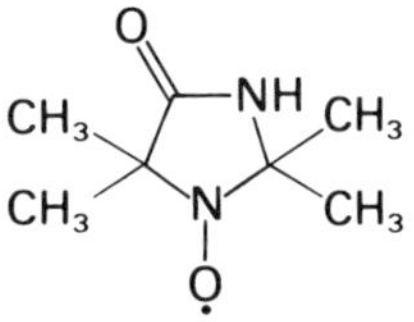 **(6.117)**	226	—	600
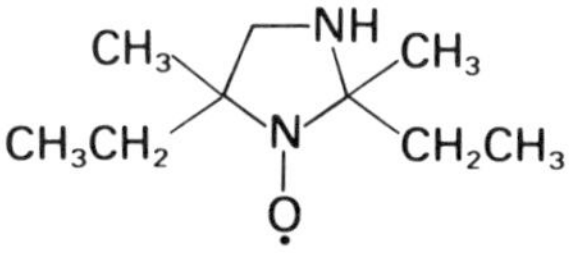 **(6.118)**	115	—	320
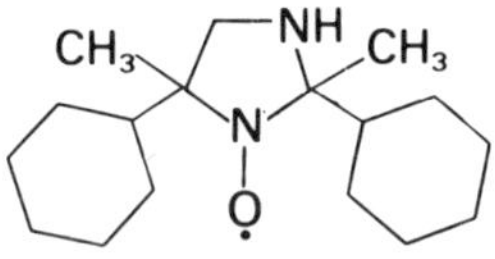 **(6.119)**	175	—	260
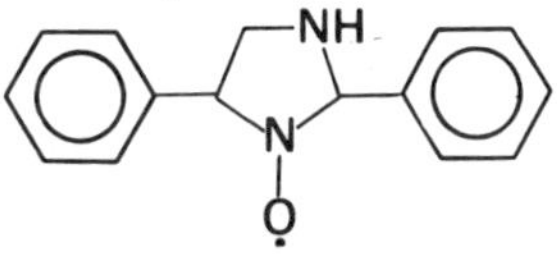 **(6.120)**	228	—	320
None	—	—	40

—15-benzyl-7,15-diazadispiro[5.1.5.3]hexadecane-14,16-dioxoamine **(6.122)**;[294]

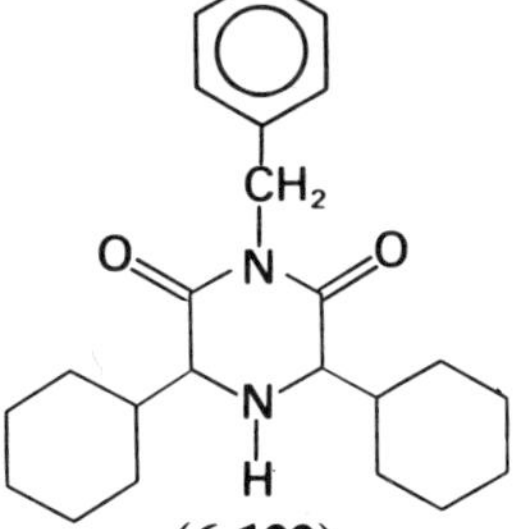

(6.122)

—15-benzyl-7,15-diazadispiro[5.1.5.3]hexadecane-14,16-dion-15-yl)-3,6-dioxaoctane (**6.123**).[294]

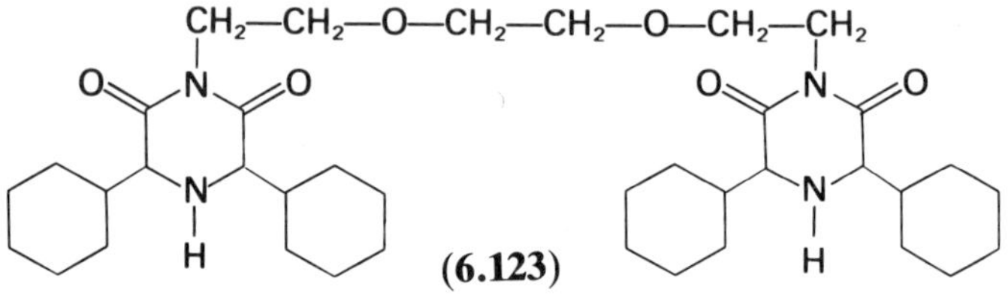

(**6.123**)

During oxidation, they form stable partially hindered nitroxy radicals (six-line ESR spectra), which are further oxidized to new, stable, fully hindered nitroxy radicals (three-line ESR spectra) (Fig. 6.20)[689,700] Field desorption mass spectroscopy of the oxidation of these amines shows that the primary nitroxy radicals, which form first, lose two hydrogen atoms and add an oxygen atom according to the following mechanism:

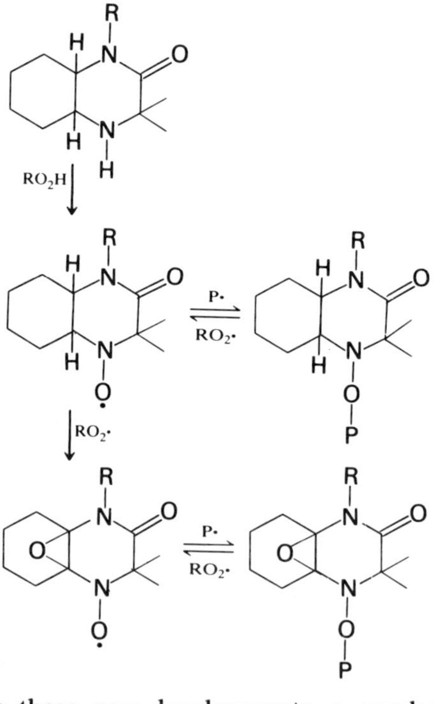

(6.99)

(6.100)

(6.101)

(6.102)

In addition to these new developments, a number of new nitroxy radicals have been synthesized and tested for their antioxidant properties (Table 6.16).

7

Polymer-Bound Additives

7.1 PROPERTIES OF POLYMER-BOUND ADDITIVES

The following methods have been used to produce polymer-bound additives (polymeric antioxidants or polymeric photostabilizers):

(i) Homopolymerization or copolymerization of vinyl additive monomers.
(ii) Reactions of functional groups in additives with functional groups in polymers.
(iii) Thermal or photo- grafting of vinyl additive monomers on to polymers, which can be performed in two different ways, i.e. by bulk or surface grafting, depending on the location of the grafted chains containing the additive groups. Bulk or surface grafting of the monomeric additives may be achieved by appropriate selection of grafting technique and conditions.

Polymer-bound additives are generally melt-miscible with the polymer to which they are added. A polymer-bound additive uniformly distributed will always have the limited mobility associated with the 'crank-shaft' movement of the polymer chain in the amorphous region of the polymer.

The non-uniform distribution of additives in the amorphous part of semicrystalline polymer implies that a small region will contain all the additive whereas the rest will have none. In this case the mobility (diffusion) of additive molecules in a polymer matrix is important to achieve their stabilizing activity.

The polarity of grafted chains may attract the polar groups in stabilizers, thus reducing their migratory tendency and improving their resistance to extraction and volatilization. Often graft copolymers may exhibit synergistic activity in combination additives.

Since the surface of polymers is considered to be the region that is most susceptible towards photo-oxidative degradation, the surface grafting of stabilizers seems to be the most effective method of protection of polymers against degradation.

Polymer-bound additives are especially useful in polymeric fibres which are widely subjected to water, aqueous detergents and dry-cleaning solvents.

Polymer-bound antioxidants have wide application in rubbers due to their great sensitivity to oxidation (particularly at high temperatures) and photo-oxidation.

The migration of polymer additives is especially important in food packaging, in toys for children, in medical applications where polymers are in contact with body fluids, and in organ replacement surgery. For that reason polymer bound-additives have special and important significance in the design and manufacture of medical equipment and highly durable medical replacements.

The disadvantages in the application of polymer-bound additives (antioxidants or photostabilizers) arise from:

(i) difficulties in the synthesis of vinyl monomers;
(ii) purification of reactive multifunctional stabilizers;
(iii) homopolymerization;
(iv) reduction in efficiency of the stabilizer system.

A number of review articles on polymeric antioxidants[820,1015,1021] and polymeric photostabilizers[150,820,1101,1102,1128–1131,1134] have been published.

7.2 POLYMERIC ANTIOXIDANTS

Polymeric hindered phenols (**7.1**) and their stable aryloxy polyradicals (**7.2**) can be synthesized by the following reactions:[207,549]

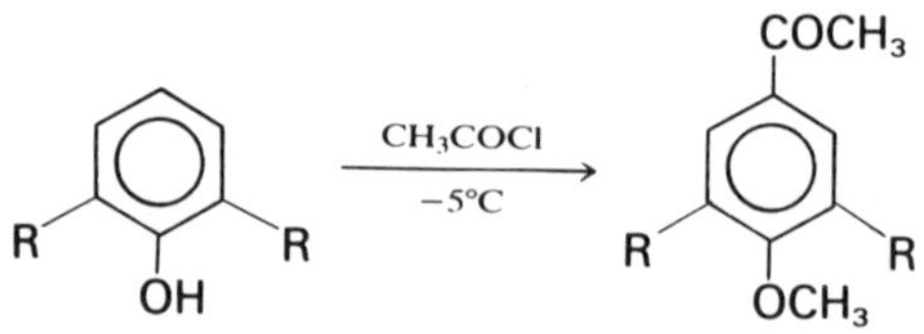

(7.1)

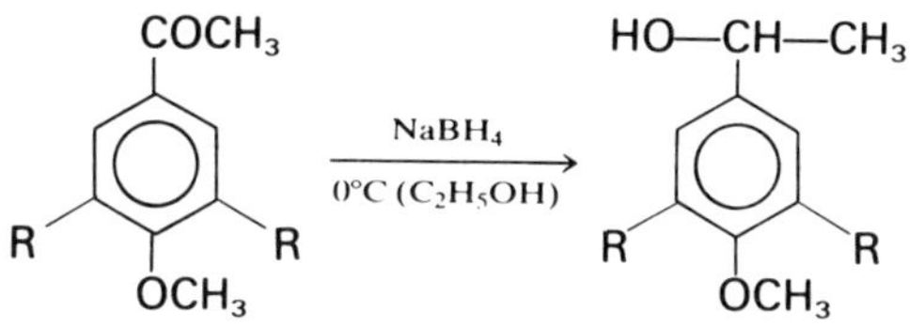

(7.2)

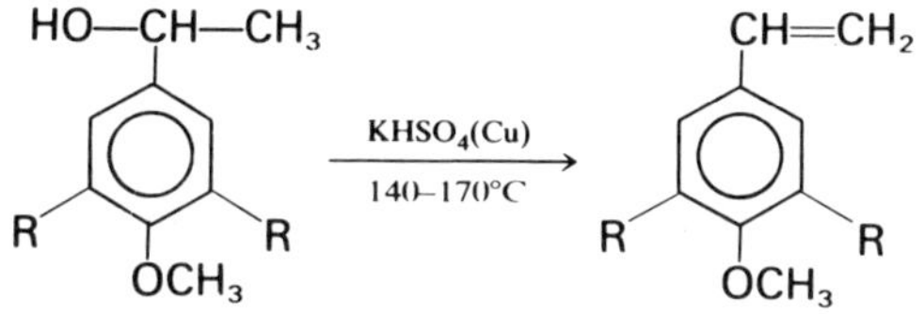

(7.3)

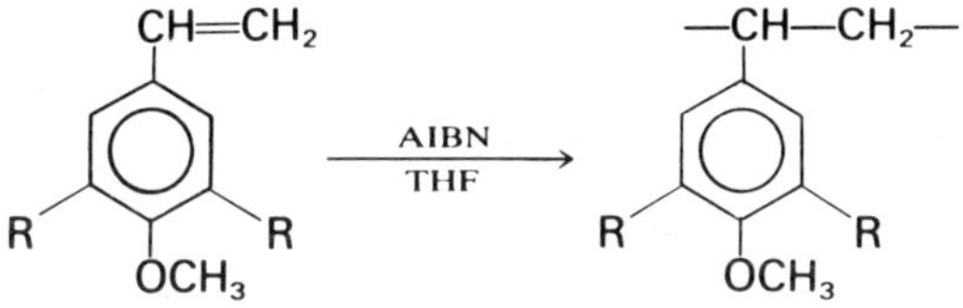

(7.4)

(AIBN = azo-bis-isobutyronitric)

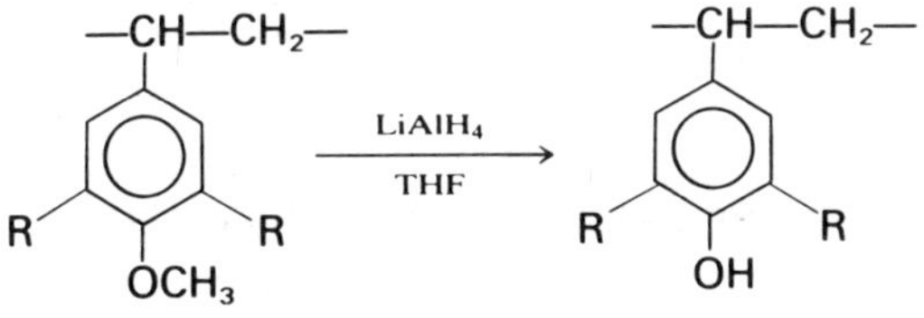

(7.5)

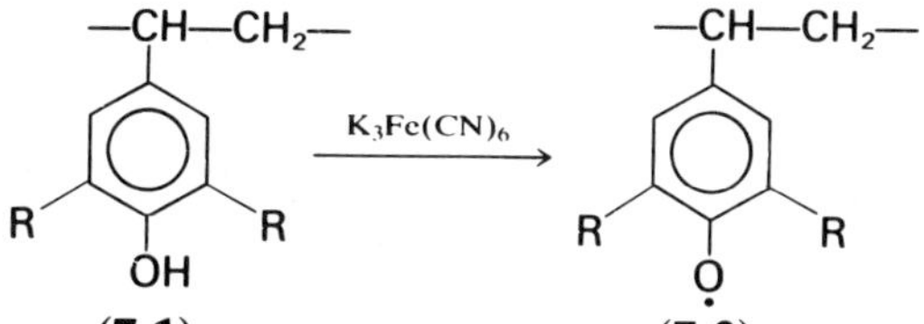

(7.6)

(7.1) **(7.2)**

In another method, α,β-unsaturated acid chlorides (**7.3**) were used to prepare polymerizable hindered phenolic monomers by acylation of 2,6-di-*tert.*-butylhydroquinone, 2,6-di-*tert.*-butyl-4-aminophenol, 3,5-di-*tert.*-butyl-4-hydroxyphenyl alcohols (**7.4**), and 3,5-di-*tert.*-butyl-4-hydroxyphenylalkylamines:[651]

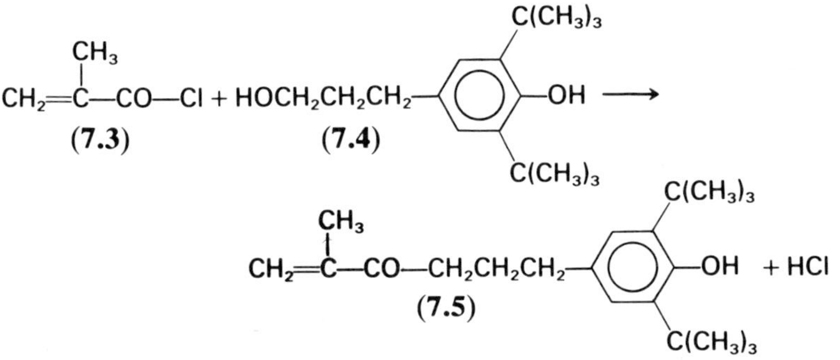

(7.7)

Vinyl monomers of hindered phenols such as 2,6-di-*tert.*-butyl-4-vinylphenol (**7.6**) and 2,6-di-*tert.*-butyl-4-isopropenylphenol (**7.5**) have been copolymerized with styrene, isoprene or butadiene.[8,497]

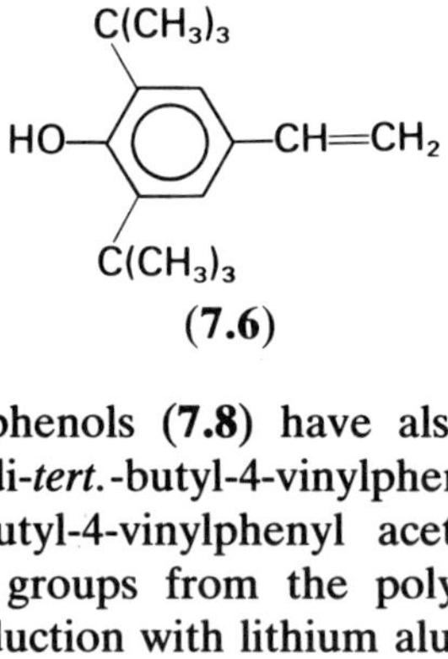

(**7.6**)

Polymeric hindered phenols (**7.8**) have also been synthesized by polymerization of 2,6-di-*tert.*-butyl-4-vinylphenyl trimethylsilyl ether (**7.7**) and 2,6-di-*tert.*-butyl-4-vinylphenyl acetate (**7.9**) and further splitting off protecting groups from the polymers by hydrolysis in acidic medium or by reduction with lithium aluminium hydride:[207]

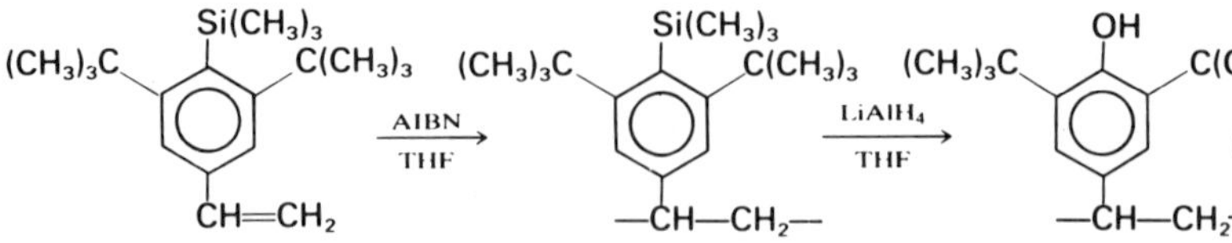

(7.8)

(**7.7**) (**7.8**)

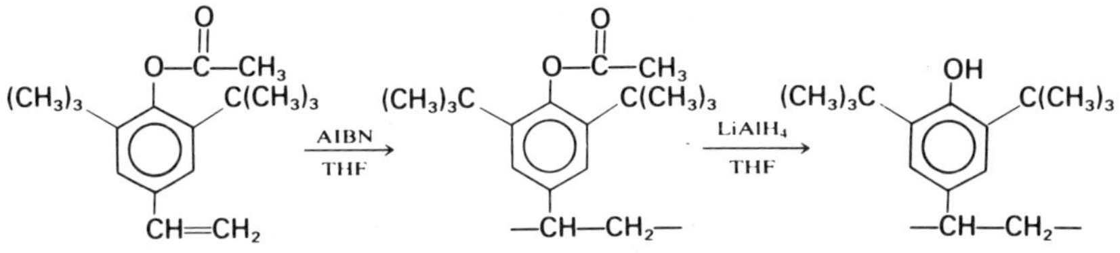

(7.8) (7.9)

(7.9)

Commercially produced nitrile rubber (Chemigum HR 665) contains small amounts (<2%) of a copolymerized vinyl monomer of *sec*-alkylaminodiphenylamine (**7.10**) which is the best rubber antidegradant:[566,651,787]

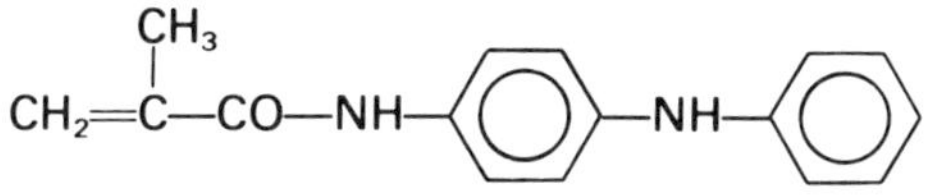

(**7.10**)

This monomer(**7.10**) can be obtained by acylation of *p*-aminodiphenylamine (**7.11**) and *p*-hydroxydiphenylamine with chlorides of α,β-unsaturated acids (**7.3**):[651]

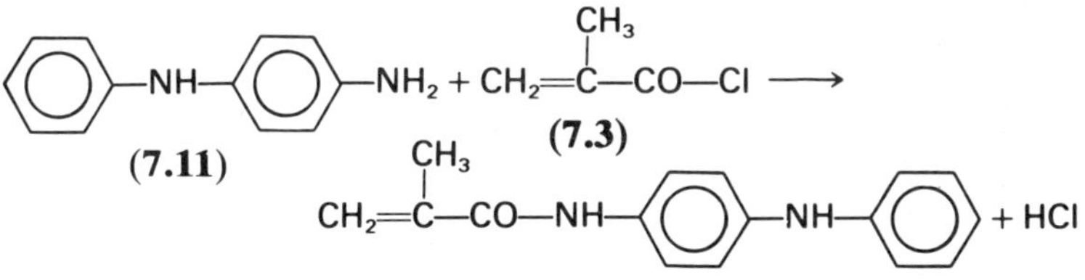

(7.10)

(**7.10**)

The *p*-nitrosodiphenylamine (**7.11**) or *N,N*-diethyl-*p*-nitrosoaniline can be bound to the rubber during vulcanization:[223]

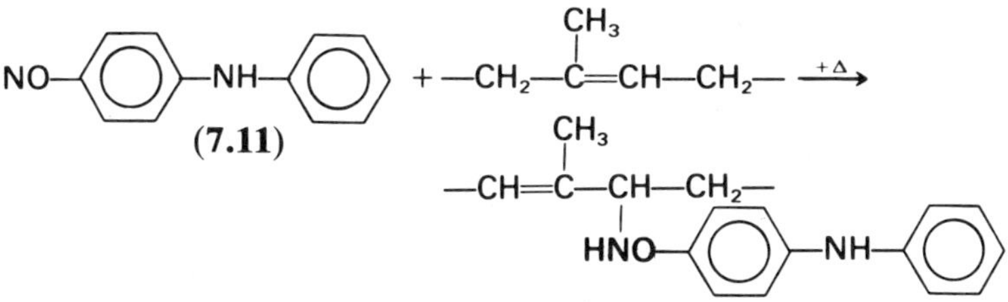

(7.11)

Polymeric antioxidants can also be obtained by polycondensation of hindered bisphenols with acid bischlorides.[344] In this way polyesters of moderate weight are formed:

$$HO{-}R{-}OH + Cl{-}\overset{O}{\overset{\|}{C}}{-}(CH_2)_8{-}\overset{O}{\overset{\|}{C}}{-}Cl \longrightarrow \left[{-}O{-}R{-}O{-}\overset{O}{\overset{\|}{C}}{-}(CH_2)_8{-}\overset{O}{\overset{\|}{C}}{-}\right]_n \quad (7.12)$$

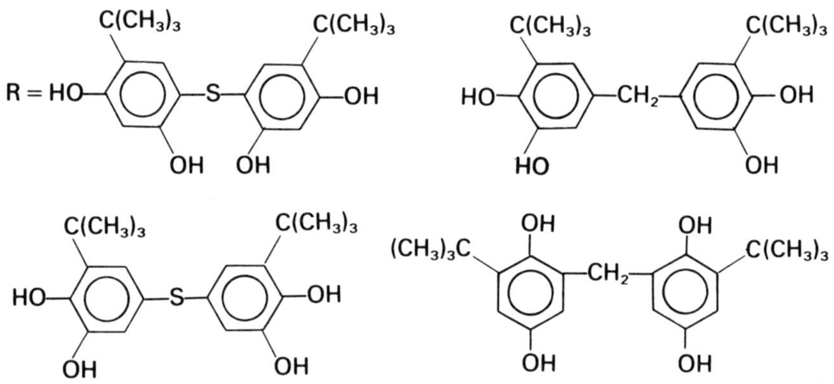

One of the most common methods for bonding antioxidants to polymer macromolecules is to employ the reactivity of epoxy groups with amine, hydroxy or carboxy groups. In this way polypropylene was grafted with glycidyl methacrylate (Table 7.1) or methacrylic acid (Table 7.2) and then reacted with antioxidant containing carboxy or epoxy groups, respectively:[820]

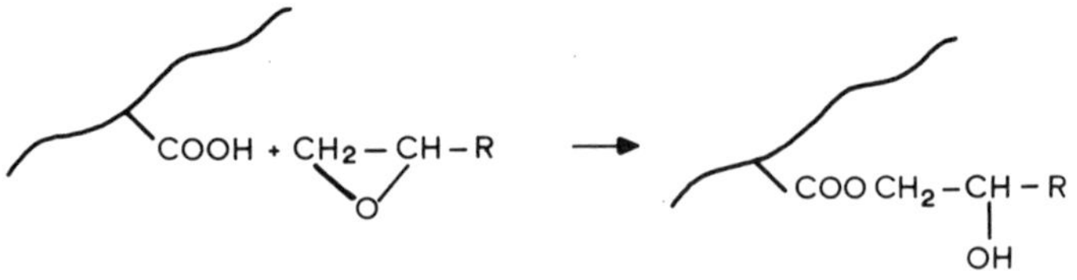

(7.13)

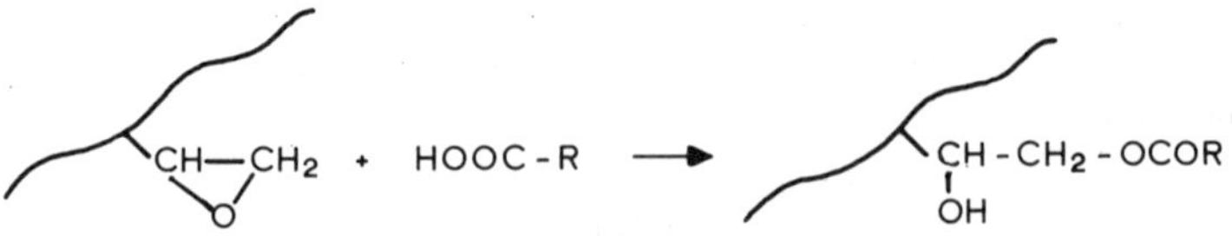

(7.14)

Reactions between the antioxidant and the grafted polypropylene

Table 7.1
Different antioxidants chemically bonded to polypropylene grafted with 3% glycidyl methacrylate

Compound	*Structure*	*Reference*
3-Butylthio-1-propionic acid	$n\text{-}C_4H_9\text{—S—}CH_2\text{—}CH_2\text{—COOH}$ **(7.12)**	831
3-Dodecylthio-1-propionic acid	$n\text{-}C_{12}H_{25}\text{—S—}CH_2\text{—}CH_2\text{—COOH}$ **(7.13)**	830
3-Octadecylthio-1-propionic acid	$n\text{-}C_{18}H_{37}\text{—S—}CH_2\text{—}CH_2\text{—COOH}$ **(7.14)**	832
4-[3′,5′-Di-*tert.*-butyl-4′-hydroxyphenyl-acetamido]benzoic acid	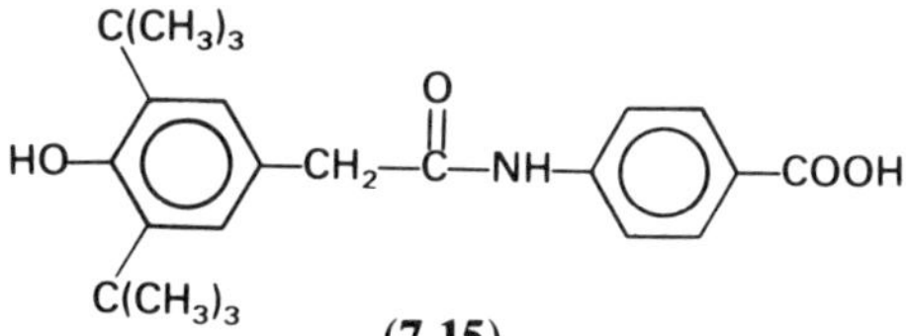 **(7.15)**	833
4-[3′,5′-Di-*tert.*-butyl-4′-hydroxyphenyl-β-propionamido] salicylic acid	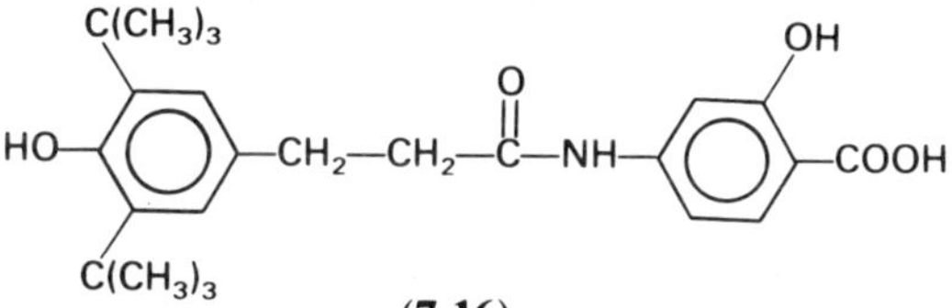 **(7.16)**	833
4-[3′,5′-Di-*tert.*-butyl-4′-hydroxyphenyl-β-propionamido]phenol	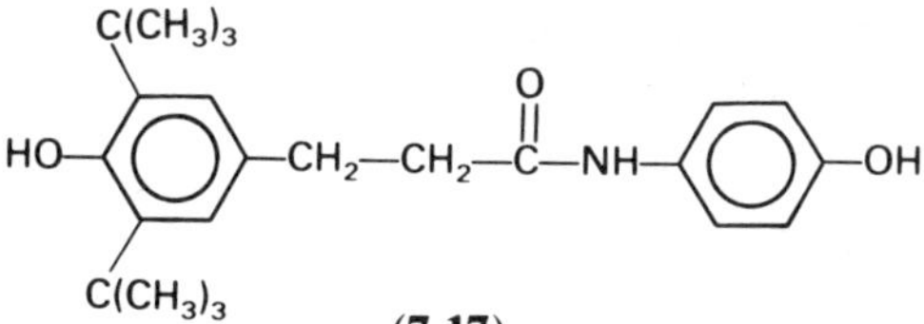 **(7.17)**	829
3,5-Di-*tert.*-butyl-4-hydroxyphenyl--β-propionic acid	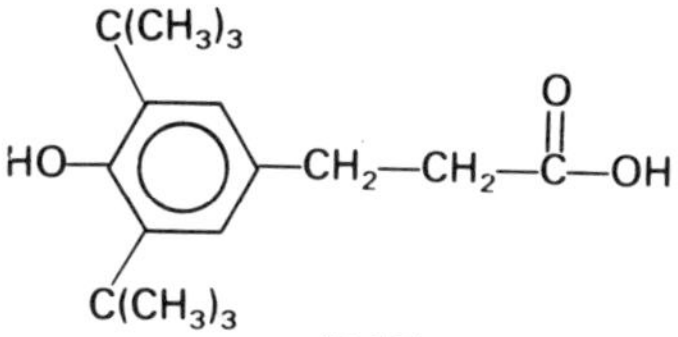 **(7.18)**	829, 836

Table 7.2
Different antioxidants chemically bonded to polypropylene grafted with 4·7% methacrylic acid

Compound	*Structure*	*Reference*
Glycidyl 3-[3,5-di-*tert.*-butyl-4-hydroxyphenyl]propionate	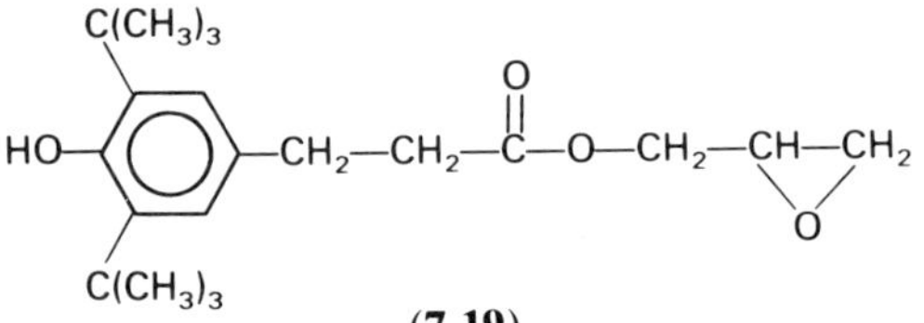 (7.19)	834, 837 839
3-*tert.*-Butyl-4-hydroxyphenyl glycidyl ether	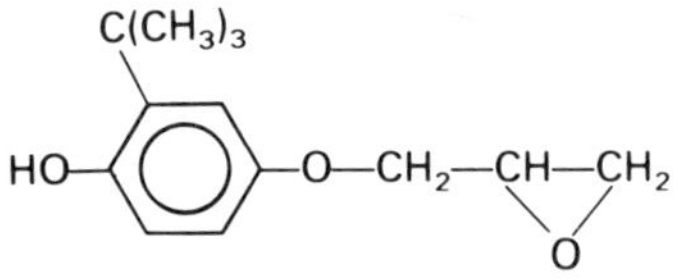 (7.20)	835
Dodecyl glycidyl thioether	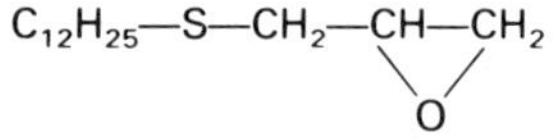 (7.21)	839
Dodecylthioethyl glycidyl ether	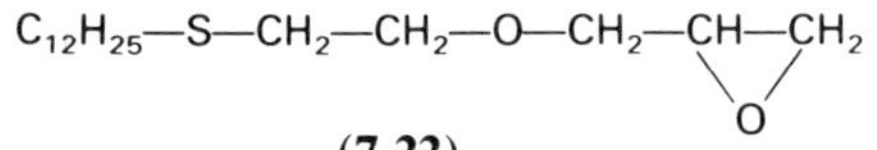 (7.22)	836
Glycidyl 3-dodecyl-thiopropionate	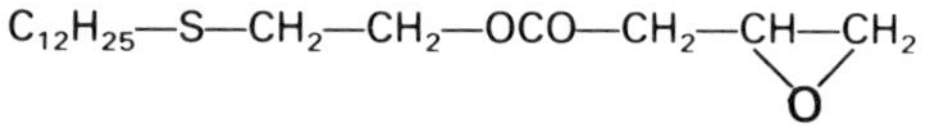 (7.23)	838

occurred in the melt so that bulk stabilization could be achieved. The grafted polypropylene containing 3–5% monomer was blended with 0·5% antioxidant, the mixture was milled and then it was melt-spun and stretched to obtain stabilized polypropylene yarn.

Masterbatches containing 17% bound antioxidant were obtained in the same way, starting with polypropylene having a higher content of grafted monomer, e.g. 8%. The masterbatch was then diluted with polypropylene to obtain the concentration of the bound antioxidant, i.e. about 0·5%. Sometimes, mixtures of polypropylene and grafted polypropylene were melt-spun and the fibres were immersed in a hot solution of antioxidant. Under these conditions the reaction between the antioxidant and the reactive groups of the grafted chains takes place mainly at the surface of the fibres.[586,840]

At high processing temperatures, the stabilizers evaporate from the polymer, and thus their concentration and efficiency in the polymer product decrease. The contamination of the technological equipment is a very important factor, since it causes difficulties in spinning as well as a deterioration of the work environment.

Polydiene rubbers can be easily expoxidized and epoxide group (**7.24**) reacts with *sec*-alkylaminodiphenylamine (**7.25**):[638]

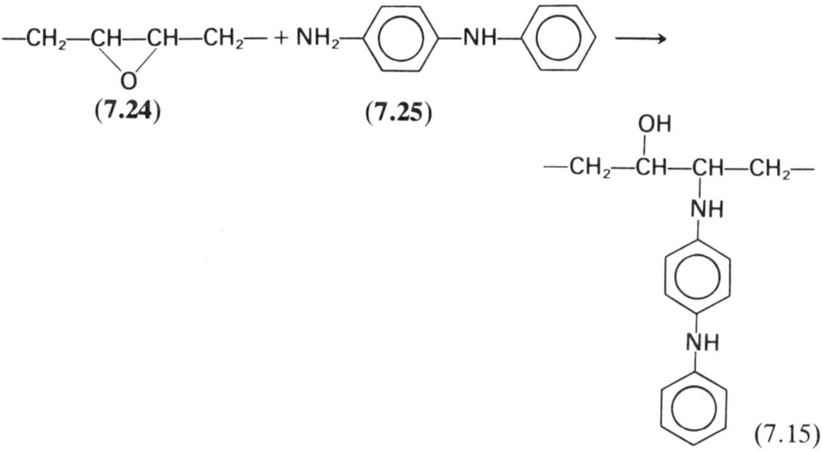

(**7.24**) (**7.25**)

(7.15)

A variety of additives such as antioxidants, e.g. 3,5-di-*tert.*-butyl-4-hydroxybenzyl mercaptan (**7.26**) or 4-mercaptoacetamidodiphenylamine (**7.27**), and photostabilizers, e.g. 4-benzoyl-3-hydroxyphenoxyethyl thioglycolate (**7.28**), can be added to polyolefins,[19,20,403,673,674,1013,1023,1024,1030,1035,1036] natural rubber, polybutadiene, nitrile rubber, styrene–butadiene rubber, EPDM (ethylene–propylene–diene terpolymers)[20,458,674,1024,1025,1027–1032] and poly(vinyl chloride)[318–320] through reaction with double bonds (present or formed in mechanical shear) in mechanical processing [eqn (7.16)].[1014,1017,1021]

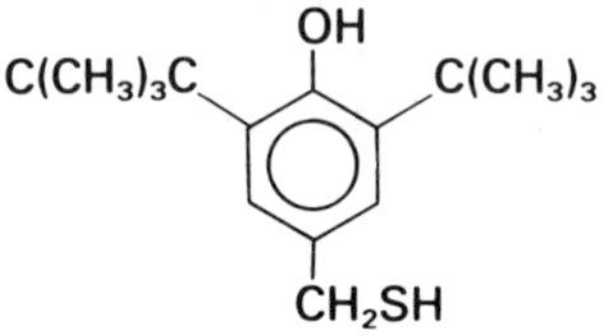

(7.26)

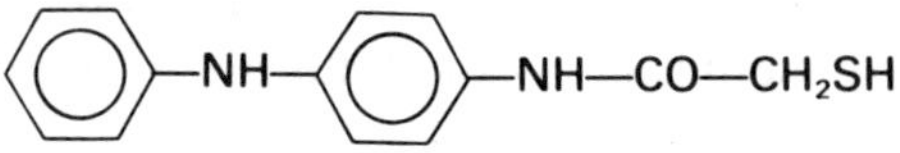

(7.27)

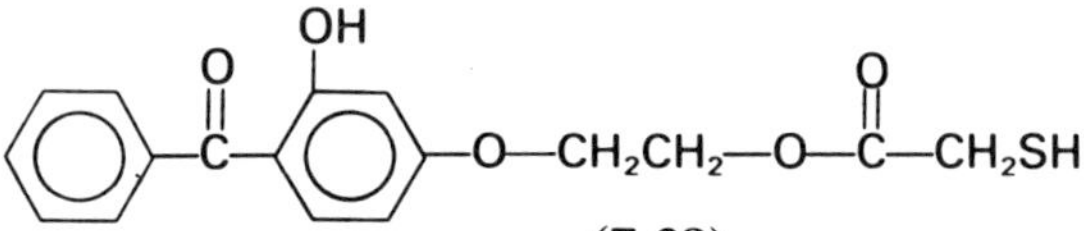

(7.28)

$$\text{—CH=CH—} + \text{RSH} \longrightarrow \text{—CH}_2\text{—}\underset{\substack{|\\ \text{SR}}}{\text{CH}}\text{—} \qquad (7.16)$$

The extent of mechanochemical modification of polymers is strongly dependent upon a number of interrelated parameters, such as:

(i) The intrinsic viscosity of the polymer at the temperature of the reaction.
(ii) Absence of oxygen in mixer, because it competes with thiyl radicals (RS•). No reactions occur in an open mixer.[20,1031]
(iii) The concentration of hydroperoxide groups in the polymer sample. In the presence of a molar excess of hydroperoxide, the thiol becomes an effective antioxidant in its own right and inhibits the completion of adduct formation by a small amount of residual hydroperoxide.
(iv) The concentration of thiol added. The decrease in yield of adduct at very high thiol concentrations is due to their plasticizing effect in the polymer matrix which leads to a reduction in the mechanical shearing forces acting on the polymer chains.[20]

Antioxidants and photostabilizers bound to rubbers through sulphur show a powerful synergistic effect in the photostabilization of this polymer which cannot be matched by synergistic mixtures of conventional additives.[458,459] Conventional additives are dispersed throughout the polymer matrix in rubber-modified plastics, whereas polymer-bound antioxidants are located almost exclusively in the rubber phase, which undergoes the most rapid photo-oxidation, and hence the most sensitive phase will be selectively stabilized.[1011,1015]

Hindered phenols can be grafted on to polyolefins and polyoxymethylene by the following reactions:[627]

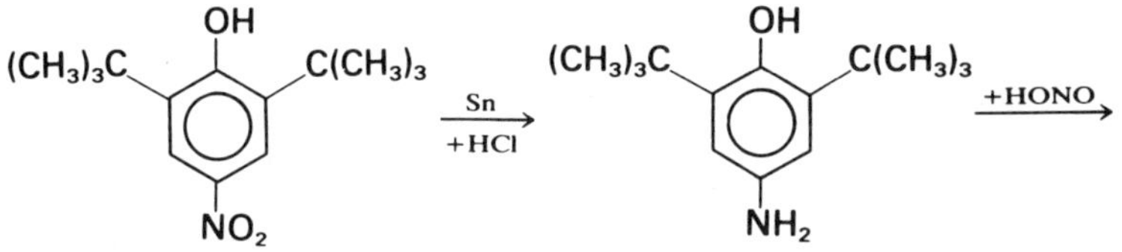

(7.17)

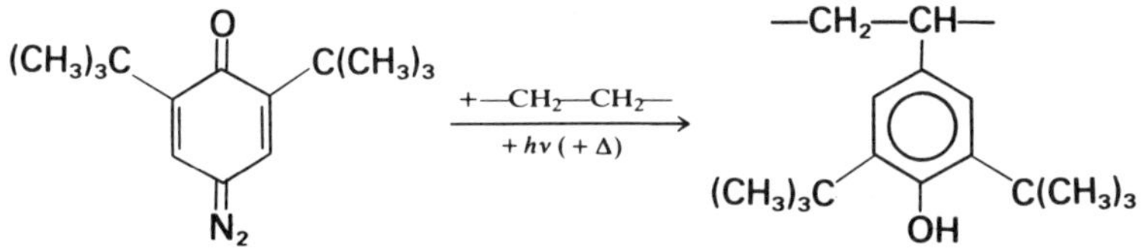

(7.18)

A common method of bonding antioxidants to polymers is by thermal or photo- grafting reactions. In this way the following monomeric antioxidants were photografted on to polyolefins[382,821,825] and rubber:[1009] 3,5-di-*tert.*-butyl-4-hydroxybenzyl methacrylate (**7.29**), 2,6-di-*tert.*-butyl-4-vinylphenol (**7.30**), *trans*-3,5-di-*tert.*-butyl-4-hydroxycinnamic acid (**7.31**), and *N*-[3,5-di-*tert.*-butyl-4-hydroxybenzyl]maleimide (**7.32**). The films were immersed in a benzene solution containing monomer and benzophenonone as photoinitiator. The grafting reaction, initiated by UV radiation, depends on such

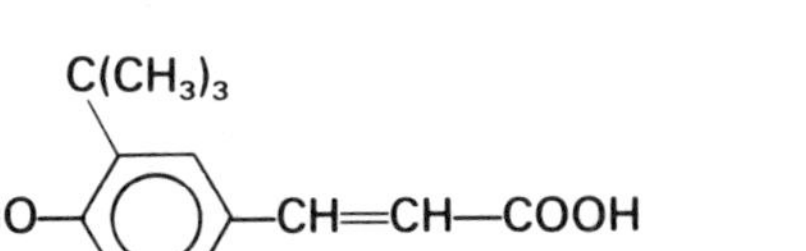

(**7.29**)

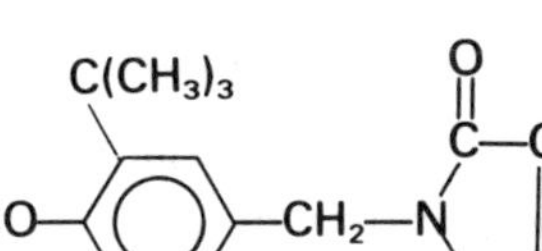

(**7.30**)

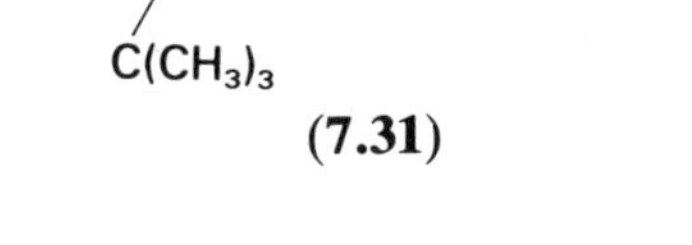

(**7.31**)

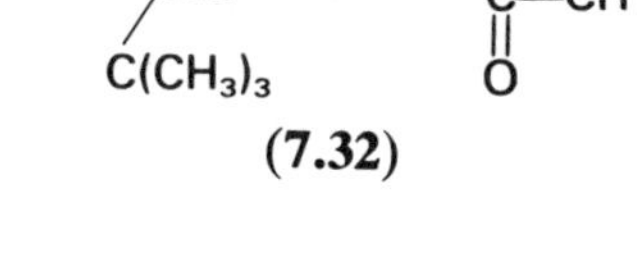

(**7.32**)

parameters as the ratio of monomer to benzophenone, irradiation time and the presence or absence of air. After extraction with acetone the surface-grafted antioxidant was much more effective than the antioxidants with the same stabilizing group incorporated in polypropylene at a similar concentration by conventional melt-compounding. The oxidative stability of the grafted film was much higher than that of polypropylene containing the monomeric antioxidant **7.29**, an antioxidant with a long alkyl chain, and even the homopolymer of the monomeric antioxidant **7.29**.

A lower grafting frequency, i.e. shorter and more frequent antioxidant chains attached to the polyolefin backbone, is more effective than very few but long grafted chains.[382]

7.3 POLYMERIC PHOTOSTABILIZERS

Most common low-molecular-weight photostabilizers have been transformed to polymeric form. The difficult step is still to obtain polymerizable vinyl or acrylate monomers with photostabilizing bonded groups. Examples of such syntheses, which are sometimes complicated, are given in the following sections.

Polymerizable UV absorbers are recommended for applications in thin polymeric films which should have a service life of five or more years.

The advantage of using polymeric UV stabilizers for the protection of polymers is most evident in samples with high surface area to volume ratio where low-molecular-weight stabilizers are easily lost.

7.3.1 Vinylcyanophenyl Cinnamate Monomer and Polymer

The polymerizable monomer of ethyl-4-vinyl-α-cyano-β-phenyl cinnamate (**7.33**) has been synthesized by the following route:[1084,1085,1099]

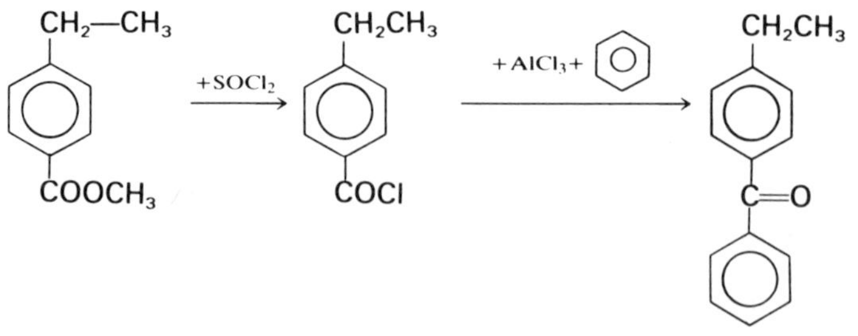

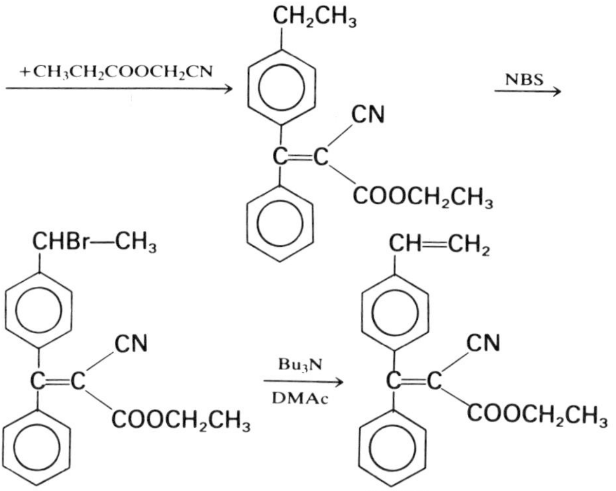

(7.19)*

(7.33)

This monomer was homopolymerized and copolymerized with styrene and methyl methacrylate with radical initiators.

7.3.2 Salicylate Monomers and Polymers

The following salicylate monomers have been synthesized:

(i) 3-Vinylsalicylate monomers (**7.34** to **7.37**).[607]

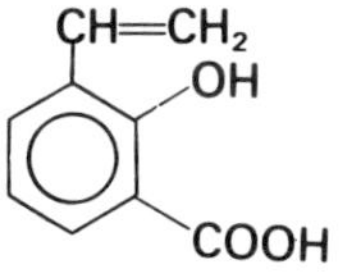

(7.34)
3-Vinylsalicylic acid

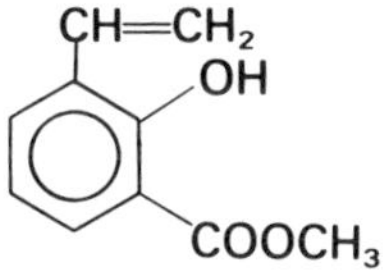

(7.35)
Methyl 3-vinylsalicylate

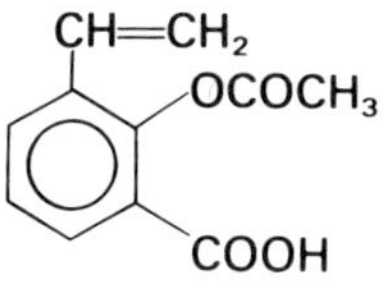

(7.36)
3-Vinylsalicylic acid

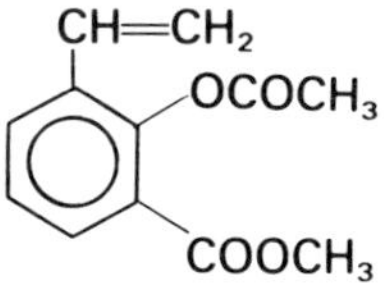

(7.37)
3-Vinylacetylsalicylic acid

* Abbreviations used in schemes (7.19)–(7.22) NBS, *N*-bromosuccinimide; Ac_2O, acetic anhydride; Bu_3N, tributylamine; Et_3N, triethylamine; DmAc, *N,N*-dimethylacetamide; P_4, phosphor.

(ii) 4-Vinylsalicylate monomers (**7.38** to **7.41**).[1099,1103]

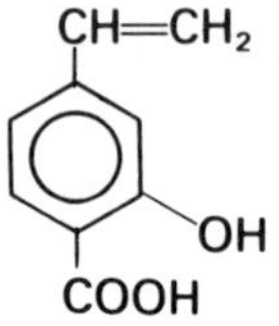

(7.38)
4-Vinylsalicylic acid

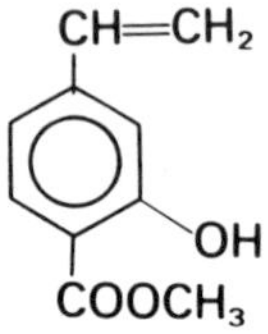

(7.39)
Methyl 4-vinyl salicylate

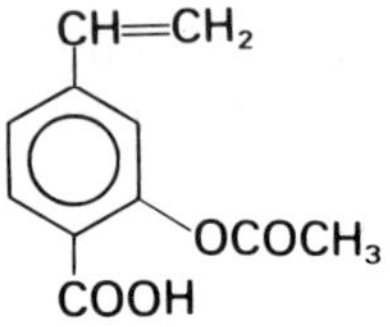

(7.40)
4-Vinylsalicylic acid

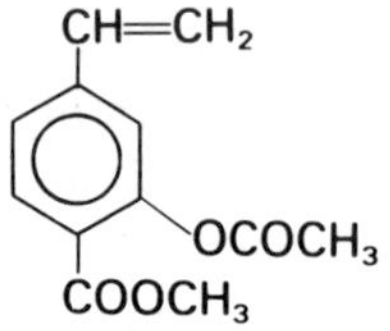

(7.41)
4-Vinylacetylsalicylic acid

(iii) 5-Vinylsalicylate monomers (**7.42** to **7.45**).[149]

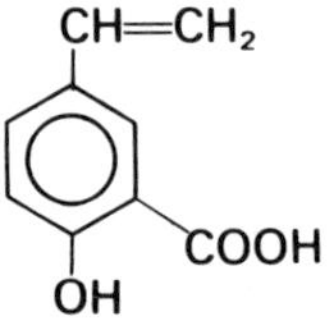

(7.42)
5-Vinylsalicylic acid

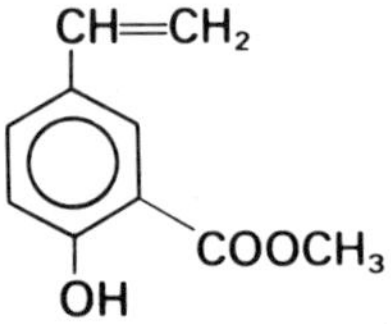

(7.43)
Methyl 5-vinyl salicylate

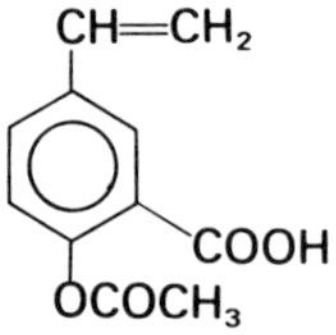

(7.44)
5-Vinylsalicylic acid

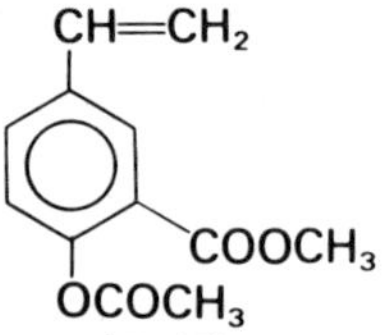

(7.45)
5-Vinylacetylsalicylic acid

These monomers have been homopolymerized and copolymerized with acrylic acid and methacrylic acid.[21,149,607,1103] Methyl 5-vinylsalicylate (**7.43**) has been grafted on to *cis*-1,4-polybutadiene and 1,2-polybutadiene.[645,646] Polymeric salicylic acid derivatives have been tested as effective UV-absorbing agents in skin protection formulations.[21]

(iv) Phenyl 5-acryloxy(or methacryloxy)methylsalicylates (**7.46**).[404]

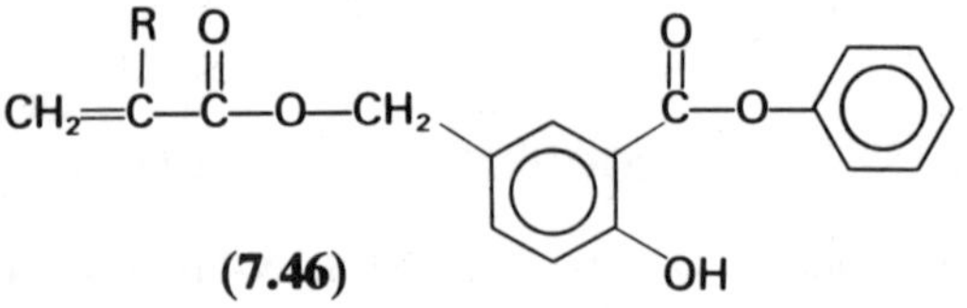

$R = H, CH_3$

(7.46)

These monomers may be readily homopolymerized and copolymerized with a wide range of other monomers, giving very good yields of polymer with high molecular weight.

The following synthetic routes were used for the preparation of vinylsalicylate monomers.

(i) 3-Vinylsalicylate monomers:[607]

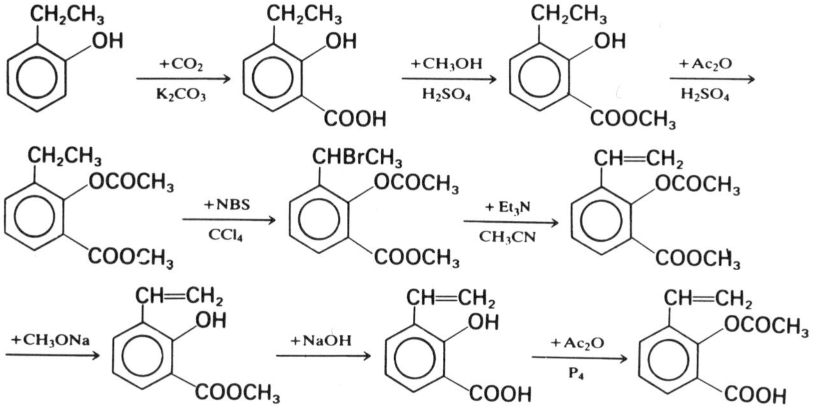

(7.20)*

(ii) 4-Vinylsalicylate monomers:[1103]

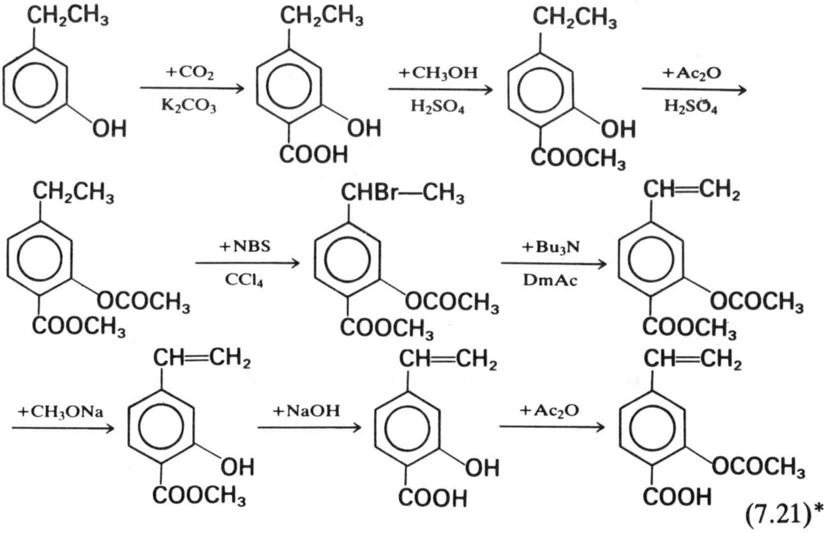

(7.21)*

* Abbreviations used in schemes (7.19)–(7.22) NBS, *N*-bromosuccinimide; Ac_2O, acetic anhydride; Bu_3N, tributylamine; Et_3N, triethylamine; DmAc, *N,N*-dimethylacetamide; P_4, phosphor.

(iii) 5-Vinylsalicylate monomers:[149]

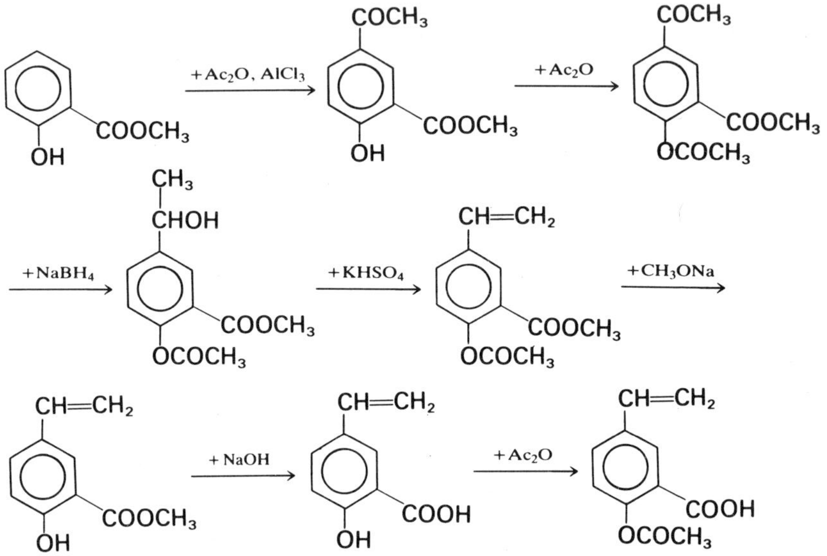

(7.22)*

(iv) Phenyl 5-acryloxy(or methacryloxy)methylsalicylates:[404]

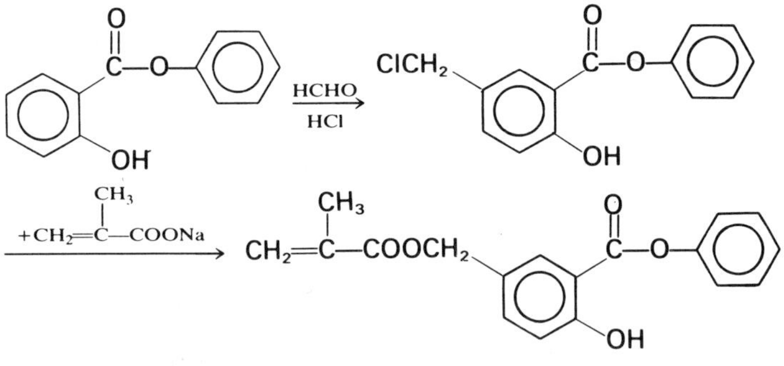

(7.23)

7.3.3 Hydroxybenzophenone Monomers and Polymers

The following hydroxybenzophenone monomers have been synthesized: 2-hydroxy-4′-vinylbenzophenone (**7.47**);[553] 2,4-dihydroxy-4′-

* Abbreviations used in schemes (7.19)–(7.22) NBS, *N*-bromosuccinimide; Ac_2O, acetic anhydride; Bu_3N, tributylamine; Et_3N, triethylamine; DmAc, *N,N*-dimethylacetamide; P_4, phosphor.

vinylbenzophenone (**7.48**);[1099–1101] 2-hydroxy-4-acryloxy(or methacryloxy)benzophenone (**7.49**);[822,872,873,1050,1108] 2-hydroxy-4-acryloxyethoxybenzophenone [Cyasorb UV 2126 (American Cyanamid), polymeric form; **7.50**];[632,735] 2-hydroxy-4-[3′-acryloxy(or methacryloxy)-2′-hydroxypropoxy]benzophenone (**7.51**);[221,405,776,822,872,873,1050] and 2-hydroxy-3-[3′-methacryloxy-2′-hydroxypropoxy]benzophenone (Permasorb MA, National Starch and Chemical Company, USA; **7.52**).

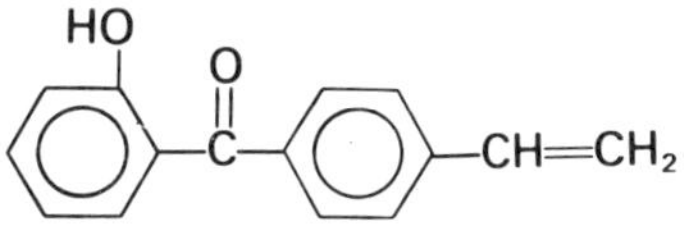

(**7.47**)

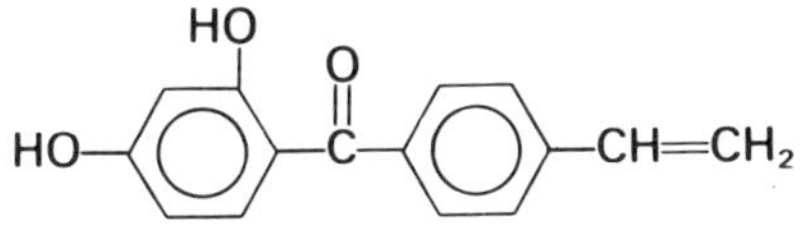

(**7.48**)

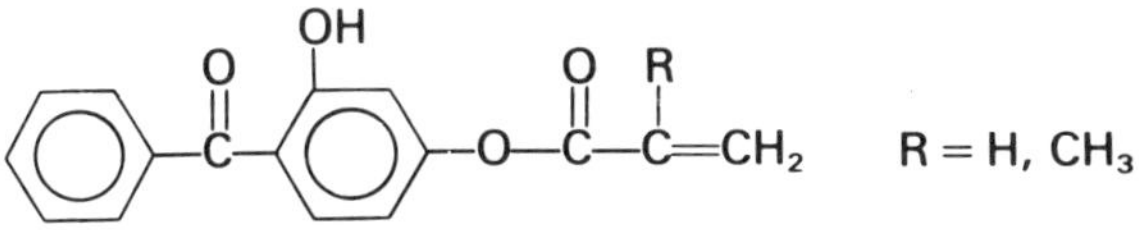

(**7.49**)

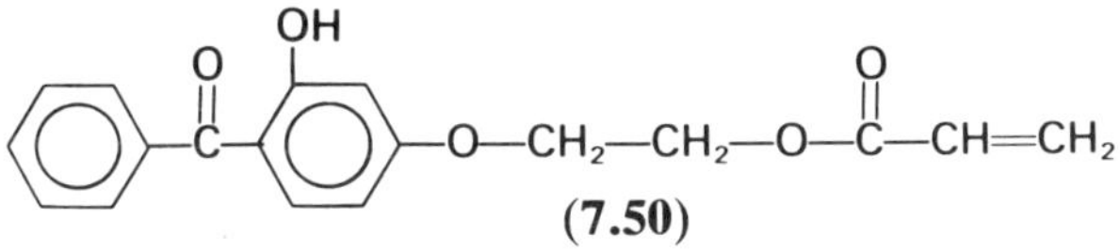

(**7.50**)

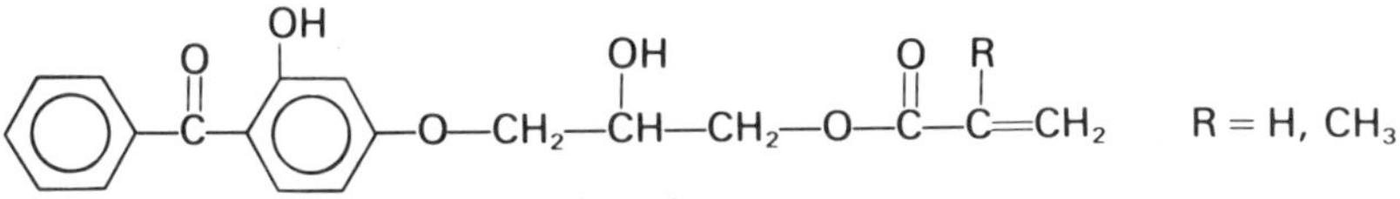

(**7.51**)

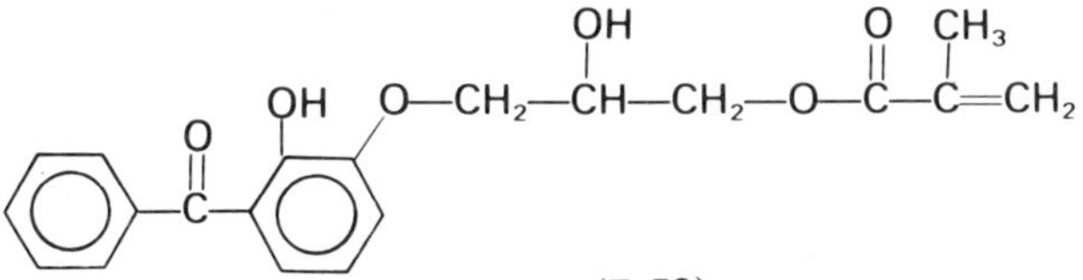

(**7.52**)

Monomers **7.47**, **7.48**, **7.49** and **7.50** were homopolymerized and copolymerized with styrene, methyl methacrylate, vinyl chloride or vinylidene chloride[148,405,515,553,796] or grafted onto polyethylene (LDPE, HDPE), polypropylene and polystyrene in a screw extruder at a temperature of 130–300°C under a nitrogen atmosphere in the presence of organic peroxides as initiators.[822,872,873,1050] However, only 3–17% of the added monomer has been grafted, the rest being unreacted and homopolymerized. The grafting efficiency depends on polyolefin type and grafting temperature and it is higher in the case of methacryloxy derivatives. Comparative photodegradation studies of chemically grafted and physically dispersed monomers showed a longer lifetime for the grafted polymers in polyolefins.

The more convenient method is radiation-induced grafting. For example, 2-hydroxy-4-[(3′-methacryloxy-2′-hydroxypropoxy)]benzophenone (**7.51**) has been grafted on to polyolefins with gamma-radiation.[220,938] Grafting was made by immersion of films into solutions of monomer (10–20 wt%) in methanol, *n*-propanol, 2-methoxyethanol, benzene or tetrahydrofuran.

The following synthetic routes were used for the preparation of vinylhydroxybenzophenone monomers.[553,893]

—2-Hydroxy-4′-vinylbenzophenone (**7.47**), by reaction scheme (7.24).[553,893]

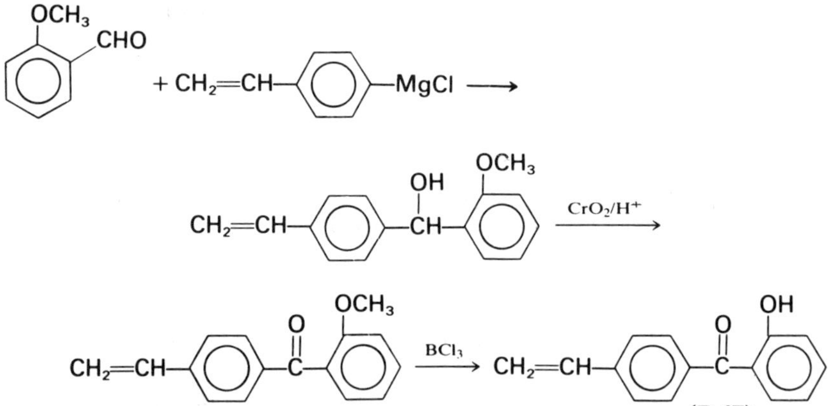

(7.47)

(7.24)

—2,4-Dihydroxy-4′-vinylbenzophenone (**7.48**), by reaction scheme (7.25).[148,1100]

The monomer **7.48** has also been copolymerized with styrene and methyl methacrylate.

—2-Hydroxy-4-acryloxy (or methacryloxy)ethoxybenzophenone (**7.50**), [by reaction sequence (7.26)].[632,735]

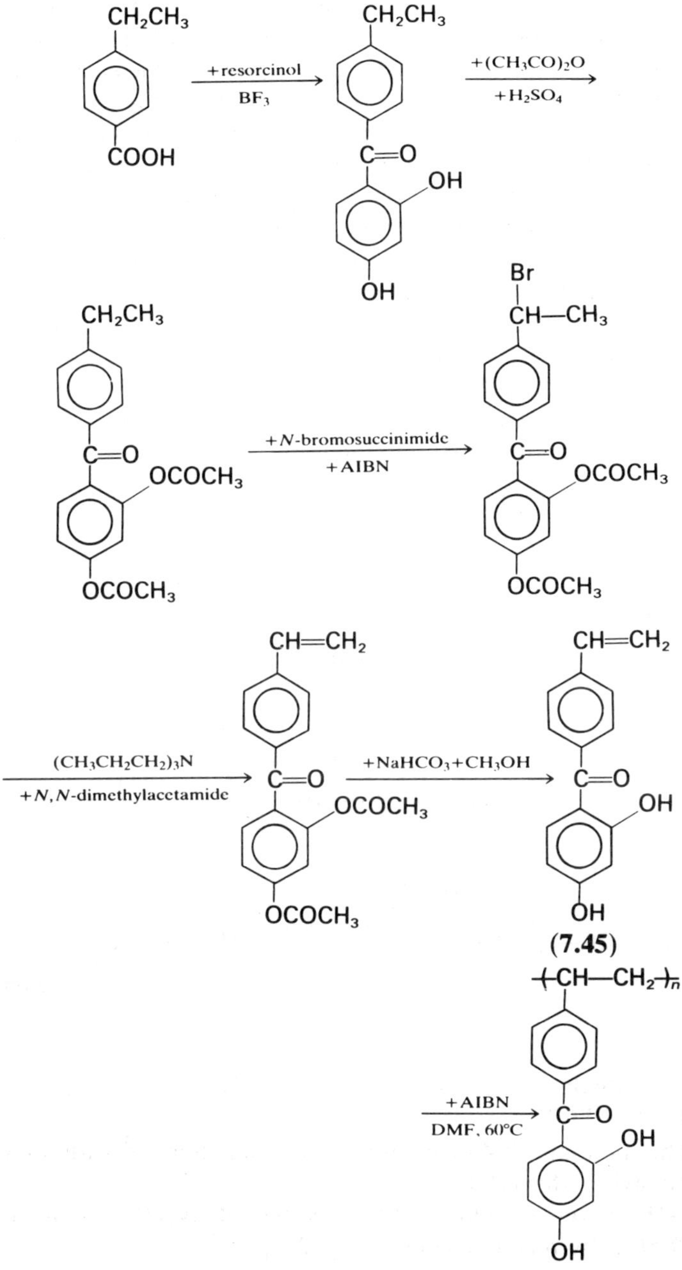
CH2CH3
COOH
+resorcinol
BF3
C=O
OH
+(CH3CO)2O
+H2SO4
OCOCH3
+N-bromosuccinimide
+AIBN
Br
CH—CH3
CH=CH2
(CH3CH2CH2)3N
+N,N-dimethylacetamide
+NaHCO3+CH3OH
+AIBN
DMF, 60°C

(7.45)

(7.25)

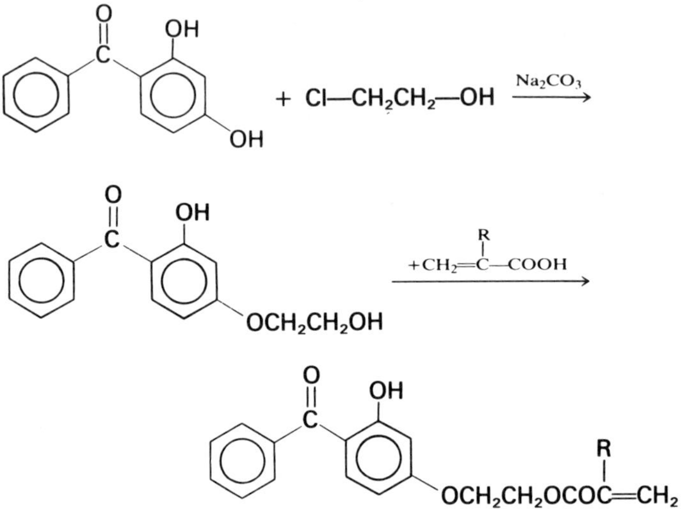

(7.50) R = H, CH3

(7.26)

—2-Hydroxy-4-[3′-acryloxy (or methacryloxy)-2′-hydroxypropoxy]-benzophenone (**7.51**), by reaction sequence (7.27).[405,776]

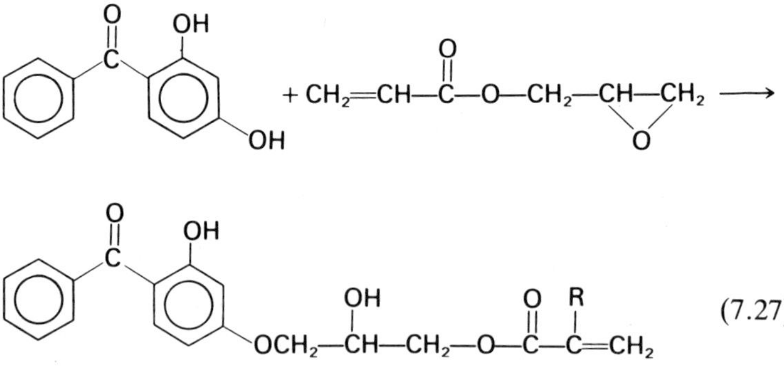

(7.27)

(**7.51**) R = H, CH3

These monomers were homopolymerized and copolymerized with vinyl chloride, vinylidene chloride, styrene and methyl methacrylate.

Quite different polymerizable monomeric photostabilizers (**7.55**) have been synthesized by reaction of 2,4-dihydroxybenzophenone

(7.53) with *N*-hydroxymethylacrylamide **(7.54)**:[625]

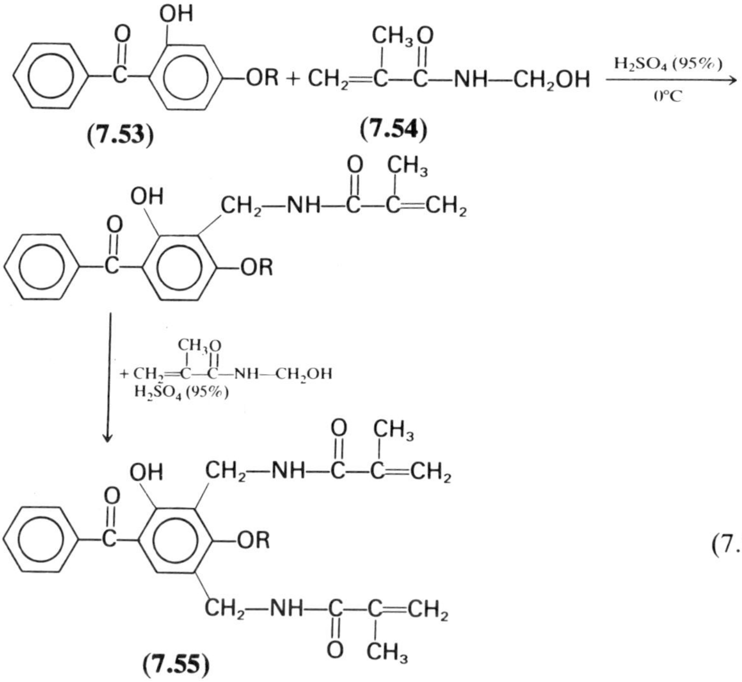

(7.53) **(7.54)**

(7.55)

$R = H, CH_3, nC_8H_{17}$

(7.28)

These monomers undergo free-radical polymerizations. Polymers absorb in the UV region from 250 to 350 nm as strongly as corresponding monomers and conventional 2,4-dihydroxybenzophenone derivatives.

The well-known reactivity of epoxy groups has been employed to synthesize 2-hydroxy-4-(2,3-epoxypropoxy)benzophenone **(7.56)**:[756]

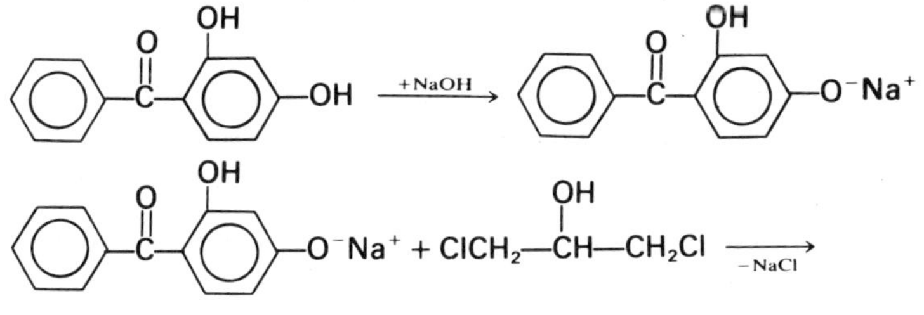

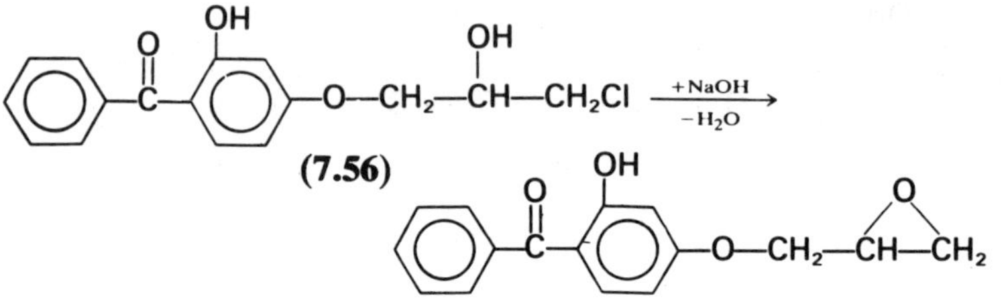

(7.56)

(7.29)

This monomer can be polymerized by tertiary amines[741] or tin(IV) chloride:[755]

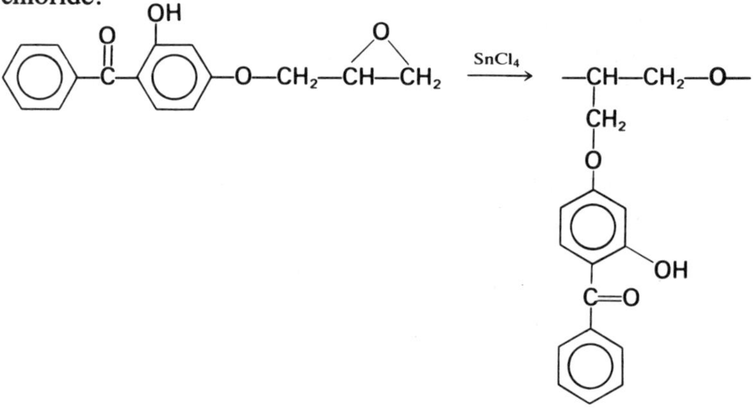

(7.30)

Copolymerization of the monomer **7.56** with cyclic anhydrides of dicarboxylic acids gives polymeric photostabilizers of the polyester type.[739,743]

An interesting group of polymeric photostabilizers based on 2-hydroxybenzophenone attached to polyesters (**7.57**) or polyepoxides

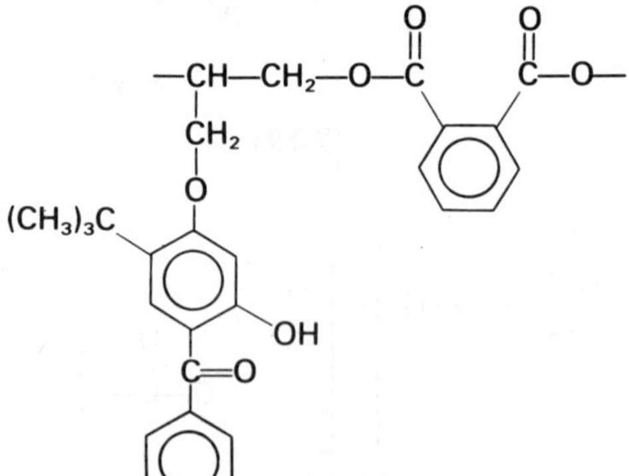

(7.57)

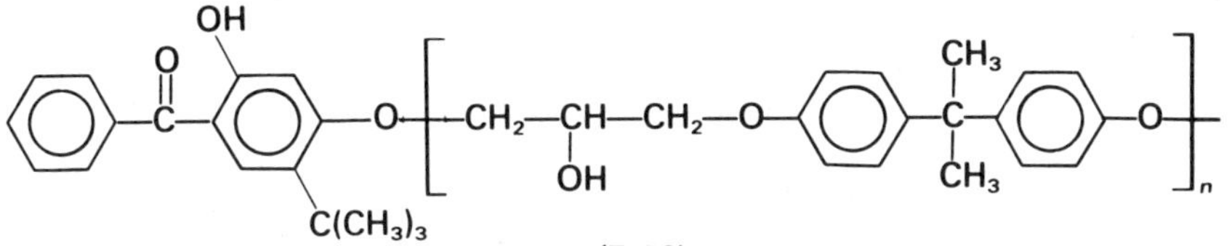

(7.58)

(**7.58**) has been reported.[740,742] These polymeric photostabilizers have high melting temperatures (**7.57**, ~148°C and **7.58**, ~200°C), and thermal stability above 400°C. They are especially useful in the spinning processing of polypropylene which occurs at 330°C.[740] They are also very stable against aqueous detergent and petrol washing of polypropylene fibres.[742]

Polyesters with structures **7.59** and **7.60** have excellent properties as UV-barrier coatings for protecting different polymers and non-polymeric materials (Table 7.3).[312]

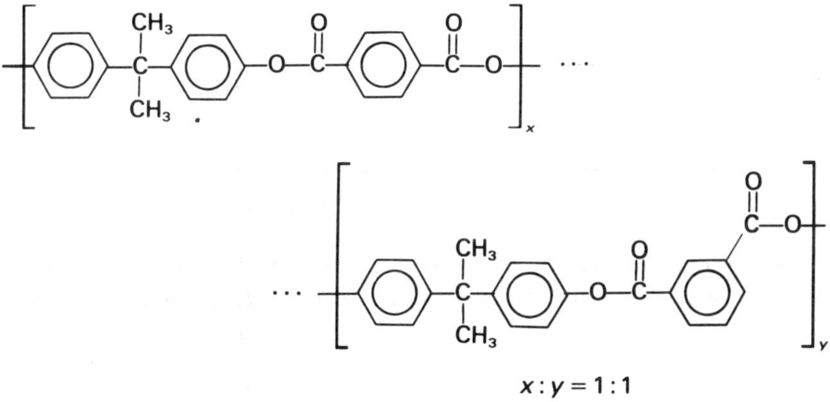

(7.59)

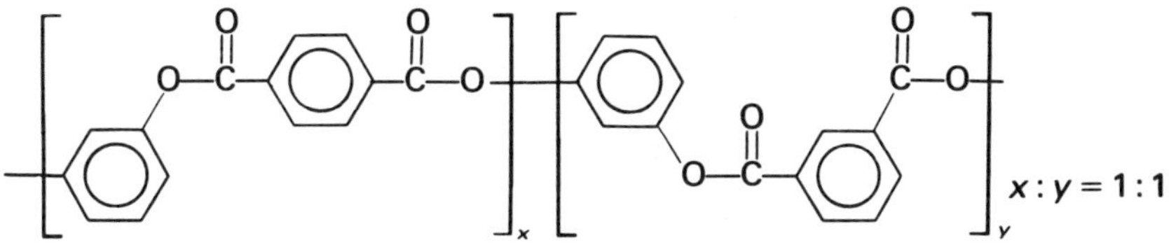

(7.60)

Table 7.3

Protection of polymers coated with UV-barrier resins during accelerated irradiation by various UV sources[312]

Substrate	*Type of exposure*	*Time (h)*	*Unirradiated (initial)*	*Irradiated*	
				Protected (coated)	*Unprotected (uncoated)*
Poly(vinyl chloride)					
Translucent, neutral colour, rigid sheet	Weatherometer	1120	'Pearl' colour, slight gloss	Very slight yellowing where irregular in coating thickness; physical properties maintained	Complete blackening or dark brown; brittle
Translucent, neutral colour, rigid sheet	450 W UV lamp	180	'Pearl' colour, slight gloss	No change	Sample turned a deep black–blue colour
Plasticized sheet, transport	Westinghouse sun lamp	2106	Polyester film laminated to PVC; some haze due to interfacial polymer mixing	Virtually unchanged; no visible yellowing; retained transparency	Yellow–tan colour developed, with loss of plasticizer; frosty appearance of surface
Cellulose acetate butyrate	450 W UV lamp	232	Transparent sheet with 0% yellow	1·5% yellow measured; no visible changes	Definite yellow measured as 16%; slight frosting of surface
Cellulose nitrate	450 W UV lamp	300	Transparent sheet with 1·5% yellow initially	An increase of 4% in yellowing occurred; retained excellent transparency	A dark tan colour developed, with loss in transparency at about 120 h

Poly(vinyl alcohol)	450 W UV lamp	96	Colourless, transparent and flexible	No visible change	Very yellow, brittle
Nylon 66	450 W UV lamp	96	White, opaque and flexible	No visible yellowing; still flexible	Straw colour developed with loss of flexibility
Polycarbonate of bisphenol-A	450 W UV lamp	96	Transparent, flexible	Retained transparency and flexibility	Deep yellow colour formed with loss in transparency
Poly(ethylene terephthalate)	450 W UV lamp	96	Clear, tough and flexible	No visible changes; properties retained	Very brittle with loss of transparency; very slight colour change
Poly(vinyl butyral)	450 W UV lamp	96	Transparent, colourless and flexible	No change	Tan colour present with loss of flexibility and some frosting of surface
Poly(vinyl acetate)	450 W UV lamp	96	Colourless and transparent	No visible change	Yellow colour with loss in flexibility
Poly(methyl methacrylate)	450 W UV lamp	96	Transparent, colourless and rigid	No visible change	Slight yellow colour; slight frosting of surface
Polyacrylonitrile	450 W UV lamp	96	Opaque, white	No visible change or loss in properties	Brown colour with loss in tensile strength
Polyethylene	450 W UV lamp	96	Translucent, flexible	No change noticeable	Very brittle; many cracks throughout sample
Polystyrene	450 W UV lamp	96	Transparent, colourless	No visible change	Very deep yellow; very brittle

(continued)

Table 7.3—*contd.*

Substrate	*Type of exposure*	*Time (h)*	*Unirradiated (initial)*	*Irradiated*	
				Protected (coated)	*Unprotected (uncoated)*
Poly(vinylidene chloride-co-vinyl chloride) (Saran; Dow Chemical)	450 W UV lamp	96	Transparent, flexible and colourless	No visible change	Dark brown in colour and very brittle
Polypropylene	450 W UV lamp	96	Opaque, white, rigid	No visible change	Sample disintegrated to a powder
Cellulose	450 W UV lamp	96	Opaque, in the form of undyed fibres	No loss in colour or physical properties	Brown colour developed
Cellulose	450 W UV lamp	96	In the form of red coloured fibres	No loss in physical properties; some bleaching of colour noticeable	Colour completely bleached away; brown colour developed

These polyesters under UV irradiation undergo a *photo-Fries rearrangement* to give polymeric *ortho*-hydroxybenzophenones:[919,934]

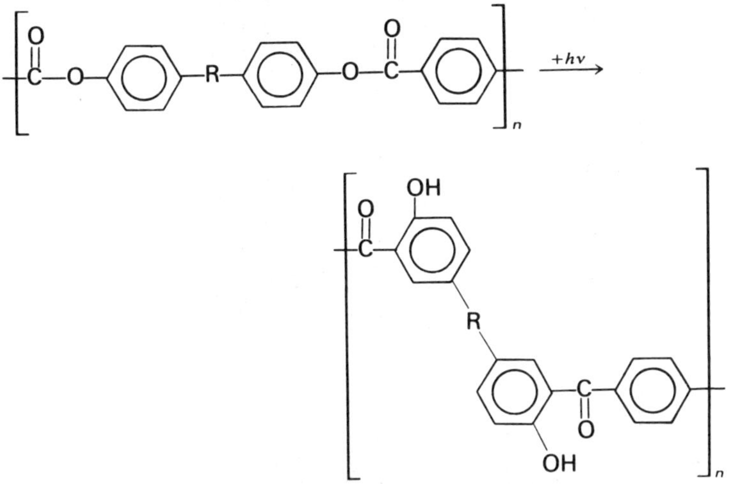

(7.31)

7.3.4 Hydroxybenzotriazole Monomers and Polymers

The following hydroxybenzotriazole monomers have been synthesized.

(i) 2-(2-Hydroxy-5-vinylphenyl)-2*H*-benzotriazole (**7.61**).[1099,1130,1172,1173]

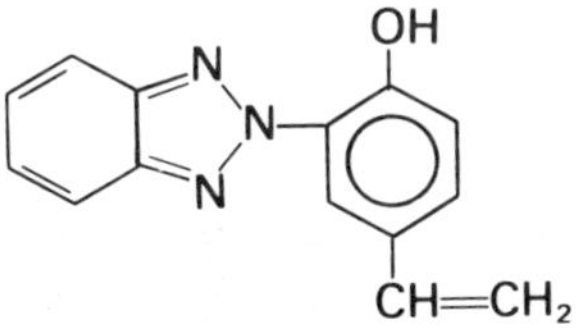

(7.61)

(ii) 2-(2-Hydroxy-3-vinyl-5-methyl)-2*H*-benzotriazole (**7.62**).[945]

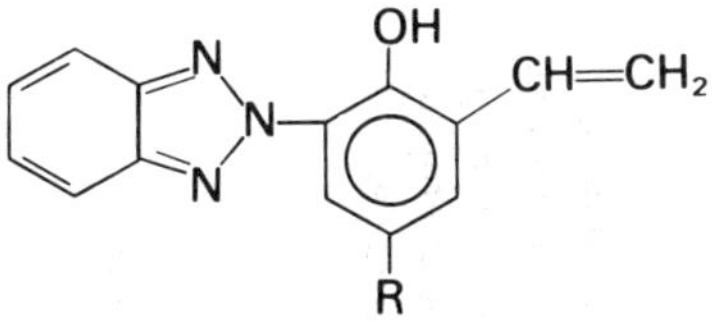

(7.62)

(iii) 2-(2-Hydroxy-5-isopropenylphenyl)-2*H*-benzotriazole (**7.63**).[516,851,1130,1172,1173]

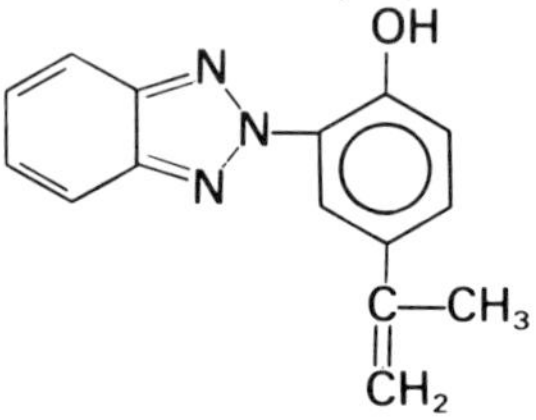

(7.63)

(iv) 2-(2-Hydroxy-5-methylphenyl)-2*H*-4-vinylbenzotriazole (**7.64**).[1171]

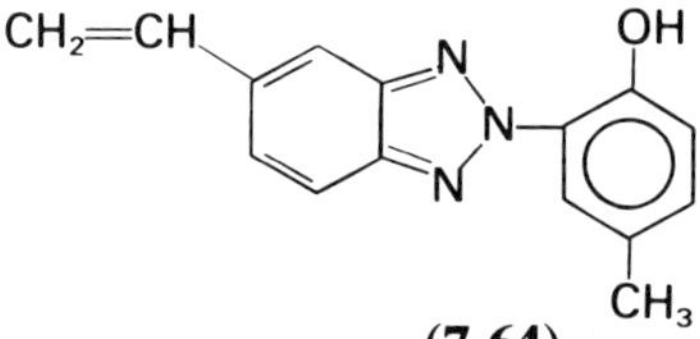

(7.64)

(v) 2-(2-Hydroxy-4-acryloxy (or methacryloxy)phenyl)-2*H*-benzotriazole (**7.65**).[712]

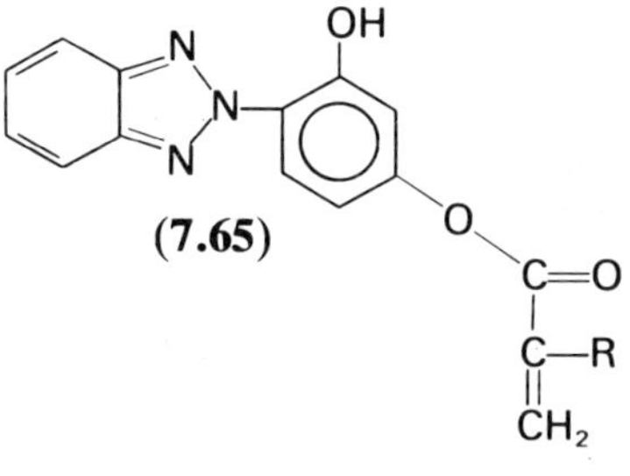

(7.65)

R = H, CH_3

(vi) 1,3-[2-Hydroxy-4-acryloxy (or methacryloxy)phenyl]di(2*H*-benzotriazole) (**7.66**).[713]

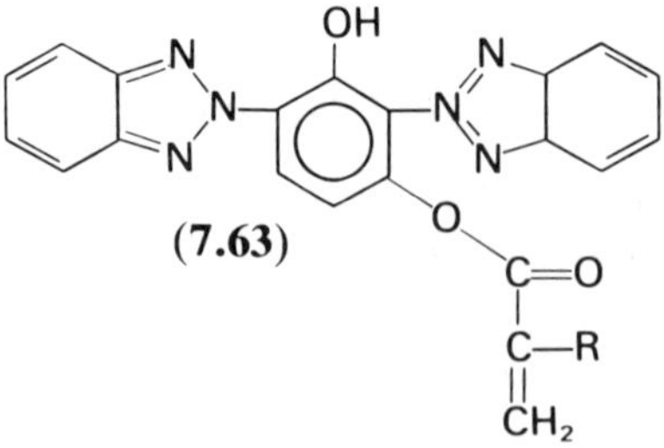

(7.63)

R = H, CH_3

(vii) 1,3-[2,6-Dihydroxy-4-acryloxy (or methacryloxy)phenyl]-di-(2*H*-benzotriazole) (**7.67**).[713]

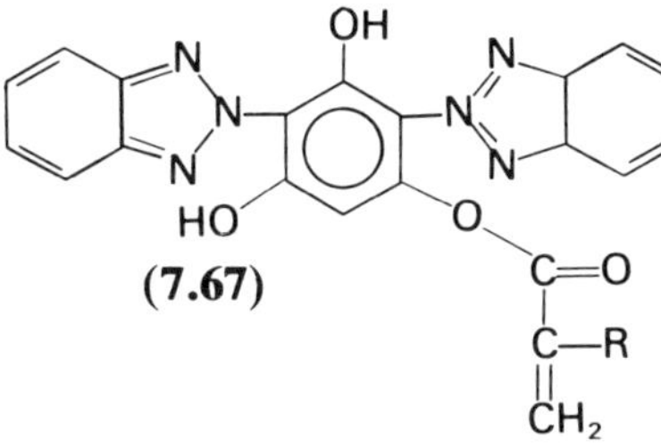

(**7.67**) R = H, CH_3

(viii) 2-(2-Hydroxy-5[2-hydroxyethylphenyl])-2*H*-benzotriazole-2-isocyanatoethyl methacrylate (**7.68**).[142]

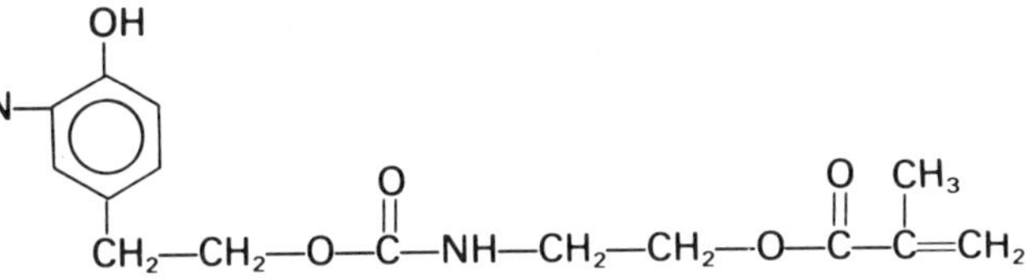

(**7.68**)

Vinylhydroxybenzotriazole monomers were polymerized and copolymerized with styrene, methyl methacrylate and *n*-butyl acrylate in an oxygen-free atmosphere, using azobisisobutyronitrile as initiator at 150°C (Table 7.4).[515,1130]

Table 7.4
Copolymerization parameters for vinyl-substituted 2-(2-hydroxyphenyl)-2*H*-benzotriazoles (M_1) and vinyl monomers (M_2) at 50°C using azobisisobutyronitrile as initiator

M_1	M_2	r_1	r_2	*Solubility in monomer at 50°C (mol %)*
2-(2-Hydroxy-5-vinylphenyl)-2*H*-benzotriazole (**7.61**)	Styrene	0·74 ± 0·06	0·10 ± 0·30	20
	Methyl methacrylate	0·45 ± 0·01	1·18 ± 0·30	19
	n-Butyl acrylate	0·21 ± 0·00	2·10 ± 0·28	14
2-(2-Hydroxy-5-isopropenylphenyl)-2*H*-benzotriazole (**7.63**)	Styrene	1·07 ± 0·08	0·0	11
	Methyl methacrylate	0·41 ± 0·01	0·0	10
	n-Butyl acrylate	0·12 ± 0·00	0·76 ± 0·26	11

Reprinted with permission from Ref 1131, © 1985 Butterworth & Co. (Publishers) Ltd.

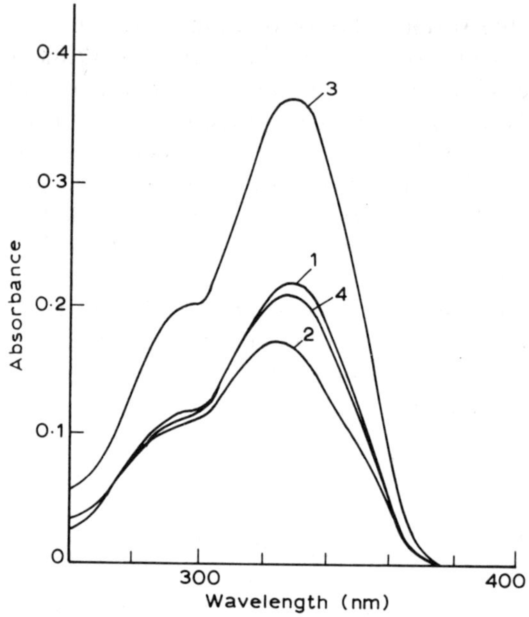

Fig. 7.1. Absorption spectra of (1) 2-(2-hydroxy-4-acryloxyphenyl)-2*H*-benzotriazole (**7.65**), (2) its polymeric form, (3) its copolymer with polystyrene and (4) its copolymer with poly(methyl methacrylate).[721]

They exhibit two absorption bands in the region 300–350 nm (Fig. 7.1).

Photophysical studies show that incorporation of small amounts of 2-(2-hydroxy-5-vinylphenyl)-2*H*-benzotriazole into polystyrene results in efficient quenching of the polystyrene monomer and excimer fluorescence in the solid state.[322,514,856]

Simulated accelerated weathering tests of copolymers with 2-(2-hydroxy-5-vinylphenyl)-2*H*-benzotriazole and 2-(2-hydroxy-5-isopropenylphenyl)-2*H*-benzotriazole showed them to be effective photostabilizers for acrylic polymers and polydienes.[721] The lifetime to failure for unstabilized samples (6–8 years) has been exceeded by 20 years for the polymeric photostabilizer-containing copolymers.[513,517]

2-(2-Hydroxy-5-vinylphenyl)-2*H*-benzotriazole (**7.61**) has been grafted on to atactic polypropylene, ethylene/propylene copolymer, ethylene/vinyl acetate copolymer, poly(methyl acrylate), and poly(methyl methacrylate) (all the grafting had high efficiency, 68–75%). Reactions were carried out in chlorobenzene at 150°C in the presence of di(*tert.*-butyl) peroxide as initiator.[322,516,912]

Studies on the surface photo-oxidation of styrene copolymers with 2-(2-hydroxy-3-vinyl-5-methyl)-2*H*-benzotriazole (**7.62**) indicated that the bulk material was photostabilized by comparison with unstabilized polystyrene; however, the surface oxygen uptake was enhanced.[813,814] After an initial induction period the subsequent rate of oxygen uptake appeared to be directly related to the surface concentration of **7.62**. These results and also others studied by ESCA[815] show that poly[2-(2-hydroxy-3-vinylmethyl)-2*H*-benzotriazole] is extremely susceptible to surface photo-oxidation.

Photophysical studies on copolymers of 2-(2-hydroxy-4-methacryloxyphenyl)-2*H*-benzotriazole (**7.65**) have been carried out recently.[465]

The 2-(2-hydroxy-5-vinylphenyl)-2*H*-benzotriazole (**7.61**) monomer was also grafted on to *cis*-1,4-polybutadiene and 1,2-polybutadiene.[645,646]

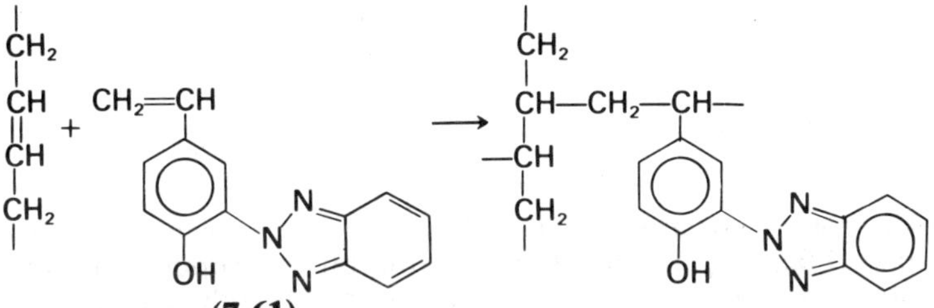

(**7.61**)

(7.32)

and unsaturated polyesters (made from maleic anhydride, phthalic anhydride and 1,3-propanediol) (70 wt%) and styrene (30 wt%):[194,1072]

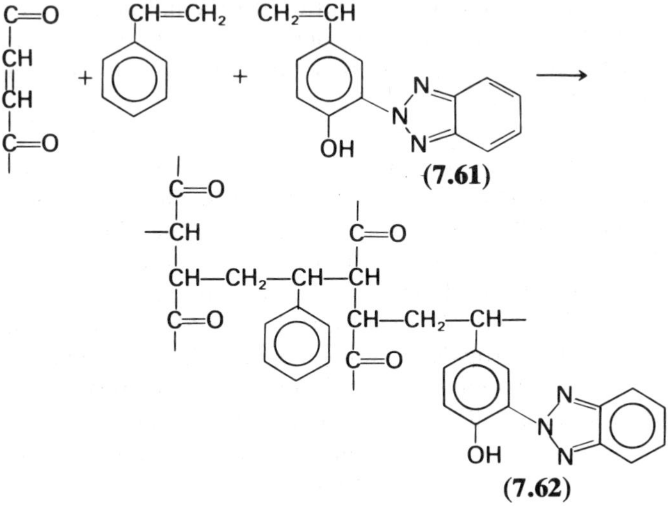

(**7.61**)

(**7.62**)

(7.33)

The following synthetic routes were used for the preparation of vinylhydroxybenzotriazole monomers

(i) 2-(2-Hydroxy-5-vinylphenyl)-2*H*-benzotriazole (**7.61**), by reaction sequence (7.34).[1130,1172,1173]

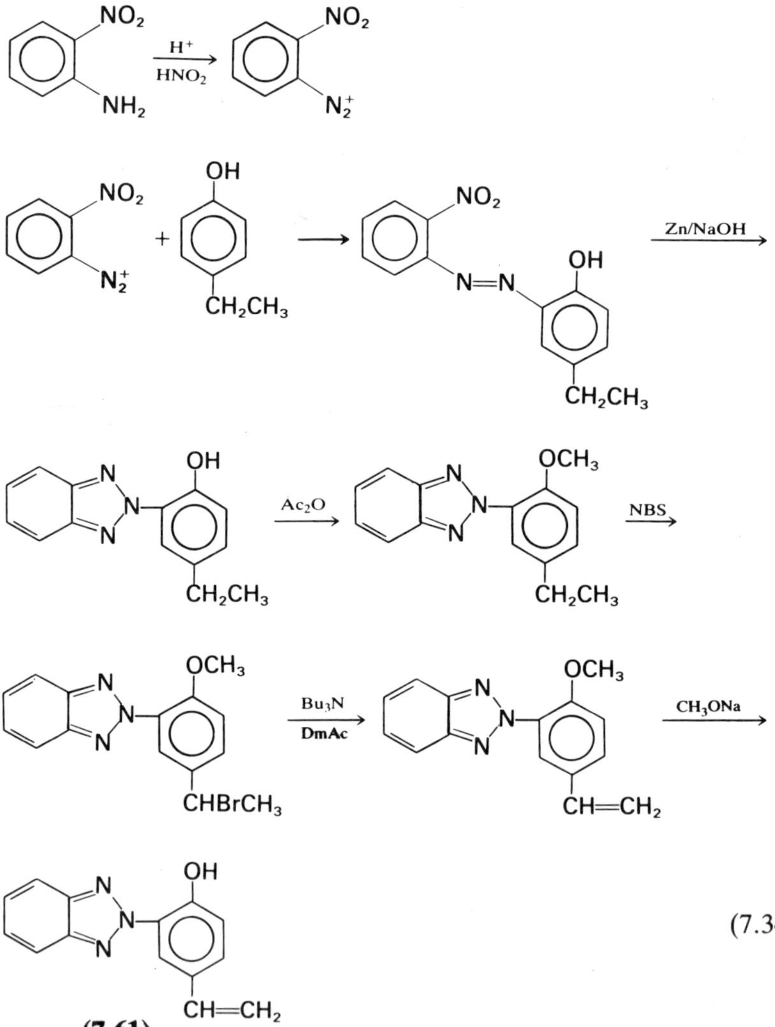

(7.34)

(7.61)

Abbreviations: Ac_2O, acetic anhydride; NBS, *N*-bromosuccinimide; Bu_3N, tributylamine.

(ii) 2-(2-Hydroxy-5-isopropenyl)-2*H*-benzotriazole (**7.63**), by reaction sequence (7.35).[516,1130,1172,1173]

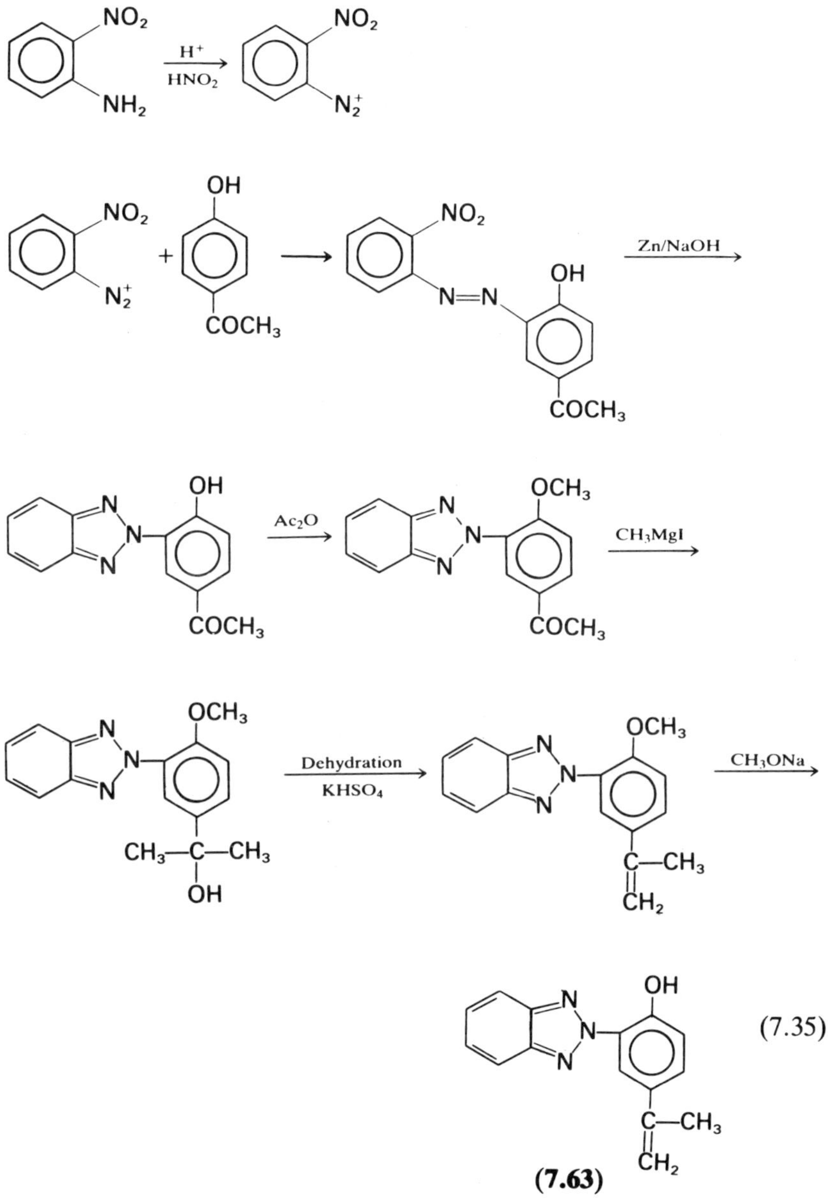

(7.35)

(**7.63**)

(iii) 2-(2-Hydroxy-4-acryloxy (or methacryloxy)phenyl)-2*H*-benzotriazole (**7.65**), by reaction sequence (7.36).[712]

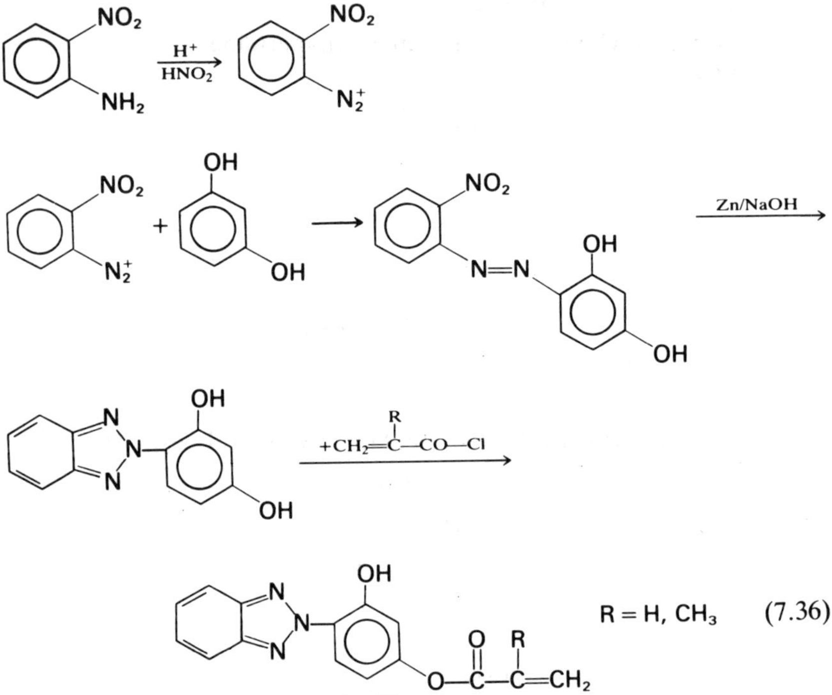

(7.36)

(7.65)

The reactivity of epoxy groups in glycidyl methacrylate (**7.69**) has been employed for grafting of 2-(2,4-dihydroxyphenyl)-2*H*-benzotriazole (**7.70**) on to this polymer:[1087]

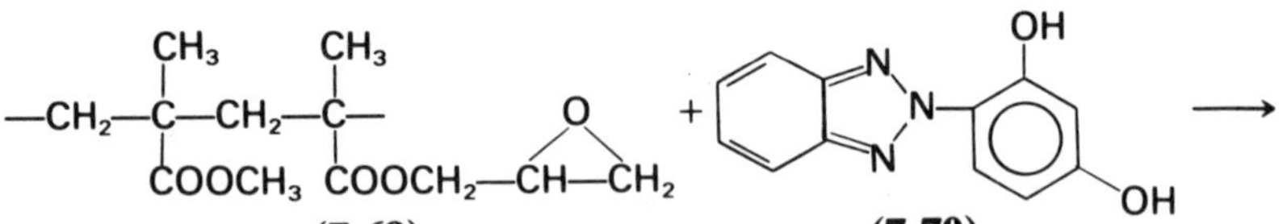

(7.69) **(7.70)**

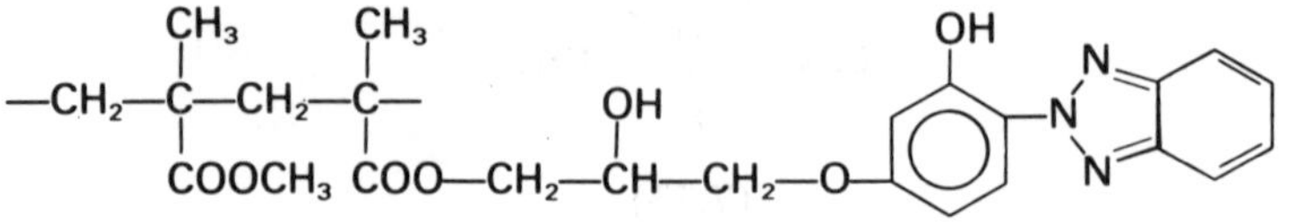

(7.37)

7.4 HINDERED AMINE (HALS) MONOMERS AND POLYMERS

A number of piperidine (and piperidinoxy radicals) monomers and polymers have been synthesized.

(i) 4-Methacryloxy-2,2,6,6-tetramethylpiperidine (**7.71**).[295,296,631,675,793]

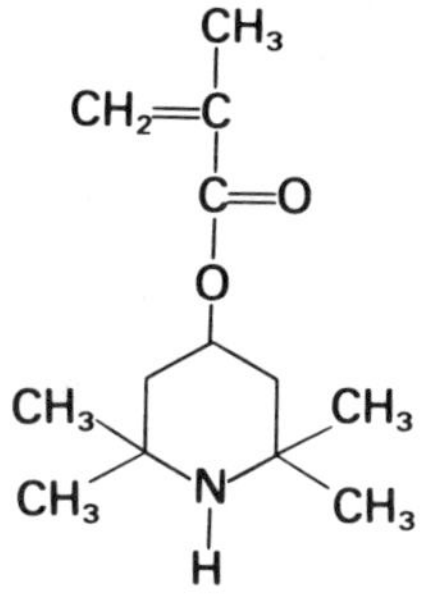

(7.71)

(ii) 4-Methacryloxy-1,2,2,6,6,-pentamethylpiperidine (**7.72**).[295,296]

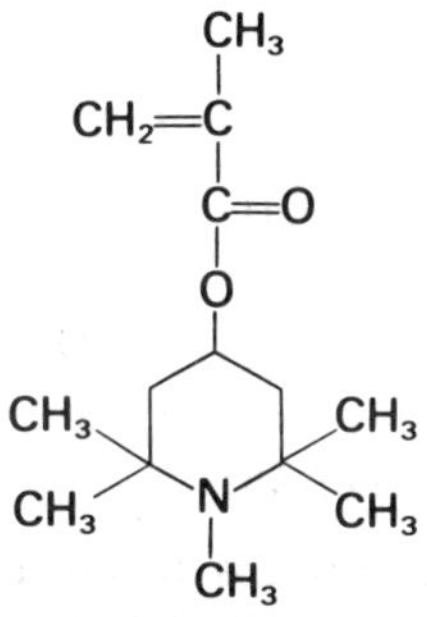

(7.72)

(iii) 4-Methacrylamino-2,2,6,6-tetramethylpiperidine (**7.73**).[294–296]

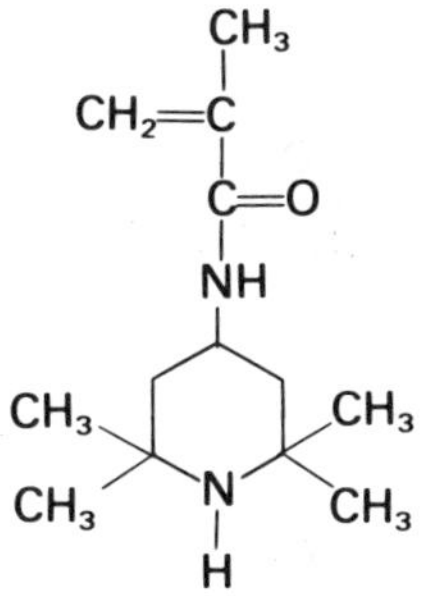

(7.73)

These monomers were homopolymerized or copolymerized with methacrylates and styrene[295,296] and thermally[879] or photochemically (in the presence of benzophenone in acetone)[793] grafted on to polypropylene.

Poly(methyl methacrylate) containing a small amount of pendant 2,2,6,6-tetramethylpiperidine groups is readily miscible with other copolymers such as poly(styrene-co-acrylonitrile) and poly(α-methylstryene-co-acrylonitrile).[475]

The higher efficiency of monomeric derivatives compared with those bound by copolymerization indicates that the mobility of the stabilizer plays an important role in the overall mechanism of stabilization by hindered amines. Homopolymers also have lower efficiency, which can be slightly improved by optimization of the molecular weight.[556] For example, the optimal molecular weight is 2700 for poly(1,2,2,6,6,-pentamethyl-4-piperidyl acrylate) used as a light stabilizer in polypropylene.[499]

The efficiency of the copolymerized stabilizer is in some cases higher than that of the homopolymer, whose ability to migrate is higher. This can be explained by better distribution of stabilizer in the case of copolymer compared with the added homopolymer.[296]

The hindered amine (HALS) monomer **(7.73)** is easy to obtain by the following route:[294]

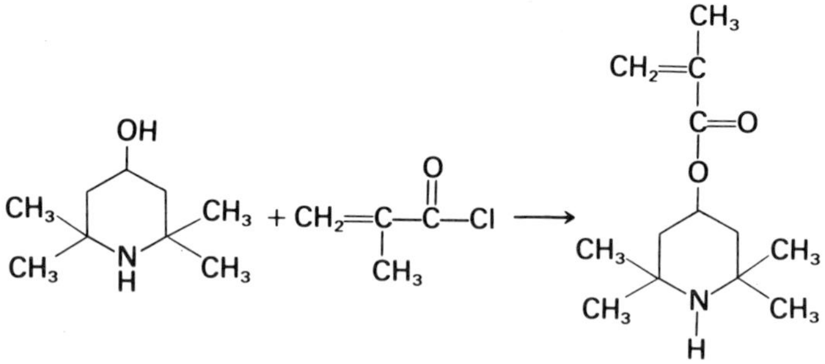

(7.38)

(7.73)

Mononeric piperidinoxy radicals **(7.74)**[294] can be further obtained by oxidation of monomers **(7.73)** with hydrogen peroxide in the presence of the sodium salt of tungstic acid (Na_2WO_4).[752]

Another type of hindered amine monomer is bis(2,2,6,6-tetramethylpiperidin-4-yl) maleate **(7.75)**, which can react with polypropylene in

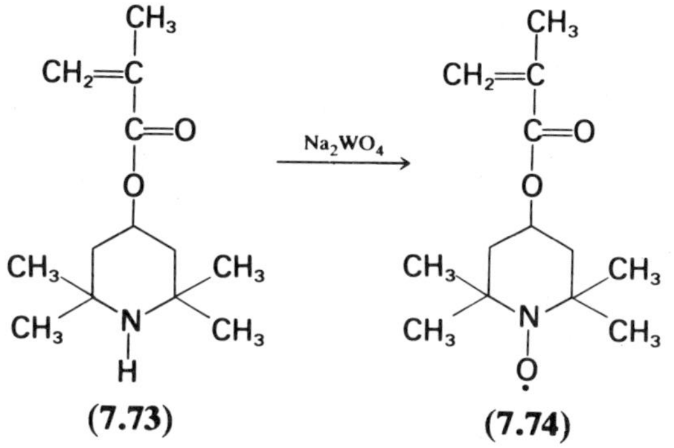

(7.39)

the melt to a high level (90%) in the presence of dicumyl peroxide. However this polymer-bound HALS has a similar level of photoantioxidant activity to a commercial hindered piperidine (Tinuvin 770; **6.3**) used as an additive at the same concentration.[113]

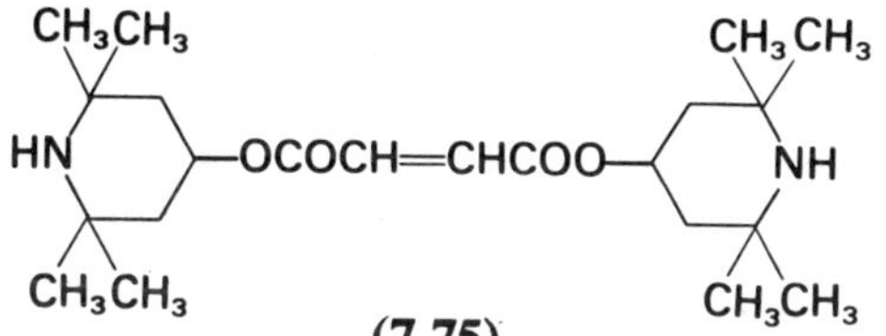

(7.75)

7.5 POLYMERS WITH HINDERED N-OXY GROUP IN THE MAIN CHAIN

Hindered bisnitroxy radicals (**7.76**) react with *p*-xylene to a copolymer:[423]

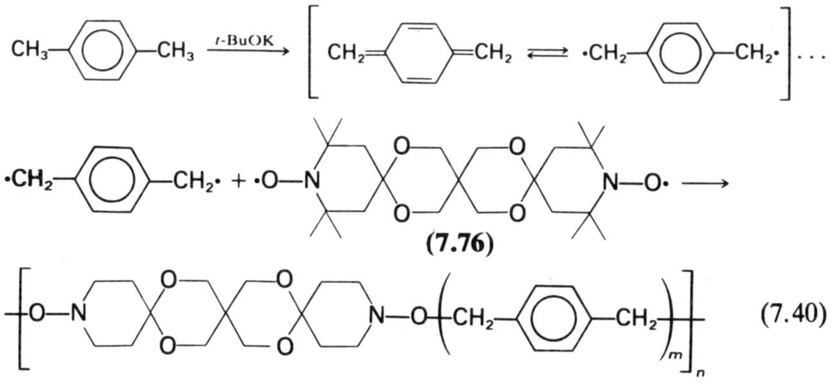

(7.40)

Table 7.5
Stabilizing effectiveness of polymers with N-oxy groups in the main chain on photo-oxidation of polypropylene 100 μm thick[423]

Copolymer at 0·25 *wt%*	*Melting point (°C)*	*Molecular weight* M_n/M_w	*Improvement factor, IF*
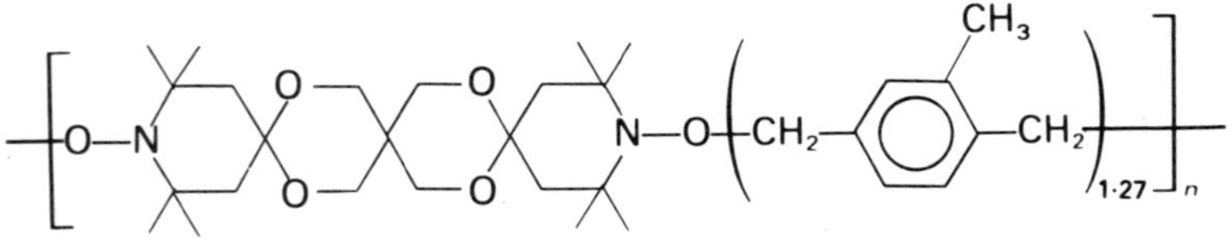 **(7.77)**	174–180	3 000 3 600	8·6
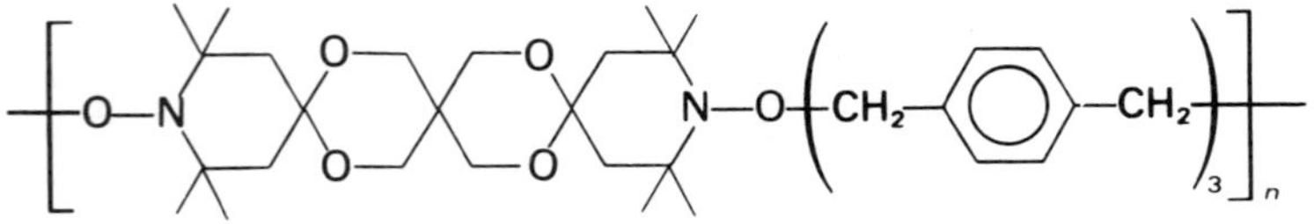 **(7.78)**	180–200	8 500 27 000	4·5
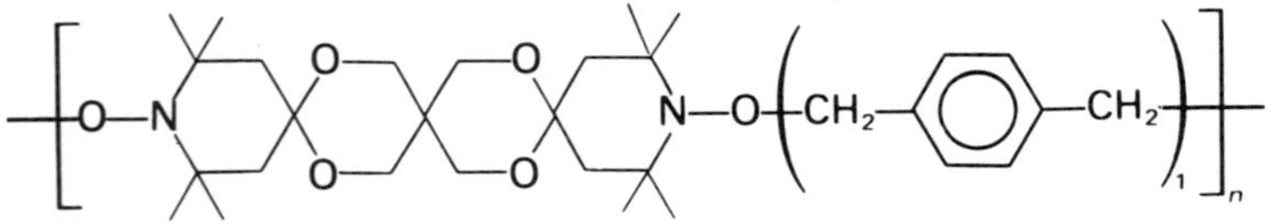 **(7.79)**	183–211	2 300 2 600	5·9
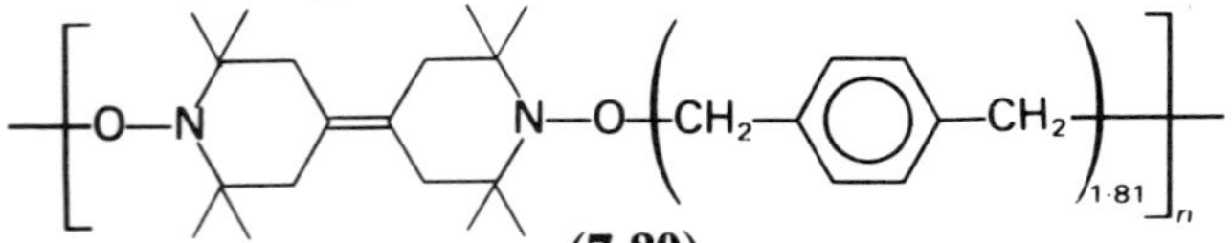 **(7.80)**	215–221	2 100 3 500	2·6

In a similar way other copolymers were obtained (Table 7.5) which exhibit almost the same improvement factor, defined here as:

$$\text{IF(improvement factor)} = \frac{\text{Half-lifetime of polymer with stabilizer}}{\text{Half-lifetime of polymer without stabilizer}} \tag{7.41}$$

where half-lifetime = time (in hours) to 50% retention of original elongation.

8

Industrial Aspects of Polymer Stabilization

8.1 INDUSTRIAL METHODS OF INCORPORATING ADDITIVES

There are four methods of incorporating additives.[428,765,783,820,1076]

(i) Additives are added to the monomer, i.e. before or during the production of the polymer (e.g. the stabilization of impact-resistant polystyrene and the prestabilization of poly(vinyl chloride) by suspension polymerization; in both of these cases, suitable phenolic antioxidants are added to the monomers.

(ii) Additives are added during the working-up of the polymer (e.g. stabilization of modified styrene polymers such as ABS latex; the additives must satisfy specific requirements determined by the process, such as emulsifiability, solubility in certain solvents, low volatility or stability to hydrolysis). Polyolefins prepared by solution, slurry or gas-phase processes are frequently provided with antioxidants at this stage in order to keep oxidative damage during isolation and drying at a low level.

(iii) Additives are added in a separate manufacturing step [polymer compounding: powder mixing, paste production, agglomerate production for poly(vinyl chloride) and granule production for polyolefins, polystyrenes, polyamides].

(iv) Additives are incorporated directly before or during moulding of a plastic to a semi-finished product or the finished article.

A frequent technological application problem is the determination of the optimum amount of additives (photostabilizers and antioxidants) for a plastics compound with respect to the processability, properties of the finished article and cost of formulation. Optimization calculations have to be applied to such cases.

8.2 TOXICITY OF PHOTOSTABILIZERS

The rapid development of plastics as packaging materials for food, pharmaceuticals, textiles and children's toys gives the possibility that incorporated photostabilizers and antioxidants may migrate in small quantities to the surface of the article and into the surrounding medium and so directly or indirectly into the human body. Very often this will result in a subchronic or chronic intoxication. For that reason, the health authorities in different countries (e.g. the Food and Drug Administration and Federal Health Office in the USA and the Bundesgeundheitsamt in the FRG) have made strict regulations for the approval of additives to plastics which may have contact with the human body.

The producer of a packaging material has to show what quantities of the additive migrate into the packaged food. He has to carry out the necessary test. The appropriate tests are standardized in most countries and they differ only slightly.[705]

A few descriptions of toxicity tests are given below as examples.[1198]

—*Photostabilizer*: *2-hydroxy-4-methoxybenzophenone* (**5.28**; *Cyasorb UV* 9; *Cyanamid*) is not expected to be injurious to the health by any route of exposure under ordinary circumstances of handling. the acute oral LD_{50} (defined at the end of this section) of the products for rats is greater than 5 g kg^{-1} and rats tolerated a level of 1% in the feed for 27 days without apparent harmful effect. Dosages of 16 g kg^{-1} maintained in continuous contact with the clipped skin of rabbits for 24 h caused no significant irritation and no signs of systematic toxicity. Installation of a small amount of the dry product into the eyes of rabbits produced only a slight inflammatory reaction of short duration.

—*Photostabilizer*: *2,2'-dihydroxy-4-methoxybenzophenone* (**5.29**; *Cyasorb UV 24*; *Cyanamid*) is considered to be practically non-toxic in single doses by mouth as the acute oral LD_{50} is greater than 10 g kg^{-1} for rats. It is not irritating in the rabbit eye nor on rabbit skin. Dosages of 10 g held in contact with the clipped skin of rabbits for 24 h produced no primary irritation or signs of systematic toxicity. The produoct was incorporated into the diet of groups of rats at concentration of 2·5, 5·0, and 10% for 36 days. Average dosages were 2·67, 4·83 and 8·95 g kg^{-1}. There were no deaths in any group during this time, but mean food intake and mean body weight gained were significantly lower in the test group than in the control group. Behaviour and appearance of the animals at the 2·5% level were normal, but those at the two higher levels appeared depressed and in relatively poor condition.

—*Photostabilizer*: *2-hydroxy-4-*n*-octoxybenzophenone* (**5.30**; *Cyasorb UV 531*; *Cyanamid*) had a lower order of acute toxicity for animals. The single oral dose LD_{50} for rats is greater than $10\,g\,kg^{-1}$, and the product is considered, therefore, to be practically nontoxic by swallowing in single doses. It is not irritating to rabbit skin or eyes, and dosages of $10\,g\,kg^{-1}$ were maintained in contact with the clipped skin of rabbits for 24 h with no signs of systematic toxicity resulting.

8.2.1 United States Food and Drug Administration Approval

Cyasorb UV 9 is permitted by Section 121.2591 of the Food Additives Regulations as a stabilizer in semirigid acrylic and modified acrylic plastics intended for use in contact with food.

Cyasorb UV 531 is permitted by Section 121.2566 of the Food Additives Regulations for use as a stabilizer in olefin polymers at concentrations not exceeding 0·5% by weight of the polymer. The finished articles are to be used in contact only with foods of high water content.

Approval for an additive is granted in most countries only under condition that a 90-day feeding test in rats has been carried out. The

Table 8.1
Classification of the acute lethal dose (LD_{50}) in rats[444]

LD_{50} by single application in rats ($mg\,kg^{-1}$)	*Designation in use*
$\leqslant 1$	Extremely toxic
1 to 50	Highly toxic
50 to 500	Moderately toxic
500 to 5000	Lightly toxic
5000 to 15 000	Practically nontoxic
>15 000	Relatively harmless

cost of toxicological investigation of an additive, depending on the size of the test, is approximately \$50 000–100 000.

Note: The *acute oral toxicity* is the toxicity of a substance after single application by mouth (per os). This value is called the LD_{50} (*lethal dose*) or DL_{50} (*dosis lethalis*) and means the amount of substance per body weight of the test animal ($mg\,kg^{-1}$) which kills 50% of the animals after a single oral application. These values are relative ones and can only be classified in comparison with other substances to give useful information.[705] To estimate the degree of toxicity of a substance based on the LD_{50} value, the classification shown in Table 8.1 is used.

8.3 'WORLD INDEX OF POLYOLEFINE STABILIZERS'

This book is a special index which contains 4000 records on polyolefin stabilizers.[544] Each entry printed in it comprises:

(i) Trade name.
(ii) Stabilizer class.
(iii) Supplier.
(iv) Chemical identity.
(v) Formula code.

Stabilizers in this index have been divided according to the chemical type:

(i) Antioxidants: phenols;
(ii) Antioxidants: amines;
(iii) Antioxidants: synergistic;
(iv) Antioxidants: mixed functions;
(v) UV stabilizers;
(vi) Metal deactivators;
(vii) Prodegradants;
(viii) Mixtures (different chemical types);
(ix) Mixtures (same chemical type);
(x) Unknown.

Antioxidants (i)–(iv) have been subdivided into classes of distinct structural chemical types:

A Monophenols
B Bisphenols
C Dihydroxybenzenes
D Trisphenols
E Polyphenols
F Monoamines
G Diamines
H Heterocyclic amines
J Miscellaneous amines
K Sulphides
L Phosphites
M Thiophosphites, etc.
N Miscellaneous phosphorus compounds
P Phenolic sulphides
Q Amine sulphides, thiocarbamates, etc.
R Phenolic phosphites, etc.
S Benzophenones, etc.
T Esters
U Benzotriazoles, etc.
V Metal quenchers
W Hindered amines

This index gives addresses of additive suppliers (161 companies) from different countries (Table 8.2).

Table 8.2
Statistical comparison of the number of companies producing additives for polyolefins[544]

County	*No. of companies*
USA	80
UK	25
Japan	14
France	13
Germany (FRG)	10
Italy	4
Netherlands	4
Switzerland	4
Czechoslovakia	1
Austria	1
Belgium	1
Canada	1
Germany (GDR)	1
India	1
Spain	1
Hungary	—
Poland	—
USSR	—
Total	161

8.4 INDUSTRIAL APPLICATION OF ANTIOXIDANTS

Polyolefins (polyethylene and polypropylene) are extremely susceptible to heat-initiated oxidative degradation:

—Low-density polyethylene (LDPE), which is mostly used for the manufacture of films, during processing around 200°C is partially oxidized and crosslinked with formation of gels; such particles are visible in the films as agglomerates, known to the processor as 'fish-eyes' or gels.
—High-density polyethylene (HDPE) exhibits different behaviour depending on the catalyst and technology used for its manufacture. In many cases, cross linking predominates over chain scission, resulting in a decrease of the Melt Flow Index (cf. Section 9.18.3). Chain degradation is the main degradation route in the Ziegler–Natta catalysed slurry process. Crosslinking, however, predominates in several of the newer HDPE resins.
—Polypropylene is mainly degraded during processing and increase of the Melt Flow Index is observed.

Unless protected against typically high processing temperatures (180–220°C) and oxidation resulting from ambient temperature exposure, polyolefins cannot realize their full performance potential. The useful lifetime of unstabilized polyethylene is predicted to vary from a few months in hot weather (30–40°C) to almost two years in cool weather (0–10°C), which correlate well with experimental results[560,564] and outdoor weathering (cf. Section 9.20).

Practically all final products made from polyolefins contain antioxidants to protect them against thermal and oxidative degradation during processing and storage, and photostabilizers against photodegradation and photo-oxidation in many outdoor applications. For that reason the polyolefin industry consumes *c.* 60% of the antioxidants *c.* 70% of the photostabilizers produced for thermoplastic polymers.[484,504,505,523]

A number of reviews discuss polyolefin stabilization based on research in both academic and industrial laboratories. Many efforts have been directed towards understanding polyolefin photodegradation, photo-oxidation and photostabilization mechanisms and developing new classes of phosostabilizers and antioxidants.[125,126,179,241,435,1079,1125]

Addition of stabilizers (antioxidants) minimizes degradation pro-

Table 8.3
Oxygen uptake at 140°C of low-density polyethylene (LDPE)[1198]

Antioxidant	*Time to absorb 10 cm³ O₂/g (h)*	*Induction period (h)*
—Control (unstabilized)	2	2
0·01% 2,2′-Methylenebis(6-*tert.*-butyl-4-methylphenol) (Plastanox 2246; **3.36**) (structure in Table 3.2)	55	58
0·01% 4,4′-Thiobis(6-*tert.*-butyl-*m*-cresol)	43	46
0·01% 2,6-Di-*tert.*-butyl-*p*-cresol	5	6
0·1% Plastanox 2246	260	265
0·1% 2,6-Di-*tert.*-butyl-*p*-cresol	14	15

cesses such as chain scission, crosslinking and oxidation and it minimizes changes in melt flow.

It is well known that high-density polyethylene (HDPE) requires lower concentrations of antioxidants than polypropylene. Processing stability is achieved here by the use of a hindered phenolic antioxidant alone. For example, 2,2′-methylenebis(6-*tert.*-butyl-4-methylphenol) (**3.36**; Plastanox 2246; Cyanamid) has the ability to retard oxygen uptake and inhibit carbonyl group formation, as shown in Tables 8.3 and 8.4, respectively. This antioxidant remarkably improves processing stability (Table 8.5).

Copolymers of vinyl–ethylene acetate have very low heat stability. When they are exposed to temperatures above 120°C for any length of time, the polymer yellows and suffers a loss in tensile strength and

Table 8.4
Carbonyl formation in low-density polyethylene after oven ageing at 140°C[1198]

Antioxidant	*Induction period (h)*
—Control (unstabilized)	4
0·1% 2,2′-Methylenebis(6-*tert.*-butyl-4-methylphenol) (Plastanox 2246; **3.36**—structure in Table 3.2)	48
0·1% 4,4′-Thiobis(6-*tert.*-butyl-*m*-cresol)	32
0·1% 4,4′-Methylenebis(2,6-di-*tert.*-butylphenol)	13
0·1% 2,6-Di-*tert.*-butyl-*p*-cresol	5

Table 8.5
Physical properties of low-density polyethylene (LDPE) after mill ageing at 165°C[1198]

Antioxidant	*Milling time (min)*	*Carbonyl (%)*	*Melt index*	*Elongation retained (%)*
—Control (unstabilized)	5	—	2·05	—
	15	0·034	2·25	80
	30	0·050	2·37	—
	45	0·204	>5·00	—
	60	0·324	>10·00	—
0·1% 2,2′-Methylenebis-(6-*tert.*-butyl-4-methyl-phenol) (Plastanox 2246;—**3.36**—structure in Table 3.2)	5	—	1·69	100
	15	—	1·91	100
	30	—	1·81	100
	60	—	1·91	100
	75	—	1·99	100
	90	—	2·14	100
	120	0·063	>5·00	48

elongation capacity. This thermal oxidative degradation of co(vinyl–ethylene acetate)s can be dramatically improved by addition of antioxidants, e.g. hindered phenols (Table 8.6).

Sulphide anitoxidants such as dilauryl thiodipropionate (**3.56**) (Cyanox LTDP; Cyanamid) and distearyl thiodipropionate (**3.59**) (Cyanox STDP; Cyanamid) are particularly effective for thermal stabilization of polyolefins and co(acrylonitrile–butadiene–styrene) (ABS) (Figs 8.1 and 8.2).[1198] At a concentration of 0·2 wt% exhibit both excellent

Table 8.6
Oven ageing of ethylene vinyl acetate at 150°C[1198]

Antioxidant	*Time for a yellow-index change of 15 (h)*
—Control (unstabilized)	40
0·1% 2,2′-Methylenebis(6-*tert.*-butyl-4-methylphenol) (Plastanox 2246; **3.36**—structure in Table 3.2)	100
0·1% Plastanox 2246 + 0·1% distearyl thiodipropionate (Plastanox STDP; **3.59**—structure in Table 3.7)	250
0·1% Plastanox 2246 + 0·2% STDP	600

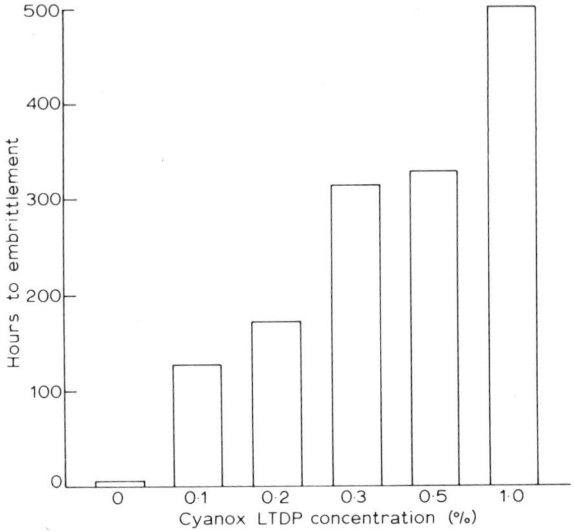

Fig. 8.1. Oven ageing of polypropylene films (at 140°C) in the presence of different concentrations of Cyanox LTDP (dilauryl thiodipropionate; **3.56**).[1198]

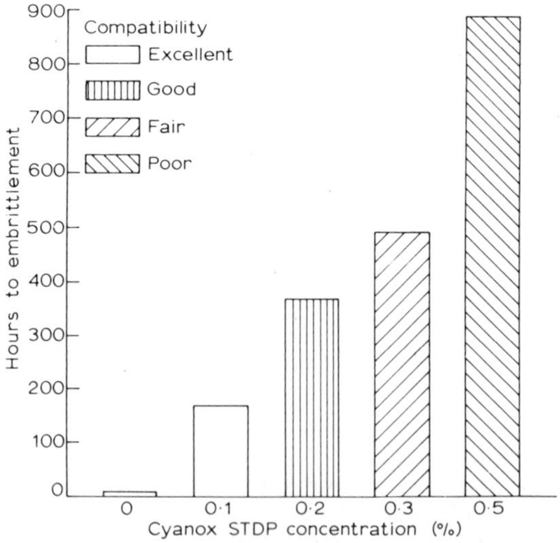

Fig. 8.2. Oven ageing of polypropylene films (at 140°C) in the presence of different concentrations of Cyanox STDP (distearyl thiodipropionate; **3.59**).[1198]

Table 8.7
Oven ageing of polypropylene at 140°C[1198]

Antioxidant	*Time to embrittlement (h)*
—Control (unstabilized)	4
0·2% Dilauryl thiodipropionate (LTDP; **3.56**—structure in Table 3.7)	170
0·2% 2,2′-Methylenebis(6-*tert.*-butyl-4-methylphenol) (Plastanox 2246; **3.36**—structure in Table 3.2)	35
0·1% 4,4′-Butylidenebis(2-*tert.*-butyl-5-methylphenol) (AO-I)	23
0·2% 1,1,3-Tri(5-*tert.*-butyl-4-hydroxy-2-methylphenyl)butane (AO-II)	115
0·2% Cyanox LTDP + 0·2% Plastanox 2246	415
0·2% Cyanox LTDP + 0·1% AO-I	455
0·2% Cyanox LTDP + 0·2% AO-II	335

stabilizing activity and good compatibility. Above this concentration some exudation of the antioxidants occurs.

They are also very effective in combination with phenolic antioxidants (Tables 8.7 and 8.8). The thermal stability obtained with a mixture of these antioxidants is essentially double that which would be expected if an additive effect rather than a synergistic effect occurred.

Phenolic antioxidants are used in mixtures with phosphite/

Table 8.8
Oven ageing of polypropylene at 140°C[1198]

Antioxidant	*Time to embrittlement (h)*
—Control (unstabilized)	4
0·2% Distearyl thiopropionate (STDP; **3.59**—structure in Table 3.7)	375
0·2% 2,2′-Methylenebis(6-*tert.*-butyl-4-methylphenol) (Plastanox 2246; **3.36**—structure in Table 3.2)	35
0·1% 4,4′-Butylidenebis(2-*tert.*-butyl-5-methylphenol) (AO-I)	23
0·2% 1,1,3-Tri(5-*tert.*-butyl-4-hydroxy-2-methylphenyl)butane (AO-II)	115
0·2% Cyanox STDP + 0·2% Plastanox 2246	730
0·2% Cyanox STDP + 0·1% AO-I	715
0·2% Cyanox STDP + 0·2% AO-II	765

Table 8.9
Synergistic blends of phenolic and phosphite/phosphonite antioxidants[1195]

Irganox blends	*Phenolic antioxidant*	*Chemical structure*	*Phosphite/ phosphonite costabilizer*	*Chemical structure*	*Other*	*Ratio*	*Key applications*
B 225	Irganox 1010 (**3.32**)	See Table 3.2	Irgafos 168 (**3.105**)	See Table 3.9	—	1:1	PP/HDPE Emphasis on long-term heat stability
B 215	Irganox 1010		Irgafos 168		—	1:2	PP/HDPE Good balance of long-term heat and processing stability
B 220	Irganox 1010		Irgafos 168		—	1:3	PP/HDPE Applications requiring improved processing stability
B 561	Irganox 1010		Irgafos 168		—	1:4	PP/HDPE/LDPE Applications requiring superior processing stability
B 921 FF	Irganox 1076 (**3.21**)	See Table 3.2	Irgafos 168		—	1:2	HDPE/LDPE Good balance of long-term heat and processing stability

B 900	Irganox 1076		Irgafos 168		—	1:4	HDPE/LDPE Applications requiring superior processing stability
B 1411	Irganox 3114 (**3.31**)	See Table 3.2	Irgafos 168		—	1:1	Filled PP and PP fibres
B 1412	Irganox 3114		Irgafos 168		—	1:2	Filled PP and PP fibres Emphasis on processing stability
B 501 W	Irganox 1425 (**3.34**)	See Table 3.2	Irgafos 168		Polyethylene wax	1:2:1	PP fibres
B 936 W	Irganox 1425		Irgafos 168		Polyethylene wax	1:1:1	PP fibres
B 712 W	Irganox 1010		Irgafos P-EPQ (**3.106**)	See Table 3.9	—	1:2	PP/HDPE/LDPE Good balance of processing and long-term stability, colour improvement
B 311	Irganox 1330 (**3.30**)	See Table 3.2	Irgafos 168		—	1:1	PP, mainly films and tapes Emphasis on long-term heat stability
B 313	Irganox 1330		Irgafos 168		—	1:2	PP, mainly films and tapes Good balance of long-term heat and processing stability

PP, Polypropylene; HDPE, high density polyethylene; LDPE, low density polyethylene.

phosphonite antioxidants (Table 8.9), because of a very strong synergistic effect between these compounds.

8.5 PHOTOSTABILIZATION OF LOW-DENSITY POLYETHYLENE FOR AGRICULTURAL APPLICATIONS

Low-density polyethylene film has found application for greenhouse covering, tunnels and mulch. The advantages of this material are low cost, optical clarity, heat retention, wind resistance, and easy replaceability. However, polyethylene is degraded by UV radiation and addition of photostabilizer is necessary to achieve optimum outdoor life. Testing of a number of photostabilizers showed that extended photostability has been obtained with 2-hydroxy-4-*n*-octoxybenzophenone (Cyasorb UV-531; **5.30**) and 2,2′-thiobis(4-*tert.*-octylphenolato)-*n*-butylamine nickel(II) (Cyasorb UV-1084; **5.63**) (Fig. 8.3). When both photostabilizers were added in a 1 : 1 combination, a synergistic effect was observed.

A typical formulation of the agricultural polyethylene film is given in Table 8.10. However, mixing, extruding and other processing steps

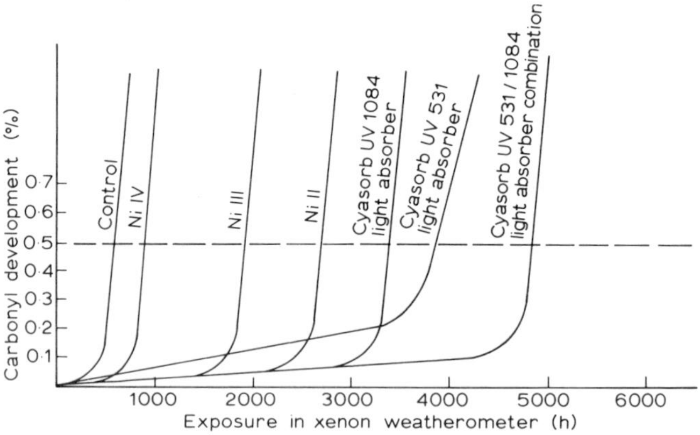

Fig. 8.3. Development of carbonyl index during photoageing of low-density polyethylene (LDPE) in the presence of different photostabilizers: Cyasorb UV 531 (2-hydroxy-4-*n*-octoxybenzophenone; **5.30**); Cyasorb UV 1084 [2,2′-thiobis(4-*tert.*-octylphenolato)-*n*-butylamine nickel(II); **5.63**]; Ni II [nickel(II) bis(2-hydroxy-4-methylphenylundecyl) oxime]; Ni III [dibutyldithiocarbamate; nickel(II), **5.66**]; Ni IV [bis(*o*-butyl-3,5-*tert.*-butyl-4-hydroxybenzylphosphonate) nickel(II); Irgastab 2002; **5.67**].[1198]

Table 8.10
A typical formulation of the agricultural polyethylene film proposed by the American Cyanamid Company[1198]

Ingredient	*Commercial name*	*Chemical name and structure*	*Parts by weight in formulation*
Low-density polyethylene			100
Primary antioxidant: phenolic–phosphite composition	Cyanox 1735	—	0·3
Secondary antioxidant	Cyanox LTDP (**3.56**)	See Table 3.7	0·25
Photostabilizer I	Cyasorb UV 531 (**5.30**)	See Table 5.5	0·25
Photostabilizer II	Cyasorb UV 1084 (**5.86**)	See Table 5.10	1·0

have an influence on the final commercially available film's properties and photostability.

8.6 COMMERCIALLY PRODUCED POLYMERIC HALS

A number of commercially produced polymeric HALS (Table 8.11) are available on the market. They are used for photostabilization of polyolefins. For example, Tinuvin 622 (**6.5**) and Chimassorb 944 (**6.6**) are used for stabilization of polypropylene fibres (cf. Section 8.7). They offer excellent performance as light and heat stabilizers at moderate use concentrations (0·15–0·3 wt%) (Table 8.12). They also show outstanding antioxidant activity on 200 μm blown low-density polyethylene (LDPE) (Table 8.13). In LDPE, as in polypropylene fibre, Chimassorb 944 was found superior to Tinuvin 622.[718,811,1116]

Comparison of the stabilizing activity between two polymeric HALS photostabilizers such as Chimassorb 944 (**6.6**) and Cyasorb UV 3346 (**8.1**) in low-density polyethylene (LDPE) and polypropylene (Tables 8.14 and 8.15) shows the following interesting effects:[73,74]

(i) Both polymeric HALS **6.6** and **8.1** are quite effective oven-ageing antioxidants when compared with the control films, the latter being more efficient than the former. This is interesting

Table 8.11
Examples of commercially available polymeric hindered amine light stabilizers (polymeric HALS)

Chemical name	*Commercial name*	*Chemical structure*
Polyester of succinnic acid with *N*-β-hydroxy-ethyl-2,2,6,6-tetramethyl-4-hydroxypiperidine	Tinuvin 622 (Ciba–Geigy)	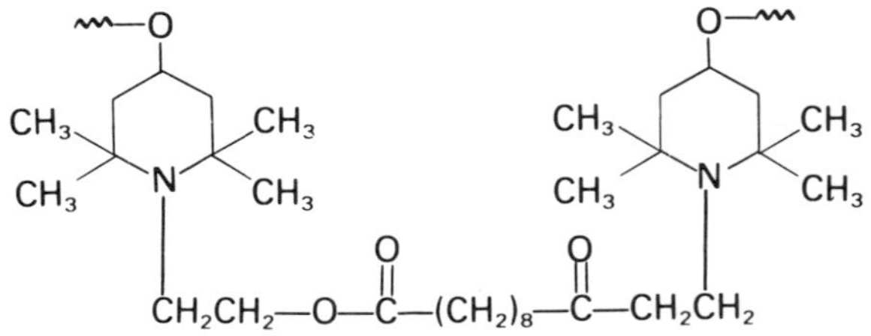 **(6.5)**
CAS name: Poly[[6-[(1,1,3,3-tetramethylbutyl) amino]-1,3,5-triazine-2,4-diyl][(2,2,6,6-tetramethyl-4-piperidinyl)imino]-1,6-hexanediyl[(2,2,6,6-tetramethyl-4-piperidinyl)imino]]	Chimassorb 944 (Ciba–Geigy)	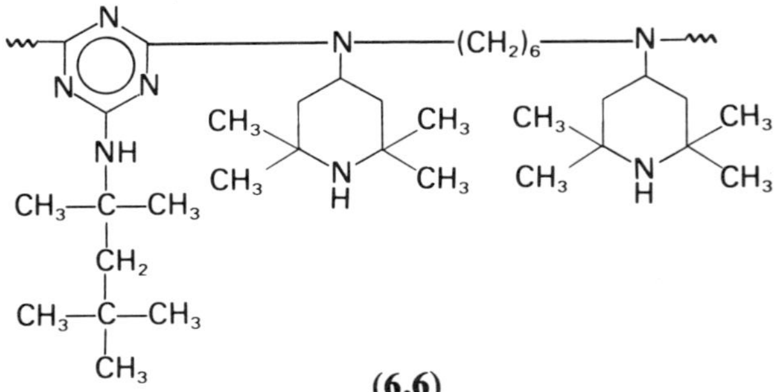 **(6.6)**

Poly[2-*N,N'*-di(2,2,6,6-tetramethyl-4-piperidinyl)hexanediamine-4-(*N*-morpholine)sym-triazine] *CAS name*: Poly(6-morpholine)-1,3,5-triazine-2,4-diyl-[(2,2,6,6-tetramethyl-4-piperidinyl)-imino]-1,6-hexanediyl[(2,2,6,6-tetramethyl-4-piperidinyl)imino]	Cyasorb UV 3346 (Cyanamid)	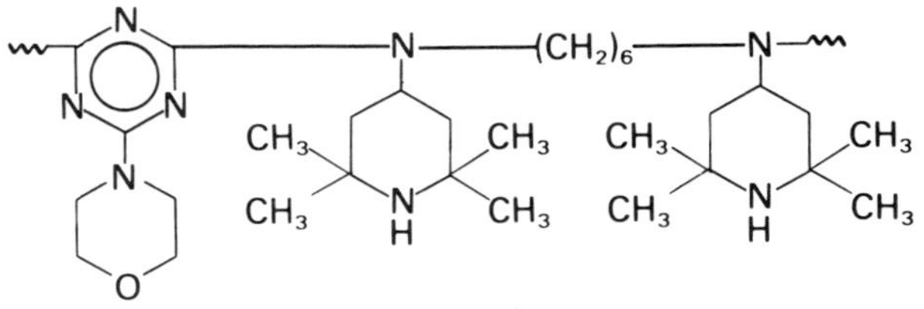 **(8.1)**
CAS name: Poly[(2,2,6,6-tetramethyl-4-piperidinyl)imino]-1,6-hexanediyl[(2,2,6,6-tetramethyl-4-piperidinyl)imino]	Spinuvex A-36 (Borg–Werner)	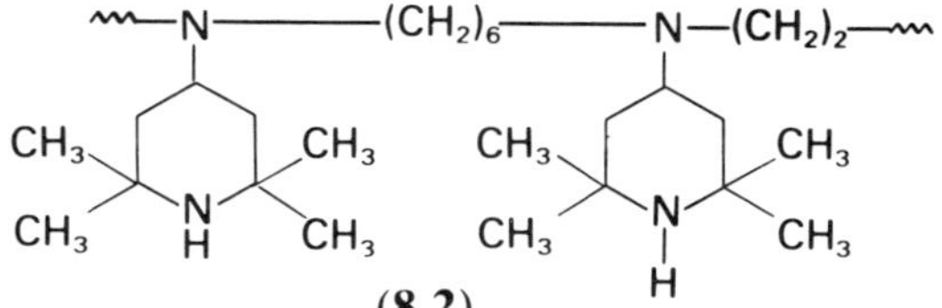 **(8.2)**

Table 8.12
Photostabilizing performance of polymeric: HALS-A (Tinuvin 622 LD; **6.5**)[a] and HALS-B (Chimassorb 944 LD; **6.6**[a]) in polypropylene multi-filaments (130/37 DEN), upon irradiation in Xenotest 1200[b]

HALS	*Half-lifetime*[c] *(h)*	*Time to brittleness in oven ageing test at 110°C (days)*
—Control (nonstabilized)	650	16
0·15% Polymeric HALS-A	3 200	70
0·3% Polymeric HALS-A	4 600	86
0·15% Polymeric HALS-B	3 600	134
0·3% Polymeric HALS-B	4 500	158

[a] Structures in Table 6.1.
[b] Reprinted with permission from Ref. 811, © 1985 American Chemical Society.
[c] Defined as time to 50% retention of tenacity.

since monomeric HALS are known to be ineffective processing antioxidants.

(ii) Both polymeric HALS are good photostabilizers and compound **8.1** is more effective when compared with **6.6**. The overall greater efficiency of **8.1** is evidently associated with the cyclic

Table 8.13
Antioxidant activity of polymeric HALS-A (Tinuvin 622 LD; **6.5**[a]) and HALS-B (Chimassorb 944 LD; **6.6**[a]) in low-density polyethylene (LDPE)[b]

HALS	*Time to failure*[c] *(days)*
—(Base stabilization)	
0·15% Polymeric HALS-A	190
0·3% Polymeric HALS-A	790
0·075% Polymeric HALS-B	1 050
0·15% Polymeric HALS-B	730
	>920 (100%)

[a] Structures in Table 6.1.
[b] Reprinted with permission from Ref. 811, © 1985 American Chemical Society.
[c] Defined as time at 100°C to 50% retention of elongation.

Table 8.14
Thermal and photo- oxidative stabilities of polypropylene and low-density polyethylene (LDPE) films containing polymeric HALS-A (Cyasorb UV 3346; **8.1**[a]) and HALS-B (Chimassorb 944 LD; **6.6**[b])

Additive (0·1% w/w)	*Embrittlement time (h)*			
	Polyethylene (100°C)		*Polypropylene (130°C)*	
	Thermal	*Photo-*	*Thermal*	*Photo-*
—Control (nonstabilized)	200	150	10	100
HALS-A	850	900	170	1 300
HALS-B	600	750	140	920

[a] Structure in Table 8.11.
[b] Structure in Table 6.1.

morpholino group which appears to impart better antioxidant properties than the (1,1,3,3,-tetramethylbutyl)amino groups.

(iii) Polymeric HALS (Tinuvin 622; **6.3**) and Chimassorb 944 (**6.6**) are less effective in polypropylene than low-molecular-weight HALS, due to their poor compatibility. They are also strongly

Table 8.15
Influence of prior thermal oxidation at 130°C on photostability of polymeric HALS-A (Cyasorb UV 3346; **8.1**[a]), and HALS-B (Chimassorb 944 LD; **6.6**[b]) in polypropylene films

Heating time (h)	*UV embrittlement time (h)*	
	HALS-A	*HALS-B*
0	1 300	920
10	2 270	1 110
25	1 920	1 200
50	1 860	1 160

[a] Structure in Table 8.11.
[b] Structure in Table 6.1.

antagonized by the presence of phenolic antioxidant (Goodrite 3125) (Table 6.12).[67]

(iv) Thermal and photo-oxidative stabilization mechanisms are probably very similar.

ESR spectroscopy shows that relative nitroxy radical concentration in polypropylene for both polymeric HALS are higher than those in low-density polyethylene (Table 8.16).[73] This conforms with the known behaviour of hydroperoxides in the two polymers and also stoichiometric reaction between hydroperoxides and HALS during thermal oxidation.[94,126,267] In polypropylene much higher concentration levels of tertiary hydroperoxides are produced than in polyethylene during thermal oxidation.

Table 8.16
Effect of thermal oxidation on nitroxy radical concentration in polypropylene and low-density polyethylene (LDPE) films containing polymeric HALS-A (Cyasorb UV 3346; **8.1**[a]) and HALS-B (Chimassorb 944LD; **6.6**[b])

Heating time (h)	*Relative nitroxy concentration, by ESR*			
	Polypropylene (130°C)		*Polyethylene (100°C)*	
	HALS-A	*HALS-B*	*HALS-A*	*HALS-B*
0	1·0	0·5	1·0	0·5
10	11·0	11·0	1·5	1·0
25	19·5	10·5	5·0	1·5
50	15·0	10·7	15·0	4·0
100	—	—	8·0	4·0

[a] Structure in Table 8.11.
[b] Structure in Table 6.1.
Reprinted with permission from Ref. 73, © Butterworth & Co. (Publishers) Ltd.

In addition, ESR measurements show much lower relative nitroxy radical concentrations are produced for polymeric HALS **8.1** compared with those for **6.6**. These results confirm the important properties of the morpholino group in compound **8.1**, which operates as an effective oxygen and radical scavenger, thereby extending the life of the piperidine group for further reactions.

8.7 PHOTOSTABILIZATION OF POLYPROPYLENE TAPES AND FIBRES

Polypropylene products with developed surfaces such as tapes or fibres are especially susceptible to photo-oxidative degradation and must be carefully stabilized.

Table 8.17 illustrates the superiority of HALS photostabilizers to classical UV absorbers such as 2-hydroxy-4-*n*-octoxybenzophenone (**5.30**; Chimmasorb 81) used at a threefold concentration. The low-molecular weight HALS such as Tinuvin 770 (**6.3**) gives better UV protection than polymeric HALS such as Chimmasorb 944 (**6.6**) or Tinuvin 622 (**6.5**), but cannot be used for food contact applications.

The influence of polymeric HALS (Chimassorb 944; **6.6**) concentration on the photostability of polypropylene tapes is shown in Table 8.18.

Optimum photostability is obtained with a combination of low-molecular-weight HALS (Tinuvin 770) and polymeric HALS (Chimassorb 944) (Table 8.19).

Table 8.17
Light stability of polypropylene tapes (50 μm) protected by different types of HALS photostabilizers[1194]

Type of photostabilizer	*Commercial name*	*Chemical name and structure*	*Concentration (%)*	*Solar energy ($kJ\ cm^{-2}$) in Florida required for 50% decrease in tensile strength*
Low-mol.-wt HALS	Tinuvin 770 (**6.3**)	See Table 6.1	0·1	9·6
Polymeric HALS	Chimassorb 944LD (**6.6**)	See Table 6.1	0·1	4·1
Polymeric HALS	Tinuvin 622 LD (**6.5**)	See Table 6.1	0·1	3·2
UV absorber (hydroxy-benzophenone type)	Chimassorb 81 (**5.30**)	See Table 5.5	0·3	2·2
Without photostabilizer	—	—	—	1·4

Table 8.18
Light stability of polypropylene tapes (50 μm) protected by combinations of low-molecular-weight HALS [bis(2,2,6,6-tetramethyl-4-piperidinyl) sebacate; **6.3**; Tinuvin 770[a]] and polymeric HALS (Chimassorb 944; **6.6**[a])[1194]

Combination of photostabilizers	*Concentration (%)*	*Solar energy (kJ cm^{-2}) in Florida required for 50% decrease in tensile strength*	*Time in Xenotest (h) (type 1200) required for 50% decrease in tensile strength*
Control (nonstabilized)		0·6	800
Tinuvin 770 DF Chimassorb 944 LD	0·05 0·05	6·4	4 000
Tinuvin 770 DF Chimassorb 944 LD	0·1 0·1	13·0	9 750
Tinuvin 770 DF Chimassorb 944 LD	0·2 0·2	15·9	17 000

[a] Structures in Table 6.1.

Polypropylene fibres place special requirements on the stabilizer system due to the high surface area, the severe processing conditions, and the use of various colourants and post-spinning treatments such as tentering, laundering and dry-cleaning. Oligomeric HALS like Tinuvin 292, Tinuvin 144, Tinuvin 622 and Chimassorb 944 (for chemical structures see Table 6.1) meet the required demands. In particular, Chimassorb 944 (**6.6**) has superior properties.[1116]

Table 8.19
Influence of polymeric HALS (Chimassorb 944; **6.6**[a]) concentration on the stability of polypropylene tapes (50 μm)[1194]

Concentration of polymeric HALS (%)	*Solar energy (kJ cm^{-2}) in Florida required for 50% decrease in tensile strength*
Control (nonstabilized)	1·59
0·05	3·2
0·10	4·6
0·20	7·6
0·40	12·4

[a] Structure in Table 6.1.

8.8 PHOTOSTABILIZATION OF RUBBER-MODIFIED PLASTICS

Rubber-modified plastics, such as polyblends of acrylonitrile–butadiene–styrene (ABS), styrene–butadiene–styrene (SBS), methyl methacrylate–butadiene–styrene (MBS), which are high-impact-resistant products, are very susceptible towards photo-oxidative degradation.[460–462] This has been attributed to the presence of the unsaturated segment of butadiene in the molecule which, due to its selective peroxidation, sensitizes to auto-oxidation both the copolymers and polyblends containing it.[460–462]

Conventional photostabilizers and antioxidants are relatively ineffective in the long-term stabilization of these plastics for high-temperature (automotive) and outdoor application (Table 8.20). This problem has been partially solved by application of polymer-bound photostabilizers and antioxidants (cf. Chapter 7) to the rubber phase.[403,662,674]

8.9 PHOTOSTABILIZATION OF POLY(VINYL CHLORIDE) BY SURFACE PHOTOGRAFTING

Epoxy–acrylate coatings which absorb UV radiation of wavelength below 300 nm have been photografted on the poly(vinyl chloride) and act as effective UV filters. Photografting was done in the presence of α-hydroxyacetophenone photoinitiators according to the reactions:[342]

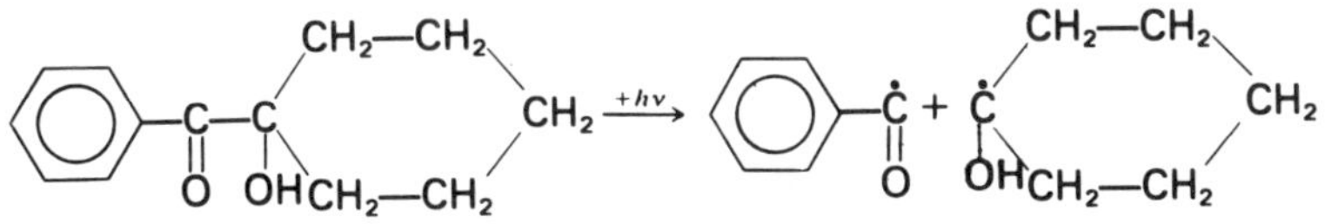

(8.1)

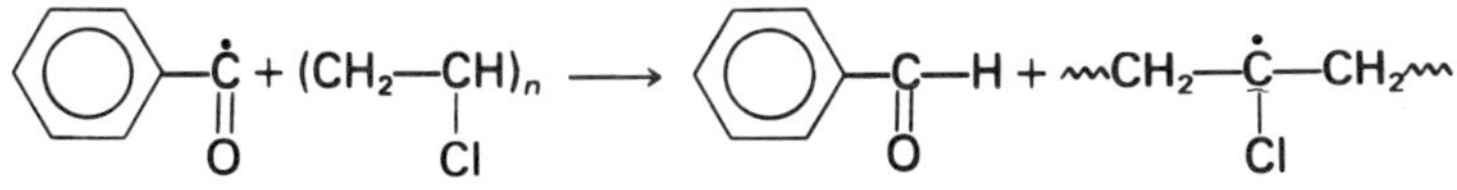

(8.2)

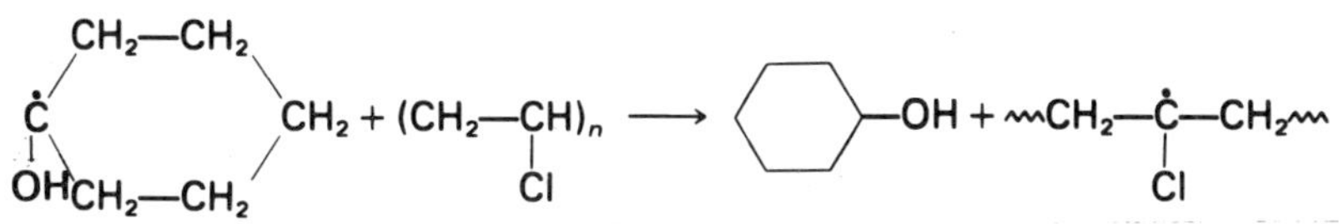

(8.3)

Table 8.20
Photostabilizing effectiveness of some of commercially available antioxidants and photostabilizers in acryl–butadiene–styrene (ABS) resin[a]

Additive	*Chemical structure*	*Trade name*	*Concentration in polymer sample (10^3 mol/100g)*	*Carbonyl index period during photo-oxidation (h)*	*Embrittlement time during photo-oxidation (h)*
Control	—	—	—	3	22
2-Hydroxy-4-*n*-octoxy-benzophenone	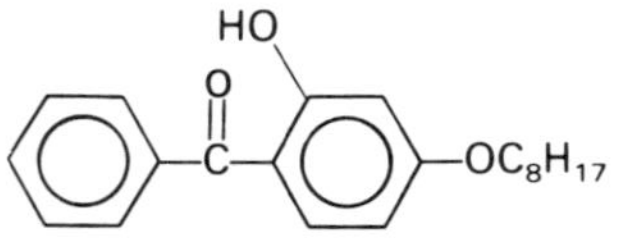 (5.30)	Cyasorb UV 531	3	11	35
2-(2-Hydroxy-5-methyl-phenyl)-2*H*-benzo-triazole	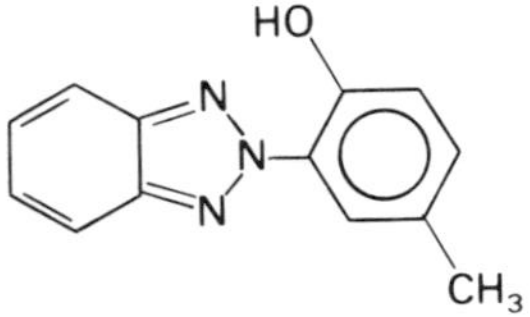 (5.43)	Tinuvin P	3	12	40
Bis(2,2,6,6-tetra-methyl-4-piper-idinyl) sebacate	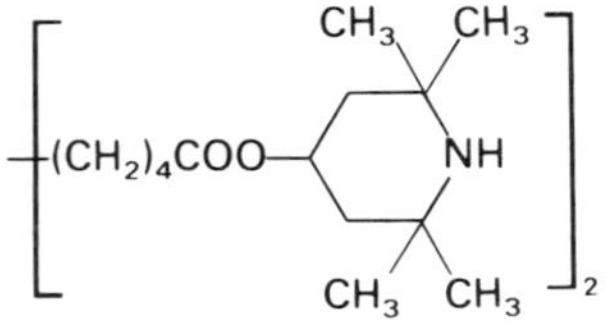 (6.3)	Tinuvin 770	3	16	37

2,6-Di-*tert.*-butyl-4-methylphenol	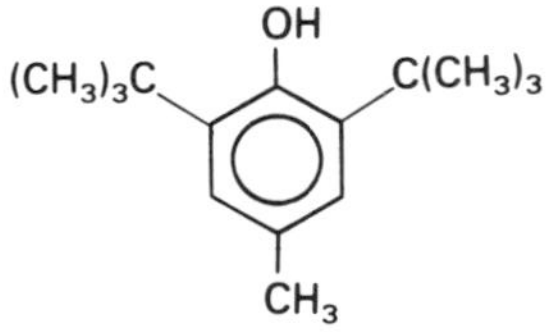 **(8.3)**	Topanol OC	4·9	10	34
Octadecyl 3-(3,5-di-*tert.*-butyl-4-hydroxyphenyl)-propionate	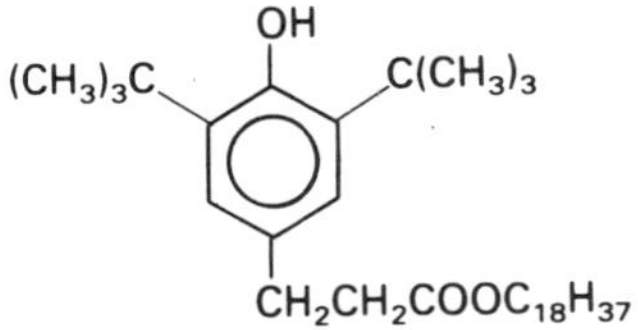 **(3.21)**	Irganox 1076	4·9	10	34
2,2′-Thiobis(4-methyl-6-*tert.*-butylphenol)	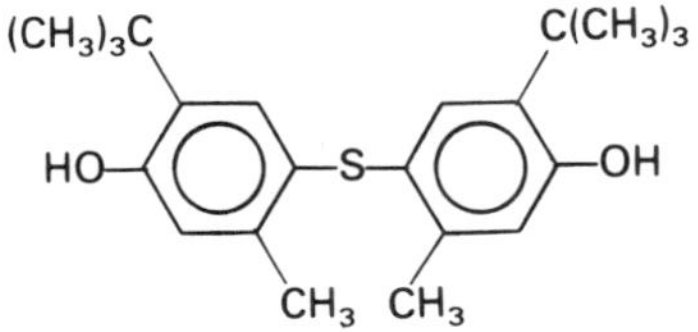 **(3.29)**	Irganox 1081	4·9	10	42
Distearyl thiodi-propionate	$(C_{18}H_{37}OCOCH_2CH_2)_2S$ **(3.59)**	Cyanox STDP	1·5	5	25
Distearyl pentaery-thrityl diphosphate	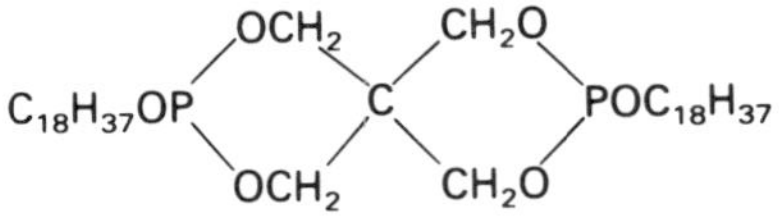 **(3.113)**	Weston 618	2·5	10	40

(continued)

Table 8.20—*contd.*

Additive	*Chemical structure*	*Trade name*	*Concentration in polymer sample* $(10^3\ mol/100g)$	*Carbonyl index period during photo-oxidation (h)*	*Embrittlement time during photo-oxidation (h)*
Zinc(II) dibutyldithio-carbamate	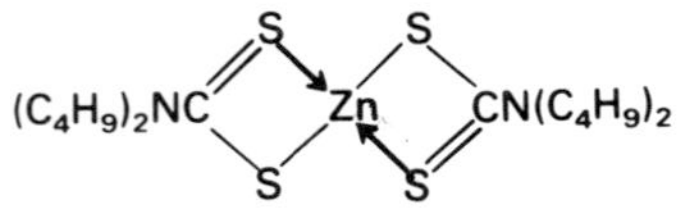 (3.85)	ZnDBC	2·0	7	30
Nickel(II) dibutyldi-thiocarbamate	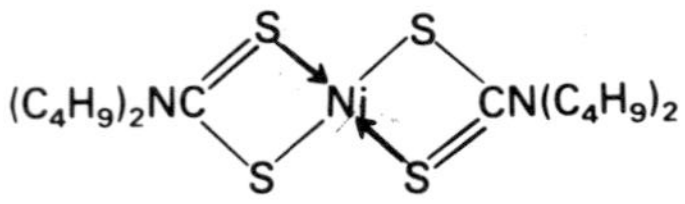 (3.85)	NiDBC	2·0	12	44

[a] Reprinted with permission from Ref. 403, © 1980 Pergamon Press PLC.

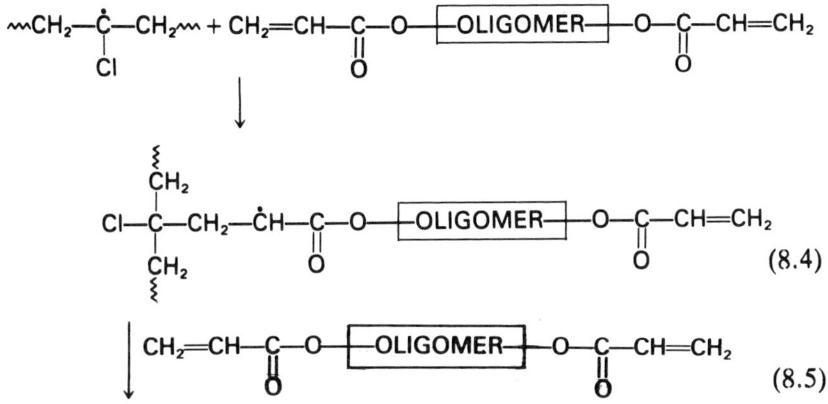

(8.4)

(8.5)

Epoxy–acrylate network grafted on PVC

The best results were obtained by incorporating photoinitiator into the top layer of the PVC sheet (film). Highly-crosslinked epoxy–acrylate coatings are very stable towards UV radiation, and by their light-screening effect they provide an excellent protection against photodegradation (Fig. 8.4).

Polymers stabilized by grafting are not subject to physical loss of stabilizers due to volatility because the stabilizing groups are chemically bonded to polymer chains. However, the thermal stability of the

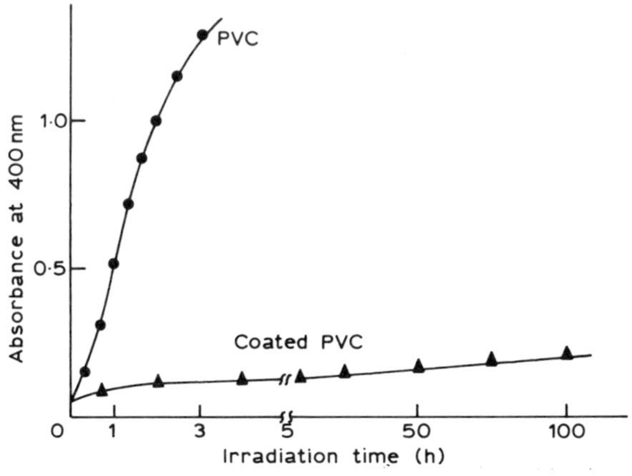

Fig. 8.4. Kinetics of the discoloration of a 100 μm poly(vinyl chloride) film (●) and of an epoxyacrylate coated poly(vinyl chloride) film (▲) upon exposure to UV radiation.[342]

grafted chains containing the stabilizing groups should be sufficiently high in order to survive the processing conditions.

8.10 PHOTOSTABILIZATION OF COATINGS

Relatively few studies have been made on the mechanism of photo-oxidative degradation of organic coatings.[574–577,668] However, the degradation chemistry and ageing of acrylic/amine coatings which are commonly used as automotive topcoats have been extensively studied.[159–162,379,445,447–449,451,452,642]

Photostabilization of surface coatings is a very important industrial problem. On exterior surfaces, numerous coatings check, crack, peel, blister and generally deteriorate because of harmful photo-oxidative processes. Photoageing of surface coatings may be postponed or even prevented entirely by incorporating a UV-absorbing compound into the clear coating. The UV-absorber will protect the coating and produce a UV barrier which will protect the coated surface. Pigmented surface coatings also suffer from UV degradation but to a lesser extent than clear coatings. The inclusion of a UV-absorber can increase the useful life of such pigmented coatings.

The choice of photostabilizer for coatings depends on many factors:

(i) the type of coating,
(ii) the coating thickness,
(iii) the type and level of pigmentation,
(iv) the cure schedule, and
(v) the type and degree of protection desired.

The photostabilizers most commonly employed for coating protection are hydroxybenzophenones and hindered amine light stabilizers (HALS).

Hydroxybenzophenones may be incorporated in a variety of surface coatings to produce a clear film largely opaque to UV radiation. The resulting screen can be applied directly to UV-sensitive substrates such as wood, to protect them efficiently and economically from photo-degradation. In addition, these coatings containing photostabilizers can be applied to window glass to set up a UV barrier which will protect display merchandise from sunlight.

Hydroxybenzophenones such as 2-hydroxy-4-methoxybenzophenone (Cyasorb UV 9; Cyanamid; **5.28**), 2,2′-dihydroxy-4-methoxybenzophenone (Cyasorb UV 24; Cyanamid; **5.29**) and 2-hydroxy-4-*n*-octoxybenzophenone (Cyasorb UV 531; Cyanamid; **5.30**) in a concentration of 0·5 wt% are widely used for the protection of vinyl, phenolic, alkyd and polyurethane coatings. Cyasorb 24 is especially useful in typical marine, tung-oil phenolic resins. Similar results have been obtained with alkyd, linseed, isophthalic, phenolic and a variety of other varnishes.[1198]

Hydroxybenzophenones are more dependent upon coating thickness than HALS. Thick films (>1·0 μm) typically require lower hydroxybenzophenones concentration than thin films (<1·0 μm) which often require a higher concentration. Table 8.21 gives some useful information on the proper choice of photostabilizers and recommended ranges for evaluation purposes.

During the degradation of coatings, gloss loss is observed (cf. Section 9.17). In order to prevent this phenomenon, combinations of hydroxybenzophenone and HALS photostabilizers are the most effective (especially for systems which contain a low degree of pigmentation).

During extensive degradation of coatings cracking and loss of coating integrity are observed as a result of chain scission and/or crosslinking. Also observed are increases in film hardness, embrittle-

Table 8.21
Guidelines for the proper choice(s) of photostabilizer (UVA)[1193]

Condition	*Recommended type of photostabilizer (UVA)*	*Recommended concentrations (%)*
Clear coating on non-UV-sensitive substrate	UVA + HALS	1·0 + 1·0–2·0
	HALS alone	0·5–3·0
Clear coating on UV-sensitive substrate	UVA + HALS	1·0–3·0 + 0·5–2·0
	UVA alone	0·5–3·0
Low-pvc[a] or metallic coatings	UVA + HALS	1·0–3·0 + 0·5–2·0
High-pvc, with light-stable pigments	HALS alone	1·0–3·0
	UVA + HALS	1·0 + 1·0–2·0
High-pvc, with light-sensitive pigments	UVA + HALS	1·0–3·0 + 1·0–3·0
Overbake yellowing control or heat exposure	Antioxidant	0·1–1·0
	HALS	0·25–1·5

[a] Abbreviations: pvc, pigment volume concentration; UVA, UV absorber.

ment, and loss of flexibility. Once film integrity is lost, corrosion or rotting of the substrate is more likely.

HALS stabilizers provide the greatest degree of protection in preventing or delaying the formation of cracks while maintaining film flexibility and desired hardness. Synergism between hydroxybenzophenone photostabilizers and HALS has a special effect in systems of low pigmentation such as clear or metallic finishes.[1193]

Loss of pigment intensity and changes in shade (i.e. darkening) is a result of pigment (or dye) photolysis (or rearrangement) into colourless or different colour moieties. Stabilization of the loss of pigment intensity is a very complex problem because of very complex photochemical reactions, which occur and the diversity of pigments and dyes used in coating formulations.

If the pigmented coating is underneath a clear topcoat, the use of hydroxybenzophenone photostabilizers in the topcoat is highly recommended to prevent fading or colour change. For single-coat pigmented systems, combinations of hydroxybenzophenone and HALS photostabilizers are generally preferred. HALS alone may be sufficient if pigment or dye breakdown is primarily due to attack by free radicals generated by binder degradation. If the pigment or dye undergoes direct photolytic breakdown, a combination of hydroxybenzophenone and HALS photostabilizers is recommended. Particularly unstable pigments or dyes may require high levels (3·0 wt%) of photostabilizers to maintain their desired shade and intensity for extended exposures.[1193]

It is well known that polyurethane coatings are susceptible to deterioration when exposed to sunlight for prolonged periods of time. The coatings discolour noticeably, initially turning a light yellow, later tan and eventually dark brown. Polyurethane coatings also undergo a severe breakdown as evidenced by lifting and peeling of the coating from the substrate. The best photostabilizers for protecting polyurethane coatings are hydroxybenzophenones.[209] in combination with HALS and hindered phenolic antioxidants (Fig. 8.5).

The use of combinations of photostabilizers has an additional effect for improving properties of coatings against delamination, peeling and chalking.

Chalking is caused by polymer erosion at the coating surface which exposes pigment particles to the degree that they can be washed or rubbed out. This can be a first sign of increased water permeability of the coating. Chalking has been purposely used by some formulators to

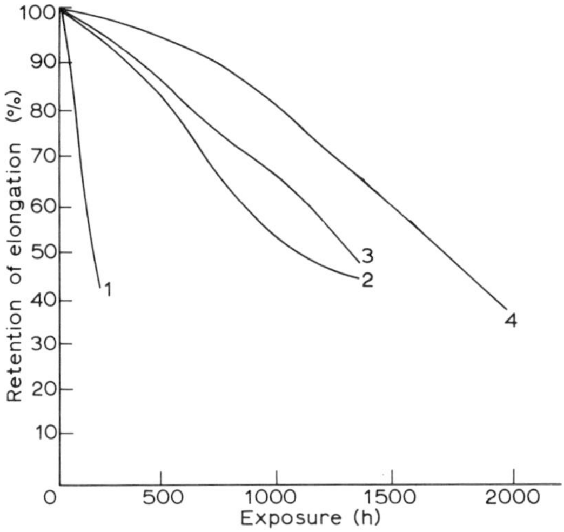

Fig. 8.5. UV degradation of clear thermoplastic urethane coatings: 1, unstabilized sample; 2–4, in the presence of (2) 0·25% Tinuvin 144 [bis(1,2,2,6,6-pentamethyl-4-piperidinyl)-2-*n*-butyl-(3,5-di-*tert.*-butyl-4-hydroxybenzyl)malonate; **6.7**] + 0·25% Irganox 1010 [pentaerithrityl tetrakis [3-(3′,5′-di-*tert.*-butyl-4′-hydroxyphenyl)propionate]; **3.32**], (3) 0·50% Tinuvin 144 and (4) 0·50% Tinuvin 144 + 0·50% Irganox 1010.[1193]

maintain dirt-free white surfaces. However, chalking may be uneven and can also discolour adjoining coloured surfaces. HALS are the most effective class of photostabilizers for delaying significantly the onset of chalking. Optimal protection against chalking may be obtained by the combination of hydroxybenzophenones with HALS, especially as the level of pigmentation decreases.

The amount of radiation absorbed by a coating is a function of film thickness and absorber concentration. In order to express the interrelation of these variables in the simplest way, the *CL value* is often used. It is the product of the absorber concentration as a percentage (C) and film thickness in μm (L):[1198]

$$CL \text{ value} = \text{Concentration } (\%) \times \text{thickness } (\mu\text{m}) \qquad (8.6)$$

A specific CL value can thus be obtained from an infinite number of combinations of absorber concentration and film thickness. For example, a CL value of 10 for a photostabilizer such as 2-hydroxy-4-methoxybenzophenone (**5.28**; Cyasorb UV 9; Cyanamid) in plasticized vinyl could be obtained from any of the following combinations:

CL value	=	Cyasorb UV 9 light-absorber concentration (%)	×	film thickness (μm)	(8.7)
10	=	1·0	×	10	
10	=	2·0	×	5	
10	=	10·0	×	1	

For each UV-absorber a specific *CL* value will produce a given spectral transmission curve denoting the absorption of definite amount of UV radiation, so any of the above or similar combinations of UV absorber concentration and film thickness would give the same *CL* value and therefore the same spectral transmission curve. In Figs 8.6–8.8 are shown the spectral transmission curves for different UV absorbers (Cyasorb UV 9, Cyasorb UV 24 and Cyasorb UV 531) in flexible poly(vinyl chloride) films. Because of the different absorption characteristics of each absorber, the *CL* value curves for each UV absorber will be different. The *CL* value of 5 cuts off all radiation from 351, 376 and 345 nm for UV 9, UV 24 and UV 531, respectively.

The proper manipulation of *CL* values makes possible the most practical, economical selection of a UV-opaque screen to protect a given substrate. The curves presented in Figs 8.6–8.8 for flexible poly(vinyl chloride) are quite similar to curves which would be obtained in acrylics, alkyds and any number of other coatings.

One of the practical uses for a UV screen is in window coatings. Merchandise displayed in store windows loses its value because of

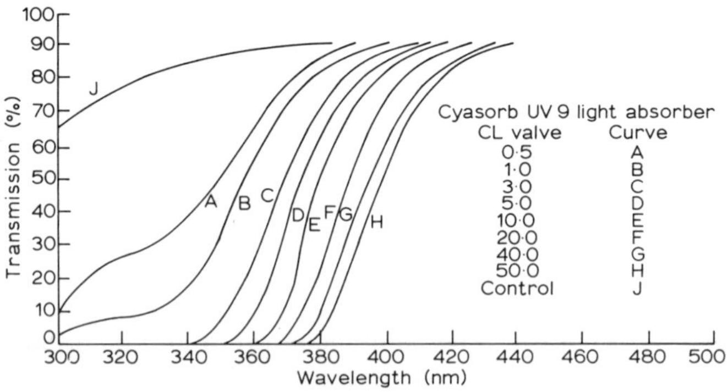

Fig. 8.6. Transmission curves for Cyasorb UV 9 (2-hydroxy-4-methoxybenzophenone; **5.28**) in flexible poly(vinyl chloride) films.[1198]

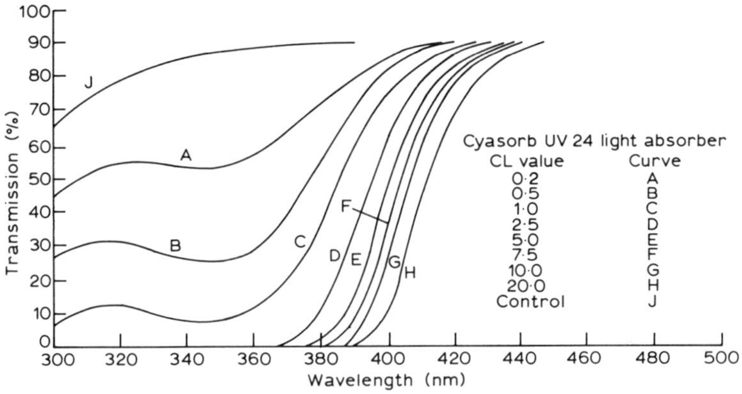

Fig. 8.7. Transmission curves for Cyasorb UV 24 (2,2′-dihydroxy-4-methoxybenzophenone; **5.29**) in flexible poly(vinyl chloride) films.[1198]

degradation from sunlight. This degradation can be minimized by applying, to the display window, a surface coating containing a UV absorber. Both alkyd- and acrylic-type coatings containing *o*-hydroxybenzophenone photostabilizers are used in this manner.

In Table 8.22 are collected data of the retarding of polystyrene plaque photodegradation in display windows by application of acrylic coating containing 2,2′-dihydroxy-4-methoxybenzophenone (**5.29**; Cyasorb UV 24; Cyanamid). The coating containing UV absorber is

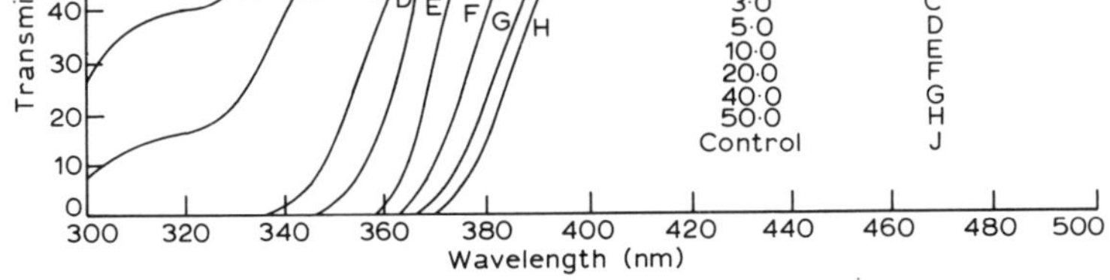

Fig. 8.8. Transmission curves for Cyasorb UV 531 (2-hydroxy-4-*n*-octoxybenzophenone; **5.30**) in flexible poly(vinyl chloride) films.[1198]

Table 8.22
Retarding photodegradation in display windows by applying acrylic coating containing photostabilizer 2,2′-dihydroxy-4-methoxybenzophenone (Cyasorb UV 24; **5.29**[a]) during irradiation in the Fadeometer[1194]

Formulation of sample	*Exposure time*			
	0h	*100h*	*200h*	*400h*
	Yellow index			
Control—acrylic coating	6	8	14	27
Acrylic coating + Cyasorb UV 24	6	6	7	9

[a] Structure in Table 5.5.

represented by curve E in Fig. 8.7, *CL* value 5; approximately 90% of the UV radiation between 300 and 400 nm is screened out. The rate of discoloration for the polystyrene plaque was much slower than that for a similar plaque exposed behind a coating with no absorber.[1198]

Many woods are extremely sensitive to UV radiation. This is manifested by severe discoloration during prolonged exposure to sunlight. The presence of hydroxybenzophenone photostabilizers in furniture lacquers can effectively screen out the harmful radiation and sometimes retard and even eliminate this deterioration.

Outdoor exposure is generally accepted as the most reliable method for the determination of coating durability.[383] The great disadvantage of this method is the long time required to differentiate between good and very good coatings.

Accelerated tests in Weatherometers (cf. Section 9.20.3) shorten test time. However, in such a test, it is impossible to guarantee that all photodegradation and photostabilization processes are accelerated to the same extent. The gloss loss results (cf. Section 9.17) in accelerated tests do not always correlate with outdoor exposure results.[495]

8.11 ADDITIVES IN A SPACE ENVIRONMENT

There is increasing interest in the application of polymers and plastics to rockets, satelites, and other space vehicles. Uses may include nose cones, surface coatings, windows, and coverings for solar batteries and measuring instruments. The space environment imposes conditions upon materials which are vastly different from those encountered in terrestrial conditions, for example the following.

(i) High vacuum exists, in which additives evaporate from the surfaces.
(ii) Photochemical action of sunlight is considerably more detrimental to polymers and plastics due to the lack of an absorbing atmosphere to filter out the extreme (10–200 nm) and far (200–300 nm) ultraviolet radiation.
(iii) Solar radiation prevails continuously rather than periodically.
(iv) Control of photodegraded surfaces is very limited.

The number of material available in published form in this field is very limited.[438,960,984,985]

9

Experimental Methods in Polymer Degradation and Stabilization

9.1 PREPARATION OF A TEST SPECIMEN

The preparation of polymer samples for testing the light stability should be carried out with special care and in very standardized conditions. Additives such as thermal stabilizers, antioxidants and photostabilizers should be incorporated into the polymer by wet blending with solvent (e.g. alcohol). The resulting mixture should then be preheated in order to remove solvent. For example a typical preparation of polypropylene test specimens is as follows.[1110] A mixture of 100 parts of polypropylene powder, 0·2 part of an antioxidant and 0·2–0·4 part of the photostabilizer, after wet blending from methanol, is preheated to 215°C for 2 min, then plasticized in a Brabender Plasticorder for 10 min at 200°C in order to produce a homogeneous material, which is further pressed for 6 min at 260°C under 15 t pressure and then rapidly cooled in cold water. The sheets, of thickness 0·5 mm, are further cut into test pieces. As a control, unstabilized polymer sheets can be prepared in a similar way.

When it is extruded partial orientation of the polymer film can be obtained by drawing.

The polymer film samples can be further exposed to irradiation in standard conditions, e.g. in a Weatherometer. A decrease in tensile strength, expressed as tenacity, over the tensile strength of the same formulation before irradiation, is a measure of the deterioration of the physical properties of the polymer. '*Failure*' in such a test is defined as a loss of 50% or more of the initial sample tenacity.

9.1.1 Microtomy

Microtomy is a useful technique to obtain shavings for various depths of an aged polymer specimen. The thickness of the microtomed sections can be determined using a micrometer. The important factors in the cutting of samples are the feed, the cutting stroke, the return stroke and the retraction of the knife after cutting. In modern commercially available microtomes, cutting is electronically controlled. Cryo-ultramicrotomy is essential for cutting sections of polymers and elastomers including plastics, rubbers, fibres, films, latex or paints and textiles. Cryo-sectioning is a rapid method when compared with other preparative techniques such as fixation, embedding, swelling or particle isolation. Microprocessor-controlled cryo-microtomes provide highly accurate and reproducible cutting of polymer samples.[543,981]

9.2 DETERMINATION OF THE ADDITIVE CONTENT AFTER POLYMER PHOTO-OXIDATIVE DEGRADATION

The most common method used for the determination of additive content in a polymer sample after light irradiation is extraction in a Soxhlet apparatus for 24 h. Extracts should be concentrated in a vacuum evaporator. Products of polymer degradation such as oligomers can be precipitated with a suitable solvent and removed by filtration. The filtrate should next be dried in a vacuum evaporator and dissolved on an appropriate volume of solvent. The additive concentration can be determined by liquid chromatography with a UV/VIS spectrophotometer detector.

9.3 QUANTITATIVE DETERMINATION OF ANTIOXIDANTS

In a number of analytical methods which can be applied for the quantitative determination of additives, one is of special interest. This method is based on oxidation of phenolic antioxidants (AO) under controlled conditions using ferric ions:[662,786]

$$AO + Fe^{3+} \longrightarrow AO + Fe^{2+} \qquad (9.1)$$

In this reaction, the Fe^{2+} ions produced react with 2,2-bipyridyl to form coloured complexes (λ_{max} 520 nm) whose intensity is proportional to the concentration of the antioxidant. The concentration of the antioxidant in the solution can be obtained from a calibration curve using antioxidant solutions of known concentrations.

9.4 DETERMINATION OF EFFECTIVENESS OF ANTIOXIDANTS

The usual method for assessing the effectiveness of antioxidants under processing conditions is the multiple extrusion test, in which changes of melt flow and colour of the extruded material are determined, commonly after the first, third and fifth pass. For high-molecular-weight HDPE, Brabender plastograms give a good indication of the stability of the polymer melt.

9.5 MEASUREMENT OF EVAPORATION OF ADDITIVES FROM POLYMER SAMPLES

Measurement of evaporation of additives from polymer samples can be made with the very simple apparatus shown in Fig. 9.1.[537] A 0·2 g strip of polymer is inserted into the evaporation sample holder, made of 10 mm o.d. Pyrex tubing. After being flushed with nitrogen, the sample holder is immersed in a temperature-controlled bath (or oven). The evaporation rate and additives content can be measured by means of gas chromatography.

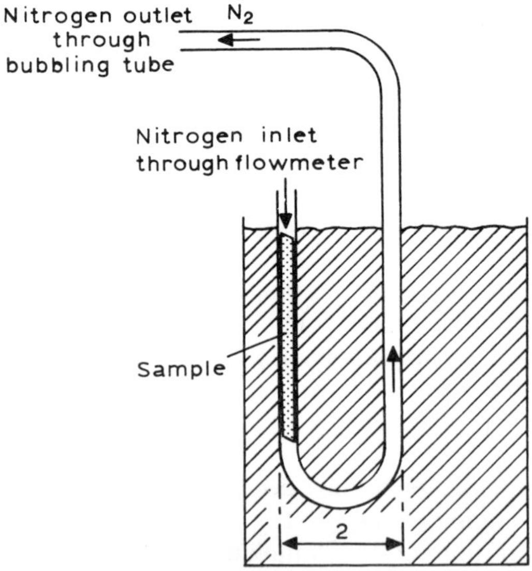

Fig. 9.1. Apparatus used to study the evaporation of antioxidants from polyethylene.[537]

The volatility of solid additives can also be studied from their rate of evaporation (*in vacuo*) from a free surface (polymer film). Vapour pressure is calculated from the equation:

$$P = K\sqrt{2\pi RT/M} \tag{9.2}$$

where P = equilibrium vapour pressure (in mm Hg);
K = rate of evaporation ($g\,cm^{-2}\,s^{-1}$);
T = absolute temperature (K);
R = molar gas constant;
M = molecular weight of the vapour.

The apparatus for determining rates of evaporation is shown in Fig. 9.2. It consists of a thermostatically controlled copper sample holder which faces a concave collecting surface that is cooled by liquid

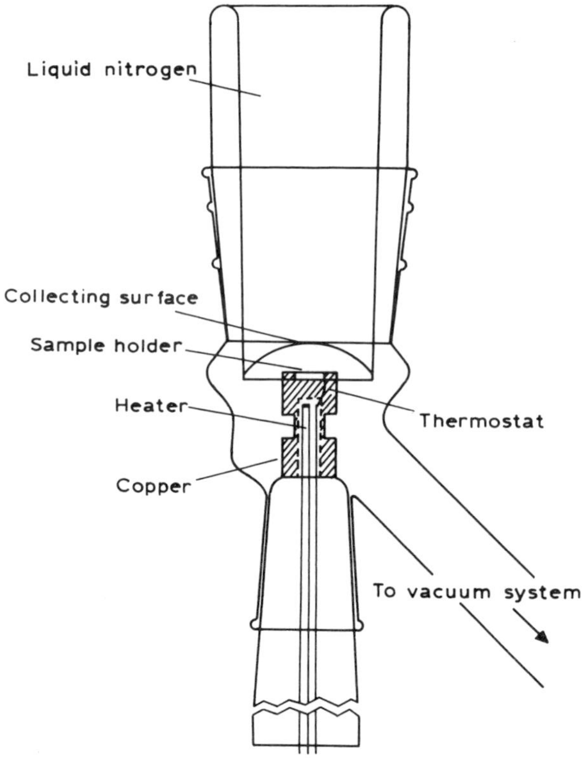

Fig. 9.2. Apparatus used for the determination rates of evaporation.[984]

nitrogen. The distance between the sample and the collecting surface is approximately 1 cm. Both the sample holder and the collecting surface are attached to the vacuum system with ground-glass joints to permit easy removal. This apparatus is connected to an oil diffusion pump via a liquid nitrogen trap.

9.6 ABSORPTION SPECTROSCOPY

The detection limit by UV/VIS absorption spectroscopy depends upon the molecular absorptivity (also called the molecular extinction coefficient), which is normally in the range 100–20 000 for stabilizers and 30 000–100 000 for optical brightening agents.

Absorption spectroscopy is a fundamental method for the determination of absorption spectra of antioxidants, photostabilizers and other additives, and for following their thermolysis and/or photolysis. Absorption UV/VIS spectroscopy together with IR spectroscopy are extremely useful methods for:

(i) identification of extracted additives from commercial polymers;[527]
(ii) study of polymer thermal and/or photo- oxidation mechanisms and kinetics;
(iii) study of antagonistic and synergistic effects between additives.

Polymer films can be cast on quartz slides (for UV and ESR studies) or on NaCl or KRS-5 salt plates (for IR studies).[915]

For UV/VIS and IR studies, samples should be mounted so that the same spot on the plate is always measured. In this way it is possible to obtain information on changes in film thickness as well as chemical composition on degradation.

An especially useful technique is derivative spectroscopy, which has improved resolution over that of the normal zero-order spectroscopic method and allows analysis of the original additive separately from that of any otherwise interfering products of oxidation.[48,57,68] UV/VIS derivative spectroscopy has been found to be very useful for studying photostabilization.[44,46,48,68,91]

9.6.1 Application of IR Spectroscopy to the Study of Polymer Oxidation

Infrared spectroscopy has been an indispensable method for the study of polymer degradation and stabilization. The technique has three important characteristics that contribute to its usefulness.

(i) No two dissimilar molecules can have the same IR spectrum. This enables all changes which occur during degradation and stabilization processes to be identified; however, interpretation of a given absorption peak is sometimes very difficult, or even impossible.
(ii) Infrared spectra of mixtures are additive and absorption is proportional to concentration. Thus, IR spectroscopy can be used for the study of the kinetics of any process involved in the degradation and stabilization.
(iii) Infrared spectra can be obtained non-destructively on any type of polymer or additive sample—solid, liquid, or even gas.

Infrared spectroscopy is still a fundamental method for the study of kinetics of polymer oxidation kinetics.

The carbonyl (CO) and hydroxy(OH)/hydroxyperoxide (OOH) frequencies are the best known of the characteristic infrared absorption bands formed upon photo-oxidative and/or thermo-oxidative degradation of polymers. These strong bands fall in regions of the spectra where few other absorptions are found.

9.6.1.1 Carbonyl Groups

The carbonyl group bands occur within 60 cm^{-1} of 1700 cm^{-1}. Variation from the normal position (1710–1720 cm^{-1}) is caused by various effects, such as:

(i) *Geometric effects,* which change the vibrational frequency, since some motions of the carbon atom are involved in the C═O stretching vibration. Deformation of the adjacent CH_2 absorbs at 1405–1440 cm^{-1}.
(ii) *Force constant effects,* which affect the vibrational frequency via the electron density within the C═O bond (i.e. bond order). Substituent groups on the carbon which reinforce the tendency of the oxygen to attract electrons in the structure ($\rangle\overset{+}{C}-\overset{-}{O}$) will decrease the force constant and lower the frequency:

$$\rangle C{=}O \rightleftharpoons \rangle\overset{+}{C}-\overset{-}{O} \qquad (9.3)$$

Conversely, electronegative substituents, which compete with the oxygen atom for electrons, decrease the contribution of the $\rangle\overset{+}{C}-\overset{-}{O}$ and increase the frequency.

An electronegative group, e.g. halogen in poly(vinyl chloride), gives two effects: a rise in the C═O frequency because of the inductive effect, and a doublet structure which results from the existence of rotational isomers.

For substituents on the α-carbon, the increase in frequency is as follows: 15–20 cm^{-1} per F, 15–20 cm^{-1} per Cl, 6 cm^{-1} per Br and 10 cm^{-1} per O—R.

(iii) *Resonance and/or mesomeric effects* affect the carbonyl frequency by reducing its electron population through interaction with adjacent double bonds, thereby lowering the frequency:

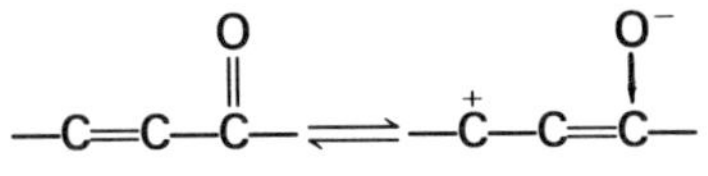

(9.4)

Conjugation with one aromatic ring or olefinic group lowers the C=O frequency by 30–40 cm^{-1}.

(iv) *Hydrogen bonding* to the carbonyl group lowers the carbonyl frequency. Hydrogen bonding decreases the normal frequency 5–10 cm^{-1}; if combined with conjugated double bonds, it can lower the frequency by 30–40 cm^{-1}.

(v) *Intermolecular effects* (*solvent effects*) affect the carbonyl frequency (it is recommended to use an inert solvent to minimize band shifts if necessary).

The overtone of the C=O frequency usually appears at slightly less than twice the fundamental frequency.

9.6.1.2 Aldehyde Groups

Aliphatic aldehyde (carbonyl) groups absorb strongly at 1725–1730 cm^{-1}, whereas aromatic aldehyde (carbonyl) groups absorb at 1690–1710 cm^{-1}. Aldehyde groups absorb at frequencies ~10 cm^{-1} higher than do the corresponding ketone groups, and are influenced in the same way by electronic effects.

Aldehyde groups have a distinctive CH absorption near 2740 and 2830 cm^{-1}; however a single band can appear only at 2830–2870 cm^{-1}.

9.6.1.3 Carboxylic Groups

Carboxylic acid groups have an absorption at 1750 cm^{-1}; however, hydrogen bonding can shift absorption towards 1700–1720 cm^{-1}. The broad absorption band of the OH group with centres in the region 3000–3400 cm^{-1} is also characteristic of the presence of carboxylic groups.

Carboxylic salts have two carbonyl groups, each equivalent to one-and-a-half bonds,

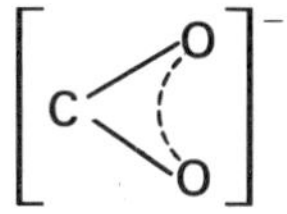

The frequency is therefore lower than that of the corresponding carboxylic acid group. A characteristic strong and broad absorption falls at 1540–1650 cm^{-1} (the asymmetric stretch) or 1360–1450 cm^{-1} (weaker, the symmetric stretch).

9.6.1.4 Mixed Carbonyl Groups

During the photo-oxidation of polymers (e.g. polyethylene) a complex mixture of carbonyl groups is formed (Table 9.1) resulting from initial and subsequent free-radical reactions.[134,250,353,440]

Identification of polymer oxidation products by IR spectroscopy[11,941] can be supported by NMR studies.[281,610] High-resolution NMR spectroscopy is capable of detecting oxidation products down to a 0·03 mol kg^{-1} level, which is easily attained by FTIR (Fourier Transform Infrared) spectroscopy.

It is easy to differentiate between carbonyl groups which belong to ketone and carboxylic groups. When a photo-oxidized polymer film is treated with 0·5–1·0 M-sodium hydroxide solution, a decrease of the absorption at 1715–1720 cm^{-1} is observed whereas a new band appears at 1580 cm^{-1} typical of carboxylate ions. This is evidence for the presence of carboxylic acids and what remains from the initial absorption at 1715 cm^{-1} is the contribution of the ketones.[454]

Table 9.1
Absorption IR bands for carbonyl groups in photo-oxidized polyethylene[440]

Group	*Wavenumber* (cm^{-1})
R—COOH	1 710
R—CO—R′	1 720
R—CO—CH_3	1 725
R—CHO	1 735
R—COO—R	1 750
R—COO—OH/R—COO—OR	1 785

9.6.1.5 Carbonyl Index

The *carbonyl index* (measured at 1710–1720 cm^{-1}) is commonly used for the determination of *embrittlement time* (mainly in polyolefins), i.e. the time (in hours) to reach 0·06 carbonyl units from the plot of absorbance of the carbonyl groups versus irradiation time:[24,25,31,32, 47,68]

$$\text{Carbonyl index} = [(\log I_0/I_t)/d]100 \qquad (9.5)$$

where I_0 = intensity of incident light,
I = intensity of transmitted light,
d = film thickness (μm).

By extrapolation of the steep portion of the curve to the time axis at zero carbonyl, the induction period (mainly in thermal oxidative degradation) is determined.

9.6.1.6 Differentiation of Hydroperoxide Groups by Reactions with Nitric Oxide

Nitric oxide (NO) does not react with most commercially available polymers; however, hydroperoxide (OOH) and hydroxyl (OH) groups react with NO to produce organo-nitrates and nitrites, respectively.[320,1052,1095]

During the exposure of oxidized samples to nitric oxide (NO), the clearly visible absorption band at 3400 cm^{-1} (OOH/OH) disappears. The NO reaction time required for complete reaction is from 16 h for thin films (25–50 μm) and about 30 h for thicker films (up to 150 μm). Photo-oxidized polyolefin films (up to 150 μm) require even 150 h of exposure.

Direct use of nitric oxide (NO) reactions to identify and quantify hydroxy (OH) and/or hydroperoxy (OOH) groups is complicated by the formation of equimolar mixtures of nitrates (with absorption bands at ~1630, 1302, 1290, 1278, 860 cm^{-1}) and nitrites (with strong absorption bands at ~1645, and weak bands at 760–780 cm^{-1}):

$$\text{—}\overset{|}{\underset{|}{C}}\text{—OOH} + NO \longrightarrow \text{—}\overset{|}{\underset{|}{C}}OO\cdot + HNO \qquad (9.6)$$

$$\text{—}\overset{|}{\underset{|}{C}}\text{—OO}\cdot + NO \longrightarrow \text{—}\overset{|}{\underset{|}{C}}OONO \longrightarrow \text{—}\overset{|}{\underset{|}{C}}\text{—}ONO_2 \qquad (9.7)$$

$$\text{—}\overset{|}{\underset{|}{C}}\text{—OH} + 2\,NO \longrightarrow \text{—}\overset{|}{\underset{|}{C}}\text{—ONO} + HNO \qquad (9.8)$$

In Table 9.2 IR data are presented for model nitrates and nitrites.

The use of NO and sulphur tetrafluoride (SF_4) gas treatments allows more precise identification and quantification of hydroxy (OH) and hydroperoxy (OOH) groups (as nitrites and nitrates, respectively, after NO reaction) and carboxylic acids (after SF_4 treatment to produce acid fluorides) and ketones (after removal of the overlapping acid absorptions and hydrogen-bonding effects with —OH groups by SF_4 reactions).[232,538] The NO reaction products are particularly informative because of their intense absorptions (up to 4–7 times stronger than those of the original OH species) and because primary, secondary and tertiary products have differing IR absorptions.[232]

Reactions of nitrogen dioxide NO_2 with oxidized polyolefins give a complex mixture of products, together with a large increase in absorption in the IR carbonyl region. This indicates direct attack on the polyolefins, as well as extensive decomposition of the hydroperoxide groups. Strong absorption at 1700, 1300 and 925 cm^{-1} develops, which can be attributed to the formation of nitric acid.[232]

9.6.1.7 Hydroxy/Hydroperoxy Groups

Identification of hydroxy (OH)/hydroperoxy (OOH) groups is based on their combined absorption in the region 3300–3600 cm^{-1}. Sharp absorption at 3550 cm^{-1} can be attributed to non-hydrogen-bonded OH groups, whereas a broad absorption from 3450 to 3300 cm^{-1} may be due to hydrogen-bonded OH groups. Intramolecular hydrogen bonding occurs at lower frequencies, generally 3200–3050 cm^{-1}.

There is generally good correlation between the iodometric and spectroscopic methods of hydroperoxide determination (Fig. 9.3) so that the two methods can be used interchangeably.[261]

9.6.1.8 Differentiation of Hydroperoxide Groups by Reaction with Sulphur Dioxide

In order to differentiate hydroperoxide (OOH) groups from hydroxyl groups (OH), oxidized polymer samples can be exposed to sulphur dioxide (SO_2), which reacts only with OOH groups, and new absorption bands at 900, 1040 and 1196 cm^{-1} are formed, attributed to the formation of C—O—O—C linkages:[302,387]

$$POOH + SO_2 \longrightarrow POSO_2OH \quad (9.9)$$

$$2\,POSO_2OH \longrightarrow POSO_2OP + H_2SO_4 \quad (9.10)$$

$$POOH + POSO_2OP \longrightarrow POOP + POSO_2OH \quad (9.11)$$

$$POOH + POSO_2OH \longrightarrow POOP + H_2SO_4 \quad (9.12)$$

Table 9.2
IR spectroscopic data for model nitrates and nitrites and for polymers (low-density polyethylene (LDPE) and isotactic polypropylene (iPP)) containing NO group[232]

Group	*Model compound diluted in hexane*	*Absorbance maximum (cm^{-1}) (Extinction coefficient, litre mol^{-1} cm^{-1})*[a]			
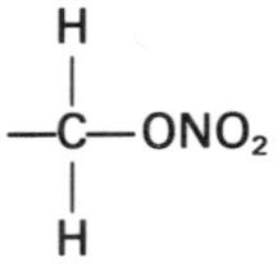	$CH_3(CH_2)_{10}ONO_2$	1 642 [2 200]	1 279 [1 210]	860 [421]	
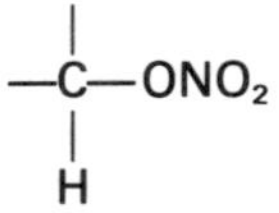	2,6,8-Trimethyl-4-nonyl nitrate	1 633 [1 933]	1 277 [660]	867 [544]	
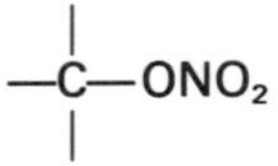	*tert.*-Butyl nitrate	1 630 [422]	1 300 [260]	860 [140]	
	1,1,3,3-Tetramethyl-1-butyl nitrate	1 628 [420]	1 292 [408]	865 [140]	
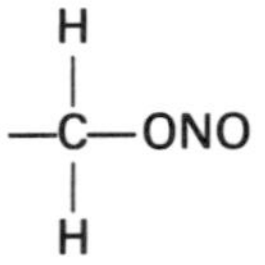	$CH_3(CH_2)_3ONO$	1 657 [470]			780 [245]

$-\overset{\vert}{\underset{\vert}{C}}-ONO_2$	*tert.*-Butyl nitrate	1 630 [422]	1 300 [260]	860 [140]	
	1,1,3,3-Tetramethyl-1-butyl nitrate	1 628 [420]	1 292 [408]	865 [140]	
$-\overset{H}{\underset{H}{C}}-ONO$	$CH_3(CH_2)_3ONO$	1 657 [470]			780 [245]
$-\overset{\vert}{\underset{H}{C}}-ONO$	2,6,8-Trimethyl-4-nonyl nitrite	1 648 [798]			778 [639]
$-\overset{\vert}{\underset{\vert}{C}}-ONO$	*tert.*-Butyl nitrite	1 638 [780]			760 [652]
Oxidized polymer + NO					
Oxidized LDPE		1 631 1 645 (wk)	1 276	870	778 (wk)
Oxidized LDPE after iodometry		1 645 1 657 (wk)			778
Oxidized iPP		1 629	1 302 (sh) 1 290 1 278 (sh)	865	760 (wk) 778 (wk)
Oxidized iPP after iodometry		1 638 1 646 (sh) 1 653			760 778

[a] Abbreviations: wk, weak; sh, shoulder.

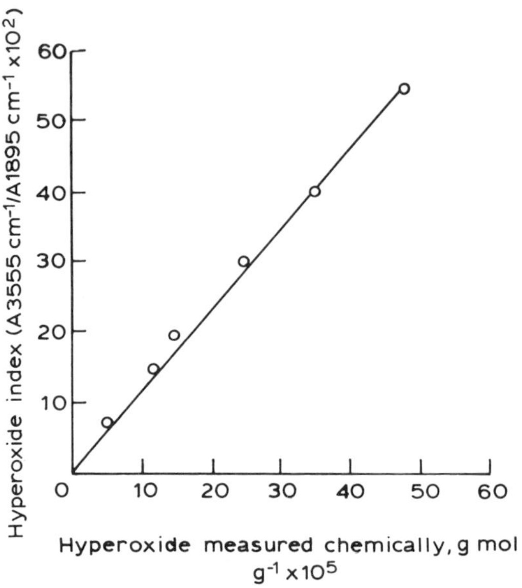

Fig. 9.3. Relationship between the iodometric and IR spectroscopy methods for determination of hydroperoxide groups in low-density polyethylene (LDPE). Reprinted with permission from Ref. 261, © 1977 Pergamon Press PLC.

Simultaneously, the decrease of the absorbance observed in the OH/OOH region at 3440 cm^{-1} is a result of disappearance of OOH groups. The concentration of OOH groups can be derived assuming a molar extinction coefficient at 3440 cm^{-1} of 70–75 mol $litre^{-1}$ cm^{-1}.[386,954]

Exposure of the SO_2-treated polyolefin films to gaseous NH_3 for a few hours leads to a change in the IR spectra. The new band positions and relative intensities at 1420 cm^{-1} and 620 cm^{-1} and the intense absorption at 1105 cm^{-1}, can be attributed to the formation of ammonium sulphate [$(NH_4)_2SO_4$] in the polymer matrix.[233]

9.6.1.9 Tagging Methods

The following tagging reactions can be utilized to distinguish keto (C=O) from aldehyde (CHO) groups and hydroxy (OH) from hydroperoxy (OOH) groups:[211,634]

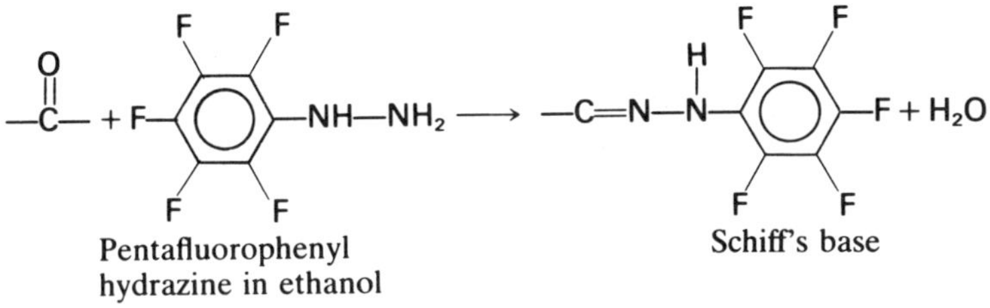

(9.13)

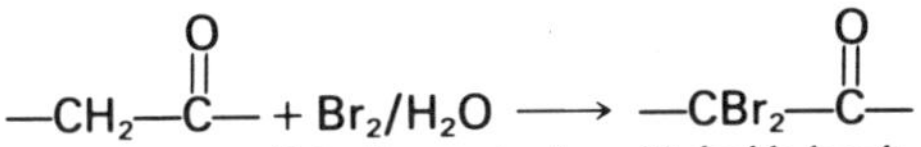

(9.14)

Note: Br_2 reacts also with double bonds.

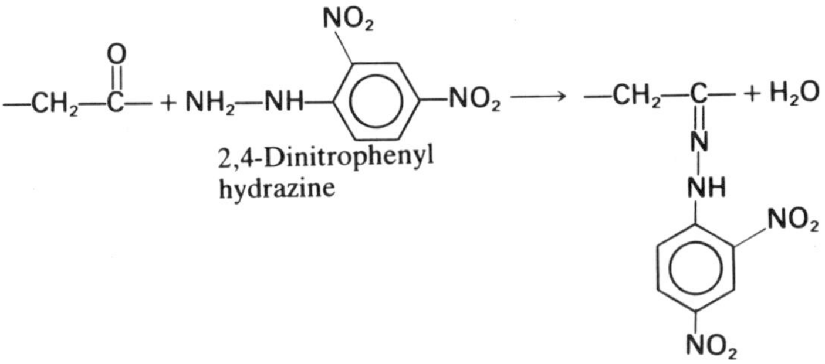

(9.15)

Note: Both ketone and aldehyde groups react.

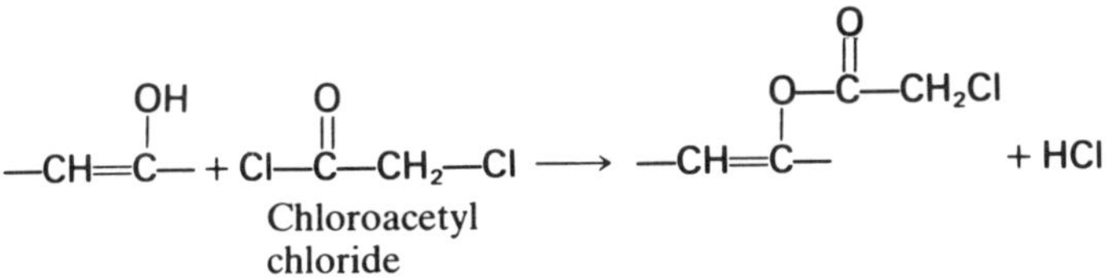

(9.16)

Note: $ClCOCH_2Cl$ reacts predominantly with enolic OH groups.

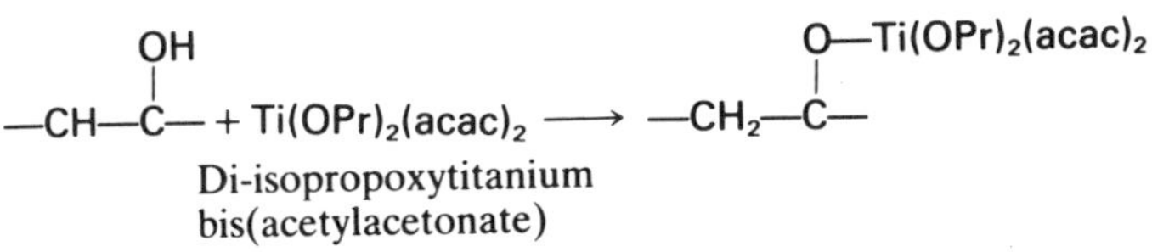

(9.17)

Note: This titanium complex reacts predominantly with alcoholic OH groups.

9.6.1.10 Determination of the Rate of Oxidation by IR Spectroscopy

The rates of photo- (and/or thermal) oxidation of polymer films are usually measured by monitoring the hydroxyl (OH/OOH) group formation (I_{OOH}) at 3400–3450 cm^{-1} and/or carbonyl (CO) group formation (I_{CO}) at 1710–1720 cm^{-1} using the following expressions, respectively:

$$I_{OOH} = \frac{\text{Absorbance at 3400–3450 cm}^{-1}}{\text{Absorbance of reference peak}} \tag{9.18}$$

$$I_{CO} = \frac{\text{Absorbance at 1710–1720 cm}^{-1}}{\text{Absorbance of reference peak}} \tag{9.19}$$

Absorbance of a reference peak compensates for changes in film thickness and should not change during oxidation time.

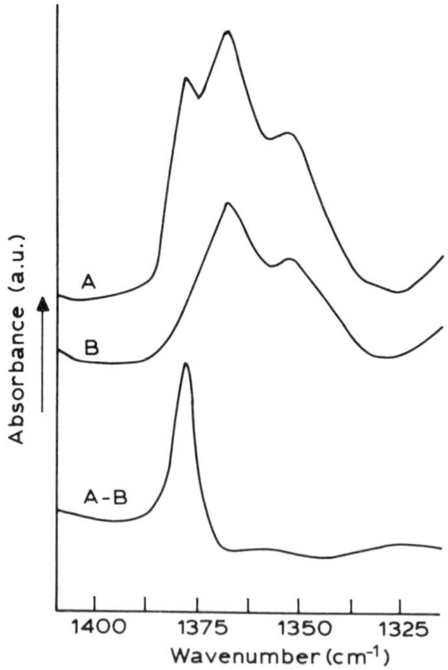

Fig. 9.4. Methylene wagging bands in given samples A and B, and difference A – B.[440]

It is very useful to measure FTIR difference absorption spectra, where the spectrometer subtracts digitally an appropriate reference spectrum from a sample spectrum.[655,656,1063] To obtain a difference spectrum in polyethylene, the methylene wagging band at 1369 cm^{-1} is used as the null wavenumber. An example of such an operation is demonstrated in Fig. 9.4.

9.6.1.11 Unsaturated Groups

During the photo- (and/or thermal) oxidation of polymers in many cases various unsaturated groups, which can be detected by IR spectroscopy (Table 9.3), are formed.[11]

Table 9.3
Absorption IR bands for unsaturated groups in photo-oxidized polyethylene[11]

Group	*Structure*	*Wavenumber* (cm^{-1})
Pendant methylene	$>C{=}CH_2$	885
End-chain vinyl	$—CH{=}CH_2$	910
trans-Vinylene	$—CH{=}CH—$	1 645

9.6.1.12 Sampling by Internal Reflection

Sampling systems based on the *multiple internal reflection* (MIR) technique, also called *attenuated total reflectance* (ATR) spectroscopy,[915] are very useful for the study of polymer surface degradation.

In this technique a beam of radiation is reflected many times, totally within an optical element made from special crystals such as KRS-5 (thallium bromide–iodide; TlBr–TlI). A portion of the energy in the IR beam extends beyond the reflecting surface at each reflection (Fig. 9.5). An absorbing sample, placed in contact with the surface, will absorb energy from the IR beam at a wavelength characteristic of the sample.

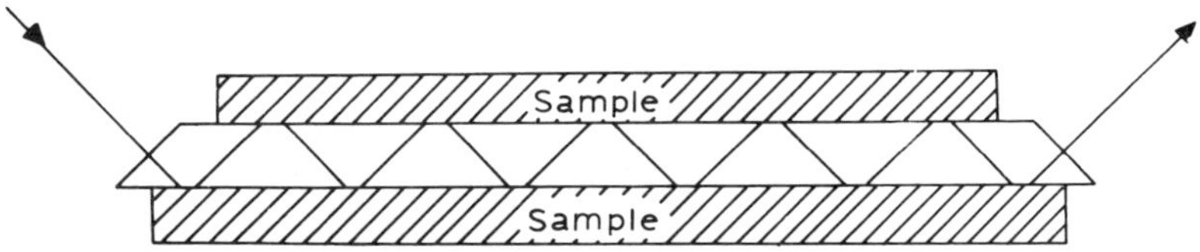

Fig. 9.5. Sample arrangement for obtaining multiple reflection.

The effective pathlength is fixed in an MIR system as long as the reflecting surface is covered completely to a thickness beyond the beam penetration (a few micrometres). Undissolved particles have little or no effect on MIR measurements. The short effective path, however, limits sensitivity for measuring minor components (less than 1% concentration). This disadvantage of the MIR sampling method makes some limitations on the study of kinetics of disappearance of additives of low concentration (<1%) during degradation of polymers. However, application of MIR method in FTIR spectrometers allows for the detection of even very weak signals, and studies on samples at very low concentration (0·2–0·5%).

A schematic representation of an IR–ATR (or IR–MIR) device is shown in Fig. 9.6. The IR radiation is successively reflected by three mirrors, M_1, M_2 and M_3. It passes into a prism (P) and is reflected on the internal face of the latter while undergoing the molecular absorption phenomena at the polymer sample surface (S). Generally, the prism (P) is made from a KRS-5 crystal with refractive index $n_1 = 2{\cdot}37–2{\cdot}38$. The crystal remains in the same position relative to the optical path irrespective of the sample thickness. A mechanism enables the selection of the IR beam angle of incidence θ to the sample surface. The angle θ must be greater than the critical angle given by

$$\theta_c = \arcsin\left(\frac{n_2}{n_1}\right) \tag{9.20}$$

IR radiation can be polarized and analysed by the pairs of polarizers: polarizer and analyser which are in the perpendicular position: the

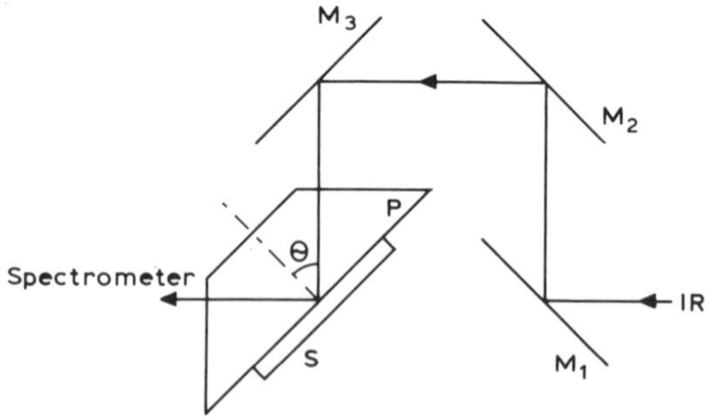

Fig. 9.6. Schematic representation of the ATR device: P, prism; S, sample; M_1, M_2, M_3, mirrors; θ, angle of incidence.

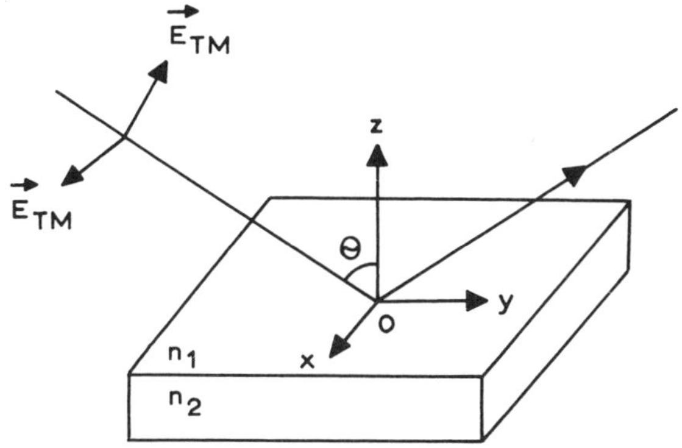

Fig. 9.7. Geometry for attenuated total reflection from polymer surfaces: n_1, n_2, refractive indexes; $\vec{E}_{TE}$, electric vector perpendicular to the plane of incidence; $\vec{E}_{TM}$, electric vector parallel to the plane of incidence; θ, angle of incidence.

electric vector $\vec{E}$ belonging to the incident plane [transverse magnetic (TM) wave] and the electric vector perpendicular to the incident plane [transverse electric (TE) wave] (Fig. 9.7).

The effective penetration depth (d_e) is given by the following equations.[526]

$$\frac{d_e}{\lambda_1}(\mathrm{TE}) = \frac{n^2 \cos\theta}{\pi(n^2-1)(n^2\sin^2\theta - 1)^{1/2}} \tag{9.21}$$

$$\frac{d_e}{\lambda_1}(\mathrm{TM}) = \frac{n^2 \cos\theta(2n^2\sin^2\theta - 1)}{\pi(n^2-1)[(n^2+1)\sin^2\theta - 1](n^2\sin^2\theta - 1)^{1/2}} \tag{9.22}$$

in which

$$\lambda_1 = \frac{\lambda}{n_1} \tag{9.23}$$

$$n = \frac{n_1}{n_2} \tag{9.24}$$

where λ = wavelength of IR radiation,
n = refractive index.

By choosing different angles θ_1 and θ_2, for TE and TM polarizations, it is possible to obtain equal effective penetration depths.[1086] The dichroic ratio is given by:

$$D = \frac{A_{TM}}{A_{TE}} \tag{9.25}$$

where A_{TM} and A_{TE} are absorbances for the TM and TE polarizations, respectively, and

$$A = \log_{10}\frac{I_0}{I} \tag{9.26}$$

where I_0 and I are transmissions at the baseline and the absorption band maximum, respectively.

IR interference ripple observed in IR spectra of polymeric films can be minimized by Harrick's method.[525]

9.7 CHEMILUMINESCENCE MEASUREMENTS

Chemiluminescence is the light emitted during chemical processes, e.g. oxidation. Generally it is a weak light which can be detected by commercial photomultipliers. The amount of light emitted can be increased by adding a *fluorescer* such as dibromo- or diphenylanthracene to the system. The fluorescer acts as an acceptor of radiation energy from the emitting substance (donor) in an energy transfer mechanism and does not have any effect on the reaction mechanism. The efficiency of light emission can be increased by a factor as much as 10^4.

Chemiluminescence measurement is based on the detection of the chemiluminescence light intensity (at constant temperature) versus time (Fig. 9.8). Such a curve provides important information:

(i) The maximum light-intensity emission (I_{max}), at the point at which steady-state oxidation is reached.[767]
(ii) The induction period ($t_{1/2}$), i.e. the time required for the light intensity to reach one-half of the maximum light-intensity emission (I_{max}).
(iii) The total time (t_t) is a direct measure of the oxidation rate of the material at that particular temperature. This value is obtained by the oxygen-drop technique. After a given period of oxidation, the oxygen is turned off.
(iv) The time required to consume all of the oxygen dissolved in the

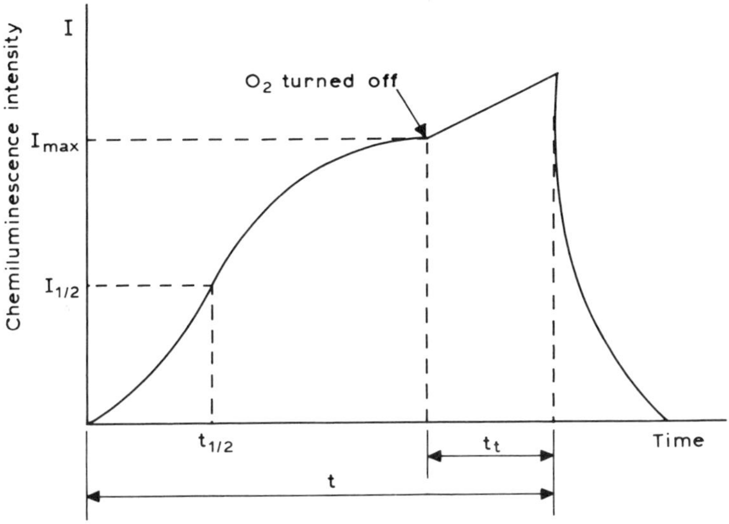

Fig. 9.8. Typical chemiluminescence curve.[767]

sample, during which the chemiluminescence intensity rises, is *t*. Oxidation continues until the oxygen concentration is essentially zero and the peak chemiluminescence intensity is reached. At that point, chemiluminescence drops as oxidation stops.

These data can be used for calculation of the activation energies of the oxidation processes:

(i) An Arrhenius treatment of I_{max} data from intensity–time curves obtained at a series of temperatures gives the activation energy for the propagation and temination of the oxidation processes.
(ii) A similar treatment of a series of induction period ($t_{1/2}$) values gives the activation energy for the hydroperoxide decomposition step, corresponding to the initiation step.

In addition, comparative intensity–time curves for different polymer samples (e.g. in the presence of antioxidants) can yield information on their ability in oxidative processes. A number of publications have reported application of the chemiluminescence method to the study of polymer thermal oxidation.[782,1184,1186]

It has also been reported the apparently pure polymer extracted with an organic solvent mixture exhibits decreased chemiluminescence intensity, which indicates the presence of oxidizable organic impurities that were removed slowly by solvent extraction.

Chemiluminescence measurement techniques can be applied to study:

(i) Effects of structural differences on a polymer's stability.
(ii) Effects of additives on polymer stabilization.[443,1186]
(iii) Processing conditions.
(iv) Environmental conditions and polymer stability.[1073]
(v) Storage of polymers for a long period.[874]
(vi) Polymer oxidation[442,770,976,1186] and photo-oxidation.[260,443,868,871,975]
(vii) Antioxidant activity.[767,976]
(viii) Volatility of stabilizers for foils.[767]

In Fig. 9.9 is shown the course of chemiluminescence with time for unstabilized and stabilized polypropylene.[767] The intensity of chemiluminescence increases with successive consumption of stabilizer and its dependence on time shows a sigmoidal form with a point of inflection at $I = 0{\cdot}5\ I_{max}$, where I_{max} is the maximum intensity for non-stabilized

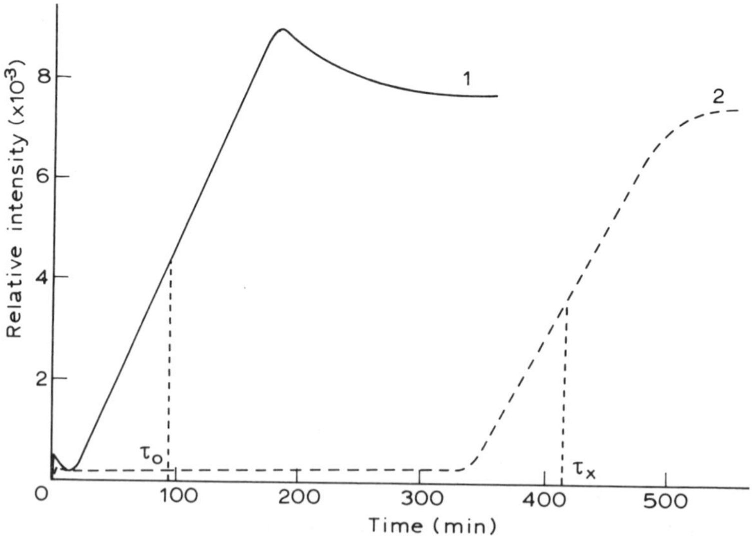

Fig. 9.9. Comparison of the course of chemiluminescence with time for unstabilized polypropylene (curve 1) and polypropylene with 2,6-di-*tert.*-butylphenol (curve 2) at 168°C.[767]

polypropylene. The *antioxidative activity* can be estimated according to the equation:

$$A = \frac{\tau_x - \tau_0}{\tau_s - \tau_0} \tag{9.27}$$

where τ_x = time necessary for attaining the point of inflection on the curves depicting chemiluminescence intensity versus time for a given sample;
τ_s = corresponding time for the stabilized polymer;
τ_0 = corresponding time for non-stabilized polymer.

The presence of antioxidant in a polymer reduces the intensity of the first maximum. The relative depression of the intensity of chemiluminescence can be estimated by the following equation:

$$I_{\text{chemilum}} = \frac{I_0 - I_x}{I_0 - I_s} \tag{9.28}$$

where I_x = intensity at the point of inflection;
I_s = intensity for the stabilized polymer;
I_0 = intensity for the non-stabilized polymer.

A sigmoidal change in the isothermal chemiluminescence intensity with time during oxidation (Fig. 9.9) has been observed for a variety of materials such as polyolefins, rubbers, acrylo–butadiene–styrene, Nylon, acrylics, etc.[1182–1186]

The *induction time,* $\ln([A]_0/[B]_0)$ and *oxidation rate* ($k_{ox}[A]_0$) values can be obtained from the S-shaped autoxidation curve (Fig. 9.10):[1186]

$$\ln\left(\frac{I_t}{I_{max} - I_t}\right) = \ln\frac{[B]_0}{[A]_0} + k_{ox}[A]_0 t \tag{9.29}$$

where I_t and I_{max} are current and maximum chemiluminescence intensities, and $[B]_0$ and $[A]_0$ are the initial concentrations of hydroperoxide and polymer, respectively.

The activation energy of oxidation can be calculated from Fig. 9.11.

There are a number of good descriptions of apparatus for chemiluminescence measurements.[442,781,860,982] For sensitive measurements, a photon counting system should be used in conjunction with a thermoelectrically refrigerated photomultiplier (Fig. 9.12). This provides optimum improvement in the signal-to-noise ratio at low levels and leaves the system gain sensitive to variations in photomultiplier high voltage. The photon counter allows the pulse height spectrum

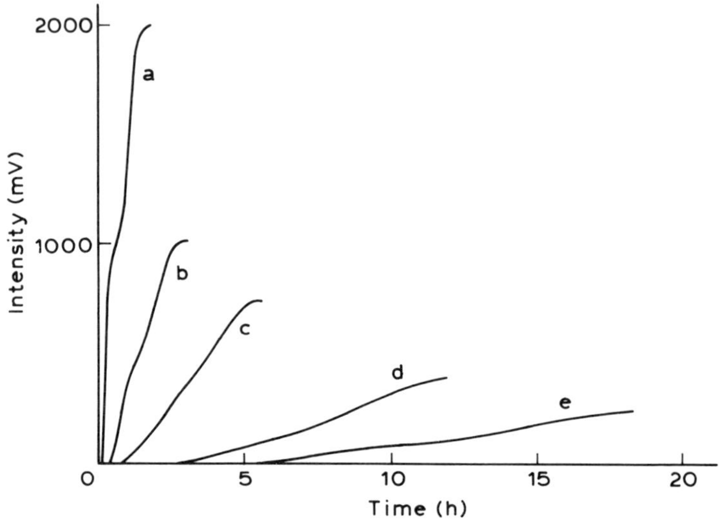

Fig. 9.10. Chemiluminescence curves for low-density polyethylene at different temperatures: (a) 190°C; (b) 180°C; (c) 170°C; (d) 160°C; (e) 150°C.[1186]

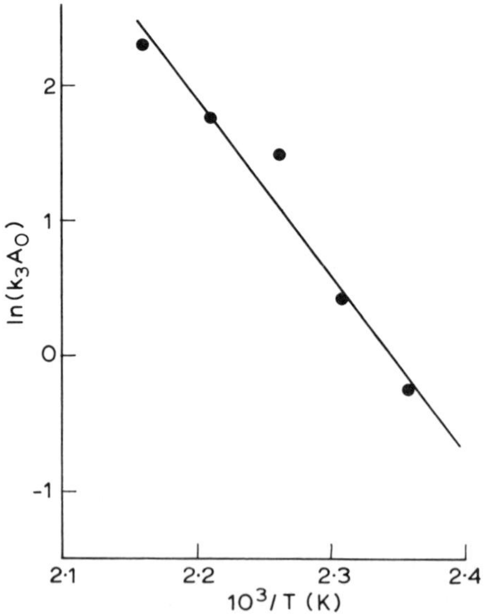

Fig. 9.11. Evaluation of the activation energy for low-density polyethylene (LDPE).[1186]

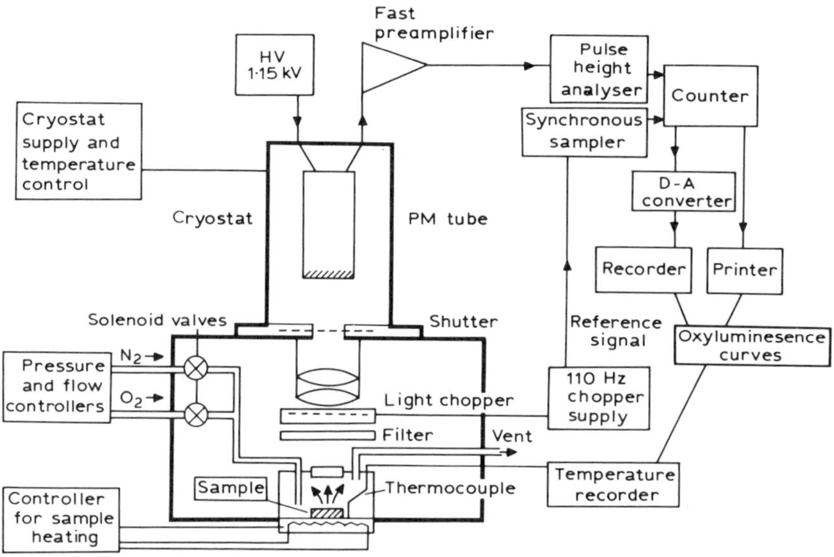

Fig. 9.12. Schematic diagram of the photon-counting oxyluminescence apparatus.[442] HV, High voltage supplier; PM, photomultiplier.

from the photomultiplier to be displayed to set discrimination levels and high voltage consistent with long-term stability. By chopping the emitted light and using synchronous sampling, real-time subtraction of the dark count from the tube can be performed. The number of discriminated pulses counted in each channel in the sampling period (typically 10 to 1000 s) is subtracted and printed out. For high emission intensities, synchronous sampling is not required and the refrigerated photomultiplier has a stable dark count at 0°C that is subtracted from the count rate. For sampling periods of 1 s or less the digital-to-analogue converter provides a display on the chart recorder. Solenoid valves provide the precise gas control required for non-stationary kinetic studies and the gas flow rate and inlet pressure are controlled and monitored. The sample temperature is proportionally controlled to better than ±0·2°C and is continuously monitored.[442]

9.8 PHOTOACOUSTIC SPECTROSCOPY

Photoacoustic spectroscopy is a useful method[963] for the study of polymer degradation, weathering and stabilization.[5,6,135,716] It has a number of advantages over conventional spectroscopic methods, such as:

(i) Sample preparation is unnecessary.
(ii) It is usually faster, more sensitive to sample changes, and yields more reproducible results than most techniques.
(iii) It is non-destructive.
(iv) It provides an optical depth profile of the absorption spectra, a unique analytical procedure.
(v) It can provide accurate information on the thermal properties of the samples.
(vi) It is not sensitive to light scattering or light reflection.
(vii) Completely opaque samples can be investigated without difficulty.

In this method a sample is placed inside a specially designed cell containing a sensitive microphone. The sample is them illuminated with chopped radiation from a strong xenon lamp (1000 W) or laser radiation (Fig. 9.13). Light absorbed by the sample is converted in part into heat by non-radiative deactivation processes within the sample. The resulting periodic heat flow from the sample to the surroundings creates pressure fluctuations in the cell. These pressure fluctuations are then detected by the microphone as a signal which is in the phase coherent with the chopping frequency. The magnitude of the resulting photoacoustic signal is directly related to the amount of light absorbed by the sample.

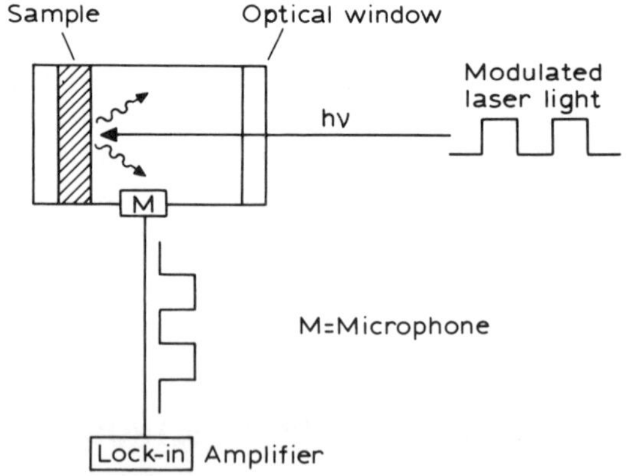

Fig. 9.13. Schematic diagram of photoacoustic set-up. Reprinted with permission from Ref. 716, © 1983 American Chemical Society.

9.9 ELECTRON SPIN RESONANCE SPECTROSCOPY

All free-radical processes involved in photo-oxidation of polymers and photostabilization may be directly investigated by electron spin resonance (ESR) spectroscopy which indicates species with unpaired electrons.[935] However, the concentrations of the unpaired spins are very low, sometimes at the threshold sensitivity level of the ESR spectrometer (10^{11} spins/0·1 ml). Modern computerized ESR spectrometers are designed to increase their sensitivity and speed of signal registration of paramagnetic species at their steady-state concentration by spin accumulation in combination with rapid scan techniques in the millisecond range. Spectral resolution can be increased in some cases, when the ESR spectra can be saturated with microwave power by applying combined electron–nuclear double (ENDOR) or triple (TRIPLENDOR) resonance techniques.

A number of transient free radicals, which are very short-lived, can be detected by coupling ESR spectroscopy with flash photolysis. The advantages of this method can be listed as follows.[192]

(i) The technique responds specifically to the free-radical intermediates in the system.
(ii) Where the free radicals are formed in solution, the ESR spectra usually contain very sharp lines and overlapping of spectra from different free radicals is not usually a problem.
(iii) ESR detection is not usually affected by samples which are optically opaque or highly scattering. Hence sample absorbance can be varied over a wide range to produce the optimum signal-to-noise ratio.
(iv) Each free radical possesses a characteristic *g*-factor and hyperfine pattern; therefore identification of transient species by ESR is usually straightforward.
(v) If the ESR spectra of two or more free radicals do overlap, the resulting spectra can often be resolved by varying instrumental conditions such as the modulation amplitude and the incident microwave power. Since the linewidth and spin relaxation characteristics differ for each free-radical type, overmodulation or microwave power saturation can be effective in isolating the ESR spectrum of one of the free components.
(vi) Absolute concentrations of free-radical intermediates can be determined relatively easily by comparison of the area under

the absorption spectrum with that of a spin concentration standard. Thus it is possible to determine kinetic rate constants which are other than first-order.

Some of the disadvantages of ESR detection are as follows:

(i) Transient intermediates with lifetimes less than about 1 μs may be difficult to observe using ESR spectroscopy. This limitation (in the absence of significant spin polarization) is imposed by the spin–lattice relaxation times characteristic of free radicals in solution ($\sim 10^{-6}$–10^{-7} s). However, since some spin polarization is always present, the relaxation criterion is not the primary limitation on the technique in this case.

(ii) Although presently available ESR spectrometers are capable of detecting about 10^{11} spins per gauss of linewidth, sensitivity considerations are very critical in transient ESR spectroscopy. Since the amplitude of a spin resonance signal is approximately proportional to the inverse square of the peak-to-peak linewidth, transient radicals with wide lines may be difficult to observe with adequate signal-to-noise, even with extensive signal averaging.

(iii) For free radicals with short lifetimes, the bandwidth of the detection system must be sufficiently broad. However, since the sensitivity is inversely proportional to the square root of the bandwidth (in inverse seconds), an increase in the bandwidth by a factor of 100 will decrease the spectrometer sensitivity by a factor of 10. Although some sensitivity can be regained by the use of a computer to average transients; however, some loss of sensitivity for short-lived free radicals still remains.

9.10 ELECTRON SPECTROSCOPY FOR CHEMICAL ANALYSIS (ESCA)

Electron spectroscopy for chemical analysis (ESCA) [also called X-ray photoelectron spectroscopy (XPS)], is an ultra-high vacuum surface analysis technique, which involves the measurement of binding energies of electrons ejected by interactions of a molecule with a monoenergetic beam of soft X-rays.[300,301,303]

In practice it is sensitive to all elements except hydrogen (H) and helium (He) and with a typical information depth of 2–5 nm,

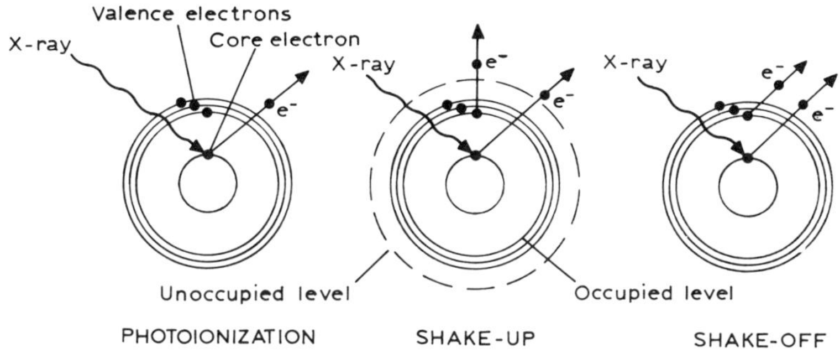

Fig. 9.14. Interaction of soft X-rays with electron in a molecule.

depending upon the density and the chemical nature of the elements in the surfaces. When a soft X-ray beam interacts with core electrons the following processes occur (Fig. 9.14).

(i) *Photoionization,* which occurs with the removal of a core electron (emission of a photoelectron);
(ii) *Shake-up,* which is excitation of a valence electron from an occupied to an unoccupied level simultaneously with (i);
(iii) *Shake-off,* which is ionization of a valence electron accompanying (i).

The last two processes (ii) and (iii) give rise to satellite peaks on the low kinetic energy side of the main photoionization peak (Fig. 9.15).

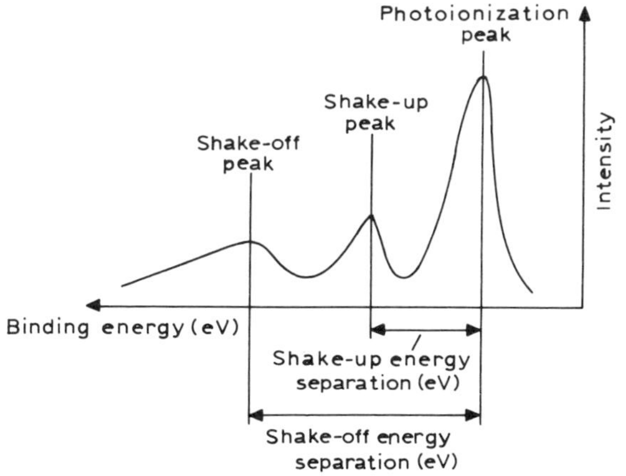

Fig. 9.15. Electron core-level spectra for a given sample.

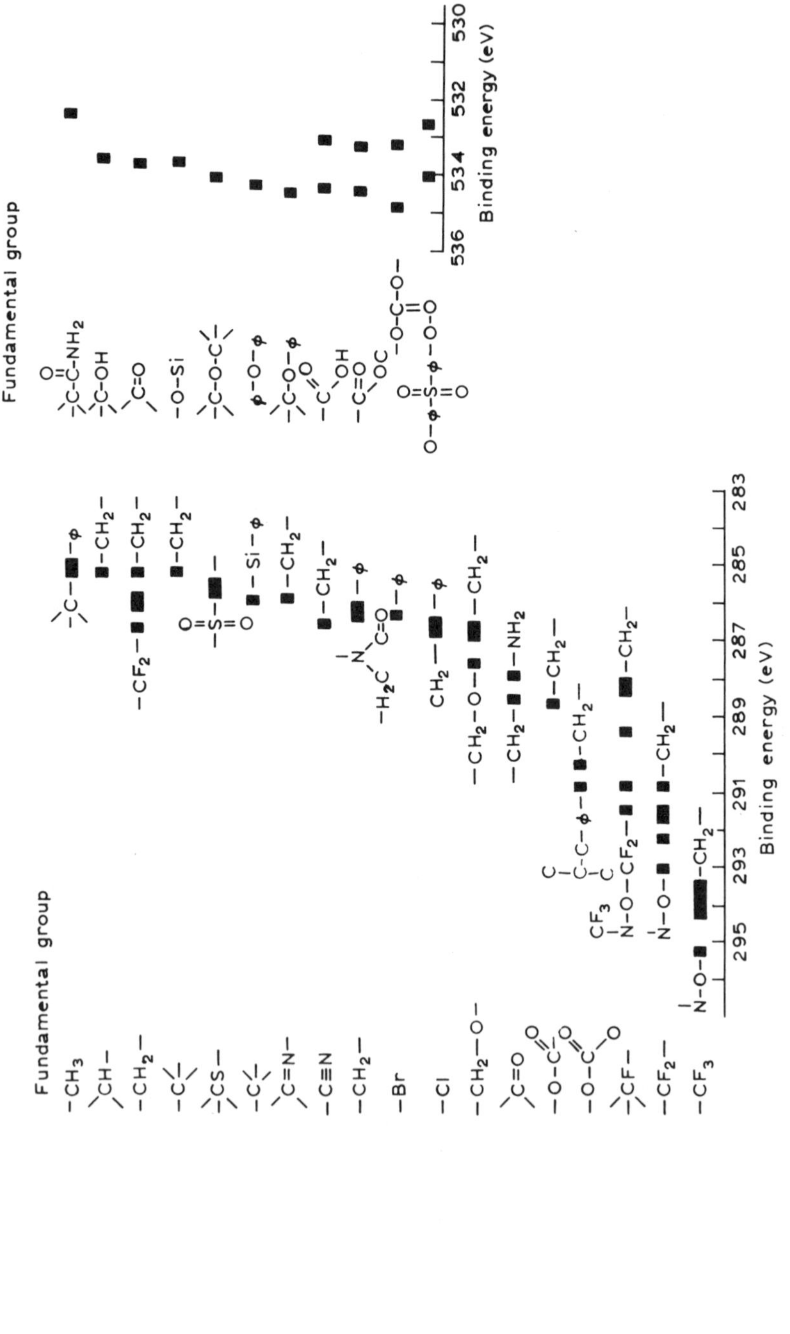

Fig. 9.16. Tabulation of experimental chemical shifts for C_{1s} and O_{1s} levels.[311]

Aromatic molecules, in particular, show large probabilities for $\pi \rightarrow \pi^*$ shake-up satellites.

ESCA analysis implies measurement of the binding energies (E_B) of the core electrons ejected by bombardment of the surface with a monoenergetic beam of X-rays. E_B is given by the following equation:

$$E_B = h\nu - E_i - \phi - C \qquad \text{(eV)} \tag{9.30}$$

where $h\nu$ = energy of the X-ray photon;
E_i = kinetic energy of the ejected electron;
ϕ = spectrometer work function;
C = the energy (positive or negative) associated with the charging up.

A systematic study of a large number of homopolymers provides a compilation of substituent effects on the C_{1s}, N_{1s}, O_{1s}, F_{1s}, Si_{2p}, P_{2p}, S_{2p} and Cl_{2p} levels.[311] In Fig. 9.16 some of the data pertaining to substituent effects on C_{1s} levels in polymers are shown.

The O_{1s} region of the spectrum is composed of two peaks of equal intensities at 532·7 and 534·3 eV. On photo-oxidation, the intensities of both peaks increase, that at lower binding energy increasing more rapidly. The lower binding energy component can originate in doubly bonded oxygen and in singly bonded oxygen in alcohol, ether and peroxide groups. The greater intensity of this peak reflects the increased concentration of both the surface carbonyl groups giving rise to the C_{1s} peak at 287·7 eV and hydroperoxide, hydroxide, etc. groups contributing to the component at 287·5 eV. The increase in carboxyl groups evidenced in the C_{1s} peak at 289·1 eV gives rise to an equal increase in intensity for both O_{1s} components.[889] This selected example shows the large number of information levels available from a single experiment.

ESCA spectroscopy has also been used for the study of polymers oxidized by singlet oxygen ($^1O_2$).[350] In Fig. 9.17 O_{1s} and C_{1s} core level spectra of polyethylene and polystyrene exposed to $^1O_2$ generated in the microwave unit are shown.

ESCA spectroscopy can be extended by the application of chemical tagging, i.e. use of organic derivatizing agents, which[384]

(i) specifically react with one unique functional group on the surface (Table 9.4), and
(ii) contain atom(s) that are easily detected by its (their) ESCA signals.

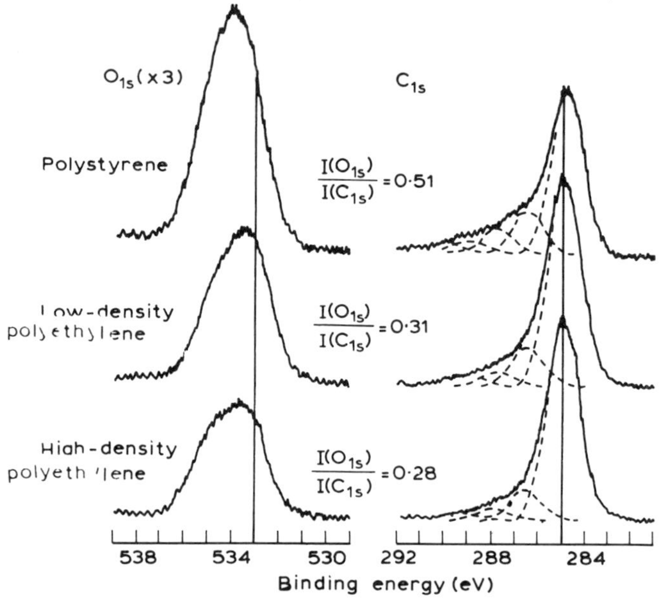

Fig. 9.17. The O_{1s} and C_{1s} core-level ESCA spectra of high-density polyethylene (HDPE), low-density polyethylene (LDPE) and polystyrene after 1 h exposure to the effluent of the singlet oxygen ($^1O_2$).[350]

This method is especially useful for the study of photo-oxidized surfaces where different oxy groups, e.g. C—O and C—O—O, have similar shifts or may overlap due to secondary substituent effects.

Typical ESCA spectra obtained during photo-oxidation, e.g. poly(methyl methacrylate), are shown in Fig. 9.18. From these spectra it is possible to evaluate the rate of polymer oxidation, by the comparison of the intensity of C_{1s} and O_{1s} peaks (Fig. 9.19).[889]

The C_{1s} peak at 285·0 eV arises from carbon with no directly bonded oxygen, that at ~286·5 eV from carbon singly bonded to one oxygen, while those carbons singly bonded to two oxygens or doubly bonded to one oxygen appear at 287·7 eV. The component at 298·0 eV is due to carbon with three bonds to oxygen, e.g.

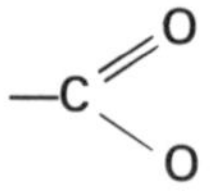

The precise depth to which the surface reactions penetrate is not

Table 9.4
Examples of possible tagging reactions for complementary functional group analysis in ESCA spectroscopy

Functional group	*Reagent*	*Product*	*Species analysed*
$>C{=}C<$	Br_2/CCl_4	$>C(Br){-}C(Br)<$	Br_{3p}
$-CH_2-OH$	$CF_3(CF_2)_2COCl$	$-CH_2-O-CO-(CF_2)_2CF_3$	F_{1s}
	CBr_3COOH } $C_6H_{11}NCNC_6H_{11}$ }	$-CH_2OCOBr_3$	$Br_{3p^{3/2}}$
$>C{=}C-OH$	$ClCH_2COCl$	$>C{=}C-OCOCH_2Cl$	Cl_{2p}
$-CH_2-C\leqslant$	Br_2	$-CBr_2-C\leqslant$	Br_{3p}
$>C{=}O$	$C_6F_5NH-NH_2$	$>C{=}N-NHC_6F_5$	F_{1s}
$-COOH$	$TiOC_2H_5$	$-COO^-Tl^+$	Tl_{4s}
	$BaCl_2$	$-(COO^-)_2Ba^{2+}$	Ba_{3s}
$-C-OOH$	SO_2	$-C-OSO_2OH$	S_{2p}
$-\overset{O}{C-C}-$	HCl	$>C(OH){-}C(Cl)<$	Cl_{2p}
$-NH_2$	C_6F_5CHO	$-N{=}CHC_6F_5$	F_{1s}
	$C_2H_5S-COCF_3$	$-NH-COCF_3$	F_{1s}
	CS_2	$-NHCSSH$	S_{2p}
$-SH$	$AgNO_3$	$-SAg$	Ag_{3d}, S_{2p}
	$[C_6H_3(NO_2)_2(COONa)-S]_2$	$-S-S-C_6H_3(NO_2)_2(COONa)$	S_{2p}, Na_{1s}

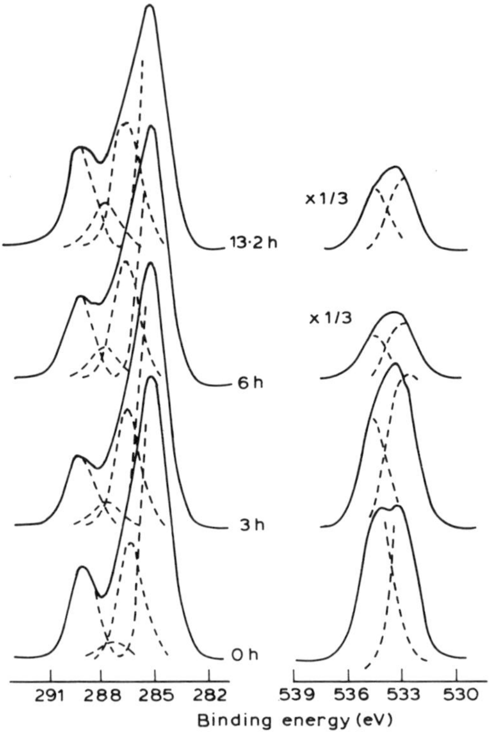

Fig. 9.18. The C_{1s} and O_{1s} regions of the ESCA spectra of poly(methyl methacrylate) before and after photo-oxidation at various times.[889]

clear at present. They may be two types of 'surface' reaction, one associated with the ESCA sampling depth (20–50 Å) and the other with reflection IR (ATR) (1–5 μm). One of the main features from ESCA studies of surface photo-oxidation of aromatic polymers at wavelengths <290 nm is the apparent oxidation of the aromatic rings, resulting in loss of 'aromaticity'.[307,813,888] Evidence for this arises from the loss of intensity in the $\pi \rightarrow \pi^*$ shake-up satellites in the corresponding C_{1s} spectra, the extent of oxygen uptake and the nature of the photo-oxidation of polymers containing no aliphatic carbons. IR (ATR) spectroscopy does not indicate similar phenomena during irradiation at wavelengths >290 nm. This suggests that the nature of the reactions observed by ESCA spectroscopy is very surface-specific and that, on going further into the material, they change to those

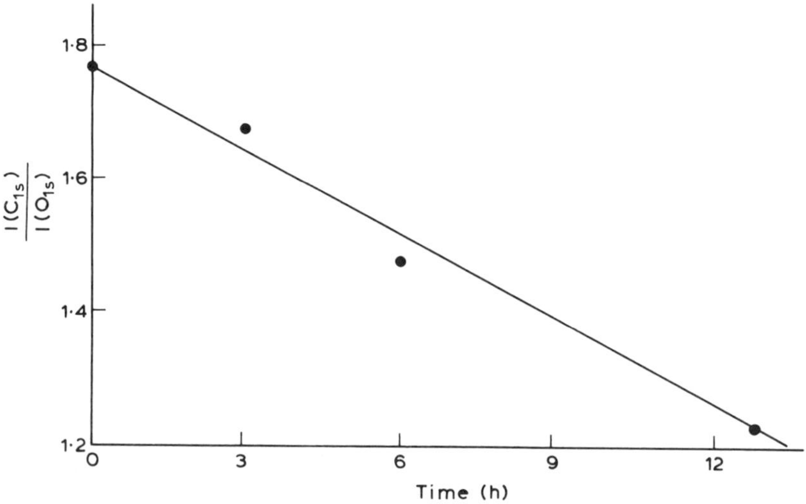

Fig. 9.19. The ratio of the intensities of the C_{1s} and O_{1s} bands for poly(methyl methacrylate) plotted as a function of oxidation time.[889]

normally associated with 'bulk' chemistry. The depth at which this changeover occurs, i.e. from predominantly surface to predominantly 'bulk' reaction, has yet to be determined.[813]

ESCA spectroscopy has a number of advantages as a spectroscopic method in the study of polymers:

(i) It is a non-destructive technique.
(ii) Study is possible *in situ* in the working environment, with a minimum of preparation.
(iii) A large amount of information is available from a single measurement.
(iv) The method provides the capability to study surface, subsurface and bulk depth profiling.
(v) The data are often complementary to those obtained by other spectroscopic methods.

ESCA spectroscopy has been widely used for the study of surface photo-oxidation of polyolefins,[350,889] polymethacrylates,[889] polystyrene,[305,307,308,350,818,888] polyetherketones,[819] polycarbonates,[304–306,812] polysulphones,[816,817] cured polyesters,[613] cellulose nitrate,[310] and for the study of surface photo-oxidation in the presence of stabilizers or grafted stabilizers.[194,813–815,1072] An interesting ESCA paper has been devoted to the study of pigments.[1113]

9.11 ANALYTICAL METHODS FOR THE DETERMINATION OF HYDROPEROXIDE GROUPS

Hydroperoxide (OOH) group concentration can be determined by a number of analytical methods, such as:

(i) the iodometric method,
(ii) the thiocyanate method, and
(iii) the triphenylphosphine method.

The advantages and limits of these methods in solid polymers have been reviewed.[431] The iodometric method can be used only when the hydroperoxides are thermally stable up to 80°C. On the other hand, the thiocyanate method is particularly well adapted to unstable hydroperoxides; however, this method requires the complete solubilization of the polymer. If a polymer sample is not soluble it is recommended to swell it in a given solvent for a minimum of one day. In Table 9.5 a comparison is given of different methods of titration with the determination of hydroperoxide groups by a spectroscopic method in which SO_2 was used (cf. Section 9.6.1.8). The sulphur

Table 9.5
Comparison of titration methods (I_3^- and Fe^{2+}) with the IR spectroscopic method (SO_2) for the determination of OOH groups[431]

Polymer analysed	*ROOH* (*mol kg*$^{-1}$)	*Method*			*Agreement of the methods*
		I_3^-	Fe^{2+}	SO_2	
Low-density polyethylene (LDPE)	0·01	+	+	+	+
Polypropylene	0·1	+	−	+	+
EPDM[a]	0·3	−	−	+	
Nylon 6	0·01	−	+	+	
Nylon 11	0·02	−	+	+	+
Nylon 12	0·04	−	+	+	+
Poly(butylene terephthalate)	0·1	+	+	+	+
Thermoplastic polyurethane elastomers	0·1	+	−	+	+
Copoly(ether-esters)s	0·4	+	−	+	+

[a] Ethylene/propylene/1,4-hexadicne copolymer.

dioxide (SO_2) method can be carried out easily, but with some polymers secondary effects are observed which perturb the measurements.

9.11.1 Iodometric Method

Hydroperoxide (OOH) groups react with sodium iodide (NaI) by the following reaction.[431,471,753]

$$\text{R—OOH} + 3\text{I}^- + 2\text{H}^+ \longrightarrow \text{I}_3^- + \text{H}_2\text{O} + \text{R—OH} \qquad (9.31)$$

The iodine, generated as I_3^-, is measured spectrophotometrically at a wavelength of 362 nm (the molar absorptivity, i.e. molar extinction coefficient of I_3^- at 362 nm is 25 280 $\text{M}^{-1}\,\text{cm}^{-1}$). The concentration of hydroperoxide can be read off a calibration curve set up using cumene hydroperoxide as a standard. In the case when there is interference at 362 nm, e.g. from additives, it is recommended that absorption be measured at a longer wavelength, i.e. 425 nm

The concentration of iodine can also be determined by titration with standard 0·01 N-sodium thiosulphite.

In the iodometric method of OOH concentration determination, a known weight of polymer sample (1–2 g) should be cut into the smallest pieces possible and refluxed for 30 min in a mixture of 2-propanol (8 ml), glacial acetic acid (2 ml) and sodium iodide (0·5 g). Sometimes it is recommended that the cut pieces of polymer be allowed to swell 18–24 h in a given solvent, e.g. chloroform, before adding glacial acetic acid and sodium iodide. Instead of the 0·5 g solid sodium iodide, 1–2 ml of a freshly prepared deaerated 5% solution of sodium iodide in methanol is recommended to be used. A probe prepared in this way should be stored 4–6 h in the dark, for complete reaction under a blanket of nitrogen.[247,261,439]

The application of this method for the determination of OOH concentration in the presence of hindered amine light stabilizers and nitroxy radicals is limited because both react with I_3^-.[396,1111,1167]

The iodometric method determines the total concentration of hydroperoxides, peroxides and peracids. In order to determine peracids in the presence of hydroperoxides and peroxides, an equal volume of 7-tetradecene solution (2×10^{-2} M) is added to a given volume used for analysis. Peracids (R—CO—OOH) react with olefins forming a corresponding epoxide, according to the reaction:

$$\text{R—CO—OOH} + {>}\text{C}{=}\text{C}{<} \longrightarrow \text{R—CO—OH} + {>}\overset{\text{O}}{\text{C—C}}{<} \qquad (9.32)$$

After 16 h standing in the dark, the remaining peroxides and hydroperoxides can be analysed by the iodometric method and determined from the equation:[396]

$$C_{peracids} = C_{total} - C_{remaining} \tag{9.33}$$

9.11.2 Thiocyanate Method

Hydroperoxide (OOH) groups can be determined colorimetrically with thiocyanate [NH_4SCN—$FeSO_4(NH_4)_2SO_4 \cdot 6H_2O$] reagent:[431,758,891,892,1067,1178]

$$R{-}OOH + 2\,Fe^{2+} + 2H^+ \longrightarrow 2\,Fe^{3+} + H_2O + R{-}OH \tag{9.34}$$

The application of colorimetric methods for the analysis of hydroperoxides in polymers requires that the sample has to be dissolved, or at least swollen, in a solvent (the reactants have to diffuse into the amorphous regions in the polymer matrix). The polymer sample should be in contact with reagent for a minimum of 30 min (in the dark). As a solvent for polyolefins or poly(vinyl chloride), a mixture of benzene–methanol (95 : 4 or 92 : 8) can be used.[892,1178] The colorimetric measurement is carried out at $\lambda = 512{\cdot}5$ nm (λ_{max} of the ferric thiocyanate complex). The concentration of hydroperoxide in the polymer sample is calculated according to the expression:

$$C_{ROOH} = \frac{2{\cdot}5 \times 10^7 A}{e \varepsilon w} \quad (\text{mmol kg}^{-1}) \tag{9.35}$$

where A = absorbance of a sample in a 1 cm cell as measured against the blank solution (thiocyanate reagent in pure benzene),
e = stoichiometric coefficient of hydroperoxide interaction with the reagent ($e \sim 2$),
w = weight of polymer sample,
ε = molar absorptivity (molar extinction coefficient) $= 1{\cdot}68 \times 10^4$ litre mol^{-1} at 509 nm.

This method can be applied for the determination of the OOH concentration in the presence of phenolic and sulphur antioxidants, and photostabilizers such as benzotriazoles. However, its application for the determination of hydroperoxide (OOH) groups in the presence of hindered amine light stabilizers (HALS) is limited, because instead of a purple thiocyanate complex a yellow coloration is formed probably due to the formation of a coloured complex between the HALS and transition metal ions (Fe^{2+}) (cf. Section 6.2).[204]

9.11.3 Triphenylphosphine Method

Hydroperoxide (OOH) groups react with triphenylphosphine (**9.1**) by the following reaction:[158,358]

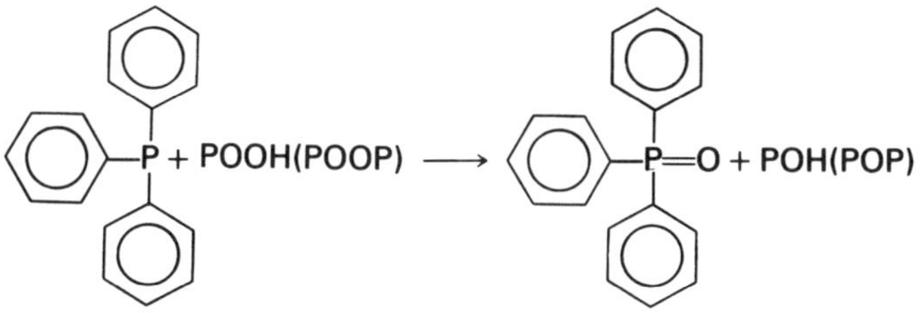

(9.36)

(9.1) **(9.2)**

The triphenylphosphine reagent consists of 1 mg of triphenylphosphine and 0·1 mg of antioxidant (di-*tert.*-butyl-*p*-cresol) in 1 ml of isopropyl alcohol. This is conveniently prepared by weighing 25 mg of triphenylphosphine into a 25 ml volumetric flask. The antioxidant, 2·5 ml of an isopropyl alcohol solution, containing 1 mg ml^{-1} of di-*tert.*-butyl-*p*-cresol, is added and the solution made up to volume with isopropyl alcohol. A peroxide sample containing about 2×10^{-3} meq of peroxide is weighed into a 10 ml flask and dissolved in 1 ml of isopropyl alcohol. After addition of 1 ml of triphenylphosphine reagent, the flask is stoppered, swirled to mix, and inverted once to rinse the neck. A secondary dilution of 1 ml to 10 ml is made for spectrophotometric reading. The absorbance of the reaction mixture is measured at 260 nm against isopropyl alcohol. A sample blank composed of a peroxide sample in isopropyl alcohol is measured at 260 nm to correct for the initial absorbance of the sample. A reagent blank consisting of 1 ml of triphenylphosphine reagent and 1 ml of isopropyl alcohol and subsequently treated as a sample is prepared with each series of samples. The decrease in absorbance of the reaction mixture at 260 nm is used as a measure of the hydroperoxide originally present. Since oxidation of triphenylphosphine (**9.1**) yields triphenylphosphine oxide (**9.2**) (reaction 9.36), which also has a significant absorbance at this wavelength, allowance must be made for the contribution of triphenylphosphine oxide to the absorbance of the total reaction mixture.

$$\text{Peroxide value (meq kg}^{-1}) = \frac{A_r + A_s - A}{\text{sample weight}} \times 21{\cdot}1 \qquad (9.37)$$

where A_r = absorbance of reagent blank,
A_s = absorbance of the sample blank calculated for the weight of sample in the reaction mixture,
A = absorbance of reaction mixture.

The constant, 21·1, is a quotient obtained by dividing 2×10^5 (to convert to meq kg^{-1} for the dilution specified) by the difference between the molar absorptivity of triphenylphosphine ($1{\cdot}10 \times 10^4$) and triphenylphosphine oxide ($1{\cdot}52 \times 10^3$) at 260 nm.

This method is sensitive to the presence of air, which slowly oxidizes phosphine solutions. On the other hand, this method requires a comparable concentration of phosphine with respect to the hydroperoxides (or peroxides) determined because the accuracy of determination decreases when a large excess of phosphine is used. The determination of hydroperoxide content by this method (coupled with quantitative analysis of the resulting alcohols by gas–liquid chromatography) is not affected by the presence of phenolic antioxidants, hindered piperidines (HALS) and piperidinoxy radicals.[1037]

9.12 STUDIES OF THE KINETICS OF OXIDATION BY OXYGEN UPTAKE MEASUREMENTS

Oxygen uptake measurement is a very common method used for studying the kinetics of thermal (or photo-) oxidation and the activity of antioxidants, photostabilizers and other additives.[117,181,205,349,486,487,915]

For example, the antioxidant activity can be determined by direct measurement of oxygen absorption in a forced-draught oven (Fig. 9.20). The test involves exposing polymer samples (e.g. polyethylene films) at high temperatures (140 ± 0·5°C) in an atmosphere of pure oxygen. After substantial oxygen uptake (75–100 ml), the amount consumed is converted to ml/g of sample exposed and plotted against time. From the resulting curve, the time required for the absorption of 10 ml of oxygen/g of polymer is determined. By extrapolating the steep portion of the curve (indicative of rapid autocatalytic oxidation) to the time axis at zero oxygen absorption, the induction period is obtained.

The *induction period* is defined as the time after which the polymer begins to reach the end of the induction period. Therefore, the longer the time required to reach the induction period, the more effective the antioxidant.

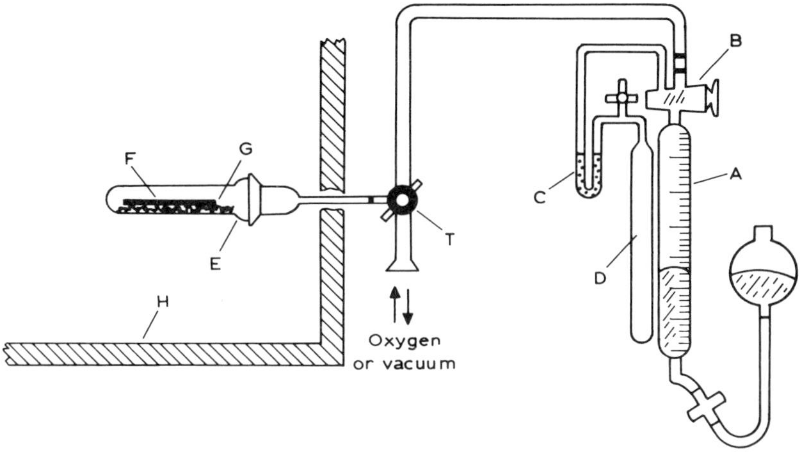

Fig. 9.20. Diagram of oxygen uptake apparatus: A, gas volumetric burette, B, two-way stopcock; C, manometer; D, compensating tube; E, 25/30 ball and socket; F, aluminium sample dish; G, molecular sieve; H, circulating air oven; T, stopcock.[533]

The photo-oxidation kinetics can be followed by the oxygen uptake in a device shown in Fig. 9.21.[489] It contains a specially constructed oxygen pressure monitor with a pressure transducer (Vallidyne Type P10) mounted in a metal cell covered with a quartz window. Polymer samples of 2–3 cm diameter have first to be evacuated at 10^{-5} torr, and

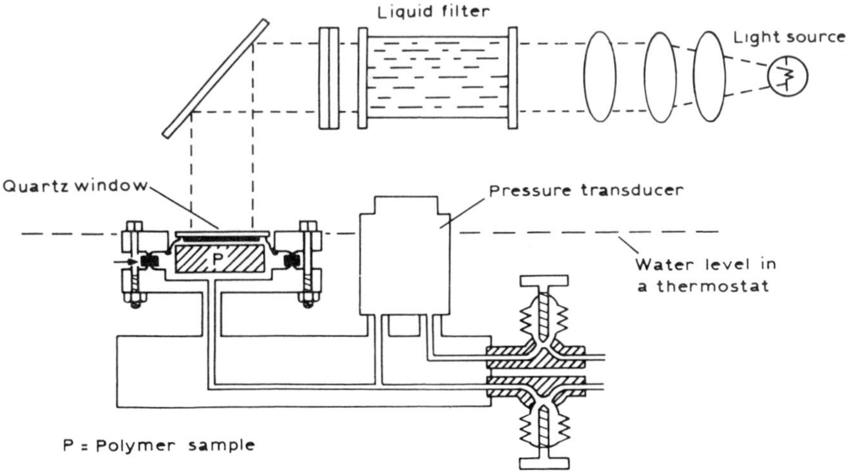

Fig. 9.21. Oxygen uptake device used in studies of photo-oxidation kinetics.[489]

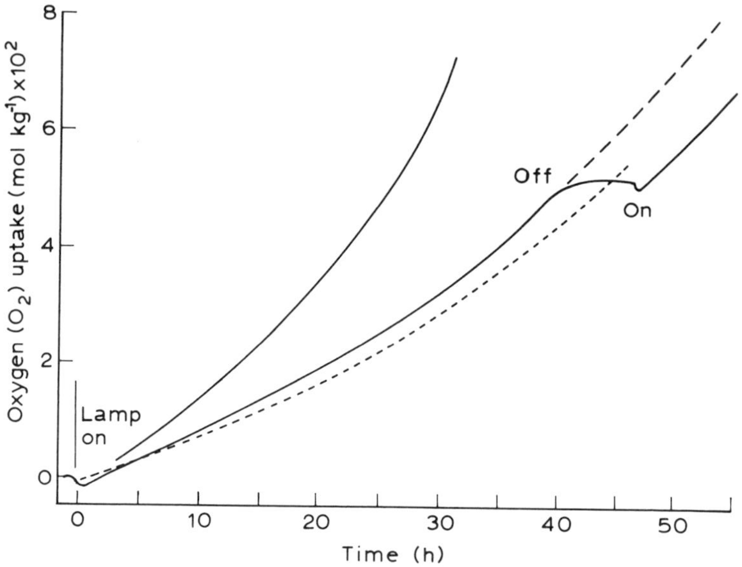

Fig. 9.22. Typical oxygen uptake curves for polypropylene films.[489]

the cell is further filled with oxygen at 1 atm. The oxygen uptake is monitored with a pressure transducer and *XY* recorder.

Figure 9.22 shows examples of oxygen uptake curves for polypropylene, obtained with the oxygen pressure monitor represented in Fig. 9.21.

More sophisticated equipment for the measurement of oxygen uptake has been described recently.[175] This photo-oxidation apparatus (Fig. 9.23) consists of two cells of equal volume which are connected to the opposite sides of a differential pressure transducer (DPT). The sample cell is a circular depression in a stainless-steel block. An optical quartz window is sealed to the block by a butyl rubber O-ring and a clamping mechanism. A shutter prevents light from reaching the sample while thermal equilibrium is being attained. Gas inlet ports are drilled through the block to connect the sample cell, reference cell, differential pressure transducer, and external plumbing. The apparatus is immersed in a water bath which thermostatically controls the temperature of photo-oxidation from ambient (25°C) to 70°C. A comprehensive vacuum line system enables the introduction of pure gases or gas mixtures into the sample at known pressures and compositions.

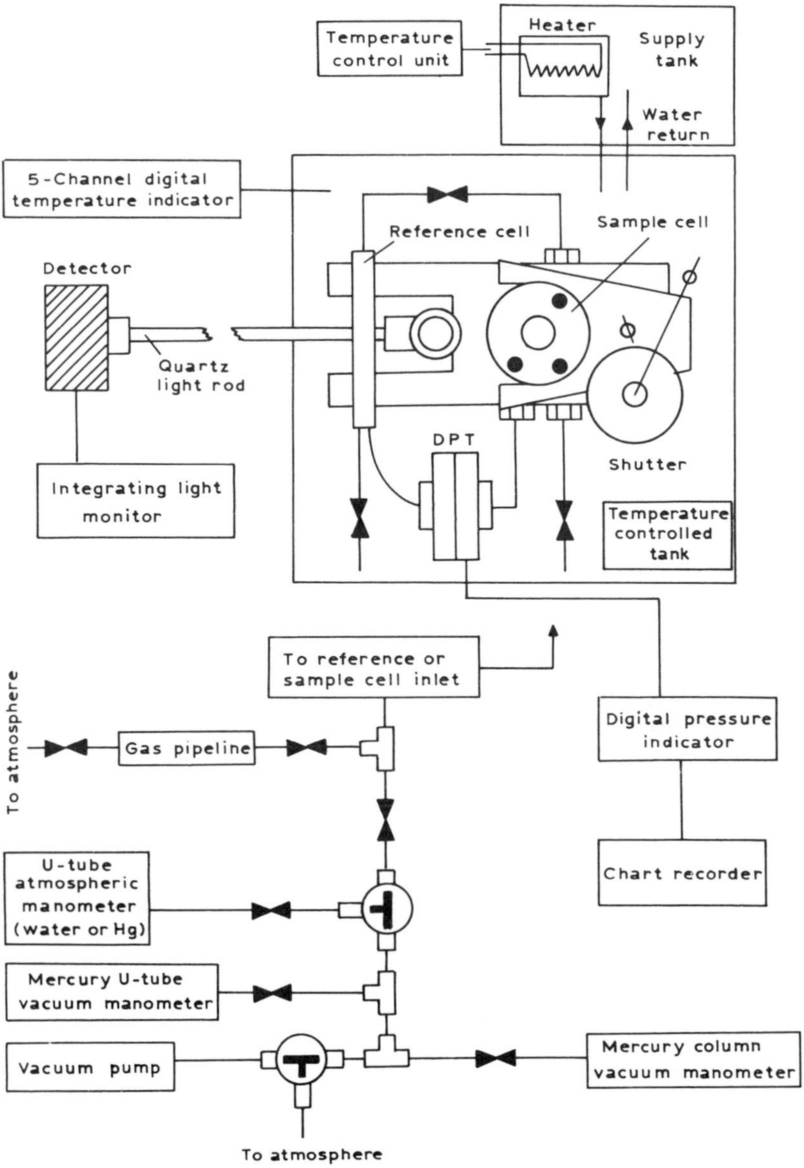

Fig. 9.23. Schematic representation of the photo-oxidation apparatus and vacuum line system.[175]

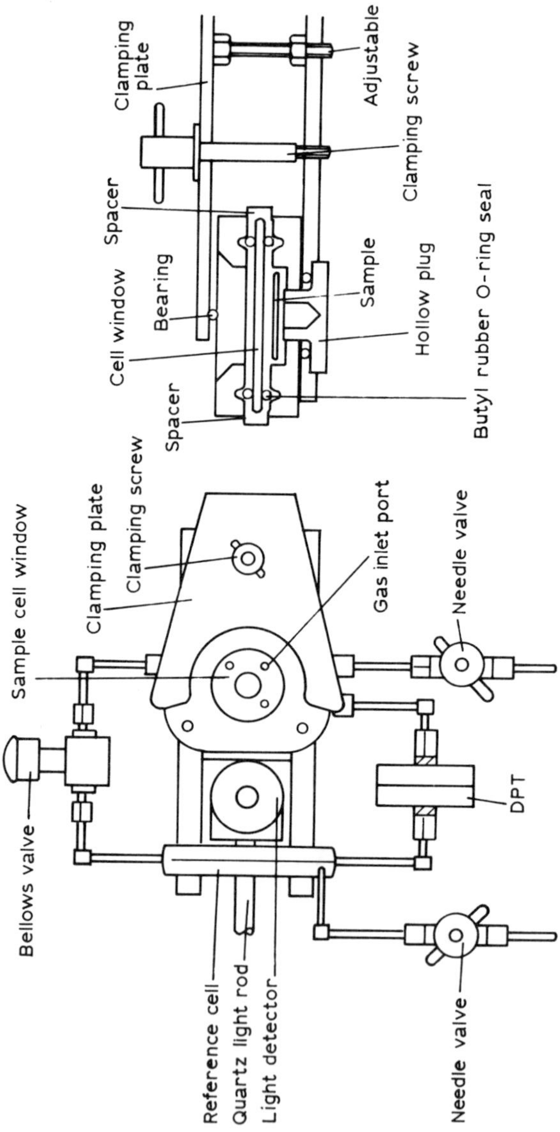

Fig. 9.24. Plan and side elevation of the photo-oxidation cell.[175]

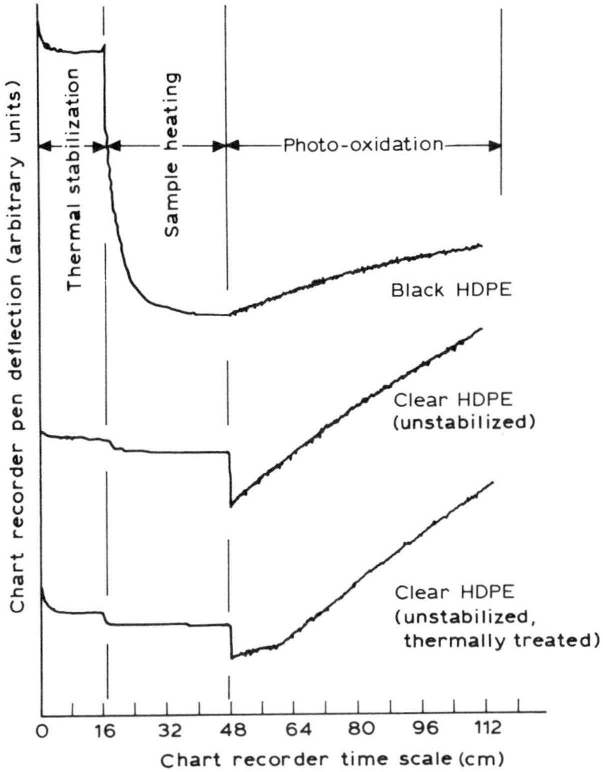

Fig. 9.25. Recorder traces of the photo-oxidation of high-density polyethylene (HDPE) films.[175]

The details of the construction of the photo-oxidation cell are shown in Fig. 9.24. Typical photo-oxidation profiles obtained using this apparatus are shown in Fig. 9.25. The first portion of each curve shows the differential pressure versus time during the thermal equilibration of the apparatus. A zero pressure change observed for an extended period of time indicates that the apparatus is adequately sealed and that thermal equilibrium has been achieved. The pressure in the sample cell decreases as photo-oxidation proceeds.

The main problem in all types of this apparatus is that gases formed during photo-oxidation, eg. CO, CO_2, CH_4, water vapour, etc., change the pressure inside the measuring cell, and measurement can be inaccurate. On the other hand, photo-oxidation cells are very

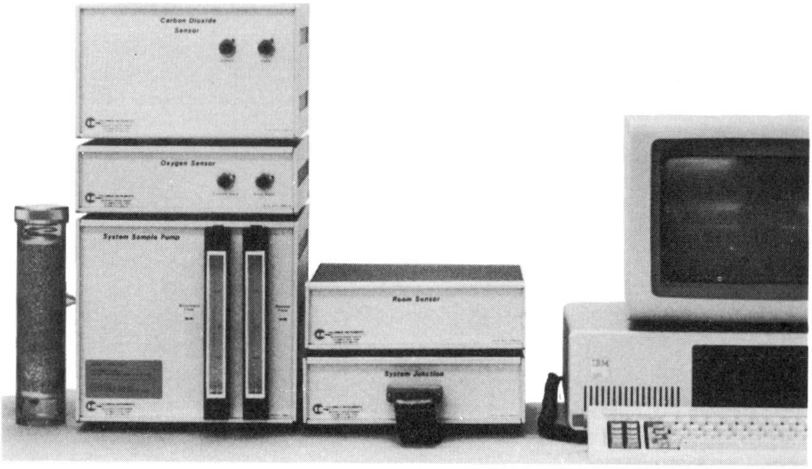

Fig. 9.26. Micro-Oxymac computerized oxygen uptake instrument produced by Columbus Instrument International Corporation, Columbus, Ohio, USA.

sensitive to changes of temperature (they change volume with changing temperature) and they have to be very well thermostated. Coloured samples can warm up under irradiation by absorbing light, and they then emit heat to the cell, which will change in volume. All these disadvantages have been solved by the Columbus Instruments International Corporation, Columbus, Ohio, USA, which has designed and produced a unique Micro-Oxymac computerized oxygen uptake instrument (Fig. 9.26) which has the capability to measure a volumetric change of 2×10^{-7} litre h^{-1} of oxygen. This instrument can also measure production of CO_2 as a result of oxidation reactions at the same volumetric change rate of 2×10^{-7} litre h^{-1}. This instrument can be very valuable for studies of polymer photo-oxidation where oxygen is consumed and CO_2 is evolved simultaneously.[197]

9.13 DETERMINATION OF THE QUENCHING OF SINGLET OXYGEN

The most effective singlet oxygen ($^1O_2$) quencher is rubrene (9,10,11,12-tetraphenylnaphthacene) (**9.3**) (orange), which is oxidized to the 9,10-endoperoxide (**9.4**) (colourless).[1077]

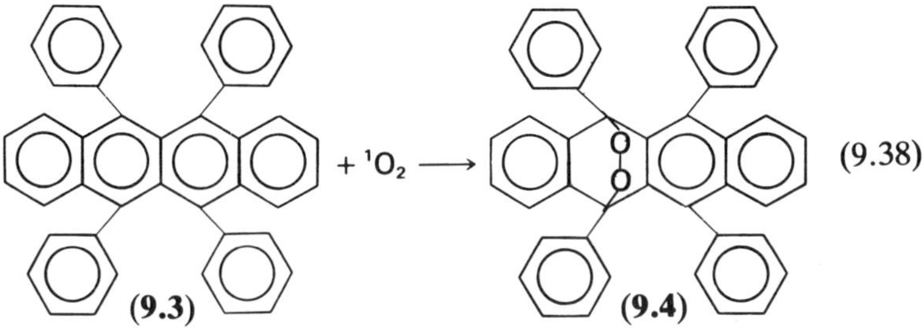

The rate constant of quenching (k_q) of singlet oxygen ($^1O_2$) by a quencher (Q) in the presence of rubrene (Ru) is given by eqn (9.39):[243,245,246,512,570,803,1187]

$$k_q = (k_{ox}[R_u] + k_d)K \tag{9.39}$$

where k_{ox} = rate constant for oxidation of rubrene = $7{\cdot}3 \times 10^{-7}$ litre $mol^{-1}\,s^{-1}$,[243]

k_d = rate of the radiationless deactivation of singlet oxygen ($^1O_2$) in a given solvent (i.e. relative lifetime, Table 9.6),

$[R_u]$ = initial concentration of rubrene (usually 5×10^{-6} M)

Table 9.6
Relative lifetime of singlet oxygen ($^1O_2$), τ, in different solvents[919]

Solvent or solvent mixture	τ (*s*)
H_2O	$2{\cdot}0 \times 10^{-6}$
$H_2O + CH_3OH$ (1:1)	$3{\cdot}5 \times 10^{-6}$
D_2O	$2{\cdot}0 \times 10^{-5}$
CH_3OH	$5{\cdot}0 \times 10^{-6}$
CH_3CH_2OH	$5{\cdot}6 \times 10^{-6}$
Benzene	$1{\cdot}25 \times 10^{-5}$
Cyclohexane	$1{\cdot}7 \times 10^{-5}$
Toluene	$2{\cdot}0 \times 10^{-5}$
Iso-octane	$2{\cdot}0 \times 10^{-5}$
Dioxane	$3{\cdot}2 \times 10^{-5}$
$CHCl_3$	$6{\cdot}0 \times 10^{-5}$
$CHCl_3 + CH_3OH$ (9:1)	$2{\cdot}6 \times 10^{-5}$
Chlorobenzene	$9{\cdot}1 \times 10^{-5}$
CH_2Cl_2	$1{\cdot}05 \times 10^{-4}$
CS_2	$2{\cdot}0 \times 10^{-4}$
CCl_4	$7{\cdot}0 \times 10^{-4}$
CF_3Cl (Freon 11)	$1{\cdot}0 \times 10^{-3}$
CCl_2FCClF_2 (Freon 113)	$2{\cdot}2 \times 10^{-3}$

and

$$K = \frac{\dfrac{d([Ru]/dt)}{d([Ru]_{(q)}/dt)} - 1}{[Q]} \qquad (9.40)$$

Here d[Ru]/dt and d$[Ru]_{(q)}$/dt are the initial rates of bleaching of rubrene in the absence and in the presence of quencher (Q), respectively, and [Q] is the concentration of the quencher (usually 10^{-2}–10^{-3} M). The bleaching of rubrene (decreasing absorption) is monitored at 520 nm.

9.14 RADIOACTIVE LABELLING TECHNIQUES

Radioactive labelling of polymers or additives is one of the most sensitive detecting methods, and allows the concentration to be measured down to 10^{-6}%.[546] The most commonly used radioactive elements are ^{14}C, ^{3}H (tritium) and ^{35}S, which are all β-radiation emitters and have half-lives in the range 0·1–6000 years (Table 9.7). There are no radioisotopes of oxygen or nitrogen of sufficient stability for practical use.

In planning radioactive labelling experiments, the following factors have to be taken into consideration:

(i) The isotope to be used in the detection equipment available.
(ii) The concentration of the isotope (its specific activity), which

Table 9.7
The β-emitters for tracer work[546]

Isotope	*Half-life (years)*	*Energy (MeV)*	*Toxicity*
^{3}H	12·26	0·02	Low
^{14}C	5 760	0·16	Low
^{32}P	0·04	1·71	Moderate
^{35}S	0·24	0·17	Moderate
^{36}Cl	300 000	0·71	Moderate
^{13}N	<0·01	1·2[a]	—
^{15}O	<0·01	1·7[a]	—

[a] Emits β^{+} (positrons).

Table 9.8
Use of radiolabelled stabilizers in polymer stabilization problems[546]

Analysis	Calibration of methods
	Optimization of extraction techniques
	Correlation with differential thermal analysis
Mobility	Water extraction
	Migration into foodstuffs
	Volatility
	Diffusion
	Adsorption on to fillers/pigments
Chemistry	'Binding' into polymers
	Losses by chemical transformations
	Kinetics

depends upon the commercially available precursors and experimental yields in the preparations.

(iii) The position of label, which depends upon the availability/cost of precursors and the chemistry of the system: ideally, loss of the label should equal loss of the labelled compound.

(iv) Isotope half-life (cf. Table 9.7).

Once a material has been labelled, estimation of the label would give an accurate analysis of the whole compound. Conversely, loss of the label would indicate loss of the reactive species. Table 9.8 shows possible applications of radioactive labelling techniques to the study of different problems in polymer stabilization.

A number of papers have been published on application of radioactive labelled stabilizers to the study of their diffusion and migration through polymer matrices.[406,528,544,546,618,1053]

Neutron activation analysis is extremely sensitive for metal-containing antioxidants and photostabilizers.[546]

9.15 VISCOSITY MEASUREMENTS

Viscosity measurements are commonly used to monitor rate of change in molecular weight during the thermal and photo- oxidation of polymer samples. Flow times of the polymer solutions (*c.* 1 wt%) are measured using a viscometer. All the solutions should be filtered through glass wool prior to measurements in order to remove microgel.

The specific viscosity (η_{sp}) (for a dilute solution) is given by:

$$\eta_{sp} = \frac{t - t_0}{t_0} \tag{9.41}$$

where t_0 and t are the flow times for pure solvent and polymer solution, respectively.

The viscosity number (reduced specific viscosity, η_{red}) is given by:

$$\eta_{red} = \frac{\eta_{sp}}{c} \tag{9.42}$$

where c is the concentration of polymer sample in g ml^{-1}.

The percentage decrease in viscosity number is given by the relationship:

$$\%\eta_{red} = \left[1 - \left(\frac{\eta_{red(0)} - \eta_{red(t)}}{\eta_{red(0)}}\right)\right] \times 100 \tag{9.43}$$

where $\eta_{red(0)}$ and $\eta_{red(t)}$ are the viscosity numbers of the polymer sample prior to degradation and at time t.

9.16 DETERMINATION OF CHAIN SCISSION

The extent of the degradation (chain scission only) can be expressed as the *number of main-chain scission* (S) by the equation:

$$S = \frac{w}{M_{n(0)}}\left(\frac{M_{n(0)}}{M_{n(t)}} - 1\right) \tag{9.44}$$

where w = weight of the irradiated sample (g),
$M_{n(0)}, M_{n(t)}$ = initial and final number-average molecular weight of a polymer after irradiation time t.

The *number of main-chain scissions* (S) can be determined from the viscometric measurements according to eqn (9.45).

$$S = \left(\frac{[\eta_0]}{[\eta_t]}\right)^{1/\alpha} - 1 \tag{9.45}$$

where $[\eta_0]$ and $[\eta_t]$ are the limiting viscosity numbers of polymer before and after irradiation, and α is the Mark–Houwink coefficient.

The *quantum yield for the main-chain scissions* (ϕ_s), which is the number of scissions occurring per quantum absorbed, is given by the following equation:

$$\phi_s = \frac{c}{M_{n(0)}}\left(\frac{d(M_{n(0)}/M_{n(t)} - 1)}{d(I_a t)}\right) \tag{9.46}$$

where c = concentration of the polymer (in solution),
I_a = intensity of light absorbed by the polymer (einsteins g^{-1}),
t = time of irradiation.

In practice ϕ_s is calculated from the equation:

$$\phi_s = \frac{c}{M_{n(0)}}\frac{S}{I_a t} \tag{9.47}$$

The addition of a photostabilizer affects ϕ_s in the following way:

$$\frac{\phi_{s(q)}}{\phi_s} = \frac{S}{S_{(q)}}\frac{I_{a(q)}}{I_a} \tag{9.48}$$

where $\phi_{s(q)}$ = quantum yield for the main-chain scission in the presence of photostabilizer,
S, $S_{(q)}$ = numbers of main-chain scissions without and in the presence of photostabilizer, respectively,
I_a, $I_{a(q)}$ = intensities of light absorbed by a polymer without and in the presence of photostabilizer, respectively.

The ratio $I_a/I_{a(q)}$ is equal to one in the case where photostabilizer does not absorb at the wavelength of irradiation. If the photostabilizer absorbs at this wavelength, the ratio $I_a/I_{a(q)}$ is given by eqn (9.49).

$$\frac{I_a}{I_{a(q)}} = \frac{1 - 10_{(q)}^{-A}}{A_{(q)}}\frac{A}{1 - 10^{-A}} \tag{9.49}$$

where A, $A_{(q)}$ = absorbance of the polymer as a film or in solution without and in the presence of photostabilizer, respectively.

9.17 THE GLOSS LOSS TEST

Estimation of photodegradation and photostabilization of coatings is mainly based on measurement of the rates of loss of physical properties such as gloss.[371]

Gloss loss is caused by the degradation and subsequent erosion or pitting of the coating polymer at the surface. This leads to the exposure of pigment or filler particles at the surface, resulting in light scattering. In the case of clear or metallic finishes, microcracking at the surface affects gloss while microcracking in the film interior appears to affect the depth of image by diffracting incoming and reflecting light.

The *gloss loss* is given by the following equation:[168,371,1064]

$$\text{Gloss} = 100 \exp\left(-\frac{4\pi\sigma\cos\theta}{\lambda}\right)^2 \tag{9.50}$$

where θ = angle of reflection,
λ = wavelength of radiation used,
σ = surface roughness.

The *reduced surface roughness* $\langle S\rangle$ can be calculated from eqn (9.51).

$$\langle S\rangle = A\sigma = [\ln(\text{gloss}/100)]^{1/2} \tag{9.51}$$

where A is an instrument constant.

The gloss data can be fitted to the following expression:[448]

$$\text{Gloss}(t) = \text{Gloss}\,(0)\exp(-k_{\text{gloss}}\,t) \tag{9.52}$$

where k_{gloss} is the rate constant for the gloss loss, which depends on the composition of coating, the level of pigment, and the harshness of the exposure. Rate constant k_{gloss} is proportional to the slope in Fig. 9.27.

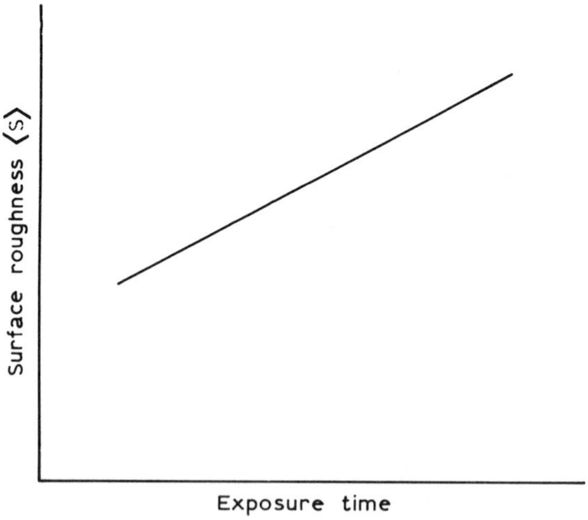

Fig. 9.27. Change of the reduced surface roughness $\langle S\rangle$ as a function of UV irradiation time.

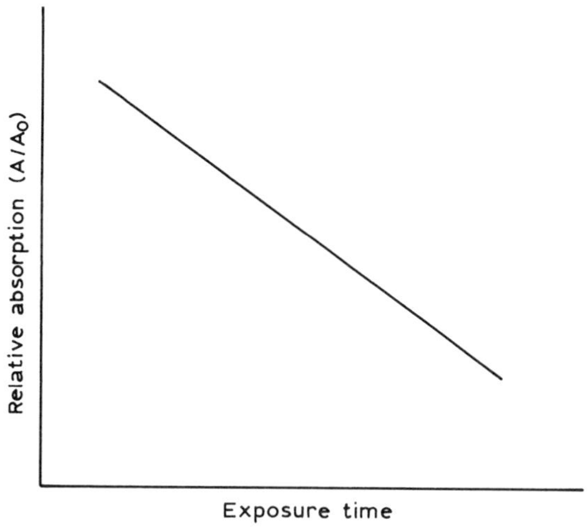

Fig. 9.28. Change of the absorbance as a function of UV irradiation time.

The relationship between gloss loss and resin loss can be determined by comparing gloss loss rate constants to rates of resin loss. Resin loss can be estimated in thin (5–10 μm) unpigmented coating films by following the rate of decrease of absorption of the C—H stretching band in the IR. The change in absorbance with time can be given by

$$A(t) = A(0)\exp(-k_e t) \tag{9.53}$$

where k_e is the rate constant for resin loss (erosion), which depends on resin composition and exposure harness. Rate constant k_e is proportional to the slope in Fig. 9.28.

9.18 TESTING METHODS OF THERMO-OXIDATIVE DEGRADATION AND STABILITY OF POLYMERS AND PLASTICS

Thermo-oxidative degradation of polymers and plastics can be determined by measurement of a property of a sample as a function of temperature and oxygen concentration.[317,503]

The following types of thermal analytical techniques are used:

(i) Static methods associated with weight change such as isobaric and isothermal weight-change determinations.

(ii) Dynamic methods associated with weight change, such as thermogravimetry and derivative thermogravimetry.
(iii) Methods associated with energy change, such as differential thermal analysis (DTA), long-term differential thermal analysis (ILDTA) and differential scanning calorimetry (DSC).
(iv) Methods associated with evolved volatiles, such as thermal volatilization analysis (TVA).
(v) Methods associated with change of mechanical properties, such as thermal mechanical analysis (TMA) and torsional braid analysis (TBA).

All of these methods are described in books on experimental methods in polymer chemistry[314,915] and thermal analysis of polymers.[314,521,635,747,748,1145,1163] Thermal analysis is especially useful in the study of the thermal oxidative stability of polymers and plastics.[178,331,503,667,1075,1078] For industrial purposes the oven ageing technique is very commonly used.[418,500]

9.18.1 Thermogravimetric Analysis

Thermogravimetric analysis (TGA) is a useful method for studying thermal oxidative degradation of polymers. The kinetics of thermogravimetric weight loss can be described by the equation:[421,915]

$$-\frac{dW}{dt} = kW^n \qquad (9.54)$$

where W = sample weight at time t,
k = rate constant,
n = reaction order.

The rate constant (k) can be expressed by the Arrhenius equation:

$$k = A \exp(-E/RT) \qquad (9.55)$$

where E is the apparent activation energy of the thermal degradation.

Since the heating rate in TGA is constant, i.e.

$$B = \frac{dT}{dt} = \text{constant} \qquad (9.56)$$

$$-\frac{dW}{dT} = (A/B)\exp(-E/RT)W^n \qquad (9.57)$$

and

$$\Delta \ln\left(-\frac{dW}{dt}\right) = (-E/R)\Delta(1/T) + n\Delta \ln W \quad (9.58)$$

and

$$X(T) = \frac{\Delta(1/T)}{\Delta \ln W} \quad (9.59)$$

$$Y(T) = \frac{\Delta \ln(-dW/dT)}{\Delta \ln W} \quad (9.60)$$

A linear regression program can be developed to calculate the kinetic parameters E and n from a linear least-squares fit of data in a plot of $Y(T)$ versus $X(T)$.[714] The thermal properties of photostabilizers have been studied by this method.[740]

9.18.2 Oxygen Induction Time Measurements

Oxygen induction time measurements are used as an accelerated thermal ageing test for polymers (mainly for polyolefins). Oxygen induction times are measured by differential scanning calorimetry (DSC) or differential thermal analysis (DTA), usually at temperatures above 200°C with the polymer in the molten state. This method produces rapid results which, however, do not correlate with the corresponding oven-ageing data gathered from the polymer in the solid state. The poor correlation between oxygen induction time values and oven-ageing data is shown in Tables 9.9 and 9.10.

Table 9.9

Comparison of test methods in high-density polyethylene (HDPE); DSC–OIT (differential scanning calorimetry–oxygen induction time test) and oven ageing[1195]

Stabilizer	*Chemical structure*	*DSC–OIT at 200°C (min)*	*Oven ageing at 120°C (days)*
Control (nonstabilized)		3·5	15
0·1% Irganox 1010 (**3.32**)	See Table 3.2	45	109
0·1% Irganox 1035 (**3.27**)	See Table 3.2	48	226
0·05% Irganox 1010 + 0·1% Irganox PS 800 (**3.56**)	See Table 3.7	26	223
0·1% 4,4′-Thiobis(2-*tert.*-butyl-5-methylphenol)		57	111
0·1% Tinuvin 622LD (**6.5**)	See Table 6.1	4	217
0·1% Chimassorb 944LD (**6.6**)	See Table 6.1	6	199

Table 9.10
Comparison of test methods in low-density polyethylene (LDPE): DSC–OIT and oven ageing[1195]

Stabilizer	*Chemical structure*	*DSC–OIT at 190°C (min)*	*Oven ageing at 100°C (days)*
Control (unstabilized)		<1	17
0·05% Irganox 1010 (**3.32**)	See Table 3.2	28	>240
0·05% Irganox 1076 (**3.21**)	See Table 3.2	7	190
0·05% Irganox 1035 (**3.27**)	See Table 3.2	18	220
0·05% 4,4′-Thiobis(2-*tert.*-butyl-5-methylphenol)		38	65
0·05% Tinuvin 622LD (**6.5**)	See Table 6.1	<1	>240
0·05% Chimassorb 944LD (**6.6**)	See Table 6.1	1·5	>240

Despite this lack of correlation, many national and international standards, intended to indicate thermal behaviour in the solid phase, are still based on oxygen induction time measurements. Accelerated tests in the melt can be useful for quality control purposes (constant polymer/stabilizer system). Oxygen induction time measurements do not permit selection of a stabilizer formulation with respect to its performance under actual end-use conditions.

9.18.3 Melt Flow Index

Measurement of the *Melt Flow Index* is a very common method for control of the change of the original molecular weight and molecular weight distribution of a polymer sample through use of melt viscosity measurements. Such tests typically measure the rate of flow of molten polymer through a standard orifice at a specified temperature and pressure. The results are usually expressed as grams per 10 minutes' flow. Depending on the testing standard used, results from the same test are referred to as:

(i) Melt Index;
(ii) Melt Flow;
(iii) Melt Flow Index;
(iv) Melt Flow Rate;
(v) Flow Rate.

The test apparatus commonly used is described in ISO 1333, DIN 53734 and ASTM 1238 standards.

Conventional Melt Flow Index tests are carried out at 190°C for polyethylene and 230°C for polypropylene; they are subject to an error of about 5–10%.[521]

9.18.4 Temperature Mill Ageing

In this test a two-roll rubber mill is used as the ageing environment. The polymer sample (control or with additives; 100–200 g) is milled at the required temperature (100–200°C). Samples of 10 or 20 g each are then removed at 15-min intervals. These specimens are subsequently compression-moulded into 15-μm films and tested for carbonyl content, change in melt index and percentage elongation retained.

Oxidized polyolefins are unstable and can continue to oxidize thermally during storage at room temperature.[237] To minimize this post-oxidation, oxidized samples should be analysed immediately, or stored at −15°C until analysis is possible.

9.19 MECHANICAL TESTING OF POLYMER (PLASTIC) SAMPLES

The deterioration of mechanical properties is normally assessed by measurement of fracture resistance by pendulum impact test methods or slow-speed flexural strength measurements and ensuring that the surface of the specimens subjected to tensile stresses corresponds to the face exposed to the UV (or weathering) source. These tests are generally more meaningful than non-destructive tests, such as modulus measurements, since the chemical changes and the creation of fine cracks (often invisible to the naked eye) may involve only a very fine layer on the surface of the exposed samples.

Tensile test pieces are usually in the form of dumb-bells (Fig. 9.29). The international standard ISO 37 specifies three sizes of dumb-bell, the smallest being intended for use when only small amounts of material are available (see also ASTM D412 and BS 903 Part 2). Dumb-bells are sometimes difficult to grip and the measurement of elongation cannot be taken from grip separation as the strain along the whole test piece is not uniform.

Measurement of the force on the stretched test pieces has been much improved by eliminating the heavy pendulum dynamometer and

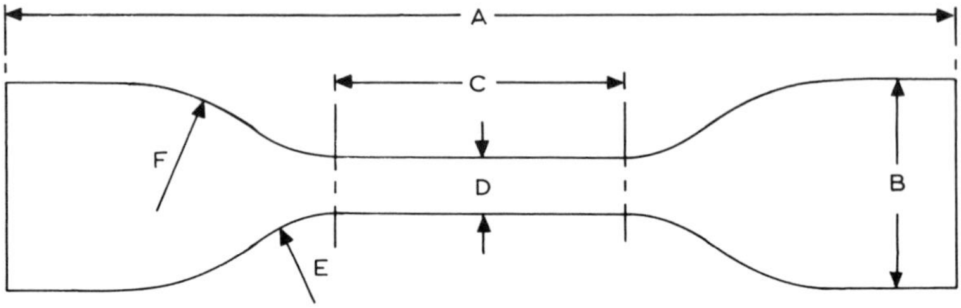

Fig. 9.29. Dumb-bell test piece (ISO Committee TC 45):

Dimensions	*Type 1 (mm)*	*Type 2 (mm)*	*Type 3 (mm)*
A	115 minimum	75 minimum	35 minimum
B	25 ± 1	$12{\cdot}5 \pm 1$	$6 \pm 0{\cdot}5$
C	33 ± 2	25 ± 1	$12 \pm 0{\cdot}5$
D	$6{\cdot}0 + 0{\cdot}4$ $- 0{\cdot}0$	$4 \pm 0{\cdot}1$	$2 \pm 0{\cdot}1$
E	14 ± 1	$8 \pm 0{\cdot}5$	$3 \pm 0{\cdot}1$
F	25 ± 2	$12{\cdot}5 \pm 1{\cdot}0$	$3 \pm 0{\cdot}1$

employing inertialess transducers which convert the force into an electrical signal. In all standard procedures tensile strength and modulus are expressed as force per unit area of the initial cross-section before stretching.[214]

The *retention of tensile strength* (for yield, elongation, breaking) (F) can be calculated from the equation:

$$F = \frac{F_t}{F_0} \times 100 \qquad (\%) \tag{9.61}$$

where F_t and F_0 are tensile strength after exposure and of the original specimen, respectively.

9.20 TESTING OF WEATHERING IN POLYMERS

Weatherability of polymer (plastic) can be generally defined as the ability to withstand complex and variable atmospheric conditions similar to those prevailing in nature.[623]

Exposure during the weathering test is influenced by a whole range of factors:

(i) Exposure to sunlight.
(ii) Exposure to heat (day and night temperatures).
(iii) Humidity (moisture, rainfall, acidity of rain).
(iv) Atmospheric contamination (ozone, smog, sulphur dioxide, etc.).[937]
(v) Mechanical corrosion (gusts of wind loaded with dust, impact of raindrops or of hail, freezing in winter, etc.).

The effectiveness of these factors depends upon the geographical location of the testing place and its elevation above sea level. Weather resistance can only be determined with complete reliability in outdoor exposure tests. The average frequency and amplitude of climatic changes are, however, fairly constant for prolonged periods, e.g. from a few to a score of years. Therefore, in order to obtain a certain average effect of climatic factors, it is necessary to carry out a test during no less an interval than 1–5 (even 10) years. Such prolonged periods of testing are usually required because the changes of physico-chemical and mechanical properties of most polymeric materials occur very slowly. Samples for testing are usually taken at one or two months, or one-year intervals.

The test data should include the exact period of the test (not only the duration), the location (where possible stating climatic data) and information on the changes observed. The first page of a test requisition for outdoor exposure evaluation is given in Fig. 9.30. The result of testing should not be given in relation to the accumulated time of weathering but to the summarized activity of external factors such as the number of hours of exposure to sunlight, mean humidity, temperature, etc. These values are measured by meteorological institutes, presented in the form of special reports and recorded in computer memory systems. Such investigations are carried out simultaneously in several stations located in different climatic regions.

Measurement of incident sunlight energy is made with pyroheliometers (ASTM E903) and expressed as joules (J) cm^{-2} (SI system) (previously Langley, $1\ Ly = 1\ cal\ cm^{-2} = 3{\cdot}187\ J\ cm^{-2}$). This measurement gives the global energy radiation from the UV to the infrared region, but it does not inform on the UV radiation intensity, which is responsible for polymer photodegradation.

In order to obtain reproducible results the tests should be carried

Ⓐrmstrong RESEARCH & DEVELOPMENT CENTER
CARD COMPANY

JOB NO. ____
NO. OF SAMPLES ____
EXPOSURE NO. ____
DATE REC'D ____

OUTDOOR EXPOSURE TEST REQUISITION

To PHYSICAL STANDARDS DEPARTMENT
(Three Copies Required)

PURPOSE ____

REQUESTED BY ____ DATE ____ PROJECT ____

NOTEBOOK NO. ____ PAGE(S) ____ NAME ____

COPIES TO ____

MATERIAL ____ TRADE NAME ____

COMPANY ☐ Ⓐ ☐ ____ ☐ Competitive ☐ Experimental ☐ Factory Test ☐ Production

END USE: ☐ Flooring, Sheet ☐ Flooring, Tile ☐ Roofing ☐ Siding, Exterior ☐ Wall Covering ☐ ____

PROCESSING: ☐ Calendered ☐ Cast ☐ Coated ☐ Dry Blend ☐ Embossed ☐ Extruded ☐ Laminated ☐ Milled ☐ Molded ☐ Pressed ☐ Sprayed ☐ ____ ☐ ____ ☐ ____ ☐ ____ ☐ ____

FORM:	☐ Adhesive	☐ Foam	☐ Slab	☐ ________
	☐ Board	☐ Foil	☐ Tape	
	☐ Coating (Not Paint)	☐ Laminate	☐ Tile	☐ ________
	☐ Corrugated	☐ Rod	☐ Tube	
	☐ Film	☐ Sheet (+0·010″)	☐ Size ________	☐ ________
EXPOSURE:	☐ Arizona	☐ Florida	☐ Lancaster	☐ ________
	☐ 0° ☐ 45°	☐ 90°	☐ N ☐ S	☐ D ☐ UG ☐ ________
	Months (circle)—3, 6, 12, 18, 24, 36, 48, 60,			________
VISUAL EXAMINATION:	☐ Blistering	☐ Delamination	☐ Staining/Bleeding	
	☐ Chalking	☐ Dirt Retention	☐ Substrate Swell	
	☐ Checking	☐ Flaking	☐ ________	
	☐ Color Change	☐ Mildew	☐ ________	
	☐ Cracking	☐ Peeling	☐ ________	
PHYSICAL TESTS:	☐ Dimensional Change	☐ Stiffness	☐ 60° Gloss	
	☐ Elongation	☐ Tensile	☐ ________	
	☐ Hardness	☐ Thickness	☐ ________	
REMARKS:				

(For PSD Use)
RACK NO. ________

LOCATION	DATE EXPOSED	DATE TERMINATED
Arizona		
Florida		
Lancaster		

Form 41038 7-66,

Fig. 9.30. The first page of a test requisition for outdoor exposure evaluation.[775]

out with films (or sheets) and not with other finished products. In a more complex article additional physical factors due to the processing, e.g. molecular orientation, internal stress and crystallinity, may also be involved. For that reason samples for this kind of testing should be pressed or cast to an even thickness and not cut out from rolled or calendered materials. In the investigation of industrial products it is necessary to know and to consider the processing conditions, especially the possibilities of internal stress and anisotropy of the samples.

The samples are mounted on racks at an angle of 30–45° to the vertical in special frames facing south (equatorial mounts). It is also important to take into account the immediate environment in which the frames are placed because of the different reflectance of UV radiation and infrared absorption by earth, rock, concrete and asphalt. The height at which the frames are placed above ground level is also important.

The South Florida (high solar radiation coupled with high humidity) and the Sonora Desert in Arizona (high solar radiation with a low humidity) exposures were taken as outdoor weathering tests. The present direct natural exposure in Miami (Florida) is regarded as 'real time' exposure, against which the results obtained with various accelerated weathering devices may be compared.

An analysis of natural weathering in the USA has indicated that the reason for the good correlation found between the exposure results obtained in aggressive environments, such as Florida and Arizona, and the more temperate climates, is the fact that maximum UV, water and temperature stresses are similar everywhere.[495,496] Elimination of annual, seasonal and location rate differences by mathematical treatment of results (rank correlation procedure) has made it possible to predict consistently the relative merits of materials by natural accelerated weathering. Correlation of results obtained in various kinds of weathering and their relation to practical durability are the criteria which determine the value of the individual test methods.

Weathering of polymers has been the subject of a special monograph.[339] A number of papers are devoted to weathering of polymers, for example polyolefins,[240,694,757,1044–1047,1097] poly(vinyl chloride),[5,15,16,202,473,660] polymethacrylates,[620] polyacetals,[983] polycarbonates,[185,703,953] polyurethanes,[6] polysulphones,[338] two-phase systems (PVC–PIB, PMM–PIB),[657,661] coatings,[371,450,451] rubbers,[4,7,335–337,970] polymer blends[402,659] and wood.[395,563]

9.20.1 Effect of Temperature on Weathering

For outdoor applications, the variation of temperature has an important effect on the useful lifetimes of hydrocarbon polymers.[1069] Figure 9.31 shows the variation of time to failure (5% oxidation) with temperature. The decrease in lifetime of unstabilized samples ranges from a few months in hot tropical weather (30–40°C), to almost two years in cool weather (0–10°C).

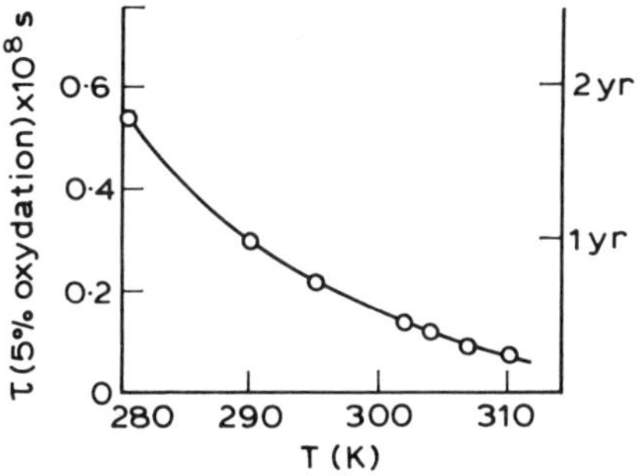

Fig. 9.31. Time to failure (5% oxidation) as a function of temperature. Reprinted with permission from Ref. 1069, © 1985 American Chemical Society.

The Arrhenius plot (Fig. 9.32) shows a apparent activation energy of 10–16 kcal mol^{-1} near atmospheric temperatures.

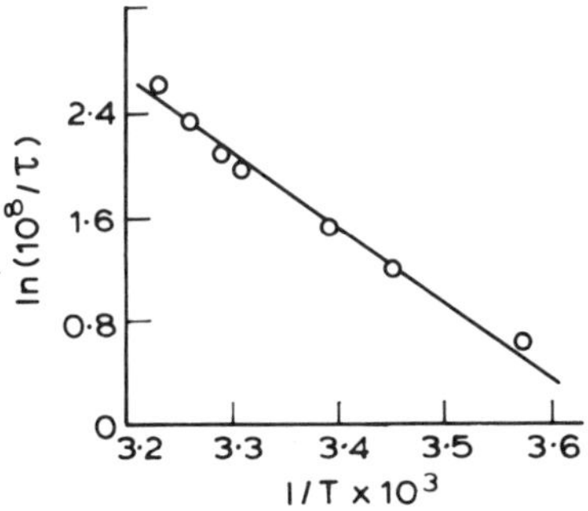

Fig. 9.32. Arrhenius plot of the rate of oxidation (*k* vs 1/*T*). Reprinted with permission from Ref. 1069, © 1985 American Chemical Society.

9.20.2 Accelerated Outdoor Testing Methods

The plastics industries require a quick answer on product ageing in environmental and/or aggressive conditions. They continuously demand the development of accelerated ageing tests which allow prediction of the behaviour of materials in working conditions. In spite of developing numerous accelerated tests and testing standards, there is a widespread criticism of existing methods and results evaluations.

The most common accelerated outdoor testing devices are sun-following devices such as:[254,255,372,373,380,916,1179]

—EMMA (Equatorial Mount with Mirrors for Acceleration) is designed to simulate arid conditions (ASTM D 1435 and ASTM E 838; cf. Table 9.11).
—EMMAQUA (EMMA in which the samples are sprayed with water for 8 min during each hour of operation) is designed to simulate wet climates (ASTM D 4141, ASTM D 4364 and ASTM G 90; cf. Table 9.11).

These devices automatically follow the path of the sun in the sky. The EMMA contains 10 highly polished aluminium mirrors precisely

Table 9.11
ASTM standards for the outdoor weathering testing of polymers and laboratory testing devices

ASTM D 1006	Outdoor weathering; conducting exterior exposure test for house/trim paints; practice.
ASTM D 1014	Outdoor weathering; exterior exposure of paint on steel; method.
ASTM D 1435	Outdoor weathering; plastics; practice for exposure.
ASTM D 2565	Exposure tests—xenon arc, light/water apparatus, plastics; practice for operating.
ASTM D 4141	Accelerated outdoor exposure test of coatings (applied to metal substrates); practice.
ASTM D 4364	Accelerated outdoor weathering of plastics using concentrated natural sunlight; practice.
ASTM D 4459	Exposure tests—xenon arc, light/water apparatus; exposure of plastics for indoor applications; practice.
ASTM E 782	Exposure to natural weathering (under conditions simulating operational mode); practice.
ASTM E 838	Accelerated outdoor exposure testing (of various materials) using concentrated natural sunlight and a Fresnel reflector; practice.
ASTM E 881	Exposure to natural weathering (in conditions simulating stagnation); practice.
ASTM G 90	Performing accelerated outdoor weathering of non-metallic materials using concentrated sunlight; practice.

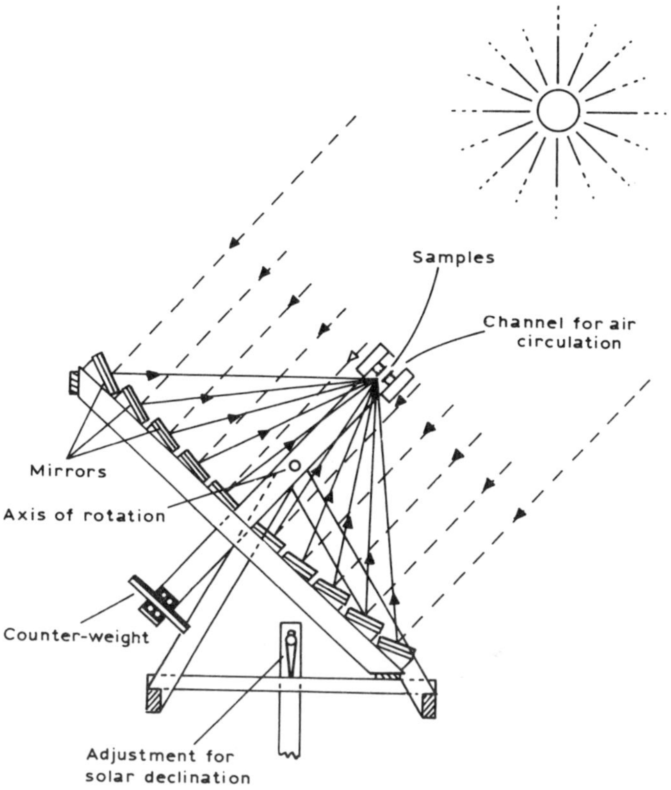

Fig. 9.33. Schematic representation of the EMMA device.

placed on a rack which is turned and tilted at computed angles in order to focus the sunlight on the surface of the sample (3 cm × 13 cm) (Figs 9.33, 9.34). The mirrors reflect 70–78% of the UV radiation of sunlight and consequently the intensity of irradiation is about 8–10 times higher than in the case of a fixed equatorial mounting. On account of the high temperature (150°C) at the spot at which the sunlight is focused, it is necessary to employ an intense cooling system using air circulation.

The EMMA devices are successfuly used by many manufacturers, particularly in the surface-coating industry, who wish to evaluate rapidly the durability of their products outdoors. However, these devices are also used in the study of accelerated weathering of almost all polymeric materials and plastics. One of the biggest testing stations

Fig. 9.34. The EMMA device in the field.

in the world is DSET (Desert) Laboratories Inc., Phoenix, Arizona 85029 (Black Canyon Stage 1), USA; it employs hundreds of EMMA devices.

The objectives of accelerated testing will vary according to the person or institution interested in the results, who may be, for example:

—producers of coating materials;

—users of paints (car manufacturers, or in the machine and electric industries);
—plastics industries;
—textile industries;
—manufacturers of testing devices, designers and/or operators of accelerated test methods.

Certain industrial paint users specify exposure for 1000 h in a Weatherometer, and several motor manufacturers use indirect outdoor exposure on EMMAQUA devices (in Arizona, USA) for 160 000 Ly. This amount of irradiation is obtained, in about 5 weeks, and furnishes results comparable with those from 6 months' direct exposure in South Florida.[383]

A number of companies producing plastics and additives have developed and used their own tests; this causes complexity in comparison of products and in a standard prediction of their lifetimes.[503] It is difficult to find correlation between accelerated test methods and outdoor ageing in service conditions.[1989] The main reason for this is lack of correlation between physical and mechanical properties of different polymers and formulations of plastics materials, measured by different methods and in different experimental conditions. Still there is no mathematical model which can describe accelerated and outdoor ageing taking into account all the parameters which influence these ageing processes.

There are certain points which may be helpful when selecting an accelerated weathering test for a project:[383]

(i) Samples have to be prepared with the utmost care and in the best possible manner. Where possible, duplicate panels should be used for each test.
(ii) Uniform and precise operation and maintenance of artificial weathering devices are essential.
(iii) Whenever new products are tested, a standard product based on a similar formulation and of known performance must be included in a control.
(iv) If time and circumstances permit, the samples should be exposed to at least two different methods of accelerated weathering (e.g. on an EMMAQUA device and in the weathering laboratory testing devices (QUV—'quick ultraviolet' cyclic testers) (cf. Section 9.20.3). The first method takes slightly longer but, being an acceleration of natural outdoor conditions,

it will tend to give results which correlate better with long-term natural exposure and real-life performance. The QUV test gives fast results, but tends to exaggerate some minor weaknesses of the product disproportionately.

(v) When evaluating results, the relative importance of any particular property with respect to the overall efficiency of the product should be borne in mind, unless the sole objective of testing is to compare one specific component or a characteristic property of a complex system.

Accelerated outdoor weathering and artificial weathering have been discussed in a number of books[339,622,934] whereas Weatherometers are described elsewhere.[916]

The standardization of outdoor weathering testing of polymers, plastics and paints can be accomplished by using established standards which are distributed by the following organizations.

(i) Association Française de Normalisation, Tour Europe, Cedex 7, 92080 Paris, La Defence, France.
(ii) DIN Deutsches Institut für Normung, Burggrafenstrasse 4–10, Postfach 1107, D-1000, Berlin 30.
(iii) British Standards Institution, 2 Park Street, London W1, UK.
(iv) US National Bureau of Standards, Washington, DC 20034, USA.
(v) American Society of Testing and Materials (ASTM), 1916 Race St, Philadelphia, PA 19103, USA (ASTM Standards, cf. Table 9.11).
(vi) USSR State Committee for Standards, Leninsky Prospekt 8, Moskva 117049, USSR.
(vii) Japanese Industrial Standards Committee, c/o Standards Department, Agency of Industrial Science and Technology, Ministry of International Trade and Industry, 1-3-1 Kusumigaseki, Chiyoda-ku, Tokyo 100, Japan.

Studies of polymer photostabilization on polymers directly throw light on overall phenomenological aspects under conditions relevant to practice. However it is extremely difficult to draw unambiguous mechanistic conclusions from this kind of investigation.[745,762]

9.20.3 Weathering Laboratory Testing Devices

There are several commercially available laboratory testing devices for artificial ageing and weathering of polymers and plastic materials for example:[916]

—*UVCON*® (Fig. 9.35): trade mark registered by the Atlas Electric Devices Co., USA. A device for exposing materials to alternate cycles of fluorescent UV radiation and visible light and condensation.[1200]

—*Fadeometer*®: trade mark registered by the Atlas Electric Devices Co., USA. An instrument in which materials are tested for lightfastness only.

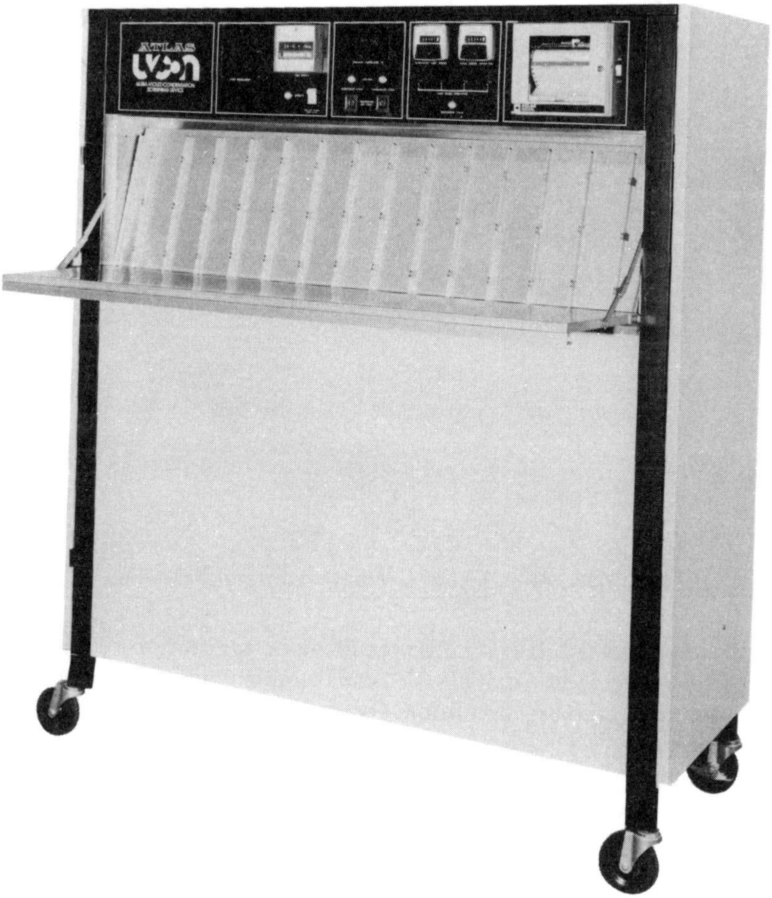

Fig. 9.35. Atlas UVCON® device.

Fig. 9.36. Atlas series C Weatherometer® device.

—*Weatherometer*® (Fig. 9.36): trade mark registered by the Atlas Electric Devices Co., USA. Apparatus in which materials are tested for climatic conditions corresponding to those found in nature.[1199]

—*Xenotests*®: trade mark registered by Heraeus Hanau GmbH, FRG, which produces a series of devices similar in action to Weatherometers (Fig. 9.37).

In all of these devices materials are tested for climatic conditions corresponding to those found in nature. The following parameters can be simulated and carefully controlled.

(i) Optical radiation.
(ii) Specimen temperature.
(iii) Humidity conditions.
(iv) Concentration of atmospheric contaminants.
(v) Sample wetting.
(vi) Thermal shock.

Weatherometers and Xenotests are usually equipped with:

(i) Light sources (xenon lamp or carbon arc lamp).
(ii) A sample exposure rack.

(a)

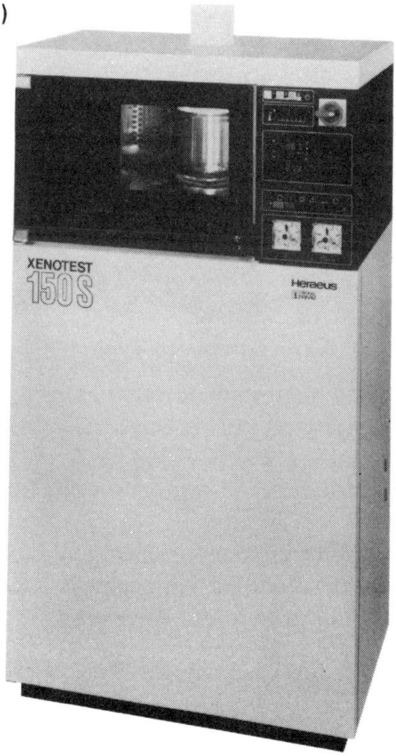

Fig. 9.37. Series of Xenotests®: (a) type 150S.

(b)

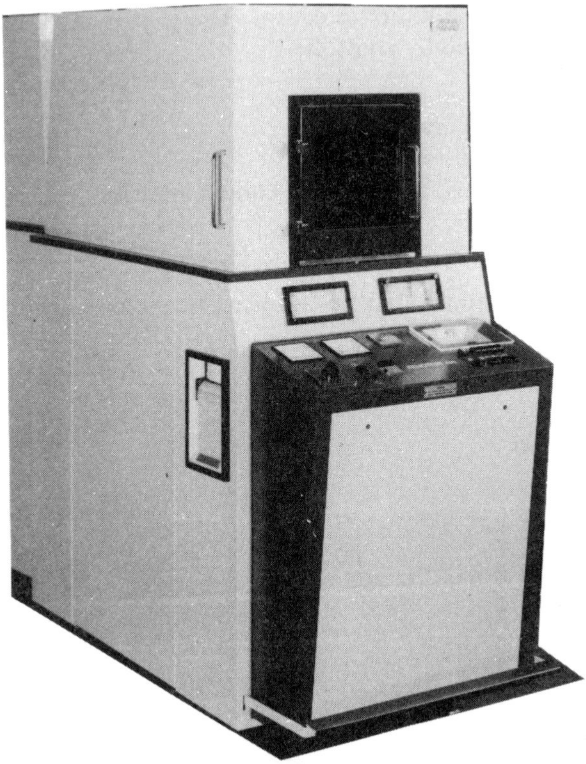

Fig. 9.37—*contd.* (b) Type 450.

(iii) Digital controls to maintain automatically preset temperature and humidity conditions during day (light-on) and night (light-off) cycles.
(iv) Independently programmed water sprays for wetting of test samples.
(v) Gas (ozone, sulphur dioxide or nitrogen dioxide) controls to simulate the effects of adding atmospheric contaminants.
(vi) Heating and cooling programmed systems.

Products such as textiles, paints, inks, plastics, coatings and colorants may be tested in such devices under controlled laboratory conditions on a repeatable basis.

The procedures to be used in such devices are all covered by National Specifications and international standardization in the course of preparation by the ISO Technical Committee 35.

(c)

Fig. 9.37—*contd.* (*c*) *Type* 1200.

In light-stability testing of polymer samples the following data are usually recorded:

(1) Hours to initial change of gloss.
(2) Hours to initial change of surface cracking.
(3) Hours to initial flex break.
(4) Hours to initial colour change.
(5) Other observations.

An important factor to determine which tests are chosen is how the equipment will be used—for technical service, production control or basic research and development?

Correlation between devices of the same type, either in the same laboratory or at different locations, can be very good if they are operated under the same test conditions. The results between devices with different sources of radiation (xenon lamp or carbon arc) will vary according to spectral emission characteristics of sources and spectral absorption characteristics of the material being tested. Thus it is possible that some materials may fail at an equal rate in models having different types of radiation sources; however, other materials may react quite differently. For this reason, correlation is only valid when exposures between devices having the same type of radiation source operating under the same test conditions are rated. Publications on artificial weathering indicate that scientific and industrial organizations involved in comparative weathering studies tend to use several types of equipment.

In view of the still-existing lack of knowledge of the fundamental reactions of the weathering processes it has been necessary, in devising accelerated weathering cycles, to examine the incidence of the discernible process variables in the breakdown process and to endeavour to obtain the correct balance of these variables by trial and error. When considering correlation of results in natural and artificial weathering studies, from among the hundreds of publications on the topic, it is difficult to find even a few views which fully agree in this respect.[501,761,1189]

9.21 COMPUTER MODELLING OF PHOTO-OXIDATION AND PHOTOSTABILIZATION

A computer model of polymer photo-oxidation and stabilization has been developed.[1069] This model can generate realistic concentration versus time profiles of the chemical species formed during photo-oxidation of hydrocarbon polymers using as input data a set of elementary reactions with corresponding rate constants and initial conditions. A numerical integration procedure on the computer provides the predicted values of the concentration terms of all species over time.

Modelling of photo-oxidation of polyethylene shows that the C—H bonds become oxidized slowly at first and then more rapidly later on.[1069] A 5% C—H bond (1% monomer units) oxidation level has been used to assign a point of failure (marked change in mechanical properties). Under typical conditions, the time to failure (τ_f) of unstabilized polyethylene would be three to four months in temperate climates and shorter in regions of high solar intensity.

Neither can the lifetime of a polymeric material be predicted nor can an optimum method of stabilization be determined, unless the effect of stress on the kinetics and mechanism of ageing under stress has been assessed and taken into account.[940]

The kinetic processes involved in photo-oxidative degradation are complex and their mathematical expression is still far from accurate and consequently cannot yet be applied for predicting long-term properties.

References

1. Abdel-Bary, E. M., Sarhan, A. A. & Abdel-Razik, E. A. (1987) *Polym. Degrad. Stabil.*, **18,** 145.
2. Abed-Ali, S., Ahmed, A. K., McWhinnie, W. R. & Scott, G. (1986) *J. Appl. Polym. Sci.*, **32,** 4857.
3. Abed-Ali, S. S., McWhinnie, W. R. & Scott, G. (1988) *Polym. Degrad. Stabil.*, **21,** 211.
4. Abu-Zeid, M. E. (1986) *J. Appl. Polym. Sci.*, **32,** 2875.
5. Abu-Zeid, M. E., Nofal, E. E., Marafi, M. A., Tahseen, L. A., Abdoul-Rasoul, F. A. & Ledwith, A. (1984) *J. Appl. Polym. Sci.*, **29,** 3431.
6. Abu-Zeid, M. E., Nofal, E. E., Tahseen, L. A., Abdoul-Rasoul, F. A. & Ledwith, A., (1984) *J. Appl. Polym. Sci.*, **29,** 2443.
7. Abu-Zeid, M. E., Youssef, Y. A. & Abdul-Rasoul, F. A. (1986) *J. Appl. Polym. Sci.*, **32,** 3345.
8. Adam, D. & Braun, D. (1980) *J. Polym. Sci.*, **18,** 629.
9. Adamczyk, A. & Wilkinson, F. (1974) *J. Appl. Polym. Sci.*, **18,** 1225.
10. Adamic, K., Bowman, D. F. & Ingold, K. U. (1970) *J. Amer. Oil Chem. Soc.*, **47,** 109.
11. Adams, J. H. (1970) *J. Polym. Sci., A1,* **8,** 1077.
12. Addis, R. R. Jr., Ghos, A. K. & Walkiem, F. G. (1968) *Appl. Phys. Lett.*, **12,** 397.
13. Adeniyi, J. B., Al-Malaika, S. & Scott, G. (1986) *J. Appl. Polym. Sci.*, **32,** 6063.
14. Adeniyi, J. B., Al-Malaika, S. & Scott, G. (1987) *J. Appl. Polym. Sci.*, **34,** 2673.
15. Adeniyi, J. B. & Scott, G. (1987) *J. Polym. Mater.* **4,** 207.
16. Adeniyi, J. B. & Scott, G. (1987) *Polym. Degrad. Stabil.*, **17,** 117.
17. Agster, A. (1954) *Melliand Textilber.*, **35,** 1209.
18. Ahmadulina, A. G., Yarkova, E. G. & Mukmeneva, N. A. (1973) *Zh. Prikl. Khim.*, **46,** 1291.
19. Ajiboye, O. & Scott, G. (1982) *Polym. Degrad. Stabil.*, **4,** 397.
20. Ajiboye, O. & Scott, G. (1982) *Polym. Degrad. Stabil.*, **4,** 415.
21. Albertsson, A. C., Donarauma, L. G., Salamone, A., Matsui, C.,

Konishi, H., Wada, K., Yoshida, S. & Vogl. O. (1984) *J. Macromol. Sci.,* **A21,** 77.
22. Allan, M., Bally, T., Haselbach, E. & Suppan, P. (1986) *Polym. Degrad. Stabil.,* **15,** 311.
23. Allara, D. L. & White, C. W. (1978) *Amer. Chem. Soc. Adv. Chem. Ser.,* **169,** 273.
24. Allen, N. S. (1980) *Makromol. Chem., Rapid Commun.,* **1,** 235.
25. Allen, N. S. (1980) *Makromol. Chem.,* **181,** 2413.
26. Allen, N. S. (1980) *Polym. Degrad. Stabil.,* **2,** 129.
27. Allen, N. S. (1980) *Polym. Degrad. Stabil.,* **2,** 155.
28. Allen, N. S. (1980) *Polym. Degrad. Stabil.,* **2,** 179.
29. Allen, N. S. (1980) *Polym. Degrad. Stabil.,* **2,** 269.
30. Allen, N. S. (1980) *Polym. Degrad. Stabil.,* **3,** 73.
31. Allen, N. S. (1981) *Polym. Photochem.,* **1,** 243.
32. Allen, N. S. (1981) *Developments in Polymer Photochemistry—2* (Allen, N. S. ed.), Applied Science Publishers, London, p. 239.
33. Allen, N. S. (1983) *Degradation and Stabilisation of Polyolefins* (Allen, N. S., ed.), Applied Science Publishers, London, p. 337.
34. Allen, N. S. (1983) *Polym. Photochem.,* **3,** 167.
35. Allen, N. S. (1986) *Chem. Soc. Rev.,* **15,** 373.
36. Allen, N. S. (1989) *Comprehensive Polymer Science,* Vol. 6 (Allen, G. & Bevington, J. C., eds) Pergamon Press, Oxford, p. 579.
37. Allen, N. S., Binkley, J. P., Parsons, B. J., Phillips, G. O. & Tennet, N. H. (1983) *Dyes and Pigments,* **4,** 11.
38. Allen, N. S., Bullen, D. J. & McKellar, J. F. (1977) *Chem. Ind.,* 797.
39. Allen, N. S., Bullen, D. J. & McKellar, J. F. (1979) *J. Mater. Sci.,* **14,** 1941.
40. Allen, N. S., Chirinos-Padron, A. & Appleyard, J. H. (1982) *Polym. Degrad. Stabil.,* **4,** 223.
41. Allen, N. S., Chirinos-Padron, A. & Appleyard, J. H. (1983) *Polym. Degrad. Stabil.,* **5,** 29.
42. Allen, N. S., Chirinos-Padron, A. & Appleyard, J. H. (1983) *Polym. Degrad. Stabil.,* **5,** 55.
43. Allen, N. S., Chirinos-Padron, A. & Appleyard, J. H. (1983) *Polym. Degrad. Stabil.,* **5,** 323.
44. Allen, N. S., Chirinos-Padron, A. & Appleyard, J. H. (1984) *Europ. Polym. J.,* **20,** 433.
45. Allen, N. S., Chirinos-Padron, A. & Appleyard, J. H. (1984) *Polym. Degrad. Stabil.,* **6,** 31.
46. Allen, N. S., Chirinos-Padron, A. & Appleyard, J. H. (1984) *Polym. Degrad. Stabil.,* **6,** 149.
47. Allen, N. S., Chirinos-Padron, A. & Appleyard, J. H. (1984) *Polym. Degrad. Stabil.,* **9,** 15.
48. Allen, N. S., Chirinos-Padron, A. & Appleyard, J. H. (1984) *Polym. Photochem.,* **5,** 333.
49. Allen, N. S., Chirinos-Padron, A. & Appleyard, J. H. (1985) *Europ. Polym. J.,* **21,** 101.

50. Allen, N. S., Chirinos-Padron, A., Appleyard, J. H. & Henman, T. J. (1983) *Polym. Degrad. Stabil.*, **5,** 105.
51. Allen, N. S., Chirinos-Padron, A., Appleyard, J. H. & Henman, T. J. (1985) *Polym. Degrad. Stabil.*, **12,** 149.
52. Allen, N. S., Chirinos-Padron, A. & Henman, T. J. (1985) *Polym. Degrad. Stabil.*, **13,** 31.
53. Allen, N. S., Chirinos-Padron, A. & Taylor, C. (1984) *Europ. Polym. J.*, **20,** 1031.
54. Allen, N. S. & Fatinikun, K. O. (1981) *Polym. Degrad. Stabil.*, **3,** 327.
55. Allen, N. S., Fatinikun, K. O., Gardette, J. L. & Lemaire, J. (1981) *Polym. Degrad. Stabil.*, **3,** 243.
56. Allen, N. S., Fatinikun, K. O., Gardette, J. L. & Lemaire, J. (1982) *Polym. Degrad. Stabil.*, **4,** 95.
57. Allen, N. S., Fatinikun, K. O. & Henman, T. J. (1981) *Chem. Ind.*, 119.
58. Allen, N. S., Fatinikun, K. O. & Henman, T. J. (1982) *Polym. Degrad. Stabil.*, **4,** 59.
59. Allen, N. S., Fatinikun, K. O. & Henman, T. J. (1983) *Europ. Polym. J.*, **19,** 551.
60. Allen, N. S., Fatinikun, K. O. & Henman, T. J. (1987) *Polym. Degrad. Stabil.*, **17,** 81.
61. Allen, N. S., Gardette, J. L. & Lemaire, J. (1981) *Polym. Degrad. Stabil.*, **3,** 199.
62. Allen, N. S., Gardette, J. L. & Lemaire, J. (1981) *Poly. Photochem.*, **1,** 111.
63. Allen, N. S., Gardette, J. L. & Lemaire, J. (1982) *Dyes and Pigments*, **3,** 295.
64. Allen, N. S., Gardette, J. L. & Lemaire, J. (1982) *J. Appl. Polym. Sci.* **27,** 2761.
65. Allen, N. S., Gardette, J. L. & Lemaire, J. (1982) *Polym. Degrad. Stabil.*, **4,** 261.
66. Allen, N. S., Gardette, J. L. & Lemaire, J. (1983) *Polym. Photochem.*, **3,** 251.
67. Allen, N. S., Gardette, L. J. & Lemaire, J. (1984) *Polym. Degrad. Stabil.*, **8,** 133.
68. Allen, N. S., Hamidi, A., Loffelman, F. F., MacDonald, P., Rauhut, M. & Susi, P. V. (1985) *Plast. Rubber Process. Appl.*, **5,** 259.
69. Allen, N. S., Harrison, M. J. & Follows, G. W. (1988) *Polym. Degrad. Stabil.*, **21,** 251.
70. Allen, N. S., Harrison, M. J., Follows, G. W. & Matthews, V. (1987) *Polym. Degrad. Stabil.*, **19,** 77.
71. Allen, N. S., Homer, J. & McKellar, J. F. (1978) *J. Appl. Polym. Sci.*, **22,** 611.
72. Allen, N. S., Homer, J. & McKellar, J. F. (1978) *Makromol. Chem.*, **179,** 1575.
73. Allen, N. S., Kotecha, J. L., Gardette, J. L. & Lemaire, J. (1984) *Polym. Commun.*, **25,** 235.

74. Allen, N. S., Kotecha, J. L., Gardette, J. L. & Lemaire, J. (1985) *Polym. Degrad. Stabil.*, **11,** 181.
75. Allen, N. S., Kotecha, J. L., Parkinson, A., Lofelman, F. F., Rauhut, M. M. & Susi, P. V. (1985) *Polym. Degrad. Stabil.*, **10,** 1.
76. Allen, N. S. & McKellar, J. F. (1975) *Chem. Soc. Rev.*, **4,** 553.
77. Allen, N. S. & McKellar, J. F. (1977) *Brit. Polym. J.*, **9,** 302.
78. Allen, N. S. & McKellar, J. F. (1978) *J. Appl. Polym. Sci.*, **22,** 3277.
79. Allen, N. S. & McKellar, J. F. (1978) *Macromol. Rev.*, **13,** 241.
80. Allen, N. S. & McKellar, J. F. (1979) *Photochemistry of Man-Made Polymers,* Applied Science Publishers, London.
81. Allen, N. S. & McKellar, J. F. (1979) *Plast. Rubb. Mater. Appl.*, **4,** 170.
82. Allen, N. S. & McKellar, J. F. (1980) *Europ. Polym. J.*, **16,** 533.
83. Allen, N. S. & McKellar, J. F. (1980) *Photochemistry of Dyed and Pigmented Polymers* (Allen, N. S. & McKellar, J. F., eds), Applied Science Publishers, London.
84. Allen, N. S. & McKellar, J. F. (1980) *Photochemistry of Dyed and Pigmented Polymers* (Allen, N. S. & McKellar, J. F., eds), Applied Science Publishers, London p. 247.
85. Allen, N. S., McKellar, J. F., Phillips, G. O. & Chapman, C. B. (1974) *J. Polym. Sci., B,* **12,** 723.
86. Allen, N. S., McKellar, J. F., Phillips, G. O. & Wood, D. G. M. (1974) *J. Polym. Sci., B,* **12,** 241.
87. Allen, N. S., McKellar, J. F. & Wilson, D. (1978) *Chem. Ind.*, 237.
88. Allen, N. S., McKellar, J. F. & Wilson, D. (1979) *Polym. Degrad. Stabil.*, **1,** 205.
89. Allen, N. S., McKellar, J. F. & Wood, G. M. (1975) *J. Polym. Sci., A1,* **13,** 2319.
90. Allen, N. S., Mudher, M. & Green, P. (1984) *Polym. Degrad. Stabil.*, **7,** 83.
91. Allen, N. S., Mudher, M. & Green, P. (1984) *Polym. Degrad. Stabil.*, **9,** 145.
92. Allen, N. S. & Parkinson, A. (1982) *Polym. Degrad. Stabil.*, **4,** 161.
93. Allen, N. S. & Parkinson, A. (1983) *Polym. Degrad. Stabil.*, **5,** 189.
94. Allen, N. S., Parkinson, A., Gardette, J. L. & Lemaire, J. (1983) *Polym. Degrad. Stabil.*, **5,** 135.
95. Allen, N. S., Parkinson, A., Loffelman, F. F., MacDonald, P., Rauhut, M. M. & Susi, P. V. (1985) *Polym. Degrad. Stabil.*, **12,** 363.
96. Allen, N. S., Parkinson, A., Loffelman, F. F., Rauhut, M. M. & Susi, P. V. (1984) *Polym. Degrad. Stabil.*, **7,** 153.
97. Allen, N. S., Parkinson, A., Loffelman, F. F. & Susi, P. V. (1983) *Polym. Degrad. Stabil.*, **5,** 241 (1983).
98. Allen, N. S., Parkinson, A., Loffelman, F. F. & Susi, P. V. (1983) *Angew. Makromol. Chem.*, **116,** 203.
99. Allen, N. S., Parkinson, A., Loffelman, F. F. & Susi, P. V. (1983) *Polym. Degrad. Stabil.*, **5,** 403.
100. Allen, N. S., Parkinson, A., Loffelman, F. F. & Susi, P. V. (1984) *Polym. Degrad. Stabil.*, **6,** 65.

101. Al-Malaika, S. (1984) *British Polym. J.*, **16,** 301.
102. Al-Malaika, S. (1987) *History of Polymer Composites* (Seymour, R. B. & Deanin, R. D., eds), VNU Science Press BV, Netherlands, p. 225.
103. Al-Malaika, S. (1988) *Polym. Plast. Technol. Eng.*, **27,** 261.
104. Al-Malaika, S. (1989) *Comprehensive Polymer Science* (Allen, G. & Bevington, J. C., eds), Pergamon Press, Oxford, p. 539.
105. Al-Malaika, S., Chakraborty, K. B. & Scott, G. (1983) *Developments in Polymer Stabilization* (Scott, G., ed.), Applied Science Publishers, London, p. 73.
106. Al-Malaika, S., Chakraborty, K. B., Scott, G. & Tao, Z. B. (1985) *Polym. Degrad. Stabil.*, **10,** 55.
107. Al-Malaika, S., Chakraborthy, K. B., Scott, G. & Tao, Z. B. (1985) *Polym. Degrad. Stabil.*, **13,** 261.
108. Al-Malaika, S., Coker, M. & Scott, G. (1985) *Polym. Degrad. Stabil.*, **10,** 173.
109. Al-Malaika, S., Coker, M. & Scott, G. (1988) *Polym. Degrad. Stabil.*, **22,** 147.
110. Al-Malaika, S., Desai, P. & Scott, G. (1985) *Plast. Rubber Process. Appl.*, **5,** 16.
111. Al-Malaika, S., Hoggokusumo, S. & Scott, G. (1986) *Polym. Degrad. Stabil.*, **16,** 25.
112. Al-Malaika, S., Huczkowski, P. & Scott, G. (1984) *Polymer,* **25,** 1006.
113. Al-Malaika, S., Ibrahim, A. Q. & Scott, G. (1988) *Polym. Degrad. Stabil.*, **22,** 233.
114. Al-Malaika, S., Kok, L. & Scott, G. (1981) *Polym. Commun.*, **25,** 233.
115. Al-Malaika, S., Marogi, A. & Scott, G. (1985) *J. Appl. Polym. Sci.*, **30,** 789.
116. Al-Malaika, S., Marogi, A. & Scott, G. (1985) *Polym. Degrad. Stabil.*, **10,** 237.
117. Al-Malaika, S., Marogi, A. & Scott, G. (1986) *J. Appl. Polym. Sci.*, **31,** 685.
118. Al-Malaika, S., Marogi, A. & Scott, G. (1987) *J. Appl. Polym. Sci.*, **33,** 1455.
119. Al-Malaika, S., Marogi, A. & Scott, G. (1987) *Polym. Degrad. Stabil.*, **18,** 89.
120. Al-Malaika, S., Omikorede, E. A. & Scott, G. (1986) *Polym. Commun.*, **27,** 173.
121. Al-Malaika, S., Omikorede, E. O. & Scott, G. (1987) *J. Appl. Polym. Sci.*, **33,** 703.
122. Al-Malaika, S. & Scott, G. (1980) *Europ. Polym. J.*, **16,** 503.
123. Al-Malaika, S. & Scott, G. (1980) *Europ. Polym. J.*, **16,** 709.
124. Al-Malaika, S. & Scott, G. (1982) *Polym. Commun.*, **23,** 1711.
125. Al-Malaika, S. & Scott, G. (1983) *Degradation and Stabilization of Polyolefins* (Allen, N. S., ed.), Applied Science Publishers, London, p. 247.
126. Al-Malaika, S. & Scott, G. (1983) *Degradation and Stabilization of Polyolefins* (Allen, N. S., ed.), Applied Science Publishers, London, p. 283.

127. Al-Malaika, S. & Scott, G. (1983) *Europ. Polym. J.*, **19**, 235.
128. Al-Malaika, S. & Scott, G. (1983) *Europ. Polym. J.*, **19**, 241.
129. Al-Malaika, S. & Scott, G. (1983) *Polym. Commun.*, **24**, 24.
130. Al-Malaika, S. & Scott, G. (1983) *Polym. Degrad. Stabil.*, **5**, 415.
131. Al-Malaika, S., Scott, G. & Huczkowski, P. (1984) *Polym. Degrad. Stabil.*, **7**, 95.
132. Altwicker, E. R. (1967) *Chem. Ind.*, **67**, 475.
133. Amin, M. U. & Scott, G. (1974) *Europ. Polym. J.*, **10**, 1019.
134. Amin, M. U., Scott, G. & Tillekeratne, L. M. K. (1975) *Europ. Polym. J.*, **11**, 85.
135. Anani, A., Mobasher, A. & Rasould, F. A. (1984) *J. Appl. Polym. Sci.*, **29**, 1491.
136. Andrady, A. L. & Schultz, A. R. (1987) *J. Appl. Polym. Sci.*, **33**, 1389.
137. Arakawa, K., Seguchi, T., Watanabe, Y. & Hayakawa, N. (1982) *J. Polym. Sci., A1*, **20**, 2681.
138. Arct, J., Dul, M., Rabek, J. F. & Rånby, B. (1981) *Europ. Polym. J.*, **17**, 1041.
139. Arnaud, R. & Lemaire, J. (1979) *Developments in Polymer Degradation—2* (Grassie, N., ed.), Applied Science Publishers, London, p. 159.
140. Arnaud, R. & Lemaire, J. (1981) *Developments in Polymer Photochemistry—2* (Allen, N. S., ed.), Applied Science Publishers, London, p. 135.
141. Arnaud, R., Lemaire, J. & Penot, G. (1982) *Polym. Photochem.*, **2**, 39.
142. Arnold, C. Jr & Ranscler, C. L. (1988) *Polym. Commun.*, **29**, 95.
143. Bagheri, R., Chakraborthy, K. B. & Scott, G. (1980) *Chem. Ind.*, 865.
144. Bagheri, R., Chakraborthy, K. B. & Scott, G. (1982) *Polym. Degrad. Stabil.*, **4**, 1.
145. Bagheri, R., Chakraborthy, K. B. & Scott, G. (1983) *Polym. Degrad. Stabil.*, **5**, 145.
146. Bagheri, R., Chakraborthy, K. B. & Scott, G. (1983) *Polym. Degrad. Stabil.*, **5**, 197.
147. Bagheri, R., Chakraborthy, K. B. & Scott, G. (1985) *Polym. Degrad. Stabil.*, **11**, 1.
148. Bailey, D., Tirrell, D., Pinazzi, C. & Vogl, O. (1978) *Macromolecules*, **11**, 312.
149. Bailey, D., Tirrell, D. & Vogl, O. (1976) *J. Polym. Sci., A1*, **14**, 2725.
150. Bailey, D. & Vogl, O. (1976) *J. Macromol. Sci., Rev. Macromol. Chem.*, **C14**, 267.
151. Bálint, G., Kelen, T., Rehak, A. & Tüdos, F. (1976) *React. Kinet. Catal. Lett.*, **4**, 467.
152. Bálint, G., Kelen, T., Tüdos, F. & Rehak, A. (1979) *Polym. Bull.*, **1**, 647.
153. Bálint, G., Rockenbauer, A., Kelen, T., Tüdos, F. & Jókay, L. (1981) *Polym. Photochem.*, **1**, 139.
154. Ball, S. & Bruice, T. C. (1980) *J. Amer. Chem. Soc.*, **102**, 6498.
155. Ballardini, R., Beggiato, G., Bortolus, P., Fucitano, A., Buttafava, A. & Gratani, F. (1984) *Polym. Degrad. Stabil.*, **7**, 41.

156. Barashkova, I. I., Vasserman, A. M. & Rapport N. Ya., *Vysokomol. Soedin,* **21A,** 1683.
157. Barksdale, J. (1966) *Titanium, Its Occurrence, Chemistry and Technology,* Ronald Press, New York, 2nd edn.
158. Barnard, D. & Wong, K. C. (1976) *Anal. Chim. Acta,* **84,** 355.
159. Bauer, D. R. (1982) *J. Appl. Polym. Sci.,* **27,** 3651.
160. Bauer, D. R. (1987) *Polym. Degrad. Stabil.,* **19,** 97.
161. Bauer, D. R. & Briggs, L. M. (1984) *Amer. Chem. Soc. Symp. Ser.,* **243,** 271.
162. Bauer, D. R. & Gerlock, J. L. (1986) *Polym. Degrad. Stabil.,* **14,** 97.
163. Bauer, R. H. & Coppinger, G. M. (1973) *Tetrahedron,* **19,** 1201.
164. Baum, B. & Deanin, R. D. (1973) *Polym. Plast. Technol. Eng.,* **2,** 1.
165. Baum, B., Margosiak, S. A., Llompart, A. & Deanin, R. (1971) *British Plastics,* **44,** 79.
166. Beavan, S. W., Hackett, P. A. & Phillips, D. (1974) *Europ. Polym. J.,* **10,** 925.
167. Beckett, A. & Porter, G. (1963) *Trans. Faraday Soc.,* **59,** 2051.
168. Beckman, P. & Spizzichino, A. (1963) *The Scattering of Electromagnetic Waves from Rough Surfaces,* Pergamon, Oxford, Ch. 5.
169. Belluš, D., Lind, H. & Wyatt, J. F. (1972) *J. Chem. Soc., Chem. Commun.,* 1199.
170. Benbow, A. W., Cullis, C. F. & Laver, H. S. (1978) *Polymer,* **19,** 824.
171. Benson, S. W. (1964) *J. Chem. Phys.,* **40,** 1007.
172. Benson, S. W. (1965) *J. Chem. Educ.,* **42,** 501.
173. Berliner, L. J. (ed.) (1974) *Spin Labelling—Theory and Practice,* Academic Press, New York.
174. Bickley, R. I. & Jayanty, R. K. M. (1974) *Disc. Faraday Soc.,* **58,** 194.
175. Bigger, S. W. & Delatycki, O. (1987) *J. Polym. Sci.,* **25,** 3311.
176. Bigger, S. W. & Delatycki, O. (1988) *Polymer,* **29,** 1277.
177. Bilek, J. G., Kollonitsch, V. & Kline, C. H. (1966) *Ind. Eng. Chem.,* **58,** 31.
178. Billingham, N. C., Bott, D. C. & Manke, A. S. (1981) *Developments in Polymer Degradation—3* (Grassie, N., ed.), Applied Science Publishers, London, p. 63.
179. Billingham, N. C. & Calvert, P. D. (1980) *Developments in Polymer Stabilization—3* (Scott, G., ed.), Applied Science Publishers, London, p. 139.
180. Billingham, N. C., Calvert, P. D. & Manke, A. S. (1981) *J. Appl. Polym. Sci.,* **26,** 3543.
181. Billingham, N. C. & Peebles, I. H. (1975) *J. Polym. Sci., A1,* **13,** 1209.
182. Billingham, N. C., Prentice, P. & Walker, T. J. (1976) *J. Polym. Sci., Polym. Symp.,* **57,** 287.
183. Billingham, N. C. & Walker, T. J. (1975) *J. Polym. Sci., A1,* **13,** 1209.
184. Black, J. F. (1978) *J. Amer. Chem. Soc.,* **100,** 527.
185. Blaga, A. & Yamasaki, R. S. (1976) *J. Mater. Sci.,* **11,** 1513.
186. Blair, H. E. (1973) *Polym. Eng. Sci.,* **13,** 435.
187. Bocian, D. F., Huston, A. L. & Scott, G. (1983) *J. Chem. Phys.,* **79,** 5802.

188. Boehm, H. P. (1966) *Angew. Chem.*, **76,** 617.
189. Boehm, H. P. & Hermann, M. (1967) *Z. Anorg. Allg. Chem.*, **352,** 156.
190. Boehm, H. P. & Hermann, M. (1967) *Z. Anorg. Allg. Chem.*, **368,** 73.
191. Bolsman, T. A. B. M., Block, A. B. & Frijns, J. H. G. (1978) *Rec. Trav. Chim. Pays Bas,* **97,** 310.
192. Bolton, J. R. & Warden, J. T. (1974) *Creation and Detection of the Excited State,* Vol. 2 (Ware, W. R., ed.) Decker, New York, p. 63.
193. Boonstra, A. H. & Mutsaers, C. A. H. A. (1975) *J. Phys. Chem.*, **79,** 1694.
194. Borsig, E., Gupta, A., Rånby, B. & Vogl, O. (1984) *Polym. Bull.*, **12,** 245.
195. Boss, C. R. & Chien, J. C. W. (1966) *J. Polym. Sci., A1,* **4,** 1543.
196. Bousquet, J. A. & Fouassier, J. P. (1982) *J. Photochem.*, **20,** 197.
197. Bousquet, J. A. & Fouassier, J. P. (1983) *Polym. Degrad. Stabil.*, **5,** 113.
198. Bousquet, J. A. & Fouassier, J. P. (1984) *Europ. Polym. J.*, **20,** 985.
199. Bousquet, J. A. & Fouassier, J. P. (1984) *J. Polym. Sci., A1*, **22,** 3865.
200. Bousquet, J. A. & Fouassier, J. P. (1987) *Europ. Polym. J.*, **23,** 367.
201. Bousquet, J. A. & Fouassier, J. P. (1987) *Polym. Degrad. Stabil.*, **18,** 163.
202. Bowley, H. J., Gerrard, D. L. & Biggin, I. S. (1988) *Polym. Degrad. Stabil.*, **20,** 257.
203. Bradon, D. V. & Falkehag, S. I. (1976) US Patent 3 984 801.
204. Bragg, L. & Claringbull, G. F. (1965) *Crystal Structure of Minerals,* Vol. 4, Bell, G. and Sons, London.
205. Brandrup, J. & Peebles, L. H., Jr (1968) *Macromolecules,* **1,** 64.
206. Brauar, M., Mlinac, M. & Rolich, J. (1976) *J. Polym. Sci., Polym. Symp.*, **57,** 161.
207. Braun, D. & Wittig, W. (1980) *Makromol. Chem.*, **181,** 557.
208. Breck, A. K., Taylor, C. L., Russel, K. E. & Wan, J. K. S. (1974) *J. Polym. Sci.*, **12,** 1505.
209. Breskman, E. L. & Pappas, S. P. (1976) *J. Coating Technol.*, **48,** 34.
210. Bridgewater, A. J. & Sexton, M. D. (1978) *J. Chem. Soc., Perkin II,* 384.
211. Briggs, D. & Kendall, C. R. (1982) *Int. J. Adhesives,* **199,** 13.
212. Briggs, P. J. & McKellar, J. F. (1967) *Chem. Ind.*, 622.
213. Briggs, P. J. & McKellar, J. F. (1968) *J. Appl. Polym. Sci.*, **12,** 1825.
214. Brown, R. P. (1986) *Physical Testing of Rubber,* Elsevier Applied Science Publishers, London.
215. Brunton, B., Gilbert, C. & Mawby, R. J. (1976) *J. Chem. Soc., Perkin II,* 1267.
216. Brydson, J. A. (1982) *Plastic Materials,* Butterworths, London.
217. Buben, I. & Pospišil, J. (1976) *J. Polym. Sci., Polym. Symp.* **57,** 261.
218. Buchachenko, A. L. (1965) *Stable Radicals,* Consultants Bureau, New York.
219. Buchachenko, A. L. & Vasserman, A. M. (1973) *Free Stable Radicals,* Khimia, Moscow (in Russian).

220. Burchill, P. J. & Pinkerton, D. M. (1976) *J. Polym. Sci., Polym. Symp.*, **55,** 185.
221. Burchill, P. J. & Pinkerton, D. M. (1980) *Polym. Degrad. Stabil.*, **2,** 239.
222. Buss, A. D., Malati, M. A. & Atkinson, R. (1976) *J. Oil Colour Chem. Assoc.*, **59,** 369.
223. Cain, M. E., Knight, G. T., Lewis, P. M. & Saville, B. (1969) *J. Rubber Res. Inst. Malaya,* **22,** 289.
224. Caldwell, R. A. & Schwertzel, R. E. (1972) *J. Amer. Chem. Soc.*, **94,** 1035.
225. Calvert, P. D. & Billingham, N. C. (1979) *J. Appl. Polym. Sci.*, **24,** 357.
226. Calvert, P. D. & Ryan, T. G. (1978) *Polymer,* **19,** 611.
227. Cameron, G. G. & Bullock, A. T. (1982) *Developments in Polymer Characterisation—2* (Dawkins, J. V., ed.), Applied Science Publishers, London, p. 107.
228. Campbell, T. W. & Coppinger, G. M. (1952) *J. Amer. Chem. Soc.*, **74,** 1469.
229. Carlson, D. P. (1963) US Patent 3 081 280.
230. Carlson, D. P. & Esayian, M. (1964) French Patent 1 381 743.
231. Carlsson, D. J., Bazan, G., Chmiela, S., Wiles, D. M. & Russell, K. E., (1987) *Polym. Degrad. Stabil.*, **19,** 195.
232. Carlsson, D. J., Brousseau, R., Can Zhang & Wiles, D. M. (1987) *Polym. Degrad. Stabil.*, **17,** 303.
233. Carlsson, D. J., Brousseau, R. & Wiles, D. M. (1986) *Polym. Degrad. Stabil.*, **15,** 67.
234. Carlsson, D. J., Chan, K. H., Durmis, J. & Wiles, D. M. (1982) *J. Polym. Sci., A1*, **20,** 575.
235. Carlsson, D. J., Chan, K. H., Garton, A. & Wiles, D. M. (1980) *Pure Appl. Chem.*, **52,** 389.
236. Carlsson, D. J., Chan, K. H. & Wiles, D. M. (1981) *Amer. Chem. Soc. Symp. Ser.* **151,** 51.
237. Carlsson, D. J., Dobbin, C. J. B., Jensen, J. P. T. & Wiles, D. M. (1985) *Amer. Chem. Soc. Symp. Ser.* **280,** 359.
238. Carlsson, D. J., Dobbin, C. J. B. & Wiles, D. M. (1985) *Macromolecules,* **18,** 1791.
239. Carlsson, D. J., Garton, A. & Wiles, D. M. (1976) *Macromolecules,* **9,** 695.
240. Carlsson, D. J., Garton, A. & Wiles, D. M. (1977) *J. Appl. Polym. Sci.*, **21,** 2963.
241. Carlsson, D. J., Garton, A. & Wiles, D. M. (1979) *Developments in Polymer Stabilization—1* (Scott, G., ed.), Applied Science Publishers, London, p. 219.
242. Carlsson, D. J., Grattan, D. W., Suprunchuk, T. & Wiles, D. M. (1978) *J. Appl. Polym. Sci.*, **22,** 2217.
243. Carlsson, D. J., Mendenhall, G. D., Suprunchuk, T. & Wiles, D. M. (1972) *J. Amer. Chem. Soc.*, **94,** 8960.
244. Carlsson, D. J., Sproule, D. E. & Wiles, D. M. (1972) *Macromolecules,* **5,** 659.

245. Carlsson, D. J., Suprunchuk, T. & Wiles, D. M. (1973) *J. Polym. Sci., B,* **11,** 61.
246. Carlsson, D. J., Suprunchuk, T. & Wiles, D. M. (1974) *Can. J. Chem.,* **52,** 3728.
247. Carlsson, D. J. & Wiles, D. M. (1969) *Macromolecules,* **2,** 587.
248. Carlsson, D. J. & Wiles, D. M. (1974) *J. Polym. Sci., A1,* **12,** 2217.
249. Carlsson, D. J. & Wiles, D. M. (1974) *Macromolecules,* **7,** 259.
250. Carlsson, D. J. & Wiles, D. M. (1976) *J. Macromol. Sci., Rev. Macromol. Chem.,* **C14,** 65.
251. Carlsson, D. J. & Wiles, D. M. (1984) *Polym. Degrad. Stabil.,* **6,** 1.
252. Carr, D. S., Baum, B., Margosiak, S. A. & Llompart, A. (1970) *Modern Plastics,* **47,** 114.
253. Carr, D. S., Baum, B., Margosiak, S. A. & Llompart, A. (1971) *Modern Plastics,* **48,** 160.
254. Caryl, C. R. & Helmick, W. E. (1960) US Patent 2 945 417.
255. Caryl, C. R. & Rheineck, A. E. (1965) *Off. Dig.,* **37,** 129.
256. Casale, A. & Porter, R. S. (1978) *Polymer Stress Reaction,* Academic Press, New York.
257. Castor, W. S., Jr & Manassion, J. A. (1978) *Additives For Plastics,* Vol. 1 (Seymour, R. B., ed.) Academic Press, New York, p. 233.
258. Catignani, G. L. & Carter, M. E. (1982) *J. Food Sci.,* **47,** 1745.
259. Cegera, J., Ribe, J. & Miro, P. (1972) *J. Soc. Dyers Colourists,* **88,** 293.
260. Cen, J. F., Konoma, F. & Osawa, Z. (1986) *Polym. Photochem.,* **7,** 469.
261. Chakraborthy, K. B. & Scott, G. (1977) *Europ. Polym. J.,* **13,** 731.
262. Chakraborthy, K. B. & Scott, G. (1977) *Europ. Polym. J.,* **13,** 1007.
263. Chakraborthy, K. B. & Scott, G. (1977) *Polymer,* **18,** 98.
264. Chakraborthy, K. B. & Scott, G. (1978) *Chem. Ind.,* 237.
265. Chakraborthy, K. B. & Scott, G. (1979) *Europ. Polym. J.,* **15,** 35.
266. Chakraborthy, K. B. & Scott, G. (1979) *Polym. Degrad. Stabil.,* **1,** 37.
267. Chakraborthy, K. B. & Scott, G. (1980) *Polym. Commun.,* **21,** 252.
268. Chakraborthy, K. B. & Scott, G. (1984) *J. Polym. Sci., B,* **22,** 553.
269. Chakraborthy, K. B. & Scott, G. (1986) *Polym. Degrad. Stabil.,* **16,** 1.
270. Chakraborthy, K. B., Scott, G. & Poyner, W. R. (1983) *Plast. Rubber Process. Appl.,* **3,** 59.
271. Chakraborthy, K. B., Scott, G. & Poyner, W. R. (1984) *Polym. Degrad. Stabil.,* **8,** 1.
272. Chakraborthy, K. B., Scott, G. & Yaghmour, H. (1985) *Polym. Degrad. Stabil.,* **10,** 221.
273. Chan, K. H., Carlsson, D. J. & Wiles, D. M. (1980) *J. Polym. Sci., B,* **18,** 607.
274. Chandra, R. (1983) *Polym. Photochem.,* **3,** 367.
275. Chandra, R. (1984) *Makromol. Chem.,* **185,** 103.
276. Chandra, R. & Singh, B. P. (1982) *Europ. Polym. J.,* **18,** 199.
277. Chasar, D. W. (1980) *Polym. Degrad. Stabil.,* **3,** 121.
278. Chattopadhyay, S. K., Das, P. K. & Hug, G. L. (1983) *J. Amer. Chem. Soc.,* **105,** 6205.
279. Chaudet, J. H., Newland, G. C., Patton, H. W. & Tamblyn, J. W. (1961) *SPE Trans.,* Jan., 26.

280. Chaudet, J. H. & Tamblyn, J. W. (1961) *SPE Trans.* Jan., 57.
281. Cheng, H. N., Stredonsky, E. R. & Whitesides, G. M. (1976) *Macromolecules,* **9,** 363.
282. Cheu, E. L. & Osawa, Z. (1975) *J. Appl. Polym. Sci.,* **19,** 2947.
283. Chevassus, F. & DeBroutelles, R. (1963) *The Stabilization of Polyvinyl Chloride,* Arnold, London.
284. Chew, C. H., Gan, L. M. & Scott, G. (1977) *Europ. Polym. J.,* **13,** 361.
285. Chien, J. C. W. (1965) *J. Phys. Chem.,* **69,** 4317.
286. Chien, J. C. W. & Boss, C. R. (1967) *J. Polym. Sci., A1,* **5,** 1683.
287. Chien, J. C. W. & Boss, C. R. (1972) *J. Polym. Sci., A1,* **10,** 1579.
288. Chien, J. C. W. & Conner, W. P. (1968) *J. Amer. Chem. Soc.,* **90,** 1001.
289. Chien, J. C. W., Vanderberger, E. J. & Jabloner, H. (1968) *J. Polym. Sci., A1,* **6,** 381.
290. Chirinos-Padrón, A. J., Hernández, P. H., Allen, N. S., Vasilon, C., Marchall, G. P. & DePoortere, M. (1987) *Polym. Degrad. Stabil.,* **19,** 177.
291. Chirinos-Padrón, A. J., Hernández, P. H. Chevaes, E., Allen, N. S., Vasilon, C. & DePoortere, M. (1987) *Europ. Polym. J.,* **23,** 935.
292. Chirinos-Padrón, A. J., Hernández, P. H. & Suarez, F. A. (1988) *Polym. Degrad. Stabil.,* **20,** 237.
293. Chirinos-Padrón, A. J., Suarez, F. A. & Berroterán, H. (1986) *Polym. Degrad. Stabil.,* **14,** 295.
294. Chmelǎ, S. & Hrdlovič, P. (1984) *Chem. Zvesti,* **38,** 199.
295. Chmelǎ, S. & Hrdlovič, P. (1985) *Polym. Degrad. Stabil.,* **11,** 339.
296. Chmelǎ, S., Hrdlovič, P. & Manašek, Z. (1985) *Polym. Degrad. Stabil.,* **11,** 233.
297. Cholvad, V., Staško, A., Tkač, A., Buchachenko, A. L. & Malik, M. (1981) *Collect. Czech. Chem. Commun.,* **46,** 823.
298. Ciceri, L. & Pavioli, F. (1964) Italian Patent 671 214.
299. Cichetti, O. (1970) *Adv. Polym. Sci.,* **7,** 70.
300. Clark, D. T. (1977) *Adv. Polym. Sci.,* **24,** 125.
301. Clark, D. T. (1977) *Characterisation of Metal and Polymer Surfaces* (Lee, L. H., ed.), Academic Press, p. 5.
302. Clark, D. T. (1982) *Pure Appl. Chem.,* **54,** 415.
303. Clark, D. T., Fox, Y. T. C. & Roberts, G. G. (1981) *J. Electron. Spectrosc. Relat. Phenom.,* **22,** 173.
304. Clark, D. T. & Munro, H. S. (1982) *Polym. Degrad. Stabil.,* **4,** 441.
305. Clark, D. T. & Munro, H. S. (1983) *Polym. Degrad. Stabil.,* **5,** 227.
306. Clark, D. T. & Munro, H. S. (1984) *Polym. Degrad. Stabil.,* **8,** 195.
307. Clark, D. T. & Munro, H. S. (1984) *Polym. Degrad. Stabil.,* **8,** 213.
308. Clark, D. T. & Munro, H. S. (1984) *Polym. Degrad. Stabil.,* **9,** 63.
309. Clark, D. T. & Peeling, J. (1981) *Polym. Degrad. Stabil.,* **3,** 177.
310. Clark, D. T. & Stephenson, P. J. (1982) *Polym. Degrad. Stabil.,* **4,** 185.
311. Clark, D. T. & Thomas, H. R. (1978) *J. Polym. Sci., A1,* **16,** 791.
312. Cohen, S. M., Young, R. H. & Markhart, A. H. (1971) *J. Polym. Sci., A1,* **9,** 3263.

313. Coleman, R. A. & Weicksel, J. A. (1959) *Modern Plastics,* (8), 117.
314. Collins, E. A., Bares, J. & Billmeyer, F. W. (1973) *Experiments in Polymer Science,* Wiley–Interscience, New York.
315. Comyn, J. (ed.) (1985) *Polymer Permeability,* Elsevier Applied Science Publishers, London.
316. Comyn, J. (1985) *Polymer Permeability* (Comyn, J., ed.) Elsevier Applied Science Publishers, London, p. 1.
317. Conley, R. T. (1970) *Thermal Stability of Polymers,* Dekker, New York.
318. Cooray, B. B. & Scott, G. (1980) *Europ. Polym. J.,* **16,** 1145.
319. Cooray, B. B. & Scott, G. (1981) *Europ. Polym. J.,* **17,** 379.
320. Cooray, B. B. & Scott, G. (1981) *Europ. Polym. J.,* **17,** 385.
321. Coppinger, G. M. & Bell, E. R. (1966) *J. Phys. Chem.,* **70,** 3479.
322. Coulter, D. R., Gupta, A., Yavrouian, A., Scott, G. W., O'Connor, D., Vogl, O. & Li, S. C. (1986) *Macromolecules,* **19,** 1227.
323. Crank, J. (1975) *The Mathematics of Diffusion,* 2nd edn, Oxford University Press, London.
324. Crank, J. & Park, G. S. (1968) *Diffusion in Polymers,* Academic Press, New York.
325. Cronemeyer, D. G. (1952) *Phys. Rev.,* **87,** 876.
326. Cundall, R. B., Rudham, R. & Salim, M. (1976) *J. Chem. Soc., Faraday I,* **72,** 1642.
327. Cunliffe, A. V. & Davis, A. (1982) *Polym. Degrad. Stabil.,* **4,** 17.
328. Dagonneau, M., Ivanov, V. B., Rozantsev, E. G., Sholle, V. D. & Kagan, E. Sh. (1982) *J. Macromol. Sci., Rev. Macromol. Chem. Phys., C,* **22,** 169.
329. Dai, G., Wu, S. K., Sustic, A. & Vogl, O. (1988) *Polym. Bull.,* **20,** 67.
330. Dan, E. & Guillet, J. E. (1973) *Macromolecules,* **6,** 230.
331. Daniels, T. (1977) *Thermal Analysis,* Kogan Page, London.
332. Darmanyan, A. P. & Tatikolov, A. S. (1986) *J. Photochem.,* **32,** 157.
333. Davidson, R. S. & Meek, R. R. (1982) *Polym. Photochem.,* **2,** 1.
334. Davies, A. G., Griller, D. & Roberts, B. P. (1972) *J. Chem. Soc., Perkin Trans. II,* 933.
335. Davis, A. & Gordon, D. (1974) *J. Appl. Polym. Sci.,* **18,** 1159.
336. Davis, A. & Gordon, D. (1974) *J. Appl. Polym. Sci.,* **18,** 1173.
337. Davis, A. & Gordon, D. (1974) *J. Appl. Polym. Sci.,* **18,** 1181.
338. Davis, A., Howes, B. V., Ledbury, K. J. & Pearce, P. J. (1979) *Polym. Degrad. Stabil.,* **1,** 121.
339. Davis, A. & Sims, D. (1983) *Weathering of Polymers,* Applied Science Publishers, London.
340. Deanin, R. D., Margosiak, S. A. & Llompart, A. (1969) *Modern Plastics,* **46,** 114.
341. Deanin, R. D., Orroth, S. A., Eliasen, R. W. & Greer, T. N. (1970) *Polym. Eng. Sci.,* **10,** 228.
342. Decker, C. (1983) *J. Appl. Polym. Sci.,* **28,** 97.
343. Decker, C., Mayo, F. R. & Richardson, H. (1973) *J. Polym. Sci., A1,* **11,** 2879.
344. Deits, W. & Vogl, O. (1981) *J. Macromol. Sci., Chem.,* **A16,** 1145.

345. De Jonge, C. R. H. I., Van der Maeden, F. P. B., Biemond, A. E. F., Huysmans, W. G. B. & Mijs, W. J. (1976) *J. Polym. Sci., Polym. Symp.*, **57,** 197.
346. Denisov, E. T. (1982) *Developments in Polymer Stabilization—5* (Scott, G., ed.), Applied Science Publishers, London, p. 23.
347. DePaoli, M. A. & Furlan, L. T. (1985) *Polym. Degrad. Stabil.*, **11,** 327.
348. DePaoli, M. A. & Furlan, L. T. (1985) *Polym. Degrad. Stabil.*, **13,** 129.
349. DePaoli, M. A. & Geuskens, G. (1988) *Polym. Degrad. Stabil.*, **21,** 277.
350. Dilks, A. (1981) *J. Polym. Sci., A1,* **19,** 1319.
351. Dills, W. L. & Reeve, T. B. (1970) *Plast. Technol.*, **16,** 50.
352. Dobrescu, V., Andrei, C. & Andrei, G. (1988) *Europ. Polym. J.*, **24,** 289.
353. Domke, W. D. & Steinke, H. (1986) *J. Polym. Sci., A1,* **24,** 2701.
354. Donnet, J. B. & Voet, A. (1976) *Carbon Black: Physics, Chemistry and Elastomer Reinforcement,* Dekker, New York.
355. Dubinskij, V. Z., Roginkij, V. A., Miller, V. & Shlapnikova, I. A. (1975) *Izv. Akad. Nauk SSSR,* 1180.
356. Dulog, L. & Blecher, R. (1982) *Makromol. Chem., Rapid Commun.*, **3,** 153.
357. Dulog, L. & Blecher, R. (1986) *Makromol. Chem.*, **187,** 2357.
358. Dulog, L. & Burg, K. H. (1964) *Z. Anal. Chem.*, **203,** 184.
359. Dunn, J. L. & Heffner, M. H. (1973) *Soc. Plast. Eng. Techn. Paper,* **19,** 483.
360. Dunn, T. S., Williams, E. E. & Williams, J. L. (1982) *J. Polym. Sci., A1*, **20,** 1599.
361. Duran, N. & Mansilla, H. (1981) *J. Macromol. Sci.—Chem.* **A21,** 1467.
362. Durmis, J., Carlsson, D. J., Chan, K. H. & Wiles, D. M. (1981) *J. Polym. Sci., B,* **19,** 549.
363. Durmis, J., Karvaš, M., Čaučik, P. & Holčik, J. (1975) *Europ. Polym. J.*, **11,** 219.
364. Edgar, O. B., Ritchie, C. F., Lawrance, D. R., Hughes, W. & Lederer, G. (1967) UK Patent 1 162 799.
365. Efimov, A. A., Ivanov, V. B., Kutimova, G. V., Lozovskaya, E. L. & Shlyapintokh, V. Ya. (1983) *Polym. Photochem.*, **3,** 231.
366. Efremkin, A. F. & Ivanov, V. B. (1981) *Polym. Photochem.*, **4,** 179.
367. Efremkin, A. F. & Ivanov, V. B. (1982) *Vysokomol. Soedin,* **B24,** 662.
368. Efremkin, A. F., Ivanov, V. B. & Shlyapintokh, V. Ya. (1985) *Europ. Polym. J.*, **21,** 769.
369. Egerton, G. S. & Shah, K. M. (1961) *Nature (London)*, **202,** 81.
370. El-Azmirly, M. A., Mosri, S. E., Issa, R. M. & Barsoum, S. (1975) *Europ. Polym. J.*, **11,** 95.
371. Ellinger, M. L. (1977) *Progr. Org. Coatings,* **5,** 21.
372. Ellinger, M. L. (1979) *J. Oil. Colour Chem. Assoc.*, **62,** 136.
373. Ellinger, M. L. (1979) *Polym. Paints Colour J.*, 112.
374. Ellis, A. B., Kaiser, S. W. & Wrighton, M. S. (1976) *J. Amer. Chem. Soc.*, **98,** 6855.

375. Emanuel, N. M. & Buchachenko, A. L. (1987) *Chemical Physics of Polymer Degradation and Stabilization,* VNU Science Press, Utrecht.
376. Emanuel, N. M., Denisov, E. T. & Maizus, Z. K. (1967) *Liquid Phase Oxidation of Hydrocarbons,* Plenum Press, New York.
377. Emanuel, N. M., Roginsky, B. A. & Buchachenko, A. L. (1982) *Uspekh. Khim.,* **51,** 361.
378. Emanuel, N. M., Zaikov, G. E. & Maizus, Z. K. (1984) *Oxidation of Organic Compounds,* Pergamon Press, Oxford.
379. English, A. D. & Spinelli, H. J. (1984) *Amer. Chem. Soc. Symp. Ser.,* **243,** 257.
380. Ershov, Yu. A. & Gladyshev, G. P. (1977) *Vysokomol. Soedin,* **A19,** 1267.
381. Evans, A. (1979) *Aldrichim. Acta,* **12,** 23.
382. Evans, B. W. & Scott, G. (1974) *Europ. Polym. J.,* **10,** 453.
383. Evans, N. A., Rosevear, J., Waters, P. J. & Wilshire, J. F. K. (1986) *Polym. Degrad. Stabil.,* **14,** 263.
384. Everhart, D. S. & Reilley, C. N. (1981) *Surface Interface Anal.,* **3,** 126.
385. Evleth, E. M. (1976) *J. Amer. Chem. Soc.,* **93,** 1637.
386. Factor, A. & Chu, M. E. (1980) *Polym. Degrad. Stabil.,* **2,** 203.
387. Fairgrieve, S. P. & MacCallum, J. R. (1984) *Polym. Commun.,* **25,** 44.
388. Fairgrieve, S. P. & MacCallum, J. R. (1984) *Polym. Degrad. Stabil.,* **8,** 107.
389. Fairgrieve, S. P. & MacCallum, J. R. (1985) *Polym. Degrad. Stabil.,* **11,** 251.
390. Fairgrieve, S. P. & MacCallum, J. R. (1986) *Polym. Degrad. Stabil.,* **15,** 81.
391. Falkehag, S. I. & Bradon, D. V. (1971) US Patent 3 789 043.
392. Fanton, E., Pellerau, B., Arnaud, R. & Lemaire, J. (1985) *Polym. Photochem.,* **6,** 437.
393. Faucitano, A., Buttafava, A., Martinotti, F. & Comincioli, V. (1986) *Polym. Photochem.,* **7,** 491.
394. Faucitano, A., Buttafava, A., Martinotti, F., Granti, F. & Bortolus, P. (1985) *Polym. Photochem.,* **6,** 259.
395. Feist, W. C. & Hon, D. N. S. (1981) *Amer. Chem. Soc. Symp. Ser.,* **207,** 401.
396. Felder, B. (1985) *Amer. Chem. Soc. Symp. Ser.,* **280,** 69.
397. Felder, B. & Schumacher, R. (1973) *Angew. Makromol. Chem.,* **31,** 35.
398. Felder, B., Schumacher, R. & Sitek, F. (1980) *Chem. Ind.,* 155.
399. Felder, B., Schumacher, R. & Sitek, F. (1980) *Helv. Chim. Acta,* **63,** 132.
400. Felder, B., Schumacher, R. & Sitek, F. (1980) *Org. Coat. Plast. Chem.,* **42,** 561.
401. Felder, B., Schumacher, R. & Sitek, F. (1981) *Amer. Chem. Soc. Symp. Ser.,* **151,** 65.
402. Feldman, D. (1981) *J. Appl. Polym. Sci.,* **26,** 3493.
403. Fernando, W. S. E. & Scott, G. (1980) *Europ. Polym. J.,* **16,** 971.
404. Fertig, J., Goldberg, A. I. & Skoultchi, M. (1965) *J. Appl. Polym. Sci.,* **9,** 903.

405. Fertig, J., Goldberg, A. I. & Skoultchi, M. (1966) *J. Appl. Polym. Sci.*, **10,** 663.
406. Figge, K., Kocha, J. & Freytag, W. (1978) *Food Cosmet. Toxic.* **16,** 135.
407. Fitton, S. L., Haward, R. N. & Williamson, G. R. (1970) *Brit. Polym. J.*, **2,** 217.
408. Fitzky, H. G., Kämpf, G., Klaeren, A. & Völtz, H. G. (1980) *Amer. Chem. Soc., Div. Org. Coat. Plast. Chem., Preprints,* **42,** 660.
409. Fitzky, H. G., Kämpf, G., & Völtz, H. G. (1974) *Progr. Org. Coatings,* **3,** 223.
410. Flockhart, B. D., Ivin, K. J., Pink, R. C. & Sharman, B. D. (1971) *Chem. Commun.*, 339.
411. Flom, S. R. & Barbara, P. F. (1983) *Chem. Phys. Lett.*, **94,** 488.
412. Flood, J., Russell, K. E. & Wan, J. K. S. (1973) *Macromolecules,* **6,** 669.
413. Foote, C. F., Change, Y. C. & Denny, R. W. (1970) *J. Amer. Chem. Soc.*, **92,** 5216.
414. Foote, C. F. & Denny, R. W. (1968) *J. Amer. Chem. Soc.*, **90,** 6233.
415. Foote, C. S., Denny, R. W., Weaver, L., Chang, Y. & Peters, J. (1970) *Ann. N.Y. Acad. Sci.*, **171,** 139.
416. Fornes, R. E., Gilbert, R. D. & Stowe, B. S. (1974) *Polym. Plast. Technol.*, **3,** 159.
417. Forrester, A. K., Hay, G. M. & Thomson, R. H. (1967) *Organic Chemistry of Stable Free Radicals,* Academic Press, London, p. 405.
418. Forsman, J. P. (1964) *Soc. Plast. Eng. Tech. Papers,* **10,** 8-2.
419. Frank, H. P. (1976) *J. Polym. Sci. Polym. Symp.*, **57,** 311.
420. Frank, H. P. & Lehner, H. (1970) *J. Polym. Sci., Polym. Symp.*, **31,** 193.
421. Freeman, E. S. & Caroll, B. (1958) *J. Phys. Chem.*, **62,** 394.
422. Fujishima, A. & Honda, K. (1972) *Nature (London),* **238,** 37.
423. Fujita, T., Kurumada, T., Toda, T. & Yoshika, T. (1981) *J. Polym. Sci., B,* **19,** 609.
424. Furimsky, E. & Howard, J. A. (1973) *J. Amer. Chem. Soc.*, **95,** 369.
425. Furlan, L. T., Rodrigues, M. A. & dePaoli, M. A. (1985) *Polym. Degrad. Stabil.*, **13,** 337.
426. Furneaux, G. C., Ledbury, K. J. & Davis, A. (1980) *Polym. Degrad. Stabil.*, **3,** 431.
427. Furue, H. & Russell, K. E. (1978) *Can. J. Chem.*, **56,** 1598.
428. Gächter, R. & Müller, H. (eds) (1985) *Plastic Additives Handbook,* Hanser, Munich.
429. Ganem, B. (1975) *J. Org. Chem.*, **40,** 1998.
430. Gantz, G. M. & Summer, G. W. (1957) *Text. Res. J.*, **24,** 244.
431. Gardette, J. L. & Lemaire, J. (1986) *Polym. Photochem.*, **7,** 409.
432. Garner, A. & Wilkinson, F. (1978) *Singlet Oxygen* (Rånby, B. & Rabek, J. F. eds) Wiley, Chichester, p. 251.
433. Garton, A., Carlsson, D. J. & Wiles, D. M. (1978) *J. Polym. Sci., A1,* **16,** 33.

434. Garton, A., Carlsson, D. J. & Wiles, D. M. (1979) *Macromolecules,* **12,** 1071.
435. Garton, A., Carlsson, D. J. & Wiles, D. M. (1980) *Developments in Polymer Photochemistry—1* (Allen, N. S., ed.), Applied Science Publishers, London, p. 93.
436. Garton, A., Carlsson, D. J. & Wiles, D. M. (1980) *J. Polym. Sci., A1,* **18,** 3245.
437. Garton, A., Carlsson, D. J. & Wiles, D. M. (1980) *Macromol. Chem.,* **181,** 1841.
438. Garton, A., McLean, P. D., Wiebe, W. & Densley, R. J. (1986) *J. Appl. Polym. Sci.,* **32,** 3941.
439. Geddes, W. C. (1964) *Europ. Polym. J.,* **3,** 733.
440. Geetha, R., Torikai, A., Nagaya, S. & Fueki, K. (1987) *Polym. Degrad. Stabil.,* **19,** 279.
441. George, G. A. (1974) *J. Appl. Polym. Sci.,* **18,** 117.
442. George, G. A. (1979) *Polym. Degrad. Stabil.,* **1,** 217.
443. George, G. A. (1981) *Developments in Polymer Degradation—3* (Grassie, N., ed.), Applied Science Publishers, London, p. 173.
444. Gerarde, H. W. (1960) *Toxicology and Biochemistry of Aromatic Compounds,* Elsevier, London.
445. Gerlock, J. L. (1983) *Anal. Chem.,* **54,** 1529.
446. Gerlock, J. L. & Bauer, D. H. (1984) *J. Polym. Sci., A1,* **22,** 447.
447. Gerlock, J. L., Bauer, D. R. & Briggs, L. M. (1984) *Amer. Chem. Soc. Symp. Ser.,* **243,** 285.
448. Gerlock, J. L., Bauer, D. R. & Briggs, L. M. (1984) *Amer. Chem. Soc. Symp. Ser.,* **280,** 119.
449. Gerlock, J. L., Bauer, D. R. & Briggs, L. M. (1985) *Polym. Degrad. Stabil.,* **14,** 53.
450. Gerlock, J. L., Mielewski, D. F. & Nauer, D. R. (1988) *Polym. Degrad. Stabil.,* **20,** 123.
451. Gerlock, J. L., Riley, T. & Bauer, D. R. (1985) *Polym. Degrad. Stabil.,* **14,** 73.
452. Gerlock, J. L., Van Oene, H. & Bauer, D. R. (1983) *Europ. Polym. J.,* **19,** 11.
453. Geuskens, G., Debie, F., Kabamba, M. S. & Nedelkos, G. (1984) *Polym. Photochem.,* **5,** 313.
454. Geuskens, G. & Kamba, M. S. (1982) *Polym. Degrad. Stabil.,* **4,** 69.
455. Geuskens, G. & Kamba, M. S. (1983) *Polym. Degrad. Stabil.,* **5,** 399.
456. Geuskens, G. & Kamba, M. S. (1987) *Polym. Degrad. Stabil.,* **19,** 315.
457. Geuskens, G. & Nedelkos, G. (1987) *Polym. Degrad. Stabil.,* **19,** 365.
458. Ghaemy, M. & Scott, G. (1980) *Polym. Degrad. Stabil.,* **3,** 253.
459. Ghaemy, M. & Scott, G. (1980) *Polym. Degrad. Stabil.,* **3,** 405.
460. Ghaffar, A., Scott, A. & Scott, G. (1976) *Europ. Polym. J.,* **12,** 615.
461. Ghaffar, A., Scott, A. & Scott, G. (1977) *Europ. Polym. J.,* **13,** 83.
462. Ghaffar, A., Scott, A. & Scott, G. (1977) *Europ. Polym. J.,* **13,** 89.
463. Ghatge, N. D. & Vernekar, S. P. (1971) *Angew. Makromol. Chem.,* **20,** 165.

464. Ghatge, N. D. & Vernekar, S. P. (1971) *Angew. Makromol. Chem.*, **20,** 175.
465. Ghiggino, K. P., Scully, A. D., Bigger, S. W., Yandell, M. D. & Vogl, O. (1988) *J. Polym. Sci., C,* **26,** 505.
466. Giesse, R. & DePaoli, M. A. (1988) *Polym. Degrad. Stabil.*, **21,** 181.
467. Gijzemann, O. L. J., Kaufman, F. & Porter, G. (1973) *J. Chem. Soc., Faraday Trans. II,* **69,** 727.
468. Gilbert, B. C., Gill, B. & Ramsden, M. J. (1979) *Chem. Ind.*, 283.
469. Gilbert, B. C., Gill, B. & Sexton, M. D. (1978) *J. Chem. Soc., Perkin II,* 530.
470. Gilead, D. & Scott, G. (1982) *Developments in Polymer Stabilization—5* (Scott, G., ed.), Applied Science Publishers, London, p. 71.
471. Glavind, J. & Hartmann, S. (1955) *Acta Chem. Scand.*, **9,** 497.
472. Gleria, M., Minto, F., Flamigni, L. & Bortolus, P. (1988) *Polym. Degrad. Stabil.*, **22,** 125.
473. Glück, L. & Poschet, G. (1982) *Kunststoffe,* **72,** 353.
474. Goh, S. H. (1984) *Polym. Degrad. Stabil.*, **8,** 123.
475. Goh, S. H., Lee, S. Y. & Siow, K. S. (1986) *J. Appl. Polym. Sci.*, **31,** 2055.
476. Goh, S. H. & Lim, Y. B. (1979) *Thermochim. Acta,* **32,** 81.
477. Goh, S. H., Ng, A. S. & Kok, P. S. (1982) *J. Rubber Res. Inst. Malaysia,* **30,** 45.
478. Goldberg, V. M. & Zaikov, G. E. (1987) *Polym. Degrad. Stabil.*, **19,** 221.
479. Golemba, F. J. & Guillet, J. E. (1969) *J. Paint Technol.*, **41,** 315.
480. Golubev, V. A., Zhdanov, R. I., Gida, V. M. & Rozantsev, E. G. (1971) *Izv. Akad. Nauk SSSR, Ser. Khim.*, 853.
481. Gordon, D. A. (ed.) (1970) *Enyclopedia of Chemical Technology,* Vol. 21, Interscience, New York, p. 115.
482. Gordon, D. A. (1986) *Amer. Chem. Soc. Adv. Chem. Ser.*, **85,** 224.
483. Gordon, G. Ya. (1964) *Stabilization of Synthetic High-Polymers,* Israel Program for Scientific Translations, Jerusalem.
484. Gould, R. W., Henman, T. J. & Billingham, N. C. (1984) *Brit. Polym. J.*, **16,** 284.
485. Grassie, N. & Scott, G. (1985) *Polymer Degradation and Stabilisation,* Cambridge University Press, Cambridge.
486. Grassie, N. & Weir, N. A. (1965) *J. Appl. Polym. Sci.*, **9,** 963.
487. Grassie, N. & Weir, N. A. (1965) *J. Appl. Polym. Sci.*, **9,** 975.
488. Grattan, D. W., Carlsson, D. J., Howard, J. A. & Wiles, D. M. (1979) *Can. J. Chem.*, **57,** 2834.
489. Grattan, D. W., Carlsson, D. J. & Wiles, D. M. (1978) *Chem. Ind.*, 228.
490. Grattan, D. W., Carlsson, D. J. & Wiles, D. M. (1979) *Polym. Degrad. Stabil.*, **1,** 69.
491. Grattan, D. W., Reddoh, A. H., Carlsson, D. J. & Wiles, D. M. (1978) *J. Appl. Polym. Sci., B,* **16,** 143.
492. Green, J. A., II & Singer, L. A. (1974) *J. Amer. Chem. Soc.*, **96,** 2730.

493. Green, J. A. II, Singer, L. A. & Parks, J. H. (1973) *J. Chem. Phys.*, **58,** 2690.
494. Griva, A. P. & Denisov, E. T. (1976) *J. Polym. Sci., A1,* **14,** 1051.
495. Grossman, G. W. (1977) *J. Coating Technol.*, **49,** 45.
496. Grossman, G. W. (1977) *J. Coating Technol.*, **49,** 633.
497. Grossman, G. W., Stolarczyk, A. & Vogl, O. (1981) *Monatsh. Chem.*, **112,** 1279.
498. Gugumus, F. (1979) *Developments in Polymer Stabilization—1* (Scott, G., ed.), Applied Science Publishers, London, p. 261.
499. Gugumus, F. (1981) *Res. Discl.*, **209,** 357.
500. Gugumus, F. (1985) Paper presented at the 17th Colloquium of Danubian Countries on *Natural and Artificial Ageing of Plastics,* Basel (Switzerland), 4–6 June.
501. Gugumus, F. (1985) Paper presented at the Symposium *Polymer Degradation and Stabilisation: Problems, Techniques and Applications,* Manchester, 18–25 September.
502. Gugumus, F. (1985) *Plastic Additives* (Gächter, A. & Muller, H., eds), Hanser, Munich, p. 97.
503. Gugumus, F. (1987) *Developments in Polymer Stabilisation—8* (Scott, G., ed.), Elsevier Applied Science Publishers, London, p. 239.
504. Gugumus, F. (1987) *Kunststoffe,* **77,** 1065.
505. Gugumus, F. (1987) *Kunststoffe,* **77,** 1070.
506. Guillet, J. E. (1977) *Pure Appl. Chem.*, **49,** 249.
507. Guillet, J. E. (1978) *Amer. Chem. Soc. Symp. Ser.*, **169,** 1.
508. Guillet, J. E. (1985) *Polymer Photophysics and Photochemistry,* Cambridge University Press, Cambridge.
509. Guillory, J. P. & Becker, H. S. (1974) *J. Polym. Sci., A1,* **12,** 993.
510. Guillory, J. P. & Cook, C. F. (1971) *J. Polym. Sci., A1,* **9,** 1529.
511. Guillory, J. P. & Cook, C. F. (1973) *J. Amer. Chem. Soc.*, **95,** 4885.
512. Guillory, J. P. & Cook, C. F. (1973) *J. Polym. Sci., A1,* **11,** 1927.
513. Gupta, A., Coulter, D., Miskowski, V. (1986) *Amer. Chem. Soc., Div. Polym. Chem., Polym. Preprints,* **27** (2), 372.
514. Gupta, A., Liang, R., Coulter, D., Vogl, O. & Scott, G. (1984) *Amer. Chem. Soc., Div. Polym. Chem., Polym. Preprints,* **25** (1), 54.
515. Gupta, A., Scott, G. & Kliger, D. (1981) *Amer. Chem. Soc. Symp. Ser.*, **151,** 27.
516. Gupta, A., Scott, G., Kliger, D. & Vogl, O. (1983) *Amer. Chem. Soc. Symp. Ser.*, **220,** 293.
517. Gupta, A., Scott, G. & Vogl, O. (1982) *Amer. Chem. Soc., Div. Polym. Chem., Polym. Preprints,* **23** (1), 219.
518. Gupta, B. D., Jiráčkova-Audouin, L. & Verdu, J. (1988) *Europ. Polym. J.*, **24,** 947.
519. Gysling, H. & Heller, H. J. (1961) *Kunststoffe,* **51,** 13.
520. Hansen, R. H., Russell, C. A., DeBenedictis, T., Martin, W. M. & Pascale, J. Y. (1964) *J. Polym. Sci., A1,* **2,** 587.
521. Hanson, D. E. (1978) ICI Technical Service Note PP132.
522. Harbour, J. R. & Hair, M. L. (1977) *J. Phys. Chem.*, **81,** 1791.

523. Hardy, W. B. (1980) *Developments in Polymer Photochemistry—3* (Allen, N. S., ed.), Applied Science Publishers, London, p. 287.
524. Harper, D. J. & McKellar, J. F. (1974) *J. Appl. Polym. Sci.,* **18,** 1233.
525. Harrick, N. J. (1977) *Appl. Spektr.,* **31,** 548.
526. Harrick, N. J. (1979) *Internal Reflection Spectroscopy,* Harrick Pressing, New York.
527. Haslam, J., Willis, H. A. & Squirrel, D. C. M. (1980) *Identification and Analysis of Plastics,* Heyden, London.
528. Hauserman, R. G. & Johnson, M. (1976) *J. Appl. Polym. Sci.,* **20,** 2533.
529. Hawkins, W. L. (1963) *Rubber Plast. Weekly,* **142,** 291.
530. Hawkins, W. L. (1972) *Polymer Stabilization,* Wiley–Interscience, New York.
531. Hawkins, W. L. (1976) *J. Polym. Sci., Polym. Symp.,* **57,** 319.
532. Hawkins, W. L. (1984) *Polymer Degradation and Stabilization,* Springer Verlag, Berlin.
533. Hawkins, W. L., Hansen, R. H., Matreyek, W. & Winslow, F. W. (1959) *J. Appl. Polym. Sci.,* **1,** 37.
534. Hawkins, W. L., Lauza, V. L., Loeffer, B. B., Matreyek, W. & Winslow, F. H., *J. Appl. Polym. Sci.,* **1,** 43.
535. Hawkins, W. L. & Sautter, H. (1963) *J. Polym. Sci., A1,* **1,** 3499.
536. Hawkins, W. L. & Winslow, F. H. (1961) *Plast. Inst. Trans. J.,* **29,** 82.
537. Hawkins, W. L., Worthington, M. A. & Matreyek, W. (1960) *J. Appl. Polym. Sci.,* **3,** 277.
538. Heacock, J. F. (1963) *J. Appl. Polym. Sci.,* **7,** 2319.
539. Heller, H. J. (1969) *Europ. Polym. J. Supplement,* 105.
540. Heller, H. J. & Blattmann, H. R. (1972) *Pure Appl. Chem.,* **30,** 145.
541. Heller, H. J. & Blattmann, H. R. (1973) *Pure Appl. Chem.,* **36,** 141.
542. Heller, H. J. & Wasserman, E. (1965) *J. Chem. Phys.* **42,** 949.
543. Hemsley, D. A. (1984) *The Light Microscopy of Synthetic Polymers,* Oxford University Press and Royal Microscopical Society, Oxford.
544. Henman, T. J. (1982) World Index of Polyolefine Stabilizers, Kogan Page, London.
545. Henman, T. J. (1985) *Developments in Polymer Stabilization—6* (Scott, G., ed.), Applied Science Publishers, London, p. 107.
546. Henman, T. J. & Oldland, S. R. (1984) *Developments in Polymer Stabilization—7* (Scott, G., ed.), Applied Science Publishers, London, p. 275.
547. Hermann, J. M., Disdier, J. & Pichat, P. (1979) *J. Catal.,* **60,** 369.
548. Heskins, M. & Guillet, J. E. (1968) *Macromolecules,* **1,** 97.
549. Hewgill, F. R. & Smith, W. T. (1976) *J. Polym. Sci., B,* **14,** 463.
550. Heyward, I. P., Chan, M. G. & Ludwick, A. G. (1985) *Amer. Chem. Soc. Symp. Ser.,* **280,** 299.
551. Hilado, C. J. (1976) *Pyrolysis of Polymers,* Technomic Publications, Westport.
552. Hirt, R. C., Searle, N. Z. & Schmitt, R. G. (1961) *Soc. Plast. Eng. Trans.,* 21.
553. Hodgeman, D. K. C. (1979) *Polym. Degrad. Stabil.,* **1,** 155.

554. Hodgeman, D. K. C. (1980) *J. Polym. Sci., A1,* **118,** 533.
555. Hodgemen, D. K. C. (1981) *J. Polym. Sci., A1,* **19,** 807.
556. Hodgeman, D. K. C. (1982) *Developments in Polymer Degradation—4* (Grassie, N., ed.), Applied Science Publishers, London, p. 189.
557. Holčik, J., Karvaš, M., Kaššovicova, D. J. & Durmis, J. (1976) *Europ. Polym. J.,* **12,** 173.
558. Holčik, J., Koenig, J. L. & Shelton, J. R. (1983) *Polym. Degrad. Stabil.,* **5,** 373.
559. Holdsworth, J. D., Scott, G. & Williams, D. (1964) *J. Chem. Soc.,* 4692.
560. Holmström, A. & Sörvik, E. M. (1978) *J. Polym. Sci., A1,* **16,** 2555.
561. Holtzen, D. W. (1976) *Proc. Soc. Plast. Eng. ANTEC,* **22,** 4888.
562. Hon, D. N. S. (1981) *Developments in Polymer Degradation—3* (Grassie, N., ed.), Applied Science Publishers, London, p. 229.
563. Hon, D. N. S. (1983) *J. Appl. Polym. Sci., Appl. Polym. Symp.,* **37,** 845.
564. Horng, P. L. & Klemchuk, P. O. (1985) *Amer. Chem. Soc. Symp. Ser.,* **280,** 235.
565. Horswill, E. C. & Ingold, K. U. (1966) *Can. J. Chem.,* **44,** 263.
566. Horvath, J. W., Grimm, D. C. & Stevčik, J. A. (1975) *Rubber Chen. Technol.,* **48,** 337.
567. Howard, J. A. & Thait, J. C. (1978) *J. Org. Chem.,* **43,** 4279.
568. Howell, B. F., McCrackin, F. L. & Wang, F. W. (1985) *Polymer,* **26,** 433.
569. Hoytink, G. (1969) *Account Chem. Res.,* **2,** 114.
570. Hrdlovič, P., Daněček, J., Karvaš, M. & Durmis, J. (1974) *Chem. Zvesti,* **28,** 792.
571. Hrdlovič, P., Lukač, I. & Maňásek, Z. (1972) *Chem. Zvesti,* **26,** 433.
572. Hubart, F. J. & Rowley, G. N. (1972) *J. Appl. Polym. Sci.,* **16,** 715.
573. Hughes, W. (ed.) (1970) *Tenth FATIPEC Congress,* Verlag Chemie, Weinheim, p. 67.
574. Hult, A. & Rånby, B. (1984) *Polym. Degrad. Stabil.,* **8,** 75.
575. Hult, A. & Rånby, B. (1984) *Polym. Degrad. Stabil.,* **8,** 89.
576. Hult, A. & Rånby, B. (1984) *Polym. Degrad. Stabil.,* **9,** 1.
577. Hult, A., Yuan, Y. Y. & Rånby, B. (1984) *Polym. Degrad. Stabil.,* **8,** 241.
578. Humphris, K. J. & Scott, G. (1973) *J. Chem. Soc., Perkin II,* 826.
579. Humphris, K. J. & Scott, G. (1973) *Pure Appl. Chem.,* **36,** 163.
580. Husband, M. J. & Scott, G. (1979) *Europ. Polym. J.,* **15,** 249.
581. Huston, A. L. & Scott, G. (1982) *Proc. SPIE Int. Soc. Optical Engineering,* **332,** 215.
582. Huston, A. L. & Scott, G. (1987) *J. Phys. Chem.,* **91,** 1408.
583. Huston, A. L., Scott, G. & Gupta, A. (1982) *J. Chem. Phys.,* **76,** 4978.
584. Hutson, G. V. & Scott, G. (1972) *Chem. Ind.,* 725.
585. Huvet, A., Philippe, J. & Verdu, J. (1978) *Europ. Polym. J.,* **14,** 709.
586. Ide, F., Oseki, T., Itoh, K. & Kimura, K. (1968) Japanese Patent 68 16397.
587. Ingold, K. U. (1956) *Can. J. Chem.,* **34,** 600.

588. Ingold, K. U. (1963) *Can. J. Chem.*, **41,** 2816.
589. Irick, G. Jr. (1972) *J. Appl. Polym. Sci.*, **16,** 2387.
590. Irick, G. Jr., Newland, G. C. & Wang, R. H. S. (1980) *Amer. Chem. Soc., Div. Coat. Plastics, Preprints,* **42,** 653.
591. Irick, G. Jr., Newland, G. C. & Wang, R. H. S. (1981) *Amer. Chem. Soc. Symp. Ser.*, **151,** 147.
592. Iring, M., Foldes, E., Barabsa, K., Kelen, T. & Tüdos, F. (1986) *Polym. Degrad. Stabil.*, **14,** 319.
593. Ito, M. & Porter, R. S. (1982) *J. Appl. Polym. Sci.*, **27,** 4471.
594. Ivanov, S. K. & Kateva, J. D. (1976) *J. Polym. Sci., Polym. Symp.*, **37,** 237.
595. Ivanov, V. B. & Anisimova, O. M. (1980) *Dokl. Akad. Nauk SSSR,* **253,** 1401.
596. Ivanov, V. B., Burkova, S. G., Morozov, Yu. L. & Shlyapintokh, V. Ya. (1977) *Vysokomol. Soedin,* **19B,** 359.
597. Ivanov, V. B., Burlatsky, S. F., Rozenboym, N. A. & Shlyapintokh, V. Ya. (1980) *Europ. Polym. J.*, **16,** 65.
598. Ivanov, V. B., Efremkin, A. F., Rozenboym, N. A. & Shlyapintokh, V. Ya. (1983) *Vysokomol. Soedin,* **A25,** 1209.
599. Ivanov, V. B., Lozovskaya, E. L., Efremkin, A. F. & Shlyapintokh, V. Ya. (1983) *Angew. Makromol. Chem.*, **114,** 35.
600. Ivanov, V. B., Lozovskaya, E. L. & Shlyapintokh, V. Ya. (1982) *Polym. Photochem.*, **2,** 55.
601. Ivanov, V. B., Rozenboym, N. A. & Angert, L. G. (1982) *Vysokomol. Soedin,* **B24,** 234.
602. Ivanov, V. B., Rozenboym, N. A., Angert, L. G. & Shlyapintokh, V. Ya. (1978) *Dokl. Akad. Nauk SSSR,* **241,** 609.
603. Ivanov, V. B. & Shlyapintokh, V. Ya. (1987) *Developments in Polymer Stabilization—8* (Scott, G., ed.), Applied Science Publishers, London, p. 29.
604. Ivanov, V. B., Shlyapintokh, V. Ya., Khvostach, O. M., Shapiro, A. B. & Rozantsev, E. G. (1974) *Izv. Akad. Nauk SSSR,* Ser Khim, 1916.
605. Ivanov, V. B., Shlyapintokh, V. Ya., Khvostach, O. M., Shapiro, A. B. & Rozantsev, E. G. (1975) *J. Photochem.*, **4,** 313.
606. Ivanov, V. B. & Zhuravlev, M. A. (1986) *Polym. Photochem.*, **7,** 55.
607. Iwasaki, M., Tirrell, D. & Vogl, O. (1980) *J. Polym. Sci.*, **18,** 2755.
608. Jackson, R. A., Oldland, S. R. D. & Pajaczkowski, A. (1968) *J. Appl. Polym. Sci.*, **12,** 1297.
609. Jaeger, W., Schulke, U., Dietrick, K. & Reinish, G. (1970) British Patent 1 311 526.
610. Jelinski, L. W., Dumais, J. J., Luongo, J. P. & Cholli, A. L. (1984) *Macromolecules,* **17,** 1650.
611. Jellinek, H. H. G. (1978) *Aspects of Degradation and Stabilization of Polymers,* Elsevier, Amsterdam.
612. Jellinek, H. H. G. & Lipovac, S. N. (1970) *Macromolecules,* **3,** 237.
613. Jian, S. Z., Lucki, J., Rabek, J. F. & Rånby, B. (1985) *Amer. Chem. Soc. Symp. Ser.*, **280,** 353.

614. Jiráčkova, L., Jelinkova, T., Totschova, J. & Pospišil, J. (1979) *Chem. Ind.*, 384.
615. Jiráčkova, L. & Pospišil, J. (1972) *Europ. Polym. J.*, **8**, 75.
616. Jiráčkova, L. & Pospišil, J. (1978) *Angew. Makromol. Chem.*, **66**, 95.
617. Jiráčkova-Audouin, L., Verdu, J. & Pospilšil, J. (1984) *Polym. Degrad. Stabil.*, **7**, 67.
618. Johnson, M. & Hauserman, R. G. (1977) *J. Appl. Polym. Sci.*, **21**, 3457.
619. Jones, H. C. (1972) *Modern Plastics*, **49**(1), 90.
620. Jordan, J. M., McIllroy, R. E. & Pearce, E. M. (1967) *J. Appl. Polym. Sci., Appl. Polym. Symp.*, **4**, 205.
621. Joschek, H. J. & Miller, S. J. (1966) *J. Amer. Chem. Soc.*, **88**, 14.
622. Kamal, M. R. (ed.) (1967) *Weatherability of Plastic Materials J. Appl. Polym. Sci., Appl. Polym. Symp.* No. 4, Interscience Publishers, New York.
623. Kamal, M. R. (1970) *Polym. Eng. Sci.*, **10**, 108.
624. Kamiya, Y. (1972) *J. Catal.*, **24**, 69.
625. Kamogawa, H., Nanasawa, M. & Uehara, Y. (1977) *J. Polym. Sci., B*, **15**, 675.
626. Kämpf, G., Völz, H. G. & Klaeren, A. (1978) *XIV FATIPEC Congress*, Budapest, 4–9 June, p. 291.
627. Kaplan, M. L., Kelleher, P. G., Bebbington, G. H. & Hartless, R. L. (1973) *J. Polym. Sci., B*, **11**, 357.
628. Karcher, W., Haesen, G. & Schwarze, A. (1979) *Chem. Ind.*, 342.
629. Karpukhin, O. N. (1976) *Mater. Plast. Elast.*, **3**, 116.
630. Karpukhin, O. N. & Slobodetskaya, E. M. (1979) *J. Polym. Sci., A1*, **17**, 3687.
631. Karrer, F. E. (1980) *Makromol. Chem.*, **181**, 595.
632. Karvaš, M., Jexová, E., Holčik, J. & Balog, A. (1968). *Chemicky Prum*, **18**, 427.
633. Katbab, A. A., Ogunbanjo, A. & Scott, G. (1985) *Polym. Degrad. Stabil.*, **12**, 333.
634. Kato, K. (1974) *J. Appl. Polym. Sci.*, **18**, 2449.
635. Ke, B. (1964) *Thermal Analysis of High Polymers*, Wiley–Interscience, New York.
636. Keana, I. F. W. (1978) *Chem. Rev.*, **78**, 37.
637. Kelen, T., Tüdos, F., Balint, G. & Rockenbauer, A. (1985) *Amer. Chem. Soc. Symp. Ser.*, **280**, 109.
638. Kenpicher, V. P., Yajubischi, L. K. & Malglysh, C. N. (1970) *Rubber Chem. Technol.*, **43**, 1228.
639. Kesting, R. E. (1971) *Synthetic Polymeric Membranes*, McGraw–Hill, New York.
640. Khalifa, W. M., El-Azmerly, M. A., Morsi, S. E. & Barsoum, S. (1979) *J. Appl. Polym. Sci.*, **23**, 2225.
641. Kikkawa, K., Nakahara, Y. & Ohkatsu, Y. (1987) *Polym. Degrad. Stabil.*, **18**, 237.
642. Killgoar, P. C., Jr & van Oene, H. (1976) *Amer. Chem. Soc. Symp. Ser.*, **25**, 407.

643. Kiryushkin, S. G., Yakimchenko, O. E., Shlyapintokh, Yu. A., Parisky, G. B., Toptygin, D. Ya. & Lebedev, Ya. S. (1975) *Vysokomol. Soedin,* **17B,** 385.
644. Kishore, K., Ravindran, K. & Nagarajan, R. (1984) *J. Polym. Sci., B,* **22,** 539.
645. Kitayama, M. & Vogl, O. (1982) *J. Macromol. Sci—Chem.,* **19,** 375.
646. Kitayama, M. & Vogl, O. (1982) *Polym. J.,* **14,** 537.
647. Klein, A. J., Yu, H. & Yen, W. M. (1981) *J. Appl. Polym. Sci.,* **26,** 2381.
648. Klemchuk, P. P. (1983) *Polym. Photochem.,* **3,** 1.
649. Klemchuk, P. P. (1985) *Amer. Chem. Soc. Symp. Ser.,* **280,** 1.
650. Klemchuk, P. P. & Gande, M. E. (1988) *Polym. Degrad. Stabil.,* **22,** 141.
651. Kline, R. H. & Miller, P. (1973) *Rubber Chem. Technol.,* **46,** 96.
652. Klöpffer, W. (1970) *Kunststoffe,* **60,** 385.
653. Klöpffer, W. (1976) *J. Polym. Sci., Polym. Symp.* **57,** 205.
654. Knoll, H. & Kuhnold, U. (1967) *Angew. Chem. Int. Ed. Engl.,* **6,** 978.
655. Koenig, J. L. (1975) *Appl. Spectrosc.,* **29,** 293.
656. Koenig, J. L. & Anton, M. K. (1978) *Appl. Opt.,* **17,** 1374.
657. Kolawole, E. G. (1984) *Europ. Polym. J.,* **20,** 629.
658. Kolawole, E. G. & Adeniyi, J. B. (1982) *Europ. Polym. J.,* **18,** 469.
659. Kolawole, E. G. & Agboola, M. O. (1982) *J. Appl. Polym. Sci.,* **27,** 2317.
660. Kolawole, E. G., Mbamali, G. E. & Olayemi, J. Y. (1980) *J. Appl. Polym. Sci.,* **25,** 2133.
661. Kolawole, E. G. & Olugbeni, P. O. (1985) *Europ. Polym. J.,* **21,** 183.
662. Kolawole, E. G. & Scott, G. (1981) *J. Appl. Polym. Sci.,* **26,** 2581.
663. König, T., Männel, D., Habicher, W. D. & Schwetlick, K. (1988) *Polym. Degrad. Stabil.,* **22,** 137.
664. König, T., Schwetlick, K., Kudelka, I. & Pospišil, J. (1986) *Polym. Degrad. Stabil.,* **15,** 151.
665. Kotulak, L., Rotschová, J. & Pospišil, J. (1983) *Collect. Czech. Chem. Comm.,* **48,** 3384.
666. Kovács, E. & Wolkober, Z. (1976) *J. Polym. Sci., Polym. Symp.,* **57,** 171.
667. Kramer, E. & Koppenhamn, J. (1986) *Polym. Degrad. Stabil.,* **16,** 261.
668. Krejcar, E. & Kolar, O. (1972) *Progr. Org. Coatings,* **1,** 249.
669. Kresta, J., Majer, J. & Vessely, K. (1968) *J. Polym. Sci., Polym. Symp.,* **22,** 329.
670. Kúdelka, I., Brodilová, J., Prusiková, M. & Pospišil, J. (1987) *Polym. Degrad. Stabil.,* **17,** 287.
671. Kúdelka, I., Misro, P. K., Pospišil, J., Korbanka, H., Riedel, T. & Pfahler, G. (1985) *Polym. Degrad. Stabil.,* **12,** 303.
672. Kuhnhold, U. (1962) Uber die Photochemische Bildung von Titanperoxid, Dissertation, Freien University, Berlin.
673. Kularatne, K. W. S. & Scott, G. (1978) *Europ. Polym. J.,* **14,** 835.
674. Kularatne, K. W. S. & Scott, G. (1979) *Europ. Polym. J.,* **15,** 827.

675. Kurosaki, T., Lee, K. W. & Okawara, M. (1972) *J. Polym. Sci., A1,* **10,** 3295.
676. Kurumada, T., Ohsawa, H., Fujita, T. & Toda, T. (1982) *J. Polym. Sci., A1,* **22,** 277.
677. Kurumada, T., Oshawa, H., Fujita, T., Toda, T. & Yoshioka, T. (1985) *J. Polym. Sci., A1,* **23,** 2747.
678. Kurumada, T., Ohsawa, H., Oda, O., Fujita, T., Toda, T. & Yoshioka, T. (1985) *J. Polym. Sci., A1,* **23,** 1477.
679. Kurumada, T., Ohsawa, H. & Yamazaki, T. (1987) *Polym. Degrad. Stabil.,* **19,** 263.
680. Küster, B. & Herlinger, H. (1974) *Angew. Makromol. Chem.,* **40,** 265.
681. Küster, B., Tschang, C. J. & Herlinger, H. (1976) *Angew. Makromol. Chem.,* **54,** 55.
682. Kuzmin, V. A. & Tatikolov, A. S. (1977) *Chem. Phys. Lett.,* **51,** 45.
683. Kuzmin, V. A. & Tatikolov, A. S. (1978) *Chem. Phys. Lett.,* **53,** 606.
684. Kuzminsky, A. S. (1981) *Developments in Polymer Stabilization—4* (Scott, G., ed.), Applied Science Publishers, London, p. 71.
685. Kuzminski, A. S. (1971) *The Ageing and Stabilization of Polymers,* Elsevier Publishing Company, Amsterdam.
686. Kuznetsov, A. N. (1976) *Methods of Spin Probes,* Nauka, Moscow (in Russian).
687. Lacoste, J., Arnaud, B. & Lemaire, J. (1984) *J. Polym. Sci., A1,* **22,** 3885.
688. Lacoste, J., Arnaud, R., Singh, R. P. & Lemaire, J. (1988) *Makromol. Chem.,* **189,** 651.
689. Lai, J. T., Son, P. N. & Jennings, E. (1985) *Amer. Chem. Soc. Symp. Ser.,* **280,** 91.
690. Lala, D. & Rabek, J. F. (1980) *Polym. Degrad. Stabil.,* **3,** 383.
691. Lala, D. & Rabek, J. F. (1981) *Europ. Polym. J.,* **17,** 7.
692. Lala, D., Rabek, J. F. & Rånby, B. (1980) *Europ. Polym. J.,* **16,** 735.
693. La Mantia, F. P. (1986) *Polym. Degrad. Stabil.,* **15,** 283.
694. La Mantia, F. P. & Schifani, R. (1985) *Polym. Degrad. Stabil.,* **10,** 67.
695. Lamola, A. A. & Sharp, L. J. (1966) *J. Phys. Chem.,* **70,** 2634.
696. Land, E. J. & Porter, G. (1963) *Trans. Faraday Soc.,* **59,** 2016.
697. Lappin, G. R. (1971) *Encyclopedia of Polymer Science and Technology,* (Mark, H. F., Gaylord, N. G. & Bikales, M. M., eds.), Vol. 14, Interscience, New York, p. 125.
698. Lattimer, R. P., Hooser, E. R., Diem, H. E., Layer, R. W. & Rhee, C. K. (1980) *Rubber Chem. Technol.,* **53,** 1170.
699. Laver, H. S. (1979) *Developments in Polymer Stabilization—1* (Scott, G., ed.), Applied Science Publishers, London, p. 167.
700. Layer, R. W., Lai, J. T., Lattimer, R. P. & Westfahl, J. C. (1985) *Amer. Chem. Soc. Symp. Ser.,* **280,** 99.
701. Leaver, I. H., Waters, P. J. & Evans, N. A. (1979) *J. Polym. Sci., A1,* **17,** 1531.
702. Lebedeva, L. P. & Levin, P. I. (1982) *Vysokomol. Soedin,* **B24,** 379.
703. Ledbury, K. J. (1973) *Plast. Polym.,* **41,** 162.
704. Lee, H. & Cava, M. P. (1981) *J. Chem. Soc., Chem. Commun.,* 277.

705. Leimgruber, R. (1985) *Plastic Additives* (Gächter, R. & Müller, H., eds.), Hanser, Munich, p. 685.
706. Lemaire, J. (1982) *Pure Appl. Chem.*, **54,** 1667.
707. Lemaire, J. & Arnaud, R. (1984) *Polym. Photochem.*, **5,** 243.
708. Lemaire, J., Lopez, D. & Baule, P. (1980) Research Report, Université Clermont Ferrand, p. 19.
709. Lerchová, J., Nikoforov, C. A. & Pospišil, J. (1976) *J. Polym. Sci, Polym. Symp.*, **57,** 249.
710. Levin, P. I., Mikhailov, V. V. & Medvedev, A. I. (1970) *Inhibition of Polymer Oxidation with Stabilizing Mixtures,* NITEKhim, Moscow (in Russian).
711. Levis, P. M. (1986) *Polym. Degrad. Stabil.*, **15,** 37.
712. Li, S., Albertsson, A. C., Gupta, A., Bassett, W., Jr & Vogl, O. (1984) *Monat. Chem.*, **115,** 853.
713. Li, S., Gupta, A., Albertsson, A. C., Bassett, W., Jr & Vogl, O. (1984) *Polym. Bull.*, **12,** 237.
714. Li, W., Shi, L., Shen, D. & Luo, B. (1988) *Polym. Degrad. Stabil.*, **22,** 375.
715. Li, X. B., Winnik, M. A. & Guillet, J. E. (1983) *J. Polym. Sci., A1,* **21,** 1263.
716. Liang, R. H., Coulter, D. R., Dao, C. & Gupta, A. (1983) *Amer. Chem. Soc. Symp. Ser.*, **220,** 265.
717. Lichtenstein, G. I. (1974) *Methods of Spin Labels in Molecular Biology,* Nauka, Moscow (in Russian).
718. Linhart, H., Gugumus, F. & Pedrazzetti, E. (1980) Optimum light stabilization of polyolefins. Paper presented at the Plastics Institute of Australia Inc. Residential Technical Seminar, 23–25 October.
719. Lisa, E., Kotulak, L., Petranek, J. & Pospišil, J. (1972) *Europ. Polym. J.*, **8,** 501.
720. Liu, R. X., Lin, F., Wu, S. K. & Rabek, J. F. (1988) *J. Polym. Sci., B,* **26,** 17.
721. Liu, R. X., Wu, S. K., Li, S., Xi, F. & Vogl, O. (1988) *Polym. Bull.*, **20,** 59.
722. Lock, L. M. & Frank, G. C. (1973) *Text. Res. J.*, **43,** 502.
723. Low, H., Peterson, J., Tedder, J. M. & Walton, J. (1977) *J. Chem. Soc., Chem. Commun.*, 171.
724. Lozovskaya, E. L., Ivanov, V. B. & Shlyapintokh, V. Ya. (1985) *Vysokomol. Soedin,* **A27,** 1589.
725. Lozovskaya, E. L. & Shlyapintokh, V. Ya. (1983) *Polym. Photochem.*, **3,** 235.
726. Lucas, P. C. & Porter, R. G. (1985) *Polym. Degrad. Stabil.*, **13,** 287.
727. Lucas, P. C. & Porter, R. G. (1988) *Polym. Degrad. Stabil.*, **22,** 175.
728. Lucki, J. (1985) *Polym. Degrad. Stabil.*, **11,** 75.
729. Lucki, J. (1985) *Polym. Photochem.*, **6,** 273.
730. Lucki, J., Jian, S. Z., Rabek, J. F. & Rånby, B. (1986) *Polym. Photochem.*, **7,** 27.
731. Lucki, J., Rabek, J. F. & Rånby, B. (1984) *Amer. Chem. Soc. Symp. Preprints,* **25**(1), 38.

732. Lucki, J., Rabek, J. F. & Rånby, B. (1984) *Polym. Photochem.*, **5,** 351.
733. Lucki, J., Rabek, J. F., Rånby, B. & Dai, G. S. (1984) *Polym. Photochem.*, **5,** 385.
734. Lucki, J., Rabek, J. F., Rånby, B. & Watanabe, Y. (1988) *J. Appl. Polym. Sci.*, **36,** 1067.
735. Lukač, I., Hrdlovič, P., Maňásek, Z. & Belluš, D. (1969) *Europ. Polym. J. Supplement,* 523.
736. Lunak, S. & Veprek-Siska, J. (1981) *Z. Naturforsch.*, **36B,** 654.
737. Lunsford, J. & Wong, N. W. (1972) *J. Chem. Phys.*, **56,** 2664.
738. Lustoň, J. (1980) *Developments in Polymer Stabilization—2* (Scott, G., ed.), p. 185.
739. Lustoň, J., Guniš, J. & Maňásek, Z. (1973) *J. Macromol. Sci.—Chem.*, **A7,** 587.
740. Lustoň, J. & Maňásek, Z. (1977) *J. Appl. Polym. Sci.*, **21,** 915.
741. Lustoň, J., Maňasek, Z. & Palovčik, R. (1975) *J. Macromol. Sci—Chem.*, **A9,** 1413.
742. Lustoň, J., Maňasek, Z. & Samuhelova, Z. (1979) *J. Appl. Polym. Sci.*, **24,** 2005.
743. Lustoň, J., Schubertova, N. & Maňasek, Z. (1973) *J. Polym. Sci., Polym. Symp.*, **40,** 33.
744. MacCallum, J. R. (1977) *Developments in Polymer Degradation—1* (Grassie, N., ed.), Applied Science Publishers, London, p. 237.
745. MacCallum, J. R. (1985) *Developments in Polymer Degradation—6* (Grassie, N., ed.), Applied Science Publishers, London, p. 191.
746. MacCallum, J. R. & Rudkin, A. L. (1978) *Europ. Polym. J.*, **14,** 655.
747. Mackenzie, R. (1970/1972) *Differential Thermal Analysis,* Vols 1 and 2, Academic Press, New York.
748. Madorski, S. I. (1964) *Thermal Degradation of Organic Polymers,* Wiley–Interscience, New York.
749. Maeda, K., Nakane, A. & Tsubomura, H. (1975) *Bull. Chem. Soc. Japan,* **48,** 2448.
750. Magell, O. L. & Sheppard, C. S. (1970) *Organic Peroxides* (Swern, D., ed.), Wiley–Interscience, New York, p. 19.
751. Mahoney, L. R. (1967) *Angew. Chem.*, **81,** 555.
752. Mailey, E. A. & Doy, A. R. (1957) *J. Org. Chem.*, **22,** 1061.
753. Mair, R. D. & Graupner, A. J. (1964) *Anal. Chem.*, **36,** 194.
754. Malati, M. A. & Seager, N. J. (1981) *J. Oil Colour Chem. Assoc.*, **64,** 231.
755. Maňásek, Z., Lustoň, J. & Vašš, F. (1982) *J. Macromol. Sci.—Chem.*, **A17,** 653.
756. Maňásek, Z., Zwari, I. & Lustoň, J. (1976) *Zhurn. Prikl. Khim.*, **49,** 1129.
757. Manassio, J. A. & Castor, W. S., Jr (1978) *Additives for Plastics,* Vol. 2 (Seymour, R. B., ed.), Academic Press, New York, p. 93.
758. Marchall, J. (1980) *J. Radiat. Phys. Chem.*, **16,** 27.
759. Margolin, A. L., Velichko, V. A., Sorokina, A. V., Postnikov, L. M., Levin, V. S., Zabara, A. Ya. & Shlyapintokh, V. Ya., *Vysokomol. Soedin.*, **27A,** 1313.

760. Marom, G. (1985) *Polymer Permeability* (Comyn, J., ed.), Elsevier Applied Science Publishers, London, p. 341.
761. Marshall, G. P. (1982) *Plast. Rubber Int.*, **7,** 146.
762. Martin, J. W. (1984) *J. Appl. Polym. Sci.*, **29,** 777.
763. Martin, K. G. (1972) *Brit. Polym. J.*, **4,** 53.
764. Maruglio, P., Sharrock, R. F. & Kennedy, R. J. (1978) *J. Oil Colour Chem. Assoc.*, **61,** 79.
765. Mascia, L. (1974) *The Role of Additives in Plastics,* Arnold Publishers, London.
766. Maslov, S. A. & Zaikov, G. E. (1987) *Developments in Polymer Stabilization—8* (Scott, G., ed.), Applied Science Publishers, London, p. 1.
767. Matisová-Rychlá, L., Ambrovič, P., Kuličková, N. & Rychlý, J. (1976) *J. Polym. Sci., Polym. Symp.*, **57,** 181.
768. Matisová-Rychlá, L., Rychlý, J., Ambrovič, P., Csomorova, K., Mogel, L. & Schulz, M. (1986) *Polym. Degrad. Stabil.*, **14,** 147.
769. Matisová-Rychlá, L., Rychlý, J., Meske, M. & Schulz, M. (1988) *Polym. Degrad. Stabil.*, **21,** 323.
770. Matisová-Rychlá, L., Rychlý, J. & Vavrekova, M. (1978) *Europ. Polym. J.*, **14,** 1033.
771. Matsumoto, S. (1983) *J. Polym. Sci., A1,* **21,** 557.
772. Matsuura, T., Omura, K. & Nakashima, R. (1965) *Bull. Chem. Soc. Japan,* **38,** 1358.
773. Matsuura, T., Yoshimura, N., Nishinaga, A. & Saito, I. (1969) *Tetrahedron Lett.*, 1669.
774. Matsuura, T., Yoshimura, N., Nishinaga, A. & Saito, I. (1972) *Tetrahedron,* **28,** 4933.
775. McGarvey, A. R., Gavan, F. M. & Zabawsky, Z. (1967) *J. Appl. Polym. Sci., Appl. Polym. Symp.*, **4,** 263.
776. McIntyre, N. S., Thompson, K. R. & Weltner, W., Jr (1971) *J. Phys. Chem.*, **75,** 3243.
777. McKellar, J. F. & Allen, N. S. (1979) *Photochemistry of Man-Made Polymers,* Applied Science Publishers, London.
778. Meares, P. (1954) *J. Amer. Chem. Soc.*, **76,** 3415.
779. Mellor, D. C., Moir, A. B. & Scott, G. (1973) *Europ. Polym. J.*, **9,** 219.
780. Mendelson, M. A., Navish, F. W., Jr, Luck, R. W. & Yeoman, F. A. (1983) *Amer. Chem. Soc. Symp. Ser.*, **220,** 39.
781. Mendenhall, G. D. (1977) *Angew. Chem. Int. Ed. Engl.*, **16,** 225.
782. Mendenhall, G. D., Agarwaj, H. K., Cooke, J. M. & Dziemianowicz, T. S. (1985) *Amer. Chem. Soc. Symp. Ser.*, **280,** 373.
783. Menzel, G. (1985) *Plastic Additives* (Gächter, R. & Muller, H., eds), Hanser, Munich, p. 715.
784. Menzel, G. (1974) *Plastverarbeiter,* **25,** 137.
785. Merritt, C., Scott, G., Gupta, A. & Yavrouian, A. (1980) *Chem. Phys. Lett.*, **69,** 169.
786. Metcalf, K. & Tomlinson, R. F. (1960) *Plastics,* **25,** 319.
787. Meyer, G. D., Kavchok, R. W. & Naples, J. F. (1973) *Rubber Chem. Technol.*, **46,** 106.

788. Michaels, A. S. & Bixler, J. F. (1961) *J. Polym. Sci.*, **50,** 393.
789. Michaels, A. S. & Bixler, J. F. (1961) *J. Polym. Sci.*, **50,** 413.
790. Michaels, A. S. & Parker, R. B. (1959) *J. Polym. Sci.*, **41,** 53.
791. Milligan, B. & Holt, L. A. (1983) *Polym. Degrad. Stabil.*, **5,** 339.
792. Milligan, B. & Holt, L. A. (1985) *Polym. Degrad. Stabil.*, **10,** 335.
793. Mingbo, H. & Xingzhou, H. (1987) *Polym. Degrad. Stabil.*, **18,** 321.
794. Minsker, K. S. & Fedoseyeva, G. T. (1979) *Destruction and Stabilization of Poly(vinyl chloride)*, Khimia, Moscow, p. 503 (in Russian).
795. Mitchell, J., Jr & Perkins, L. R. (1967) *J. Appl. Polym. Sci., Appl. Polym. Symp.*, **4,** 167.
796. Mizutani, Y. & Kusumoto, K. (1975) *J. Appl. Polym. Sci.*, **19,** 713.
797. Mlinac, M., Rolich, J. & Bravar, M. (1976) *J. Polym. Sci., Polym. Symp.*, **57,** 161.
798. Moad, G., Rozzardo, E. & Solomon, D. (1981) *Tetrahedron Lett.*, **22,** 1165.
799. Moisan, J. Y. (1980) *Europ. Polym. J.*, **16,** 979.
800. Moisan, J. Y. (1981) *Europ. Polym. J.*, **17,** 857.
801. Moisan, J. Y. (1985) *Polymer Permeability* (Comyn, J., ed.), Elsevier Applied Science Publishers, London, p. 119.
802. Moisan, J. Y. & Lever, R. (1982) *Europ. Polym. J.*, **18,** 407.
803. Monroe, B. M. (1977) *J. Phys. Chem.*, **81,** 1861.
804. Monroe, B. M. & Mrowca, J. J. (1979) *J. Phys. Chem.*, **83,** 591.
805. Moore, R. F. (1963) *Polymer,* **4,** 493.
806. Morterra, C., Chiorno, A. & Bouizzi, F. (1981) *Z. Phys. Chem. Neiue Folge,* **124,** 211.
807. Morterra, C., Chiorno, A. & Zecchina, A. (1979) *Gazz. Chim. Ital.*, **109,** 683.
808. Morterra, C., Chiorno, A. & Zecchina, A. (1979) *Gazz. Chim. Ital.*, **109,** 691.
809. Morterra, C., Ghiotti, G., Garrone, E. & Bouizzi, F. (1980) *J. Chem. Soc., Faraday Trans. I.*, **76,** 2102.
810. Müller, H. (1983) *Plastic Additives* (Gächter, R. & Müller, H., eds.), Hanser, Munich, p. 75.
811. Müller, H. K. (1985) *Amer. Chem. Soc. Symp. Ser.*, **280,** 55.
812. Munro, H. S. & Allaker, R. S. (1985) *Polym. Degrad. Stabil.*, **11,** 349.
813. Munro, H. S., Banks, J., Bottino, F. A., Pollicino, A. & Recca, A. (1987) *Polym. Degrad. Stabil.*, **17,** 185.
814. Munro, H. S., Banks, J., Recca, A., Bottino, F. A. & Pollicino, A. (1986) *Polym. Degrad. Stabil.*, **15,** 161.
815. Munro, H. S., Bottino, F. A., Pollicino, A. & Recca, A. (1988) *Polym. Degrad. Stabil.*, **23,** 19.
816. Munro, H. S. & Clark, D. T. (1985) *Polym. Degrad. Stabil.*, **11,** 225.
817. Munro, H. S. & Clark, D. T. (1987) *Polym. Degrad. Stabil.*, **17,** 319.
818. Munro, H. S., Clark, D. T. & Peeling, J. (1984) *Polym. Degrad. Stabil.*, **9,** 185.
819. Munro, H. S., Clark, D. T. & Recca, A. (1987) *Polym. Degrad. Stabil.*, **19,** 353.
820. Munteanu, D. (1987) *Developments in Polymer Stabilization—8* (Scott, G., ed.), Applied Science Publishers, London, p. 179.

821. Munteanu, D., Mracec, M., Tincul, I. & Csunderlik, C. (1985) *Polym. Bull.*, **13,** 77.
822. Munteanu, D. & Nanu, I. (1977) *Rev. Roum. Chem.*, **22,** 923.
823. Munteanu, D. & Tincul, I. (1981) *Mater. Plast. (Bucharest)*, **21,** 79.
824. Munteanu, D., Tincul, I. & Chirila, T. (1981) *Mater. Plast. (Bucharest)*, **18,** 147.
825. Munteanu, D., Tincul, I., Scuderlik, C. & Iorga, I. (1983) *Proc. 29th IUPAC Symposium on Macromolecules, Bucharest,* p. 185.
826. Murayama, K., Morimura, S. & Yoshioka, T. (1969) *Bull. Chem. Soc. Japan,* **12,** 1640.
827. Murayama, K. & Yoshioka, T. (1969) *Bull. Chem. Soc. Japan,* **42,** 1942.
828. Myers, A. W., Stannett, V. & Szwarc, M. (1959) *J. Polym. Sci.*, **35,** 285.
829. Nakatsuna, K., Ide, F. & Itoh, K. (1968) Japanese Patent 68 16392.
830. Nakatsuna, K., Ide, F. & Itoh, K. (1968) Japanese Patent 68 16393.
831. Nakatsuna, K., Ide, F. & Itoh, K. (1969) Japanese Patent 69 07345.
832. Nakatsuna, K., Ide, F. & Itoh, K. (1969) Japanese Patent 69 07346.
833. Nakatsuna, K., Ide, F. & Itoh, K. (1969) Japanese Patent 69 07347.
834. Nakatsuma, K., Ide, F., Itoh, K., Nakagawa, O. & Kawakami, S. (1969) Japanese Patent 69 02714.
835. Nakatsuma, K., Ide, F., Itoh, K., Nakagawa, O. & Kawakami, S. (1969) Japanese Patent 69 02715.
836. Nakatsuma, K., Ide, F., Itoh, K., Nakagawa, O. & Kawakami, S. (1969) Japanese Patent 69 02716.
837. Nakatsuma, K., Ide, F., Itoh, K., Nakagawa, O. & Kawakami, S. (1969) Japanese Patent 69 02717.
838. Nakatsuma, K., Ide, F., Itoh, K., Nakagawa, O. & Kawakami, S. (1969) Japanese Patent 69 02718.
839. Nakatsuma, K., Ide, F., Itoh, K., Nakagawa, O. & Kawakami, S. (1969) Japanese Patent 69 02719.
840. Nakatsuma, K., Ide, F., Kamat, K. & Hasegawa, A. (1968) Japanese Patent 68 16396.
841. Neiman, M. B. (1964) *Uspekh. Khim.*, **33,** 28.
842. Neiman, M. B. (1965) *Ageing and Stabilization of Polymers,* Consultants Bureau, New York.
843. Neiman, M. B. & Rozantsev, E. G. (1964) *Izv. Akad. Nauk SSSR, Ser. Khim.*, **7,** 1178.
844. Neville, G. H. J. (1969) *Kirk-Othmer Encyclopedia of Chemical Technology* (Mark, H., McKetta, J. J. & Othmer, D. E., eds.), Vol. 2, 2nd edn Interscience–Wiley, New York, p. 380.
845. Newland, G. C. & Tamblyn, J. W. (1964) *J. Appl. Polym. Sci.*, **8,** 1949.
846. Newland, G. C. & Tamblyn, J. W. (1968) US Patent 3 396 143.
847. Ng, H. G. & Guillet, J. E. (1978) *Macromolecules,* **11,** 929.
848. Ninagawa, A., Iseki, T. & Matsuda, H. (1983) *Makromol. Chem.*, **184,** 71.
849. Ninagawa, A., Nakamura, M. & Matsuda, H. (1982) *Makromol. Chem.*, **183,** 1969.

850. Ninagawa, A., Tokunaga, Y. & Matsuda, H. (1979) *Makromol. Chem.*, **180,** 625.
851. Nir, Z., Vogl, O. & Gupta, A. (1982) *J. Polym. Sci.*, **20,** 2725.
852. Nowakowska, M. (1983) *Polym. Photochem.*, **3,** 221.
853. Nowakowska, M. (1983) *Polym. Photochem.*, **3,** 243.
854. Nowakowska, M. & Kowal, J. (1980) *Europ. Polym. J.*, **16,** 615.
855. O'Connell, E. J., Jr (1968) *J. Amer. Chem. Soc.* **90,** 6550.
856. O'Connor, D. B., Scott, G. W., Coulter, R., Gupta, A., Webb, S. P., Yen, S. W. & Clark, J. H. (1985) *Chem. Phys. Lett.*, **121,** 417.
857. Ogiwara, Y., Kubota, H. & Kimura, Y. (1978) *J. Polym. Sci., A1,* **16,** 2865.
858. Ogubanjo, A. & Scott, G. (1985) *Europ. Polym. J.*, **21,** 541.
859. Ohashi, K., Nakai, K., Kitayama, S. & Kawaguchi, T. (1988) *Polym. Degrad. Stabil.*, **22,** 223.
860. O'Keefe, E. S. & Billingham, N. C. (1985) *Polym. Degrad. Stabil.*, **10,** 137.
861. Olson, D. R. (1980) *Org. Plast. Prep.*, **43,** 808.
862. Olson, D. R. (1983) *J. Appl. Polym. Sci.*, **28,** 1159.
863. Olson, D. R. & Schroeter, S. H. (1978) *J. Appl. Polym. Sci.*, **22,** 2165.
864. Osawa, Z. (1984) *Developments in Polymer Stabilization—7* (Scott, G., ed.), Applied Science Publishers, London, p. 193.
865. Osawa, Z. (1988) *Polym. Degrad. Stabil.*, **20,** 203.
866. Osawa, Z. & Aiba, M. (1982) *Polym. Photochem.*, **2,** 447.
867. Osawa, Z., Kobayashi, K. & Kayano, E. (1985) *Polym. Degrad. Stabil.*, **11,** 63.
868. Osawa, Z., Konoma, F., Wu, S. & Cen, J. F. (1986) *Polym. Photochem.*, **7,** 337.
869. Osawa, Z., Kurishu, N. & Kuroda, H. (1980) *Polym. Degrad. Stabil.*, **3,** 265.
870. Osawa, Z., Kurishu, N., Nagashima, K. & Nakano, K. (1979) *J. Appl. Polym. Sci.*, **23,** 2583.
871. Osawa, Z., Kuroda, N. & Kobayashi, Y. (1984) *J. Appl. Polym. Sci.*, **29,** 2843.
872. Osawa, Z., Matsui, K. & Ogiwara, Y. (1967) *J. Macromol. Sci.—Chem.*, **A1,** 581.
873. Osawa, Z., Suzuki, Y., Ogiwara, Y., Sugaya, I., Hirama, K. & Kasuga, T. (1971) *J. Macromol. Sci.—Chem.*, **A5,** 275.
874. Osawa, Z., Wu, S. & Konoma, F. (1988) *Polym. Degrad. Stabil.*, **22,** 97.
875. Otterstedt, J. E. A. (1973) *J. Chem. Phys.*, **58,** 5715.
876. Owen, E. D. (1984) *Degradation and Stabilization of Poly(vinyl chloride)* (Owen, E. D., ed.), Elsevier Applied Science Publishers, London, p. 197.
877. Padron, A. J. C., Colmenares, M. A., Rubinztain, Z. & Albornoz, L. A., (1987) *Europ. Polym. J.*, **23,** 723.
878. Padron, A. J. C., Rubinztain, Z. & Colmenares, M. A. (1987) *Europ. Polym. J.*, **23,** 729.
879. Pan, J., Zhang, C. & Ma, Z. (1987) *Chinese J. Polym. Sci.*, **5,** 366.

880. Pappas, S. P. & Fischer, R. M. (1974) *J. Paint Technol.*, **46**, 65.
881. Pappas, S. P. & Kuhhirt, W. (1975) *J. Paint Technol.*, **47**, 42.
882. Pappas, S. P. & Winslow, F. H. (eds) (1981) *Photodegradation and Photostabilization of Coatings, Amer. Chem. Soc. Symp. Ser.*, **151**.
883. Parrish, M. A. (1977) *J. Oil Colour Chem. Assoc.*, **60**, 474.
884. Pastor, S. A. & Odorisio, P. (1987) US Patent 46 17336.
885. Patel, A. R. & Usilton, A. R. (1978) *Amer. Chem. Soc. Symp. Ser.*, **169**.
886. Patel, J. & Troth, H. G. (1967) Stabilization of polyolefins with ZnO synergistic combinations. Paper presented at International Symposium *Weathering of Plastics and Rubber,* London.
887. Patton, T. C. (ed.) (1973) *Pigment Handbook,* Wiley–Interscience, New York.
888. Peeling, J. & Clark, D. T. (1981) *Polym. Degrad. Stabil.*, **3**, 97.
889. Peeling, J. & Clark, D. T. (1981) *Polym. Degrad. Stabil.*, **3**, 177.
890. Penot, G., Arnaud, R. & Lemaire, J. (1982) *Polym. Photochem.*, **2**, 39.
891. Petruj, J., Zehnacker, S., Sédlař, J. & Marchal, J. (1986) *Analyst,* **111**, 671.
892. Petruj, J., Zehnacker, S., Sédlař, J. & Marchal, J. (1986) *Polym. Degrad. Stabil.*, **15**, 193.
893. Pinnazzi, C. P. & Fernandez, A. (1973) *Makromol. Chem.*, **167**, 147.
894. Piotrovsky, K. B. & Tarasova, Z. N. (1980) *Ageing and Stabilization of Synthetic Rubbers and Vulcanized Rubbers,* Khimia, Moscow (in Russian).
895. Pivovarov, A. P., Ershov, Yu. A. & Lukovnikov, A. F. (1974) *Vysokomol. Soedin,* **2**, 220.
896. Pivovarov, A. P., Pivovarova, T. S. & Lukovnikov, A. F. (1973) *Vysokomol. Soedin,* **A15**, 661.
897. Plant, M. A. & Scott, G. (1971) *Europ. Polym. J.*, **7**, 1173.
898. Pobiedimskii, D. G. & Kirpichnikov, P. A. (1980) *J. Polym. Sci., A1,* **18**, 815.
899. Pobiedimskii, D. G., Mukmeneva, N. A. & Kirpichnikov, P. A. (1980) *Developments in Polymer Stabilization—2* (Scott, G., ed.), Applied Science Publishers, London, p. 125.
900. Popov, A. A., Rapoport, N. Ya. & Zaikov, G. E. (1987) *Oxidation of Polyolefins under Pressure,* Khimia, Moscow (in Russian).
901. Porubská, M., Antoš, K. & Hodúl, P. (1988) *Polym. Degrad. Stabil.*, **21**, 139.
902. Porubská, M., Krb, R. & Welnitz, L. (1988) *Polym. Degrad. Stabil.*, **21**, 191.
903. Porubská, M. & Krkošková, M. (1988) *Polym. Degrad. Stabil.*, **21**, 121.
904. Porubská, M., Zahradnickova, A. & Sedlar, J. (1988) *Polym. Degrad. Stabil.*, **21**, 29.
905. Pospišil, J. (1968) *Antioxidanty,* Academia, Prague (in Czechoslovak).
906. Pospišil, J. (1979) *Developments in Polymer Stabilization—1* (Scott, G., ed.), Applied Science Publishers, London, p. 1.
907. Pospišil, J. (1980) *Adv. Polym. Sci.*, **36**, 69.

908. Pospišil, J. (1981) *Developments in Polymer Photochemistry—2* (Allen, N. S., ed.), Applied Science Publishers, London, p. 53.
909. Pospišil, J. (1984) *Developments in Polymer Stabilization—7* (Scott, G., ed.), Applied Science Publishers, London, p. 1.
910. Pospišil, J. (1985) *Amer. Chem. Soc. Symp. Ser.*, **280**, 157.
911. Pospišil, J. (1988) *Polym. Degrad. Stabil.*, **20**, 181.
912. Pradellok, W., Gupta, A. & Vogl, O. (1981) *J. Polym. Sci., A1,* **19**, 3307.
913. Prušilova, M., Kudelka, I., Brodilova, J. & Pospišil, J. (1986) *Polym. Degrad. Stabil.*, **16**, 277.
914. Rabek, J. F. (1975) *Chemical Kinetics,* Vol. 14 (Bamford, C. H. & Tipper, C. F. H. eds), Elsevier Scientific, Amsterdam, p. 425.
915. Rabek, J. F. (1980) *Experimental Methods in Polymer Chemistry,* Wiley, Chichester.
916. Rabek, J. F. (1982) *Experimental Methods in Photochemistry and Photophysics,* Wiley, Chichester.
917. Rabek, J. F. (1982) *Org. Coatings Appl. Polym. Sci. Proc.*, **46**, 391.
918. Rabek, J. F. (1985) *Singlet Oxygen,* Vol. 4 (Frimer, A. A., ed.), CRC Press, Boca Raton, USA, p. 1.
919. Rabek, J. F. (1987) *Mechanisms of Photophysical Processes and Photochemical Reactions in Polymers,* Wiley, Chichester.
920. Rabek, J. F. & Lala, D. (1980) *J. Polym. Sci., B,* **18**, 427.
921. Rabek, J. F. & Rånby, B. (1974) *J. Polym. Sci., A1,* **12**, 273.
922. Rabek, J. F. & Rånby, B. (1978) *Photochem. Photobiol.*, **28**, 557.
923. Rabek, J. F., Rånby, B. & Arct, J. (1983) *Polym. Degrad. Stabil.*, **5**, 65.
924. Rabek, J. F., Rånby, B., Arct, J. & Golubski, Z. (1982) *Europ. Polym. J.*, **18**, 81.
925. Rabek, J. F., Rånby, B., Arct, J. & Liu, R. (1984) *J. Photochem.*, **25**, 519.
926. Rabek, J. F. & Sanetra, J. (1986) *Macromolecules,* **19**, 1679.
927. Rabek, J. F., Sanetra, J. & Rånby, B. (1968) *Macromolecules,* **19**, 1674.
928. Raciszewski, Z. & Mullins, D. H. (1970) *J. Appl. Polym. Sci.*, **14**, 967.
929. Ramey, C. E. & Rostek, C. J. (1985) *Amer. Chem. Soc. Symp. Ser.*, **280**, 149.
930. Ranaweera, R. P. R. & Scott, G. (1974) *Chem. Ind.*, 771.
931. Ranaweera, R. P. R. & Scott, G. (1975) *J. Polym. Sci., B,* **13**, 71.
932. Ranaweera, R. P. R. & Scott, G. (1976) *Europ. Polym. J.*, **12**, 591.
933. Ranaweera, R. P. R. & Scott, G. (1976) *Europ. Polym. J.*, **12**, 825.
934. Rånby, B. & Rabek, J. F. (1975) *Photodegradation, Photooxidation and Photostabilization of Polymers,* Wiley, London.
935. Rånby, B. & Rabek, J. F. (1977) *ESR Spectroscopy in Polymer Research,* Springer Verlag, Berlin.
936. Rånby, B. & Rabek, J. F. (1979) *J. Appl. Polym. Sci., Appl. Polym. Symp.*, **35**, 243.
937. Rånby, B. & Rabek, J. F. (1983) *Amer. Chem. Soc. Symp. Ser.*, **220**, 291.

938. Ranogajec, F., Mlinac, M. & Dvornik, I. (1981) *Radiat. Phys. Chem.* **18,** 511.
939. Rapoport, N. Ya., Shlyapintokh, Yu. A., Gromov, B. A. & Dubinsky, V. Zh. (1972) *Vysokomol. Soedin,* **14A,** 1540.
940. Rapoport, N. Ya. & Zaikov, G. E. (1985) *Developments in Polymer Degradation—6* (Grassie, N., ed.), Applied Science Publishers, London, p. 207.
941. Rasmusen, J. R., Stedronsky, E. R. & Whitesides, G. M. (1977) *J. Amer. Chem. Soc.,* **99,** 4736.
942. Rasti, F. & Scott, G. (1980) *Europ. Polym. J.,* **16,** 1153.
943. Razumovskii, S. D. & Zaikov, G. E. (1983) *Developments in Polymer Stabilisation—6* (Scott, G., ed.), Applied Science Publishers, London, p. 239.
944. Razumovskii, S. D. & Zaikov, G. E. (1984) *Ozone and Its Reactions with Organic Compounds,* Elsevier, Amsterdam.
945. Recca, A., Libertini, E., Finocchiaro, P., Munro, H. S. & Clark, D. T., *Macromolecules,* **21,** 2641.
946. Renschler, C. L. (1985) *High Temp. Sci.,* **19,** 197.
947. Richters, P. (1970) *Macromolecules,* **3,** 262.
948. Rickatson, W. & Stevens, T. S. (1963) *J. Chem. Soc.,* 3960.
949. Riediger, W., Schulz, M., Stamphel, G. & Wegwart, H. (1976) *J. Polym. Sci., Polym. Symp.,* **57,** 329.
950. Rieker, A. & Kessler, H. (1966) *Z. Naturforsch.,* **21b,** 939.
951. Rieker, A., Muller, E. & Becker, W. (1962) *Z. Naturforsch.,* **17b,** 718.
952. Rigaudy, J. & Klesney, S. P. (1979) *Nomenclature of Organic Chemistry,* Pergamon Press, Oxford.
953. Rivaton, A., Gardette, J. L. & Lemaire, J. (1985) *Caoutchoucs et Plastiques,* **651,** 81.
954. Rivaton, A. & Lemaire, J. (1988) *Polym. Degrad. Stabil.,* **23,** 51.
955. Rode, R. (1975) *VDI-Gesellschaft Kunstofftechnik: Einfarben von Kunststoffen,* VDI Verlag, Dusseldorf.
956. Roe, R. J., Bair, H. E. & Gniewski, C. (1971) *J. Appl. Polym. Sci.,* **18,** 843.
957. Roe, R. J., Bair, H. E. & Gniewski, C. (1974) *J. Appl. Polym. Sci.,* **18,** 843.
958. Roger, A., Sallet, D., Lacoste, J. & Lemaire, J. (1986) *Makromol. Chem.,* **187,** 1891.
959. Rogers, C. E. (1985) *Polymer Permeability* (Comyn, J., ed.), Elsevier Applied Science Publishers, London, p. 11.
960. Rosato, D. V. (1969) *Environmental Effects on Polymeric Materials* (Rosato, D. V. & Schwartz, R. T., eds), Interscience–Wiley, New York, p. 733.
961. Rosato, D. V. & Schwartz, R. T. (eds) (1968) *Environmental Effects on Polymeric Materials,* Vols 1–2, Interscience–Wiley, New York.
962. Rose, W. C., Walden, M. K. & Moore, J. E. (1961) *Text. Res. J.,* **31,** 495.
963. Rosencweig, A. (1973) *Anal. Chem.,* **47,** 595A.
964. Rozantsev, E. G. (1970) *Free Nitroxyl Radicals,* Plenum Press, New York.

965. Rozantsev, E. G., Kagan, E. Sh. & Sholle, V. D. (1985) *Paramagnetic Models of Drugs and Biochemicals* (Zdanov, R. I., ed.), CRC Press, Boca Raton, USA.
966. Rozantsev, E. G., Kagan, E. Sh., Sholle, V. D., Ivanov, V. B. & Smirnov, V. A. (1985) *Amer. Chem. Soc. Symp. Ser.*, **280**, 11.
967. Rozantsev, E. G. & Sholle, V. D. (1971) *Russian Chem. Rev.*, **40**, 233.
968. Rozantsev, E. G. & Sholle, V. D. (1979) *Organic Chemistry of Free Radicals*, Khimia, Moscow (in Russian).
969. Rozenboyem, N. A., Angert, L. G., Ivanov, V. B. & Shlyapintokh, V. Ya. (1980) *Kautschuk i Rezina*, 31.
970. Ruhnke, G. & Biritz, L. F. (1972) *Kunststoffe*, **62**, 250.
971. Russel, G. A. & Pascale, J. V. (1963) *J. Appl. Polym. Sci.*, **7**, 959.
972. Russel, R. M. (1969) *Brit. Polym. J.*, **1**, 53.
973. Ryan, T. G. & Calvert, P. D. (1982) *Polymer*, **23**, 877.
974. Ryan, T. G., Calvert, P. D. & Billingham, N. C. (1978) *Amer. Chem. Soc. Symp. Ser.*, **169**, 261.
975. Rychlá, L. M., Fodor, Z. & Rychlý, J. (1981) *Polym. Degrad. Stabil.*, **3**, 371.
976. Rychlá, L., Rychlý, J. & Krivostek, I. (1988) *Polym. Degrad. Stabil.*, **20**, 325.
977. Sabaa, M. W., Madkour, T. M. & Yassin, A. A. (1988) *Polym. Degrad. Stabil.*, **22**, 195.
978. Sabaa, M. W., Madkour, T. M. & Yassin, A. A. (1988) *Polym. Degrad. Stabil.*, **22**, 205.
979. Sadrmohaghegh, C. & Scott, G. (1980) *Polym. Degrad. Stabil.*, **3**, 333.
980. Savides, C. (1973) *Soc. Plastics Eng. J.*, 38.
981. Sawyer, L. C. & Grubb, D. I. (1987) *Polymer Microscopy*, Chapman and Hall, London.
982. Schard, M. P. & Russell, C. A. (1964) *J. Appl. Polym. Sci.*, **8**, 985, 977.
983. Schmidt, H. & Wolters, E. (1972) *Kunststoffe*, **61**, 261.
984. Schmitt, R. G. & Hirt, R. C. (1960) *J. Polym. Sci.*, **45**, 35.
985. Schmitt, R. G. & Hirt, R. C. (1962) *J. Polym. Sci.*, **61**, 361.
986. Schnabel, W. (1981) *Polymer Degradation; Principles and Practical Applications*, Hanser, München.
987. Schrauzer, G. N. & Mayweg, V. P. (1965) *J. Amer. Chem. Soc.*, **87**, 1483.
988. Schrauzer, G. N., Mayweg, V. P. & Heinrich, W. (1966) *J. Amer. Chem. Soc.*, **88**, 5174.
989. Schulz, M., Mogel, L. & Siffarth, K. (1984) *Plaste Kautschuk*, **31**, 92.
990. Schulz, M., Schilimper, R. & Wegwart, H. (1977) *Plaste Kautschuk*, **24**, 401.
991. Schwarzenbach, K. (1985) *Plastics Additives* (Gächter, R. & Müller, H., eds), Hanser, Munich, p. 1.
992. Schwertzel, R. E. & Caldwell, R. A. (1973) *J. Amer. Chem. Soc.*, **95**, 1382.
993. Schwetlick, K. (1983) *Pure Appl. Chem.*, **55**, 1629.
994. Schwetlick, K., König, T., Pionteck, J., Sasse, D. & Habicher, W. D. (1988) *Polym. Degrad. Stabil.*, **22**, 353.

995. Schwetlick, K., König, T., Rüger, C. & Pionteck, J. (1986) *Z. Chem.*, **26,** 360.
996. Schwetlick, K., König, T., Rüger, C., Pionteck, J. & Habicher, W. D. (1986) *Polym. Degrad. Stabil.*, **15,** 97.
997. Schwetlick, K., Pionteck, J., König, T. & Habicher, W. D. (1987) *Europ. Polym. J.*, **23,** 383.
998. Schwindt, R. (1978) *Defazet*, **32,** 108.
999. Scott, A. C., Tedder, J. M., Walton, J. C. & Mhatre, S. (1980) *J. Chem. Soc., Perkin Trans. II*, 260.
1000. Scott, G. (1965) *Atmospheric Oxidation and Antioxidants*, Elsevier, Amsterdam.
1001. Scott, G. (1969) *Europ. Polym. J. Supplement*, 189.
1002. Scott, G. (1971) *Brit. Polym. J.*, **3,** 24.
1003. Scott, G. (1972) *Pure Appl. Chem.*, **30,** 267.
1004. Scott, G. (1973) *Macromol. Chem.*, **8,** 319.
1005. Scott, G. (1973) *Polymers and Ecological Problems* (Guillet, J. E., ed.), Plenum Publishing Corp., New York, p. 24.
1006. Scott, G. (1974) British Patent 1 356 107.
1007. Scott, G. (1976) *Amer. Chem. Soc. Symp. Ser.*, **25,** 340.
1008. Scott, G. (1976) *J. Polym. Sci., Polym. Symp.* **57,** 357.
1009. Scott, G. (1977) *Plast. Rubber Proc.*, (June), 41.
1010. Scott, G. (1978) *Polym. Plast. Technol. Eng.*, **11,** 1.
1011. Scott, G. (1979) *Developments in Polymer Stabilization—1* (Scott, G., ed.), Applied Science Publishers, London, p. 309.
1012. Scott, G. (1981) *Developments in Polymer Stabilization—4* (Scott, G., ed.), Applied Science Publishers, London, p. 1.
1013. Scott, G. (1981) *Developments in Polymer Stabilization—4* (Scott, G., ed.), Applied Science Publishers, London, p. 181.
1014. Scott, G. (1983) *Pure Appl. Chem.*, **55,** 128.
1015. Scott, G. (1984) *Amer. Chem. Soc. Symp. Ser.*, **280,** 173.
1016. Scott, G. (1984) *Developments in Polymer Stabilization—7* (Scott, G., ed.), Applied Science Publishers, London, p. 65.
1017. Scott, G. (1984) *Polym. Eng. Sci.*, **24,** 1007.
1018. Scott, G. (1985) *Chem. in Brit.*, **21,** 648.
1019. Scott, G. (1985) *Polym. Degrad. Stabil.*, **10,** 97.
1020. Scott, G. (1987) *Chem. Ind.*, 841.
1021. Scott, G. (1987) *Developments in Polymer Stabilization—8* (Scott, G., ed.), Applied Science Publishers, London, p. 209.
1022. Scott, G. (1988) *Bull. Chem. Soc. Japan.*, **61,** 165.
1023. Scott, G. & Setoudeh, E. (1983) *Polym. Degrad. Stabil.*, **5,** 1.
1024. Scott, G. & Setoudeh, E. (1983) *Polym. Degrad. Stabil.*, **5,** 11.
1025. Scott, G. & Setoudeh, E. (1983) *Polym. Degrad. Stabil.*, **5,** 81.
1026. Scott, G., Shenna, H. H. & Harriman, A. M. (1978) *Europ. Polym. J.*, **14,** 1071.
1027. Scott, G. & Suharto, R. (1984) *Europ. Polym. J.*, **20,** 139.
1028. Scott, G. & Suharto, R. (1985) *Europ. Polym. J.*, **21,** 765.
1029. Scott, G. & Tavakoli, S. M. (1982) *Polym. Degrad. Stabil.*, **4,** 267.
1030. Scott, G. & Tavakoli, S. M. (1982) *Polym. Degrad. Stabil.*, **4,** 279.

1031. Scott, G. & Tavakoli, S. M. (1982) *Polym. Degrad. Stabil.*, **4**, 343.
1032. Scott, G. & Tavakoli, S. M. (1988) *Plast. Rubber Process. Appl.*, **9**, 59.
1033. Scott, G. & Yusoff, M. F. (1980) *Europ. Polym. J.*, **16**, 497.
1034. Scott, G. & Yusoff, M. F. (1980) *Polym. Degrad. Stabil.*, **2**, 309.
1035. Scott, G. & Yusoff, M. F. (1980) *Polym. Degrad. Stabil.*, **3**, 13.
1036. Scott, G. & Yusoff, M. F. (1980) *Polym. Degrad. Stabil.*, **3**, 48.
1037. Sedlař, J. & Marchal, J. (1987) *Polym. Degrad. Stabil.*, **19**, 251.
1038. Sedlař, J., Marchal, J. & Petrúj, J. (1982) *Polym. Photochem.*, **2**, 175.
1039. Sedlař, J., Petrúj, J., Pac, J. & Navratil, M. (1980) *Polymer*, **21**, 5.
1040. Sedlař, J., Petrúj, J., Pac, J. & Zahradničková, A. (1980) *Europ. Polym. J.*, **16**, 659.
1041. Sedlař, J., Petrúj, J., Pac, J. & Zahradničková, A. (1980) *Europ. Polym. J.*, **16**, 663.
1042. Searle, N. Z. & Hirt, R. C. (1961) *Soc. Plast. Eng. Trans.*, **1**, 1.
1043. Sereda, P. J. & Litvan, G. G. (1980) *Durability of Building Materials*, American Society of Testing and Materials, Philadelphia, USA.
1044. Severini, F., Gallo, R. & Ipsale, S. (1986) *Polym. Degrad. Stabil.*, **14**, 341.
1045. Severini, F., Gallo, R. & Ipsale, S. (1987) *Polym. Degrad. Stabil.*, **17**, 57.
1046. Severini, F., Gallo, R. & Ipsale, S. (1988) *Polym. Degrad. Stabil.*, **22**, 53.
1047. Severini, F., Gallo, R. & Ipsale, S. (1988) *Polym. Degrad. Stabil.*, **22**, 285.
1048. Seymour, R. B. & Tsang, H. S. (1971) *Texas J. Sci.*, **23**, 187.
1049. Sezi, R. & Springer, J. (1981) *Colloid Polym. Sci.*, **259**, 1170.
1050. Sharma, Y. N., Naqvi, M. K., Gawande, P. S. & Bhardwaj, I. S. (1982) *J. Appl. Polym. Sci.*, **27**, 2605.
1051. Sharples, K. B., Gondon, K. M., Laner, K. M., Patrick, S. W., Singer, S. P. & Young, M. W. (1975) *Chem. Sci.*, **84**, 9.
1052. Shelton, J. R. & Kopczewski, R. F. (1967) *J. Org. Chem.*, **32**, 2908.
1053. Shepherd, M. J. & Gilbert, J. (1981) *Europ. Polym. J.*, **17**, 285.
1054. Shilov, Yu. B. & Denisov, E. T. (1972) *Vysokomol. Soedin*, **A14**, 2385.
1055. Shilov, Yu. B. & Denisov, E. T. (1974) *Vysokomol. Soedin*, **A16**, 2313.
1056. Shlyapintokh, V. Ya. (1982) *Developments in Polymer Photochemistry—2* (Allen, N. S., ed.), Applied Science Publishers, London, p. 215.
1057. Shlyapintokh, V. Ya. (1985) *Photochemical Conversions and Polymer Stabilization*, Hanser, München.
1058. Shlyapintokh, V. Ya. & Goldenberg, V. I. (1974) *Europ. Polym. J.*, **10**, 679.
1059. Shlyapintokh, V. Ya. & Ivanov, V. B. (1982) *Developments in Polymer Stabilization* (Scott, G., ed.), Applied Science Publishers, London, **5**, 41.
1060. Shlyapintokh, V. Ya., Ivanov, V. B., Khvostach, V. B., Chapiro, O. M. & Rozantsev, A. B. (1975) *Dokl. Akad. Nauk SSSR*, **225**, 1132.
1061. Shlyapnikova, I. A., Miller, V. B. & Shlyapnikov, Yu. A. (1977) *Europ. Polym. J.*, **13**, 331.

1062. Shvets, V. A. & Kazansky, V. B. (1972) *J. Catal.*, **25**, 123.
1063. Siesler, H. W. (1980) *J. Mol. Spectr.*, **59**, 694.
1064. Simpson, L. A. (1978) *Progr. Org. Coat.*, **6**, 1.
1065. Singer, L. A., Davis, G. A. & Muralidharan, V. P. (1969) *J. Amer. Chem. Soc.*, **91**, 897.
1066. Singh, R. P. & Chandra, R. (1982) *Europ. Polym. J.*, **18**, 289.
1067. Singh, R. P., Lacoste, J., Arnaud, R. & Lemaire, J. (1988) *Polym. Degrad. Stabil.*, **20**, 49.
1068. Skowronski, T. A., Rabek, J. F. & Rånby, B. (1984) *Polym. Degrad. Stabil.*, **8**, 37.
1069. Sommersall, A. C. & Guillet, J. E. (1985) *Amer. Chem. Soc. Symp. Ser.*, **280**, 211.
1070. Son, P. N. (1980) *Polym. Degrad. Stabil.*, **2**, 295.
1071. Son, T. T., Rotschova, J., Prusnikova, M. & Pospišil, J. (1983) *Polym. Degrad. Stabil.*, **5**, 205.
1072. Song, Z., Rånby, B., Gupta, A., Borsig, E. & Vogl, O. (1984) *Polym. Bull.*, **12**, 245.
1073. Soria, V., Celda, B., Tejero, R. & Figueruelo, J. E. (1985) *Polym. Degrad. Stabil.*, **10**, 287.
1074. Spacht, R. B., Hollingshead, W. S., Bullard, H. L. & Wills, D. C. (1964) *Rubber Chem. Technol.*, **37**, 210.
1075. Steiner, G. & Koppelmann, J. (1987) *Polym. Degrad. Stabil.*, **19**, 307.
1076. Štěpek, J. & Daoust, H. (1983) *Additives for Plastics*, Springer Verlag, Berlin.
1077. Stevens, B. & Ors, J. A. (1976) *J. Phys. Chem.*, **80**, 2164.
1078. Still, R. H. (1977) *Developments in Polymer Degradation—1* (Grassie, N., ed.), Applied Science Publishers, London, p. 1.
1079. Stivala, S. S., Kimura, J. & Gabby, S. M. (1983) *Degradation and Stabilization of Polyolefins* (Allen, N. S., ed.), Elsevier Applied Science Publishers, London, p. 63.
1080. Strobel, A. F. & Catino, S. C. (1962) *Ind. Eng. Chem. Prod. Res. Dev.*, **1**, 241.
1081. Strobel, A. F. & Catino, S. C. (1962) Belgium Patent 619 809.
1082. Stempe, J. & Schwetlick, K. (1987) *Polym. Degrad. Stabil.*, **17**, 103.
1083. Sulivan, W. F. (1972) *Prog. Org. Coat.*, **1**, 157.
1084. Sumida, Y. & Vogl, O. (1981) *Polym. J.*, **13**, 521.
1085. Sumida, Y., Yoshida, S. & Vogl, O. (1980) *Polym. Prepr. Amer. Chem. Soc. Div. Polym. Chem.*, **21**(1), 201.
1086. Sung, C. S. P. & Hobbs, J. P. (1984) *Chem. Eng. Commun.*, **30**, 229.
1087. Sustic, A., Albertsson, A. C. & Vogl, O. (1987) *Proc. Amer. Chem. Soc. Div. Polym. Mater. Sci. Eng.*, **56**(2), 1.
1088. Tabata, M. & Sohma, J. (1979) *Polym. Degrad. Stabil.*, **1**, 139.
1089. Taimr, L. & Pospišil, J. (1974) *Angew. Makromol. Chem.*, **39**, 189.
1090. Taimr, L. & Pospišil, J. (1976) *Angew. Makromol. Chem.*, **52**, 31.
1091. Taimr, L. & Pospišil, J. (1976) *J. Polym. Sci., Polym. Symp.*, **57**, 213.
1092. Taimr, L. & Pospišil, J. (1984) *Polym. Degrad. Stabil.*, **8**, 23.
1093. Taimr, L. & Pospišil, J. (1984) *Polym. Degrad. Stabil.*, **8**, 67.

1094. Tamblyn, J. W., Newland, G. S. & Watson, M. T. (1958) *Plast. Technol.*, **4**, 427, 455.
1095. Tarte, P. (1952) *J. Chem. Phys.*, **20**, 1560.
1096. Taylor, H. A., Tincher, W. C. & Hamner, W. F. (1970) *J. Appl. Polym. Sci.*, **14**, 141.
1097. Taylor, L. J. & Tobias, J. W. (1981) *J. Appl. Polym. Sci.*, **26**, 2917.
1098. Thomas, M. J. & Foote, C. S. (1978) *Photochem. Photobiol.*, **27**, 683.
1099. Tirrell, D. (1981) *Amer. Chem. Soc., Symp. Ser.*, **151**, 43.
1100. Tirrell, D., Bailey, D. B., Pinazzi, C. & Vogl, O. (1978) *Macromolecules*, **11**, 213.
1101. Tirrell, D., Bailey, D. B. & Vogl, O. (1977) *Polym. Prepr. Amer. Chem. Soc. Div. Polym. Chem.*, **18**(1), 542.
1102. Tirrell, D., Bailey, D. B. & Vogl, O. (1978) *Polymeric Drugs* (Donaruma, G. & Vogl, O., eds), Academic Press, New York, p. 263.
1103. Tirrell, D. & Vogl, O. (1980) *Makromol. Chem.*, **181**, 2097.
1104. Tkac, A. (1987) *Developments in Polymer Stabilization—8* (Scott, G., ed.), Applied Science Publishers, London, p. 61.
1105. Tkač, A. & Omelka, L. (1973) *J. Polym. Sci., Polym. Symp.*, **40**, 119.
1106. Tkač, A., Omelka, L. & Holčik, J. (1973) *J. Polym. Sci., Polym. Symp.*, **40**, 105.
1107. Tobolsky, A. V. & Mesrobian, R. B. (eds) (1954) *Organic Peroxides*, Interscience, New York.
1108. Tocker, S. (1967) *Macromol. Chem.*, **101**, 23.
1109. Toda, T. & Kurumada, T. (1983) *Sanko Kenkyusho Nempo*, **35**, 1.
1110. Toda, T., Kurumada, T. & Murayama, K. (1985) *Amer. Chem. Soc. Symp. Ser.*, **280**, 37.
1111. Toda, T., Mori, E. & Murayama, K. (1972) *Bull. Chem. Soc. Japan*, **45**, 1904.
1112. Torikai, A., Geetha, R., Nagaya, S. & Fueki, K. (1986) *Polym. Degrad. Stabil.*, **16**, 199.
1113. Toussaint, A. (1978) *XIV FATIPEC Congress, Budapest, 4–9 June*, p. 657.
1114. Tozzi, A., Cantatore, G. & Masina, F. (1978) *Text. Res. J.*, **48**, 433.
1115. Tsurugi, J., Murakami, S. & Goda, K. (1971) *Rubber Chem. Technol.*, **44**, 857.
1116. Tucker, R. J. & Susi, P. V. (1985) *Amer. Chem. Soc. Symp. Ser.*, **280**, 137.
1117. Tüdos, F., Balint, G. & Kelen, T. (1983) *Developments in Polymer Stabilization—6* (Scott, G., ed.), Applied Science Publishers, London, p. 121.
1118. Uri, N. (1970) *Israel, J. Chem.*, **8**, 125.
1119. Uzelmeier, C. (1970) *Soc. Plastics Eng. J.*, **26**, 69.
1120. Van Amerongen, G. J. (1950) *J. Polym. Sci.*, **5**, 307.
1121. Van Amerongen, G. J. (1964) *Rubber Chem. Technol.*, **37**, 1965.
1122. Van Hoof, J. H. C. (1968) *J. Catal.*, **11**, 277.
1123. Verizhnikov, L. V., Kirpichnikov, P. A., Kolyubakina, N. S., Mukmeneva, N. A. & Karpova, S. I. (1971) *Vysokomol. Soedin*, **A13**, 714.

1124. Vink, P. (1980) *Developments in Polymer Stabilization—3* (Scott, G., ed.), Applied Science Publishers, London, p. 117.
1125. Vink, P. (1983) *Degradation and Stabilization of Polyolefins* (Allen, N. S., ed.), Elsevier Applied Science Publishers, London, p. 213.
1126. Vink, P. & van Veen, Th. J. (1978) *Europ. Polym. J.*, **14,** 533.
1127. Virt, J., Rosik, L., Kovarova, J. & Pospišil, J. (1980) *Europ. Polym. J.*, **16,** 247.
1128. Vogl, O. (1979) *Pure Appl. Chem.*, **51,** 2409.
1129. Vogl, O., Albertsson, A. C., Bansleben, D. A., Borsig, E., Grosso, P., Janovic, C., Kitayama, M., Li, S. J., Mugge, J., Nir, C., Okhata, Y., Pradellok, W., Purgett, M. & Xi, F. (1984) *Makromol. Chem. Supplement* **7,** 1.
1130. Vogl, O., Albertsson, A. C. & Janovic, Z. (1985) *Amer. Chem. Soc. Symp. Ser.*, **280,** 197.
1131. Vogl, O., Albertsson, A. C. & Janovic, Z. (1985) *Polymer,* **26,** 1288.
1132. Vogl, O., Loeffler, P., Bensleben, D. & Mugge, J. (1983) *Polymeric Science and Technology,* Vol. 19, *Coordination Polymerization* (Price, E. C. & Vandenberg, E., eds), Plenum Press, New York, p. 95.
1133. Vogl, O., Mugge, J. & Bansleben, D. (1980) *Polym. J.*, **12,** 677.
1134. Vogl, O. & Yoshida, S. (1980) *Rev. Roum. Chim.*, **25,** 1123.
1135. Voigt, J. (1966) *Die Stabiliserung der Kunststoffe gegen Licht und Wärme,* Springer Verlag, Berlin.
1136. Völz, H. G., Kämpf, G. & Fitzky, H. G. (1973) *Progr. Polym. Coat.*, **1,** 1.
1137. Völz, H. G., Kämpf, G., Fitzky, H. G. & Klaeren, A. (1980) *Amer. Chem. Soc. Div. Org. Coat. Plast. Chem., Preprints,* **42,** 660.
1138. Völz, H. G., Kämpf, G., Fitzky, H. G. & Klaeren, A. (1981) *Amer. Chem. Soc. Symp. Ser.*, **151,** 163.
1139. Völz, H. G., Kämpf, G. & Klaeren, A. (1976) *Farbe u. Lack,* **82,** 805.
1140. Wallder, V. T., Clarke, W. J., DeCoste, J. B. & Howard, J. B. (1950) *Ind. Eng. Chem.*, **42,** 2320.
1141. Waters, P. J. & Evans, N. A. (1978) *Textile Res. J.*, **48,** 251.
1142. Waters, P. J. & Evans, N. A. (1983) *J. Textile Inst.*, **74,** 99.
1143. Watkins, A. R. (1974) *Chem. Phys. Lett.*, **29,** 526.
1144. Watkins, A. R. (1980) *Chem. Phys. Lett.*, **70,** 262.
1145. Wendland, W. W. & Collins, L. W. (1977) *Thermal Analysis,* Wiley–Interscience, New York.
1146. Werner, T. (1979) *J. Phys. Chem.*, **83,** 320.
1147. Werner, T., Kramer, H. E. A., Kuster, B. & Herlinger, H. (1976) *Angew. Makromol. Chem.*, **54,** 15.
1148. Werner, T., Woessner, G. & Kramer, H. E. A. (1981) *Amer. Chem. Soc. Symp. Ser.*, **151,** 1.
1149. Westfahl, J. C. (1973) *Rubber Chem. Technol.*, **46,** 1134.
1150. Westfahl, J. C., Carman, C. J. & Layer, R. W. (1972) *Rubber Chem. Technol.*, **45,** 402.
1151. Westlake, J. F. & Johnson, M. (1975) *J. Appl. Polym. Sci.*, **19,** 319.
1152. Wicks, Z. W., Jr & Kuhhirt, W. (1975) *J. Paint Technol.*, **47,** 49.
1153. Wicks, Z. W., Jr & Pappas, S. P. (1978) *UV Curing: Science and*

Technology (Pappas, S. P., ed.), Technology Marketing, Stamford, USA, p. 79.
1154. Wiles, D. M. (1978) *Pure Appl. Chem.*, **50,** 291.
1155. Wiles, D. M. & Carlsson, D. J. (1980) *Polym. Degrad. Stabil.*, **3,** 61.
1156. Wiles, D. M., Tovborg Jensen, J. P. & Carlsson, D. J. (1983) *Pure Appl. Chem.*, **55,** 1651.
1157. Williams, R. F., Jordan, M. W. & Dannenberg, E. M. (1965) *J. Appl. Polym. Sci.*, **9,** 861.
1158. Wilson, J. E. (1954) *J. Chem. Phys.*, **22,** 334.
1159. Windle, A. H. (1985) *Polymer Permeability* (Comyn, J., ed.), Elsevier Applied Science Publishers, London, p. 75.
1160. Winslow, F. H. & Matreyek, W. (1975) *Polym. Prepr. Amer. Chem. Soc., Div. Polym. Chem.*, **16,** 606.
1161. Wismontski-Knittel, T. & Kilp, T. (1983) *J. Polym. Sci., A1,* **21,** 3209.
1162. Wössner, G., Goeller, G., Kollat, P., Stezowski, J., Hauser, M., Klein, U. K. A. & Kramer, H. E. A. (1984) *J. Phys. Chem.*, **88,** 5544.
1163. Wrasidlo, W. (1974) *Advances in Polymer Science Series,* Vol. 13, *Thermal Analysis of Polymers,* Springer Verlag, Berlin.
1164. Wu, S. K., Dai, G. S., Liu, L. S. & Chang, J. K. (1986) *Polym. Degrad. Stabil.*, **16,** 169.
1165. Yakimtchenko, O. E., Gaponova, I. S., Goldberg, V. M., Parisky, G. B., Toptygin, D. Ya. & Lebedev, Ya. S. (1974) *Izv. Akad. Nauk SSSR, Ser. Khim.*, 352.
1166. Yakimtchenko, O. E., Kiryushkin, S. G., Parisky, G. B., Toptygin, D. Ya., Shlyapintokh, D. Ya. & Lebedev, Ya. S. (1975) *Izv. Akad. Nauk SSSR, Ser. Khim.*, 2235.
1167. Yang, X. Z., Chen, Y., Dickinson, L. C. & Chien, J. C. W. (1988) *Polym. Degrad. Stabil.*, **20,** 1.
1168. Yang, Y. Y., Lucki, J., Rabek, J. F. & Rånby, B. (1983) *Polym. Photochem.*, **3,** 47.
1169. Yang, Y. Y., Lucki, J., Rabek, J. F. & Rånby, B. (1983) *Polym. Photochem.*, **3,** 97.
1170. Yassin, A. A. & Sabaa, M. W. (1982) *Polym. Degrad. Stabil.*, **4,** 313.
1171. Yoshida, S., Lillya, C. P. & Vogl, O. (1982) *J. Polym. Sci., A1,* **20,** 2215.
1172. Yoshida, S. & Vogl, O. (1980) *Polym. Prepr. Amer. Chem. Soc. Div. Polym. Chem.*, **21**(1), 203.
1173. Yoshida, S. & Vogl, O. (1982) *Makromol. Chem.*, **183,** 259.
1174. Yoshioka, T., Hagashida, S., Morimura, S. & Murayama, K. (1971) *Bull. Chem. Soc. Japan,* **44,** 2207.
1175. Zahradničkova, A., Petruj, J. & Sedlář, J. (1983) *Polym. Photochem.*, **3,** 295.
1176. Zahradničkova, A., Sedlář, J., Kovařova, J. & Pospišil, J. (1982) *Polym. Photochem.*, **2,** 349.
1177. Zannucci, J. S. & Lappin, G. R. (1974) *Macromolecules,* **7,** 393.
1178. Zeppenfeld, G. (1967) *Makromol. Chem.*, **90,** 439.
1179. Zerlaut, G. A. (1974) *Proc. Inst. Environm. Sci.*, 153.

1180. Zhang, C., Carlsson, D. J. & Wiles, D. M. (1986) *J. Polym. Sci., B,* **24,** 453.
1181. Zhdanov, R. I. (1981) *Paramagnetic Models of Biological Active Compounds,* Nauka, Moscow (in Russian).
1182. Zlatkevich, L. (1983) *J. Polym. Sci., B,* **21,** 57.
1183. Zlatkevich, L. (1984) *Polym. Eng. Sci.,* **24,** 1421.
1184. Zlatkevich, L. (1985) *Amer. Chem. Soc. Symp. Ser.,* **280,** 387.
1185. Zlatkevich, L. (1985) *J. Polym. Sci., A2,* **23,** 1691.
1186. Zlatkevich, L. (1987) *Polym. Degrad. Stabil.,* **19,** 51.
1187. Zweig, A. & Henderson, W. A., Jr, (1975) *J. Polym. Sci., A1,* **13,** 717.
1188. Zweig, A. & Henderson, W. A., Jr, (1975) *J. Polym. Sci., A1,* **13,** 993.
1189. Anon. (1966) *The Correlation of Natural Weathering with Accelerated Testing,* ASTM Committee G-3 Report, ASTM, Philadelphia, USA.
1190. Anon. (1967) Titangesellshaft Chart GmbH, Dutch Patent 6 610 998.
1191. Anon. (1975) *Plastic Technol.,* **21**(10), 11.
1192. Anon. (1976) *Guide to Plastics Property and Specification Charts,* McGraw–Hill, New York.
1193. Anon. (1985) *Light and Heat Stabilization of Coatings,* Ciba–Geigy A-288A5M, USA.
1194. Anon., Ciba–Geigy Publ. No. 28/472/e.
1195. Anon. (1988) Ciba–Geigy Publ. No. 28622e.
1196. Anon. (1988) *Modern Plastics Encyclopedia* McGraw–Hill, New York.
1197. Anon. (1987) *Catalogue Handbook of Fine Chemicals,* Aldrich-Chemie, Steinheim, FRG.
1198. Anon. (1984–87) Cynamide International Publications ORG No. 23, 28, 27, 79, 115, 127, 228, 231 and other short publications.
1199. Anon (1979–1988) *Atlas Series C Weather-Ometers, Bulletins and Catalogues.*
1200. Anon (1986–1988) *Atlas UVCON Bulletins.*

Appendices

(Appendix 1: pp. 545–571
Appendix 2: pp. 572–579)

APPENDIX 1: COMMERCIALLY AVAILABLE ANTIOXIDANTS[1196]

Trade name or designation	Supplier	Recommended for:									FDA sanction and type	General characteristics
		ABS	*Cellulosic*	*Nylon*	*Polyester*	*PE*	*PP*	*PS*	*PVC*	*Other plastics*		
Alkylated phenols and bisphenols												
Antioxidant MBP-5P, 5T	Aceto	X	X	X	X	X	X	X	X	Acetal, urethane	For polyolefins, 0·1% maximum	Nondiscolouring, nonstaining
Antioxidant SP	Harwick					X	X	X	X			Nondiscolouring, nonstaining
BHA	Eastman, UOP		X			X	X	X	X		Indirect food additive	White, waxy tablets or flakes
Butylated hydroxytoluene (BHT)	Aceto, Borg–Warner, ICI Americas, Mobay, Neville–Synthese, Stanchem, UOP, PMC, Plastics & Chem., USI, Uniroyal	X	X		X	X	X	X	X	Polyols, elastomers, polyurethane	Indirect and direct food additive	Nonstaining, nondiscolouring crystal
Bisphenol A	Shell								X			Nonstaining

CAO-1, CAO-3	PMC		X	X	X	X	X	X	X	Acetal, urethane	Indirect food additive	Nondiscolouring, nonstaining
CAO-5	PMC	X	X	X	X	X	X	X	X	Acetal, urethane	For adhesives; for polyole-fins, 0·1% max.	Nondiscolouring, nonstaining
CAO-41	PMC		X			X						Nonstaining, liquid
CAO-94	PMC		X			X	X	X	X			Nonstaining, molten
DBPC	Neville–Synthese	X	X			X	X	X	X	Polyols, elastomers		Nonstaining, nondiscolouring flake or crystal
Escoflex A-122, A-123	Plastics & Chem.	X	X	X	X	X	X	X	X	EPDM, olefin copolymers		
Ethanox 376	Ethyl	X	X	X	X	X	X	X	X	EPDM, olefin copolymers, polyurethane	Polyolefins, ABS, PS, impact PS, 178.20.10; adhesives, 175.105	Nondiscolouring, nonstaining, colourless, odourless
Ethanox 702	Ethyl	X	X		X	X	X	X		Coumarone indene	Food packaging adhesives, food wrap for nonfat foods	Slightly discolouring
Ethanox 703	Ethyl		X		X	X				Coumarone indene		Slightly discolouring

(continued)

APPENDIX 1—*contd.*

Trade name or designation	Supplier	Recommended for:									FDA sanction and type	General characteristics
		ABS	Cellulosic	Nylon	Polyester	PE	PP	PS	PVC	Other plastics		
Irganox 100	Ciba–Geigy			X	X	X	X	X	X	Ethylene, SBR latex, EPDM	For olefins (excluding EPDM), ethylene, isobutylene, polystyrene, polyoxy-methylene, poly-l-butene 178.2010	Low volatility, odourless, tasteless, extraction-resistant, thermally stable
Irganox 259	Ciba–Geigy	X	X	X	X	X	X	X	X	EPDM, polyurethane		Nondiscolouring, nonstaining, colourless, odourless
Irganox 1035	Ciba–Geigy, USI	X	X	X		X	X	X		EPDM, polyurethane	Adhesives, 175.105	Nondiscolouring, nonstaining, colourless, odourless
Irganox 1076	Ciba–Geigy, USI	X	X	X	X	X	X	X	X	EPDM, olefin copolymers, polyurethane	For polyolefins, ABS, PS, impact PS, 178.2010; for adhesives, 175.105	Nondiscolouring, nonstaining, colourless, odourless

Irganox 1098	Ciba–Geigy			X						Acetal, polyurethane		Nondiscolouring, nonstaining, colourless, odourless
Irganox MD-1024	Ciba–Geigy		X	X	X	X	X	X	X	EP, EPDM, olefin copolymers		Nondiscolouring, nonstaining, metal deactivator (e.g., copper)
Irganox 1425 WL C-9	Ciba–Geigy				X	X	X					
Isonox 129	Schenectady	X			X	X	X	X	X	Polyurethane	For polyolefins in contact with all foods under all current use conditions, 178.2010: for adhesives, 175.105	Nonstaining crystalline powder
Isonox 132	Schenectady								X	Polyurethane	For adhesives, 175.105	Nonstaining liquid
MO-14	Aceto	X	X		X	X	X	X		Coumarone indene	Food packaging adhesives, food wrap for nonfat foods	Slightly discolouring
Naugard 76	Uniroyal	X	X	X	X	X	X	X	X	EPDM, LLDPE, HIPS	Polyolefins, ABS, HIPS, 178.2010; adhesives, 175.105	Nondiscolouring solid, powder, granules, or flake

(*continued*)

APPENDIX 1—*contd.*

Trade name or designation	*Supplier*	*Recommended for:*									*FDA sanction and type*	*General characteristics*
		ABS	*Cellulosic*	*Nylon*	*Polyester*	*PE*	*PP*	*PS*	*PVC*	*Other plastics*		
Naugard 431	Uniroyal	X	X	X	X	X	X	X	X	Urethane	Adhesives, polyolefins, PS	Nondiscolouring, nonstaining
Naugard BHT tech. BHT food grade	Uniroyal	X	X	X	X	X	X	X	X		Indirect and direct food additive	Nondiscolouring, nonstaining
Naugard SP	Uniroyal		X	X		X	X	X	X	Urethane	Adhesives, 175.105	Nonstaining, liquid
Nevastain A & B, Nevastain 21	Neville					X	X	X			For adhesives, 175.105	Nonstaining, light amber
Nonox WSP	ICI Americas, Stanchem					X	X				In sealing closures, max. 1%; PE packaging for foodstuffs, max. 0·2%	Nontoxic
Permanax WSL	Stanchem	X						X				Nontoxic
Permanax WSO	Stanchem					X	X					Nondiscolouring, nonstaining
Prodox 120	PMC	X				X	X	X	X		For adhesives, 175.105	Nonstaining, amber liquid

Prodox 148, 148B	PMC	X										
Prodox 340, 341	PMC	X			X	X	X	X				
Santech A/O 15-1 Conc.	Santech	X	X	X	X	X	X	X	X	EPDM, polyurethane		Nondiscolouring, nonstaining, colourless, odourless
Santech A/O 18-271, 18-122 Conc.	Santech					X	X			EPDM		Nondiscolouring
Santech A/O 18-121, 18-123 Conc.	Santech					X	X			EPDM		Nonstaining, colourless, odourless
Stabilite SP-49-464	Reichhold	X								Urethane, EPDM		
Stabilite 49-467, 49-470	Reichhold	X				X	X	X	X	Urethane, EPDM	Adhesives, 175.105	Nondiscolouring, nonstaining, low volatility
Sustane BHA	UOP		X			X	X	X	X		Indirect and direct food additive	White, waxy tablets or flakes
Sustane BHT	UOP		X		X	X	X	X	X		Indirect and direct food additive	Nonstaining, nondis-colouring crystal
Sustane TBHQ	UOP				X						Food packag-ing, 21CFR 177.2420	White to light tan crystals
TBHQ	Aceto				X							White to tan crystals
TBHQ	Eastman, UOP				X						Food packag-ing, 21CFR 177.2420	White to light tan crystals
Topanol O	Stanchem		X		X	X	X	X	X			Nondiscolouring, nonstaining
Topanol OC	ICI Americas, Stanchem	X	X		X	X	X	X	X	Polyols, elastomers, polyurethane	Indirect and direct food additive	Nondiscolouring, nonstaining

(continued)

APPENDIX 1—*contd.*

Trade name or designation	Supplier	Recommended for:									FDA sanction and type	General characteristics
		ABS	Cellulosic	Nylon	Polyester	PE	PP	PS	PVC	Other plastics		
Ultranox 226	Borg–Warner	X	X			X	X	X	X		Indirect food additive	Nonstaining powder
Ultranox 246	Borg–Warner	X				X	X	X		Acetals	For polyolefins, 178.2010; for adhesives, 175.105	Nonstaining powder
Ultranox 276	Borg–Warner	X	X	X	X	X	X	X	X	Elastomers, EPDM	For polyolefins, HIPS, ABS, 178.2010; for adhesives, 175.105	Nonstaining, nondiscolouring
Uvi-Nox 1494	GAF		X	X	X	X	X	X	X			Nondiscolouring, nonstaining, low volatility
Vanox PC	Vanderbilt	X		X	X	X	X	X	X	Polyurethane	Indirect food additive	Nonstaining granules or flakes
Vanox SKT	Vanderbilt	X	X	X	X	X	X	X	X	EPDM, polyurethane	For polyolefins in contact with all types of food under all current use conditions, 178.2010	Nondiscolouring, good high-temperature stability

Vanox 102	Vanderbilt			X		X	X	X	X	Urethane		Nonstaining liquid
Vanox 1003	Vanderbilt					X	X		X	Polyurethane		Amber liquid
Vanox 1290, 1320	Vanderbilt	X			X	X	X	X	X	Polyurethane		Nondiscolouring
Vanox 2246	Vanderbilt	X			X	X	X	X	X			Nonstaining, nondiscolouring
Vulkanox BKF	Mobay	X	X	X	X	X	X	X	X	Acetal, urethane	For adhesives; for polyolefins, 0·1% max.	Nondiscolouring, nonstaining
Vulkanox DS	Mobay	X				X	X	X	X			Nondiscolouring, nonstaining
Vulkanox KB	Mobay		X			X	X	X	X			Nondiscolouring, nonstaining
Wingstay C	Goodyear					X	X	X				Nondiscolouring, nonstaining
Wingstay S & T	Goodyear					X	X	X	X		175.105, 177.2600	Nondiscolouring, nonstaining amber liquid
Alkylidene bis-, tris- and polyphenols												
Cyanox 425	Am. Cyanamid	X		X		X	X		X		For adhesives, 176.250; for ABS contacting foods, not to exceed 0·6% wt., 178.2010	Cream coloured to white free-flowing powder
Cyanox 1735	Am. Cyanamid	X				X	X			EVA		Nonstaining, pale yellow liquid

(continued)

APPENDIX 1—*contd.*

Trade name or designation	*Supplier*	*Recommended for:*									*FDA sanction and type*	*General characteristics*
		ABS	*Cellulosic*	*Nylon*	*Polyester*	*PE*	*PP*	*PS*	*PVC*	*Other plastics*		
Cyanox 1790	Am. Cyanamid	X	X	X	X	X	X	X	X	Acetal, urethane, EPDM, olefin copolymers	For polyolefins in contact with all foods under all current use conditions, PS, impact PS, 178.2010	Free-flowing white powder, nondiscolouring, nonstaining, low gas yellowing
Cyanox 2246	Am. Cyanamid	X				X	X			EVA	For polyolefins, 178.2010; for adhesives, 176.250	Off-white free-flowing powder
Cyanox 2777 Blend	Am. Cyanamid	X			X	X	X	X		Polyolefins, polyurethane	For polyolefins, PS and impact PS	Nondiscolouring
Ethanox 330	Ethyl	X	X	X	X	X	X	X	X	EPDM	Up to 0·5% wt. in polymers, 178.2010	Nondiscolouring, nonstaining, odourless, crystalline powder
Hostanox 03	Hoechst Celanese	X	X	X	X	X	X	X		Polyolefins, polyurethane	For polyolefins, 178.2010	Nondiscolouring, colourless, odourless powder or granulate

Irganox 1010	Ciba–Geigy, USI	X	X	X	X	X	X	X	X	Acetals, EPDM, olefin co-polymers, polyurethane	For polyolefins and other polymers, 178.2010; for adhesives, 175.105	Nondiscolouring, nonstaining, colourless, odourless
Irganox B-Blends	Ciba–Geigy	X	X	X	X	X	X	X	X	HDPE	For polyolefins, 178.2010	Odourless, tasteless, low volatility
Mixxim AO-30	Fairmount	X	X	X	X	X	X	X	X	Acetals, EPDM, olefin co-polymers, polyurethane	For polyolefins and other polymers, 178.2010; for adhesives, 175.105	Nondiscolouring, nonstaining, colourless, odourless
Naugard XL-1	Uniroyal	X	X	X	X	X	X	X	X	EPDM, polyurethane, EVA	For polyolefins, ABS, PS, impact PS, 178.2010; for adhesives, 175.105	Antioxidant and metal deactivator
Naugawhite	Uniroyal	X				X	X			Urethane, EVA	Adhesives, 175.105; ABS, 178.2010	Nonstaining, amber liquid
Oxi-Chek 114	Ferro	X				X	X			EVA	For polyolefins, 178.2010; for adhesives, 176.250	Off-white free-flowing powder

(continued)

APPENDIX 1—*contd.*

Trade name or designation	*Supplier*	*Recommended for:*									*FDA sanction and type*	*General characteristics*
		ABS	*Cellulosic*	*Nylon*	*Polyester*	*PE*	*PP*	*PS*	*PVC*	*Other plastics*		
Oxi-Chek 116	Ferro	X	X	X	X	X	X	X	X	EPDM, olefin copolymers, polyurethane	For polyolefins, ABS, PS, impact PS, 178.2010; for adhesives, 175.105	Nondiscolouring, nonstaining, colourless, odourless
Oxi-Chek 414	Ferro	X	X	X	X	X	X	X	X	Acetals, EVA	For polyolefins, 178.2010; for adhesives, 175.105	Nonstaining, clear, synergistic light-stable
Santowhite powder	Harwick, Monsanto	X	X	X	X	X	X	X	X	Acetals, EVA	For adhesives, 175.105; for polyolefins, 178.2010	Clear, nonstaining, light-stable, synergistic
Topanol, CA	ICI Americas, Stanchem	X	X	X	X	X	X	X	X	Acetal, olefin copolymers	For use with all thermoplastic resins	Nontoxic, nondiscolouring, nonstaining
Vanox GT	Vanderbilt	X	X	X	X	X	X	X	X	EPDM, polyurethane	For polyolefins in contact with fatty and non-fatty foods, 178.2010	Nondiscolouring, improves UV resistance, highly effective in conjunction with secondary antioxidants

Wytox PAP	Olin	X	X	X	X	X	X	X	X	EPDM		Nonstaining, amber liquid
Thio and dithiobis-, tris-, and poly-alkylated phenols												
Antioxidant TBM-6P, 6T	Aceto	X	X	X	X	X	X	X	X		For polyolefins, 178.2010	Clear, nonstaining, light-stable
Irganox 1035	Ciba–Geigy	X	X	X	X	X	X			Elastomers	Adhesives, 175.105	Nondiscolouring, nonstaining, white crystalline powder
Santonox, Santonox R	Monsanto, USI	X	X	X	X	X	X	X	X		For adhesives, 175.105; for polyolefins, 178.2010	Clear, nonstaining, light-stable, synergistic
Ultranox 236	Borg–Warner	X	X	X	X	X	X	X	X		For poly-ethylene, 178.2010; for adhesives, 175.105	Nonstaining crystalline powder
Phenol condensation products												
Mixxim AO-30	Fairmount	X	X	X	X	X	X	X	X		For use with all thermoplastic resins	Nonstaining, nonblooming, nontoxic
Topanol CA	ICI Americas, Stanchem	X	X	X	X	X	X	X	X	Acetal, olefin copolymers	For use with all thermoplastic resins	Nonstaining, nonblooming, nontoxic

(continued)

APPENDIX 1—*contd.*

Trade name or designation	*Supplier*	*Recommended for:*									*FDA sanction and type*	*General characteristics*
		ABS	*Cellulosic*	*Nylon*	*Polyester*	*PE*	*PP*	*PS*	*PVC*	*Other plastics*		
Wingstay L	Goodyear	X	X	X	X	X	X	X	X	Acetals, acrylates	178.2010, 175.105, 177.2600	Nondiscolouring, nonstaining light tan powder or flake
Wytox PMW	Olin	X	X	X	X	X	X	X	X	Acetals	175.105, 177.2600	Nondiscolouring, nonstaining light tan powder
Amines												
Additin 30	Mobay	X	X	X	X	X	X	X	X			Discolouring, brownish-violet flakes
Goodrite 3120, 3140	Goodrich					X	X					Slight degree of staining or discolouration
Irganox LO-6[2]	Ciba–Geigy											Light tan powder
Naugard A, BG	Uniroyal			X							Hot-melt adhesives	Discolouring, secondary amine
Naugard J	Uniroyal	X				X	X				Adhesives, 175.105	Generally discolouring
Naugard 445	Uniroyal	X	X	X	X	X	X	X	X	Polyacetal, polyacrylate, polyurethane polyols	Adhesives, 175.105; polyester elastomers, 177.1590; side-seam cements-nylon, 175.300	Generally nondiscolouring or nonstaining

Naugard O	Uniroyal									Crosslinked PE		Tan powder
Polylite	Uniroyal	X										Discolouring, liquid
Vanox 1080, 1081	Vanderbilt	X	X	X	X	X	X	X		Polyurethane polyols, hot melts	Adhesives, 175.105	Slight discolouration non-skinning, viscosity stabilizer
Vulkanox 4010 NA	Mobay					X	X			Acrylics	177.2600	Discolouring, light grey flakes
Vulkanox 4020 F	Mobay					X	X			Acrylics		Discolouring, brownish-violet flakes
Vulkanox 4030	Mobay					X	X			Acrylics		Discolouring, dark-red liquid
Wingstay 29	Goodyear		X	X	X	X	X	X		Acetal, acrylate, urethane		Nondiscolouring, nonstaining
Wytox ADP	Olin	X	X	X	X	X	X	X		Polyurethane		Slight degree of stain and discolouration
Esters												
Argus DLTDP, DSTDP	Argus	X				X	X		X	EPDM, EVA, elastomers	Food packaging materials	Clear, nonstaining
Argus DMTDP	Argus	X	X			X	X		X	EPDM, EVA, elastomers	Food packaging	Clear
Argus DTDTDP	Argus	X				X	X		X			Clear, nonstaining
Carstab DLTDP	Morton Thiokol	X	X			X	X	X	X	EPDM, EVA	Food packaging materials	Clear
Carstab DMTDP, DSTDP	Morton Thiokol	X	X			X	X		X	EPDM, EVA	Food packaging materials	Clear
Cyanox LTDP, STDP, MTDP, 1212	Am. Cyanamid	X	X			X	X			EVA	Food packaging materials	White crystalline flakes and powder
Cyanox 711	Am. Cyanamid	X										Colourless to slightly yellow clear liquid

(continued)

APPENDIX 1—*contd.*

Trade name or designation	*Supplier*	*Recommended for:*									*FDA sanction and type*	*General characteristics*
		ABS	*Cellulosic*	*Nylon*	*Polyester*	*PE*	*PP*	*PS*	*PVC*	*Other plastics*		
Dilauryl thiodipropionate; distearyl thiodipropionate	Aceto, Stanchem	X				X	X		X		Food packaging materials	Clear, nonstaining
Dilauryl thiodipropionate; distearyl thiodipropionate; dimyristyl thiodipropionate	Argus, Evans	X				X	X		X			Synergistic
Dimethyl thiodipropionate	Argus									Elastomers		
Ditridecyl thiodipropionate	Argus, Evans	X				X	X		X			Clear, nonstaining
Evanstab 13	Evans	X	X			X	X		X			Clear, nonstaining
Evanstab 12, 14, 18	Evans	X	X			X	X		X		Food packaging materials	Clear, nonstaining
Hostanox SE 10	Hoechst Celanese	X	X			X	X					Clear, synergistic, specially for Hostanox 03
Hostanox SE 2	Hoechst Celanese	X	X			X	X		X	EPDM, EVA	Food packaging materials	Clear, synergistic
Mark 2140	Argus	X			X	X	X	X	X	Elastomers, polyethers, urethane		
Mark 5089	Argus	X				X	X	X	X	Elastomers, urethane		
Mark 5095	Argus	X	X			X	X			EVA	Food packaging materials	White crystalline flakes
Poly TDP 2000	Eastman	X	X			X	X		X	EPDM, EVA		Clear

Seenox 412S (Pentaerythrityl tetrakis-[β-laurylthiopropionate])	Argus	X	X		X	X	X		X	EPDM, EVA, elastomers, PC-polyesters		White crystalline powder
Vandox DSTDP	Vanderbilt	X	X			X	X		X	EPDM, EVA	Food packaging materials	Clear, nonstaining flakes
Organic phosphites and phosphates												
Bis-tridecyl phosphite	Albright & Wilson, Borg–Warner, Plastics & Chem.						X					
Dibutyl phosphite	Albright & Wilson, Stauffer						X					Clear, nonstaining
Didecyl phosphite	Argus, Borg–Warner					X	X					Clear, nonstaining
Dilauryl phosphite	Albright & Wilson, Argus, Borg–Warner		X	X	X	X	X		X			Clear, nonstaining
Dimethyl phosphite	Albright & Wilson				X							Metal chelator, stabilizer
Dioctyl phosphite	Albright & Wilson, Argus, Borg–Warner						X					Clear, nonstaining
Dioleyl phosphite	Albright & Wilson		X	X	X		X		X			Clear, nonstaining
Diphenyldecyl phosphite	Akzo, Argus, Borg–Warner, Plastics & Chem.								X			
Ditetradecyl phosphite	Plastics & Chem.						X					Clear, nonstaining

(continued)

APPENDIX 1—*contd.*

Trade name or designation	*Supplier*	*Recommended for:*									*FDA sanction and type*	*General characteristics*
		ABS	*Cellulosic*	*Nylon*	*Polyester*	*PE*	*PP*	*PS*	*PVC*	*Other plastics*		
Hostanox PAR 24	Hoechst Celanese			X	X	X	X	X		Polycarbonate		Solid phosphite to improve processing stability, good hydrolysis resistance
Interstab CH-55, CH-55R	Akzo	X	X		X	X	X	X	X	Epoxy	Food packaging materials	Clear, nonstaining
Interstab CH-90, CH-301	Akzo	X					X		X			Clear, nonstaining
Interstab CH-300	Akzo			X	X	X			X	Epoxy, polyacrylates		Light stability
Mark 329	Argus	X					X		X		178.2010	Clear, nonstaining
Mark C	Argus	X					X		X			Clear, nonstaining
Mark 1178, 1178B	Argus	X	X		X	X	X	X	X		178.2010	Nonstaining; heat, colour, and processing stabilizer
Mark 1500	Argus	X							X		178.2010	Nonstaining; heat, colour, and processing stabilizer
Mark 2112	Argus	X	X		X	X	X	X	X	Polycarbonate	178.2010	Nondiscolouring, synergistic, good hydrolysis resistance
Mark 5050, 5060	Argus				X	X	X	X	X		178.2010	Solid, nondiscolouring, synergistic
Mark 5070	Argus	X		X	X	X	X	X	X		178.2010	Solid, nondiscolouring, synergistic

Mark 5082	Argus					X	X			Elastomers, olefin polymers	178.2010	Hydrolytically stable
Mixxim P-11	Fairmount				X	X	X	X	X			Solid, nondiscolouring, synergistic
Mixxim P-21	Fairmount	X		X	X	X	X	X	X			Solid, good hydrolysis resistance
Naugard P	Uniroyal	X	X	X	X	X	X	X	X	EVA, polycarbonate	178.2010	Clear, nonstaining, nondiscolouring liquid
Naugard 512	Uniroyal	X	X	X	X	X	X	X	X	EVA, polycarbonate	178.2010	Liquid, good hydrolysis resistance, nondiscolouring
Naugard 524	Uniroyal	X	X		X	X	X	X	X		Polyolefins, 178.2010	Good hydrolytic stability, synergistic, white crystalline solid
Octyl diphenyl phosphite	Argus, Borg–Warner, Plastics & Chem.	X					X		X			
Phenyldidecyl phosphite	Akzo, Argus, Borg–Warner, Plastics & Chem.								X			
Polygard	Uniroyal	X						X		SAN	Food packaging materials	Clear, nonstaining, nondiscolouring, liquid
Polygard HR								X				
Sandostab P-EPQ	Sandoz	X		X	X	X	X	X	X	Polycarbonate	177.1520, 177.1580, 177.1640	High-temperature stabilizer, nonstaining, nondiscolouring
TPP	Borg–Warner									Butyrates		
Thermolite 187	M & T	X					X		X			Clear, low plate-out, synergistic
Tributyl phosphite	Albright & Wilson, Stauffer		X	X	X		X		X			Clear, nonstaining

(continued)

APPENDIX 1—*contd.*

Trade name or designation	*Supplier*	*Recommended for:*									*FDA sanction and type*	*General characteristics*
		ABS	*Cellulosic*	*Nylon*	*Polyester*	*PE*	*PP*	*PS*	*PVC*	*Other plastics*		
Tridecyl phosphite	Akzo, Argus, Borg–Warner, Plastics & Chem.								X			
Tri-isopropyl phosphite	Albright & Wilson				X							Metal chelator, stabilizer
Trimethyl phosphite	Albright & Wilson, Stauffer				X							Metal chelator, stabilizer
Trioctyl phosphite	Albright & Wilson, Argus, Borg–Warner		X	X	X		X		X			Clear, nonstaining
Triphenyl phosphite	Akzo, Argus, Borg–Warner, Morton Thiokol, Plastics & Chem.	X			X		X	X	X	Epoxy		Light stability
Trisnonylphenyl phosphite	Akzo, Argus, Borg–Warner, Olin, Plastics & Chem., Uniroyal	X		X	X	X	X	X	X	Epoxy	Food packaging materials	Clear, nonstaining, synergistic
Tris(2-chloroethyl phosphite)	Albright & Wilson		X	X	X		X		X			Clear, nonstaining

Ultranox 624	Borg–Warner								X	Polycarbonate	For PP, HDPE, 178.2010; for PE copolymers, 175.105	White powder
Ultranox 626	Borg–Warner	X	X	X	X	X	X	X	X	Polycarbonate	For PP, HDPE, 178.2010; for PE copolymers, 175.105	White powder
Vanox 1005	Vanderbilt				X	X	X	X	X	EVA, polyurethane	178.2010, 175.105	Improves colour and stability
Weston DPDP, PDDP	Borg–Warner											
Weston ODPP	Borg–Warner	X				X	X	X		SAN		
Weston TLTTP	Borg–Warner					X	X					
Weston TNPP, 399	Borg–Warner	X				X	X	X		Polyurethane, SBR	Food packaging materials	Clear, nonstaining
Weston TPP	Borg–Warner									Epoxy		
Weston 600	Borg–Warner				X					Butyrates		
Weston 618	Borg–Warner				X	X	X	X	X		177.1520, 176.170	Solid, nondiscolouring, synergistic
Weston 626	Borg–Warner	X		X	X	X	X	X	X		Selected food packaging materials	Solid, good hydrolysis resistance
Weston 430, 474, 491, 494, DHOP, PTP, PNPG, TDP, THOP, TIOP, TLP, TSP	Borg–Warner			X	X	X	X		X	Polyurethane		
Weston DOPI, DPP, DSP	Borg–Warner						X					
Weston EGTPP	Borg–Warner						X			Epoxy, polyurethane		
Wytox 345	Olin	X		X	X	X	X	X	X		177.2600, 175.105	Clear, nonstaining, synergistic

(continued)

APPENDIX 1—*contd.*

Trade name or designation	*Supplier*	*Recommended for:*									*FDA sanction and type*	*General characteristics*
		ABS	*Cellulosic*	*Nylon*	*Polyester*	*PE*	*PP*	*PS*	*PVC*	*Other plastics*		
Wytox 540	Olin										177.2600, 175.105	Clear, nonstaining, synergistic
Wytox 312, 320	Olin	X		X	X	X	X	X	X	Polyurethane, SBR	Food packaging materials	Clear, nonstaining, synergistic
Miscellaneous												
Antioxidant 451	Uniroyal											Nondiscolouring, nonstaining
p-Benzoquinone	Eastman			X	X							Yellow crystalline solid
Colour suppressant 190	M & T	X			X	X	X	X	X		Food packaging materials	Clear, nonstaining, synergistic
Colour suppressant 195	M & T	X			X				X			
Eastman Inhibitor OABH	Eastman					X	X					White crystalline solid
Hostanox VP OSP 1	Hoechst Celanese					X	X				Adhesives, 178.2010	Metal deactivator (e.g., copper)
Hostanox System VN 23681	Hoechst Celanese						X					Concentrate, carbon black synergist
Hostanox System VP 01961	Hoechst Celanese					X				Polyolefins	Food packaging	Concentrate
Hydroquinone	Eastman		X		X					Acrylic, acrylate		White or light grey crystals
Hydroquinone monomethyl ether	Eastman				X					Acrylic, acrylate		White crystals, nonstaining
Mono-*tert*.-butylhydroquinone	Eastman				X							White to tan crystals

2,5-Di-*tert.*-butylhydroquinone	Eastman				X	X	X					Buff powder
Toluhydroquinone	Eastman				X							Pink to white crystals
Mark 135	Argus		X		X					Polycarbonate, polyethers		Clear, nonstaining
Mark 158	Argus						X					
Mark 328	Argus					X	X				178.2010	Nonstaining; good heat, colour, and processing stabilizers
Mark 522	Argus						X			EVA		
Mark 1220	Argus	X								IPS	181.22, 178.2010	
Mark 1409	Argus					X	X	X			178.2010	
Mark 1475	Argus				X	X	X	X	X	EPDM, olefin copolymers, thermoplastic elastomers		Nonstaining, metal deactivator (e.g. copper)
Mark 1589	Argus		X	X	X	X	X	X	X	EPDM, olefin copolymers, thermoplastic elastomers	178.2010	White powder
Mark 1589B	Argus		X	X	X	X	X	X			178.2010	White powder
Mark 292, 1900	Argus	X							X			
Mark 2100, 2100A	Argus					X	X			Olefin copolymers		For halogenated flame retardants
Mark 2180	Argus					X	X			Olefin copolymers		White powder clarifier stabilizer
Mark 5090A, B, C, D	Argus					X	X	X		Olefin copolymers	178.2010	
MiRaStab 706-K	M-R-S			X	X	X	X		X			
Naugard 492	Uniroyal	X	X	X	X	X	X	X	X	Urethane	178.2010	Liquid, nonstaining, phenolic phosphite antioxidant
Naugard XL-517	Uniroyal	X	X	X	X	X	X	X	X		178.2010, 175.105	Solid phenolic phosphite antioxidant/metal deactivator

(*continued*)

APPENDIX 1—*contd.*

Trade name or designation	*Supplier*	*Recommended for:*									*FDA sanction and type*	*General characteristics*
		ABS	*Cellulosic*	*Nylon*	*Polyester*	*PE*	*PP*	*PS*	*PVC*	*Other plastics*		
Nonox DLTDP	ICI Americas, Stanchem	X				X	X	X	X		In contact with fatty foods, 0·1%; non-fatty, 0·25% except in PS, ABS (applies to Nonox DLTDP and Nonox DSTDP)	
Nonox DSTDP	ICI Americas, Stanchem					X	X					White powder, synergistic
Stabaxol P	Mobay									Urethane, EVA		Hydrolysis agent for millable urethane, heat-resistant stabilizers for curable EVA
Stabaxol M	Mobay									Urethane		Hydrolysis agent for polyester-based cast urethane
Stabilizer B	Harwick									CO, ECO, EPDM		Nondiscolouring, metal deactivator

Thermolite 20, 31, 35	M & T	X							X	Acrylic, chlorinated polymers		Low usage, clear
Thiodipropionic acid	Evans								X		Food packaging materials	Clear, nonstaining
Vanox 1030	Vanderbilt	X					X	X				
Vanox 3240, 3245	Vanderbilt						X					Mineral deactivator
Vanstay SC, SD, SG, SH	Vanderbilt								X			Improves colour and heat stability
Voidox 100%	Guardian	X				X	X	X		Polyolefins	Food additive, 0·02%; food packaging materials, 0·33%	Waxy flake, lubricant
Vulkanox MB-2/MGC	Mobay		X	X	X	X	X	X	X			Nonstaining, yellowish-white powder
Vulkanox ZMB-2/C5	Mobay		X	X	X	X	X	X	X			Nonstaining, white powder

Names and Addresses of Suppliers Listed in Appendix 1

Aceto Corp., 126-02 Northern Blvd., Flushing, NY 11368 (718-898-2300)

Akzo Chemie America, Interstab Chemicals, 500 Jersey Ave., P.O. Box 638, New Brunswick, NJ 08903 (201-247-2202)

Albright & Wilson Inc., P.O. Box 26229, Richmond, VA 23260-6229 (804-752-6118)

American Cyanamid Co., Polymer Additives Dept., P.O. Box 6885, Bridgewater, NJ 08807 (201-356-2600)

Argus Div., Witco Corp., 633 Court St., Brooklyn, NY 11231 (718-858-5678)

Borg–Warner Chemicals, Borg–Warner Corp., International Center, Parkersburg, WV 26101 (304-424-5411)

Ciba–Geigy Corp., Additives Dept., 3 Skyline Dr., Hawthorne, NY 10532 (914-347-4700)

Eastman Chemical Products Inc., sub. Eastman Kodak Co., P.O. Box 431, Kingsport, TN 37662 (615-229-2000)

Ethyl Corp., Chemicals Group, Ethyl Tower, 451 Florida Blvd., Baton Rouge, LA 70801 (504-388-7353)

Evans Chemitics, Organic Chemicals Div., W. R. Grace & Co., 90 Tokeneke Rd., Darien, CT 06820 (203-655-8741)

Fairmount Chemical Co., 117 Blanchard St., Newark, NJ 07105 (201-344-5790)

Ferro Corp., Chemical Div., 7050 Krick Rd., Bedford, OH 44146 (216-641-8580)

GAF Corp., 1361 Alps Rd., Wayne, NJ 07470 (201-628-3000)

Goodrich, B. F., Chemical Group, 6100 Oak Tree Blvd., Cleveland, OH 44131 (216-447-6000)

Goodyear Tire & Rubber Co., 1485 E. Archwood Ave., Akron, OH 44316 (216-796-6400)

Guardian Chemical, Div. United-Guardian Inc., P.O. Box 2500, Smithtown, NY 11787 (516-273-0900)

Harwick Chemical Corp., 60 S. Seiberling St., P.O. Box 9360, Akron, OH 44305 (216-796-9300)

Hoechst Celanese Corp., Specialty Products Group, Rte. 202-206 N., Somerville, NJ 08876 (201-231-2073)

ICI Americas Inc., Concord Pike & New Murphy Rd., Wilmington, DE 19897 (302-575-3000)

M & T Chemicals Inc., Woodbridge Ave., P.O. Box 1104, Rahway, NJ 07065 (201-499-2457)

M-R-S Chemicals Inc., 2498 Adie Rd., Maryland Heights, MO 63043 (314-872-8750)

Mobay Chemical Corp., Mobay Rd., Pittsburgh, PA 15205 (412-777-2000)

Monsanto Co., 800 N. Lindbergh Blvd., St. Louis, MO 63167 (314-694-1000)

Morton Thiokol, Carstab Products, 2000 West St., Cincinnati, OH 45215 (513-733-2100)

Neville Chemical Co., Neville Island, Pittsburgh, PA 15225 (412-331-4200)

Neville-Synthese Organics Inc., Neville Island, Pittsburgh, PA 15225 (412-331-4200)

Olin Chemical Co., Polychem Products, 120 Long Ridge Rd., Stamford, CT 06904 (203-356-2000)

PMC Specialties, Group Div. PMC Inc., 20525 Center Ridge Rd., Rocky River, OH 44116 (216-356-0700)

Plastics & Chemicals Inc., P.O. Box 160, Cedar Grove, NJ 07009 (201-221-0002)

Richhold Chemicals Inc., Resins & Binders Div., P.O. Box 1433, Pensacola, FL 32596 (904-433-7621)

Sandoz Chemicals Corp., 4000 Monroe Rd., Charlotte, NC 28205 (704-372-0210)

Santech Inc., 150 Norseman St., Toronto, Ont. M8Z 5M4 Canada (416-233-1234)

Schenectady Chemicals Inc., P.O. Box 1046, Schenectady, NY 12301 (518-370-4200)

Shell Chemical Co., 1 Shell Plaza, P.O. Box 2463, Houston, TX 77002 (713-241-6161)

Stanchem, a bus. unit of CIL, Inc., 90 Sheppard Ave. E., Box 200, Sta. A. N. York, ON, Canada M2N 6H2 (416-229-7000)

Stauffer Chemical Co., Specialty Chemical Div., Westport, CT 06881 (203-222-3000)

UOP Inc., Biological & Food Products, 25 E. Algonquin Rd., Des Plaines, IL 60017-5017 (312-391-3093)

USI Chemicals, 11500 Northlake Dr., Cincinnati, OH 45249 (513-530-6500)

Uniroyal Inc., Chemical Group, Middlebury, CT 06749 (203-573-2000)

Vanderbilt, R. T., Co. Inc., Plastics Dept., 30 Winfield St., E. Norwalk, CT 06855 (203-853-1400)

APPENDIX 2: COMMERCIALLY AVAILABLE PHOTOSTABILIZERS[1196]

Name	Trade names, designations, and suppliers[a]	Recommended for:										Form	Melting point (°C)
		ABS	Cellulosics	Epoxy	Polyesters	Polyethylene	Polypropylene	Polystyrene	Flexible PVC	Rigid PVC	Vinylidene chloride		
Benzophenones													
2,4-Dihydroxybenzophenone	Uvinul 400 (BASF); Mixxim LS-24 (Fairmount); Syntase 100 (Neville); Uvasorb 20H (3-V Chem.)		X	X	X			X				Powder	136–149
2-Hydroxy-4-acryloxyethoxybenzophenone (polymer)	Cyasorb UV2126 (Am. Cyanamid)					X	X			X		Powder	—
2-Hydroxy-4-methoxybenzophenone	Acetorb A (Aceto); Cyasorb UV9 (Am. Cyanamid); Uvinul M-40 (BASF); UV-Absorber 325 (Rhein-Chemie); Syntase 62 (Neville); Uvasorb MET (3-V Chem.)	X	X		X			X	X[b]	X	X	Powder	62–63·5
Ethyl-2-cyano-3,3-diphenyl acrylate	Uvinul N-35 (BASF)									X	X	Powder	96
2-Ethylhexyl-2-cyano-3,3-diphenyl acrylate	Uvinul N-539 (BASF)								X[b]		X	Liquid	−10
2,2′-Dihydroxy-4-methoxybenzophenone	Cyasorb UV 24 (Am. Cyanamid)								X[b]			Powder	68
2,2′-Dihydroxy-4,4′-dimethoxybenzophenone	Uvinul D-49 (BASF)			X	X							Powder	125

2-Hydroxy-4-*n*-octoxybenzophenone[d]	Advastab 46 (Aceto); Cyasorb UV 531 (Am. Cyanamid); Hostavin ARO 8 (Hoechst Celanese); 10057 MB (Ampacet); Mark 1413 (Argus); Uvinul 408 (BASF); UV Chek AM-300, -301 (Ferro); A-19459EF (Polycom Huntsman); Syntase 800, 805 (Neville); C-PPR-8000, -2498, -1689, C-PE-1746, -2612, -2388, -1747 (Reed); Uvasorb 3C (3-V Chem.); Spectrathene (USI)	X	X	X	X	X	X	X	X[b]	X	X	Powder	48–49
2,2′4,4′-Terahydroxybenzophenone	Uvinul D-50 (BASF)			X	X							Powder	195
4-Dodecyloxy-2-hydroxybenzophenone	Eastman Inhibitor DOBP (Eastman); UV Chek AM-320 (Ferro)	X	X	X	X	X	X	X	X[b]	X	X	Powder	52
Proprietary	Uvinul 490 (BASF)			X	X							Powder	Mixture
Proprietary	Mark 1535 (Argus)	X	X	X	X	X	X	X	X[b]	X	X	Liquid	—
Proprietary	Permyl B 100 (Ferro)		X	X	X	X		X	X[b]	X		Powder	—
Benzotriazoles													
2-(2′-Hydroxy-5′-methylphenyl)benzotriazole	Viosorb 520 (Aceto); Tinuvin P (Ciba–Giegy); Mixxim LS-14 (Fairmount); Uvasorb SV (3-V Chem.); Vanox UV-1 (Vanderbilt)	X	X	X	X			X	X[b]	X	X	Powder	129
2(3′,5′-Di-*t*-butyl-2′-hydroxyphenyl)-5-chlorobenzotriazole	Tinuvin 327 (Ciba–Geigy)	X	X	X	X	X	X		X[b]	X	X	Powder	151
2-(3′-*t*-Butyl-2′-hydroxy-5′-methylphenyl)-5-chlorobenzotriazole	Tinuvin 326 (Ciba–Geigy)	X	X	X	X	X	X		X[b]	X	X	Powder	140
2-(2′-Hydroxy-3′,5′-di-*t*-butylphenyl)benzotriazole	Tinuvin 320 (Ciba–Geigy)	X	X	X	X	X			X[b]	X	X	Powder	152
2-(2′-Hydroxy-3′,5′-di-*t*-amylphenyl)benzotriazole	Tinuvin 328 (Ciba–Geigy)	X	X	X	X	X	X	X	X[b]	X	X	Powder	81
2-(2′-Hydroxy-5-*t*-octylphenyl)benzotriazole	Cyasorb UV 5411 (Am. Cyanamid); Syntase UV 746 (Neville)	X	X	X	X	X	X	X	X[b]	X	X	Powder	101–105
Proprietary	Mark 446, 446B (Argus)				X	X		X	X[b]	X	X	Powder	—

(continued)

APPENDIX 2—*contd.*

Name	Trade names, designations, and suppliers[a]	Recommended for:										Form	Melting point (°C)
		ABS	Cellulosics	Epoxy	Polyesters	Polyethylene	Polypropylene	Polystyrene	Flexible PVC	Rigid PVC	Vinylidene chloride		
Salicylates													
Phenyl salicylate	Salol (Dow)		X		X	X			X[c]		X	Solid	62
t-Butyl phenyl salicylate	Anti UVK (Aceto); C-HDPE-2723, -2301 (Reed)	X	X		X	X	X	X			X	Crystal	65–66
Proprietary	10316 COP MB (Ampacet)					X						Pellets	—
Nickel organic													
Nickel bis (octyl phenol sulphide)	UV Chek Am-101, AM-105 (Ferro)					X	X					Flakes	130
[2,2′-Thiobis(4-*t*-octylphenolato)]-*n*-butylamine nickel	Cyasorb UV 1084 (Am. Cyanamid); 10071 MB (Ampacet); C-PPR-8008 (Reed); Uvasorb NI (3-V Chem); Spectrathene (USI)	X				X	X					Powder	258–281
Nickel bis [*O*-ethyl(3,5-di-*t*-butyl-4-hydroxybenzyl)]phosphonate	Irgastab 2002 (Ciba–Geigy); C-PPR-8006, -8007 (Reed)					X	X					Powder	—
Nickel dibutyldithiocarbamate	Rylex NBC (Du Pont); UV Chek AM-104 (Ferro); C-PPR-8002, -8003 (Reed); Naugard NBC (Uniroyal); Vanox NBC (Vanderbilt)					X	X					Flakes	88
Nickel di-isobutyldithiocarbamate	ISO Butyl Niclate (Vanderbilt)					X	X					Powder	173–181
Nickel dimethyldithiocarbamate	Methyl Niclate (Vanderbilt)					X	X					Powder	>290
Proprietary	UV Chek Am-205 (Ferro)					X	X					Flakes	140

Proprietary	A-23773EF (Polycom Huntsman)					X	X					Pellets	—
Proprietary	Mark 1306A (Argus)					X	X					Powder	—
Proprietary	10071 PE MB (Ampacet)					X						Pellets	—
Proprietary	10% Nickel Hex-Cem (Mooney)				X					X	X	Liquid	—
Proprietary	Hostavin VP NiCsi (Hoechst Celanese)					X	X					Powder	
Monobenzoates													
Resorcinol monobenzoate	Eastman Inhibitor RMB (Eastman)		X					X	X[c]		X	Powder	133
Oxalanilide													
Proprietary	Sanduvor EPU (Sandoz)					X	X					Powder	116–124
Proprietary	Sanduvor VSU (Sandoz)				X			X	X[b]	X		Powder	127
Others													
2,4-Di-*t*-butylphenyl 3,5-di-*t*-butyl-4-hydroxybenzoate	UV Chek AM-340 (Ferro); C-PPR-8001 (Reed)	X				X	X	X				Powder	196
3,5-Di-*t*-butyl-*p*-hydroxybenzoic acid	Aceto (Aceto)					X	X					Powder	200–210
n-Hexadecyl 3,5-di-*t*-butyl-4-hydroxybenzoate[d]	Cyasorb UV 2908 (Am. Cyanamid)	X			X	X	X					Powder	60
Hindered amine[d]	Cyasorb UV 3346 (Am. Cyanamid)	X				X	X	X				Powder	110–130
Hindered amine	40407 PP MB (Ampacet)						X					Pellets	—
Hindered amine	Spinuvex A-36 (Borg–Warner)	X	X	X		X	X	X				Powder	80
Hindered amine	Tinuvin 144 & 770 (Ciba–Geigy); C-PPR-8004, -8005, C-HDPE-2764 (Reed)	X	X	X	X	X	X	X				Powder	80–150
Hindered amine[d]	Chimassorb 944FL (Ciba–Geigy)					X	X					Powder	155–125
Hindered amine	Tinuvin 765 Ciba–Geigy)											Liquid	<20
Hindered amine[d]	Tinuvin 622LD (Ciba–Geigy)				X	X	X		X[b]			Powder	55–70
Hindered amine	Mixxim Hals 57 (Fairmount)		X	X	X	X	X	X		X		Powder	130–140

(*continued*)

APPENDIX 2—*contd.*

Name	*Trade names, designations, and suppliers*[a]	*Recommended for:*										*Form*	*Melting point (°C)*
		ABS	*Cellulosics*	*Epoxy*	*Polyesters*	*Polyethylene*	*Polypropylene*	*Polystyrene*	*Flexible PVC*	*Rigid PVC*	*Vinylidene chloride*		
Hindered amine	Mixxim Hals 62, 67 (Fairmount)											Liquid	—
Hindered amine	Mixxim Hals 63 (Fairmount)	X	X	X	X	X	X		X[c]			Powder	80–90
Hindered amine	Mixxim Hals 68 (Fairmount)					X	X					Powder	70–80
Hindered amine	A-25904EF (Polycom Huntsman)					X	X					Pellets	—
Hindered amine	Sunshade 17-132, 17-163 (Santech)	X			X	X	X	X	X[c]	X		Pellets	120–140
Proprietary	10154 PE MB, 10478 PE MB (Ampacet)					X						Pellets	—
Proprietary	Hostavin System L-1920 (Hoechst Celanese); 10057, 10071, 10325, 10476 (Ampacet)					X	X					Pellets	130–225
Proprietary	Hostavin Systems VN28721, VN30373, 22193, L1922, VN 25027, VN26402 (Hoechst Celanese)				X	X	X					Pellets	—
Proprietary	Vanstay L (Vanderbilt)								X[b]			Powder	—
Proprietary	AST-1001, Merix (Merix)				X	X		X	X[b]	X	X	Liquid	—
Proprietary	Parabolix-100 (Merix)	X			X		X	X				Liquid	—

Proprietary	UV Chek AM-595 (Ferro)	X							X[b]	X	X	Powder	—
Proprietary (indol derivative)	UV-Absorber 340 (Rhein Chemie)		X		X			X	X[b]	X	X	Powder	97–99
Proprietary	Sunshade 20-76, 20-77 (Santech)								X	X		Powder	—
Steric hindered amine	Hostavin N 20 (Hoechst Celanese)	X	X			X	X					Powder	225
Zinc oxide	Azo-77, -77S (Asarco); Horse Head, Kadox 15, Protox-169 (NJ Zinc)	X	X	X	X	X	X	X				Powder	>1800
Zinc oxide/synergist	Sunshade 17-166, 17-221 (Santech)	X			X	X	X		X[c]	X		Pellets	120–140
	Sunshade 17-222 (Santech)								X[c]			Flakes	—
	Sunshade 18-90 (Santech)									X		Paste	—
	Sunshade 18-87, 18-107 (Santech)				X	X	X		X[c]			Flakes	—
	Sunshade 18-98 (Santech)							X				Chips	—
	Sunshade 18-99, 18-90 (Santech)								X[c]			Paste	—
	Sunshade 18-92, 18-95, 18-104 (Santech)						X					Pellets	—
	Sunshade 18-88 (Santech)					X	X					Flakes	—
	Sunshade 18-89, 18-94, 18-96, 19-37, 19-38, 19-39 (Santech)					X						Pellets	—
	Sunshade 18-81 (Santech)	X										Granules	—
	Sunshade 18-91 (Santech)									X		Granules	—

[a] For full names and addresses of suppliers (in parentheses), see p. 578.
[b] Includes semirigid.
[c] Does not include semirigid.
[d] FDA cleared for polyolefins.

Names and Addresses of Suppliers Listed in Appendix 2

Aceto Chemical Co. Inc., 126-02 Northern Blvd., Flushing, NY 11368 (718-898-2300)

American Cyanamid Co., Polymer Additives Dept., P.O. Box 6885, Bridgewater, NJ 08807 (201-356-2600)

Ampacet Corp., 250 S. Terrace Ave., Mount Vernon, NY 10550 (914-699-9100)

Argus Div., Witco Corp., 633 Court St., Brooklyn, NY 11231-2193 (718-858-5678)

Asarco Inc., P.O. Box 327, Columbus, OH 43216 (614-294-5566)

BASF Corp., Chemicals Div., 100 Cherry Hill Rd., Parsippany, NJ 07054 (201-263-3400)

Borg–Warner Chemicals, Borg–Warner Corp., International Center, Parkersburg, WV 26101 (304-424-5411)

Ciba–Geigy Corp., Additives Dept., 3 Skyline Dr., Hawthorne, NY 10532 (914-347-4700)

Dow Chemical Co., 2020 Dow Center, Midland, MI 48640 (517-636-1000)

DuPont de Nemours, E. I., & Co. Inc., Chemicals and Pigments Dept., 1007 Market St., Wilmington, DE 19898 (302-774-2421)

Eastman Chemical Products Inc., Eastman Kodak Co., P.O. Box 431, Kingsport, TN 37662 (615-229-2000)

Fairmount Chemical Co., 117 Blanchard St., Newark, NJ 07105 (201-344-5790)

Ferro Corp., Chemical Div., 7050 Krick Rd., Bedford, OH 44146 (216-641-8580)

Hoechst Celanese Corp., Specialty Products Group, Rte. 202-206 N., Somerville, NJ 08876 (201-231-2073)

Merix Chemical Co., 2234 E. 75th St., Chicago, IL 60649 (312-221-8242)

Mooney Chemicals Inc., 2301 Scranton Rd., Cleveland, OH 44113 (216-781-8383)

Neville-Synthese Organics Inc., Neville Island, Pittsburgh, PA 15225 (412-331-4200)

New Jersey Zinc Co. Inc. Palmerton, PA 18071 (215-826-2111)

Polycom Huntsman, Inc., 90 W. Chestnut St., Washington, PA 15301 (412-225-2220)

Reed Plastics Corp. Holden Industrial Park, Holden, MA 01520 (617-829-6321)

Rhein Chemie, 3200 Gilchrist Rd., Mogadore, OH 44260 (216-784-1029)

Sandoz Chemicals Corp., 4000 Monroe Rd., Charlotte, NC 28205 (704-372-0210)

Santech Inc., 150 Norseman St., Toronto, Ont. M8Z 5M4 Canada (416-233-1234)

3-V Chemical Corp., 7422

Carmel Executive Park, Charlotte, NC 28226 (704-543-1000)

USI Chemicals, 11500 Northlake Dr., Cincinnati, OH 45249 (513-530-6500)

Uniroyal Chemical Co. Inc., Middlebury, CT 06749 (203-573-2000)

Vanderbilt, R. T., Co., Vanstay Dept., 30 Winfield St., E. Norwalk, CT 06855 (203-853-1400)

Index

Numbers in italic type indicate those pages on which topics are given in full.